Lecture Notes in Computer Science 944

Edited by G. Goos, J. Hartmanis and J. van Leeuwen

Springer
Berlin
Heidelberg
New York
Barcelona
Budapest
Hong Kong
London
Milan
Paris
Tokyo

Zoltán Fülöp Ferenc Gécseg (Eds.)

Automata, Languages and Programming

22nd International Colloquium, ICALP 95
Szeged, Hungary, July 10-14, 1995
Proceedings

 Springer

Series Editors

Gerhard Goos
Universität Karlsruhe
Vincenz-Priessnitz-Straße 3, D-76128 Karlsruhe, Germany

Juris Hartmanis
Department of Computer Science, Cornell University
4130 Upson Hall, Ithaca, NY 14853, USA

Jan van Leeuwen
Department of Computer Science, Utrecht University
Padualaan 14, 3584 CH Utrecht, The Netherlands

Volume Editors

Zoltán Fülöp
Ferenc Gésceg
Department of Computer Science, Attila József University
Aradi vértanúk tere 1, H-6720 Szeged, Hungary

CIP data applied for

Die Deutsche Bibliothek - CIP-Einheitsaufnahme

Automata, languages, and programming : 22nd international
colloquium ; proceedings / ICALP 95, Szeged, Hungary, July
1995. Zoltán Fülöp ; Ference Gécseg (ed.). - Berlin ; Heidelberg
; New York : Springer, 1995
  (Lecture notes in computer science ; Vol. 944)
  ISBN 3-540-60084-1
NE: Fülöp, Zoltán [Hrsg.]; ICALP <22, Szeged>; GT

CR Subject Classification (1991): F, D.1, E.1, E.3, G.2, I.3.5

ISBN 3-540-60084-1 Springer-Verlag Berlin Heidelberg New York

© Springer-Verlag Berlin Heidelberg 1995
Printed in Germany

Typesetting: Camera-ready by author
SPIN 10486389   06/3142 – 5 4 3 2 1 0   Printed on acid-free paper

Foreword

The International Colloquium on Automata, Languages and Programming (ICALP) is an annual conference series sponsored by the European Association for Theoretical Computer Science (EATCS). It is intended to cover all important areas of theoretical computer science such as computability, automata, formal languages, term rewriting, analysis of algorithms, computational geometry, computational complexity, symbolic and algebraic computation, cryptography, data types and data structures, theory of data base and knowledge bases, semantics of programming languages, program specification, transformation and verification, foundations of logic programming, theory of logical design and layout, parallel and distributed computation, theory of concurrency, theory of robotics. ICALP 95 was held in Szeged, Hungary, from July 10 to July 14, 1995. Previous colloquia were held in Jerusalem (1994), Lund (1993), Wien (1992), Madrid (1991), Warwick (1990), Stresa (1989), Tampere (1988), Karlsruhe (1987), Rennes (1986), Nafplion (1985), Antwerp (1984), Barcelona (1983), Aarhus (1982), Haifa (1981), Amsterdam (1980), Graz (1979), Udine (1978), Turku (1977), Edinburgh (1976), Saarbrücken (1974) and Paris (1972). ICALP 96 will be held in Paderborn from July 8 to July 12, 1996.

The number of papers submitted was 111. Each submitted paper was sent to at least four Programme Committee members, who were often assisted by their referees. The Programme Committee meeting took place at the József Attila University in Szeged on the 3rd and 4th of February 1995. This volume contains the 53 papers selected at the meeting plus the four invited papers. We would like to thank all the Programme Committee members and the referees who assisted them in their work. The list of referees is as complete as we can achieve and we apologize for any omissions and errors.

We would like to thank the members of the Organizing Committee and the members of our department for their contribution throughout the preparation.

We also gratefully acknowledge support from

Academy of Finland
Hungarian Academy of Sciences
József Attila University
Human Resources Development Project (FEFA)
National Scientific Research Fund (OTKA)
ZENON Biotechnology Ltd.
ZENON Computer Engineering and Trading Ltd.

April 1995   Zoltán Fülöp and Ferenc Gécseg

Invited Lecturers

E. Best, Hildesheim
R. Freivalds, Riga
G. Paun, Bucharest
G. Rozenberg, Leiden

Programme Committee

S. Abiteboul, Paris
A. Apostolico, Padova and West Lafayette
S. Arnborg, Stockholm
C. Calude, Auckland
E. Clarke, Pittsburgh
J. Diaz, Barcelona
Z. Fülöp, Szeged
F. Gécseg, Szeged (Chairman)
R. Gorrieri, Bologna
J.-P. Jouannaud, Paris
J. van Leeuwen, Utrecht
B. Mahr, Berlin
M. Nivat, Paris
E.-R. Olderog, Oldenburg
B. Rovan, Bratislava
A. Salomaa, Turku
E. Shamir, Jerusalem
U. Vishkin, Maryland and Tel Aviv
D. Wood, London, Ontario

Organizing Committee

Tibor Csendes
János Csirik (Chairman)
Károly Dévényi
Tamás Gaizer
Ferenc Gécseg
Balázs Imreh

List of referees

Alt H.
Amir A.
Andersson A.
Asperti A.
Atallah M.
Attali I.
Baeza-Yates R.
Balcázar J.L.
Baldamus M.
Bartha M.
Beauquier D.
Ben-Or M.
Berg C.
Boadlender H.L.
Boasson L.
Boucheron S.
Busi N.
Campos S.
Carlsson S.
Chazelle B.
Chen J.-S.
Corradini F.
Courcelle B.
Crescenzi P.
Csuhaj-Varjú E.
Dányi G.
Dessmark A.
Devienne P.
Ding C.
Dinic Y.
Dinitz Y.
Dombi J.
Dor D.
Duriš P.
ElMaftouhi A.
Ésik Z.
Farinas del Cerro L.
Felsner S.
Ferrari G.
Focardi R.
Fortnow L.

Gilleron R.
Glas R.
Goldmann M.
Golumbic M.C.
Gruska D.
Guessarian I.
Gyenizse P.
Håstad J.
Hajnal P.
Harju T.
He X.
Heintze N.
Hofbauer D.
Honkala J.
Horváth T.
Horváth S.
Horváth Gy.
Hunyadvári L.
Imreh B.
Italiano G.F.
Jacquemart F.
Jha S.
Jonsson B.
Jonsson H.
Kann V.
Kannan S.
Karger D.
Karhumäki J.
Karlsson R.
Katona E.
Khuller S.
Kirchner C.
Korec I.
Kriznac D.
Lagergren J.
Lamma E.
Laneve C.
Levcopoulos C.
Lingas A.
Linial N.
De Luca A.

Mauri G.
Mello P.
Miller G.
Minea M.
Montanari U.
Mount D.
Muthukrisnan S.
Näslund M.
Nickelsen A.
De Nicola R.
Nicolescu R.
Nicollin X.
Niepel Ľ.
Nilsson B.J.
Older S.
Olejár D.
Păun G.
Pacholski L.
Pagli L.
Palamidessi C.
Panconesi A.
Parnas M.
Parra A.
Peron A.
Petit A.
Pfenning F.
Pietracaprina A.
Pnueli A.
Pollák Gy.
Pooyan-Weihs L.
Prívara I.
Pucci G.
Rangah P.
Rao S.
Rauch-H. M.
Reisig W.
Rensink A.
Renvall A.
Restivo A.
Roccetti M.
Rossi F.

Sahihalp S.C.
Salomaa K.
Sangiorgi D.
Santha M.
Satta G.
Scheffler P.
Schieferdecker I.
Schrettner L.
Schuster A.
Segala R.
Seidl H.
Senizergues G.
Serna M.
Shamir R.
Sifakis J.
Škoviera M.
Sleator D.
Slobodová A.
Sotteau D.
Statman R.
Ştefănescu G.
Steinby M.
Šturc J.
Swanson K.
Swierstra D.
Szíjártó M.
Thomas W.
Thurimella R.
Tishby N.
Toczki J.
Tolksdorf R.
Toran J.
Tsay Y.-K.
Turán Gy.
Tyugy E.
Vágvölgyi S.
Varga L.
Vereshchagin N.K.
Vogler H.
Voisin F.
Vrťo I.

Fraenkel A.	Luccio F.	de Rougemont M.	Walters R.F.C.
Gabarró J.	Manoussakis Y.	Ružička P.	Warhow T.
Gasarch W.	Marche C.	Séébold P.	Wiebrock S.
Gavalda R.	Marchetti-S. A.	Sýkora O.	Zaroliagis C.D.
Giachini L.-A.	Margenstern M.	Sabadini N.	Zlatuška J.
Giancarlo R.	Martínez C.	Sacks E.	Zwick U.
Gibbons J.	Mateescu A.		

Table of Contents

Concurrency I

Theory of 2–Structures
A. Ehrenfeucht, T. Harju, G. Rozenberg (Invited Speaker) 1

A Domain for Concurrent Termination: A Generalization of
Mazurkiewicz Traces
V. Diekert, P. Gastin 15

Nonfinite Axiomatizability of the Equational Theory of Shuffle
Z. Ésik, M. Bertol 27

Automata and Formal Languages I

The Algebraic Equivalent of AFL Theory
W. Kuich 39

Finite State Transformations of Images
K. Culik, J. Kari 51

Post Correspondence Problem: Words Possible as Primitive Solutions
M. Lipponen 63

Computing the Closure of Sets of Words under Partial Commutations
Y. Métivier, G. Richomme, P.-A. Wacrenier 75

Algorithms I

Intervalizing k-Colored Graphs
H. L. Bodlaender, B. de Fluiter 87

NC Algorithms for Finding a Maximal Set of Paths with Application to
Compressing Strings
Z.-Z. Chen 99

On the Construction of Classes of Suffix Trees for Square Matrices:
Algorithms and Applications
R. Giancarlo, R. Grossi 111

How to Use the Minimal Separators of a Graph for its Chordal
Triangulation
A. Parra, P. Scheffler 123

Communication Protocols

Fast Gossiping by Short Messages
J.-C. Bermond, L. Gargano, A. A. Rescigno, U. Vaccaro 135

Break Finite Automata Public Key Cryptosystem
F. Bao, Y. Igarashi 147

Short Memory in Stochastic Graphs: Fully Dynamic Connectivity in
Poly-Log Expected Time
S. Nikoletseas, J. Reif, P. Spirakis, M. Yung 159

On the Number of Random Bits in Totally Private Computation
C. Blundo, A. De Santis, G. Persiano, U. Vaccaro 171

Computational Complexity I

Lower Time Bounds for Randomized Computation
R. Freivalds (Invited Speaker), M. Karpinski 183

New Collapse Consequences of NP Having Small Circuits
J. Köbler, O. Watanabe 196

The Complexity of Searching Succinctly Represented Graphs
J. L. Balcázar 208

Algorithms II

Optimal Shooting: Characterizations and Applications
F. Bauernöppel, E. Kranakis, D. Krizanc, A. Maheshwari, M. Noy,
J.-R. Sack, J. Urrutia 220

Placing Resources in a Tree: Dynamic and Static Algorithms
V. Auletta, D. Parente, G. Persiano 232

Shortest Path Queries in Digraphs of Small Treewidth
S. Chaudhuri, C. D. Zaroliagis 244

A Dynamic Programming Algorithm for Constructing Optimal
Prefix-Free Codes for Unequal Letter Costs
M. J. Golin, G. Rote 256

Algorithms III

Parallel Algorithms with Optimal Speedup for Bounded Treewidth
H. L. Bodlaender, T. Hagerup 268

Approximating Minimum Cuts under Insertions
M. Rauch Henzinger 280

Linear Time Algorithms for Dominating Pairs in Asteroidal
Triple-Free Graphs
D. G. Corneil, S. Olariu, L. Stewart 292

On-line Resource Management with Applications to Routing and
Scheduling
S. Leonardi, A. Marchetti-Spaccamela 303

Automata and Formal Languages II

Alternation in Simple Devices
H. Petersen 315

Hybrid Automata with Finite Bisimulations
T. A. Henzinger 324

Generalized Sturmian Languages
L.-M. Lopez, P. Narbel 336

Polynomial Closure and Unambiguous Product
J.-E. Pin, P. Weil 348

Computational Complexity II

Lower Bounds on Algebraic Random Access Machines
A. M. Ben-Amram, Z. Galil 360

Improved Deterministic PRAM Simulation on the Mesh
A. Pietracaprina, G. Pucci 372

On Optimal Polynomial Time Approximations: P-Levelability
vs. Δ-Levelability
K. Ambos-Spies 384

Computability

Weakly Useful Sequences
S. A. Fenner, J. H. Lutz, E. Mayordomo 393

Graph Connectivity, Monadic NP and Built-in Relations of Moderate
Degree
T. Schwentick 405

The Expressive Power of Clocks
T. A. Henzinger, P. W. Kopke, H. Wong-Toi 417

Automata and Formal Languages III

Grammar Systems: A Grammatical Approach to Distribution
and Cooperation
G. Păun (Invited Speaker) 429

Compactness of Systems of Equations in Semigroups
T. Harju, J. Karhumäki, W. Plandowski 444

Sensing versus Nonsensing Automata
P. Ďuriš, Z. Galil 455

Algorithms IV

New Upper Bounds for Generalized Intersection Searching Problems
P. Bozanis, N. Kitsios, C. Makris, A. Tsakalidis 464

OKFDDs versus OBDDs and OFDDs
B. Becker, R. Drechsler, M. Theobald 475

Bicriteria Network Design Problems
M. V. Marathe, R. Ravi, R. Sundaram, S. S. Ravi, D. J. Rosenkrantz,
H. B. Hunt III 487

On Determining Optimal Strategies in Pursuit Games in the Plane
N.-M. Lê 499

Foundations of Programming

Extension Orderings
A. Rubio 511

The PushDown Method to Optimize Chain Logic Programs
S. Greco, D. Saccà, C. Zaniolo 523

Automatic Synthesis of Real Time Systems
J. H. Andersen, K. J. Kristoffersen, K. G. Larsen, J. Niedermann 535

Self-Correcting for Function Fields of Finite Transcendental Degree
M. Blum, B. Codenotti, P. Gemmell, T. Shahoumian 547

Learning, Coding, Robotics

Measure, Category and Learning Theory
L. Fortnow, R. Freivalds, W. I. Gasarch, M. Kummer, S. A. Kurtz,
C. Smith, F. Stephan 558

A Characterization of the Existence of Energies for Neural Networks
M. Cosnard, E. Goles 570

Variable-Length Codes for Error Correction
H. Jürgensen, S. Konstantinidis 581

Graphbots: Mobility in Discrete Spaces
S. Khuller, E. Rivlin, A. Rosenfeld 593

Semantics

Solving Recursive Net Equations
E. Best (Invited Speaker), M. Koutny 605

Implicit Definability and Infinitary Logic in Finite Model Theory
A. Dawar, L. Hella, P. G. Kolaitis 624

Concurrency II

The Limit of $Split_n$–Language Equivalence
W. Vogler 636

Divergence and Fair Testing
V. Natarajan, R. Cleaveland 648

Causality for Mobile Processes
P. Degano, C. Priami 660

Internal Mobility and Agent-Passing Calculi
D. Sangiorgi 672

Author Index 685

Theory of 2-Structures

A. Ehrenfeucht

Department of Computer Science, University of Colorado at Boulder
Boulder, Co 80309, U.S.A.

T. Harju

Department of Mathematics, University of Turku
FIN-20500 Turku, Finland

G. Rozenberg

Department of Computer Science, Leiden University
P.O.Box 9512, 2300 RA Leiden, The Netherlands

and

Department of Computer Science, University of Colorado at Boulder
Boulder, Co 80309, U.S.A.

1. Introduction

In this paper we review some basic notions and results of the theory of 2-structures as initiated in [8]. For a more detailed introduction to the topic we refer to [6].

A 2-structure $g = (D, R)$ consists of a finite domain D together with an equivalence relation R on the ordered pairs $(x, y) \in D \times D$ with $x \neq y$. Hence a 2-structure can be considered as a complete directed graph with an 'abstract colouring' of the edges.

A colouring of the edges can be made concrete if one adds a labeling function to a 2-structure obtaining in this way a labeled 2-structure. In the litterature the term '2-structure' is used both as a technical term as defined above and as a generic term refering to the theory of labeled and 'unlabeled' 2-structures.

These structures can be used to study various combinatorial properties of systems such as graphs, partially ordered sets and communication networks, see [2], [10] and [17] for some of the intuitions and motivations.

We cover both the 'static' part of 2-structures concerned mainly with the hierarchical representations, and the 'dynamic' part of the theory concerned with the local transformations of 2-structures.

The main results on the static properties of 2-structures are presented in Sections 2 and 3, where we are interested in the *clan decomposition* of these structures. The clan decomposition of 2-structures is closely related to the *modular decomposition* of graphs and directed graphs, see [16].

In the clan decomposition a 2-structure g is divided into substructures, and a quotient 2-structure is constructed to indicate the necessary relationships between the substructures.

The results on the local transformations are presented in Sections 4. Here one assumes that the labels used by a labeled 2-structure form a group. One uses then the group operations to induce transformations of the labels of the edges – the domain, *i.e.*, the set of nodes, does not change in these transformations. In this way a single labeled 2-structure generates a whole set of labeled 2-structures, which is refered to as a *dynamic labeled 2-structure*.

Also here the notion of a clan remains central. As a matter of fact one of the main themes investigated in the theory of dynamic labeled 2-structures is the behaviour of the clan decomposition under the group induced by the local transformations.

2. 2-Structures and Their Clans

2.1. Definition of a 2-structure

Let D be a finite nonempty set, and let $E_2(D) = \{(x,y) \mid x \in D, y \in D, x \neq y\}$ be the set of all (*directed*) *edges* between the elements of D. For an edge $e = (x,y)$, we denote by $e^{-1} = (y,x)$ the *reverse edge* of e.

A *2-structure* $g = (D,R)$ consists of a finite nonempty set D, called its *domain*, and an equivalence relation $R \subseteq E_2(D) \times E_2(D)$ on its *edges*. The domain of g consists of *nodes* and it is denoted by $\mathrm{dom}(g)$. For an edge $e \in E_2(D)$ let $eR = \{e' \mid eRe'\}$ be the equivalence class (or *edge class*) of R containing e.

A 2-structure $g = (D,R)$ can be represented as a complete directed graph with nodes D and directed edges $E_2(D)$ together with a *colouring* $\alpha\colon E_2(D) \to K$ of the edges corresponding to the edge classes such that $e_1 R e_2$ iff $\alpha(e_1) = \alpha(e_2)$. Notice that such a colouring α is by no means unique, since the choice of the colours is *arbitrary* in the representation.

An edge e of g is said to be *symmetric*, if eRe^{-1}, and an edge class a is *symmetric*, if all edges $e \in a$ are symmetric, *i.e.*, $e \in a$ iff $e^{-1} \in a$.

Example 2.1. Let $G = (D,E)$ be an undirected graph (where E is a subset of 2-element subsets $\{x,y\}$ of D with $x \neq y$). The graph G can be *represented* as a 2-structure $g(G) = (D,R)$ with two symmetric edge classes by defining $e_1 R e_2$ for all $e_1, e_2 \in E$ and for all $e_1, e_2 \notin E$. Notice that the representing 2-structure $g(G)$ cannot distinguish between G and its complement graph.

A 2-structure $g = (D,R)$ is *reversible* when for all edges $e_1, e_2 \in E_2(D)$, $e_1 R e_2$ iff $e_1^{-1} R e_2^{-1}$. Equivalently, g is reversible, if for each edge class a there corresponds a unique edge class a^{-1}, the *reverse class* of a, such that $e \in a$ iff $e^{-1} \in a^{-1}$. If $a = a^{-1}$, then the edge class a is symmetric.

We simplify the drawings of reversible 2-structures by (1) omitting the reverse edge classes of the chosen edge classes, (2) drawing a line segment between the nodes x and y, if the edge (x,y) is symmetric, and (3) often omitting one symmetric edge class.

Example 2.2. The 2-structure g in Fig. 1 has two symmetric edge classes c and d. The edges in the edge class d have not been drawn, *e.g.*, the edge (x_1, x_3) and its reverse edge are in the edge class d. The 2-structure g has four nonsymmetric edge classes a, a^{-1}, b and b^{-1} of which we have drawn the classes a and b only.

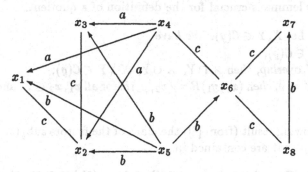

Figure 1: A simplified representation of a reversible 2-structure g

2.2. Clans

Let X be a nonempty subset of the domain D of a 2-structure $g = (D, R)$. Then the *substructure* induced by X is defined as the 2-structure $\text{sub}_g(X) = (X, R \cap (E_2(X) \times E_2(X)))$, which is obtained from g by restricting the relation R to the subset $E_2(X)$ of $E_2(D)$.

We call a subset $X \subseteq \text{dom}(g)$ a *clan*, if

$$(y, x_1)R(y, x_2) \quad \text{and} \quad (x_1, y)R(x_2, y) \quad \text{for all} \quad x_1, x_2 \in X, \ y \notin X.$$

The set of all clans of g is denoted by $\mathcal{C}(g)$. A clan $X \in \mathcal{C}(g)$ is said to be *proper*, if X is a proper subset of the domain of g. It follows immediately that $\emptyset$, $\text{dom}(g)$ and the singletons $\{x\}$ ($x \in \text{dom}(g)$) are clans. These sets are the *trivial clans* of g. A 2-structure g that has only the trivial clans is called *primitive*.

A substructure $\text{sub}_g(X)$ induced by a nonempty $X \in \mathcal{C}(g)$ is a *factor* of g.

Example 2.3. Consider the 2-structure $g = (D, R)$ of Example 2.2, see Fig. 1. The substructures $\text{sub}_g(X)$ and $\text{sub}_g(Y)$ are factors of g for $X = \{x_1, x_2, x_3\}$ and $Y = \{x_7, x_8\}$.

The next result shows that restricting ourselves to reversible 2-structures does not imply a loss of generality as far as the clans are concerned.

Theorem 2.4. *For each $g = (D, R)$ there exists a reversible 2-structure $h = (D, S)$ such that for all $X \subseteq D$, $\mathcal{C}(\text{sub}_g(X)) = \mathcal{C}(\text{sub}_h(X))$. In particular, $\mathcal{C}(g) = \mathcal{C}(h)$.*

Reversible 2-structures are simpler to handle than the general 2-structures. As an example, we have the following one-way condition for clans: A subset $X \subseteq \text{dom}(g)$ of a reversible g is a clan iff for all $y \notin X$ and $x_1, x_2 \in X$, $(y, x_1)R = (y, x_2)R$.

We take the advantage of Theorem 2.4, and *consider only reversible 2-structures in the rest of the paper.*

Let $g = (D, R)$ be a (reversible) 2-structure. We say that two subsets X and Y of the domain D *overlap*, if $X \cap Y \neq \emptyset$, $X \setminus Y \neq \emptyset$ and $Y \setminus X \neq \emptyset$. The next rather obvious lemma is crucial for the definition of a quotient.

Lemma 2.5. *Let $X, Y \in \mathcal{C}(g)$. We have*

(1) $X \cap Y \in \mathcal{C}(g)$;

(2) *if X, Y overlap, then $X \cap Y$, $X \cup Y$, $X \setminus Y \in \mathcal{C}(g)$;*

(3) *if $X \cap Y = \emptyset$, then $(x_1, y_1)R = (x_2, y_2)R$ for all $x_1, x_2 \in X$ and $y_1, y_2 \in Y$.*

By the following result (from [8]) the clans of the factors $\mathrm{sub}_g(X)$ are exactly those clans of g that are contained in X.

Theorem 2.6. *For a factor $h = \mathrm{sub}_g(X)$, $\mathcal{C}(h) = \{Y \mid Y \subseteq X, Y \in \mathcal{C}(g)\}$.*

3. Decompositions of 2-Structures

3.1. Prime clans and quotients

A nonempty clan $P \in \mathcal{C}(g)$ is a *prime clan*, if it does not overlap with any clan of g. We denote by $\mathcal{P}(g)$ the set of all prime clans of g. For a prime clan P $\mathrm{sub}_g(P)$ is called a *prime factor* of g.

Prime clans of factors are characterized in the following result from [8].

Theorem 3.1. *For a factor $h = \mathrm{sub}_g(X)$, $\mathcal{P}(h) = \{P \mid P \in \mathcal{P}(g), P \subseteq X\} \cup \{X\}$.*

Let $g = (D, R)$ be a 2-structure. A partition $\mathcal{X} = \{X_1, X_2, \ldots, X_k\}$ of D into nonempty clans is called a *factorization* of g. The *quotient* of g by a factorization $\mathcal{X}$ is defined as the 2-structure $g/\mathcal{X} = (\mathcal{X}, R_{\mathcal{X}})$, where

$$(X_1, Y_1) R_{\mathcal{X}} (X_2, Y_2) \text{ iff } (x_1, y_1) R (x_2, y_2) \text{ for some } x_i \in X_i, y_i \in Y_i, (X_i, Y_i \in \mathcal{X}).$$

Thus a quotient $g/\mathcal{X}$ is obtained from g by contracting each clan $X \in \mathcal{X}$ into a single node, and then inheriting the edge classes from g. By Lemma 2.5(3), the quotient $g/\mathcal{X}$ is well-defined, *i.e.*, it is independent of the choice of the representatives.

Example 3.2. The 2-structure g from Example 2.2 (see Fig. 1) has a factorization $\mathcal{X} = \{X, Y, \{x_4\}, \{x_5\}, \{x_6\}\}$. The quotient $g/\mathcal{X}$ is now obtained by contracting the factors into nodes as in Fig. 2.

A *decomposition* $(\mathrm{sub}_g(X_1), \ldots, \mathrm{sub}_g(X_k); g/\mathcal{X})$ of a 2-structure g consists of the factors $\mathrm{sub}_g(X_i)$ with respect to a factorization $\mathcal{X} = \{X_1, \ldots, X_k\}$ together with the quotient $g/\mathcal{X}$ that gives the relationships between the factors.

In the general case of 2-structures (that we have been studying now in contrast to the *labeled* 2-structures) we may lose some information in the formation

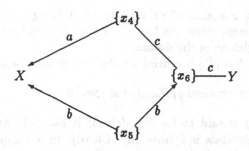

Figure 2: $g/\mathcal{X}$

of a decomposition. Indeed, two different 2-structures may possess a common decomposition, see *f.g.* [6].

Next we shall study the clans of the quotients $g/\mathcal{X}$ for factorizations $\mathcal{X} \subseteq \mathcal{C}(g)$. For this we adopt the following notations. If $\mathcal{A}$ is a family of sets, then let

$$\cup \mathcal{A} = \bigcup_{A \in \mathcal{A}} A, \quad \text{and} \quad \cap \mathcal{A} = \bigcap_{A \in \mathcal{A}} A.$$

Theorem 3.3. *Let* $g = (D, R)$ *and let* $\mathcal{X}$ *be a factorization of* g.
(1) *If* $\mathcal{Z} \in \mathcal{C}(g/\mathcal{X})$, *then* $\cup \mathcal{Z} \in \mathcal{C}(g)$.
(2) *If* $\mathcal{Z} \in \mathcal{P}(g/\mathcal{X})$ *is not a singleton, then* $\cup \mathcal{Z} \in \mathcal{P}(g)$.
(3) *If* $\mathcal{X} \subseteq \mathcal{P}(g)$, *then for each prime clan* $\mathcal{Z} \in \mathcal{P}(g/\mathcal{X})$, *also* $\cup \mathcal{Z} \in \mathcal{P}(g)$.

3.2. The clan decomposition theorem

A prime clan $P \in \mathcal{P}(g)$ is a *maximal prime clan*, if it is maximal with respect to inclusion among the proper prime clans of g. We denote by $\mathcal{P}_{\max}(g)$ the set of maximal prime clans of g. For $P \in \mathcal{P}_{\max}(g)$ the factor $\text{sub}_g(P)$ is called a *maximal prime factor* of g.

We observe that all 2-structures $g = (D, R)$ with $|D| \geq 2$ have maximal prime clans, because every singleton is a prime clan. By definition of maximality, this is not true in the trivial case, where $|D| = 1$. For this reason we set $\mathcal{P}_{\max}(g) = \{D\}$, if $|D| = 1$.

Theorem 3.4. *The maximal prime clans* $\mathcal{P}_{\max}(g)$ *of* g *form a partition of* $\text{dom}(g)$, *and therefore the quotient* $g/\mathcal{P}_{\max}(g)$ *is well-defined.*

Since each singleton is a prime clan, it follows that every $Y \in \mathcal{C}(g)$ is the union of the prime clans $P \in \mathcal{P}(g)$ that are contained in Y. By Theorem 3.1, the prime clans P of the factor $\text{sub}_g(Y)$ are exactly the prime clans of g that are contained in Y. Therefore a nonsingleton clan $Y \in \mathcal{C}(g)$ is the union of the prime clans $P \in \mathcal{P}(g)$ that are the maximal prime clans of the factor $\text{sub}_g(Y)$.

We generalize the notion of primitivity by defining a 2-structure g to be *special*, if all its prime clans are trivial. Hence if g is special, then $\mathcal{P}_{\max}(g)$ consists of the singletons of the domain.

The next result from [8] is crusial for the clan decomposition theorem.

Theorem 3.5. *The quotient $g/\mathcal{P}_{\max}(g)$ is special.*

A 2-structure g is said to be *complete*, if it has only one edge class. In particular, this edge class is symmetric. Clearly, in a complete 2-structure g each subset of the domain is a clan, and therefore a complete 2-structure is special.

A 2-structure g is *linear*, if g has a nonsymmetric edge class a that linearly orders the domain D of g, i.e., D has an ordering $x_1, x_2, \ldots, x_n$ such that $(x_i, x_j) \in a$ if and only if $i < j$. In this case the nonempty clans of g are exactly the segments $\{x_i, x_{i+1}, \ldots, x_{i+k}\}$ for $1 \leq i \leq i + k \leq n$. Consequently, a linear 2-structure is special.

It is also clear that a primitive 2-structure is special. A 2-structure g with at most two nodes is always primitive and at the same time either complete or linear. For this reason we say that a primitive g is *truly primitive*, if it has at least three nodes.

As stated in the following theorem (from [8]) there are only three types of special 2-structures.

Theorem 3.6. *A 2-structure g is special iff it is either linear, or complete, or truly primitive.*

We have thus obtained the following characterization result of [8].

Theorem 3.7 (Clan decomposition theorem). *For each 2-structure g, the quotient $g/\mathcal{P}_{\max}(g)$ is either linear, or complete, or truly primitive.*

For a proof of Theorem 3.7 we refer to [8], [14], [15] or [6].

3.3. The shape of a 2-structure

By the clan decomposition theorem, each g can be decomposed into its maximal prime factors so that the resulting quotient is special. Also, the maximal prime factors (as substructures of g) can be decomposed themselves unless they are already special. Therefore using the clan decomposition theorem iteratively we find a hierarchical representation (the shape) of a 2-structure in a form of a tree. For a more formal definiton we notice first that the partially ordered set $T(g) = (\mathcal{P}(g), \subseteq)$ of prime clans forms a rooted tree, where the domain $\mathrm{dom}(g)$ is the root and the singletons are the leaves. We call the tree $T(g)$ the *prime tree* of g.

The *shape* of a 2-structure g is defined as a pair

$$\mathrm{shape}(g) = (T(g), \Psi_g),$$

where $\Psi_g(P) = \mathrm{sub}_g(P)/\mathcal{P}_{\max}(\mathrm{sub}_g(P))$ is a function, which associates the special quotient to each $P \in \mathcal{P}(g)$.

The shape of a 2-structure g can conveniently be drawn by contracting each prime factor $\mathrm{sub}_g(P)$ to the corresponding quotient $\Psi_g(P)$, and then drawing a line down from a node Q of $\Psi_g(P)$ to the corresponding quotient $\Psi_g(Q)$ of Q.

Example 3.8. Let g be the 2-structure of Example 2.2 (see, Fig. 1). The shape of g is given in Fig. 3. In this figure we have not identified the inner nodes (circles) of the quotients (rectangles). Indeed, each node $Q \in \Psi_g(P)$ is the prime clan consisting of the singletons x_i that are descendants of this node.

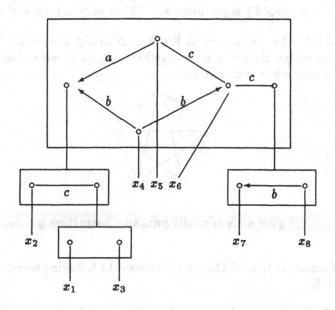

Figure 3: The shape of g

3.4. Hereditary properties of primitivity

Of the three special 2-structures linear and complete are determined by (an ordering of) their domains, and therefore the only interesting special 2-structures are the truly primitive ones.

We shall state some properties of the primitive 2-structures. The first of these is from [9].

Theorem 3.9 (Downward hereditarity). *In each truly primitive* $g = (D, R)$ *there are nodes* x *and* y *(possibly* $x = y$*) such that* $\mathrm{sub}_g(D \setminus \{x, y\})$ *is primitive.*

In particular, in a truly primitive g there is always a small truly primitive substructure:

Lemma 3.10. *Every truly primitive 2-structure g has a primitive substructure consisting of three or four nodes.*

A primitive 2-structure $g = (D, R)$ is *critically primitive*, if g has no primitive substructures $\mathrm{sub}_g(D \setminus \{x\})$ with $x \in D$. The critically primitive 2-structures were characterized in [1] and [17]. In particular, the following result was proved in [17].

Theorem 3.11. *Let $g = (D, R)$ be critically primitive with $|D| = n$. If the substructure $h = \mathrm{sub}_g(X)$ is primitive with $|X| = m \geq 3$, then $n \equiv m \pmod 2$.*

Example 3.12. The 2-structure in Fig. 4 is critically primitive. Notice that g has only two edge classes, a class together with its reverse class. In graph theoretical terms, g is a tournament.

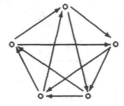

Figure 4: A critically primitive 2-structure g

A proof similar to that of Theorem 3.9 shows the following hereditrary result, see [17] or [15]

Theorem 3.13 (Upward hereditarity). *Let $g = (D, R)$ be primitive, and let $h = \mathrm{sub}_g(X)$ be a proper truly primitive substructure of g. Then there are nodes $x, y \in D \setminus X$ (possibly $x = y$) such that $\mathrm{sub}_g(X \cup \{x, y\})$ is primitive.*

As shown in Lemma 3.10 a truly primitive 2-structure contains a primitive substructure with 3 or 4 nodes. A structure $g = (D, R)$ is called *angular* in [11], if $\mathrm{sub}_g(X)$ is nonprimitive for every subset X with $|X| = 3$. These structures were further studied in [12] (see also [7]), where the notion of *text* is defined using T-structures (angular 2-structures without symmetric edge classes). A text is a structured word $\tau = (\lambda, \rho_w, \rho_s)$, where $\lambda \colon D \to A$ is a function and ρ_w, ρ_s are two linear orders on D. Here $\lambda \rho_w$ can be considered as an ordinary word over the alphabet A, for which ρ_s gives a grammatical structure.

4. Dynamic Labeled 2-Structures

4.1. Labeled 2-structures

As noted in Section 3 the decompositions of 2-structures are not unique in the sense that two different 2-structures may possess the same decomposition. In *labeled* 2-structures this ambiguity does not occur.

Labeled 2-structures as defined in [8] are obtained from 2-structures by setting different labels for the edge class. Equivalently (see *f.g.* [6]) we can define a *labeled 2-structure* or an *ℓ2-structure*, for short, as a function $g: E_2(D) \to \Delta$ to a set Δ of *labels*. Again, we assume that our *ℓ2*-structures are *reversible*, i.e., there is a permutation $\delta: \Delta \to \Delta$ of order two on the labels that satisfies the condition: $g(e^{-1}) = \delta(g(e))$ for all $e \in E_2(D)$. Later the set Δ will be a group, and the reversibility function $\delta: \Delta \to \Delta$ will be a group involution.

For each *ℓ2*-structure $g: E_2(D) \to \Delta$ we define the *underlying* unlabeled 2-structure as $g' = (D, R)$, where $e_1 R e_2$ iff $g(e_1) = g(e_2)$.

We say that $g_1: E_2(D) \to \Delta_1$ and $g_2: E_2(D) \to \Delta_2$ are *strictly isomorphic*, if there exists a bijection $\psi: \Delta_1 \to \Delta_2$ such that for all $e \in E_2(D)$, $g_2(e) = \psi(g_1(e))$.

The next result gives a connection between the unlabeled and the labeled 2-structures.

Theorem 4.1. *Two ℓ2-structures g and h are strictly isomorphic iff they have the same underlying (unlabeled) 2-structure.*

We shall now modify some of the basic definitions and results of the previous sections for labeled 2-structures.

Let $g: E_2(D) \to \Delta$. The *substructure* $\mathrm{sub}_g(X)$ induced by a subset $X \subseteq D$ is the restriction of g onto $E_2(X)$, i.e., $\mathrm{sub}_g(X): E_2(X) \to \Delta$ satisfies $\mathrm{sub}_g(X)(e) = g(e)$ for all $e \in E_2(X)$. Clearly, a substructure $\mathrm{sub}_g(X)$ has the same reversibility function $\delta: \Delta \to \Delta$ as g.

A subset $X \subseteq D$ is a *clan* of $g: E_2(D) \to \Delta$, if for all $x, y \in X$ and all $z \notin X$, $g(z, x) = g(z, y)$ (and hence $g(x, z) = g(y, z)$ by the reversibility condition). Again, let $\mathcal{C}(g)$ denote the family of clans of g. By the following lemma the basic results of Sections 2 and 3 hold for $\mathcal{C}(g)$.

Lemma 4.2. *Let $g: E_2(D) \to \Delta$ be an ℓ2-structure and $g' = (D, R)$ the underlying 2-structure. Then for each $X \subseteq D$, $\mathrm{sub}_{g'}(X)$ is the underlying 2-structure of $\mathrm{sub}_g(X)$, and $\mathcal{C}(g) = \mathcal{C}(g')$.*

We define *quotients* of *ℓ2*-structures as for the unlabeled 2-structures. A quotient $g/\mathcal{X}$ (by a factorization $\mathcal{X}$) is defined by: for all $X, Y \in \mathcal{X}$ (with $X \neq Y$), $(g/\mathcal{X})(X, Y) = g(x, y)$, whenever $x \in X$ and $y \in Y$.

Unlike for an unlabeled 2-structure the decompositions of a labeled 2-structure g are characteristic to g: Two *ℓ2*-structures g_1 and g_2 have a common decomposition iff $g_1 = g_2$.

4.2. Motivation of dynamic labeled 2-structures

In our treatment of dynamic labeled 2-structures we shall follow [13], where these systems were initiated.

Often in a valuation of objects of a mathematical structure the values come from a *structured set*. Frequently such a set is an algebraic structure, where the elements are bound together by operations *f.g.*, a finite automaton may be considered as a directed graph, where the edges have values from a free monoid. In graph theory the values (the labels of edges or nodes) are usually taken from an algebraic structure, *e.g.*, in many optimality problems the edges are labeled by elements of the field $(\mathbb{R}, +, \cdot)$ of real numbers.

We study $\ell2$-structures that have a group structure on their sets of labels. The group structure of the labels gives then a method of locally transforming $\ell2$-structures into each other, and this leads to dynamic $\ell2$-structures.

Since we are mostly interested in *reversible* $\ell2$-structures, a group is a natural candidate for the set labels of edges. The dynamic $\ell2$-structures were also given a more concrete motivation in [13] by evolution of networks and similar processes. In this motivation groups are not only natural but even necessary.

Consider a network of processors D, where each $x \in D$ is connected by a channel to each $y \in D$ for $y \neq x$. A channel (x, y) may assume certain value or state from a set Δ at a specific time. The *concurrent* activities of the processors then modify the states of the channels. The activity of $x \in D$ is described by two sets of actions, the *output activities* O_x and the *input activities* I_x, by which x changes the values of the channels from and to x, respectively. These actions are thus mappings on Δ.

For each channel there are two processors (x and y) that change the value of (x, y) concurrently; x changes it by a mapping from O_x and y by a mapping from I_y. In order to accomodate this concurrency the mappings from O_x and I_y should commute with each other. Also, to avoid unnecessary sequencing of actions the composition of actions from O_x (I_x, resp.) should be an action. Further, to assure a minimal freedom for each x we assume that for each $a, b \in \Delta$ and $x \in D$ there are $\varphi \in O_x$ and $\gamma \in I_x$ such that $\varphi(a) = b$ and $\gamma(a) = b$.

These assumptions simplify the situation in the network considerably. Indeed, as shown in [13] (see, also [3]), if $|D| \geq 3$, then there are two isomorphic groups O and I of permutations on Δ such that for each $x \in D$, $O_x = O$ and $I_x = I$. Thus the actions come from two groups O and I, which are independent of the processors. Moreover, there is a group operation on Δ so that Δ is isomorphic to O and I. In fact, now the groups O and I become defined by left and (involutive) right multiplication of the group Δ.

These observations lead to the definition of a dynamic $\ell2$-structure in terms of *selectors*. A selector σ captures the action of each processor $x \in D$ at a specific stage of the network, and it will be simply a function $\sigma : D \to \Delta$.

A global state of a network is now represented by an $\ell2$-structure with Δ as its set of labels. An evolution is presented as a set of $\ell2$-structures that represent the possible global states of the network. A transition from one global state to another is becomes defined by a selector.

4.3. Basics of dynamic labeled 2-structures

We assume that the set Δ of labels forms a (possibly infinite) *group*. The identity element of Δ is usually denoted by 1_Δ (or by 0, if Δ is an abelian group).

A mapping $\delta : \Delta \to \Delta$ is an *involution* of Δ, if it is a permutation such that $\delta(ab) = \delta(b)\delta(a)$ for all $a, b \in \Delta$ and $\delta^2(a) = a$ for all $a \in \Delta$.

Example 4.3. (1) For any group Δ the most obvious involution is the inverse function: $\delta(a) = a^{-1}$ for each $a \in \Delta$.

(2) If Δ is the group of nonsingular $n \times n$-matrices (over $\mathbb{Q}$, say), then transposition, $\delta(M) = M^T$, is an involution.

As before we will be dealing only with reversible $\ell 2$-structures; now only the involution will determine reversibility. Let δ be an involution of the group Δ. Then $g : E_2(D) \to \Delta$ is a *δ-reversible $\ell 2$-structure* or a *$\delta\ell$-structure*, for short, if δ is its reversibility function: $g(e^{-1}) = \delta(g(e))$ for all edges $e \in E_2(D)$.

We shall now give the definition of a dynamic 2-structure from [13]. Let g be a $\delta\ell$-structure for a group Δ and its involution δ. A function $\sigma : D \to \Delta$ is a *selector*. Define for a selector σ the $\ell 2$-structure g^σ by

$$g^\sigma(x, y) = \sigma(x) \cdot g(x, y) \cdot \delta(\sigma(y))$$

for all $(x, y) \in E_2(D)$. The family $[g] = \{g^\sigma \mid \sigma \text{ a selector}\}$ is a (*single axiom*) *dynamic $\delta\ell$-structure* (*generated by* g).

Example 4.4. Let $\Delta = (\mathbb{Z}_2, +)$ be the cyclic group of order two. In this group the involution δ is necessarily the identity mapping. One can consider a $\delta\ell$-structure g as an undirected graph, where $g(e) = 1$ means that e is an edge of g, and $g(e) = 0$ means that e is not an edge of g. Let $D = \{x_1, x_2, x_3, x_4\}$ and consider a $\delta\ell$-structure g as in Fig. 5, where a line denotes value 1.

Figure 5: g with labels in $\mathbb{Z}_2$

There are now $2^{|D|} = 16$ selectors $\sigma : D \to \Delta$, but some of them have the same image g^σ. In fact, we have only 8 different images g^σ as depicted in Fig. 6, where the nodes are labeled by the values of a selector σ which applied to g yields g^σ.

The *product* of selectors $\sigma_1, \sigma_2 : D \to \Delta$ is defined by $\sigma_2\sigma_1(x) = \sigma_2(x)\sigma_1(x)$ ($x \in D$). In particular, $\sigma_2\sigma_1$ is a selector. It can be shown that the image g^σ of a $\delta\ell$-structure g is also δ-reversible, after which it is immediate that *the selectors form a group*. The inverse σ^{-1} becomes defined by $\sigma^{-1}(x) = \sigma(x)^{-1}$ ($x \in D$).

From these observations we obtain

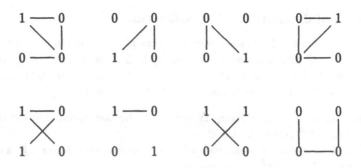

Figure 6: The images g^σ

Lemma 4.5. *Let g be a $\delta\ell$-structure. Then $[g^\sigma] = [g]$ for all $\sigma : D \to \Delta$.*

4.4. Clans of a dynamic $\delta\ell$-structure

Let $[g]$ be a dynamic $\delta\ell$-structure. A clan of each individual $h \in [g]$ is called a *clan* of $[g]$. We shall write

$$C[g] = \bigcup_{h \in [g]} C(h).$$

Example 4.6. Recall that the intersection of two clans $X, Y \in C(g)$ of an $\ell 2$-structure g is always a clan. This does not hold in general for the dynamic $\delta\ell$-structures. Indeed, consider $\Delta = \mathbb{Z}_2$, and the $\ell 2$-structure g from Fig. 7 together with its image g^σ.

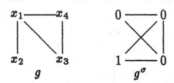

Figure 7: A $\delta\ell$-structure g and its image g^σ.

Here $X = \{x_2, x_3, x_4\}$ is a clan of g, and $Y = \{x_1, x_2, x_3\}$ is a clan of g^σ, and hence both of these are in $C[g]$, but $X \cap Y = \{x_2, x_3\}$ is not in $C[g]$ as can be easily verified.

Theorem 4.7. *The set $C[g]$ of clans of a dynamic $\delta\ell$-structure $[g]$ is closed under complements.*

We let $\overline{X}$ denote the complement $D \setminus X$ of $X \subseteq D$. A clan $X \in C(g)$ is said to be *isolated*, if $g(x, y) = 1_\Delta$ for all $x \in X$ and $y \in \overline{X}$. Hence, if X is an isolated clan, then its complement $\overline{X}$ is also a clan of g.

4.5. Horizons and quotients

By Theorem 4.7, for each clan $X \in C[g]$ there is an $h \in [g]$ such that X and $\overline{X}$ are both clans of h. In particular, this holds for the singleton clans $\{x\}$ of $[g]$, and therefore for each $x \in D$, $D \setminus \{x\} \in C[g]$. We call a node $x \in D$ a *horizon* of $g : E_2(D) \to \Delta$, if $g(x, y) = 1_\Delta$ for all $y \neq x$. Now, for each g and each $x \in \mathrm{dom}(g)$ there is an image g^σ of g such that x is a horizon of g^σ.

The next lemma states that the creation of a clan $X \in C[g]$ in a dynamic $[g]$ depends only on X. A selector $\sigma : D \to \Delta$ is said to be *constant* on a subset $X \subseteq D$, if $\sigma(x_1) = \sigma(x_2)$ for all $x_1, x_2 \in X$.

Lemma 4.8. *Let X be a proper clan of a $\delta\ell$-structure g and let σ be a selector. Then $X \in C(g^\sigma)$ iff σ is constant on X.*

As a corollary, see [13], we obtain that for each clan X of a dynamic $\delta\ell$-structure $[g]$ either X or its complement is a clan of any $h \in [g]$ which has a horizon.

Theorem 4.9. *Let $[g]$ be a dynamic $\delta\ell$-structure, and let $h \in [g]$ have a horizon. Then $C[g] = \{X \mid X \in C(h) \text{ or } D \setminus X \in C(h)\}$.*

We define now the factorizations and quotients of $[g]$. These constructions are not quite so straightforward as for 2-structures, because the clans forming a partition $\mathcal{X} \subseteq C[g]$ of the domain D may come from different images g^σ of g.

A partition $\mathcal{X}$ of the domain D into clans is called a *factorization* of $[g]$. The *quotient* of $[g]$ by $\mathcal{X}$ is the family

$$[g]/\mathcal{X} = \{h/\mathcal{X} \mid h \in [g], \mathcal{X} \text{ a factorization of } h\}.$$

The following theorem was proved in [5].

Theorem 4.10. *If $\mathcal{X}$ is a factorization of a dynamic $[g]$, then there is a selector σ such that $[g]/\mathcal{X} = [g^\sigma/\mathcal{X}]$.*

In general, two given g_1 and g_2 in $[g]$ can have drastically different clan structures. Nevertheless, it is shown in [5] that the ℓ2-structures of a dynamic $[g]$ do have strong structural relationship. This relationship is expressed by the planes of $[g]$, which are defined as follows.

Let $h \in [g]$ have a horizon x. The substructure $\pi_x(g) = \mathrm{sub}_g(D \setminus \{x\})$ is called the *x-plane* of G.

The following result from [5] shows that the x-planes in $[g]$ are quite similar to each other.

Theorem 4.11. *Let $[g]$ be a dynamic $\delta\ell$-structure. The x-planes of $[g]$ are strictly isomorphic.*

Using this result it can be shown that for any two x-planes h_1 and h_2 the special quotients $\Psi_{h_1}(X)$ and $\Psi_{h_2}(X)$ (for prime clans X) are of the same type, *i.e.* they are both primitive or both complete or both linear. Therefore, we can define a tree structure for dynamic $\delta\ell$-structures, which resembles the shape of an ℓ2-structure. For further details we refer to [5].

References

[1] Bonizzoni, P., Primitive 2-structures with the $(n-2)$-property, *Theoret. Comput. Sci.* **132** (1994), 151 – 178.

[2] Buer, H. and R.H. Möhring, A fast algorithm for the decomposition of graphs and posets, *Math. Oper. Res.* **8** (1983), 170 – 184.

[3] Ehrenfeucht, A., T. Harju and G. Rozenberg, Permuting transformation monoids, *Semigroup Forum* **47** (1993), 123 – 125.

[4] Ehrenfeucht, A., T. Harju and G. Rozenberg, Invariants of 2-structures on groups of labels, Manuscript (1994).

[5] Ehrenfeucht, A., T. Harju and G. Rozenberg, Quotients and plane trees of group labeled 2-structures, Leiden University, Department of Computer Science, Technical Report No. 03, 1994.

[6] Ehrenfeucht, A., T. Harju and G. Rozenberg, 2-structures, Manuscript, 1995.

[7] Ehrenfeucht, A., H.J. Hoogeboom, P. ten Pas, and G. Rozenberg, An introduction to context-free text grammars, in *Developments in Language Theory*, G. Rozenberg and A. Salomaa, eds., World Scientific Publishing, 1994.

[8] Ehrenfeucht, A. and G. Rozenberg, Theory of 2-structures, Parts I and II, *Theoret. Comput. Sci.* **70** (1990), 277 – 303 and 305 – 342.

[9] Ehrenfeucht, A. and G. Rozenberg, Primitivity is hereditary for 2-structures *Theoret. Comput. Sci.* **70** (1990), 343 – 358.

[10] Ehrenfeucht, A. and G. Rozenberg, Partial (set) 2-structures; part II: State spaces of concurrent systems, *Acta Informatica* **27** (1990), 343 – 368.

[11] Ehrenfeucht, A. and G. Rozenberg, Angular 2-structures, *Theoret. Comput. Sci.* **92** (1992), 227 – 248.

[12] Ehrenfeucht, A. and G. Rozenberg, T-structures, T-functions, and texts, *Theoret. Comput. Sci.* **116** (1993), 227 – 290.

[13] Ehrenfeucht, A. and G. Rozenberg, Dynamic labeled 2-structures, *Mathematical Structures in Computer Science*, to appear.

[14] Engelfriet, J., T. Harju, A. Proskurowski and G. Rozenberg, Characterization and Complexity of Uniformly Non-Primitive Labeled 2-Structures, *Theoret. Comput. Sci.*, to appear.

[15] Harju, T. and G. Rozenberg, Decomposition of infinite labeled 2-structures, *Lecture Notes in Computer Science* **812** (1994), 145 – 158.

[16] Muller J.H. and J. Spinrad, Incremental Modular Decomposition, *J. of the ACM* **36** (1989), 1 – 19.

[17] Schmerl, J. H. and W. T. Trotter, Critically indecomposable partially ordered sets, graphs, tournaments and other binary relational structures, *Discrete Math.* **113** (1993), 191 – 205.

A Domain for Concurrent Termination
A Generalization of Mazurkiewicz Traces

(Extended Abstract)[*]

Volker Diekert[1] and Paul Gastin[2]

[1] Universität Stuttgart, Institut für Informatik
Breitwiesenstr. 20-22, D-70565 Stuttgart
[2] Institut Blaise Pascal, L.I.T.P.
4, place Jussieu, F-75252 Paris Cedex 05

Abstract. This paper generalizes the concept of Mazurkiewicz traces to a description of a concurrent process, where a known prefix is given as a trace in a first component and a second alphabetic component yields some information about future actions. This allows to define a good semantic domain where the concatenation is continuous with respect to the Scott- and to the Lawson topology. For this we define the notion of $\alpha-$ and of δ-trace. We show various mathematical results proving thereby the soundness of our approach. Our theory is a proper generalization of the theory of finite and infinite words (with explicit termination) and of the theory of finite and infinite (real and complex) traces. We make use of trace theory, domain theory, and topology.

1 Introduction

The theory of Mazurkiewicz traces has been recognized as an important tool for investigating concurrent systems, see [3] for the state of the art. The underlying idea is that for a given alphabet Σ of actions (or events) we fix an irreflexive and symmetric independence relation $I \subseteq \Sigma \times \Sigma$. The semantics is that independent actions can be executed concurrently; and finite processes are described by elements of the quotient monoid $\mathbb{M}(\Sigma, D) = \Sigma^*/\{ab = ba \mid (a, b) \in I\}$. The complement $D = \Sigma \times \Sigma \setminus I$ is called the *dependence relation*. The elements of $\mathbb{M}(\Sigma, D)$ are called *traces*. The concatenation of traces combines by a single operation sequential and parallel composition. If $D = \Sigma \times \Sigma$ is full, then the concatenation is the sequential composition of words. If s, t are traces such that the actions of s and t are pairwise independent, then st can be viewed as the parallel composition $s \| t$.

Another important situation is given when the dependency D is transitive. Then we have $\mathbb{M}(\Sigma, D) = \Sigma_1^* \times \cdots \times \Sigma_k^*$ for some partition $\Sigma = \Sigma_1 \mathbin{\dot\cup} \cdots \mathbin{\dot\cup} \Sigma_k$. In this case we describe a system with k independent components, where each component behaves sequentially. Let us continue with this example. A process p

---

[*] This research has been supported by the ESPRIT Basic Research Actions No. 6317 ASMICS II and No. 6067 CALIBAN.

is then described as a tuple of words $p = (v_1, \ldots, v_k)$. Assume that p' is a prefix of p describing the part of p which is accomplished, and let $q = (w_1, \ldots, w_k)$ be a second process, which should be executed after p. Assume that $p' = (v'_1, \ldots, v'_k)$ has some components which are terminated, $v'_i = v_i$, and some components, where $v'_i < v_i$ is a proper prefix. Following a usual convention, we use the symbol $\perp$ for non-termination, we may assume $p' = (v'_1 \perp, \ldots, v'_i \perp, v_{i+1}, \ldots, v_k)$. We want the following law:

$$(v'_1 \perp, \ldots, v'_i \perp, v_{i+1}, \ldots, v_k)(w_1, \ldots, w_k) = (v'_1 \perp, \ldots, v'_i \perp, v_{i+1} w_{i+1}, \ldots, v_k w_k)$$

This law expresses the fact that if we know termination on a subset of components $K \subseteq \{1, \ldots, k\}$ (e.g., as above $K = \{i+1, \ldots, k\}$), then the execution of $q = (w_1, \ldots, w_k)$ can be started on these components without waiting for global termination. On the other components we have no further information than $v'_i \perp$.

This basic idea generalizes perfectly as follows in the frame of Mazurkiewicz traces. We describe a process p as a pair (t, A) where $t \in \mathbb{IM}(\Sigma, D)$ is trace and $A \subseteq \Sigma$ is a subset of actions which may still be performed before termination. If $A = \emptyset$ is the empty set then this explicitly means global termination. One should note the difference with failure (or ready) semantics in process algebra like CCS or CSP [8, 11], where usually the behavior in the next step is specified. Here the finite alphabetic information A is given for the whole future. It should be noted that defining composition laws using dependencies from the alphabetic information about the future has been done in process algebra too. Let us refer to [12, 14].

If we are uncertain about the future, then we have possibly to allow everything. Thus in our formalism (t, Σ) is always an approximation of (t, A) for all $A \subseteq \Sigma$. More precisely, if $A \subseteq B$ then (t, A) imposes more restrictions on the process than (t, B) does. Hence for $A \subseteq B$, we view (t, A) as a better approximation than (t, B) and therefore $(t, B) \sqsubseteq (t, A)$. In the following we will call these pairs α-traces. In many cases, it is convenient to abstract from the second component to the set $D(A) = \{b \in \Sigma \mid (a, b) \in D\}$, since, as we will see, this is the only information we need for concatenation. This leads to the notion of δ-trace, which identifies (t, A) with (t, A'), if $(t, D(A)) = (t, D(A'))$. Although our motivation stems from the description of finite processes, we are able to cope with infinite behaviors, too. In fact, it is convenient to deal with finite and infinite behaviors simultaneously since, if the process runs on different independent components, then it is very natural that on some components there is termination whereas on other components it runs infinitely. The precise mathematical formalism is given below. We will develop a theory which results in an algebraic cpo (Scott domain), where the concatenation is continuous. In particular, it is monotone with respect to both components. Moreover, we will define a natural metric on the space of α-traces (δ-traces resp.) such that the induced topological space becomes compact (by the Lawson topology [10]). The concatenation is in fact uniformly continuous in this topology. Our theory generalizes the theory of finite and infinite words (with explicit termination) and the theory of infinite (real and complex) traces. Some of the following ideas were proposed in the invited lecture [1]. For lack of space, some proofs had to be omitted. The full version of this paper will be published elsewhere.

2 Preliminaries from poset theory

Let $(Z, \leq)$ be a poset, i.e., a partially ordered set. A subset $Y \subseteq Z$ is *coherent* if for all $x, y \in Y$, there exists $z \in Z$ such that $x \leq z$ and $y \leq z$. A subset $Y \subseteq Z$ is *directed* if it is non-empty and for all $x, y \in Y$, there exists $z \in Y$ such that $x \leq z$ and $y \leq z$. A subset $Y \subseteq Z$ is *bounded* if there exists $z \in Z$ such that $x \leq z$ for all $x \in Y$. The poset $(Z, \leq)$ is a *coherently complete partial order* (a *ccpo* for short) if every coherent subset of Z has a least upper bound in Z. Similarly, the poset $(Z, \leq)$ is a *directed complete partial order* (a *dcpo* for short) if every directed subset of Z has a least upper bound in Z. Finally, the poset $(Z, \leq)$ is a *bounded complete partial order* (a *bcpo* for short) if every bounded subset of Z has a least upper bound in Z. Note that a ccpo is indeed both a dcpo and a bcpo. Whenever they exist, the least upper bound and the greatest lower bound of a subset $Y \subseteq Z$ will be denoted by $\bigsqcup Y$ and $\sqcap Y$ respectively.

Let $(Z, \leq)$ be a ccpo. An element $x \in Z$ is *prime* if for all *coherent* subset $Y \subseteq Z$, $x \leq \bigsqcup Y$ implies $x \leq y$ for some $y \in Y$. For all $z \in Z$, we denote by $P(z) = \{x \in Z \mid x \leq z \text{ and } x \text{ is prime}\}$ the set of prime elements of Z less than or equal to z. A ccpo $(Z, \leq)$ is *prime algebraic* if $z = \bigsqcup P(z)$ for all $z \in Z$.

Similarly, let $(Z, \leq)$ be a dcpo. An element $x \in Z$ is *compact* if for all *directed* subsets $Y \subseteq Z$, $x \leq \bigsqcup Y$ implies $x \leq y$ for some $y \in Y$. For $z \in Z$, we denote by $K(z) = \{x \in Z \mid x \leq z \text{ and } x \text{ is compact}\}$ the set of compact elements less than or equal to z. A dcpo $(Z, \leq)$ is *algebraic* if $K(z)$ is directed and $z = \bigsqcup K(z)$ for all $z \in Z$. Note that, if any pair of elements of Z which is bounded above admits a least upper bound (in particular if $(Z, \leq)$ is a ccpo or a bcpo) then $K(z)$ is directed for all $z \in Z$. Finally, an algebraic dcpo is called a *(Scott) domain*. For a poset $(Z, \leq)$ and $z \in Z$ we use the following standard notations:

$$\downarrow z = \{y \in Z \mid y \leq z\} \text{ and } \uparrow z = \{y \in Z \mid y \geq z\}.$$

Let $(X, \leq)$ and $(Y, \leq)$ be two posets. A mapping $f : X \longrightarrow Y$ is *monotone* if for all $x \in X$ and $y \in Y$, $x \leq y$ implies $f(x) \leq f(y)$. Let $(X, \leq)$ and $(Y, \leq)$ be two dcpos. A monotone mapping $f : X \longrightarrow Y$ is *continuous* if $f(\bigsqcup Z) = \bigsqcup f(Z)$ for all directed sets $Z \subseteq X$.

3 Preliminaries from trace theory

We assume that the reader is familiar with the basic concepts of trace theory. Let Σ be a finite alphabet with a symmetric and irreflexive independence relation $I \subseteq \Sigma \times \Sigma$. The complement $D = \Sigma \times \Sigma \setminus I$ is called the dependence relation, and (Σ, D) is called a dependence alphabet. A dependence alphabet is represented as an undirected graph where self-loops are omitted. A dependence graph (over (Σ, D)) is (an isomorphism class of) a directed acyclic node-labelled graph $[V, E, \lambda]$ where V is a countable set of vertices, $\lambda : V \longrightarrow \Sigma$ is the node-labelling, and where the edge relation $E \subseteq V \times V$ satisfies the following restriction: $E \cup E^{-1} \cup \mathrm{id}_V = \lambda^{-1}(D)$, i.e., edges are between dependent vertices, only. Furthermore, for technical reasons we demand that the partial order, which is induced by the reflexive and transitive closure E^*, is well-founded. The set of dependence graphs $\mathbb{G}(\Sigma, D)$ is a monoid with the following concatenation $[V_1, E_1, \lambda_1] \cdot [V_2, E_2, \lambda_2] = [V, E, \lambda]$ where $V = V_1 \:\dot\cup\: V_2$, $\lambda = \lambda_1 \:\dot\cup\: \lambda_2$ and $E = E_1 \:\dot\cup\: E_2 \:\dot\cup\: (V_1 \times V_2 \cap \lambda^{-1}(D))$. This means that we take the disjoint union and introduce new arcs from V_1 to V_2 whenever vertices have dependent labels. The neutral element is the empty graph, denoted by $1 = [\emptyset, \emptyset, \emptyset]$.

For $x \in V$ we denote by $\downarrow x$ the dependence graph which is induced by all elements below x, i.e., $\downarrow x$ is the restriction of $[V, E, \lambda]$ to the set $\{x' \in V \mid (x', x) \in E^*\}$. If the set V is finite, then $[V, E, \lambda]$ has a canonical identification with a finite trace of $\mathbb{M}(\Sigma, D) = \Sigma^* / \{ab = ba \mid (a, b) \in I\}$. If all vertices of V have a finite past, i.e., if $\downarrow x$ is finite for all $x \in V$, then $[V, E, \lambda]$ is called a *real trace*. The set of real traces is denoted by $\mathbb{R}(\Sigma, D)$. Note that $\mathbb{R}(\Sigma, D)$ is not a submonoid of $\mathbb{G}(\Sigma, D)$, e.g. the concatenation of a^ω with itself yields a dependence graph with a non-empty transfinite part, i.e., some vertices do not have a finite past. See [5] for a detailed presentation about the theory of infinite traces.

For dependence graphs we have a natural notion of prefix (here the assumption of well-foundedness is used). For $g, h \in \mathbb{G}(\Sigma, D)$ we write $g \leq h$ if $gf = h$ for some dependence graph f. More generally, we denote this *prefix relation* by $\leq$ for any monoid. Note that $\leq$ is not antisymmetric, in general. The subset $\mathbb{R}(\Sigma, D)$ with the induced prefix ordering forms a prime algebraic coherently complete domain, [4]. The same is *almost* true for $\mathbb{G}(\Sigma, D)$, [6]. We do not go into details what "almost" means, but the basic reason that $\mathbb{G}(\Sigma, D)$ can not form a complete partial order results from set theoretical considerations: The least upper bound of all countable ordinals is not countable anymore. For $g = [V, E, \lambda] \in \mathbb{G}(\Sigma, D)$, we let the *real part* $\mathrm{Re}(g)$ of g be the restriction of g to the set $\{x \in V \mid \downarrow x \text{ is finite}\}$. The real part $\mathrm{Re}(g) \in \mathbb{R}(\Sigma, D)$ is the maximal real prefix of g. We also define the *alphabet* and *alphabet at infinity* for $g = [V, E, \lambda] \in \mathbb{G}(\Sigma, D)$ by:

$$\mathrm{alph}(g) = \lambda(V)$$
$$\mathrm{alphinf}(g) = \lambda(\{x \in V \mid \downarrow x \text{ is infinite or } \lambda^{-1}\lambda(x) \text{ is infinite}\})$$

For $A \subseteq \Sigma$ and $g = [V, E, \lambda] \in \mathbb{G}(\Sigma, D)$, the maximal real prefix of g which is independent of A, denoted by $\mu_A(g)$, is the restriction of g to the set $\{x \in V \mid \downarrow x \text{ is finite and } \mathrm{alph}(\downarrow x) \times A \subseteq I\}$. Note that $\mu_A(g) \in \mathbb{R}(\Sigma, D)$ is a real trace with $\mathrm{alph}(\mu_A(g)) \cap D(A) = \emptyset$ and that $\mu_\emptyset(g) = \mathrm{Re}(g)$.

Since $\mu_A(g)$ is a prefix of g and $\mathbb{G}(\Sigma, D)$ is left-cancellative, we can define the dependence graph $\mu_A(g)^{-1}g$ by the equation $\mu_A(g)(\mu_A(g)^{-1}g) = g$. In the following we are only interested in the alphabet of this graph. As a matter of fact, it is more convenient to include the set $\mathrm{alphinf}(g)$ in this alphabet. Thus, we define

$$\sigma_A(g) = \mathrm{alph}(\mu_A(g)^{-1}g) \cup \mathrm{alphinf}(g)$$

4 The calculus of α- and δ-traces

Definition 1. An α-trace (δ-trace resp.) is a pair $x = (r, A)$ (a pair $x = (r, D(A))$ resp.) where $r \in \mathbb{R}(\Sigma, D)$ is a real trace, and $A \subseteq \Sigma$ is a subset of Σ such that $\mathrm{alphinf}(r) \subseteq A$. The trace r is also denoted by $\mathrm{Re}(x)$. The set A (the set $D(A)$ resp.) is denoted by $\mathrm{Im}(x)$ and it is called the *imaginary part* of x. The set of α-traces is denoted by $\mathbb{F}_\alpha(\Sigma, D)$, the set of δ-traces is denoted by $\mathbb{F}_\delta(\Sigma, D)$. Both sets are monoids with $(1, \emptyset)$ as neutral element and the following concatenation

$$(r, A) \cdot (s, B) = (r\mu_A(s), \sigma_A(s) \cup A \cup B) \qquad \text{for } \alpha\text{-traces}$$
$$(r, D(A)) \cdot (s, D(B)) = (r\mu_A(s), D(\sigma_A(s) \cup A \cup B)) \quad \text{for } \delta\text{-traces}$$

Note that the concatenation of δ-traces is well-defined since $\mu_A(g) = \mu_{A'}(g)$ and $\sigma_A(g) = \sigma_{A'}(g)$ for all $g \in G$ and $A, A' \subseteq \Sigma$ such that $D(A) = D(A')$.

Example 1. Let $(\Sigma, D) = a$ — b — c. A first try to define a concatenation of real traces might ask for the following laws: $a^\omega b^\omega = a^\omega$, $b^\omega c^\omega = b^\omega$, and $a^\omega c^\omega = (ac)^\omega$. However, then we have:

$$(a^\omega b^\omega)c^\omega = a^\omega c^\omega = (ac)^\omega \neq a^\omega = a^\omega b^\omega = a^\omega(b^\omega c^\omega) \ .$$

Since for obvious reasons (a and c are independent) we want to keep the law that $a^\omega c^\omega \neq a^\omega$, we have to change the law for $a^\omega b^\omega$ (and, of course, for $b^\omega c^\omega$) in order to have an associative operation. This is what happens in our calculus: The interpretation as an α-trace yields $x^\omega = (x^\omega, \{x\})$ for $x \in \Sigma$. Hence $a^\omega b^\omega = (a^\omega, \{a\})(b^\omega, \{b\}) = (a^\omega, \{a, b\}) \neq a^\omega$. Here are some more lines of the associative calculus of α-traces:

$$(acbac, \{c\})(aaccb, \emptyset) = (acbaaac, \{b, c\})$$
$$(acbac, \Sigma)(aaccb, \emptyset) = (acbac, \Sigma)$$
$$(acbac, \emptyset)(aaccb, \{a\}) = (acbaaacccb, \{a\})$$
$$(a^\omega, \{a\})(b^\omega, \{b\})(c^\omega, \{c\}) = (a^\omega, \Sigma)$$
$$(a^\omega, \{a\})(c^\omega, \{c\})(b^\omega, \{b\}) = ((ac)^\omega, \Sigma)$$
$$(a^\omega, \{a\})((ac)^\omega, \{a, c\}) = (a^\omega, \{a\})(c^\omega, \{c\})$$
$$= ((ac)^\omega, \{a, c\})$$

We can think of an α- or δ-trace as an approximation of a process. The first component is a real trace which describes a prefix of the actual process; the second part is an alphabetic information about possible future actions in the α-case or about possible future dependencies (which is sufficient in many cases) in the δ-case. This interpretation leads to the following definition of a natural approximation ordering.

Definition 2. For α-traces (r, A), (s, B) we define $(r, A) \sqsubseteq (s, B)$, if $r \leq s$ and $\mathrm{alph}(r^{-1}s) \cup B \subseteq A$. Similarly, for δ-traces $(r, D(A))$, $(s, D(B))$ we define $(r, D(A)) \sqsubseteq (s, D(B))$, if $r \leq s$ and $D(\mathrm{alph}(r^{-1}s) \cup B) \subseteq D(A)$.

Remark.

i)   The natural mapping $\mathbb{F}_\alpha(\Sigma, D) \longrightarrow \mathbb{F}_\delta(\Sigma, D)$, $(r, A) \longmapsto (r, D(A))$ is a surjective and monotone homomorphism.

ii)   The prefix relation $\leq$ of $\mathbb{F}_\alpha$ ($\mathbb{F}_\delta$ resp.) is antisymmetric, whence a partial order. Moreover, the prefix ordering $\leq$ and the approximation ordering $\sqsubseteq$ are disjoint in the following sense: If $(r, A) \leq (s, B)$ and $(r, A) \sqsubseteq (s, B)$ then we have $(r, A) = (s, B)$.

iii)   If $D = \Sigma \times \Sigma$ is the full dependence relation then the consideration of $\mathbb{F}_\delta$ is equivalent to the formalism of Σ^∞ together with the non-termination symbol $\bot$. In fact, for $D(A)$ we only have two possibilities: either $D(A) = \Sigma$, which means non-termination, or $D(A) = \emptyset$, which means explicit termination. A δ-trace in this special case either is a finite non-terminated word $v\bot$, with $v \in \Sigma^*$, or it is a finite terminated word $v \in \Sigma^*$, or it is an infinite word $v \in \Sigma^\omega$ (where the non-termination symbol is redundant and therefore it may be omitted). The calculus is as usual such that $v\bot$ or infinite words are right-absorbent. Furthermore, we have $u\bot \sqsubseteq v\bot \sqsubseteq v$ if and only if $u \leq v$, $u, v \in \Sigma^\infty$.

iv)   The bottom element with respect to $\sqsubseteq$ is the pair $(1, \Sigma)$. It signifies that nothing has happened and everything is possible. Note that $(1, \Sigma)$ is right-absorbent with respect to the concatenation and we have $(r, A) \cdot (1, \Sigma) = (r, \Sigma)$.

We have a sequence of canonical mappings

$$\mathbb{M}(\Sigma, D) \hookrightarrow \mathbb{R}(\Sigma, D) \hookrightarrow \mathbb{G}(\Sigma, D) \xrightarrow{\varphi} \mathbb{F}_\alpha(\Sigma, D) \twoheadrightarrow \mathbb{F}_\delta(\Sigma, D)$$

The first two mappings are inclusions, the mapping $\varphi : \mathbb{G}(\Sigma, D) \to \mathbb{F}_\alpha(\Sigma, D)$, $g \mapsto (\mathrm{Re}(g), \mathrm{alphinf}(g))$ is a homomorphism, but it is not surjective (for $\Sigma \neq \emptyset$), and $\mathbb{F}_\alpha(\Sigma, D) \to \mathbb{F}_\delta(\Sigma, D)$ is the surjective homomorphism defined above. The image of $\mathbb{G}(\Sigma, D)$ in $\mathbb{F}_\alpha(\Sigma, D)$ is in fact the monoid of α-complex traces $\mathbb{C}_\alpha(\Sigma, D)$, the image of $\mathbb{G}(\Sigma, D)$ in $\mathbb{F}_\delta(\Sigma, D)$ is the monoid of complex traces $\mathbb{C}(\Sigma, D)$; both monoids are defined in [2].

The mapping $\mathbb{R}(\Sigma, D) \to \mathbb{F}_\delta(\Sigma, D)$, $r \mapsto (r, D(\mathrm{alphinf}(r)))$ is injective, hence $\mathbb{R}(\Sigma, D) \subseteq \mathbb{C}(\Sigma, D) \subseteq \mathbb{F}_\delta(\Sigma, D)$. Similarly, the mapping $\mathbb{R}(\Sigma, D) \to \mathbb{F}_\alpha(\Sigma, D)$, $r \mapsto (r, \mathrm{alphinf}(r))$ is injective, hence $\mathbb{R}(\Sigma, D) \subseteq \mathbb{C}_\alpha(\Sigma, D) \subseteq \mathbb{F}_\alpha(\Sigma, D)$.

In the following we focus our attention to the set of α-traces $\mathbb{F}_\alpha(\Sigma, D)$. Analogous results can be obtained for the slightly more abstract model of δ-traces. This is left to the reader.

The homomorphism $\mathbb{M}(\Sigma, D) \hookrightarrow \mathbb{F}_\alpha(\Sigma, D)$, $t \mapsto (t, \emptyset)$ is monotone with respect to the prefix ordering $\leq$. It maps a finite trace t to a finite terminated process $(t, \emptyset)$. If $s \neq t$ then $(s, \emptyset)$ and $(t, \emptyset)$ are incomparable with respect to the approximation ordering $\sqsubseteq$. This is a special case of Rem. 4, ii).

The prefix ordering of the monoid $\mathbb{F}_\alpha(\Sigma, D)$ has the following nice property.

Proposition 3. *The set* $(\mathbb{F}_\alpha(\Sigma, D), \leq)$ *is a coherently complete partial order. If* $X \subseteq \mathbb{F}_\alpha(\Sigma, D)$ *is* $\leq$*-coherent, then we have* $\bigsqcup_\leq X = (r, A)$ *where* $r = \bigsqcup_\leq \mathrm{Re}(X)$ *and* $A = \mathrm{alphinf}(r) \cup \bigcup_{x \in X} \mathrm{Im}(x)$.

The proposition above allows in a convenient way to define an ω-product of α-traces $x_1 x_2 \cdots$ for $x_i \in \mathbb{F}_\alpha(\Sigma, D)$, $i \geq 1$. The ω-product is defined as the least upper bound $\bigsqcup_{i<\omega}(x_1 \ldots x_i)$. The generalization to infinite products over any ordinal is straightforward.

The main drawback of the prefix ordering is that it is not monotone with respect to the concatenation. If $x' \leq x$ and $y' \leq y$ then we cannot infer $x'y' \leq xy$. Overcoming this difficulty is a basic motivation to study $\mathbb{F}_\alpha(\Sigma, D)$ with the approximation ordering $\sqsubseteq$.

5 Poset properties of $\mathbb{F}_\alpha(\Sigma, D)$ with the approximation ordering

Proposition 4. *The concatenation of α-traces is monotone for the approximation order: Let $x, x', y, y' \in \mathbb{F}_\alpha(\Sigma, D)$ be α-traces. If $x' \sqsubseteq x$ and $y' \sqsubseteq y$ then $x' \cdot y' \sqsubseteq x \cdot y$.*

Proof. Let $x' = (r', A')$, $x = (r, A)$, $y' = (s', B')$ and $y = (s, B)$ such that $r' \leq r$, $A' \supseteq \text{alph}(r'^{-1}r) \cup A$ and $s' \leq s$, $B' \supseteq \text{alph}(s'^{-1}s) \cup B$. We have $x'y' = (r'\mu_{A'}(s'), \sigma_{A'}(s') \cup A' \cup B')$. Note that $\mu_{A'}(s') \leq \mu_A(s') \leq \mu_A(s)$. Furthermore, since $A' \supseteq \text{alph}(r'^{-1}r)$, we see that $r'\mu_{A'}(s') \leq r\mu_A(s)$. Therefore it remains to show that $\sigma_{A'}(s') \cup A' \cup B' \supseteq \sigma_A(s) \cup A \cup B \cup \text{alph}((r'\mu_{A'}(s'))^{-1}(r\mu_A(s)))$. Note that $(r'\mu_{A'}(s'))^{-1}(r\mu_A(s))(\mu_A(s)^{-1}s) = (r'^{-1}r)(\mu_{A'}(s')^{-1}s')(s'^{-1}s)$. Since $A' \supseteq A \cup \text{alph}(r'^{-1}r)$ and $B' \supseteq B \cup \text{alph}(s'^{-1}s)$, the result follows from $\sigma_{A'}(s') \supseteq \sigma_A(s)$.

The following generalization remains to the reader.

Proposition 5. *Let $(x_i)_{i \geq 1}$, $(y_i)_{i \geq 1}$ be two infinite sequences of α-traces such that $x_i \sqsubseteq y_i$ for all $i \geq 1$. Then we have for the infinite ω-product*

$$x_1 x_2 \ldots \sqsubseteq y_1 y_2 \ldots.$$

The following investigation shows that $(\mathbb{F}_\alpha(\Sigma, D), \sqsubseteq)$ is a good semantic domain with many desirable completeness properties. Moreover, based on Prop. 4, we will see that the concatenation behaves well with respect to these properties.

Theorem 6. *The poset $(\mathbb{F}_\alpha(\Sigma, D), \sqsubseteq)$ is a coherently complete partial order. If $X \subseteq \mathbb{F}_\alpha(\Sigma, D)$ is a coherent set, then we have*

$$\bigsqcup X = (\bigsqcup_{\leq} \text{Re}(X), \bigcap_{x \in X} \text{Im}(x)).$$

Proof. Clearly $\text{Re}(X)$ is a coherent subset of $(\mathbb{R}(\Sigma, D), \leq)$, and $r = \bigsqcup_{\leq} \text{Re}(X) \in \mathbb{R}(\Sigma, D)$ exists. The next observation is that if $x \in X$ then $\text{Im}(x) \supseteq \text{alphinf}(r)$. This follows from the fact that $\text{Im}(y) \supseteq \text{alphinf}(\text{Re}(y))$ for all α-traces y and that X is coherent. Once we have this, we see that $(r, \bigcap_{x \in X} \text{Im}(x))$ is an α-trace. An easy reflection then shows the relation $x \sqsubseteq (r, \bigcap_{x \in X} \text{Im}(x))$ for all $x \in X$. Finally let $x \sqsubseteq (s, B)$ for all $x \in X$. Then $r \leq s$ and $\text{Im}(x) \supseteq \text{alph}(r^{-1}s) \cup B$ for all $x \in X$. Hence we get the result.

Corollary 7. *Let* $Y \subseteq \mathbb{F}_\alpha(\Sigma, D)$ *be any non-empty set of α-traces. Then the greatest lower bound* $\sqcap Y$ *exists and we have* $\sqcap Y = (r, A)$ *where*

$$r = \sqcap \operatorname{Re}(Y) \quad and \quad A = \bigcup_{y \in Y} \left(\operatorname{Im}(y) \cup \operatorname{alph}(r^{-1} \operatorname{Re}(y)) \right)$$

Theorem 8. *An α-trace* $(t, A) \in \mathbb{F}_\alpha(\Sigma, D)$ *is compact if and only if t is a finite trace. Moreover, the poset* $(\mathbb{F}_\alpha(\Sigma, D), \sqsubseteq)$ *is an algebraic coherently complete partial order, in particular it is a domain.*

Proof. Let $x = (r, A)$ be any α-trace. It is easy to see that the set of finite α-traces

$$\{ (p, \operatorname{alph}(p^{-1}r) \cup A) \mid p \leq r \text{ and } p \text{ is finite} \}$$

is directed. Moreover for some $p \leq r$ which is large enough we have $\operatorname{alph}(p^{-1}r) = \operatorname{alphinf}(r)$. Since $\operatorname{alphinf}(r) \subseteq A$ by definition, we see from the formula in Thm. 6 and [4, Prop. 3.3] that x is the least upper bound of this directed set of finite α-traces. Hence compact elements must be finite. To see that finite α-traces are compact we consider any directed set $Y \subseteq \mathbb{F}_\alpha(\Sigma, D)$ and $x = (t, A) \sqsubseteq \sqcup Y$ such that t is finite. Since $\operatorname{Re}(Y)$ is a directed set of $(\mathbb{R}(\Sigma, D), \leq)$ and t is finite, we have $t \leq \operatorname{Re}(y)$ for some $y \in Y$. Since Y is directed we may in fact assume that $t \leq \operatorname{Re}(y)$ for all $y \in Y$. Next we replace Y by the set $Y' = \{(t, \operatorname{alph}(t^{-1} \operatorname{Re}(y)) \cup \operatorname{Im}(y)) \mid y \in Y\}$. The new set Y' is still directed and we have $x \sqsubseteq \sqcup Y'$. Since for all $y' \in Y'$ there exists some $y \in Y$ with $y' \sqsubseteq y$, it is enough to find some $y' \in Y'$ such that $x \sqsubseteq y'$. Now, the set Y' is finite, hence $\sqcup Y' \in Y'$ and the claim follows. Since we have just seen that every α-trace is approximated by finite (compact) elements, we conclude that $(\mathbb{F}_\alpha(\Sigma, D), \sqsubseteq)$ is algebraic.

For a finite trace $t \in \mathbb{M}(\Sigma, D)$ we denote by $\max(t)$ the (pairwise independent) set of labels of maximal elements in the dependence graph of t. Following Viennot [13] we call a finite trace with exactly one maximal element a *pyramid*.

The next lemma characterizes prime elements. Recall that prime elements are compact. Hence only finite α-traces can be prime.

Proposition 9.
i) *The α-trace* $(1, A) \in \mathbb{F}_\alpha(\Sigma, D)$, $A \subseteq \Sigma$ *is prime if and only if* $|\Sigma \setminus A| = 1$.
ii) *Let* $(t, A) \in \mathbb{F}_\alpha(\Sigma, D)$, $t \neq 1$, *be an α-trace. Then* (t, A) *is prime if and only if t is a pyramid (hence finite) and* $\Sigma = \max(t) \cup A$.

Proof. i) Let $x = (1, A)$ with $A \subseteq \Sigma$. If $|\Sigma \setminus A| \geq 2$, then choose $a, b \in \Sigma \setminus A$, $a \neq b$. We have $(1, A) = \sqcup \{(1, A \cup \{a\}), (1, A \cup \{b\})\}$, but neither $(1, A) \sqsubseteq (1, A \cup \{a\})$ nor $(1, A) \sqsubseteq (1, A \cup \{b\})$. Hence $(1, A)$ is not prime in this case. If $|\Sigma \setminus A| = 0$, then $A = \Sigma$ and $(1, \Sigma) = \sqcup \emptyset$ is not prime. Finally, if $|\Sigma \setminus A| = 1$ and $x \sqsubseteq \sqcup X$ for some coherent set X, then, from the formula in Thm. 6, there exists some $y \in X$ with $A \supseteq \operatorname{Im}(y)$. It follows that $x \sqsubseteq y$. Hence x is prime.
ii) Let $x = (t, A)$ with $t \in \mathbb{M}(\Sigma, D)$, $t \neq 1$ and $A \subseteq \Sigma$. Consider first the case where $A = \Sigma$. Then we have $x \sqsubseteq \sqcup X$ if and only if $t \leq \sqcup_{\leq} \operatorname{Re}(X) \in \mathbb{R}(\Sigma, D)$.

It follows (e.g. from [4]) that (t, Σ) is prime if and only if t is a pyramid. If $|\Sigma \setminus A| \geq 2$, then a similar argument as in i) applies showing that (t, A) is not prime. Therefore we may assume that $|\Sigma \setminus A| = 1$ and $A \cup \{a\} = \Sigma$ for some $a \in \Sigma$.

Now, let first t be a pyramid such that $\max(t) = \{a\}$. Consider a coherent set $X \subseteq \mathbb{F}_\alpha(\Sigma, D)$ such that $(w, E) = \bigsqcup X$ and $(t, A) \sqsubseteq (w, E)$. Knowing $a \notin E$ there has to be some $(s, C) \in X$ such that $a \notin C$. If we have $t \leq s$ then $A \supseteq \mathrm{alph}(t^{-1}w) \cup C \supseteq \mathrm{alph}(t^{-1}s) \cup C$ and hence $(t, A) \sqsubseteq (s, C)$. So assume that $a \notin C$ and that t is not a prefix of s. Applying the well-known generalization of Levi's Lemma, see e.g. [3], we can factorize $t = pu$, $s = pv$ and $w = puvq$ such that $\mathrm{alph}(u) \times \mathrm{alph}(v) \subseteq I$. In addition, since t is a pyramid, we have $a \in \mathrm{alph}(u)$. Since $(s, C) \sqsubseteq (w, E) = (suq, E)$ we have in particular $C \supseteq \mathrm{alph}(uq) \ni a$ which is a contradiction. Hence, if t is a pyramid and $A \cup \max(t) = \Sigma$, then (t, A) is prime.

There remain two cases: either t is not a pyramid or $A \cup \max(t) \neq \Sigma$. In both cases it is easy to see that (t, A) is not prime. This is left to the reader.

The following theorem summarizes the results above. It is the main theorem of this section. The proof uses the characterization of prime α-traces, Prop. 9. It is left to the reader.

Theorem 10. *The poset* $(\mathbb{F}_\alpha(\Sigma, D), \sqsubseteq)$ *is a prime algebraic coherently complete domain.*

Let us point out that δ-traces do not form a prime algebraic domain in general. We have the following counterexample

$$
\begin{array}{ccc}
a & — & b \\
| & & | \\
d & — & c
\end{array}
$$

Example 2. Let $(\Sigma, D) = d — c$ the cycle of four letters. Then the finite δ-trace $(1, D(a))$ is not prime. In fact: $(1, D(a))$ has only two approximations, $(1, \Sigma)$ and $(1, D(a))$, and none of them is prime. Indeed, $(1, \Sigma) = \bigsqcup \emptyset$ and $(1, D(a)) \sqsubseteq (1, \emptyset) = \bigsqcup \{(1, D(b)), (1, D(c))\}$. We conclude that $(1, D(a))$ is not the least upper bound of its prime approximations.

Although we have not yet been able to characterize those dependence alphabets where $\mathbb{F}_\delta(\Sigma, D)$ is prime algebraic, we can state the following theorem.

Theorem 11. *The poset* $\mathbb{F}_\delta(\Sigma, D)$ *of δ-traces is an algebraic coherently complete domain. In general it is not prime algebraic.*

The next theorem states that in both domains the concatenation is continuous with respect to the approximation order. Its proof can be based on Prop. 4. Later we will show an even stronger result, therefore the proof can be omitted.

Theorem 12. *Let* $X, Y \subseteq \mathbb{F}_\alpha(\Sigma, D)$ *(*$\mathbb{F}_\delta(\Sigma, D)$ *resp.) be two directed sets of α-traces (δ-traces resp.). Then $X \cdot Y$ is directed and we have* $\bigsqcup(X \cdot Y) = (\bigsqcup X) \cdot (\bigsqcup Y)$.

6 Topology

The Lawson topology is a refinement of the Scott topology for a domain such that on one side the topology becomes Hausdorff (thus, any two points can be separated by two open sets), on the other side (for bounded complete domains) it is still compact, see [7, Chap. III, Thm. 1.10 and Exercises].

Let us first recall the definition of the Lawson topology for a domain $(Z, \leq)$. In this case, the family

$$S = \{\uparrow k \mid k \text{ compact}\} \cup \{Z \setminus \uparrow k \mid k \text{ compact}\}$$

is a sub-basis for the Lawson topology on $(Z, \leq)$. This means the open sets are precisely the unions of finite intersections of the sub-basis.

In this section we define a natural ultrametric of $\mathbb{F}_\alpha(\Sigma, D)$ and we show that it induces the Lawson topology. The length of an α-trace $x = (t, A)$ is the length of its real part (i.e., the number of vertices in the dependence graph). It is denoted by $|x|$. The same notation is used for real and δ-traces, too. The definition below follows the following scheme. We define a function l to $\mathbb{N} \cup \{\infty\}$ and we let $d(x, y) = 2^{-l(x,y)}$ as resulting ultrametric.

Definition 13. Let x, y be real (α- resp., δ- resp.) traces. Define

$$l_{\mathbb{R}}(x, y) = \sup\{n \mid \forall p \in \mathbb{R}(\Sigma, D), |p| \leq n : p \leq x \Leftrightarrow p \leq y\}$$
$$l_\alpha(x, y) = \sup\{n \mid \forall p \in \mathbb{F}_\alpha(\Sigma, D), |p| < n : p \sqsubseteq x \Leftrightarrow p \sqsubseteq y\}$$
$$l_\delta(x, y) = \sup\{n \mid \forall p \in \mathbb{F}_\delta(\Sigma, D), |p| < n : p \sqsubseteq x \Leftrightarrow p \sqsubseteq y\}$$

The resulting ultrametrics are denoted by $d_{\mathbb{R}}$, d_α and d_δ.

Remark. The metric $d_{\mathbb{R}}$ for real traces is the same as in [9]. The metrics d_α and d_δ have been studied for α-complex and complex traces in [2].

In the following we study only d_α for α-traces; the extension to δ-traces remains to the reader. Analogous results for $d_{\mathbb{R}}$ and real traces are well-known.

Definition 14. Let $x \in \mathbb{F}_\alpha(\Sigma, D)$ be an α-trace. The n-th *open ball* $B(x, n)$ with center x is defined by

$$B(x, n) = \{y \in \mathbb{F}_\alpha(\Sigma, D) \mid l_\alpha(x, y) > n\}.$$

The n-th *approximation* $x[n]$ of x is defined by

$$x[n] = \sqcup\{y \in \mathbb{F}_\alpha(\Sigma, D) \mid y \sqsubseteq x \text{ and } |y| \leq n\}.$$

Lemma 15. *Let* $x, y \in \mathbb{F}_\alpha(\Sigma, D)$, *then we have:*
i) $B(x, n) = B(x[n], n)$
ii) $B(x, n) \cap B(y, n) \neq \emptyset$, *if and only if* $x[n] = y[n]$
iii) *The open ball* $B(x, n)$ *is closed, too.*

Proposition 16. *The set of finite α-traces $\{(t, A) \mid t \in \mathrm{IM}(\Sigma, D), A \subseteq \Sigma\}$ is an open, discrete and dense subset of $(\mathbb{F}_\alpha(\Sigma, D), d_\alpha)$.*

Proposition 17. *The metric d_α induces the Lawson topology on $\mathbb{F}_\alpha(\Sigma, D)$, which is defined with respect to the approximation order $\sqsubseteq$.*

Since $\mathbb{F}_\alpha(\Sigma, D)$ is coherently complete, it is bounded complete. Therefore, the Lawson topology yields a compact space. From general facts about the Lawson topology we deduce the following:

Theorem 18.

i) *The metric space $(\mathbb{F}_\alpha(\Sigma, D), d_\alpha)$ is compact and complete.*

ii) *Let $(x_n)_{n \geq 0} \subseteq \mathbb{F}_\alpha(\Sigma, D)$ be a $\sqsubseteq$-increasing sequence. Then we have*

$$\lim_{n \to \infty} x_n = \sqcup \{x_n \mid n \geq 0\}$$

Our final result shows that the concatenation of α-traces is uniformly continuous with respect to the metric d_α. By Prop. 4 and Thm. 18 this is a stronger result than Thm. 12.

Theorem 19. *The concatenation of α-traces is uniformly continuous. More precisely, let $x, x', y, y' \in \mathbb{F}_\alpha(\Sigma, D)$ be α-traces. Then*

$$d_\alpha(xy, x'y') \leq \max(d_\alpha(x, x'), d_\alpha(y, y'))\,.$$

Proof. For the proof it is enough to show that $(x \cdot y)[n] = (x[n] \cdot y[n])[n]$ for all $x, y \in \mathbb{F}_\alpha(\Sigma, D)$, $n \geq 0$. Let $x = (r, A)$, $y = (s, B)$, and $z = (t, C)$ with $|t| < n$. We have $xy = (r\mu_A(s), \sigma_A(s) \cup A \cup B)$.

First, let $z \sqsubseteq xy$. Then, by the generalization of Levi's Lemma (see e.g. [3]), we have $t = pv$, $r = pu$, $\mu_A(s) = vq$ such that $\mathrm{alph}(u) \times \mathrm{alph}(v) \subseteq I$, $C \supseteq \mathrm{alph}(u) \cup \mathrm{alph}(q) \cup \sigma_A(s) \cup A \cup B$. Note that $(p, \mathrm{alph}(u) \cup A) \sqsubseteq x[n]$ and $(v, \mathrm{alph}(q) \cup \sigma_A(s) \cup B) \sqsubseteq y[n]$. We obtain by the calculus of α-traces and Prop. 4 the following formula

$$z = (t, C) \sqsubseteq (t, \mathrm{alph}(u) \cup \mathrm{alph}(q) \cup \sigma_A(s) \cup A \cup B)$$
$$= (p, \mathrm{alph}(u) \cup A) \cdot (v, \mathrm{alph}(q) \cup \sigma_A(s) \cup B)$$
$$\sqsubseteq x[n] \cdot y[n]\,.$$

Since $|t| < n$, we have $z \sqsubseteq (x[n] \cdot y[n])[n]$. It follows $(x \cdot y)[n] \sqsubseteq (x[n] \cdot y[n])[n]$.

The other direction is trivial, since again by monotonicity we have $x[n] \cdot y[n] \sqsubseteq x \cdot y$ and it follows $(x[n] \cdot y[n])[n] \sqsubseteq (x \cdot y)[n]$.

Acknowledgment: We would like to thank the anonymous referees of ICALP for careful reading and Dan Teodosiu for Ex.2, which is a simplification of our original example.

References

1. V. Diekert. Complex and complex-like traces. In A. Borzyszkowski et al., editors, *Proceedings of the 18th Mathematical Foundations of Computer Science (MFCS 93), Gdansk (Polen) 1993*, number 711 in Lecture Notes in Computer Science, pages 68–82, Berlin-Heidelberg-New York, 1993. Springer. Invited Lecture.

2. V. Diekert. On the concatenation of infinite traces. *Theoretical Computer Science*, 113:35–54, 1993. Special issue STACS'91.

3. V. Diekert and G. Rozenberg, editors. *The Book of Traces*. World Scientific, Singapore, 1995.

4. P. Gastin and B. Rozoy. The poset of infinitary traces. *Theoretical Computer Science*, 120:101–121, 1993.

5. P. Gastin and A. Petit. Infinite traces. In G. Rozenberg and V. Diekert, editors, *The Book of Traces*, pages 393–486. World Scientific, Singapore, 1995.

6. P. Gastin and A. Petit. Poset properties of complex traces. In I.M. Havel and V. Koubek, editors, *Proceedings of the 17th Symposium on Mathematical Foundations of Computer Science (MFCS'92), Prague, (Czechoslovakia), 1992*, number 629 in Lecture Notes in Computer Science, pages 255–263, Berlin-Heidelberg-New York, 1992. Springer.

7. G. Gierz, K.H. Hofmann, K. Keimel, J.D. Lawson, M. Mislove, and D. Scott. *A Compendium of Continuous Lattices*. Springer, 1980.

8. C.A.R. Hoare. *Communicating Sequential Processes*. Prentice-Hall International, London, 1985.

9. M. Kwiatkowska. A metric for traces. *Information Processing Letters*, 35:129–135, 1990.

10. J.D. Lawson. The versatile continuous order. In M. Main, A. Melton, M. Mislove, and D. Schmidt, editors, *Proceedings of the Workshop on Mathematical Foundations of Programming Language Semantics*, number 298 in Lecture Notes in Computer Science, pages 134–160, Berlin-Heidelberg-New York, 1988. Springer.

11. R. Milner. *A Calculus of Communicating Systems*. Number 92 in Lecture Notes in Computer Science. Springer, Berlin-Heidelberg-New York, 1980.

12. A. Rensink and H. Wehrheim. Weak sequential composition in process algebras. In B. Jonssen and J. Parrow, editors, *Proceedings of the Fifth International Conference on Concurrency Theory CONCUR'94, Uppsala (Sweden)*, number 836 in Lecture Notes in Computer Science, pages 226–241, Berlin-Heidelberg-New York, 1994. Springer.

13. X.G. Viennot. Heaps of pieces I: Basic definitions and combinatorial lemmas. In G. Labelle et al., editors, *Proceedings Combinatoire énumerative, Montréal, Québec (Canada) 1985*, number 1234 in Lecture Notes in Mathematics, pages 321–350, Berlin-Heidelberg-New York, 1986. Springer.

14. H. Wehrheim. Parametric action refinement. In E.-R. Olderog, editor, *Proceedings of the IFIP Working Conference on Programming Concepts, Methods and Calculi (PROCOMET 94), San Miniato*, pages 247–266. Elsevier, 1994.

Nonfinite Axiomatizability of the Equational Theory of Shuffle

Zoltán Ésik*
A. József University
Department of Computer Science
Szeged, Hungary
esik@inf.u-szeged.hu

Michael Bertol**
University of Stuttgart
Department of Computer Science
Stuttgart, Germany
bertol@informatik.uni-stuttgart.de

Abstract. We consider language structures $\mathbf{L}_\Sigma = (P_\Sigma, \cdot, \otimes, +, 1, 0)$, where P_Σ consists of all subsets of the free monoid Σ^*; the binary operations $\cdot$, $\otimes$ and $+$ are concatenation, shuffle product and union, respectively, and where the constant 0 is the empty set and the constant 1 is the singleton set containing the empty word. We show that the variety **Lang** generated by the structures $\mathbf{L}_\Sigma$ has no finite axiomatization.

1 Introduction

In the interleaving or language model of concurrency, the parallel composition of two processes is given by the shuffle product, which also plays an important role in language theory. One would like to have a complete axiomatization of the equational properties of shuffle in conjunction with other operations of interest. In this paper, we consider language structures $\mathbf{L}_\Sigma = (P_\Sigma, \cdot, \otimes, +, 1, 0)$, where P_Σ consists of all subsets of the free monoid Σ^*; the binary operations $\cdot$, $\otimes$ and $+$ are concatenation, shuffle product and union, respectively, and where the constant 0 is the empty set and the constant 1 is the singleton set containing the empty word. We show that the variety **Lang** generated by the structures $\mathbf{L}_\Sigma$ has no finite axiomatization. Thus the equational properties of the shuffle product, in conjunction with the other operations and constants, cannot be captured by a finite number of axioms. In fact we establish the stronger result that **Lang** has no finite axiomatization over the variety of ordered algebras $\mathbf{Lg}_\leq$ generated by the structures $(P_\Sigma, \cdot, \otimes, 1, \subseteq)$, where $\subseteq$ is set inclusion. The fact that $\mathbf{Lg}_\leq$ has no finite axiomatization has been shown very recently, cf. [7]. (This is in contrast with the fact that $\mathbf{Lg}$, the variety generated by the reducts $(P_\Sigma, \cdot, \otimes, 1)$, does have a complete axiomatization consisting of a small finite set of simple equational axioms, cf. [6].) For completeness, we also consider the enrichment with the Kleene star and iterated shuffle operations.

* Partially supported by grant No. T7383 of the National Foundation for Scientific Research of Hungary, the Alexander von Humboldt Foundation, and by the US-Hungarian Joint Fund under grant number 351.
** Partially supported by the ESPRIT Basic Research Actions No. 6317 ASMICS II.

In our proofs, we make use of the concrete descriptions from [6] of the free algebras in the above varieties. The free algebras in the variety **Lg** are the labeled series-parallel posets. The same series-parallel posets, equipped with the trace ordering, are the free algebras in the variety **Lg$_\leq$** of ordered algebras. (The trace ordering is not the same as the subsumption order studied in [12].) Finally, "closed subsets" of series-parallel posets are the free algebras in the variety **Lang**. On the basis of these facts, it is said in [7]: "These results seem to show that two different models of parallelism are equivalent: the so called language or interleaving model and the pomset model, cf. [17]".

2 Basic notions

We assume familiarity with universal algebra. For varieties of ordered algebras the reader is referred to [4].

2.1 Bimonoids

Definition 1. A **bimonoid** $M = (M, \cdot, \otimes, 1)$ consists of a monoid $(M, \cdot, 1)$ and a commutative monoid $(M, \otimes, 1)$. Suppose M and M' are bimonoids. A **bimonoid morphism** $M \to M'$ is a function which preserves the unit and the two binary operations.

The only connection between the two monoids in a bimonoid is the common neutral element 1.

Definition 2. An **ordered bimonoid** $(M, \leq)$ is a bimonoid M equipped with a partial ordering $\leq$ such that for all $x, y, a, b \in M$,

$$x \leq a, \ y \leq b \Longrightarrow x \cdot y \leq a \cdot b \text{ and } x \otimes y \leq a \otimes b.$$

A **morphism of ordered bimonoids** is an order preserving bimonoid morphism.

The following structures are examples of bimonoids: $\mathcal{L}_\Sigma := (P_\Sigma, \cdot, \otimes, 1)$, $\mathcal{R}_\Sigma := (R_\Sigma, \cdot, \otimes, 1)$, and $\mathcal{F}_\Sigma := (F_\Sigma, \cdot, \otimes, 1)$. Here P_Σ is the collection of all subsets of the free monoid Σ^*. R_Σ consists of the regular subsets, and F_Σ is the set of the finite subsets. The operation $B \cdot C$ is (complex) concatenation and $B \otimes C$ is the shuffle product of the languages B and C:

$$B \otimes C := \{u_1 v_1 \cdots u_n v_n \mid u_1 \cdots u_n \in B, \ v_1 \cdots v_n \in C, u_i, v_i \in \Sigma^*, n \geq 0\}.$$

The constant 1 is the set consisting of the empty word. Equipped with the partial order given by set inclusion, each of the bimonoids $\mathcal{L}_\Sigma$, $\mathcal{R}_\Sigma$ and $\mathcal{F}_\Sigma$ is an ordered bimonoid.

2.2 Shuffle semirings

Definition 3. A **shuffle semiring** $(S, +, \cdot, \otimes, 0, 1)$ is a bimonoid $(S, \cdot, \otimes, 1)$ enriched with a constant 0 and a commutative, associative, idempotent addition operation $+$, such that

$$x + 0 = x$$
$$x \cdot 0 = 0 = 0 \cdot x$$
$$x \otimes 0 = 0$$
$$x \cdot (y + z) = (x \cdot y) + (x \cdot z)$$
$$(y + z) \cdot x = (y \cdot x) + (z \cdot x)$$
$$x \otimes (y + z) = (x \otimes y) + (x \otimes z),$$

for all $x, y, z \in S$. A **morphism of shuffle semirings** is a function which preserves $0, 1$ and the three binary operations $+, \cdot, \otimes$.

When writing expressions involving the operations $+$, $\cdot$ and $\otimes$, we will assume that $\cdot$ binds more closely than $\otimes$ which in turn binds more closely than $+$.

Notation. The class of all shuffle semirings is denoted **S**. The structures $\mathbf{L}_\Sigma := (P_\Sigma, +, \cdot, \otimes, 0, 1)$, $\mathbf{R}_\Sigma := (R_\Sigma, +, \cdot, \otimes, 0, 1)$, and $\mathbf{F}_\Sigma := (F_\Sigma, +, \cdot, \otimes, 0, 1)$ are shuffle semirings where the addition operation in each is given by set union and 0 is the empty set. The other operations and constants were defined above.

Suppose that S is a shuffle semiring. We define a partial order on S by

$$a \leq b \iff a + b = b \iff \exists c \, a + c = b.$$

The operations of S are monotonic with respect to the induced partial order $\leq$. Moreover, every shuffle semiring morphism is monotonic. Thus the reduct $(S, \cdot, \otimes, 1, \leq)$ is an ordered bimonoid. In the structures $\mathbf{L}_\Sigma$, the partial order $\leq$ determined by the additive structure is set inclusion.

2.3 Terms

A shuffle semiring term is a well-formed expression built in the usual way from the operation symbols $\cdot$, $\otimes$, $+$, constants $0, 1$, and variables x_i from a countable set. A bimonoid term is a shuffle semiring term not containing the symbols $+$ and 0. The set of variables which actually occur in the term t is denoted $\mathrm{var}(t)$. Sometimes we write $t(x_1, \ldots, x_n)$ to indicate that $\mathrm{var}(t) \subseteq \{x_1, \ldots, x_n\}$. When S is a shuffle semiring (bimonoid, resp.) and $t = t(x_1, \ldots, x_n)$ is a shuffle semiring term (bimonoid term, resp.), and if $a_1, \ldots, a_n \in S$, we write $t_S(a_1, \ldots, a_n)$ or just $t(a_1, \ldots, a_n)$ for the value of the term function induced by t in S on the arguments $a_1, \ldots, a_n$. We will consider equations $t = t'$ and inequations $t \leq t'$ between terms t and t'. A class $\mathcal{V}$ of shuffle semirings is a variety if there exists a set E of equations between shuffle semiring terms such that $\mathcal{V}$ is the class of all models of E:

$$\mathcal{V} = \{S \mid S \models \mathcal{E}\}.$$

Each inequation $t \leq t'$ between shuffle semiring terms can be considered as an abbreviation for the equation $t + t' = t'$.

A class $\mathcal{B}$ of bimonoids (ordered bimonoids, resp.) is a variety if there is a set E of equations (inequations, resp.) between bimonoid terms such that $\mathcal{B}$ is the class of all models of E.

3 Results

We define the following varieties: **Lg**, **Lg$_\leq$**, and **Lang**. **Lg** is the variety of bi-monoids generated by the language structures $\mathcal{L}_\Sigma$. **Lg$_\leq$** is the variety of ordered bimonoids generated by the ordered structures $(\mathcal{L}_\Sigma, \subseteq)$, and **Lang** is the variety of shuffle semirings generated by the language structures $\mathbf{L}_\Sigma$.

We note that **Lg** is also generated by the bimonoids $\mathcal{R}_\Sigma$ or $\mathcal{F}_\Sigma$, and that **Lg$_\leq$** is generated by the ordered bimonoids $(\mathcal{R}_\Sigma, \subseteq)$ or $(\mathcal{F}_\Sigma, \subseteq)$. Further, the variety **Lang** is generated by the shuffle semirings $\mathbf{R}_\Sigma$ or $\mathbf{F}_\Sigma$. It was shown in [6] that the (in)equational theory of each of the above varieties is decidable, see also [11, 16]. Our main result is the following theorem.

Theorem 4. *The variety* **Lang** *has no finite (equational) axiomatization relative to all valid inequations of* **Lg$_\leq$**.

Corollary 5. *The variety* **Lang** *is not finitely axiomatizable.*

Note that an inequation $t \leq t'$ between bimonoid terms is valid in **Lg$_\leq$** if and only if it is valid in **Lang**, in notation: **Lg$_\leq$** $\models t \leq t' \iff$ **Lang** $\models t \leq t'$.

The essence of Theorem 4 is that there exists no finite set E of shuffle semiring equations that together with all valid inequations of the variety **Lg$_\leq$** is a complete axiomatization of **Lang**. (Recall that each inequation between bimonoid terms can be considered as an equation.)

It is shown in [7] that **Lg$_\leq$** is not finitely axiomatizable. This result and Corollary 5 are independent, none of them implies the other (see also Problem 28 in Section 8). On the other hand, **Lg** has a finite axiomatization, for **Lg** is the class of all bimonoids, cf. [6] and Section 4.

The rest of the paper is organized as follows. In Section 4 we review from [6] the concrete descriptions of the free algebras in the varieties **Lg**, **Lg$_\leq$** and **Lang**. The description uses labeled posets and is utilized in the proof of Theorem 4. In Section 5, we state a general metatheorem which also provides an outline of the proof of Theorem 4, which is in turn completed in Section 6 by giving a concrete construction. A stronger form of Theorem 4 and some further results are stated in Section 7. Several open problems are given in Section 8.

4 Review

A concrete description of the free algebras in the varieties **Lg**, **Lg$_\leq$** and **Lang** was obtained in [6]. The description uses labeled posets.

Suppose that A is a set. An **A-labeled poset** $P = (P, \leq_P, \ell)$ consists of a finite poset $(P, \leq_P)$, usually written just $(P, \leq)$, and a function $\ell : V \to A$ mapping vertices in V to labels in A. Below we will usually say labeled poset, or just poset, in particular when A is understood. A morphism $f : P \to Q$ of labeled posets is a function $P \to Q$ which preserves the ordering and the labeling. We identify isomorphic posets. The operations of sequential or serial product $P \cdot Q$ and parallel or shuffle product $P \otimes Q$ are defined as follows:

$$P \cdot Q \quad := \quad (P \cup Q, \leq_{P \cdot Q}) \qquad P \otimes Q \quad := \quad (P \cup Q, \leq_{P \otimes Q}),$$

where the two sets P and Q are disjoint, and for $v, v' \in P \cup Q$,

$$v \leq_{P.Q} v' \iff v \leq_P v' \text{ or } v \leq_Q v' \text{ or } v \in P \text{ and } v' \in Q.$$
$$v \leq_{P \otimes Q} v' \iff v \leq_P v' \text{ or } v \leq_Q v'.$$

The labeling is the obvious extension of the labelings of P and Q. Labeled posets with the above operations were introduced by Pratt to model concurrent behavior of processes, cf. [17].

Definition 6. An A-labeled poset is called **series-parallel** if it is in the least class of posets containing the empty poset 1, the singleton posets a, labeled a, for each $a \in A$, closed under the operations of sequential and parallel product. We let $\mathbf{SP}(A)$ denote the collection of all A-labeled series-parallel posets.

Both $\mathbf{SP}(A)$ and the collection $\mathbf{Pos}(A)$ of all A-labeled posets are bimonoids with the operations sequential and parallel product and the constant 1.

Remark 7. A graph theoretic characterization of the series-parallel posets was found in Grabowski [13] and Valdes et al. [21]. A poset P is series-parallel if it does not have an induced subposet consisting of four vertices a, b, c, d whose only nontrivial order relations are given by:

$$a < c, \quad a < d, \quad b < d.$$

Thus any subset of a series-parallel poset, equipped with the induced partial order, is series-parallel. This fact will be used several times without mention. We identify each letter $a \in A$ with the singleton poset whose vertex is labeled a.

Theorem 8. [12] $\mathbf{SP}(A)$ *is freely generated by the set A in the variety of all bimonoids.*

Thus the variety of bimonoids is generated by both the bimonoids $\mathbf{SP}(A)$ and the bimonoids $\mathbf{Pos}(A)$.

Let $\overline{A}$ be a disjoint copy of A and suppose that the map $a \mapsto \overline{a}$ is a bijection $A \to \overline{A}$. We define

$$\Sigma(A) \quad := \quad \{a_i, \overline{a}_i \mid a \in A, \; i = 1, 2, \ldots\}.$$

The letters a_i are called open, and the letters $\overline{a}_i$ are closed. Open and closed letters have been used in several papers about action refinement, see e.g. [1, 2] and the references contained therein.

Let $h_0 : \mathbf{SP}(A) \to \mathcal{L}_{\Sigma(A)}$ be the unique bimonoid morphism determined by the map

$$h_0 : a \longmapsto \{a_1\overline{a}_1, a_2\overline{a}_2 \ldots, a_n\overline{a}_n, \ldots\}.$$

Thus, $h_0(P)$ consists of all label sequences of sequentializations of posets obtainable from P by replacing, in each possible way, each vertex labeled a by a two-point chain corresponding to the order $a_i < \overline{a}_i$, $i \geq 0$.

Theorem 9. [6] *The morphism h_0 is injective. It follows that $\mathbf{Lg}$ is the variety of all bimonoids.*

The words in Ph_0 satisfy, for $P \in \mathbf{SP}(A)$, the following conditions:

(i) $|u|_{a_i} = |u|_{\overline{a}_i}$, for each $a \in A$. ($|u|_{a_i}$ is the number of occurrences of the letter a_i in the word u.)

(ii) If w is a prefix of u, $|w|_{a_i} \geq |w|_{\overline{a}_i}$, for each $a_i \in \Sigma(A)$.

Definition 10. A monoid endomorphism φ of $(A_N \cup \overline{A}_N)^*$ is **admissible** if, for each $a \in A$, $i > 0$, there is some $j > 0$ with $a_i\varphi = a_j$, and $\overline{a}_i\varphi = \overline{a}_j$.

Definition 11. The preorder $\sqsubseteq$ on the set $\Sigma(A)^*$ is the least reflexive and transitive relation which satisfies the following conditions. $u \sqsubseteq u'$ if, for some words w, w' and distinct letters $a_i, b_j \in \Sigma(A)$,

$$u = wa_ib_jw' \text{ and } u' = wb_ja_iw'; \quad \text{or}$$
$$u = w\overline{a}_i\overline{b}_jw' \text{ and } u' = w\overline{b}_j\overline{a}_iw'; \quad \text{or}$$
$$u = w\overline{a}_ib_jw' \text{ and } u' = wb_j\overline{a}_iw'; \quad \text{or}$$
$$u = u'\varphi, \quad \text{for some admissible } \varphi.$$

For each $P \in \mathbf{SP}(A)$, the set Ph_0 is closed with respect to the preorder $\sqsubseteq$, i.e., if $u \in Ph_0$ and $v \sqsubseteq u$, then $v \in Ph_0$. Further, for each word $v \in Ph_0$ there is a maximal word $u \in Ph_0$ with $v \sqsubseteq u$. Of course, a word $u \in Ph_0$ is maximal if $u' \sqsubseteq u$ for all $u' \in Ph_0$ with $u \sqsubseteq u'$. Each letter occurs at most once in any maximal word.

Definition 12. For A-labeled posets P, Q in $\mathbf{SP}(A)$, $P \leq Q$ iff for each word $u \in Ph_0$ there is a word $u' \in Qh_0$ with $u \sqsubseteq u'$. We call the order $\leq$ the **trace ordering** on $\mathbf{SP}(A)$.

We note some easy consequences of Definition 12 and Theorem 9.

(i) For posets $P, Q \in \mathbf{SP}(A)$, $P \leq Q$ iff one of the following conditions holds:
   (a) $Ph_0 \subseteq Qh_0$.
   (b) Each maximal word in Ph_0 is in Qh_0.
   (c) For each maximal word u in Ph_0 there is a (maximal) word $v \in Qh_0$ with $u \sqsubseteq v$.

(ii) $(\mathbf{SP}(A), \leq)$ is an ordered bimonoid in $\mathbf{Lg}_\leq$.

Suppose that P is a nonempty poset in $\mathbf{SP}(A)$. We describe in brief how to construct the maximal words in Ph_0. First, replace each vertex v of P labeled a by the two-element chain $v < \overline{v}$ with v labeled a_i and $\overline{v}$ labeled $\overline{a}_i$, for some integer i, but such that each letter of $\Sigma(A)$ is used at most once. Vertex v is called an open vertex, and $\overline{v}$ is a closed vertex. Let P' denote the resulting series-parallel $\Sigma(A)$-labeled poset. A linearization of P' can be represented by a word $w = x_1 \ldots x_{2n}$ composed of the vertices x_i of P'. We can rewrite the word w as $w = s_0p_1 \ldots s_{k-1}p_k$, where each s_i is a nonempty word formed of open vertices and each p_j is a nonempty word of closed vertices. Call the linearization w maximal if whenever the open vertex v is listed in s_i and the closed vertex $\overline{u}$ is an immediate predecessor of v, then $\overline{u}$ is listed in the word p_i. Thus s_0 is a word listing all the minimal (open) vertices, and p_k is a word listing all the maximal (closed) vertices of P'. Each maximal word in Ph_0 is the label sequence corresponding to a maximal linearization of P'.

Example. Let $P = a \cdot (b \otimes c) \otimes (a \otimes a) \cdot d$. Then, up to admissible isomorphism and commutations between open or closed letters, there are two maximal words in Ph_0:

$$u_1 = a_1 a_2 a_3 \bar{a}_1 b_1 c_1 \bar{a}_2 \bar{a}_3 d_1 \bar{b}_1 \bar{c}_1 \bar{d}_1 \qquad u_2 = a_1 a_2 a_3 \bar{a}_1 \bar{a}_2 d_1 \bar{a}_3 b_1 c_1 \bar{b}_1 \bar{c}_1 \bar{d}_1.$$

Let $Q = a \cdot (a \otimes a) \otimes (a \otimes a) \cdot a$. Then, up to admissible isomorphism, there is a unique maximal word in Qh_0:

$$u_1 = a_1 a_2 a_3 \bar{a}_1 a_4 a_5 \bar{a}_2 \bar{a}_3 a_6 \bar{a}_4 \bar{a}_5 \bar{a}_6.$$

Theorem 13. [6] *For each set A, $(\mathbf{SP}(A), \leq)$ is the free ordered bimonoid generated by A in the variety of ordered bimonoids $\mathbf{Lg}_\leq$.*

We end this section by recalling from [6] the concrete description of the free shuffle semirings in the variety $\mathbf{Lang}$.

Definition 14. A subset B of $\mathbf{SP}(A)$ is closed if for all $P \in \mathbf{SP}(A)$, $Ph_0 \subseteq Bh_0$ implies $P \in B$.

In the above definition, Bh_0 denotes the union of the sets Qh_0 for Q in B. Note that Bh_0 is closed for the preorder $\sqsubseteq$. We list some easy consequences of the definition.

(i) The empty set $\emptyset$ is closed.
(ii) If B_j, $j \in J$, are closed, then so is the intersection $\cap_{j \in J} B_j$. Thus each set $B \subseteq \mathbf{SP}(A)$ is contained in a least closed set, denoted $cl(B)$.
(iii) Each closed set is an ideal; i.e., if B is closed and $P \leq Q$ and $Q \in B$, then $P \in B$.
(iv) Each principal (i.e. one-generated) ideal $(P] = \{Q \mid Q \leq P\}$ in $\mathbf{SP}(A)$ is closed. In particular, the ideal $(a] = \{a\}$ is a closed set, for each $a \in A$.
(v) If $B = cl(B_0)$ and $C = cl(C_0)$, then the inclusion $B \subseteq C$ is equivalent to either one of the following conditions:
    (a) $B_0 h_0 \subseteq C_0 h_0$, i.e., $Ph_0 \subseteq C_0 h_0$, for each $P \in B_0$.
    (b) Each word which is maximal in $B_0 h_0$ is in $C_0 h_0$.
    (c) For each (maximal) word $u \in B_0 h_0$ there is a (maximal) word $v \in C_0 h_0$ with $u \sqsubseteq v$.
(vi) If $P, Q \in \mathbf{SP}(A)$, then $P \leq Q$ iff $(P] \subseteq (Q]$.

Definition 15. Let $\mathbf{I}_\omega(A)$ denote the set of all finite closed subsets of $\mathbf{SP}(A)$:

$$\mathbf{I}_\omega(A) := \{cl(B_0) \mid B_0 \subseteq \mathbf{SP}(A), \; B_0 \text{ finite}\}.$$

Remark 16. Note that a closed subset of $\mathbf{SP}(A)$ is finite iff it is finitely generated, since the closure of a finite set is finite.

Definition 17. Suppose that $B, C \in \mathbf{I}_\omega(A)$. Then we define:

$$B + C := cl(B \cup C)$$
$$B \cdot C := cl(\{P \cdot Q \mid P \in B, \; Q \in C\})$$
$$B \otimes C := cl(\{P \otimes Q \mid P \in B, \; Q \in C\}),$$

and $0 := \emptyset$, $1 := cl(\{1\}) = \{1\}$.

If $B, C \in \mathbf{I}_\omega(A)$ then $B + C$, $B \cdot C$, $B \otimes C$ are in $\mathbf{I}_\omega(A)$. If $B = cl(B_0)$ and $C = cl(C_0)$, with B_0 and C_0 finite, then

$$B + C = cl(B_0 \cup C_0)$$
$$B \cdot C = cl(\{P \cdot Q \mid P \in B_0, \ Q \in C_0\})$$
$$B \otimes C = cl(\{P \otimes Q \mid P \in B_0, \ Q \in C_0\}).$$

Equipped with the above operations and constants, $\mathbf{I}_\omega(A)$ is a shuffle semiring in which the partial order determined by the additive structure is set inclusion. The principal ideals $(P]$, for $P \in \mathbf{SP}(A)$ determine an ordered sub bimonoid of $\mathbf{I}_\omega(A)$ isomorphic to $\mathbf{SP}(A)$.

We identify each letter $a \in A$ with the principal ideal $(a] = \{a\}$.

Theorem 18. [6] *The shuffle semiring $\mathbf{I}_\omega(A)$ is freely generated by A in* **Lang**.

We note several properties of closed sets.

Lemma 19. *Suppose that $P \in \mathbf{SP}(A)$ and $X \subseteq \mathbf{SP}(A)$. If $P \in cl(X)$ then there exists a poset $Q \in X$ with the following properties:*

(i) *For each $a \in A$, the number of vertices of P labeled a, $|P|_a$, is the same as $|Q|_a$. Thus the number of vertices of P, $|P|$, is the same as $|Q|$.*

(ii) *For each $a \in A$, the number of minimal (maximal) vertices of P labeled a is at most the number of minimal (maximal) vertices of Q labeled a.*

(iii) *The height of P is at least the height of Q.*

(iv) *The number of isolated vertices of P is at most the number of isolated vertices of Q.*

5 The metatheorem

In this section A denotes a nonempty set. By a closed ideal we mean a finite closed set in $\mathbf{I}_\omega(A)$. When t is a shuffle semiring term, the total number of occurrences of variables in t is denoted $|t|$.

Theorem 20. *Suppose that for each $n \geq 1$ we are given a poset $P_n \in \mathbf{SP}(A)$ and a closed ideal $I_n = cl(X_n) \in \mathbf{I}_\omega(A)$ with the following properties.*

(P1) *$|P_n| = |P| > n$, for all $P \in I_n$.*

(P2) *$(P_n] \subset I_n$ and $P_n \notin X_n$.*

(P3) *For all $X \subseteq \mathbf{SP}(A)$, if $(P_n] \subseteq cl(X) \subset I_n$ then $P_n \in X$.*

(P4) *I_n has no nontrivial sequential or parallel decomposition, i.e., if $I_n = I \cdot J$ or $I_n = I \otimes J$, for some $I, J \in \mathbf{I}_\omega(A)$, then $I = 1$ or $J = 1$.*

(P5) *If t is a bimonoid term and t' is a shuffle semiring term with $\mathrm{var}(t) = \mathrm{var}(t') = \{x_1, \ldots, x_k\}$, and if*

$$(P_n] \subseteq t(J_1, \ldots, J_k) \subset I_n = t'(J_1, \ldots, J_k),$$

for some $J_1, \ldots, J_k \in \mathbf{I}_\omega(A)$, then $|t| > n$.

Then for each finite set E of equations between shuffle semiring terms with **Lang** $\models E$ *there exists a finite shuffle semiring S_E and shuffle semiring terms p and p' such that* **Lang** $\models p \leq p'$, $S_E \models E$, *but $S_E \not\models p \leq p'$. Moreover, $S_E \models q \leq q'$, for all bimonoid terms q and q' with* **Lang** $\models q \leq q'$.

Remark 21. Note that for any poset $P \in \mathbf{SP}(A)$, if I is the principal ideal $(P]$, so that I is also a closed ideal, then the set $\{P\}$ is the only minimal generating set of I. Further, each generating set of a closed ideal in $\mathbf{I}_\omega(A)$ contains a minimal generating set. Thus if $(P_n] \subseteq I_n = cl(X_n)$ and $P_n \notin X_n$, then $(P_n] \subset I_n$. Thus in (P2) we can require $(P_n] \subseteq I_n$ instead of the strict inclusion $(P_n] \subset I_n$.

We briefly describe the construction of S_E. Let E be a finite set of equations between shuffle semiring terms with $\mathbf{Lang} \models E$, and let $n = \max\{|t|, |t'| \mid t = t' \in E\}$, $N = |P_n|$, so that $N > n$. First we collapse into a single point all closed ideals in $\mathbf{I}_\omega(A)$ which contain at least one "bad" poset, or two posets of different size. A poset $P \in \mathbf{SP}(A)$ is bad if $|P| > N$, or $|P| = N$ but $P \notin I_n$, or there is a letter $a \in A$ which is the label of some vertex of P, but which is not the label of any vertex of a poset in I_n. We do this so that we don't need to worry about bad posets and in order to obtain a finite model. The quotient shuffle semiring S'_E so obtained is in $\mathbf{Lang}$, and I_n is an element of S'_E. Since by (P2) $(P_n] \subset I_n$, the principal ideal $(P_n]$ is also in S'_E. We then modify the additive structure slightly to obtain S_E from S'_E. As a result, whenever $(P_n] \subseteq I \subset I_n$ holds for the closed ideal I, the two ideals I and I_n become incomparable in the induced partial order of S_E, in notation: $I \nleq I_n$.

All of the conditions (P2)–(P4) are used to establish the fact that S_E is indeed a shuffle semiring. Property (P4) proves that $S_E \models q \leq q'$ for all inequations $q \leq q'$ between the bimonoid terms q and q' with $\mathbf{Lg}_\leq \models q \leq q'$. (Recall that $\mathbf{Lg}_\leq \models q \leq q'$ iff $\mathbf{Lang} \models q \leq q'$.)

The fact that $S_E \models E$ follows by (P5). In fact we show that for all bimonoid terms t and shuffle semiring terms t' with $|t|, |t'| \leq n$, if $\mathbf{Lang} \models t \leq t'$ then $S_E \models t \leq t'$. This suffices to establish $S_E \models E$ for the following reason. First, if t and t' are shuffle semiring terms, $S_E \models t = t'$ iff $S_E \models t \leq t'$ and $S_E \models t' \leq t$, for S_E is a shuffle semiring. Consider the inequation $t \leq t'$. Using the defining equations of shuffle semirings, in particular the distributive and zero laws, we can transform t into a sum $t_1 + \ldots + t_k$, where t_i is a bimonoid term for all $i \in [k]$. (When $k = 0$ this sum is the term 0.) We have $|t_i| \leq |t|$, for all $i \in [k]$. Since S_E is a shuffle semiring, $S_E \models t \leq t'$ iff $S_E \models t_i \leq t'$, for all $i \in [k]$. But if $t = t'$ is in E, we have $\mathbf{Lang} \models t_i \leq t'$ and $|t_i|, |t'| \leq n$.

The shuffle semiring terms p and p' are found easily: We let $p = p(x_1, \ldots, x_k)$ and $p' = p'(x_1, \ldots, x_k)$ be terms with $(P_n] = p_{S_E}((a_1], \ldots, (a_k])$ and $I_n = p'_{S_E}((a_1], \ldots, (a_k])$, for some letters $a_1, \ldots, a_k \in A$ which appear as labels of some vertices of posets in I_n. Since $(P_n] \nleq I_n$ in S_E, we have $S_E \nvDash p \leq p'$. But $\mathbf{Lang} \models p \leq p'$, for it follows from our construction that $(P_n] = p_{\mathbf{I}_\omega(A)}((a_1], \ldots, (a_k])$ and $I_n = p'_{\mathbf{I}_\omega(A)}((a_1], \ldots, (a_k])$.

6   A concrete construction

In this section we give a concrete construction of the posets P_n and the closed ideals I_n satisfying the conditions (P1)–(P5) of the metatheorem, Theorem 20. For a nonnegative integer k we denote by $[k]$ the set $\{1, \ldots, k\}$.

Let A be the set

$$A \quad := \quad \{a, b_1, b_2, \ldots\}.$$

Suppose that n is a positive integer. We assign to each permutation σ : $[2n+1] \to [2n+1]$ the poset $P_\sigma \in \mathbf{SP}(A)$:

$$P_\sigma = a \cdot (b_{1\sigma} \otimes b_{2\sigma}) \otimes \left(\bigotimes_{i=2}^{n} (a \otimes a) \cdot (b_{(2i-1)\sigma} \otimes b_{(2i)\sigma}) \right) \otimes (a \otimes a) \cdot b_{(2n+1)\sigma}$$

The construction of the posets P_n and closed ideals $I_n = cl(X_n)$ is given in the following definition.

Definition 22. For each $n \geq 1$, we define X_n to be the collection of all posets P_σ, where σ is a permutation $[2n+1] \to [2n+1]$. We define

$$P_n := \bigotimes_{i=1}^{n+2} a \cdot b_i \qquad I_n := cl(X_n).$$

7 Further results

There is a stronger version of Theorem 4.

Theorem 23. *For each finite set E of shuffle semiring equations with* $\mathbf{Lang} \models E$ *there is a finite shuffle semiring S_E and an equation $p = p'$ with* $\mathbf{Lang} \models p = p'$ *between shuffle semiring terms p and p' in* **one** *variable such that $S_E \models E$, $S_E \models q \leq q'$, for all bimonoid inequations $q \leq q'$ with* $\mathbf{Lang} \models q \leq q'$, *but $S_E \not\models p = p'$.*

This result can be proved by a slight change in the concrete construction of the posets P_n and closed ideals I_n.

We briefly consider the enrichment with the Kleene star operation. For an alphabet Σ, let $\mathbf{L}_\Sigma^*$ denote the structure $(P_\Sigma, \cdot, \otimes, +, ^*, 0, 1)$. Let $\mathbf{Lang}^*$ be the variety generated by the structures $\mathbf{L}_\Sigma^*$.

Theorem 24. $\mathbf{Lang}^*$ *has no finite (equational) axiomatization over* $\mathbf{Lang}$.

Thus, in addition to all valid equations of $\mathbf{Lang}$, an infinite number of equations involving the operation * are needed to obtain a complete axiomatization of the variety $\mathbf{Lang}^*$. The same result holds if one considers iterated shuffle † defined by:

$$x^\dagger \quad := \quad 1 + x + x \otimes x + x \otimes x \otimes x + \ldots$$

But there is nothing surprising here, Theorem 24 follows from the fact that the regular sets (over the one letter alphabet) do not have a finite axiomatization, cf. [19], [9]. The proof of Theorem 24 using this fact is standard.

There are other operations of interest. In contrast with the star operation, the equational laws of reversal$^\vee$(mirror image) can be captured by the involution axioms relative to any variety considered above:

$$(x + y)^\vee = \breve{x} + \breve{y} \qquad (x^\dagger)^\vee = (\breve{x})^\dagger$$
$$(x \cdot y)^\vee = \breve{y} \cdot \breve{x} \qquad (x^*)^\vee = (\breve{x})^*$$
$$(x \otimes y)^\vee = \breve{x} \otimes \breve{y} \qquad \breve{x}^\vee = x$$

It follows from the other axioms that $0^\vee = 0$ and $1^\vee = 1$.

8 Open Problems

Problem 25. Give nontrivial infinite (in)equational axiomatizations of the varieties $\mathbf{Lg}_\leq$, $\mathbf{Lang}$, and $\mathbf{Lang}^*$. Do there exist minimal equational axiomatizations?

Problem 26. Does there exist a finitely based first order theory (e.g. universal Horn theory) whose equational part is the equational theory of $\mathbf{Lang}$, or $\mathbf{Lang}^*$, or the (in)equational theory of some other variety discussed above other than $\mathbf{Lg}$?

Problem 27. Is it possible to add some more natural operations to the collection of operations we have been considering such that the variety generated by the resulting language structures has a finite axiomatization?

Problem 28. Does the inequational theory of $\mathbf{Lg}_\leq$ have a finite axiomatization in terms of equations between shuffle semiring terms which hold in $\mathbf{Lang}$? (More precisely, the question is whether there exists a finite set E of shuffle semiring equations with $\mathbf{Lang} \models E$ such that for all inequations $t \leq t'$ between bimonoid terms t and t', if $\mathbf{Lg}_\leq \models t \leq t'$ then $E \models t \leq t'$.)

Each variety has a minimal first order axiomatization, but there are varieties with no minimal equational axiomatization. The second question of Problem 25 asks if the varieties $\mathbf{Lg}_\leq$, $\mathbf{Lang}$ and $\mathbf{Lang}^*$ studied here are such.

When one considers the regular operations $\cdot$, $+$, * on languages together with the constants 0 and 1, the corresponding variety (that we denote $\mathbf{L}$ below) has no finite axiomatization. Infinite sets of equational axioms were obtained in [15, 5], but these are not minimal. For the variety $\mathbf{L}$, positive answers to Problems 26 and 27 are given in [20, 3, 15, 14, 8, 18]. Due to Theorem 9, there is an $\mathcal{O}(n \log n)$ algorithm for deciding the equational theory of $\mathbf{Lg}$. By modifying the proof of Theorem 4 in [10], it is shown in [6] that deciding the inequational theory of the variety $\mathbf{Lg}_\leq$ is Π_2-complete in the polynomial hierarchy. The same fact holds for the equational theory of $\mathbf{Lang}$. Deciding the equational theory of the variety $\mathbf{Lang}^*$ is PSPACE-complete. There is a further difference between the two varieties $\mathbf{Lg}$ and $\mathbf{Lg}_\leq$, or $\mathbf{Lang}$, the variety $\mathbf{Lg}$ has a finite axiomatization but $\mathbf{Lg}_\leq$ and $\mathbf{Lang}$ do not. This difference is reflected by the complexity result.

Acknowledgement We thank S.L. Bloom and L. Bernátsky for carefully reading an earlier version of the paper. During the preparation of this paper, the authors exchanged visits to their respective Computer Science Departments. We thank all of our friends and colleagues in Stuttgart and Szeged who made our trips so enjoyable. Both authors would like to thank Volker Diekert for his support making our joint work possible.

References

1. L. Aceto. Full abstraction for series-parallel pomsets. In *Proceedings of TAPSOFT '91*, volume 493 of *Lecture Notes in Computer Science*, pages 1–40. Springer-Verlag, 1991.
2. L. Aceto and M. Hennessy. Towards action refinement in process algebras. *Information and Computation*, 103(2):204–269, 1993.
3. K. B. Arkhangelskii and P. V. Gorshkov. Implicational axioms for the algebra of regular languages. *Dokl. Akad. Nauk USSR Ser. A*, 10:67–69, 1987. (in Russian)
4. S. L. Bloom. Varieties of ordered algebras. *Journal of Computer and System Sciences*, 45:200–212, 1976.
5. S. L. Bloom and Z. Ésik. Equational axioms for regular sets. *Mathematical Stuctures in Computer Science*, 3:1–24, 1993.
6. S. L. Bloom and Z. Ésik. Free shuffle algebras in language varieties. Full version submitted for publication. Extended abstract to appear in the proceedings of LATIN '95.
7. S. L. Bloom and Z. Ésik. Nonfinite axiomatizability of shuffle inequalities. In: *Proceedings of TAPSOFT '95*, LNCS 915, Springer–Verlag, 1995, 318–333.
8. M. Boffa. Une remarque sur les systèmes complets d'identites rationelles. *Theoret. Inform. Appl.*, 24:419–423, 1990.
9. J. Conway. *Regular Algebra and Finite Machines*. Chapman & Hall, London, 1971.
10. J. Feigenbaum, J. A. Kahn, C. Lund. Complexity results for pomset languages. *SIAM Journal of Discrete Mathematics*, 6(3):432–444, 1993.
11. Jay Loren Gischer. *Partial orders and the axiomatic theory of shuffle*. PhD thesis, Stanford University, Computer Science Dept., 1984.
12. Jay Loren Gischer. The equational theory of pomsets. *Theoretical Computer Science*, 61:199–224, 1988.
13. Jan Grabowski. On partial languages. *Fundamenta Informatica*, IV(2):427–498, 1981.
14. D. Kozen. A completeness theorem for Kleene algebras and the algebra of regular events. Information and Computation, 110:366–390, 1994.
15. D. Krob. Complete systems of B-rational identities. *Theoretical Computer Science*, 89:207–343, 1991.
16. A. Meyer and A. Rabinovich. Private communication.
17. Vaughan Pratt. Modeling concurrency with partial orders. *International Journal of Parallel Processing*, 15(1):33–71, 1986.
18. Vaughan Pratt. Action structures and pure induction. Technical Report, Stanford University, Dept. of Computer Science, April 1991.
19. V. N. Redko. On defining relations for the algebra of regular events. *Ukrain. Mat. Z.*, 16:120–126, 1964. (in Russian)
20. Arto Salomaa. Two complete axiom systems for the algebra of regular events. *J. ACM*, 13:158–169, 1966.
21. J. Valdes, R. E. Tarjan, and E. L. Lawler. The recognition of series-parallel digraphs. *SIAM Journal of Computing*, 11(2):298–313, 1981.

The Algebraic Equivalent of AFL Theory

WERNER KUICH

Institut für Algebra und Diskrete Mathematik
Abteilung für Theoretische Informatik
Technische Universität Wien
Wiedner Hauptstraße 8-10, A-1040 Wien

Abstract. We generalize the characterization of abstract families of languages (AFLs) by abstract families of acceptors to ω-continuous semirings.

1. Introduction

This is the third in a series of ICALP-papers that generalize results of language and automata theory to ω-continuous semirings. In the first ICALP-paper [10], we showed generalizations of the Kleene Theorem and the Parikh Theorem. In the second ICALP-paper [11] we generalized the characterization of context-free languages by pushdown automata.

In this paper we generalize the characterization of AFLs by abstract families of acceptors as given in Ginsburg [3] and Kuich, Salomaa [14], Section 11.

In Section 2 we investigate how much the concept of a complete semiring morphism generalizes the concept of a representation. We prove that both concepts coincide if the basic semiring is $\mathbb{B}$, $\mathbb{N}^{\infty}$ or $\mathbb{R}_{+}^{\infty}$. Here $\mathbb{B}$ is the Boolean semiring, $\mathbb{N}^{\infty} = \mathbb{N} \cup \{\infty\}$, where $\mathbb{N}$ is the semiring of non-negative integers, and $\mathbb{R}_{+}^{\infty} = \mathbb{R}_{+} \cup \{\infty\}$, where $\mathbb{R}_{+}$ is the semiring of non-negative reals. But there exist ω-continuous semirings such that the concept of a complete semiring morphism is more general than that of a representation.

In Section 3 we generalize rational transducers. We replace the rational representation in the usual definition by a "rational" semiring morphism that is a member of a family of complete semiring morphisms closed under matricial composition. We show that the family of transductions realized by these rational transducers is closed under funtional composition.

In Section 4 we generalize the concepts of AFL and full AFL to the concept of an AFL-semiring: that are fully rationally closed semirings which are also closed under rational transductions. These AFL-semirings are then characterized by automata representing a certain "type".

We assume that the reader is familiar with semirings (see Golan [4], Hebisch, Weinert [6]), ω-continuous semirings (see Karner [7], Kuich [10], [11], [12], [13]) and formal power series (see Berstel, Reutenauer [2], Kuich, Salomaa [14], Salomaa, Soittola [15]). All notions that are not defined are from Kuich [12] or Kuich, Salomaa [14].

We now make some notational conventions: I, J and Q, possibly provided with indices, will always denote index sets. All sets Q are finite and nonempty and are subsets of some fixed countably infinite set Q_∞. By Z we denote a finite or infinite alphabet of variables. The letters Σ, Δ and Γ, possibly provided with indices, denote sets of symbols. The sets Σ are finite alphabets, the sets Δ are finite or infinite alphabets and the sets Γ are finite or infinite, possibly empty. The set Σ_∞ is a fixed infinite alphabet.

The letter A will always denote an ω-continuous semiring. The symbol $\sqsubseteq$ in connection with an ω-continuous semiring denotes the natural order: $a \sqsubseteq b$ iff there exists an element c such that $a+c=b$.

2. Representations and continuous morphims

A multiplicative morphism $\mu : \Sigma_1^* \to A^{Q \times Q}\langle\!\langle \Sigma_2^* \rangle\!\rangle$ is usually called a *representation* and can be extended to a semiring morphism

$$\mu : A\langle\!\langle \Sigma_1^* \rangle\!\rangle \to A^{Q \times Q}\langle\!\langle \Sigma_2^* \rangle\!\rangle$$

by

$$\mu(r) = \sum_{w \in \Sigma_1^*} (r,w) \otimes \mu(w), \qquad r \in A\langle\!\langle \Sigma_1^* \rangle\!\rangle,$$

if A is commutative. Here $\otimes$ denotes the Kroneckerproduct for matrices. (See Sections 4 and 6 of Kuich, Salomaa [14].)

We consider in the next theorem a generalization of representations. A semiring morphism $h : A \to A'$ is called *complete* iff for all families $(a_i | i \in I)$ in A, $h(\sum_{i \in I} a_i) = \sum_{i \in I} h(a_i)$, i.e., h is a morphism of complete semirings.

Theorem 2.1. *Let A and A' be ω-continuous semirings. Assume that A' is commutative. Consider a complete semiring morphism $h : A \to A'$ and a multiplicative monoid morphism $\mu : \Delta_1^* \to A'^{Q \times Q}\langle\!\langle \Delta_2^* \rangle\!\rangle$. Then the mapping $\mu' : A\langle\!\langle \Delta_1^* \rangle\!\rangle \to A'^{Q \times Q}\langle\!\langle \Delta_2^* \rangle\!\rangle$, defined by*

$$\mu'(r) = \sum_{w \in \Delta_1^*} h((r,w)) \otimes \mu(w), \qquad r \in A\langle\!\langle \Delta_1^* \rangle\!\rangle,$$

is a complete semiring morphism. $\qquad\square$

Corollary 2.2. *Let A be a commutative ω-continuous semiring. If $\mu : \Sigma_1^* \to A^{Q \times Q}\langle\!\langle \Sigma_2^* \rangle\!\rangle$ is a representation then the extended mapping $\mu : A\langle\!\langle \Sigma_1^* \rangle\!\rangle \to A^{Q \times Q}\langle\!\langle \Sigma_2^* \rangle\!\rangle$ is a complete semiring morphism.* $\qquad\square$

In our theory we want to replace "representation" by "complete semiring morphism". We generalize the semiring $A \langle\!\langle \Sigma_1^* \rangle\!\rangle$ to A and the semiring $A^{Q \times Q} \langle\!\langle \Sigma_2^* \rangle\!\rangle$ to $A'^{|X|}$, i. e., we consider mappings $\mu : A \to A'^{|X|}$. Our next goal is to investigate how much the concept of a complete semiring morphism $\mu : A \to A'^{|X|}$ generalizes the concept of a representation.

Given $A' \subseteq A$, we define $[A'] \subseteq A$ to be the least complete subsemiring of A that contains A'. The semiring $[A']$ is called the *complete semiring generated by A'* and is again an ω-continuous semiring. Each element a of $[A']$ can be generated from elements of A' by multiplication and summation (including "infinite summation"):

$$a \in [A'] \text{ iff } a = \sum_{i \in I} a_{i1} \ldots a_{in_i}$$

for some $a_{ij} \in A'$ and for some index set I.

An ω-continuous semiring A is called *[0,1]-semiring* iff it is the complete semiring generated by $\{a \in A | 0 \sqsubseteq a \sqsubseteq 1\}$. Observe that each element a of a [0,1]-semiring can be written in the form $a = \sum_{j \in J} a_j$, $a_j \sqsubseteq 1$, for some index set J.

Theorem 2.3. *Let A be a [0,1]-semiring and A' be an ω-continuous semiring. Consider a complete semiring morphism $\mu : A \to A'^{|X|}$ and define the mapping $h_i : A \to A'$, $i \in I$, by $h_i(a) = \mu(a)_{i,i}$ for all $a \in A$.*

Then $\mu(a)$, $a \in A$, is a diagonal matrix and h_i, $i \in I$, is a complete semiring morphism. □

Theorem 2.4. *Let A be a [0,1]-semiring and A' be an ω-continuous semiring. Consider a complete semiring morphism $\mu : A \langle\!\langle \Delta_1^* \rangle\!\rangle \to A'^{Q \times Q} \langle\!\langle \Delta_2^* \rangle\!\rangle$.*

Then there exists a complete semiring morphism $\mu_1 : A \to A'^{Q \times Q}$, where $\mu_1(a)$ is a diagonal matrix for all $a \in A$, such that

$$\mu(r) = \sum_{w \in \Delta_1^*} \mu_1((r,w)) \mu(w), \quad r \in A \langle\!\langle \Delta_1^* \rangle\!\rangle.$$ □

We now consider the [0,1]-semirings $\mathbb{B}$, $\mathbb{N}^\infty$ and $\mathbb{R}_+^\infty$. It is wellknown that a mapping $h : \mathbb{R} \to \mathbb{R}$ with $h(1) = 1$ is a ring morphism iff it is the identity. The same holds true for the semirings $\mathbb{B}$, $\mathbb{N}^\infty$ and $\mathbb{R}_+^\infty$ and can be shown by an elementary proof.

Theorem 2.5. *Let A be one of the semirings $\mathbb{B}$, $\mathbb{N}^\infty$ or $\mathbb{R}_+^\infty$. A mapping $h : A \to A$ is a semiring morphism iff it is the identity.* □

Corollary 2.6. *Let A be one of the semirings $\mathbb{B}$, $\mathbb{N}^\infty$ or $\mathbb{R}_+^\infty$. Consider a semiring morphism $\mu : A \to A^{|X|}$.*

Then $\mu(a)$, $a \in A$, is a diagonal matrix. Moreover, the mapping $h_i : A \to A$, $i \in I$, defined by $h_i(a) = \mu(a)_{i,i}$, $a \in A$, is the identity. □

Corollary 2.7. *Let A be one of the semirings $\mathbb{B}$, $\mathbb{N}^\infty$ or $\mathbb{R}_+^\infty$. Consider a semiring morphism $\mu : A \to A$. Then, for all $a \in A$, $\mu(a) = a \otimes E$.* □

Theorem 2.8. *Let A be one of the semirings* $\mathbb{B}$, $\mathbb{N}^{\infty}$ *or* $\mathbb{R}_{+}^{\infty}$. *Consider a mapping* $\mu : A\langle\langle\Sigma_1^*\rangle\rangle \to A^{Q\times Q}\langle\langle\Sigma_2^*\rangle\rangle$.

Then the following statements are equivalent:
- (i) μ *is a complete semiring morphism;*
- (ii) μ *is a representation.* $\qquad\qquad\qquad\qquad\qquad\qquad\Box$

The characterization of Theorem 2.8 is not valid in the *tropical* semiring $\langle\mathbb{N}^{\infty}, \min, +, \infty, 0\rangle$. This semiring is wellknown in theoretical computer science (see e.g. Simon [16], Krob [9]).

3. Rational Transductions

In this section we generalize rational transducers that map formal languages to formal languages (see Berstel [1]) to rational transducers that map elements of an ω-continuous semiring to elements of an ω-continuous semiring. In this section we denote by A and $\hat{A}$ ω-continuous semirings and by $A' \subseteq A$ and $\hat{A}' \subseteq \hat{A}$ subsets of A and $\hat{A}$, respectively, both containing the respective neutral elements 0 and 1. The fixed countably infinite set Q_{∞} has the following property: if $q_1, q_2 \in Q_{\infty}$ then $(q_1, q_2) \in Q_{\infty}$.

An $(\hat{A}', A')$-*rational transducer* $\mathfrak{T} = (Q, h, S, P)$ is defined by
- (i) a finite set Q of *states*,
- (ii) a semiring morphism $h : \hat{A} \to A^{Q\times Q}$, where $h(a)_{q_1, q_2} \in \mathfrak{Rat}(A')$ for all $a \in \hat{A}'$, $q_1, q_2 \in Q$,
- (iii) $S \in \mathfrak{Rat}(A')^{1\times Q}$ called the *initial state vector*,
- (iv) $P \in \mathfrak{Rat}(A')^{Q\times 1}$ called the *final state vector*.

The mapping $\|\mathfrak{T}\| : \hat{A} \to A$ *realized* by an $(\hat{A}', A')$-rational transducer $\mathfrak{T} = (Q, h, S, P)$ is defined, for all $a \in \hat{A}$, by $\|\mathfrak{T}\|(a) = Sh(a)P$. A mapping $\tau : \hat{A} \to A$ is called an $(\hat{A}', A')$-*rational transduction* iff there exists an $(\hat{A}', A')$-rational transducer $\mathfrak{T}$ such that $\tau(a) = \|\mathfrak{T}\|(a)$ for all $a \in \hat{A}$. In this case, we say that τ is realized by $\mathfrak{T}$. An (A', A')-rational transducer (in case $\hat{A} = A$ and $\hat{A}' = A'$) is called A'-*rational transducer* and an (A', A')-rational transduction is called A'-*rational transduction*.

We want to show that certain A'-rational transductions are closed under functional composition. For the proof of this result we need the application of semiring morphisms to matrices.

Let $h : \hat{A} \to A^{Q\times Q}$ be a mapping. Then we define the mapping $h' : \hat{A}^{I_1\times I_2} \to A^{(Q\times I_1)\times(Q\times I_2)}$ by

$$h'(M)_{(q_1, i_1), (q_2, i_2)} = h(M_{i_1, i_2})_{q_1, q_2}$$

for $M \in \hat{A}^{I_1\times I_2}$, $q_1, q_2 \in Q$, $i_1 \in I_1$, $i_2 \in I_2$. In the sequel we use the same notation h for the mappings h and h'.

Observe that $A^{(Q\times I_1)\times(Q\times I_2)}$ and $(A^{Q\times Q})^{I_1\times I_2}$ are isomorphic monoids (with respect to addition). If $I_1 = I_2$, they are isomorphic semirings.

The mapping $h : \hat{A} \to A^{Q \times Q}$ defines also a mapping $h' : \hat{A}^{I_1 \times I_2} \to (A^{Q \times Q})^{I_1 \times I_2}$ by

$$(h'(M)_{i_1, i_2})_{q_1, q_2} = h(M_{i_1, i_2})_{q_1, q_2}$$

for $M \in \hat{A}^{I_1 \times I_2}$, $q_1, q_2 \in Q$, $i_1 \in I_1$, $i_2 \in I_2$. Again we use in the sequel the same notation h for the mappings h and h'.

Lemma 3.1. Let $h : \hat{A} \to A^{Q \times Q}$ be a complete semiring morphism. Let $M_1 \in \hat{A}^{I_1 \times I_2}$ and $M_2 \in \hat{A}^{I_2 \times I_3}$. Then $h(M_1 M_2) = h(M_1) h(M_2)$ for $h(M_1 M_2)$ in $A^{(Q \times I_1) \times (Q \times I_3)}$ or $(A^{Q \times Q})^{I_1 \times I_3}$. □

This lemma together with some easy computations implies our next result.

Theorem 3.2. Let $h : \hat{A} \to A^{Q \times Q}$ be a complete semiring morphism. Then $h : \hat{A}^{I \times I} \to A^{(Q \times I) \times (Q \times I)}$ and $h : \hat{A}^{I \times I} \to (A^{Q \times Q})^{I \times I}$ are again complete semiring morphisms. □

Consider the family of all semiring morphisms $h : A \to A^{Q \times Q}$, Q finite. A subfamily $\mathfrak{H}$ of this family is *closed under matricial composition* iff the following conditions are satisfied for arbitrary $h : A \to A^{Q \times Q}$ and $h' : A \to A^{Q' \times Q'}$ in $\mathfrak{H}$:

(i) The functional composition $h \circ h' : A \to A^{(Q \times Q') \times (Q \times Q')}$ defined by $(h \circ h')(a) = h(h'(a))$, $a \in A$, is again in $\mathfrak{H}$.

(ii) If $Q \cap Q' = \emptyset$ then the mapping $h + h' : A \to A^{(Q \cup Q') \times (Q \cup Q')}$ defined by

$$(h + h')(a) = \begin{pmatrix} h(a) & 0 \\ 0 & h'(a) \end{pmatrix}, \qquad a \in A,$$

where the blocks are indexed by Q and Q', is again in $\mathfrak{H}$.

(iii) The identity mapping $e : A \to A$, defined by $e(a) = a$, $a \in A$, is in $\mathfrak{H}$. Clearly, the family of all complete semiring morphims $h : A \to A^{Q \times Q}$ is closed under matricial composition.

For the rest of this paper, we assume $\mathfrak{H}$ to be a family of *complete* semiring morphisms that is closed under matricial composition.

An A'-rational transducer $\mathfrak{T} = (Q, h, S, P)$ where $h \in \mathfrak{H}$ is called $\mathfrak{H}$-A'-*rational transducer*. Hence, for an $\mathfrak{H}$-A'-rational transducer $\mathfrak{T} = (Q, h, S, P)$, the morphism $h \in \mathfrak{H}$ is *complete* and $h(a) \in \mathfrak{Rat}(A')^{Q \times Q}$ for all $a \in A'$. An A'-rational transduction realized by an $\mathfrak{H}$-A'-rational transducer is called $\mathfrak{H}$-A'-*rational transduction*.

Theorem 3.3. Assume that $\mathfrak{T}$ is an $\mathfrak{H}$-A'-rational transducer and that $a \in \mathfrak{Rat}(A')$. Then $\|\mathfrak{T}\|(a) \in \mathfrak{Rat}(A')$.

Proof. Let a be the behavior of the finite A'-automaton $\mathfrak{A} = (Q, M, S, P)$. Assume that $\mathfrak{T} = (Q', h, S', P')$. We consider now the finite $\mathfrak{Rat}(A')$-auto-maton $\mathfrak{A}' = (Q' \times Q, h(M), S'h(S), h(P)P')$. By Theorem 3.6 of Kuich [12] we obtain $\|\mathfrak{A}'\| \in \mathfrak{Rat}(A')$. Since $\|\mathfrak{A}'\| = S'h(S)h(M)^*h(P)P' = S'h(SM^*P)P' = \|\mathfrak{T}\|(\|\mathfrak{A}\|)$, our theorem is proved. $\qquad\square$

Corollary 3.4. *If* $h : A \to A^{\varrho \times \varrho}$ *is a complete semiring morphism, such that* $h(a') \in \mathfrak{Rat}(A')^{\varrho \times \varrho}$ *for all* $a' \in A'$, *then* $h(a) \in \mathfrak{Rat}(A')^{\varrho \times \varrho}$ *for all* $a \in \mathfrak{Rat}(A')$. $\qquad\square$

Theorem 3.5. *The family of* $\mathfrak{H}$-A'-*rational transductions is closed under functional composition.*

Proof. Let $\mathfrak{T}_j = (Q_j, h_j, S_j, P_j)$, $j = 1, 2$, be two $\mathfrak{H}$-A'-rational transducers. We want to show that the mapping $\tau : A \to A$ defined by $\tau(a) = \|\mathfrak{T}_2\|(\|\mathfrak{T}_1\|(a))$, $a \in A$, is an $\mathfrak{H}$-A'-rational transduction.

Consider $\mathfrak{T} = (Q_2 \times Q_1, h_2 \circ h_1, S_2 h_2(S_1), h_2(P_1)P_2)$. By Theorem 3.2 and Corollary 3.4 the mapping $h_2 \circ h_1 : A \to A^{(\varrho_2 \times \varrho_1) \times (\varrho_2 \times \varrho_1)}$ is a complete semiring morphism in $\mathfrak{H}$ such that $h_2(h_1(a)) \in \mathfrak{Rat}(A')^{(\varrho_2 \times \varrho_1) \times (\varrho_2 \times \varrho_1)}$ for $a \in A'$. Furthermore, $S_2 h_2(S_1) \in \mathfrak{Rat}(A')^{1 \times (\varrho_2 \times \varrho_1)}$ and $h_2(P_1)P_2 \in \mathfrak{Rat}(A')^{(\varrho_2 \times \varrho_1) \times 1}$. Hence, $\mathfrak{T}$ is an $\mathfrak{H}$-A'-rational transducer. Now the theorem follows by

$$\|\mathfrak{T}\|(a) = S_2 h_2(S_1) h_2(h_1(a)) h_2(P_1)P_2 = S_2 h_2(\|\mathfrak{T}_1\|(a)) P_2 = \|\mathfrak{T}_2\|(\|\mathfrak{T}_1\|(a)). \quad \square$$

4. AFL-Semirings

We introduce as a generalization of abstract families of languages (briefly, AFL) the notion of an AFL-semiring.

Given $A' \subseteq A$, any subset $\mathfrak{L}$ of $[A']$ is called A'-*family of elements.*

Let $\mathfrak{T}$ be an $\mathfrak{H}$-A'-rational transducer. Then for each $a \in [A']$, we obtain $\|\mathfrak{T}\|(a) \in [A']$. Hence, the set

$$\mathfrak{M}(\mathfrak{L}) = \{\tau(a) \mid a \in \mathfrak{L}, \ \tau : A \to A \text{ is an } \mathfrak{H}\text{-}A'\text{-rational transduction}\}$$

is again an A'-family of elements for each A'-family of elements $\mathfrak{L}$. An A'-family of elements is said to be *closed under* $\mathfrak{H}$-A'-*rational transductions* iff $\mathfrak{L} = \mathfrak{M}(\mathfrak{L})$. The notation $\mathfrak{F}(\mathfrak{L})$ is used for the smallest fully rationally closed subsemiring of $[A']$ that is closed under $\mathfrak{H}$-A'-rational transductions and contains $\mathfrak{L}$. We have tried to use in our notation letters customary in AFL theory to aid the reader familiar with this theory (see Ginsburg [3]).

An A'-family of elements $\mathfrak{L}$ is called $\mathfrak{H}$-A'-*abstract family of elements* (briefly, AFE) iff $\mathfrak{L} = \mathfrak{F}(\mathfrak{L})$.

Example 4.1. Consider the case where A is the semiring of formal languages $\mathfrak{P}(\Sigma_\infty^*)$ and $A' = \{\{x\} \mid x \in \Sigma_\infty^*\} \cup \{\{\varepsilon\}, \varnothing\}$. If $\mathfrak{H}$ is the family of matrix representations then the notion of an AFE coincides with the notion of a full AFL (see Berstel [1] and Ginsburg [3]). If $\mathfrak{H}$ is the family of matrix representations corresponding to the "ε-output bounded"-transducers (Ginsburg [3]; they correspond to regulated rational transducers via the isomorphism of $\mathfrak{P}(\Sigma_\infty^*)$ and $\mathbb{B}\langle\!\langle\Sigma_\infty^*\rangle\!\rangle$) then the notion of an AFE coincides with the notion of an AFL. $\qquad\square$

It is understood that A, $A' \subseteq A$ and $\mathfrak{H}$ are fixed for the rest of this paper. Furthermore, we assume that $0,1 \in A'$ and $A = [A']$. The first condition is necessary in most of the results while the second condition will cause no loss of generality.

Example 4.1 is the reason, why we call AFEs in the sequel *AFL-semi-rings* (where the parameters A, A', $\mathfrak{H}$ are fixed). Hence, AFL-semirings constitute a generalization of both, AFLs and full AFLs.

The intersection of a family of AFL-semirings is again an AFL-semiring. The AFL-semiring $\mathfrak{F}(\mathfrak{R})$, generated by a family of elements $\mathfrak{R}$, is the intersection of all AFL-semirings containing $\mathfrak{R}$.

Theorem 4.1. *Let $\mathfrak{L}$ be an AFL-semiring. Then $\mathfrak{Rat}(A') \subseteq \mathfrak{L}$.*

Proof. Since $\mathfrak{L}$ is a semiring, we have $1 \in \mathfrak{L}$. For $a \in \mathfrak{Rat}(A')$, consider the $\mathfrak{H}$-A'-rational transducer $\mathfrak{T}_a = (\{q\}, e, 1, a)$, where $e \in \mathfrak{H}$ is the identity mapping. Given $b \in A$, we obtain $\|\mathfrak{T}_a\|(b) = ba$. Hence, $\|\mathfrak{T}_a\|(1) = a$. Since $\mathfrak{L}$ is closed under $\mathfrak{H}$-A'-rational transductions, we obtain $\|\mathfrak{T}_a\|(1) = a \in \mathfrak{L}$ for all $a \in \mathfrak{Rat}(A')$. $\qquad\square$

Let $\Delta = \Delta' \cup Z$, where $\Delta' = \{\mathbf{a} \mid a \in A'\}$ and Z is a possibly infinite alphabet of variables. Let $h : A \to A^{Q \times Q}$ be a semiring morphism in $\mathfrak{H}$. We say that h is *extended* to a monoid morphism $h : \Delta^* \to A^{Q \times Q}$ compatible with A iff $h(\mathbf{a}) = h(a)$, $\mathbf{a} \in \Delta'$, and $h(z) = 0$ for almost all variables $z \in Z$. Observe that such an extended monoid morphism $h : \Delta^* \to A^{Q \times Q}$ allows us to identify uniquely the original semiring morphism $h : A \to A^{Q \times Q}$ in $\mathfrak{H}$.

Let now $h : \Delta^* \to A^{Q \times Q}$ be a monoid morphism compatible with A. Then we extend this monoid morphism uniquely to a complete semiring morphism $h : \mathbb{N}^\infty\langle\!\langle\Delta^*\rangle\!\rangle \to A^{Q \times Q}$:

$$h(r) = h\Big(\sum_{v \in \Delta^*} (r,v)v\Big) = \sum_{v \in \Delta^*} \nu((r,v))h(v), \qquad r \in \mathbb{N}^\infty\langle\!\langle\Delta^*\rangle\!\rangle,$$

where ν is defined by $\nu(n) = \sum_{0 \leq i \leq n-1} 1$, $n \in \mathbb{N}$, and $\nu(\infty) = \sum_{i \geq 0} 1$. (See Goldstern [5], Lemma 5.6.)

We introduce now the notions of a *type T*, a *T-matrix*, a *T-auto-maton* and the *automaton representing T*. The intuitive meaning of these notions is similar to that described in Kuich, Salomaa [14], Section 11.

A *type* is a quadruple $(\Gamma_T, \Delta_T, T, \pi_T)$, where

    (i) Γ_T is the set of *storage symbols*;
    (ii) $\Delta_T \subseteq \Delta$ is the set of *instructions*;
    (iii) $T \in (\mathbb{N}^\infty\langle\Delta_T\rangle)^{\Gamma_T^* \times \Gamma_T^*}$ is the *type matrix*;
    (iv) $\pi_T \in \Gamma_T^*$ is the *initial contents of the working tape*.

In the sequel we speak of the type T if Γ_T, Δ_T and π_T are understood.

A matrix $M \in (\mathfrak{Rat}(A')^{\varrho \times \varrho})^{\Gamma_T^* \times \Gamma_T^*}$ is called a T-matrix iff there exists a monoid morphism $h : \Delta^* \to \mathfrak{Rat}(A')^{\varrho \times \varrho}$ that is compatible with A such that $M = h(T)$.

A T-automaton $\mathfrak{A} = (Q, \Gamma_T, M, S, \pi_T, P)$ is defined by

    (i) a finite set Q of *states*;
    (ii) a T-matrix M called the *transition matrix*;
    (iii) $S \in \mathfrak{Rat}(A')^{1 \times \varrho}$ called the *initial state vector*;
    (iv) $P \in \mathfrak{Rat}(A')^{\varrho \times 1}$ called the *final state vector*.

Observe that Γ_T and π_T are determined by T. The behavior of the T-automaton $\mathfrak{A}$ is given by

$$\|\mathfrak{A}\| = S(M^*)_{\pi_T, \varepsilon} P.$$

Clearly, for each such T-automaton $\mathfrak{A}$ there exists a $\mathfrak{Rat}(A')$-automaton $\mathfrak{A}' = (\Gamma_T^* \times Q, M', S', P')$ such that $\|\mathfrak{A}'\| = \|\mathfrak{A}\|$. This is achieved by choosing $M'_{(\pi_1, q_1), (\pi_2, q_2)} = (M_{\pi_1, \pi_2})_{q_1, q_2}$, $S'_{(\pi_T, q)} = S_q$, $S'_{(\pi, q)} = 0$, $\pi \neq \pi_T$, $P'_{(\varepsilon, q)} = P_q$, $P'_{(\pi, q)} = 0$, $\pi \neq \varepsilon$, $q_1, q_2, q \in Q$, $\pi_1, \pi_2, \pi \in \Gamma_T^*$.

The *automaton* $\mathfrak{A}_T$ representing a type $(\Gamma_T, \Delta_T, T, \pi_T)$ is an $\mathbb{N}^\infty \langle\!\langle \Delta_T^* \rangle\!\rangle$-automaton defined by $\mathfrak{A}_T = (\Gamma_T^*, T, S_T, P_T)$, where $(S_T)_{\pi_T} = \varepsilon$, $(S_T)_\pi = 0$, $\pi \in \Gamma_T^*$, $\pi \neq \pi_T$, $(P_T)_\varepsilon = \varepsilon$, $(P_T)_\pi = 0$, $\pi \in \Gamma_T^*$, $\pi \neq \varepsilon$. The behavior of $\mathfrak{A}_T$ is $\|\mathfrak{A}_T\| = (T^*)_{\pi_T, \varepsilon}$.

In a certain sense, $\mathfrak{A}_T$ generates an A'-family of elements by $(\Delta_T \cup \{0, \varepsilon\}, A')$-rational transductions $\tau : \mathbb{N}^\infty \langle\!\langle \Delta_T^* \rangle\!\rangle \to A$. These are realized by $(\Delta_T \cup \{0, \varepsilon\}, A')$-rational transducers $\mathfrak{T} = (Q, h, S, P)$, where $h : \Delta^* \to \mathfrak{Rat}(A')^{\varrho \times \varrho}$ is a monoid morphism compatible with A.

Given a T-automaton $\mathfrak{A} = (Q, \Gamma_T, M, S, \pi_T, P)$ where $M = h(T)$, we apply $\mathfrak{T} = (Q, h, S, P)$ to $\|\mathfrak{A}_T\|$ and obtain $\|\mathfrak{T}\|(\|\mathfrak{A}_T\|) = Sh((T^*)_{\pi_T, \varepsilon})P = S(M^*)_{\pi_T, \varepsilon} P = \|\mathfrak{A}\|$. We define now the A'-family of elements

$$\mathfrak{Rat}_T(A') = \{\|\mathfrak{A}\| \,|\, \mathfrak{A} \text{ is a } T\text{-automaton}\}.$$

Hence, $\mathfrak{Rat}_T(A')$ is the family of elements $\|\mathfrak{T}\|(\|\mathfrak{A}_T\|)$, where $\mathfrak{T} = (Q, h, S, P)$ is a $(\Delta_T \cup \{0, \varepsilon\}, A')$-rational transducer and $h : \Delta^* \to \mathfrak{Rat}(A')^{\varrho \times \varrho}$ is compatible with A.

It will turn out that $\mathfrak{Rat}_T(A')$ constitutes an AFL-semiring if T is a restart type. Here a type $(\Gamma_T, \Delta_T, T, \pi_T)$ is called a *restart type* iff $\pi_T = \varepsilon$ and the non-null entries of T satisfy the conditions $T_{\varepsilon, \varepsilon} = z_0 \in Z$, $T_{\varepsilon, \pi} \in Z - \{z_0\}$, $(T_{\pi, \pi'}, z_0) = 0$ for all $\pi', \pi \in \Gamma_T^*$, $\pi \neq \varepsilon$.

Observe that the working tape is empty at the beginning of the computation. Furthermore, only spontaneous transitions (caused by variables from Z) are possible in the case of an empty working tape.

Theorem 4.2. *If T is a restart type then* $\mathfrak{Rat}(A') \subseteq \mathfrak{Rat}_T(A')$.   □

Theorem 4.3. *If T is a restart type then* $\mathfrak{Rat}_T(A')$ *is closed under addition and multiplication.*

Proof. Assume that $\mathfrak{A}_j = (Q_j, \Gamma_T, M_j, S_j, \varepsilon, P_j)$, where $M_j = h_j(T)$, $j = 1,2$, is a T-automaton and $Q_1 \cap Q_2 = \emptyset$. We first give the construction of a T-automaton $\mathfrak{A} = (Q_1 \cup Q_2, \Gamma_T, M, S, \varepsilon, P)$ with $\|\mathfrak{A}\| = \|\mathfrak{A}_1\| + \|\mathfrak{A}_2\|$.

Let $h : \Delta^* \to A^{(Q_1 \cup Q_2) \times (Q_1 \cup Q_2)}$ be defined by

$$h(d) = \begin{pmatrix} h_1(d) & 0 \\ 0 & h_2(d) \end{pmatrix}.$$

Then $\bar{h} : A \to A^{(Q_1 \cup Q_2) \times (Q_1 \cup Q_2)}$, defined by $\bar{h}(a) = h(a)$, $a \in \Delta$, is in $\mathfrak{H}$, and h is compatible with A. Hence, $h(T)$ is a T-matrix. Consider now $M = h(T)$. By the construction of M, we obtain

$$(M^*)_{\pi_1, \pi_2} = \begin{pmatrix} (M_1^*)_{\pi_1, \pi_2} & 0 \\ 0 & (M_2^*)_{\pi_1, \pi_2} \end{pmatrix}, \quad \pi_1, \pi_2 \in \Gamma_T^*.$$

Let $S = (S_1 \ S_2)$ and $P = \begin{pmatrix} P_1 \\ P_2 \end{pmatrix}$.

Then $\|\mathfrak{A}\| = S(M^*)_{\varepsilon, \varepsilon} P = S_1 (M_1^*)_{\varepsilon, \varepsilon} P_1 + S_2 (M_2^*)_{\varepsilon, \varepsilon} P_2 = \|\mathfrak{A}_1\| + \|\mathfrak{A}_2\|$.

The construction of a T-automaton $\mathfrak{A}' = (Q_1 \cup Q_2, \Gamma_T, M', S', \varepsilon, P')$ with $\|\mathfrak{A}'\| = \|\mathfrak{A}_1\| \|\mathfrak{A}_2\|$ is as follows. Let $h' : \Delta^* \to A^{(Q_1 \cup Q_2) \times (Q_1 \cup Q_2)}$ be defined by

$$h'(z_0) = \begin{pmatrix} h_1(z_0) & P_1 S_2 \\ 0 & h_2(z_0) \end{pmatrix}, \quad h'(d) = \begin{pmatrix} h_1(d) & 0 \\ 0 & h_2(d) \end{pmatrix}, \quad d \in \Delta - \{z_0\}.$$

Define $\bar{M} \in (\mathfrak{Rat}_T(A'))^{(Q_1 \cup Q_2) \times (Q_1 \cup Q_2)) \Gamma_T^* \times \Gamma_T^*}$ by

$$\bar{M}_{\varepsilon, \varepsilon} = \begin{pmatrix} 0 & P_1 S_2 \\ 0 & 0 \end{pmatrix}, \quad \bar{M}_{\pi_1, \pi_2} = 0 \text{ for } (\pi_1, \pi_2) \neq (\varepsilon, \varepsilon).$$

Let $M' = h'(T)$. Then we have $M' = M + \bar{M}$ and $M'^* = (M^* \bar{M})^* M^*$. Hence, the $Q_1 \times Q_2$-block of $(M'^*)_{\varepsilon, \varepsilon}$ equals $(M_1^*)_{\varepsilon, \varepsilon} P_1 S_2 (M_2^*)_{\varepsilon, \varepsilon}$. Define $S' = (S_1 \ 0)$ and $P' = \begin{pmatrix} 0 \\ P_2 \end{pmatrix}$. Then $\|\mathfrak{A}'\| = S_1 (M_1^*)_{\varepsilon, \varepsilon} P_1 S_2 (M_2^*)_{\varepsilon, \varepsilon} P_2 = \|\mathfrak{A}_1\| \|\mathfrak{A}_2\|$. $\square$

Theorem 4.4. *If T is a restart type and $a \in \mathfrak{Rat}_T(A')$ then $a^* \in \mathfrak{Rat}_T(A')$.*

Proof. Assume that $\mathfrak{A} = (Q, \Gamma_T, M, S, \varepsilon, P)$, where $M = h(T)$, is a T-automaton. We give the construction of a T-automaton $\mathfrak{A}' = (Q, \Gamma_T, M', S, \varepsilon, P)$ with $\|\mathfrak{A}'\| = \|\mathfrak{A}\|^+$. Let $h' : \Delta^* \to A^{Q \times Q}$ be defined by $h'(z_0) = h(z_0) + PS$, $h'(d) = h(d)$, $d \in \Delta - \{z_0\}$. Then h' is compatible with A.

Define $\bar{M} \in (\mathfrak{Rat}_T(A')^{Q \times Q}) \Gamma_T^* \times \Gamma_T^*$ by $\bar{M}_{\varepsilon, \varepsilon} = PS$, $\bar{M}_{\pi_1, \pi_2} = 0$ for $(\pi_1, \pi_2) \neq (\varepsilon, \varepsilon)$. Let $M' = h'(T)$. Then we obtain $M' = M + \bar{M}$ and $M'^* = (M^* \bar{M})^* M^*$. We compute $(M^* \bar{M})_{\varepsilon, \pi} = 0$ for $\pi \in \Gamma_T^+$, $(M^* \bar{M})_{\varepsilon, \varepsilon} = (M^*)_{\varepsilon, \varepsilon} PS$, $((M^*)_{\varepsilon, \varepsilon} PS)^*$ and $((M^* \bar{M})^*)_{\varepsilon, \pi} = 0$, $\pi \in \Gamma_T^+$. Hence,

$$(M'^*)_{\varepsilon, \varepsilon} = ((M^* \bar{M})^*)_{\varepsilon, \varepsilon} (M^*)_{\varepsilon, \varepsilon} = ((M^*)_{\varepsilon, \varepsilon} PS)^* (M^*)_{\varepsilon, \varepsilon}$$

and

$$\|\mathfrak{A}'\| = S((M^*)_{\varepsilon, \varepsilon} PS)^* (M^*)_{\varepsilon, \varepsilon} P = \|\mathfrak{A}\|^+.$$

By Theorems 4.2 and 4.3, $\|\mathfrak{A}\|^+ + 1 = \|\mathfrak{A}\|^*$ is in $\mathfrak{Rat}_T(A')$. $\square$

This proves that $\mathfrak{Rat}_T(A')$ is a fully rationally closed semiring.

Theorem 4.5. *If T is a restart type then $\mathfrak{Rat}_T(A')$ is closed under $\mathfrak{H}$-A'-rational transductions.*

Proof. Assume that $\mathfrak{A} = (Q, \Gamma_T, M, S, \varepsilon, P)$, where $M = h(T)$, is a T-automaton and that $\mathfrak{T} = (Q', h', S', P')$ is a $\mathfrak{H}$-A'-rational transducer. Since $h : \Delta^* \to A^{Q \times Q}$ is compatible with A and $h' : A \to A^{Q' \times Q'}$ is in $\mathfrak{H}$, the monoid morphism $h' \circ h : \Delta^* \to A^{(Q' \times Q) \times (Q' \times Q)}$ is again compatible with A. We prove now that the behavior of the T-automaton $\mathfrak{A}' = (Q' \times Q, \Gamma_T, (h' \circ h)(T), S' h'(S), h'(P) P')$ is equal to $\|\mathfrak{T}\|(\|\mathfrak{A}\|)$:

$$\|\mathfrak{A}'\| = S' h'(S)(h' \circ h)(T)^* h'(P) P' = S' h'(S) h'(h(T)^*) h'(P) P' =$$
$$= S' h'(S h(T)^* P) P' = S' h'(\|\mathfrak{A}\|) P' = \|\mathfrak{T}\|(\|\mathfrak{A}\|). \qquad \square$$

Corollary 4.6. *If T is a restart type then $\mathfrak{Rat}_T(A')$ is an AFL-semiring.* $\square$

In order to get a complete characterization of AFL-semirings we need a result "converse" to Corollary 4.6.

Let $\mathfrak{L}$ be an AFL-semiring. Then we construct a restart type T such that $\mathfrak{L} = \mathfrak{Rat}_T(A')$. Assume $\mathfrak{L} = \mathfrak{F}(\mathfrak{R})$. Then for each $b \in \mathfrak{R}$ there exists an index set I_b such that $b = \sum_{i \in I_b} a_{i1} \ldots a_{in_i}$, $a_{ij} \in A'$, i.e.,

$$\mathfrak{R} = \{b \,|\, b = \sum_{i \in I_b} a_{i1} \ldots a_{in_i} \}.$$

Such a representation of $\mathfrak{L}$ is possible since $\mathfrak{L} = \mathfrak{F}(\mathfrak{L}) \subseteq [A']$. The restart type $(\Gamma_T, \Delta_T, T, \pi_T)$ is defined by

(i) $\Gamma_T = \bigcup_{b \in \mathfrak{R}} \Delta_b$, where $\Delta_b = \{a_b \,|\, a \in A'\}$ for $b \in \mathfrak{R}$;

(ii) $\Delta_T = \Delta' \cup \{z^0\} \cup \{z_b \,|\, b \in \mathfrak{R}\}$, where $\Delta' = \{a \,|\, a \in A'\}$;

(iii) $T \in (\mathbb{N}^\infty \langle \Delta_T \rangle)^{\Gamma_T^* \times \Gamma_T^*}$, where the non-null entries of T are
$T_{\varepsilon, \varepsilon} = z^0$;
$T_{\varepsilon, a_b} = z_b$ for $a_b \in \Delta_b$, $b \in \mathfrak{R}$;
$T_{\pi a_b, \pi a_b a_b'} = a$ for $\pi \in \Delta_b^*$, $a_b, a_b' \in \Delta_b$, $b \in \mathfrak{R}$;
$T_{\pi a_b, \varepsilon} = a$ for $\pi \in \Delta_b^*$, $a_b \in \Delta_b$ such that $\pi a_b = (a_{i1})_b \ldots (a_{in_i})_b$
for some $i \in I_b$, $b \in \mathfrak{R}$.

Theorem 4.7. $\mathfrak{Rat}_T(A') = \mathfrak{F}(\mathfrak{R}) = \mathfrak{L}$.

Proof. We first compute $(T^*)_{\varepsilon, \varepsilon}$. This computation is easy if we consider the blocks of T according to the partition $\{\{\varepsilon\}\} \cup \{\Delta_b^+ \,|\, b \in \mathfrak{R}\} \cup \{\Gamma\}$, where $\Gamma = \Gamma_T^+ - \bigcup_{b \in \mathfrak{R}} \Delta_b^+$. The only non-null blocks according to this partition are $\{\varepsilon\} \times \{\varepsilon\}$, $\{\varepsilon\} \times \Delta_b^+$, $\Delta_b^+ \times \{\varepsilon\}$, $\Delta_b^+ \times \Delta_b^+$, $b \in \mathfrak{R}$. Hence, by a result similar to Theorem 4.26 of Kuich, Salomaa [14], we obtain

$$(T^*)_{\varepsilon, \varepsilon} = (T(\{\varepsilon\}, \{\varepsilon\}) + \sum_{b \in \mathfrak{R}} T(\{\varepsilon\}, \Delta_b^+) T(\Delta_b^+, \Delta_b^+) T(\Delta_b^+, \{\varepsilon\}))^* =$$

$$= (z^0 + \sum_{b \in \mathfrak{R}} \sum_{i \in I_b} z_b a_{i1} \ldots a_{in_i})^*.$$

We show now $\mathfrak{L} \subseteq \mathfrak{Rat}_T(A')$. Fix a $b \in \mathfrak{R}$ and let $h : \Delta^* \to A^{2 \times 2}$ be the monoid morphism defined by

$$h(a) = \begin{pmatrix} a & 0 \\ 0 & a \end{pmatrix}, \ a \in A', \ h(z_b) = \begin{pmatrix} 0 & 1 \\ 0 & 0 \end{pmatrix}, \ h(z_{b'}) = h(z^0) = 0 \ \text{for} \ b' \in \mathfrak{R}, \ b' \neq b.$$

Clearly, h is compatible with A. We obtain

$$h((T^*)_{\varepsilon,\varepsilon}) = \Big(\sum_{i \in b} \begin{pmatrix} 0 & 1 \\ 0 & 0 \end{pmatrix} \begin{pmatrix} a_{i1} \dots a_{in_i} & 0 \\ 0 & a_{i1} \dots a_{in_i} \end{pmatrix} \Big)^* = \begin{pmatrix} 1 & b \\ 0 & 1 \end{pmatrix}$$

and infer that $b \in \mathfrak{Rat}_T(A')$. Hence, $\mathfrak{R} \subseteq \mathfrak{Rat}_T(A')$. Since $\mathfrak{Rat}_T(A')$ is an AFL-semiring, we infer $\mathfrak{F}(\mathfrak{R}) \subseteq \mathfrak{Rat}_T(A')$.

Conversely, we show now $\mathfrak{Rat}_T(A') \subseteq \mathfrak{L}$. Assume $a \in \mathfrak{Rat}_T(A')$. Then there exists a monoid morphism $h : \Delta^* \to \mathfrak{Rat}_T(A')^{Q \times Q}$ compatible with A, and $S \in \mathfrak{Rat}(A')^{1 \times Q}$, $P \in \mathfrak{Rat}(A')^{Q \times 1}$ such that $a = Sh((T^*)_{\varepsilon,\varepsilon})P$. Consider now the entries of

$$h((T^*)_{\varepsilon,\varepsilon}) = (h(z^0) + \sum_{b \in \mathfrak{R}} h(z_b)h(b))^*.$$

The entries of $h(b)$ are in $\mathfrak{F}(\mathfrak{R})$, the entries of $h(z^0)$ and $h(z_b)$ are in $\mathfrak{Rat}(A') \subseteq \mathfrak{F}(\mathfrak{R})$. Since only finitely many $h(z_b)$ are unequal to zero, the entries of $h(z^0) + \sum_{b \in \mathfrak{R}} h(z_b)h(b)$ are in $\mathfrak{F}(\mathfrak{R})$. Since $\mathfrak{F}(\mathfrak{R})$ is fully rationally closed, the entries of $h((T^*)_{\varepsilon,\varepsilon})$ are in $\mathfrak{F}(\mathfrak{R})$. This implies $a \in \mathfrak{F}(\mathfrak{R})$. $\square$

We have now achieved our main result, a complete characterization of AFL-semirings.

Corollary 4.8. *A family of elements $\mathfrak{L}$ is an AFL-semiring iff there exists a restart type T such that $\mathfrak{L} = \mathfrak{Rat}_T(A')$.* $\square$

Acknowledgement. Thanks are due to Martin Goldstern and Georg Karner for helpful discussions.

References.

[1] Berstel, J.: *Transductions and Context-Free Languages.* Teubner, 1979.

[2] Berstel, J. et Reutenauer, C.: *Les séries rationelles et leur langages.* Masson, 1984.

[3] Ginsburg, S.: *Algebraic and Automata-Theoretic Properties of Formal Languages.* North-Holland, 1975.

[4] Golan, J. S.: *The Theory of Semirings with Applications in Mathematics and Theoretical Computer Science.* Longman 1992.

[5] Goldstern, M.: *Vervollständigung von Halbringen.* Diplomarbeit Technische Universität Wien, 1985.

[6] Hebisch, U. und Weinert, H. J.: *Halbringe.* Teubner, 1993.

[7] Karner, G.: *On limits in complete semirings.* Semigroup Forum 45(1992) 148-165.

[8] Karner, G.: *On transductions of formal power series over complete semirings.* Theoretical Computer Science 98(1992) 27-39.

[9]   Krob, D.: *The equality problem for rational series with multiplicities in the tropical semiring is undecidable.* ICALP92, LNCS 623, 101–112.

[10] Kuich, W.: *The Kleene and the Parikh Theorem in Complete Semirings.* ICALP 87, LNCS 267(1987) 212–225.

[11] Kuich, W.: *ω-continuous semirings, algebraic systems and pushdown automata.* ICALP 90, LNCS 443(1990) 103–110.

[12] Kuich, W.: *Automata and languages generalized to ω-continuous semirings.* Theoretical Computer Science 79(1991) 137–150.

[13] Kuich, W.: *Automaten und Formale Sprachen.* Skriptum Technische Universität Wien, 1991.

[14] Kuich, W. and Salomaa, A.: *Semirings, Automata, Languages.* Springer, 1986.

[15] Salomaa, A. and Soittola, M.: *Automata-Theoretic Aspects of Formal Power Series.* Springer, 1978.

[16] Simon, I.: *Recognizable sets with multiplicities in the tropical semiring.* MFCS88, LNCS 324, 107–120.

Finite State Transformations of Images

Karel Culik[1] and Jarkko Kari[2]

[1] Depart. of Comp. Science, University of South Carolina,
Columbia, S.C. 29208 ***
[2] Academy of Finland

Abstract. Weighted finite transducers (WFT) are finite state devices
that serve as a powerful tool for describing and implementing a large
variety of image transformations and more generally linear operators on
real functions.
Here we show new results on WFT and demonstrate that WFT are indeed an excellent tool for image manipulation and more generally for
function transformation. We note that every WFA transformation is a
linear operator and show that most of the interesting linear operators on
real functions (on $[0,1]^2$) can be easily implemented by WFT. They include affine transformations, low-pass or high-pass filters, wavelet transform, (partial) derivatives, simple and multiple integrals.

1 Introduction

Weighted finite automata (WFA) have been introduced in [3, 4] as devices for
computing real functions on $[0,1]^n$. The main motivation has been to generate
functions on $[0,1] \times [0,1]$ interpreted as gray-tone images. In [4, 5] (see also [7])
we developed inference algorithms for WFA. Using the algorithm from [5] we can
efficiently encode any image (digitalized photograph) by a WFA. One of the best
image compression method has been developed on this basis. In [2] the generalized k-tape WFA have been introduced and in particular, the 2tape WFA called
weighted finite transducers (WFT) have been studied and shown to perform a
number of simple operations on images. When considered as only defining mappings on finite words WFT are a special case of rational transducers [10]. In [8]
the iterative WFT have been studied and shown to be strictly more powerful
image generators than MRFS [1].

Here we show new results on WFT and demonstrate that WFT are indeed an
excellent tool for image manipulation and more generally for function transformation. We note that every WFA transformation is a linear operator and show
that most of the interesting linear operators on real functions (on $[0,1]^2$) can be
easily implemented by the WFT. They include affine transformations, high-pass
or low-pass filters, wavelet transform, (partial) derivatives, simple and multiple
integrals. Since the family of WFA-functions is constructively closed under WFT
[2], an implementation of a transformation τ by a WFT is actually a proof of

*** Supported by the National Science Foundation under Grant No. CCR-9202396 the theorem stating that for each WFA A there effectively exists another WFA B that computes transformation τ of the function defined by A. In the case of integrals this has been shown in [3] for average preserving WFA. Now this result is extended to arbitrary WFA and to multiple integrals.

In Section 3 we demonstrate the surprising ability and flexibility of WFT to implement image transformations and linear function operators. We show that every piece-wise affine transformation can be implemented by WFT. The crucial, somewhat tricky example, of a WFT is the one that shifts an image by one pixel for any finite resolution. It leads us to implementation of filters, (partial) derivatives, and other linear operators.

2 Multiresolution functions, WFA and WFT

A *multiresolution function* f over alphabet Σ is a function $\Sigma^* \longrightarrow R$. It is *average preserving* if

$$\sum_{a \in \Sigma} f(ua) = p \cdot f(u), \text{ for all } u \in \Sigma^*, \tag{1}$$

where $p = |\Sigma|$ is the cardinality of the alphabet Σ.

Fig. 1. The addresses of quadrants

In image processing applications a four letter alphabet $\Sigma = \{0,1,2,3\}$ will be used. Words of length k over Σ will be interpreted as addresses of subsquares in the division of the unit square into $2^k \times 2^k$ subsquares as follows: Each letter refers to one quadrant as shown in Fig. 1. Word wa addresses the quadrant a of the subsquare addressed by w. A multiresolution function $f : \Sigma^* \longrightarrow [0,1]$ defines a sequence of gray-tone images with increasing resolutions: Its restriction to Σ^k defines an image in resolution $2^k \times 2^k$. The gray-tone intensity of a point in the subsquare addressed by $w \in \Sigma^k$ is $f(w)$. The images at different resolutions are compatible if the multiresolution function is average preserving. In this case one can easily move from higher resolution to lower one by simply computing the averages of the intensities inside each subsquare.

A multiresolution function f over $\Sigma = \{0,1,2,3\}$ defines an *infinite resolution image* $\hat{f} : [0,1]^2 \longrightarrow [0,1]$ if the sequence $f_{|\Sigma^k}$, $k = 0,1,2,\ldots$ of finite resolution images converges point-wise to $\hat{f}$. In a similar fashion, WFA over 2^m letter alphabet are used to define functions $[0,1]^m \to [0,1]$ for all $m \geq 1$.

An m-state *weighted finite automaton (WFA)* A over alphabet Σ is defined by a row vector $I^A \in R^{1 \times m}$ (called the initial distribution), a column vector

$F^A \in R^{m \times 1}$ (the final distribution), and weight matrices $W_a^A \in R^{m \times m}$ for all $a \in \Sigma$. The WFA A defines a multiresolution function f_A over Σ by

$$f_A(a_1 a_2 \ldots a_k) = I^A \cdot W_{a_1}^A \cdot W_{a_2}^A \cdot \ldots \cdot W_{a_k}^A \cdot F^A.$$

The WFA A is called average preserving if

$$\sum_{a \in \Sigma} W_a^A \cdot F^A = p \cdot F^A, \tag{2}$$

where $p = |\Sigma|$ is the cardinality of the alphabet Σ. In other words, a WFA is average preserving if its final distribution is an eigenvector of $\sum_{a \in \Sigma} W_a^A$ corresponding to its eigenvalue $|\Sigma|$. It is known (see [4]) that the multiresolution function computed by an average preserving WFA is average preserving, and that every average preserving multiresolution function computable by a WFA can be computed by an average preserving WFA.

Analogously to WFA, an n-state *weighted finite transducer* (WFT) M from alphabet Σ_1 into alphabet Σ_2 is specified by

1. weight matrices $W_{a,b} \in R^{n \times n}$ for all $a \in \Sigma_1 \cup \{\varepsilon\}$ and $b \in \Sigma_2 \cup \{\varepsilon\}$,
2. a row vector $I \in R^{1 \times n}$, called the initial distribution, and
3. a column vector $F \in R^{n \times 1}$, called the final distribution.

The WFT M is called ε-free if weight matrices $W_{\varepsilon,\varepsilon}$, $W_{a,\varepsilon}$ and $W_{\varepsilon,b}$ are zero matrices for all $a \in \Sigma_1$ and $b \in \Sigma_2$.

The WFT M defines function $f_M : \Sigma_1^* \times \Sigma_2^* \longrightarrow R$, called weighted relation between Σ_1^* and Σ_2^*, by

$$f_M(u, v) = I \cdot W_{u,v} \cdot F, \text{ for all } u \in \Sigma_1^*, v \in \Sigma_2^*,$$

where

$$W_{u,v} = \sum_{\substack{a_1 \ldots a_k = u \\ b_1 \ldots b_k = v}} W_{a_1,b_1} \cdot W_{a_2,b_2} \cdot \ldots \cdot W_{a_k,b_k}, \tag{3}$$

if the sum converges. (If the sum does not converge, $f_M(u, v)$ remains undefined.) In (3) the sum is taken over all decompositions of u an v into symbols $a_i \in \Sigma_1 \cup \{\varepsilon\}$ and $b_i \in \Sigma_2 \cup \{\varepsilon\}$, respectively.

In the special case of ε-free transducers,

$$f_M(a_1 a_2 \ldots a_k, b_1 b_2 \ldots b_k) = I \cdot W_{a_1,b_1} \cdot W_{a_2,b_2} \cdot \ldots \cdot W_{a_k,b_k} \cdot F,$$

for $a_1 a_2 \ldots a_k \in \Sigma_1^k$, $b_1 b_2 \ldots b_k \in \Sigma_2^k$, and $f_M(u, v) = 0$, if $|u| \neq |v|$.

Let $\rho : \Sigma_1^* \times \Sigma_2^* :\longrightarrow R$ be a weighted relation and $f : \Sigma_1^* \longrightarrow R$ a multiresolution function. The application of ρ to f is the multiresolution function $g = \rho(f) : \Sigma_2^* \longrightarrow R$ over Σ_2 defined by

$$g(v) = \sum_{u \in \Sigma_1^*} f(u)\rho(u, v), \text{ for all } v \in \Sigma_2^*,$$

provided the sum converges. The application $M(f)$ of WFT M to f is defined as the application of the weighted relation f_M to f, i.e. $M(f) = f_M(f)$.

If $\Sigma_1 = \Sigma_2 = \{0, 1, 2, 3\}$ the weighted relation ρ can be applied also on (integrable) infinite resolution images $\alpha : [0, 1)^2 \longrightarrow [0, 1]$. Assume there exists an (unique) average preserving multiresolution function f such that $\hat{f} = \alpha$. We define $\rho(\alpha) = \widehat{\rho(f)}$, provided $\rho(f)$ exists and converges to an infinite resolution image $\widehat{\rho(f)}$. The application of a WFT M to α is defined as the application of f_M to α.

Lemma 1. *WFT M is a linear operator $\mathbb{R}^{\Sigma_1^*} \longrightarrow \mathbb{R}^{\Sigma_2^*}$. In other words,*

$$M(r_1 f_1 + r_2 f_2) = r_1 M(f_1) + r_2 M(f_2),$$

for all $r_1, r_2 \in \mathbb{R}$ and $f_1, f_2 : \Sigma_1^* \to \mathbb{R}$. More generally, any weighted relation acts as a linear operator. □

It follows naturally from Lemma 1 that weighted relations over four letter alphabet act as linear operators of infinite resolution images.

In [2] the application of WFT M to WFA A was defined. For simplicity we define it here only for ε-free WFT. The application of an ε-free WFT M to an m-state WFA A over alphabet Σ_1 defined by initial distribution I^A, final distribution F^A and weight matrices W_a^A, $a \in \Sigma_1$, is the mn-state WFA $B = M(A)$ over alphabet Σ_2 with initial distribution $I^B = I \otimes I^A$, final distribution $F^B = F \otimes F^A$ and weight matrices

$$W_b^B = \sum_{a \in \Sigma_1} W_{a,b} \otimes W_a^A, \text{ for all } b \in \Sigma_2.$$

Here, $\otimes$ denotes the ordinary tensor product of matrices (called also Kronecker product or direct product), defined as follows: Let T and Q be matrices of sizes $s \times t$ and $p \times q$, respectively. Then their tensor product is the $st \times pq$ matrix

$$T \otimes Q = \begin{pmatrix} T_{11}Q \cdots T_{1t}Q \\ \vdots \qquad \vdots \\ T_{s1}Q \cdots T_{st}Q \end{pmatrix}$$

. Clearly, $f_B = M(f_A)$, i.e. the multiresolution function defined by B is the same as the application of the WFT M to the multiresolution function computed by WFA A.

Let the cardinalities of the alphabets be $p = |\Sigma_1|$ and $q = |\Sigma_2|$. We call the ε-free WFT *average preserving* if for all $a \in \Sigma_1$ holds

$$\sum_{b \in \Sigma_2} W_{a,b} \cdot F = \frac{q}{p} F. \tag{4}$$

In other words, the final distribution F is an eigenvector of matrices $\sum_{b \in \Sigma_2} W_{a,b}$ corresponding to eigenvalue $\frac{q}{p}$, for all $a \in \Sigma_1$.

According to the next theorem the application of an average preserving WFT to an average preserving WFA is an average preserving WFA. Moreover, application to any average preserving multiresolution function is average preserving. The following is proved in [6].

Theorem 2. *Let M be an ε-free WFT. $M(A)$ is average preserving for every average preserving WFA A if and only if M is average preserving. Moreover, if M is average preserving, $M(f)$ is average preserving for every average preserving multiresolution function f.*

The composition of weighted relations was defined in [2]. Let us recall its definition, as well as definitions of some other operations. Let $\sigma : (\Sigma_1^\sigma)^* \times (\Sigma_2^\sigma)^* \longrightarrow R$ and $\rho : (\Sigma_1^\rho)^* \times (\Sigma_2^\rho)^* \longrightarrow R$ be weighted relations. Define $\sigma \circ \rho$ (composition), $\sigma + \rho$ (sum), $r\rho$ (product with scalar $r \in R$), $\sigma \cdot \rho$ (concatenation) and ρ^+ (catenation closure) as follows:

$$(\sigma \circ \rho)(u, v) = \sum_{w \in \Sigma_2^*} \sigma(u, w)\rho(w, v),$$

$$(\sigma + \rho)(u, v) = \sigma(u, v) + \rho(u, v),$$

$$(r\rho)(u, v) = r\rho(u, v),$$

$$(\sigma \cdot \rho)(u, v) = \sum_{\substack{u = u_1 u_2 \\ v = v_1 v_2}} \sigma(u_1, v_1)\rho(u_2, v_2), \text{ and}$$

$$(\rho^+)(u, v) = \rho(u, v) + (\rho \cdot \rho)(u, v) + (\rho \cdot \rho \cdot \rho)(u, v) + \dots$$

$$= \sum_{\substack{u = u_1 \dots u_k \\ v = v_1 \dots v_k}} \rho(u_1, v_1)\rho(u_2, v_2)\dots\rho(u_k, v_k).$$

In the case of composition it is assumed that $\Sigma_2^\sigma = \Sigma_1^\rho$, and in the cases of addition and concatenation that $\Sigma_1^\sigma = \Sigma_1^\rho$ and $\Sigma_2^\sigma = \Sigma_2^\rho$. The composition is a weighted relation between $(\Sigma_1^\sigma)^*$ and $(\Sigma_2^\rho)^*$ while all others are between $(\Sigma_1^\sigma)^*$ and $(\Sigma_2^\sigma)^*$. In the definition of the catenation closure the sum is taken over all decompositions of words u and v into subwords u_i and v_i including empty words. It is possible that the sum does not converge, in which case the catenation closure is not defined. The sum in the definition of composition does not always converge either. However, if the weighted relations are defined by ε-free WFT the composition is always defined.

The following formulas for the applications of the weighted relations to multiresolution function $f : (\Sigma_1^\sigma)^* \longrightarrow R$ follow from the definitions above:

$$(\sigma \circ \rho)(f) = \rho(\sigma(f)),$$

$$(\sigma + \rho)(f) = \sigma(f) + \rho(f),$$

$$(r\rho)(f) = r\rho(f),$$

$$((\sigma \cdot \rho)(f))(w) = \sum_{w = w_1 w_2} (\sigma(f))(w_1)(\rho(f))(w_2), \text{ and}$$

(a) Original 512 × 512, 8 bpp   (b) Regenerated by a WFA, 0.07346 bpp

Fig. 2. Image Carol

$$((\rho^+)(f))\,(w) = (\rho(f))\,(w) + ((\rho \cdot \rho)(f))\,(w) + ((\rho \cdot \rho \cdot \rho))\,(f)(w) + \cdots$$
$$= \sum_{w = w_1 \ldots w_k} (\rho(f))\,(w_1) \ldots (\rho(f))\,(w_k).$$

Again, the catenation closure is defined only if the sum converges (the sum is over all decompositions of w into subwords).

The following theorems are proved in [6].

Theorem 3. *For ε-free WFT A and B there effectively exists WFT $A \circ B, A + B, rA, A \cdot B$ and A^+ such that $f_{A \circ B} = f_A \circ f_B$, $f_{A+B} = f_A + f_B$, $f_{rA} = rf_A$ for all $r \in R$, $f_{A \cdot B} = f_A \cdot f_B$ and $f_{A+} = f_A^+$. In the case of catenation closure we naturally assume that f_A^+ is defined.*

Theorem 4. *Let A and B be average preserving ε-free WFT. Then $A \circ B$, $A + B$ and rA are average preserving as well.*

3 Examples

Now, we will demonstrate that WFT can implement almost every useful linear operator. It has been shown in [2] that every affine transformation on R^2 is realized by a WFT. We give few examples of (piecewise) affine transformations restricted to $[0, 1]^2$. We will illustrate most our WFT by mapping image Carol shown in Fig. 2. The original, resolution 512 × 512 8 bits per pixel, is in Fig. 2(a). The image in Fig. 2(b) is regenerated by WFA stored in 2406 bytes (109 × compression 0.07346 bpp).

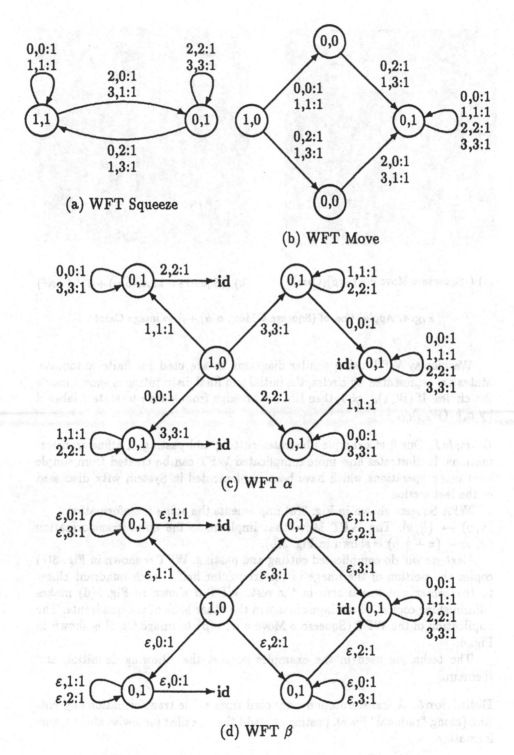

(a) WFT Squeeze

(b) WFT Move

(c) WFT α

(d) WFT β

Figure 3: Four WFT used in the next figure

 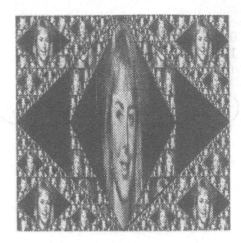

(a) $((\text{Squeeze} \circ \text{Move} \circ \alpha) + \beta)(\text{Carol})$ (b) $((\text{Squeeze} \circ \text{Move} \circ \alpha) + \beta)^2(\text{Carol})$

Fig. 4. Application of $(\text{Squeeze} \circ \text{Move} \circ \alpha) + \beta$ to image Carol

We display WFT using similar diagrams as are used for finite automata. States are represented by circles, the initial and final distribution is shown inside the circles. If $(W_{a,b})_{i,j} \neq 0$, then there is an edge from state i to state j labeled by $a, b : (W_{a,b})_{i,j}$.

Example 1. Our first example illustrates cutting and pasting of affine transformations. It illustrates how more complicated WFT can be created from simple ones using operations which have been implemented in System wftx discussed in the last section.

WFA Squeeze shown in Fig. 3(a) implements the affine transformation $(x, y) \rightarrow (\frac{x}{2}, y)$. The WFT Move that implements the affine transformation $(x, y) \rightarrow (x + \frac{1}{4}, y)$ is shown in Fig. 3(b).

Next we will do complicated cutting and pasting. WFT α shown in Fig. 3(c) copies the portion of the image in the triangular half of each quadrant closer to the center and leaves zero in the rest. WFT β shown in Fig. 3(d) makes "diminishing copies" of the input image in the other halves of the quadrants. The application of the WFT $(\text{Squeeze} \circ \text{Move} \circ \alpha) + \beta)$ to image Carol is shown in Fig. 4.

The technique used in the examples suggest the following definition and theorem.

Definition 5. A transformation composed from affine transformations by cutting (along "rational" lines), pasting and addition is called piecewise affine transformation.

shift right:

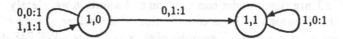

shift left:

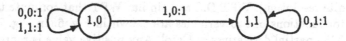

shift up in two dimensions:

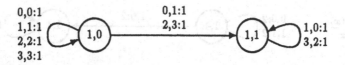

Fig. 5. WFT computing circular shifts by one pixel

Theorem 6. *Every piecewise affine transformation can be implemented by a WFT.*

Proof outline. It has been shown in [2] that every affine transformation (restricted to $[0, 1]^2$) can be implemented by a WFT. A practice of "rational" cutting, i.e. the restriction to a regular set can, clearly, be implemented by a WFT. Pasting is a special case of addition of WFT. Hence by the closure of WFT under composition and addition we can implement every piecewise affine transformation. Actually, using the operation of concatenation we can paste infinite number of copies as long as the infinite copying can be expressed by a regular set (WFA if the grayness is not uniform). □

Note that every affine transformation is implemented by an average preserving WFT. This is not necessary case for piecewise affine transformation, however, every piecewise affine transformation can be implemented by a WFT in which all states visited more than uniformly bounded number of times satisfies the average preserving condition (4).

Many interesting WFT can be designed using the technique of value shifting. In one dimension, the shift by one pixel (one step in a function table) for resolution $2^n, n \geq 1$ requires to move the value at address $w01^r$ to the address $w10^r$ for $0 \leq r \leq n - 1$ and $w \in \Sigma^{n-r-1}$. Therefore, somewhat surprisingly, we easily design a WFT with only 2 states which performs this shift. By appropriately choosing the initial distribution we can make the shift circular. WFT computing various shifts are shown in Fig. 5.

Since WFT are closed under composition and addition we, clearly, can design a WFT that computes any linear combination of the values of each cell and any fixed finite set of its neighbors. Examples of such WFT are a WFT that simulates the moves of a knight on a $2^n \times 2^n$ chess-board, WFT that implements any type of low or high pass filter or WFT Diff which for any finite resolution computes $\frac{f(x+h)-f(x)}{h}$ for table step h. Thus WFT Diff computes the derivative $\frac{df(x)}{dx}$ in the limit (infinite resolution). WFT Diff and similar WFT that compute the partial derivates for functions of two variables are shown in Fig. 6. In Fig. 7 we show the sign of the partial derivatives of Carol. Any positive value is represented by white color, negative by black.

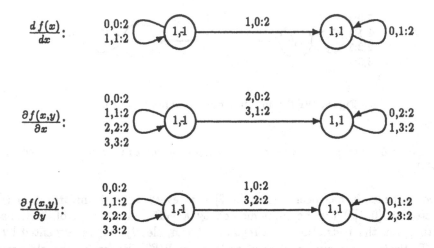

Fig. 6. WFT computing derivative and partial derivatives

Computing of integrals requires to compute linear combinations of values from an unbounded number of neighbors but that can be done, too, see [6].

In all the examples above the same small WFT computes the desired transformation for every resolution and for the infinite resolution as well. For the shifts the limiting case is the identity transformation. However, it is easy to give an average preserving WFT which does not converge to any mapping on infinite resolution. Consider one state WFT with I=F=1, and transitions (0,0:1),(1,0:1), (2,2:1) and (3,2:1).

Theorem 7. *Let* $\Sigma = \{0,1\}$ *and* A *be a WFA over* Σ, *Then we can construct WFA* B *such that*

(i) $\hat{f}_B(x) = \frac{df_A(x)}{dx}$ *for all* $x \in [0,1]$ *for which* $\frac{df_A(x)}{dx}$ *exists.*

(ii) $\hat{f}_B(x) = \int_0^x f_A(t)\,dt$ *if the Rieman integral exists.*

$$\frac{\partial \text{Carol}(x,y)}{\partial x} \qquad\qquad \frac{\partial \text{Carol}(x,y)}{\partial y} \qquad\qquad \frac{\partial \text{Carol}(x,y)}{\partial x \partial y}$$

Fig. 7. The partial derivatives of Carol

Proof. Follows from the existence of WFT that compute the operators and the closure of WFA under WFT [2].

□

Note that (ii) is an extension of a result in [3] where this result was shown for the restricted case of the average preserving WFA. However, if WFA A has n states the construction in [3] yields B with $n + 1$ states while the application of our two state integrating WFA yields WFA B with $2n$ states.

WFT Mallat shown in Fig. 8 computes the coefficients of the discrete Haar wavelet transform for any finite resolution and presents them in the Mallat form [9]. It computes the continuous Haar wavelet transform for the infinite resolution. The image *Mallat*(Carol) is shown in Fig. 8 with the contrast increased. In order to display better all the coefficients we compute them unscaled that is use wavelets which are not orthonormal. To get the orthonormal case it is sufficient just to adjust the weights of the ε transitions of our WFT.

All the examples of image transformations here were produced using a menu-driven X-windows based system wftx implemented by P.Rajcani. In wftx images are represented either in pixel form or by WFA. The conversions between these representations are implemented using the WFA inference and the WFA decoding algorithms from [5, 7]. All the operations on images, WFA and WFT, except for the concatenation closure are implemented in wftx. In particular, a WFT can be applied to an image in pixel form (resolution $2^n \times 2^n$, $n \geq 1$) and produce again an image in possibly different resolution, or a WFT can be applied to a WFA and produce a WFA. Any of the WFA computing integrals is a typical example of a WFA which has a small number of states but it is highly nondeterministic. For such a WFT, it is much more efficient to apply it to an image in the WFA representation even if it requires converting the image to the WFA representation and then decoding the resulting WFA.

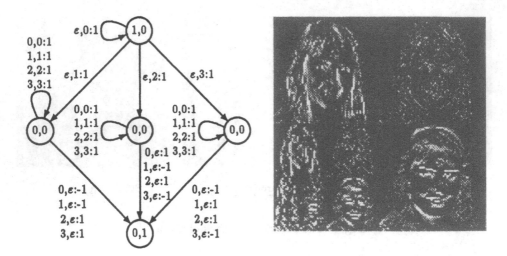

Fig. 8. WFT Mallat computing Haar wavelet coefficients in Mallat form and the image Mallat(Carol)

References

1. Culik II, K. and Dube, S., Balancing Order and Chaos in Image Generation, *Computer and Graphics* **17** 4 (1993) 465-486
2. Culik II, K. and Fris, I., Weighted Finite Transducers in Image Processing. *Discrete Applied Mathematics* (to appear)
3. Culik II, K. and Karhumäki, J., Automata Computing Real Functions, *SIAM J. on Computing* **23** 4 (1994) 789-814
4. Culik II, K. and Kari, J., Image Compression Using Weighted Finite Automata, *Computer and Graphics* **17** 3 (1993) 305-313
5. Culik II, K. and Kari, J., Image-Data Compression Using Edge-Optimizing Algorithm for WFA Inference, *Journal of Information Processing and Management* **30** 6 (1994) 829-838
6. Culik II, K. and Kari, J., Finite state transformations of images, submitted to *Discrete Applied Mathematics*
7. Culik II, K. and Kari, J., Efficient Inference Algorithm for Weighted Finite Automata, in *Fractal Image Encoding and Compression*, ed. Y.Fisher, Springer-Verlag (1994)
8. Culik II, K. and Rajcani, P., Iterative Weighted Finite Transductions, *Acta Informatica* (to appear)
9. DeVore, R. A., Jawerth, B. and Lucier, B. J., Image Compression through Wavelet Transform Coding, *IEEE Transactions of Information Theory* **38** (1992) 719-746
10. Eilenberg, S., *Automata, Languages and Machines*, Vol. A, Academic Press, New York (1974)

Post Correspondence Problem: Words Possible as Primitive Solutions *

Marjo Lipponen

Department of Mathematics, University of Turku
20500 Turku, Finland
e-mail: marlip@polaris.cc.utu.fi

Abstract. Three types of "primitive" or "prime" solutions for PCP have earlier been investigated. The sets of words belonging to one of the three types (for some instance of PCP) form an increasing hierarchy. In our main result we show that the hierarchy is proper, except the binary case. We also give a sharpened characterization of the finite type.

1 Introduction

The Post Correspondence Problem, [8], is one of the "cornerstones of undecidability". Reduction from the Post Correspondence Problem is one of the most common undecidability arguments in many areas of computer science, [9].

It is well known that the set of solutions for an instance of the Post Correspondence Problem is a star language: whenever w_1 and w_2 are solutions, then so is w_1w_2. It is natural to consider w_1 and w_2 to be "simpler" solutions than w_1w_2. There are also other ways to define what it means for a solution to be "simple", "prime" or "primitive". Such considerations have turned out to be theoretically important in various contexts. For example, in the purely morphic characterization of recursively enumerable languages given in [1], attention is restricted to solutions w such that no proper prefix of w is a solution. Because of this restriction the use of an intersecting regular language becomes unnecessary in the characterization. (The intersection is needed in the normal construction to exclude the garbage produced by unrestricted equality sets.) Solutions restricted in this way are in the sequel referred to as *F-primes*: no (nonempty) final subword can be removed in such a way that the remainder is still a solution. Similarly, a solution w for an instance of the PCP is termed an *S-prime* if no nonempty subword can be removed, and a *P-prime* if no nonempty scattered subword can be removed. These three different types of prime solutions were first introduced in [10]. In [7] the investigation was extended to concern analogous problems for the target alphabet.

In this work we take another point of view by studying words that can appear as a prime solution, for some instance of PCP. Such a study of primitive words and languages was initiated in [6].

---

* Work supported by Project 11281 of the Academy of Finland

Our main result in Section 3 studies the two crucial inclusion problems. We know, by definition, that every P-prime is an S-prime, and every S-prime is an F-prime. Thus the set of P-words is included in the set of S-words which, in turn, is included in the set of F-words. Our main result consists in showing that both of these inclusions are proper, except in the case of a binary alphabet.

Section 4 consists of a more detailed study of P-prime solutions. We consider two natural subtypes of the type of P-prime solutions and show that all type combinations are indeed possible in this framework.

The suitability of the Post Correspondence Problem for reduction arguments is due to the fact that in some sense the essence of computations is captured by PCP. Thus simple solutions of PCP mean simplifications of computations and the results contribute on an abstract level to the understanding of computations.

The study of P-words, S-words and F-words can also be viewed as a contribution to the general combinatorics on words. The property of being an S-word, for instance, is an abstract property of words that is so far rather poorly understood: of two similar-looking words, one can be an S-word, whereas the other fails to be an S-word.

2 Definitions and earlier results

Let g and h be nonerasing morphisms of Σ^* into Δ^*, where Σ and Δ are finite alphabets. The *equality set* between g and h is defined by

$$E(g, h) = \{w \in \Sigma^+ \mid g(w) = h(w)\}.$$

The pair (g, h)=PCP is also referred to as an *instance of the Post Correspondence Problem*. Words in $E(g, h)$, if any since the empty word λ is not included, are called *solutions* of PCP.

For a word w over Σ^*, we now consider the sets of words obtained from w by removing a final subword, a subword or a scattered subword, respectively. Define

$$fin\,(w) = \{v_1 \mid w = v_1 x, \text{ for some } x \in \Sigma^*\},$$
$$sub\,(w) = \{v_1 v_2 \mid w = v_1 x v_2, \text{ for some } v_1, v_2, x \in \Sigma^*\},$$
$$scatsub\,(w) = \{v_1 \cdots v_k \mid w = x_1 v_1 \cdots x_k v_k x_{k+1}, \text{ for some } x_i, v_i \in \Sigma^*\}.$$

(Observe that *fin* is not the same here as in [3].)

We can now determine three further sets, as follows:

$$F(g, h) = \{w \in E(g, h) \mid fin\,(w) \cap E(g, h) = \{w\}\},$$
$$S(g, h) = \{w \in E(g, h) \mid sub\,(w) \cap E(g, h) = \{w\}\},$$
$$P(g, h) = \{w \in E(g, h) \mid scatsub\,(w) \cap E(g, h) = \{w\}\}.$$

Words in the three sets are called *F-prime*, *S-prime* and *P-prime solutions* for the instance PCP=(g, h), respectively.

It is a direct consequence of the definitions that

$$P(g, h) \subseteq S(g, h) \subseteq F(g, h) \subseteq E(g, h).$$

One or both of the first inclusions may be strict whereas the third inclusion is always strict, provided $E(g, h)$ is nonempty. If $E(g, h)$ is nonempty then so must be the three other sets. In addition, each of the four sets is recursive.

The triple (p, s, f), where p, s and f are the cardinalities of the sets $P(g, h)$, $S(g, h)$ and $F(g, h)$, respectively, is defined to be the *primality type* of the instance PCP=(g, h). Thus p, s and f are nonnegative integers or ∞. These triples were fully characterized in [10].

Theorem 1. *A triple (p, s, f) is a primality type if and only if either $p = s = f = 0$, or else each of the following conditions (i)–(iii) holds: (i) $1 \leq p \leq s \leq f$, (ii) p is finite, (iii) if $s < f$ then $f = \infty$. An example for each possible type can be effectively constructed.*

We now consider words that can describe a prime solution, for some instance of PCP. We say that a word w over Σ is a *P-word* if, for some instance (g, h), w is in $P(g, h)$. *S-words* and *F-words* are defined similarly.

The following lemma due to [6] gives a way of constructing words that are not F-words (and thus not S- and P-words either). Here $\psi(w)$ means the Parikh vector.

Lemma 2. *If w_1 is a nontrivial prefix of w satisfying $\psi(w_1) = r\psi(w)$ for some (rational) number r, then w is not an F-word.*

The lemma gives, among others, the following examples: $ab^2a^2b^4$, $ab^6a^3b^2$.

Also the following two lemmas, originally due to [5], will be needed in the sequel.

Lemma 3. *If $uv = vz$ holds for some words u, v, z, where u is nonempty, then*

$$u = xy, \quad v = (xy)^k x, \quad z = yx,$$

for some words x, y and integer $k \geq 0$. If $uv = vu$ holds for some nonempty words u and v, then u and v are powers of the same word.

Lemma 4. *For u, v, w in Σ^+ and $m, n, p \geq 2$, if $u^m v^n = w^p$, then u, v and w are powers of the same word.*

3 Strictness of the Inclusions

We begin with an exception to the general case. The following result is due to [4]. A morphism g is *periodic* if there exists a word u over Δ such that $g(i) \in u^+$ for each $i \in \Sigma$.

Theorem 5. *Let g and h be morphisms over a binary alphabet such that at least one of them is periodic. Then*

$$P(g, h) = S(g, h).$$

Thus, under this additional assumption, in a binary alphabet P-words and S-words are the same. If g and h are both injective and $E(g, h) = \{u, v\}^+$ then u and v are S-prime solutions. However, it is conceivable that $u \in scatsub(v)$. We conjecture that this is not possible in a binary alphabet, although it is possible in bigger alphabets.

For F-words we need the following lemma, due to [2], which characterizes all the instances agreeing on a word *aabb*.

Lemma 6. *The word aabb belongs to $E(g, h)$, with $|g(a)| > |h(a)|$, iff there exist nonnegative integers t_1, t_2 and t_3 and words α and β such that*

	a	b
g	$\alpha(\beta\alpha)^{t_1}\alpha(\beta\alpha)^{t_2}\beta$	$\alpha(\beta\alpha)^{t_3}$
h	$\alpha(\beta\alpha)^{t_1}$	$\beta(\alpha\beta)^{t_2}\alpha(\alpha\beta)^{t_3}\alpha$

Theorem 7. *The word 111212 is an F-word but not an S-word.*

Proof. The word 111212 is clearly an F-prime solution for the instance (g, h) defined by the table

	1	2
g	a^2	a
h	a	a^3

In order to prove that it is not an S-word we use Lemma 6 by denoting $a = 1$ and $b = 12$. Arguing indirectly, we assume that 111212 is an S-word and, hence, in the equality set of some instance (g, h). By our notations Lemma 6 becomes applicable. Thus we have

$$g(1) = g(a) = \alpha(\beta\alpha)^{t_1}\alpha(\beta\alpha)^{t_2}\beta$$
$$g(12) = g(b) = \alpha(\beta\alpha)^{t_3}$$
$$h(1) = h(a) = \alpha(\beta\alpha)^{t_1}$$
$$h(12) = h(b) = \beta(\alpha\beta)^{t_2}\alpha(\alpha\beta)^{t_3}\alpha$$

The first two equations imply

$$\alpha(\beta\alpha)^{t_3} = \alpha(\beta\alpha)^{t_1}\alpha(\beta\alpha)^{t_2}\beta g(2)$$

and the last two

$$\beta(\alpha\beta)^{t_2}\alpha(\alpha\beta)^{t_3}\alpha = \alpha(\beta\alpha)^{t_1}h(2).$$

Therefore $\alpha\beta = \beta\alpha$, independently of the exponents t_1 and t_2. This implies that $112 \in sub(111212)$ is always a solution. $\qquad\square$

The above proof argument can be applied similarly in alphabets with more than two letters to show that the word

$$(123\ldots(k-1))^2(123\ldots k)^2 \in \{1,2,3,\ldots,k\}^*$$

is an F-word but not an S-word. Indeed, that this is an F-word is seen from the instance

$$
\begin{array}{c|cccccc}
 & 1 & 2 & 3 & \ldots & k-1 & k \\
\hline
g & a^2 & a^2 & a^2 & \ldots & a^2 & a \\
h & a & a & a & \ldots & a & a^{2k-1}
\end{array}
$$

We have the following result.

Theorem 8. *Over every alphabet Σ, $|\Sigma| \geq 2$, there exists an F-word that is not an S-word.*

The case for P- and S-words in alphabets with more than two letters now remains. The following result was originally suggested in [6] and later corrected by me by adding a new condition.

Lemma 9. *Let $w = avava$ where a is a letter and v is a word over Σ, $|\Sigma| \geq 2$, and, in addition, v begins or ends with a. If $w \in E(g,h)$ for some instance (g,h) then $g(a)$, $g(v)$, $h(a)$ and $h(v)$ are powers of the same word.*

Proof. We assume that the instance (g,h) is nontrivial, that is, $g(a) \neq h(a)$ for all $a \in \Sigma$. Without loss of generality, we denote

$$g(av) = \alpha, \quad h(av) = \beta = \alpha x, \quad |x| \geq 1,$$
$$g(a) = \alpha_1, \quad h(a) = \beta_1, \quad \alpha = \alpha_1\alpha_2, \quad \beta = \beta_1\beta_2.$$

It follows that $|x| < |\beta_2|$ and, moreover, $\alpha_1 = \beta_1 y$ where $|y| \geq 1$.

Writing the equation $g(w) = h(w)$ in the new notation

$$\beta_1 y \alpha_2 \beta_1 y \alpha_2 \beta_1 y = \beta_1\beta_2\beta_1\beta_2\beta_1 = \beta_1 y \alpha_2 x \beta_1 y \alpha_2 x \beta_1,$$

we obtain

$$\beta_1 y \alpha_2 \beta_1 y = x \beta_1 y \alpha_2 x \beta_1.$$

This implies that $|y| = 2|x|$. We write $y = y_1 y_2$ where $|y_1| = |y_2| = |x|$. Our preceding equation reads now

$$\beta_1 y_1 y_2 \alpha_2 \beta_1 y_1 y_2 = x \beta_1 y_1 y_2 \alpha_2 x \beta_1.$$

Considering subwords of the same length, we obtain

$$\beta_1 y_1 = x\beta_1, \quad y_1 = y_2, \quad \alpha_2\beta_1 y_1 y_1 = y_1\alpha_2 x\beta_1, \tag{1}$$

where we have already used the second equation in the third one. By Lemma 3 there exist words u and q such that $x = uq$, $y_1 = qu$ and $\beta_1 = (uq)^k u$. Consequently, we obtain from the last equation of (1) above

$$\alpha_2 uq = qu\alpha_2,$$

which yields that $qu = rs$, $uq = sr$ and $\alpha_2 = (rs)^l r$. Writing, without loss of generality, $q = rp$, we obtain $s = pu$ and $urp = pur$, which implies

$$p = z^i, \quad ur = z^j, \tag{2}$$

for some z. Futhermore, $uq = urp = z^{i+j}$.

Combining our knowledge, we have

$$\begin{aligned}
\alpha_1 &= (uq)^{k+2}u = (z^{i+j})^{k+2}u, \\
\alpha_2 &= r(uq)^l = r(z^{i+j})^l, \\
\beta_1 &= (uq)^k u = (z^{i+j})^k u, \\
\beta_2 &= r(uq)^{l+3} = r(z^{i+j})^{l+3}.
\end{aligned} \tag{3}$$

To complete the proof we shall use the fact that v either begins with a or ends with it. This is exhibited in two parts.

Let $v = ca$. By the last two equations in (3),

$$r(z^{i+j})^{l+3} = \beta_2 = h(v) = h(c)h(a) = h(c)\beta_1 = h(c)(z^{i+j})^k u. \tag{4}$$

We first suppose that $|u| > |z|$. Thus $u = zd$ and $u = ez$ for some d, e by the equations (2) and (4), and there exist n, o such that $e = no$, $d = on$, $z = (no)^t n$ for some t. Equation (2) now gives us

$$(no)^t nonr = [(no)^t n]^j$$
$$onr = [(no)^t n]^{j-1}.$$

Reading the words from the left, we obtain $on = no$ and thus there exists an f such that $o = f^{i_1}$ and $n = f^{i_2}$, as intended.

If, however, $t = 0$ then

$$onr = n^{j-1}$$

and either $n = om$ or $o = n\mu$. In the first case we obtain

$$omr = m(om)^{j-2},$$

which yields that $om = mo$. On the other hand, in the latter case we repeat the previous consideration for the word μ instead of o and in finitely many steps we find a word f such that $o = f^{i_1}$ and $n = f^{i_2}$, as intended. So the claim follows if $|u| > |z|$. (Obviously, it follows also in the case $|u| = |z|$.)

Next we suppose that $|u| < |z|$. In the same way as before, $z = ud$ and $z = eu$ for some d, e by the equations (2) and (4), and, further, $e = no$, $d = on$, $u = (no)^t n$. Equation (4) now gives us

$$r(z^{i+j})^{l+2}[(no)^t non]^{i+j} = h(c)(z^{i+j})^{k-1}[(no)^t non]^{i+j}(no)^t n,$$

which, by reading the words from the right, implies that $no = on$.

If, however, $k = l = 0$ then the equation (2) yields that $r = on[(no)^{t+1}n]^{j-1}$ and

$$on[(no)^{t+1}n]^{j-1} = r = \alpha_2 = g(c)\alpha_1 = g(c)z^{2i+2j}u = g(c)[(no)^{t+1}n]^{2i+2j}(no)^t n.$$

Considering the suffixes we obtain $no = on$, as intended.

The second case $v = ac$ goes in the same way. We just interchange the roles of u and r. $\square$

Theorem 10. *The word* $w = a^2b^3c^3a^4b^3c^3a^3$ *is an S-word but not a P-word.*

Proof. By denoting $v = ab^3c^3a^2$ we have $w = avava$. Further, by Lemma 9 $g(a)$, $g(v)$, $h(a)$ and $h(v)$ are powers of the same word. Let $g(a) = u^j$ and $g(v) = u^i$. Then

$$u^i = g(v) = g(a)g(b)^3g(c)^3g(a)^2 = u^jg(b)^3g(c)^3u^{2j}$$
$$u^k = g(b)^3g(c)^3, \quad k = i - 3j.$$

We denote briefly $u^k = o^3p^3$.

In the same way, we have

$$u^{k'} = h(b)^3h(c)^3 =: r^3s^3.$$

This leads to four subcases: $k = k' = 1$ or $k' > 1 = k$ or $k > 1 = k'$ or $k, k' > 1$.

The case $k = k' = 1$ yields to the instance (g, h) defined by

	a	b	c
g	u^j	o	p
h	u^i	r	s

The equations $o^3p^3 = u = r^3s^3$ and $(o^3p^3)^{9j+2} = g(w) = h(w) = (r^3s^3)^{9t+2}$ now imply that $j = t$ and, further, $g(a) = h(a)$. Thus the instance (g, h) is trivial and $w \notin P(g, h)$.

Assume now that $k = 1$ and $k' > 1$. By Lemma 4 the words u, $h(b)$ and $h(c)$ are powers of the same word. If now $u = q^l = o^3p^3$ then either $q = o^3p^3$ or again q, o and p are powers of the same word. In the latter case the instance (g, h) is periodic and we are through since $\psi(w) = 3(3,2,2)$. On the other hand, the former case yields to the instance (g, h) defined, for some $j, i_1, i_2, i_3 \in \mathbb{N}$, by

	a	b	c
g	$(o^3p^3)^j$	o	p
h	$(o^3p^3)^{i_1}$	$(o^3p^3)^{i_2}$	$(o^3p^3)^{i_3}$

Since $g(w) = (o^3p^3)^{9j+2}$ and $h(w) = (o^3p^3)^{9i_1+6i_2+6i_3}$ we must have

$$9j + 2 = 9i_1 + 6i_2 + 6i_3.$$

But this is not possible for any integers since $2 \not\equiv 0 \pmod 3$.

The last two cases are considered similarly. Thus the corresponding instance (g, h) is always either periodic or trivial and therefore w cannot be a P-word.

Nevertheless, w is an S-word for the instance (g, h) defined by

	a	b	c
g	o^{21}	o	o
h	o	o^{14}	o^{18}

This is seen immediately by considering Parikh vectors. No subword of w possesses the Parikh vectors $(3, 2, 2)$ or $2(3, 2, 2)$. □

The word presented in Theorem 10 is over three letters. One can construct S-words that are not P-words over bigger alphabets using the following obvious fact. Whenever w fails to be a P-word and φ is a nonerasing morphism, then also $\varphi(w)$ fails to be a P-word.

For instance, the morphism defined by

$$\varphi(a) = a, \quad \varphi(b) = b, \quad \varphi(c) = cd,$$

gives $\varphi(w) = a^2 b^3 (cd)^3 a^4 b^3 (cd)^3 a^3$ which is not a P-word. It is an S-word, as is easily seen from the instance

	a	b	c	d
g	o^{101}	o	o	o
h	o	o^2	o^6	o^{145}

Similarly, the morphism φ defined by

$$\varphi(a) = a, \quad \varphi(b) = b, \quad \varphi(c) = cde,$$

gives the non-P-word $a^2 b^3 (cde)^3 a^4 b^3 (cde)^3 a^3$ over five letters. An instance showing that it is an S-word is now

	a	b	c	d	e
g	o^{101}	o	o	o	o
h	o	o^2	o^9	o^{41}	o^{402}

4 Sharpening in the Finite Case

For many purposes it is useful to view the languages $F(g, h)$ and $P(g, h)$ as subsets of the equality set $E(g, h)$ obtained from $E(g, h)$ by certain operations. Consider a partial order $\leq$ on Σ^* and a language $L \subseteq \Sigma^*$. Then $MIN_\leq(L)$ is the subset of L consisting of elements minimal with respect to $\leq$, in symbols,

$$MIN_\leq(L) = \{x \in L \mid \text{whenever } y \in L \text{ satisfies } y \leq x, \text{ then } y = x\}.$$

If we consider the partial orders in Σ^*

$$uFw \text{ iff } u \text{ is in fin}(w),$$
$$uPw \text{ iff } u \text{ is in scatsub}(w),$$

we obtain

$$F(g, h) = MIN_F(E(g, h)),$$
$$P(g, h) = MIN_P(E(g, h)).$$

The following partial orders are natural from the point of view of the simplicity of solutions:

$$u \leq_\mu w \text{ iff } u = w \text{ or } |u| < |w|,$$
$$u \leq_\psi w \text{ iff } u = w \text{ or } \psi(u) < \psi(w).$$

Thus, intuitively, now we regard a solution w primitive if no word shorter than w, or no word obtained from the letters of w but not using all of them, is a solution. Define the sets of "shortest" and "Parikh-shortest" solutions, respectively, as

$$M(g, h) = MIN_\mu(E(g, h)),$$
$$\Psi(g, h) = MIN_\psi(E(g, h)),$$

and denote their cardinalities by μ and ψ, respectively. We clearly have

$$\mu \leq \psi \leq p.$$

The triple (μ, ψ, p) is defined to be the *finite primality type* of the instance PCP=(g, h).

We prepare the characterization of these finite primality types with a result concerning the pairs (ψ, p). (In the sequel the letters of the domain alphabet Σ are denoted by natural numbers.)

Lemma 11. *All the pairs* $(\psi, p) = (1, n)$, $n \geq 1$, *can be effectively constructed: for each pair* $(1, n)$, *we can effectively construct an instance* (g, h) *for which* $\psi = 1$ *and* $p = n$.

Proof. The main idea in constructing the instance (g_n, h_n) giving rise to the pair $(1, n)$ is to enlarge the alphabet Σ with two letters and the alphabet Δ with one letter at a time.

We begin with a solution 12. Next we define the morphisms for the letters 3 and 4 such that 3214 is also a solution. Clearly, $12 \notin scatsub\,(3214)$ whereas $\psi(12) = (1, 1) < (1, 1, 1, 1) = \psi(3214)$. Similarly, we define 5 and 6 such that 521436 is a solution, and so on. Thus every time we obtain one new P-prime solution whereas the number of Ψ-prime solutions remains the same.

The first four instances (g_i, h_i) are as follows.

	1	2
g_1	ab	a
h_1	a	ba

	1	2	3	4
g_2	ab	a	cba^2	c
h_2	a	ba	c	a^2bc

	1	2	3	4	5	6
g_3	ab	a	cba^2	c	dba^4bc^2	d
h_3	a	ba	c	a^2bc	d	$a^2bc^2ba^2d$

	1	2	3	4	5	6	7	8
g_4	ab	a	cba^2	c	dba^4bc^2	d	$fba^4bc^2a^2bc^2ba^2d^2$	f
h_4	a	ba	c	a^2bc	d	$a^2bc^2ba^2d$	f	$a^2bc^2ba^2d^2ba^4bc^2f$

We clearly have

$$P(g_1, h_1) = \{12\} = \Psi(g_1, h_1),$$
$$P(g_2, h_2) = \{12, 3214\},$$
$$P(g_3, h_3) = \{12, 3214, 521436\},$$
$$P(g_4, h_4) = \{12, 3214, 521436, 72143658\}.$$

Notice also the following equations

$$g_2(3) = h_2(321), \ g_2(214) = h_2(4), \ h_2(3) = g_2(4) \notin \Delta_2 = \{a, b\},$$
$$g_3(5) = h_3(52143), \ g_3(21436) = h_3(6), \ g_3(6) = h_3(5) \notin \Delta_3 = \{a, b, c\},$$
$$g_4(7) = h_4(7214365), \ g_4(2143658) = h_4(8), \ g_4(7) = h_4(8) \notin \Delta_4 = \{a, b, c, d\}.$$

Proceeding inductively, we show that $|P(g_n, h_n)| = n$. This is clearly the case for the instance (g_1, h_1). Since $|P(g_i, h_i)| = i$ the letters $\{1, \ldots, 2i\}$ cannot produce any other P-primes in the instance (g_{i+1}, h_{i+1}) either. On the other hand, if we use the letters $2i + 1$ and $2i + 2$, the only possibility is that $2i + 1$ begins and $2i + 2$ ends the solution. Together with the equations

$$g(2i + 1) = h(2i + 1)h(2)h(1)\ldots h(2i - 1)$$
$$g(2)g(1)\ldots g(2i - 1)g(2i + 2) = h(2i + 2)$$

the proof is completed. $\qquad\square$

We are now ready to state our characterization result.

Theorem 12. *The triple (μ, ψ, p) is a finite primality type iff $\mu \leq \psi \leq p$. An example for each possible type can be effectively constructed.*

Proof. The argument is divided into four subcases, depending on the properties of the given triple.

Case I $\mu = \psi = p$.

Consider the instance (g, h) defined by

	$1_1 \ldots 1_k$
g	$a_1 \ldots a_k$
h	$a_1 \ldots a_k$

We obtain

$$P(g, h) = \Psi(g, h) = M(g, h) = \{1_1, 1_2, \ldots, 1_k\}.$$

Hence the triple is (k, k, k), $k \geq 1$.

Case II $\mu < \psi = p$

Define the morphisms g and h by

	1_k	2_i	3_i
g	a_k	$b_i c_i$	b_i
h	a_k	b_i	$c_i b_i$

Now $2_i 3_i \notin M(g, h)$ since $|2_i 3_i| > |1_k|$. Thus our triple is $(k, k+i, k+i)$, $k, i \geq 1$.

Case III $\mu = \psi < p$

Let the instance (g, h) be defined by

	1_i	2_i	1	2	3	4	5	6	...
g	$a_i b_i$	a_i	ab	a	cba^2	c	dba^4bc^2	d	...
h	a_i	$b_i a_i$	a	ba	c	a^2bc	d	$a^2bc^2ba^2d$	...

By Lemma 11 we obtain the triple $(i+1, i+1, i+n)$, $i \geq 1$, $n \geq 2$, whereas the instance (g, h) without the letters 1_i and 2_i gives the triples $(1, 1, n)$, $n \geq 2$.

Case IV $\mu < \psi < p$

	1_k	2_i	3_i	1	2	3	4	5	6	...
g	a_k	$b_i c_i$	b_i	ab	a	cba^2	c	dba^4bc^2	d	...
h	a_k	b_i	$c_i b_i$	a	ba	c	a^2bc	d	$a^2bc^2ba^2d$	...

Again by Lemma 11 we obtain the triples $(k, k+i+1, k+i+n)$, $k, i \geq 1$, $n \geq 2$ and without the letters 2_i and 3_i the triples are $(k, k+1, k+n)$, $k \geq 1$, $n \geq 2$. (Observe that in the last two cases the letters 5 and 6 are not needed if $n = 2$.)

This concludes the proof for Case IV and, at the same time, the proof of Theorem 12. □

5   Conclusions: Future Work

We have shown that the set of P-words is included properly in the set of S-words which, in turn, is included properly in the set of F-words.

An important open problem, both in decidability theory and in combinatorics on words, is to find characterizations for P-, S- and F-words. At the moment, even the recursiveness of the three sets of words remains open. We have obtained some results for binary alphabets.

References

1. K. Culik II, "A Purely Homomorphic Characterization of Recursively Enumerable Sets", *J. Assoc. Comput. Mach.* 26 (1979) 345–350.
2. K. Culik II and J. Karhumäki, "On the Equality Sets for Homomorphisms on Free Monoids with Two Generators", *R.A.I.R.O. Informatique théorique* 14 (1980) 349–369.
3. M. Harrison, *Introduction to Formal Language Theory*, Addison-Wesley (1978)
4. M. Lipponen, "Primitive Words and Languages Associated to PCP", *E.A.T.C.S. Bulletin* 53 (1994) 217–226.

5. R.C. Lyndon and M.P. Schützenberger, "The Equation $a^M = b^N c^P$ in a Free Group", *Michigan Math. J.* 9 (1962) 289–298.
6. A. Mateescu and A. Salomaa, "PCP-Prime Words and Primality Types", *R.A.I.R.O. Informatique théorique* 27 (1993) 57–70.
7. A. Mateescu and A. Salomaa, "On simplest possible solutions for Post Correspondence Problems", *Acta Informatica* 30 (1993) 441–457.
8. E. Post, "A variant of a recursively unsolvable problem", *Bulletin of the American Mathematical Society* 53 (1946) 264–268.
9. G. Rozenberg and A. Salomaa, *Cornerstones of Undecidability*, Prentice Hall (1994).
10. A. Salomaa, K. Salomaa and Sheng Yu, "Primality types of instances of the Post Correspondence Problem", *E.A.T.C.S. Bulletin* 44 (1991) 226–241.

Computing the closure of sets of words under partial commutations*

Yves Métivier   Gwénaël Richomme   Pierre-André Wacrenier

LaBRI   LAMIFA
Université Bordeaux I,   Faculté de Mathématiques
351 cours de la Libération,   et d'Informatique
33405 Talence, France   33 rue Saint Leu,
{metivier,wacren}   80039 Amiens, France
@labri.u-bordeaux.fr   richomme@crihan.fr

Abstract. The aim of this paper is the study of a procedure S given in [11, 13]. We prove that this procedure can compute the closure of the star of a closed recognizable set of words if and only if this closure is also recognizable. This necessary and sufficient condition gives a semi algorithm for the Star Problem. As intermediary results, using S, we give new proofs of some known results.

In the last part, we compare the power of S with the rank notion introduced by Hashigushi [9]. Finally, we characterize the recognizability of the closure of star of recognizable closed sets of words using this rank notion.

Keywords: commutation, recognizability, rank, Star Problem, trace monoids.

1 Introduction

Free partially commutative monoids have been introduced by Cartier and Foata [2] for combinatorial purposes. These monoids, also called trace monoids, have been intensively studied since Mazurkiewicz [10] proposed them to describe the behaviour of concurrent processes.

Among these studies, many deal with the recognizability of a set of traces. The family of recognizable subsets of a free partially commutative monoid is closed under union, intersection, complementation and concatenation [3, 5]. But as soon as the trace monoid is not a free monoid, this family is not closed under the star operation.

Hence, many studies concern the decidability of the Star Problem : "*if T is a recognizable set of traces, is T^* recognizable ?*". In the general case, the decidability of the Star Problem remains open. In [16], it was proved that the Star Problem is decidable in free partially commutative monoids which does not have a sub free partially commutative monoid on the form $\{a, b\}^* \times \{c, d\}^*$. This family of free partially commutative monoids strictly contains all the previous known cases of decidability of the Star Problem [6, 7, 8, 18].

* This work has been supported by Esprit Basic Research Actions ASMICS II.

In this paper, we investigate the procedure S introduced in [13] which is defined for a given partial commutation Θ and a set of words X as follows:

$$S(X) = \left\{ w \;\middle|\; \begin{array}{l} w \in w_1(u_1 \sqcup\!\!\sqcup v_1)w_2 \ldots w_n(u_n \sqcup\!\!\sqcup v_n)w_{n+1}, \\ w_1u_1v_1w_2 \ldots w_nu_nv_nw_{n+1} \in X, \\ alph(u_i) \times alph(v_i) \subseteq \Theta \;\; \forall i \leq n \end{array} \right\}.$$

The main results of this paper are:

- the procedure S computes the closure of the star of a closed recognizable set of words X under partial commutations if and only if the closure of the star is recognizable (in this case, knowing an automaton recognizing X, we can compute an automaton recognizing $S(X^*)$),
- the rank (defined by Hashigushi in [9]) of the star of a closed recognizable set of words is finite if and only if the closure of the star is recognizable.

The paper is organised as follows. In Section 2, we recall some notions about traces and some results about the Star Problem and the Finite Power Property Problem.

In Section 3, we present the procedure S introduced in [11, 13] to compute the closure of a set of words when this closure is recognizable: this is not always possible. In the following sections, it is proved that, by iterating S, we can compute:

- the closure of the concatenation of two closed recognizable sets of words (Section 4),
- the closure of the star of a closed recognizable set containing only connected words (Section 5),
- the closure of a set of words where all iterative factors are connected (Section 6),
- the closure of a set of words in Knuth normal form (Section 8).

Finally, S gives a semi algorithm to compute the closure of X^*, when X is a closed recognizable set. Indeed, we can compute this closure iterating S if and only if this closure is recognizable (Section 7).

As a consequence of results of sections 4, 5, 6 and 8, we obtained new proofs of some known results: the concatenation of two recognizable sets of traces is recognizable [3, 5](Section 4); the star of a recognizable set of connected traces is recognizable [12] (Section 5); the closure of a recognizable set for which each iterative factor is connected is recognizable [12] (Section 6); a set T of traces is recognizable if and only if the set of Knuth normal forms of T is recognizable [14](Section 8). In [9], Hashigushi introduced the rank notion to study recognizable subsets of trace monoids. By this way he has obtained new proofs of the results on concatenation, star of connected traces, closure of a set with only connected iterative factors obtained in consequence of the studies of sections 4, 5 and 6 and detailed above. In Section 9, we recall this rank notion and compare its power with this of the procedure S.

2 Basic notions and notations

Let A be a finite set of letters. Then A^* is the set of all the words over A. Formally, A^* with the concatenation operation forms the free monoid with the set of generators A and with the empty word ε. We denote by $|x|$ the length of the word x, by $|x|_a$ the number of occurrences of the letter a in x. The notation $alph(x) = \{a \in A \mid |x|_a \neq 0\}$ is used to denote the set of letters of A actually appearing in the word x. The set $alph(x)$ is called the alphabet of the word. For a subset X of A^*, we denote $alph(X) = \bigcup_{x \in X} alph(x)$.

2.1 Free partially commutative monoid

Throughout the paper, Θ is a symmetrical and irreflexive relation over the alphabet A, called commutation (or independency relation). The complement of Θ is denoted by $\overline{\Theta}$: $\overline{\Theta} = A \times A \setminus \Theta$.

Two letters a, b such that $(a, b) \in \overline{\Theta}$ are said to be dependent. Let $\sim_\Theta$ be the congruence over A^* generated by Θ. The quotient of A^* by the congruence $\sim_\Theta$ is the free partially commutative monoid induced by the relation Θ. It is denoted by $M(A, \Theta)$. This monoid is cancellative. The elements of $M(A, \Theta)$, which are equivalence classes of words of A^* under the relation $\sim_\Theta$, are called traces. For a word x of A^*, $[x]_\Theta$ denotes the equivalence class of x under $\sim_\Theta$, and for any subset X of A^*, $[X]_\Theta$ denotes the subset of $M(A, \Theta)$ consisting of the traces generated by X: $[X]_\Theta = \{t \in M(A, \Theta) \mid \exists x \in X \text{ such that } [x]_\Theta = t\}$.

Note that for every letter a of A , the equivalence class $[a]_\Theta$ will simply be denoted by a. Similarly, the unit element of $M(A, \Theta)$ will also be denoted by ε. It is obvious that any two elements x and y of A^* such that $[x]_\Theta = [y]_\Theta$ differ only by the order in which the letters appear, therefore it is possible to define for a trace $t = [x]_\Theta$ of $M(A, \Theta)$ the length $|t| = |x|$ of t, the number $|t|_a = |x|_a$ of occurences of the letter a appearing in t, the set $alph(t) = alph(x)$ of letters appearing in t. For a subset T of $M(A, \Theta)$, let $alph(T) = \bigcup_{t \in T} alph(t)$.

For a subset T of $M(A, \Theta)$, let $\overline{T}$ be the set of all the words w of A^* such that there exists a trace $t \in T$ and $t = [w]_\Theta$.

For a subset X of A^*, we also denote by $\overline{X}$ the set of all the words equivalent to some words of X: $\overline{X} = \{y \in A^* \mid \exists x \in X \text{ such that } y \sim_\Theta x\}$.

The set $\overline{X}$ is called the closure of X. If $X = \overline{X}$, X is said to be Θ-closed.

We write $u\Theta v$ when two traces u and v of $M(A, \Theta)$ commute absolutely, i.e. $alph(u) \times alph(v) \subseteq \Theta$.

The noncommutation graph associated with any subset B of A denoted by $(B, \overline{\Theta})$ is the graph which vertices are the letters of B, and which edges are the non-commuting pairs of letters.

We will say that a trace t, or a word x, or a subset B of A are connected meaning that respectively the graph $(alph(t), \overline{\Theta})$, or the graph $(alph(x), \overline{\Theta})$, or the graph $(B, \overline{\Theta})$ is connected.

If T is a subset of $M(A, \Theta)$, we denote by $Conn(T)$ the set of all the connected traces of T, and by $NConn(T)$ the set of all the non connected traces of T.

2.2 Recognizable subsets of $M(A, \Theta)$

Let T be a subset of $M(A, \Theta)$ and let t be a trace of $M(A, \Theta)$. The left quotient of T by t is the set: $t^{-1}T = \{t' \in M(A, \Theta) \mid tt' \in T\}$.
A subset T of $M(A, \Theta)$ is recognizable if and only if the set $\{t^{-1}T \mid t \in M(A, \Theta)\}$ is finite. We have:

Proposition 1. *A subset T of $M(A, \Theta)$ is recognizable iff $\overline{T}$ is a recognizable subset of A^*.*

So a subset T of $M(A, \Theta)$ is recognizable if and only if there is an automaton which recognizes $\overline{T}$.
The family of recognizable subsets of a trace monoid is closed under union, intersection, complementation and concatenation but as soon as the alphabet contains at least two independent letters, this family is not closed under the star operation. For example, if a and b commute, $T = \{ab\}^*$ is not recognizable since $\overline{T}$ is equal to $\{x \in (a + b)^* \mid |x|_a = |x|_b\}$.
The Star Problem is the following question:

$\qquad$ *given a recognizable set T of traces, is T^* recognizable ?*

The decidability of the Star Problem is known in each free partially commutative monoid which does not have a sub free partially commutative monoid of the form $\{a, b\}^* \times \{c, d\}^*$ [16]. The question is open in other cases.

A subset X of a monoid possesses the finite power property if there exists an integer k such that:

$$X^* = \bigcup_{i=0}^{k} X^i.$$

In [17], a necessary and sufficient condition is given:

Theorem 2. *For a recognizable subset T of a free partially commutative monoid $M(A, \Theta)$, T^* is recognizable if and only if the set $Conn(T^*) \cup NConn(T)$ has the finite power property.*

3 The procedure S

In this section, we recall the procedure S introduced in [11, 13].
Given two words u and v, we call *shuffle* of u and v, the set, denoted by $u \sqcup v$, of the words $u_1 v_1 \ldots u_n v_n$ with $u = u_1 \ldots u_n$, $v = v_1 \ldots v_n$ and $u_i, v_i \in A^*$ $(1 \leq i \leq n)$. The shuffle of two sets of words X_1 and X_2, denoted by $X_1 \sqcup X_2$, is the union of the shuffles $x_1 \sqcup x_2$ with x_1 in X_1 and x_2 in X_2.

The procedure S add to X the words belonging to the sets:

$$w_1(u_1 \sqcup u_1')w_2 \ldots w_n(u_n \sqcup u_n')w_{n+1},$$

where $w_1 u_1 u_1' w_2 \ldots w_n u_n u_n' w_{n+1}$ belongs to X and for all i, $alph(u_i) \times alph(u_i') \subseteq \Theta$; i.e. let x be an element of A^*:

$$S(x) = \left\{ y \sim_\Theta x \left| \begin{array}{l} x = w_1 u_1 u_1' w_2 \ldots w_n u_n u_n' w_{n+1}, \\ alph(u_i) \times alph(u_i') \subseteq \Theta, \\ y \in w_1(u_1 \sqcup u_1')w_2 \ldots w_n(u_n \sqcup u_n')w_{n+1} \end{array} \right. \right\}$$

and $S(X) = \bigcup_{x \in X} S(x)$.

For example, let a and b be two letters such that $a\Theta b$.
We have $S(a^* b) = a^* b a^*$, $S(\{abab\}) = \{abab, baab, abba, aabb, baba\}$,
and $S^2(\{abab\}) = S(\{abab\}) \cup \{bbaa\} = \overline{\{abab\}}$.

Remark If X is recognizable then $S(X)$ is recognizable.
Indeed, let X be a recognizable subset of the free monoid A^*. We denote by $\mathcal{A} = (A, Q, q_0, F, \delta)$ the minimal automaton recognizing X: Q is the set of states of the automaton; q_0 is its initial state; F is its set of final states; δ is the transition function from $Q \times A$ into Q. The equality $X = \overline{X}$ is verified if and only if $\mathcal{A}$ is Θ-compatible (Θ-diamond) i.e. if and only if for all words x and y:

$$x \sim_\Theta y \Rightarrow \forall q \in Q, \quad \delta(q, x) = \delta(q, y).$$

Let us note that we have just to verify the previous property for words of length 2, i.e. $\mathcal{A}$ is Θ-compatible if and only if for all a, b in A:

$$(a, b) \in \Theta \Rightarrow \forall q \in Q, \quad \delta(q, ab) = \delta(q, ba).$$

We can associate to each couple of states (p, q) of $\mathcal{A}$, the recognizable subset of A^* denoted by $X_{p,q}$, and defined by the automaton $(A, Q, p, \{q\}, \delta)$. We consider the 5-tuples (p, q, r, A_1, A_2) where p, q, r are some states, and where A_1, A_2 are non empty subsets of A verifying $A_1 \times A_2 \subseteq \Theta$.
Let $\mathcal{A}_{p,q,r,A_1,A_2} = (A, Q', q_0', F', \delta')$ be the minimal automaton recognizing $(X_{p,q} \cap A_1^*) \sqcup (X_{q,r} \cap A_2^*)$. We assume that $Q \cap Q' = \emptyset$.

Let us define the diamond operation, denoted $\diamond$. Let $\mathcal{A} = (A, Q, q_0, F, \delta)$ be an automaton, let $\mathcal{A}_1 = (A, Q_1, q_1, F_1, \delta_1)$ an automaton such that $Q_1 \cap Q = \emptyset$, and let p and r two states of Q; the automaton $\mathcal{A} \diamond (p, \mathcal{A}_1, r)$ is the automaton $(A, Q \cup Q_1, q_0, F, \delta_2)$ where δ_2 is defined by:

if $s \in Q$, $\delta_2(s, a) = \delta(s, a)$,
if $s \in Q_1$, $\delta_2(s, a) = \delta_1(s, a)$,
$\delta(p, \varepsilon) = q_1$,
if $s \in F_1$, $\delta_2(s, \varepsilon) = r$.

Let us remark that for two automata $\mathcal{A}_1$ and $\mathcal{A}_2$, and for four states p_1, r_1, p_2 and r_2 of Q we have $(\mathcal{A} \diamond (p_1, \mathcal{A}_1, r_1)) \diamond (p_2, \mathcal{A}_2, r_2) = (\mathcal{A} \diamond (p_2, \mathcal{A}_2, r_2)) \diamond (p_1, \mathcal{A}_1, r_1)$.

Obviously, if X is recognized by the automaton $\mathcal{A}$ then the set $S(X)$ is recognized by the automaton obtained from $\mathcal{A}$ applying the *diamond* operation with the automata $\mathcal{A}_{p,q,r,A_1,A_2}$ for all 5-uples (p, q, r, A_1, A_2) with p, q, r some states and A_1 and A_2 some subsets of A verifying $A_1 \times A_2 \subseteq \Theta$.

Remark.   a - For all subsets X_1 and X_2 of A^*, we have $S(X_1)S(X_2) \subseteq$
$S(X_1X_2)$.
  b - For a subset X of A^*, we have $(S(X))^* \subseteq S(X^*)$.
  c - For two subsets X_1 and X_2 of A^*, if $X_1 \subseteq X_2$ then $S(X_1) \subseteq S(X_2)$.

The procedure S has been introduced to compute the closure of a recognizable
set when this closure is recognizable.
Due to the definition of S it is clear that for any set of words X that we have
$X \subseteq S(X) \subseteq \overline{X}$.
Moreover for any word $w = a_0a_1 \ldots a_n a_{n+1}$, note that S^n compute the closure
of w, we have: $\overline{w} = S(a_0.S(a_1.S(\ldots S(a_n a_{n+1}) \ldots)) \subseteq S^n(w)$.
Therefore $\overline{X} = \bigcup_{i \geq 0} S^i(X) = S^*(X)$.
Observe that a set of words X is Θ-closed if and only if X is a fixpoint of S:

Lemma 3. *The two following assertions are equivalent:*

$$\begin{cases} (i) \ \text{there exists an integer } n \text{ such that } \overline{X} = S^n(X). \\ (ii) \ \text{there exists an integer } n \text{ such that } S^{n+1}(X) = S^n(X). \end{cases}$$

The following example given by A. Arnold shows that S cannot always compute
the closure of a recognizable set when its closure is recognizable:

Example 1. Let $A = \{a, b\}$ with $a\Theta b$, and $X = (ab)^*(\{\varepsilon\} \cup a^+ \cup b^+)$. We have
$\overline{X} = A^*$, and by induction, we can prove that: $\forall n \geq 0 \qquad a^{3^n+1}b^{3^n+1} \notin S^n(X)$.
Thus, for each integer n, $S^n(X) \neq \overline{X}$.

From the previous example, we deduce that, in the general case, S does not even
provide a semi-algorithm to compute recognizable closures. The problem is now
to give sufficient conditions on the set X such that S provides an algorithm or
a semi-algorithm.

4   The procedure S and the concatenation operation

In this section, we show that S provides an algorithm in order to compute the
closure of the concatenation of two Θ-closed sets:

Proposition 4. *For all Θ-closed subsets X_1, X_2 of A^*, we have:*

$$\overline{X_1.X_2} = S^{|A|}(X_1.X_2).$$

This proposition implies in corollary that the concatenation of two recognizable
sets of traces is recognizable [3, 5]. Indeed, if T_1 and T_2 are two recognizable sets
of traces, by Proposition 1, the sets $\overline{T_1}$ and $\overline{T_2}$ are recognizable. These sets verify
the hypothesis of Proposition 4. Thus, $\overline{\overline{T_1}.\overline{T_2}}$ is recognizable. By Proposition 1,
the set $[\overline{T_1}.\overline{T_2}]_\Theta$ i.e. $T_1.T_2$ is recognizable.
A simple induction from Proposition 4 gives $\overline{X_1 \ldots X_p} = S^{(p-1)|A|}(X_1 \ldots X_p)$.
Obviously, one can give a better bound than $(p - 1)|A|$:

Corollary 5. *Let $X_1, \ldots, X_p$ be some recognizable Θ-closed subsets of A^*. We have $\overline{X_1 \ldots X_p} = S^{l|A|}(X_1 \ldots X_p)$ where l is the integer such that $2^{l-1} < p \leq 2^l$.*

We can note that *there does not exist any integer n such that for all integer p and all Θ-closed subsets $X_1, \ldots, X_p$ of A^* such that $X_i = \overline{X_i}$ ($1 \leq i \leq p$), we have $\overline{X_1 \ldots X_p} \subseteq S^n(X_1 \ldots X_p)$.*
Just take p even with $X_{2i-1} = \{a\}$ and $X_{2i} = \{b\}$ in $\{a\}^* \times \{b\}^*$.

In the next section, we will see that when all words of each X_i are connected a such integer n exists.

5 The procedure S and the star of a set of connected traces

We are going to show that, given a closed recognizable set containing only connected words, S computes the closure of X^*. Previously, we have proved this result in the particular case where all the words of the set have the same connected alphabet.

Proposition 6. *Let $M(A, \Theta)$ be a free partially commutative monoid such that $(A, \overline{\Theta})$ is connected. If $X_1, \ldots, X_p$ are some Θ-closed subsets of A^* such that for all i between 1 and p and for all x in X_i we have $alph(x) = A$ then*

$$\overline{X_1 \ldots X_p} = S^{(|A|-1)|A|}(X_1 \ldots X_p).$$

Remark that as a corollary we have $\overline{X^*} = S^{(|A|-1)|A|}(X^*)$, for any Θ-closed subset X of A^*, with (A, Θ) connected, if for all $x \in X$, $alph(x) = A$.

Let us now give the main result of this section:

Proposition 7. *Let X be a Θ-closed subset of A^*. If X contains only connected words then $\overline{X^*} = S^{(|A|+1)|A|^2}(X^*)$.*

Indeed let us assume that there exists an integer m such that for each y of $\overline{X^*}$ with $|alph(y)| < |alph(x)|$, we have $y \in S^m(X^*)$.
- If x is not connected, we can find two non empty words x_1, x_2 with $x_1 \Theta x_2$ and $x_1 \in P(x_1 x_2) \subseteq S(x_1 x_2)$. Since all elements of X are connected, x_1 and x_2 belong to X^*.
We have $|alph(x_1)| < |alph(x)|$ and $|alph(x_2)| < |alph(x)|$. By inductive hypothesis, $x_1 \in S^m(X^*)$ and $x_2 \in S^m(X^*)$ and we get we have $x \in S^{m+1}(X^*)$.
- If x is connected then we get $x \in S^{m+(|A|+1)|A|}(X^*)$ from the following lemma:

Lemma 8. *Let $M(A, \Theta)$ be a free partially commutative monoid. Let X be a Θ-closed subset of A^* and m an integer between 1 and $|A| - 1$. If there exists an integer n such that for each word y in $\overline{X^*}$ with $|alph(y)| < m$, $y \in S^n(X^*)$, then for each connected word x of $\overline{X^*}$ with $|alph(x)| = m$, $x \in S^{n+(|A|+1)|A|}(X^*)$.*

Let us note that in corollary of Proposition 7, we have: the star of a recognizable set T containing only connected traces is recognizable [12].

6 The procedure S and iterative factors

A word x is called an iterative factor of a subset of A^* if there exists two words u and v such that ux^*v is a subset of X. In [11, 13], Métivier conjectured that the procedure S gives an algorithm to compute the closure of a recognizable set of words X when all the iterative factors of X are connected. Using the notion of *star height* of a rational language and the results on S about concatenation and connected star, we have proved this conjecture:

Proposition 9. *Let X be a recognizable set of words such that all its iterative factors are connected. There exists an integer n such that $S^n(X) = \overline{X}$.*

Let us note that in corollary we have: the closure of a recognizable set of words whose all iterative factors are connected is recognizable [12].

7 The procedure S and the star of a set

Proposition 10. *Let T be a recognizable subset of $M(A, \Theta)$. The set T^* is recognizable if and only if there exists an integer n such that $S^n((\overline{T})^*) = \overline{T^*}$.*

This proposition gives a semi-algorithm to say whether the star of a recognizable set of traces is recognizable. Indeed, by Lemma 3, knowing whether $\overline{X^*}$ is recognizable (for a Θ-closed recognizable set X of words) can be done finding an integer n such that $S^n(X^*) = S^{n+1}(X^*)$.

Proof. Set $X = (\overline{T})^*$.
By Proposition 1, T^* is recognizable if and only if $\overline{X}$ is recognizable. If there exists an integer n such that $S^n(X) = \overline{X}$, then, as $\overline{T^*} = \overline{X}$, T^* is recognizable.

Let us assume by now that T^* is recognizable. Since $S^n(X) \subseteq \overline{X}$ for each integer n, we verify in the rest of the proof that there exists an integer n such that for all x in X, $\overline{x} \subseteq S^n(X)$.
Let x be an element of $\overline{X}$. If $|alph(x)| = 1$ then $\overline{x} \subseteq S^0(X)$. Let assume that there exists $m \geq 0$ such that for each word y of $\overline{X}$ with $|alph(y)| < |alph(x)|$, $\overline{y} \subseteq S^m(X)$.

• If x is connected, then the searched integer exists by Lemma 8.

• Assume that x is not connected. By Theorem 2, since T^* is recognizable, $Conn(T^*) \cup NConn(T)$ has the finite power property i.e. there exists an integer p such that $T^* = \bigcup_{i=0}^{p}(Conn(T^*) \cup NConn(T))^i$.
There exists some words $x_1, \ldots, x_j$ in $\overline{Conn(T^*) \cup NConn(T)}$ with $j \leq p$ such that $x \sim_\Theta x_1 \ldots x_j$ and then $\overline{x} = \overline{x_1 \ldots x_j}$.
If x_i belongs to $\overline{Conn(T^*)}$, then $alph(x_i) \subset alph(x)$ and thus by inductive hypothesis $\overline{x_i} \subseteq S^m(X)$. For x_i in $\overline{NConn(T)}$, we have $\overline{x_i} \subseteq \overline{T} \subseteq X \subseteq S^m(X)$.
By Corollary 5, $\overline{x} \subseteq S^{m+(l-1)|A|}(X)$.

Since $|alph(x)| \leq |A|$, the previous iteration can be done at most $|A| - 1$ times. Thus there exists an integer n bounded by $(|A| - 1)|A|sup(l - 1, |A| + 1)$ such that $S^n(X) = \overline{X}$.

8 Procedure S and Knuth normal forms

In this section, we assume that A is totally ordered by $<$. We also denote by $<$ the lexicographic order on A^* induced by the order $<$ on A. Let u and v be two words of A^*. We have $u < v$ if u is a prefix of v, or if $u = xay$ and $v = xbz$, with a, b in A and $a < b$.

For a trace t of $M(A, \Theta)$, let $Inf_\Theta(t)$ be the least representant of t for the lexicographic order i.e. $Inf_\Theta(t)$ is the least word u of A^* such that $[u]_\Theta = t$. This element is called Knuth normal form of t.

For a word w of A^*, we also denote by $Inf_\Theta(w)$, the least word for the lexicographic order equivalent by $\sim_\Theta$ to w. We have $Inf_\Theta(w) = Inf_\Theta([w]_\Theta)$.

For a set T of traces and a set X of words, we note $\begin{vmatrix} Inf_\Theta(T) = \{Inf_\Theta(t) \mid t \in T\} \\ Inf_\Theta(X) = \{Inf_\Theta(w) \mid w \in X\} \end{vmatrix}$.

In order to establish a Kleene-like Theorem for traces Ochmański proved:

Theorem 11. *Let X be a subset of A^*. The set $\overline{X}$ is recognizable iff $Inf_\Theta(X)$ is recognizable.*

Using the procedure S, we obtained a new proof of this result:

Proposition 12. *Let X be a subset of A^*.*
$$If \ Inf_\Theta(X) \subseteq X \ then \ S^{(|A|+1)|A|^2}(X) = \overline{X}.$$

This is a sufficient condition "$Inf_\Theta(X) \subseteq X$" for a recognizable set X to have a recognizable closure. This condition is not necessary as shown by this example :

Example 2. Let $A = \{a, b\}$ with $a\Theta b$ and $a < b$. Let $X = b^+.a^+$. We have $\overline{X}$ recognizable since $\overline{X} = A^* \setminus (a^* \cup b^*)$. But $Inf_\Theta(X) = a^+.b^+$ and we have not $Inf_\Theta(X) \subseteq X$.

9 S and the rank notion

In [9], Hashigushi introduced the notion of rank ($Rank_\Theta(X, u, v)$) of a decomposition (u, v) of a word in connection with a set X of words and a commutation Θ. This notion give a common formal context to prove the results about concatenation and star connected operations for recognizable trace languages.

Since the procedure S gives also a formal context to prove the same results, one can ask *for a subset X of A^*, does there exists an integer n such that $s^n(X) = S^{n+1}(X)$ if and only if X has a finite rank ?*

In subsection 9.3, we study the previous question and prove that if there exists an integer n such that $S^n(X) = S^{n+1}(X)$, then X has a finite rank. The converse is false if the considered trace monoid is not on the form $A^* \times B^*$. This results are established using a formula which indicates that the rank of a set $S(X)$ is approximately the third of the rank of X.

Then we prove that for a recognizable closed set of words X, there exists an integer n such that $S^n(X^*) = S^{n+1}(X^*)$ if and only if the rank of X^* is finite.

9.1 The rank notion

The rank of a decomposition (u, v) of a word in connection with a set X of words and a commutative relation Θ is the least integer n such that there exists a word $u_0 v_0 \ldots u_n v_n$ in X with $u \sim_\Theta u_0 \ldots u_n$, $v \sim_\Theta v_0 \ldots v_n$ and $v_i \Theta u_{i+1} \ldots u_n$ for $1 \leq i < n$:

$$Rank_\Theta(X, u, v) = min \left\{ \quad \infty, \quad n \in \mathbb{N} \quad \left| \begin{array}{l} \exists u_0 v_0 \ldots u_n v_n \in X, \\ u \sim_\Theta u_0 \ldots u_n, \\ v \sim_\Theta v_0 \ldots v_n, \\ v_i \Theta u_{i+1} \ldots u_n. \end{array} \right. \right\}$$

Proposition 13. [9] *For a subset X of A^* and two words u and v, $Rank_\Theta(X, u, v)$ is finite if and only if uv belongs to $\overline{X}$: $uv \in \overline{X} \Leftrightarrow \exists k \in \mathbb{N}, Rank_\Theta(X, u, v) = k$.*

The rank of a set of words is $Rank_\Theta(X) = max\{Rank_\Theta(X, u, v) | uv \in \overline{X}\}$.
We will say that X has a finite rank if there exists an integer k such that $Rank_\Theta(X) = k$. Hashigushi proved the two following results:

Theorem 14. *[9] If a set X of words has a finite rank, then $\overline{X}$ is recognizable.*

Proposition 15. *[9]*
– Given two recognizable closed set of words X_1 and X_2, the rank of $X_1 X_2$ is finite.
– For a recognizable closed set of connected words X, the rank of X^* is finite.
– For a recognizable set X of words for which each iterative factor is connected, the rank of X is finite.

Remark. Note that the converse of theorem 14 is false (this was a question of Hashigushi [9]). We have the following example already given by Ochmański and Wacrenier [15]: for $A = \{a, b\}$ with $a\Theta b$, the closure of the set of words $X = (ab)^*.(a^* \cup b^*)$ is recognizable since $\overline{X} = \{a, b\}^*$ but the rank of X is not finite.

9.2 S divides by 3 the rank

We denote by $\lfloor r \rfloor$ the integer part of the real r.

Lemma 16. *Let $M(A, \Theta)$ be a trace monoid, X be a subset of A^*, and u and v be two words of A^*. We have: $Rank_\Theta(S(X), u, v) = \lfloor \frac{1}{3}(Rank_\Theta(X, u, v) + 1) \rfloor$.*

Sketch of proof:
Let $n = Rank_\Theta(X, u, v)$. There exists $u_0 v_0 \ldots u_n v_n$ in X with $u \sim_\Theta u_0 \ldots u_n$, $v \sim_\Theta v_0 \ldots v_n$ and, for $1 \leq i < n$, $u_i \Theta v_{i+1} \ldots v_n$. Completing with empty words, we can consider the word $u_0 v_0 \ldots u_{3p+1} v_{3p+1}$, with $3p - 1 \leq n < 3p + 2$.
Let $w = u_0' v_0' \ldots u_p' v_p'$ with $u_0' = u_0 u_1$, $v_i' = v_{3i} v_{3i+1} v_{3i+2}$ ($0 \leq i \leq p - 1$), $u_i' = u_{3i-1} u_{3i} u_{3i+1}$ ($1 \leq i \leq p$) and $v_p' = v_{3p} v_{3p+1}$. We have $w \in S(X)$. Thus, $Rank_\Theta(S(X), u, v) \leq p \leq \lfloor \frac{n+1}{3} \rfloor$.

To establish that $Rank_\Theta(S(X), u, v) \geq \lfloor \frac{n+1}{3} \rfloor$ we have first noted that we have just to prove it in the case where $|X| = 1$ and it is easy to show that S cannot group together 4 different blocks of u_i; indeed let $M(C, \Theta)$ be a trace monoid, if $A = \{a_1, a_2, a_3, a_4\}$ and $B = \{b_1, b_2, b_3\}$ are two disjoint subsets of C then we have $S(a_1 b_1 a_2 b_2 a_3 b_3 a_4) \cap B^* A^* B^* = \emptyset$.

Since S cannot group together 4 different blocks of u_i, S divides the number of blocks by 3 or less.

9.3 Connections between S and the rank notion

Procedure S cannot compute the closure of an unbounded rank set: indeed, if $Rank_\Theta(X) = \infty$ then for each integer n $Rank_\Theta(S^n(X)) = \infty$, therefore:

Proposition 17. Let $M(A, \Theta)$ be a trace monoids and X be a subset of A^*. If $Rank_\Theta(X)$ is infinite then, for each integer n, $S^n(X) \neq \overline{X}$.

The two following examples prove that the converse of Proposition 17 is false:

Example 3. Let $A = \{a, b, c\}$ with $a\Theta b$, $a\Theta c$, $b\Theta c$. Let $X_1 = a^*(ab)^*(abc)^*$, $X_2 = a^*(ac)^*(abc)^*$, $X_3 = b^*(ba)^*(abc)^*$, $X_4 = b^*(bc)^*(abc)^*$, $X_5 = c^*(ca)^*(abc)^*$, $X_6 = c^*(cb)^*(abc)^*$. The closures $\overline{X_1}, \ldots, \overline{X_6}$ are not recognizable since for instance $\overline{X_1}$ is the set of words x such that $|x|_a \geq |x|_b \geq |x|_c$. Let $Z = X_1 \cup X_2 \cup X_3 \cup X_4 \cup X_5 \cup X_6$. The closure of Z is recognizable since $\overline{Z} = \{a, b, c\}^*$. Finally, let $X = Z.Z$. We have $\overline{X} = \{a, b, c\}^*$ and $Rank_\Theta(X) = 0$. Moreover, for each integer $n \geq 1$, $a^{4.3^n} b^{4.3^n} c^{4.3^n} \notin S^{n-1}(X)$.

Example 4. Let $A = \{a, b, c\}$ with $a\Theta b$, $a\overline{\Theta} c$, $b\overline{\Theta} c$. Let $X = (ab)^*(a^* + b^*)c(a + b)^* + (a + b)^* c(ab)^*(a^* + b^*)$. The closure $\overline{X}$ is recognizable: $\overline{X} = (a + b)^* c(a + b)^*$. We have $Rank_\Theta(X) = 0$. Moreover, for each integer $n \geq 0$, $a^{3^n} b^{3^n} ca^{3^n} b^{3^n} \notin S^{n-1}(X)$.

Now suppose that $A = B \cup C$ with $\Theta = B \times C$. Let X be a subset of A^* such that $Rank_\Theta(X) = n$, for some integer n. We have $Rank_\Theta(S^n(X)) = 0$ and then $\overline{S^n(X) \cap B^* C^*} = \overline{X}$. Thus $S^{n+1}(X) = \overline{X}$. Due to this observation and the two previous examples we get:

Proposition 18. Let $M(A, \Theta)$ be a trace monoid and X a recognizable subset of A^*. The following equivalence "There exists an integer n such that $S^n(X) = S^{n+1}(X)$ if and only if $Rank_\Theta(X)$ is finite" is true if and only if we can write $A = B \cup C$ with $\Theta = B \times C$.

9.4 Rank and Star Problem

Proposition 19. Let Θ be a commutation relation on A and X be a recognizable Θ-closed subset of A^*. The following assertions are equivalent:

- the closure $\overline{X^*}$ is recognizable,
- there exists an integer n such that $S^n(X^*) = S^{n+1}(X^*)$,
- the rank of X^* is finite.

References

1. A.V. Anisimov et D.E. Knuth, *Inhomogeneous Sorting*, Int. J. of Computer and Information Sciences, Vol 8, No 4, 1979.
2. P. Cartier et D. Foata, *Problèmes combinatoires de commutation et réarrangements*, Lecture Notes in Math. 85, 1969.
3. R. Cori and D. Perrin, *Automates et commutations partielles*, RAIRO Inform. Théor. 19, p 21-32, 1985.
4. S. Eilenberg, *Automata, languages and machines*, Academic Press, New York, 1974.
5. M. Fliess, *Matrices de Hankel*, J. Math Pures et Appl. 53, p197-224, 1974.
6. P. Gastin, E. Ochmański, A. Petit et B. Rozoy, *Decidability of the star problem in $A^* \times \{b\}^*$*, Inform. Process. Lett. 44, p65-71, 1992.
7. S. Ginsburg et E. Spanier, *Semigroups, Presburger formulas and languages*, Pacific journal of mathematics 16, p285-296, 1966.
8. S. Ginsburg et E. Spanier, *Bounded regular sets*, Proceedings of the AMS, vol. 17(5), p1043-1049, 1966.
9. K. Hashigushi, *Recognizable closures and submonoids of free partially commutative monoids*, Theoret. Comput. Sci. 86, p233-241, 1991.
10. A. Mazurkiewicz, *Concurrent program schemes and their interpretations*, Aarhus university, DAIMI rep. PB 78, 1977.
11. Y. Métivier, *Contribution à l'étude des monoïdes de commutations*, Thèse d'état, université Bordeaux I, 1987.
12. Y. Métivier, *On recognizable subsets of free partially Commutative Monoids*, Theoret. Comput. Sci. 58, p201-208, 1988.
13. Y. Métivier et B. Rozoy, *On the star operation in free partially commutative monoids*, International Journal of Foundations of Computer Science 2, p257-265, 1991.
14. E. Ochmański, *Regular behaviour of concurrent systems*, Bulletin of EATCS 27, p56-67, 1985.
15. E. Ochmański, P.-A. Wacrenier, *On Regular Compatibility of Semi-Commutations*, Proceedings of ICALP'93, LNCS 700, p445-456, 1993.
16. G. Richomme, *Some trace monoids where both the Star Problem and the Finite Power Property Problem are decidable*, Proc. of MFCS'94, LNCS 841, p577-586, 1994.
17. G. Richomme, *Equivalence decidability of the Star Problem and the Finite Power Property Problem in trace monoids*, LaBRI internal report 835.94, 1994.
18. J. Sakarovitch, *The "last" decision problem for rational trace languages*, Proceedings of LATIN'92, LNCS 583, p460-473, 1992.

Intervalizing k-Colored Graphs*

Hans L. Bodlaender and Babette de Fluiter

Department of Computer Science, Utrecht University,
P.O. Box 80.089, 3508 TB Utrecht, the Netherlands

Abstract. The problem to determine whether a given k-colored graph is a subgraph of a properly k-colored interval graph is shown to be solvable in $O(n)$ time when $k = 2$, solvable in $O(n^2)$ time when $k = 3$, and to be NP-complete for any fixed $k \geq 4$. This problem has an application in DNA physical mapping. Our algorithm for $k = 3$ is based on an extensive analysis of the precise structure of graphs of pathwidth two, dynamic programming on certain parts of the input graph, and a careful combination of the results for the different parts.

1 Introduction

In this paper, we consider the following problem.

INTERVALIZING COLORED GRAPHS [ICG]
Instance: A graph $G = (V, E)$, a coloring $c : V \to \{1, \ldots, k\}$
Question: Is there a properly colored supergraph $G' = (V, E')$ of G which is an interval graph?

The problem models a problem arising in sequence reconstruction, which appears in some investigations in molecular biology (such as protein sequencing, nucleotide sequencing and gene sequencing (see [5]). A sequence X (usually a large piece of DNA) is fragmented (or k copies of the sequence X are fragmented). For each fragment, a set of characteristics (its 'fingerprint' or 'signature') is determined, and based on respective fingerprints, an 'overlap' measure is computed. Using this overlap information, the fragments are assembled into islands of contiguous fragments (contigs). Instances of ICG model the situation where k copies of X are fragmented, and some fragments (clones) are known to overlap. Fragments of the same copy of X will not overlap. Now each vertex in V represents one fragment; the color of a vertex represents to which copy of X the fragment belongs. It can be seen that ICG helps here to predict other overlaps and to work towards reconstruction of the sequence X.

It is known that ICG for an arbitrary number of colors is NP-complete [5, 6]. However, from the application it appears that the cases where the number of

---

* This research was partially supported by the Foundation for Computer Science (S.I.O.N) of the Netherlands Organization for Scientific Research (N.W.O.) and partially by the ESPRIT Basic Research Actions of the EC under contract 7141 (project ALCOM II). E-mail: {hansb,babette}@cs.ruu.nl

colors k (= the number of copies of X that are fragmented) is some small given constant are of interest. In this paper, we resolve the complexity of this problem for all constant values k. We observe that the case $k = 2$ is easy to resolve in linear time. We show that the case $k = 3$ is solvable in $O(n^2)$ time. Finally, we show that ICG is NP-complete for four colors (and hence, for any fixed number of colors ≥ 4.)

In [5], Fellows et al. consider ICG with a bounded number of colors. They show that, although for fixed $k \geq 3$, yes-instances have bounded pathwidth (and hence bounded treewidth), standard methods for graphs with bounded treewidth will be insufficient to solve ICG, as the problem is 'not finite state'. Also, they show ICG to be hard for the complexity class $W[1]$, (which was strengthened in [1] to hardness for all classes $W[t]$, $t \in \mathbf{N}$). This result implies that it is unlikely that there exists a c, such that for any fixed number of colors k, ICG is solvable in time $O(f(k)n^c)$. Clearly, our NP-completeness result implies the fixed parameter intractability results, but is much stronger.

ICG is closely related to TRIANGULATING COLORED GRAPHS (TCG) where we look for a properly colored *triangulated* supergraph G' of a k-colored input graph G (i.e., G' does not contain a chordless cycle of length at least four). This problem is known to be NP-complete [2, 15], solvable in $O(n^{k+1})$ time for fixed k [11], and solvable in linear time for the cases $k \in \{2,3\}$ [3, 7, 8, 13]. It appears that ICG poses some additional difficulties which require more complex and time consuming algorithms. For instance, while there is an easy characterization which assures that three-colored simple cycles can be triangulated without adding edges between vertices of the same color, for ICG on three-colored simple cycles, such a simple characterization does not exist, and even this case seems to require an $O(n^2)$ algorithm, based on dynamic programming. Additionally, TCG with three colors is 'finite state', while ICG with three colors is not.

Another closely related problem is COLORED PROPER INTERVAL GRAPH COMPLETION, which asks whether a given colored graph is a subgraph of a properly colored unit interval graph. In [9, 10], it is shown that this problem is NP-complete, polynomial for a fixed number of colors, and hard for $W[1]$.

A necessary condition for a three-colored graph G to be 'intervalizable' is that the pathwidth of G is at most two [5]. Our algorithm exploits the precise structure of graphs of pathwidth two (partial two-paths). For parts of the input graphs, a dynamic programming approach is used to compute whether these parts can be intervalized, and some more information. Then, a careful case analysis is necessary to see whether all the different parts can be put together to an intervalization of the entire input graph. In Sect. 3 we analyze the structure of partial two-paths. We do this first for biconnected partial two-paths, after that for trees of pathwidth two, and finally for general partial two-paths. In Sect. 4 we consider the algorithms, again first for biconnected graphs, then for trees, and finally, we shortly discuss how information for biconnected and tree-parts of the graph can be pieced together. In Sect. 5 we discuss our NP-completeness result. There are many details to be taken care of, and for space reasons, we necessarily omit or only sketch large parts of the algorithms and proofs.

2 Preliminaries

Without loss of generality, we suppose all graphs we deal with are connected. The notion *biconnected component* denotes a non-trivial biconnected component, i.e., one with at least three vertices. An *interval graph* is a graph $G = (V, E)$ for which there is a function $\Phi : V \to \mathcal{I}$, where $\mathcal{I}$ is the set of all intervals on the real line, such that for each pair $v, w \in V$, $\Phi(v) \cap \Phi(w) \neq \emptyset \Leftrightarrow \{v, w\} \in E$. A k-coloring of a graph $G = (V, E)$ is a surjection $c : V \to \{1, \ldots, k\}$. A *proper k-coloring* is a k-coloring c such that for each edge $\{v, w\} \in E$, $c(v) \neq c(w)$. An *intervalization* of a graph $G = (V, E)$ with a k-coloring c is a supergraph $G' = (V, E')$ $(E \subseteq E')$ of G which is an interval graph and is properly colored by c. Thus ICG is the problem to determine whether there exists an intervalization for a given colored graph.

A *path decomposition* PD of a graph $G = (V, E)$ is a sequence $(V_1, \ldots, V_t)$, in which $V_i \subseteq V$ for all i, and the following conditions are satisfied:
1. For each $e \in E$, there is an i such that $e \subseteq V_i$.
2. For each $1 \leq i \leq j \leq l \leq t$, $V_i \cap V_l \subseteq V_j$.

The sets V_i are called the *nodes* of the path decomposition. The *width* of PD is $\max_i |V_i| - 1$. A graph G has *pathwidth* k if there is path decomposition of width k of G, but there is no path decomposition of width $k - 1$ of G. A graph G is called a *partial k-path* if it has pathwidth at most k. Let G be a graph, $PD = (V_1, \ldots, V_t)$ a path decomposition of G. We say that an induced, connected subgraph G' of G *occurs* in $(V_j, \ldots, V_{j'})$ if V_j and $V_{j'}$ contain an edge of G', and no node V_i on the left side of V_j (i.e. $i < j$) or on the right side of $V_{j'}$ (i.e. $i > j'$) contains an edge of G'. We say that the vertices of G' *occur* in $(V_l, \ldots, V_{l'})$ if these are the only nodes in PD containing vertices of G'. A vertex v (edge e) is an *end vertex* (*end edge*) of G' if in each path decomposition of width two of G, v (e) occurs in the leftmost or rightmost end node of the occurrence of G'.

A *proper path decomposition* of a k-colored graph G is a path decomposition $PD = (V_1, \ldots, V_t)$ in which for each node V_i and each pair $v, w \in V_i$, if $v \neq w$ then $c(v) \neq c(w)$.

Lemma 1. *Let $G = (V, E)$ be a graph, $c : V \to \{1, \ldots, k\}$ a k-coloring of G. G has an intervalization if and only if there is a proper path decomposition of G, which has width $k - 1$ at most.*

Proof. (See also [5].) For the if part, suppose $PD = (V_1, \ldots, V_t)$ is a proper path decomposition of G. Note that PD has width $\leq k - 1$. Let G' be the graph obtained from G by adding all edges between vertices which are in the same node of PD, and which are not already present. Then G' is a properly k-colored interval graph, since we can give the following function. For each $v \in V$, if v occurs in nodes $(V_j, \ldots, V_l)$, then $\Phi(v) = [j, l]$. The 'only if' part is similar. $\square$

Thus, we use the following problem, which is equivalent to ICG.

Instance: A graph $G = (V, E)$, a k-coloring $c : V \to \{1, \ldots, k\}$
Question: Is there a proper path decomposition of G?

Note that if there is a proper path decomposition of a k-colored graph G, then G is a partial $(k-1)$-path. It is easy to observe that in the case that $k = 2$, the question is equivalent to the question whether G is a properly colored partial one-path, (see Sect. 3.2 for the structure of partial one-paths). Thus, in the case $k = 2$, ICG can be solved in time linear in the number of vertices of the graph.

Theorem 2. *ICG can be solved in linear time for 2-colored graphs.*

3 The Structure of Partial Two-Paths

In this section, we first give a characterization of biconnected partial two-paths. After that, we give a characterization of trees of pathwidth two, and finally of partial two-paths in general.

3.1 Biconnected Partial Two-Paths

Given a graph $G = (V, E)$, the graph $\bar{G}$ which is obtained from G by adding all edges $\{v, w\} \notin E$ for which there are three disjoint paths from v to w in G is called the *cell completion* of G. The following lemma has been proved in [3] in the setting of partial two-trees.

Lemma 3. *Let G be a partial two-path. Each path decomposition of width two of G is a path decomposition of $\bar{G}$.*

The cell completion of a partial two-path can be found in linear time [3]. In the cell completion of a graph, each two chordless cycles have at most one edge in common. In [3], it has also been shown that the cell completion of a biconnected partial two-tree is a tree of chordless cycles. We show that the cell completion of a biconnected partial two-path is a path of chordless cycles. Figure 1 shows an example of a path of chordless cycles.

Definition 4 (Path of Chordless Cycles). A *path of chordless cycles* is a pair $(\mathcal{C}, \mathcal{S})$, where $\mathcal{C}$ is a sequence $(C_1, \ldots, C_p)$ of chordless cycles, $p \geq 1$, and $\mathcal{S}$ is a sequence $(e_1, \ldots, e_{p-1})$ of edges, such that for each i, $1 \leq i < p$, $V(C_i) \cap V(C_{i+1}) = e_i$, $E(C_i) \cap E(C_{i+1}) = \{e_i\}$ and for each i, $1 \leq i < p - 1$, if $e_i = e_{i+1}$, then $|V(C_i)| = 3$.

Lemma 5. *Let G be a biconnected graph. G is a partial two-path if and only if $\bar{G}$ can be written as a path of chordless cycles.*

Proof. We only sketch the proof. It can be shown that in a path decomposition PD of width two of G, the occurrences of different chordless cycles can overlap only in their common edge. Furthermore, for each edge in a chordless cycle, there is a node in PD containing this edge and another vertex from the cycle. This means that a chordless cycle can only have two edges in common with other chordless cycles. Furthermore, an edge of a chordless cycle C can only occur in both end nodes of the occurrence of C if $|V(C)| = 3$. □

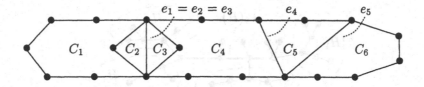

Fig. 1. A path of chordless cycles $(\mathcal{C}, \mathcal{S})$ with $\mathcal{C} = (C_1, \ldots, C_6)$, $\mathcal{S} = (e_1, \ldots, e_5)$.

In the same way as in [3], we can check whether $\bar{G}$ is a tree of chordless cycles, and make a list of all chordless cycles in linear time. After that, we can check in linear time whether the tree of chordless cycles is a path of chordless cycles.

3.2 Trees of Pathwidth Two

The following result, describing the structure of trees of pathwidth k, has form and proof similar to a result in [4].

Lemma 6. *Let H be a tree, $k \geq 1$. H is a partial k-path if and only if there is a path $P = (v_1, \ldots, v_s)$ in H such that the connected components of $H[V - V(P)]$ have pathwidth $k - 1$ at most, i.e. H consists of a path with partial $(k-1)$-paths connected to it.*

Because graphs of pathwidth one do not contain cycles, a graph of pathwidth one is a tree which consists of a path with 'sticks', vertices of degree one adjacent only to a vertex on the path ('caterpillars with hair length one').

Lemma 7. *Let H be a tree of pathwidth k, $k \geq 1$, suppose there is no vertex $v \in V(H)$ such that the components of $H[V - \{v\}]$ have pathwidth $k - 1$ or less. Then there is a unique path P in H such that the components of $H[V - V(P)]$ have pathwidth $k - 1$ or less, and P is shorter than and contained in all other paths having this property. We denote this path by $P_k(H)$.*

Proof. It can be shown that the intersecting path P'' of two paths P and P' for which $H[V - V(P)]$ and $H[V - V(P')]$ have pathwidth $k - 1$ is not empty, and that $H[V - V(P'')]$ also has pathwidth $k - 1$. □

We can show that if H has pathwidth two and there is a vertex $v \in V(H)$ such that $H[V - \{v\}]$ has pathwidth one, then there are at most seven vertices for which this holds. Furthermore, the graph induced by these vertices is connected. Figure 2 shows an example of a tree of pathwidth two.

The linear time algorithm in [4] or [12] to compute the pathwidth of a tree can also be used to find the path $P_2(H)$ if it is unique, or to find all vertices v for which $H[V - \{v\}]$ has pathwidth one.

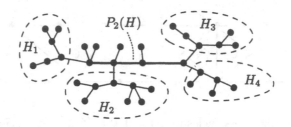

Fig. 2. A tree H of pathwidth two. $P_2(H)$ is formed by the fat edges. H_1, H_2, H_3 and H_4 are the partial one-paths with pathwidth one which are connected to $P_2(H)$.

3.3 Partial Two-Paths

A partial two-path consists of a number of biconnected components, and a number of trees of pathwidth two, which are connected to each other in a certain way. Let G be a partial two-path. By G_T we denote the graph obtained from G by deleting all edges of biconnected components of G. Clearly, each biconnected component of G can be written as a path of chordless cycles, and each component of G_T consists of a path with partial one-paths and sticks connected to it. The number of possible ways in which biconnected components and components of G_T can be connected to each other is large. A complete analysis is given in the full paper. An example of a graph of pathwidth two is given in Fig. 3. Different cases in the ways of connecting correspond to different (but often still rather similar) cases in the intervalization algorithm. We only mention the following. Let B be a biconnected component of G with path of chordless cycles $(\mathcal{C}, \mathcal{S})$, $\mathcal{C} = (C_1, \ldots, C_p)$. The vertices of B which are not in C_1 or C_p may only be adjacent to vertices $v \in V - V(B)$ which have degree one, since these vertices can only occur within the occurrence of B in any path decomposition of G of width two. Furthermore, at most two vertices of C_1 and at most two vertices of C_p may be adjacent to a vertex which does not have degree one. This follows from the next lemma.

Lemma 8. *Let $G = (V, E)$ be a partial two-path, $V' \subseteq V$, such that $G[V']$ is a connected graph. Let $PD = (V_1, \ldots, V_t)$ be a path decomposition of width two of G such that the vertices of V' occur in $(V_j, \ldots, V_{j'})$. On each side of $(V_j, \ldots, V_{j'})$, edges of at most two components of $G[V - V']$ occur.*

Proof. Suppose there are edges of at least three components of $G[V - V']$ on the left side of V_j. Let G_1, G_2, G_3 be three of these components. Let V_l, $1 \le l < j$, be the rightmost node on the left side of V_j containing an edge of one of the components G_1, G_2 and G_3, say G_1. V_l contains a vertex of G_2 and of G_3. Hence $|V_l| = 4$. □

If a vertex $v \in V(B)$ is adjacent to a vertex $w \notin V(B)$ which does not have degree one, then v is an end vertex of B.

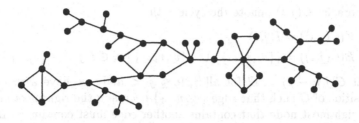

Fig. 3. A graph of pathwidth two.

Consider the set of all paths $P_2(H')$ of components H' of G_T. It can be shown that there is path P in G containing all these paths and at least one vertex of each biconnected component.

4 Intervalizing Three-Colored Graphs

In this section, we give algorithms for determining whether there is an intervalization (i.e. there is a proper path decomposition) of three-colored biconnected graphs, three-colored trees, and finally of three-colored graphs in general. The main algorithm has the following form: first the structure of G is determined, as described or indicated in Sect. 3, and then the algorithms of this section are used.

4.1 Biconnected Graphs

To make a proper path decomposition of a properly three-colored biconnected partial two-path G, we can make proper path decompositions of the chordless cycles of $\bar{G}$, thereby taking into account which edges of each chordless cycle are shared with other chordless cycles: these are the end edges of the chordless cycle. The proper path decompositions of the chordless cycles can then be concatenated in the order in which they occur in the path of chordless cycles of G, and this gives a proper path decomposition of G (of width two).

Hence we concentrate now on checking whether there exists a proper path decomposition of a chordless cycle C. Let C be a properly three-colored chordless cycle. If $|V(C)| = 3$, then there is a proper path decomposition which consists of one node. Now suppose $|V(C)| > 3$. There may be certain edges which we want to be in the leftmost node (starting edges), and certain edges which we want to be in the rightmost node node (ending edges) of the proper path decomposition. We denote the vertices and edges of C by $V(C) = \{v_0, v_1, \ldots, v_{n-1}\}$, and $E(C) = \{\{v_i, v_{i+1}\} \mid 0 \le i < n\}$ (for each i, let v_i denote $v_{i \bmod n}$). For each $j, l, 0 \le j < n$ and $l \ge 1$, let $I(j, l)$ denote the set of vertices of $V(C)$ between v_j and v_{j+l}, when going from v_j to v_{j+l} in positive direction, i.e.,

$$I(j, l) = \{v_i \mid j \le i \le j + l\}.$$

Furthermore, let $C(j, l)$ denote the cycle with

$$V(C(j, l)) = I(j, l)$$
$$E(C(j, l)) = \{\{v_j, v_{j+l}\}\} \cup \{\{v_i, v_{i+1}\} \mid v_i \in I(j, l) - \{v_{j+l}\}\}$$

Note that $C(j, n - 1) = C$ for all j, $0 \leq j < n$. If we have a proper path decomposition of C such that edge $\{v_j, v_{j-1}\}$ occurs in the rightmost end node, then the rightmost node that contains another edge must contain v_j and v_{j-1}, and either v_{j+1} or v_{j-2}. If it contains v_{j+1}, then there must be a proper path decomposition of $C(j + 1, n - 2)$ with edge $\{v_{j+1}, v_{j-1}\}$ in the rightmost node; if it contains v_{j-2}, a proper path decomposition of $C(j, n - 2)$ with $\{v_j, v_{j-2}\}$ in the rightmost node must exist. This observation is used below to obtain a dynamic programming algorithm for our problem. We define $PPW2$ as follows. Let $E_S \subseteq E(C)$ be a set of starting edges and let $0 \leq j < n$, $l \geq 1$.

$$PPW2(C, E_S, j, l) =$$
$$\begin{cases} \text{true if } \exists_{PD=(V_1,\ldots,V_t)} \; PD \text{ is a proper path decomposition of } C(j, l) \\ \qquad \land v_j, v_{j+l} \in V_t \land \exists_{e \in E_S} \; e \subseteq V_1 \\ \text{false otherwise} \end{cases}$$

$PPW2$ can be described recursively as follows. Let $E_S \subseteq E(C)$, $0 \leq j < n, l \geq 1$.

$$PPW2(C, E_S, j, l) =$$
$$\begin{cases} \{v_j, v_{j+1}\} \in E_S & \text{if } l = 1 \\ c(v_j) \neq c(v_{j+l}) \land \\ (PPW2(C, E_S, j + 1, l - 1) \lor PPW2(C, E_S, j, l - 1)) & \text{if } l \geq 2 \end{cases}$$

For a given properly three-colored cycle C and set of starting edges $E_S \subseteq E(C)$, the function $PPW2$ can be computed in $O(n^2)$ time using dynamic programming. From the values of $PPW2(C, E_S, j, n - 1)$, we can compute whether there is proper path decomposition of C. If there is an ending edge $\{v_i, v_{i-1}\}$, then there is a proper path decomposition of C if and only if $PPW2(C, E_S, i, n - 1)$ holds. If there is no ending edge, then there is a proper path decomposition of C if and only if there is a j, $0 \leq j < n$, such that $PPW2(C, E_S, j, n - 1)$ holds.

For a given three-colored biconnected graph G, the algorithm is now as follows.

1. Find the cell completion $\bar{G}$ of G and check if $\bar{G}$ is properly three-colored. If not, stop, the answer is no.
2. Check if $\bar{G}$ can be written as a path of chordless cycles. If so, construct such a path $(\mathcal{C}, \mathcal{S})$ with $\mathcal{C} = (C_1, \ldots, C_p)$ and $\mathcal{S} = (e_1, \ldots, e_{p-1})$. If not, stop, the answer is no.
3. For each chordless cycle C_i in the path, compute $PPW2$, with set of starting edges $\{e_{i-1}\}$ if $i > 1$, $E(C_i)$ otherwise, and ending edge e_i if $i < p$, no ending edge otherwise. If the computed value for $PPW2$ is true for each C_i, the answer is yes, otherwise it is no.

Step 1 and 2 run in $O(n)$ time, step 3 runs in $O(n^2)$ time, where $n = |V(G)|$.

4.2 Trees

Definition 9 (Nice Path Decomposition). Let H be a tree of pathwidth two, $PD = (V_1, \ldots, V_t)$ a path decomposition of H. We say that PD is a *nice path decomposition* of H if the following holds. If $P_2(H) = (v_1, \ldots, v_s)$ is the unique path of H, then V_1 contains a vertex $w \in V(H)$ such that $w \in V(H')$ for some partial one-path H' which is not a stick and is connected to v_1, w end point of $P_1(H')$, and V_t contains a vertex $x \in V(H)$ such that $x \in V(H'')$ for some partial one-path H'' which is not a stick and is connected to v_s, x end point of $P_1(H'')$, and $H'' \neq H'$. If $P_2(H)$ is not unique, then there is a path (v_1) for which this holds. The path P from w to x is called a *nice path* of H.

Lemma 10. *Let H be a three-colored partial two-path. If there is a proper path decomposition of H then there is a nice proper path decomposition of H.*

Proof. This is proved by starting with an arbitrary proper path decomposition PD, and transforming it into a nice proper path decomposition by using some elementary operations. □

Our algorithm for trees of pathwidth two works as follows: for certain possible nice paths of H, try if there is a nice proper path decomposition with this nice path. The number of nice paths may be large. Fortunately, we can bound the number of nice paths that has to be tried, using Lemma 8: if H is a tree of pathwidth two, $c : V \to \{1, 2, 3\}$ a proper three-coloring of H, $v \in V(H)$, then at most four components of $H[V - \{v\}]$ may have a vertex of color $c(v)$, otherwise there is no proper path decomposition of H. It can be shown that if $H[V - \{v\}]$ has four or more components which contain at least one edge, and there is a proper path decomposition of H, then there is a proper path decomposition of H in which all components of pathwidth one which do not have a vertex of color $c(v)$ can occur completely within the occurrence of v. Let $P_2(H) = (v_1, \ldots, v_s)$. Thus, for the choice of a nice path P starting with vertex w and ending with x, there are at most four (at most three if $s \geq 2$) components where w can be chosen from, and each component has pathwidth one and is connected to v_1; w is one of the two endpoints of such a path. In the same way, the number of possible choices for x is bounded.

Now, suppose we have a (possible) nice path $P = (v_1, \ldots, v_s)$. We discuss how to check whether there is a nice proper path decomposition of H with nice path P. The general method is as follows. First, for each vertex v of the nice path to which at least one partial one-path is connected, we compute some values which concern local path decompositions of partial one-paths connected to v. After that, these values are all combined into a nice proper path decomposition for the complete tree.

A useful observation for bounding the number of cases to consider is the following: if there is a nice proper path decomposition PD of H with nice path P, then there is a nice proper path decomposition PD' with the same nice path P such that no two partial one-paths of $H[V - V(P)]$ overlap, i.e. if H' and H''

are distinct partial one-paths of $H[V - V(P)]$, then PD' contains no node V_i which contains a vertex of H' and a vertex of H''.

A partial one-path H' (which is not a stick) connected to a vertex v_m on the nice path P can be 'laid out' in different ways. If H' does not contain a vertex with color $c(v_m)$, then it is 'usually' possible to have H' occur completely within the occurrence of v_m in the path decomposition to be formed. Otherwise, H' must be mapped either 'left' of v_m (there are v_j, $j < m$, with $\exists i : v_j \in V_i \wedge V(H') \cap V_i \neq \emptyset$), or 'right' of v_m (there are v_j, $j > m$, with $\exists i : v_j \in V_i \wedge V(H') \cap V_i \neq \emptyset$), or in some specific cases (which we will further not discuss), H' is mapped partially left and partially right of v_m. (It is not possible to map two partial one-paths, both connected to the same node v_m, to the same direction, since then these two components would overlap. This also helps to bound the number of different cases to consider and the total time of the algorithm.)

As an example, suppose H' is a path $(w_1, \ldots, w_t)$, with $\{v_m, w_1\} \in E(H)$. Suppose $v_{m'}$ is the first node after m ($m' > m$) with a partial one-path (not a stick) attached to it. If H' is mapped to the right, then we want to minimize $\rho_{H'} = \max\{j \mid \exists i : v_j \in V_i \wedge V(H') \cap V_i \neq \emptyset\}$ (i.e., we want H' to 'overlap' in the path decomposition with a portion of P which is as small as possible), in order to avoid overlaps with the partial one-paths attached to $v_{m'}$, which we could want to map to the left. (In 'most cases', we will have, by a 'non-overlapping argument', that $\rho_{H'} \leq m'$.) If there are no sticks connected to vertices on P between v_m and $v_{m'}$, then we can use the procedure PPW2 to compute this minimal value. All other cases (sticks adjacent to vertices on P or H', or H' connected to P in a different way) can be handled in a similar way: the case analysis becomes involved, but all cases can be handled by procedures that are essentially simple modifications of PPW2.

For each component H', not a stick, connected to a vertex on the nice path, we compute this minimal value of $\rho_{H'}$ for the case that it would be mapped to the right, and in the same way a 'maximal value for the case that it would be mapped to the left'. When we have all these values for all components connected to the nice path, one simple (linear time) scan along the nice path from left to right suffices to see whether all components can be 'laid out' together with the nice path P.

The time is quadratic, as each vertex will be involved in a constant bounded number of calls to PPW2 or a modification of PPW2.

4.3 General Graphs

Let G be a properly three-colored partial two-path. As discussed in Sect. 3.3, there is a shortest path P containing all paths $P_2(H)$ of components H of G_T and at least one vertex of each biconnected component of G. In a similar way as in the previous section, we can define a kind of nice paths for this path P, such that the number of nice paths is bounded by a constant. Then we use the same kind of algorithm as for trees. Only the computation of the values for each v_m on the nice path which has one or more partial one-path (not a stick) or biconnected component connected to it is different (but still uses a structure, similar to PPW2). Then, the combination of these values is done by the same type of left-to-right scan along the nice path as in the case of trees. Due to space limitations, we do not give a complete description of this algorithm, which is very lengthy, and is given in the full paper.

5 Intervalizing Four-Colored Graphs

For some time, it has been an open problem whether there exist polynomial time algorithms for ICG for some constant number k of colors, $k \geq 4$. Older results showed fixed parameter intractability [5, 1], but did not resolve the question. Our NP-completeness result resolves the open problem in a negative way (assuming $P \neq NP$).

Theorem 11. *ICG is NP-complete for four-colored graphs.*

We omit the proof, which is based on a transformation from 3-PARTITION. It may be interesting to note that in our reduction, only three vertices are colored with the fourth color. It may also be interesting to note that, when the maximum degree of vertices and the number of colors are bounded by constants, then ICG can be solved in polynomial time (combining techniques from e.g. [14, 11].)

6 Conclusions

In this paper, we have given (the outlines of) an $O(n^2)$ time algorithm to determine whether we can add edges to a given three-colored graph such that it becomes a properly colored interval graph. The algorithm can be modified such that it outputs an intervalization, if existing, and still uses quadratic time. To get a faster algorithm for the problem considered in this paper might well be a hard problem. It seems that even the simplest cases, e.g., when G is a simple cycle, need $O(n^2)$ time to resolve, and might well already capture the main difficulties for speed-up.

We have shown that this problem is NP-complete for four or more colors. We feel however that the graphs, arising in the reduction of this proof, will not be typical for the type of colored graphs, arising in the sequence reconstruction application. It may well be that special cases of ICG, which capture characteristics of the application data, have efficient algorithms. Further research could perhaps give new meaningful results here.

References

1. H. L. Bodlaender, M. R. Fellows, and M. Hallett. Beyond NP-completeness for problems of bounded width: Hardness for the W hierarchy. In *Proceedings of the 26th Annual Symposium on Theory of Computing*, pages 449–458, New York, 1994. ACM Press.

2. H. L. Bodlaender, M. R. Fellows, and T. J. Warnow. Two strikes against perfect phylogeny. In *Proceedings 19th International Colloquium on Automata, Languages and Programming*, pages 273–283, Berlin, 1992. Springer Verlag, Lecture Notes in Computer Science, vol. 623.

3. H. L. Bodlaender and T. Kloks. A simple linear time algorithm for triangulating three-colored graphs. *J. Algorithms*, 15:160–172, 1993.

4. J. A. Ellis, I. H. Sudborough, and J. Turner. The vertex separation and search number of a graph. *Information and Computation*, 113:50–79, 1994.

5. M. R. Fellows, M. T. Halett, and H. T. Wareham. DNA physical mapping: Three ways difficult (extended abstract). In T. Lengauer, editor, *Proceedings 1st Annual European Symposium on Algorithms ESA '93*, pages 157–168. Springer Verlag, Lecture Notes in Computer Science, vol. 726, 1993.

6. M. C. Golumbic, H. Kaplan, and R. Shamir. On the complexity of dna physical mapping. *Advances in Applied Mathematics*, 15:251–261, 1994.

7. R. Idury and A. Schaffer. Triangulating three-colored graphs in linear time and linear space. *SIAM J. Disc. Meth.*, 2:289–293, 1993.

8. S. Kannan and T. Warnow. Triangulating 3-colored graphs. *SIAM J. Disc. Meth.*, 5:249–258, 1992.

9. H. Kaplan and R. Shamir. Pathwidth, bandwidth and completion problems to proper interval graphs with small cliques. Technical Report 285/93, Inst. for Computer Science, Tel Aviv University, Tel Aviv, Israel, 1993. To appear in SIAM J. Comput.

10. H. Kaplan, R. Shamir, and R. E. Tarjan. Tractability of parameterized completion problems on chordal and interval graphs: Minimum fill-in and physical mapping. In *Proceedings of the 35th annual symposium on Foundations of Computer Science (FOCS)*, pages 780–791. IEEE Computer Science Press, 1994.

11. F. R. McMorris, T. Warnow, and T. Wimer. Triangulating vertex-colored graphs. *SIAM J. Disc. Meth.*, 7(2):296–306, 1994.

12. R. H. Möhring. Graph problems related to gate matrix layout and PLA folding. In E. Mayr, H. Noltemeier, and M. Sysło, editors, *Computational Graph Theory, Comuting Suppl. 7*, pages 17–51. Springer Verlag, 1990.

13. S.-I. Nakano, T. Oguma, and T. Nishizeki. A linear time algorithm for c-triangulating three-colored graphs. *Trans. Institute of Electronics, Information and Communication, Eng., A.*, 377-A(3):543–546, 1994. In Japanese.

14. J. B. Saxe. Dynamic programming algorithms for recognizing small-bandwidth graphs in polynomial time. *SIAM J. Alg. Disc. Meth.*, 1:363–369, 1980.

15. M. Steel. The complexity of reconstructing trees from qualitative characters and subtrees. *J. of Classification*, 9:91–116, 1992.

NC Algorithms for Finding a Maximal Set of Paths with Application to Compressing Strings

Zhi-Zhong Chen

Department of Mathematical Sciences, Tokyo Denki University
Hatoyama, Saitama 350-03, Japan. Email: chen@r.dendai.ac.jp
January 12, 1995

Abstract. It is shown that the problem of finding a maximal set of paths in a given (undirected or directed) graph is in NC. This result is then used to obtain three parallel approximation algorithms for the shortest superstring problem. The first is an **NC** algorithm achieving a compression ratio of $\frac{1}{3+\epsilon}$ for any $\epsilon > 0$. The second is an **RNC** algorithm achieving a compression ratio of $\frac{38}{63} \approx 0.603$. The third is an **RNC** algorithm achieving an approximation ratio of $2\frac{50}{63} \approx 2.793$. All the results significantly improve on the best previous ones.

1 Introduction

Let $S = \{s_1, \cdots, s_n\}$ be a set of n strings over an alphabet Σ. A *superstring* of S is a string s over Σ such that s contains each string $s_i \in S$ as a substring, i.e., s can be written as $u_i s_i v_i$ for some strings u_i and v_i over Σ. A *shortest superstring* of S is a superstring of S that has minimum length over all superstrings of S. The *shortest superstring problem* (SSP) is to find a shortest superstring of a given set of strings.

SSP has many important applications [6, 7, 9] but is unfortunately NP-hard [4]. This has motivated many researchers to find approximation algorithms with good performance guarantees for SSP [2, 3, 8, 10, 12]. To evaluate the quality of an approximation algorithm A for SSP, two measures are usually used. One is the *approximation ratio*, defined to be $max\{\frac{|A(S)|}{opt(S)} : S$ is a set of strings$\}$, where $|A(S)|$ is the length of the superstring found by algorithm A and $opt(S)$ is the length of a shortest superstring of S. The other is the *compression ratio*, defined to be $min\{\frac{|S|-|A(S)|}{|S|-opt(S)} : S$ is a set of strings$\}$, where $|S|$ is the total length of the strings in S. The best known approximation (resp., compression) ratio achieved by a polynomial-time approximation algorithm is $2\frac{3}{4}$ [1] (resp., $\frac{38}{63}$ [8]). Our objective here is to design efficient parallel approximation algorithms for SSP with a good approximation or compression ratio.

Parallel approximation algorithms for SSP were first studied by Czumaj *et al* [3]. In [3], they gave an **NC** approximation algorithm for SSP that achieves a compression ratio of $\frac{1}{4+\epsilon}$ for any $\epsilon > 0$. They also gave an **RNC** approximation algorithm for SSP that achieves an approximation ratio of $\frac{17}{6} \approx 2.833$ [3]. In this paper, we present better parallel approximation algorithms for SSP. The first is an **RNC** algorithm achieving a compression ratio of $\frac{38}{63} \approx 0.603$. This algorithm is obtained by (very) nontrivially modifying and parallelizing the approximation algorithm of Rao Kosaraju *et al* for SSP that achieves a compression ratio of $\frac{38}{63}$ [8]. At the heart of our algorithm is a slightly modified version of the path-coloring lemma (Lemma 2 in [8]) of Rao Kosaraju *et al*. The modified version states that there is a nearly optimal **NC** algorithm for partitioning the arc set of a given digraph D into two sets of vertex-disjoint directed paths provided that D satisfies the following three conditions:

(1) Each vertex in D has indegree at most 2, outdegree at most 2, and total degree at most 3;

(2) D contains no 2-cycle;

(3) D contains no arc (u, v) s.t. u has indegree 2 and v has outdegree 2.

Note that the original path-coloring lemma of Rao Kosaraju *et al* does not include the condition (3) above. To prove the modified version, we first give an NC algorithm for finding a maximal set of paths in a given undirected graph, and then use this NC algorithm to design an **NC** algorithm for partitioning the arc set of a given digraph D satisfying the above three conditions into two sets of vertex-disjoint directed paths.

Our second approximation algorithm for SSP is an **NC** algorithm achieving a compression ratio of $\frac{1}{3+\epsilon}$ for any $\epsilon > 0$. Similar to that in [3], the idea behind this algorithm is to parallelize GREEDY using the geometric grouping technique of Karmarkar and Karp [5] (with a small decrease of the compression ratio). Here, GREEDY denotes the well-known simple approximation algorithm for SSP which repeatedly merges the pair of (distinct) strings with maximum overlap until only one string remains. However, our algorithm contains a nontrivial subroutine, namely, an **NC** algorithm for finding a maximal set of directed paths in a given digraph.

In [2], Blum *et al* showed that if SSP has an approximation algorithm achieving a compression ratio of $\frac{1}{2} + \epsilon$, then it has another one achieving an *approximation* ratio of $3 - 2\epsilon$. Combining this fact with our first approximation algorithm for SSP, we obtain our final **RNC** approximation algorithm for SSP that achieves an approximation ratio of $2\frac{50}{63} \approx 2.793$.

The model of parallel computation we use is ARBITRARY CRCW PRAM. The results in this paper and their comparison with the best previous ones are shown in Table 1.

Table 1. Main results and their comparison with the best previous ones

Type of Approximations	Compression ratio		Approximation ratio	
	Previous	Ours	Previous	Ours
Polynomial-time	$\frac{38}{63}$ [8]		$2\frac{3}{4}$ [1]	
RNC	$\frac{1}{2}$ [Folkflore]	$\frac{38}{63}$	$2\frac{5}{6}$ [3]	$2\frac{50}{63}$
NC	$\frac{1}{4+\epsilon}$ [3]	$\frac{1}{3+\epsilon}$	$(2+\epsilon)\log n$ [3]	

2 Finding a maximal set of paths in parallel

In this section, we only prove that a maximal set of paths in a given undirected graph can be found in **NC**. However, at the end of this section, we will point out that this result can be easily extended to digraphs.

Throughout this section, unless stated otherwise, a graph is always undirected and simple. By a *path*, we always mean a simple path. Note that a single vertex is considered as a path (of length 0). Let $G = (V, E)$ be a graph. A set M of edges in G is called a *matching* if no two edges in M share an endpoint. A matching in G is said to be *maximal* if it is not a proper subset of another matching. For $F \subseteq E$, let $G[F]$ denote the graph (V, F). A set F of edges in G is called a *path set* if $G[F]$ is a forest in which each connected component is a path. A *maximal path set* (MPS) in G is a path set that is not properly contained in another path set.

Algorithm 1
Input: An n-vertex graph $G = (V, E)$.
Output: An MPS F in G.
1. Initialize F to be the empty set.
2. While $E \neq \emptyset$, perform the following steps:
    2.1. Construct a new graph H as follows. Corresponding to each connected component P in $G[F]$, H contains a vertex w_P. For two connected components P_1 and P_2 in $G[F]$, H contains the edge $\{w_{P_1}, w_{P_2}\}$ iff E contains some edge $\{v_1, v_2\}$ such that v_1 is contained in P_1 and v_2 is contained in P_2.
    2.2. Find a matching M in H whose size is at least a constant fraction of the size of a maximum matching in H.
    2.3. In parallel, for each edge $\{w_{P_1}, w_{P_2}\} \in M$, add to F the smallest edge $\{v_1, v_2\} \in E$ such that v_1 is contained in P_1 and v_2 is contained in P_2, and then remove the edge $\{v_1, v_2\}$ from E.
    2.4. Remove from E all edges e such that $F \cup \{e\}$ is not a path set in G.
3. Output F.

Let t be the number of executions of the while-loop in Algorithm 1. In case $t = 0$, the input graph G contains no edge and so Algorithm 1 is clearly correct and takes constant time. Thus, we may assume that $t \geq 1$. For $1 \leq i \leq t$, let E_i, F_i, H_i, and M_i denote the contents of the variables E, F, H, and M after the ith execution of the while-loop, respectively. For convenience, let $F_0 = \emptyset$ and E_0 be the edge set of the input graph G. For $1 \leq i \leq t$, let $M_i' = F_i - F_{i-1}$. Note that $E_t = \emptyset$ and $|M_i| = |M_i'|$ for $1 \leq i \leq t$. Let $0 \leq i \leq t - 1$. An *augmentation* of F_i is a set of some edges in E_i. An augmentation A of F_i is said to be *valid* if $F_i \cup A$ is a path set in G.

Lemma 1. F_t is an MPS.

Lemma 2. $t = O(\log n)$.

Proof. For $0 \leq i \leq t - 1$, let α_i be the size of a maximum valid augmentation of F_i. Let us first prove that $|M'_{i+1}|$ is at least a constant fraction of α_i. Fix an integer i with $0 \leq i \leq t - 1$. Let β be the size of a maximum matching in the graph H_{i+1}. We want to show that $\alpha_i \leq 2\beta$. Let A be a maximum valid augmentation of F_i. Since A is valid, each connected component of $G[F_i \cup A]$ must be a path. For each connected component P of $G[F_i \cup A]$ that contains at least one edge of A, we start at an endpoint of P and traverse P toward the other endpoint while labeling the edges of A on P by 0 and 1 alternately (the first edge of A on P is labeled 0). Let B be the set of those edges of A labeled 0. Clearly, $|B| \geq \frac{|A|}{2} = \frac{\alpha_i}{2}$. Moreover, it is easy to see that corresponding to B, there is a matching of size $|B|$ in H_{i+1}. This implies that $|B| \leq \beta$ and in turn that $\alpha_i \leq 2\beta$. On the other hand, by step 2.2, $|M_{i+1}| \geq c\beta$ for some constant $c > 0$. Using $\alpha_i \leq 2\beta$, we now have $|M'_{i+1}| = |M_{i+1}| \geq \frac{c}{2}\alpha_i$.

By Lemma 1, M'_{i+1} is a valid augmentation of F_i for $0 \leq i \leq t - 1$. Thus, $\alpha_{i+1} + |M'_{i+1}| \leq \alpha_i$ for $0 \leq i \leq t - 1$. Since $|M'_{i+1}| \geq \frac{c}{2}\alpha_i$, we now have $\alpha_{i+1} \leq (1 - \frac{c}{2})\alpha_i$ for $0 \leq i \leq t - 1$. Combining this with the fact that α_0 is no more than the number of edges in G, we obtain $t = O(\log n)$. $\blacksquare$

Theorem 3. Given an n-vertex m-edge graph G, an MPS of G can be found in $O(\log^3 n)$ time or in $O(\log^2 n)$ expected time with $n + m$ processors.

The following corollary will be used in Section 3.

Corollary 4. Given an n-vertex m-edge graph G and a path set F' in G, an MPS F in G with $F' \subseteq F$ can be found in $O(\log^3 n)$ time or in $O(\log^2 n)$ expected time with $n + m$ processors.

Below, we point out that Theorem 3 can be extended to digraphs. This extension will be used in Section 4. Let $D = (V, A)$ be a digraph. Hereafter, a path (resp., cycle) in D always means a simple directed path (resp., cycle) in D. If P is a path or cycle in D, then the *length* of P is the number of arcs on P and is denoted by $|P|$. For $F \subseteq A$, let $D[F]$ denote the digraph (V, F). A *directed path set* (DPS) in D is a subset B of A such that $D[B]$ is an acyclic digraph in which the indegree and outdegree of each vertex are both at most 1. Intuitively speaking, if B is a DPS in D, then $D[B]$ is a collection of vertex-disjoint paths. A *maximal directed path set* (MDPS) in D is a DPS that is not properly contained in another DPS.

Corollary 5. Given an n-vertex m-arc digraph $D = (V, A)$ and a DPS B in D, an MDPS F in D with $B \subseteq F$ can be found in $O(\log^3 n)$ time or in $O(\log^2 n)$ expected time with $n + m$ processors.

3 A parallelizable path-coloring lemma

Hereafter, a *2-path-coloring* of a digraph always mean a two-coloring of the arcs in the digraph such that the arcs in each color class form a set of vertex-disjoint paths. (Recall that a path always means a simple directed path.) In [8], Rao Kosaraju *et al* claimed the following lemma.

Lemma 6. [8] Let $D = (V, A)$ be a directed graph such that (1) each vertex has indegree at most 2, outdegree at most 2, and total degree at most 3, and (2) the graph contains no cycle of length 2. Then D has a 2-path-coloring.

Lemma 7. Same as Lemma 6 except that D is required to satisfy the following additional condition: (3) the graph contains no arc (u, v) such that u has indegree 2 and v has outdegree 2.

Lemma 7 immediately follows from Lemma 6. However, we give below an **NC** algorithm for finding a 2-path-coloring of D. Let us say that two arcs in D *conflict* if either both enter or both leave the same vertex. An arc in D is said to be *free* if it does not conflict with any other arc. Let A_f be the set of the free arcs in D and set $A_c = A - A_f$. The following lemma is very useful.

Lemma 8. Let C be a cycle in $D[A_c]$. Then, the arcs conflicting with the arcs on C must either all enter or all leave C in $D[A_c]$.

Corollary 9. All the cycles in $D[A_c]$ are vertex-disjoint. Moreover, for every cycle C in $D[A_c]$, either the vertices on C all have indegree 1 and outdegree 2 or the vertices on C all have indegree 2 and outdegree 1.

Let G_c be an undirected graph obtained from D as follows. Corresponding to each arc $e \in A_c$, there is exactly one vertex u_e in G_c. For every two arcs e and e' in A_c, the edge $\{u_e, u_{e'}\}$ is in G_c if and only if e and e' conflict in D. We call G_c the *conflict graph* of D. A cycle in G_c is said to be *even* if its length is even.

Lemma 10. Each connected component of G_c is either a path or an even cycle.

Let $P(G_c)$ denote the set of those connected components of G_c that are paths.

Lemma 11. Let C be a cycle in $D[A_c]$ and let e be an arc on C. Then, the vertex u_e has degree 1 in G_c, that is, u_e is an endpoint of some path in $P(G_c)$.

Let H be the undirected graph constructed from $P(G_c)$ as follows. Initially, set H to be $P(G_c)$. Next, for every cycle C in $D[A_c]$ and every two *consecutive* arcs e and e' on C, add the edge $\{u_e, u_{e'}\}$ to H. By the first assertion in Corollary 9, H is well-defined.

Lemma 12. Suppose S is an MPS of H such that S contains all the edges in $P(G_c)$. Then, for every cycle C in $D[A_c]$, there are at least two arcs e and e' on C such that $\{u_e, u_{e'}\}$ is an edge in S.

Lemma 13. Each vertex in $D[A_f]$ is of indegree at most 1 and outdegree at most 1. Moreover, a vertex of total degree 2 in $D[A_f]$ is not incident on any arc of A_c.

By Lemma 13, $D[A_f]$ is a collection of vertex-disjoint paths and cycles. Hereafter, we view $D[A_f]$ as a set of these paths and cycles. Now, we are ready to present our **NC** algorithm for finding a 2-path-coloring.

Algorithm 3
Input: An n-vertex digraph $D = (V, A)$ satisfying the conditions in Lemma 7.
Output: A 2-path-coloring of D.
1. Compute the set A_c of those arcs that are not free in D and find all the cycles in $D[A_c]$.
2. Construct an undirected graph G_c as follows. Corresponding to each arc $e \in A_c$, there is exactly one vertex u_e in G_c. For every two arcs e and e' in A_c, the edge $\{u_e, u_{e'}\}$ is in G_c if and only if e and e' conflict in D.
3. Compute $P(G_c)$, the set of connected components of G_c that are paths.
4. Construct an undirected graph H from $P(G_c)$ as follows. Initially, set H to be $P(G_c)$. Next, for every cycle C in $D[A_c]$ and every two consecutive arcs e and e' on C, add the edge $\{u_e, u_{e'}\}$ to H.
5. Compute an MPS S of H such that S contains all the edges in $P(G_c)$.
6. Let K be the undirected graph obtained by adding the cycles in G_c to $H[S]$. Two-color the vertices in K in a way that no two adjacent vertices receive the same color.
7. Two-color the arcs in A_c as indicated by the coloring obtained in step 6. That is, color each arc $e \in A_c$ with the color that was used to color u_e.
8. Let $A_f = A - A_c$. For each path of length ≥ 2 and each cycle in the set $D[A_f]$, choose an arbitrary arc on it, color this arc with one of the two colors, and color the rest arcs on it with the other color.
9. For each path of length 1 in the set $D[A_f]$, choose an endpoint of this path that has total degree at most 2 in D, color the unique arc on it with a color that was not used to color any arc incident to this endpoint in step 7.

Lemma 14. The cycles in $D[A_c]$ can be computed in $O(\log n)$ time with n processors.

Theorem 15. Algorithm 3 is correct and runs in $O(\log^3 n)$ time or in $O(\log^2 n)$ expected time with n processors.

4 RNC-approximation of shortest superstrings

For a string s, let $|s|$ denote the length of s. Let s and t be two strings, and let v be the longest string such that $s = uv$ and $t = vw$ for some non-empty strings u and w. $|v|$ is called the *overlap* between s and t and is denoted by $ov(s, t)$. By $s \circ t$, we denote the string uvw.

Let $S = \{s_1, s_2, \cdots, s_n\}$ be a set of strings. As in previous studies, we assume that S is *substring free*, i.e., no string in S is a substring of any other. By $opt(S)$, we denote the length of a shortest superstring of S. Define $|S| = \sum_{i=1}^{n} |s_i|$. The *overlap graph* of S is the arc-weighted digraph $OG(S) = (V, A, ov)$, where $V = \{1, 2, \cdots, n\}$, $A = \{(j, k) : 1 \leq j \neq k \leq n\}$, and $ov(j, k) = ov(s_j, s_k)$.

For a subgraph D of $OG(S)$, the *weight* of D is the total weight of the arcs in D and is denoted by $ov(D)$. Let $P = j_0, e_1, j_1, \cdots, e_l, j_l$ be a path in $OG(S)$. We call $s_{j_0} \circ s_{j_1} \circ \cdots \circ s_{j_l}$ the *string associated with* P. Note that the string associated with P is a superstring of the strings $s_{j_0}, s_{j_1}, \cdots, s_{j_l}$ and has length $\sum_{i=1}^{l} |s_{j_i}| - ov(P)$. Let P_{max} denote a maximum-weight Hamiltonian path in $OG(S)$.

Fact 1 [12] The string associated with a Hamiltonian path P in $OG(S)$ has length $|S| - ov(P)$.

By Fact 1, we can find a shortest superstring of S by computing a maximum-weight Hamiltonian path in $OG(S)$. Unfortunately, it does not seem that there is an efficient algorithm for computing a maximum-weight Hamiltonian path in $OG(S)$. In [8], Rao Kosaraju *et al* describes a polynomial-time algorithm for finding a Hamiltonian path in $OG(S)$ with weight at least $\frac{38}{63} \cdot ov(P_{max})$. Their algorithm uses the following general framework.

Algorithm 4
Input: $OG(S) = (V, A, ov)$.
1. Compute a maximum-weight cycle cover C of $OG(S)$. (Remark: A *cycle cover* of $OG(S)$ is a collection of cycles in $OG(S)$ such that each vertex is in exactly one cycle.)
2. Use C to obtain a set P of vertex-disjoint paths in $OG(S)$.
3. Construct a Hamiltonian path in $OG(S)$ by patching together the paths in P, and then output the string associated with the Hamiltonian path.

In [8], Rao Kosaraju *et al* gave three different implementations of step 2. Based on two of them, we will obtain two new implementations of step 2. Before describing our implementations, we need several definitions and notations. A cycle in C with exactly two arcs is called a *2-cycle*. The other cycles in C are called *3^+-cycles*. Let $C_1, C_2, \cdots, C_q$ be the 2-cycles in C, and let b_i (resp., c_i) be the weight of the heavier (resp., lighter) arc on C_i. A 2-cycle C_i is *unbalanced* if $b_i > 2c_i$; otherwise, it is *balanced*. Suppose that $C_1, \cdots, C_r$ are the balanced 2-cycles in C. Let $b = \sum_{i=1}^{r} b_i / ov(C)$ and $c = \sum_{i=1}^{r} c_i / ov(C)$. Clearly, $c \le b \le 2c$. Let $C' = C - \{C_1, C_2, \cdots, C_r\}$. For convenience, we give another (nonstandard) definition. A *matching* in a digraph is a set M of arcs in the digraph such that no two arcs of M are incident on a common vertex.

Our first implementation of step 2 is a slight modification of the second implementation of step 2 given in [8]. The details follow:

2.1. Construct a weighted digraph D' from $OG(S)$ as follows. For each balanced 2-cycle $C_i \in C$, redefine the weight on each arc of C_i to be $2(b_i - c_i)$. The weight on each of the rest arcs remains unchanged.
2.2. Find a maximum-weight matching M in D'.
2.3. Let D'' be the digraph $(V, M \cup C')$ in which there are two different copies of each $e \in M \cap C'$ (one in M and the other in C'). Find a subset R of the arcs in C' such that (i) $ov(R) \le \frac{1}{3} \cdot ov(C')$ and (ii) after removing the arcs

of R from D'', the resulting digraph contains no arc (j_1, j_2) such that j_1 has indegree 2 and j_2 has outdegree 2.

2.4. Let D be the digraph obtained from D'' by first removing the arcs of R and then contracting the two vertices on each balanced 2-cycle of C. (Note: Hereafter, contracting a set U of vertices in a digraph means replacing the vertices of U by a (single) new vertex without creating any self-loop.)

2.5. Find a 2-path-coloring of D and set P to be the color class with more weight, together with one arc from each balanced 2-cycle of C.

Our implementation above differs from the second implementation given by Rao Kosaraju *et al* only in step 2.3 [8]. However, this difference is essential as it enables us to parallelize Algorithm 4 without altering its compression ratio. To see this, let us first show that the subset R in step 2.3 exists. The following technical lemma is very useful.

Lemma 16. Let $G = (X, Y, E)$ be a simple bipartite undirected graph in which every vertex $x \in X$ has degree 2 and every vertex $y \in Y$ has degree at most 2. Then, $X \cup Y$ can be partitioned into three subsets U_0, U_1, and U_2 such that for each $0 \le i \le 2$, no vertex $x \in X$ has degree 2 in the subgraph of G induced by $(X \cup Y) - U_i$. Moreover, the partition can be done in $O(\log(|X + |Y|))$ time with $|X| + |Y|$ processors.

Proof. Without loss of generality, we may assume that G is connected. Then, G is either a path or an even cycle. If G is a path, a desired partition can be easily found. Thus, we may assume that G is an even cycle. Let $2p$ be the number of vertices in G. Without loss of generality, we may assume that $X = \{0, 2, \cdots, 2p-2\}$, $Y = \{1, 3, \cdots, 2p-1\}$, and the two neighbors of vertex i in G are $i-1$ (modulo $2p$) and $i+1$ (modulo $2p$). In case p is even, we set $U_0 = \{i \in Y \mid i \equiv 1 \pmod{4}\}$, $U_1 = \{i \in Y \mid i \equiv -1 \pmod{4}\}$, and $U_2 = X$. In case p is odd but $2p = 3h$ for some $h \ge 2$, we set $U_0 = \{0 \le i \le 2p - 3 \mid i \equiv 0 \pmod{3}\}$, $U_1 = \{1 \le i \le 2p - 2 \mid i \equiv 1 \pmod{3}\}$, and $U_2 = \{2 \le i \le 2p - 1 \mid i \equiv 2 \pmod{3}\}$. In case p is odd but $2p = 3h+1$ for some $h \ge 3$, we set $U_0 = \{3 \le i \le 2p-1 \mid i \equiv 0 \pmod{3}\}$, $U_1 = \{1 \le i \le 2p - 3 \mid i \equiv 1 \pmod{3}\}$, and $U_2 = \{0\} \cup \{2 \le i \le 2p - 2 \mid i \equiv 2 \pmod{3}\}$. In case p is odd but $2p = 3h + 2$ for some $h \ge 4$, we set $U_0 = \{0 \le i \le 2p - 2 \mid i \equiv 0 \pmod{3}\}$, $U_1 = \{1 \le i \le 2p - 7 \mid i \equiv 1 \pmod{3}\} \cup \{2p - 3\}$, and $U_2 = \{2 \le i \le 2p - 6 \mid i \equiv 2 \pmod{3}\} \cup \{2p - 4, 2p - 1\}$. ∎

Lemma 17. The subset R in step 2.3 exists and can be found in $O(\log n)$ time with n processors.

Proof. Consider the digraph D'' in step 2.3. First observe that each vertex in D'' has indegree at most 2, outdegree at most 2, and total degree at most 3. Let Z be the set of the lighter arcs on the unbalanced 2-cycles of C'. That is, $Z = \{e \in C_i \mid r + 1 \le i \le q \text{ and } ov(e) = c_i\}$. By the definition of an unbalanced 2-cycle, $ov(Z) < \frac{1}{3} \sum_{i=r+1}^{q} ov(C_i)$. Consider $D'' - Z$, the digraph obtained from D'' by removing the arcs of Z. An arc (j_1, j_2) in $D'' - Z$ is said to be *bad* if j_1 has indegree 2 and j_2 has outdegree 2 in $D'' - Z$; otherwise, it is said to be *good*. If $D'' - Z$ contains no bad arc, then we set $R = Z$ and we are done. So, we may assume that $D'' - Z$ contains at least one bad arc. Let $e = (j_1, j_2)$ be a bad arc in $D'' - Z$. Then, e must be an arc on a 3^+-cycle C of C' because M is a matching, C' is a collection of vertex-disjoint cycles, and $D'' - Z$ contains no 2-cycle of C'. Moreover, exactly one of the two arcs entering j_1 in $D'' - Z$ must appear on C and exactly one of the two arcs leaving j_2 in $D'' - Z$ must appear on C. We call the two arcs the *rivals* of e. Since C is a 3^+-cycle, the two rivals of e cannot be equal. Furthermore, the two rivals of e are both good arcs. Obviously, if we delete at least one of the two rivals of e from $D'' - Z$, then e will become good.

Let X be the set of the bad arcs in $D'' - Z$, and let Y be the rivals of the bad arcs in $D'' - Z$. Note that $X \cap Y = \emptyset$. Let G be the simple bipartite undirected graph (X, Y, E), where $E = \{\{x, y\} \mid x \in X, y \in Y, \text{ and } y \text{ is a rival}$ of $x\}$. Each vertex $x \in X$ must have degree 2 in G since the two rivals of x are not equal. Also observe that in $D'' - Z$, a good arc can be a rival of at most two bad arcs. From these, it is easy to see that G satisfies the condition in Lemma 16. Thus, by Lemma 16, we can partition $X \cup Y$ into three subsets U_0, U_1, and U_2 such that for each $0 \leq i \leq 2$, no vertex $x \in X$ has degree 2 in the subgraph of G induced by $(X \cup Y) - U_i$. Clearly, for each $0 \leq i \leq 2$, if we delete all the arcs in U_i from $D'' - Z$, then $D'' - Z$ will not contain any bad arc. We find a U_i $(0 \leq i \leq 2)$ with $ov(U_i) = \min\{ov(U_0), ov(U_1), ov(U_2)\}$, and set $R = Z \cup U_i$. Obviously, in the digraph $D'' - R$, there is no arc (j_1, j_2) such that j_1 has indegree 2 and j_2 has outdegree 2. Recall that none of the bad arcs and their rivals appears on an unbalanced 2-cycle of C'. From this, it is not difficult to see that $ov(R) \leq \frac{1}{3} \cdot ov(C')$. $\blacksquare$

The coloring in step 2.5 is also possible. This immediately follows from Lemma 3 in [8] but we here give a new proof, which gives an **NC** algorithm for finding a 2-path-coloring.

Lemma 18. D has a 2-path-coloring. Moreover, a 2-path-coloring of D can be found in $O(\log^3 n)$ time or in $O(\log^2 n)$ expected time with n processors.
Proof. As observed in the proof of Lemma 3 in [8], D satisfies the condition (1) in Lemma 7. By step 2.3, the digraph $D'' - R$ contains no arc (j_1, j_2) such that j_1 has indegree 2 and j_2 has outdegree 2. Also note that each vertex on a balanced 2-cycle can have total degree at most 1 in $D'' - R$. Thus, contracting the two vertices on each balanced 2-cycle does not cause $D'' - R$ to contain an arc (j_1, j_2) such that j_1 has indegree 2 and j_2 has outdegree 2. Thus, D also satisfies the condition (3) in Lemma 7.

2-cycles may exist in D. However, as shown in the proof of Lemma 3 in [8], we can use the special structure of D to construct a new graph H such that (a) H satisfies the three conditions in Lemma 7 and (b) a 2-path-coloring of D can be easily computed from a 2-path-coloring of H. Moreover, H can be easily constructed from D and does not contain more vertices or arcs than D. Note that D has n vertices and $O(n)$ arcs. Thus, by Theorem 15, a 2-path-coloring of H can be computed in $O(\log^3 n)$ time or in $O(\log^2 n)$ expected time with n processors. Given such a coloring of H, a 2-path-coloring of D can be computed in $O(1)$ time with n processors [8]. $\blacksquare$

Lemma 19. The first version of Algorithm 4 achieves a compression ratio of $\frac{7}{12} - \frac{1}{12}(b - 2c)$.

Proof. Let $d = ov(C' - R)/ov(C)$. Actually, our implementation above differs from that of Rao Kosaraju *et al* only in step 2.3. From the analysis in Section 2.2 of [8], it is not difficult to see that $ov(P) \geq \frac{1}{2}(\frac{ov(P_{max})}{2} + \frac{b-2c}{2} \cdot ov(C) + d \cdot ov(C)) + c \cdot ov(C) = \frac{ov(P_{max})}{4} + \frac{b+2d+2c}{4} \cdot ov(C)$. Noting that the weight of C' is $(1 - b - c) \cdot ov(C)$, we see that $d \geq \frac{2}{3}(1 - b - c)$ by step 2.3. Combining this with $ov(C) \geq ov(P_{max})$, we now have $ov(P) \geq (\frac{7}{12} - \frac{1}{12}(b - 2c)) \cdot ov(P_{max})$. ∎

Our second implementation of step 2 is the same as the third implementation of step 2 in [8]. To be self-contained, we include it here.

2.1'. Break each cycle C of C with $|C| > 7$ into $\lceil |C|/7 \rceil$ paths all but at most one of length 6, by deleting a set of $\lceil |C|/7 \rceil$ arcs with total weight $\leq \frac{\lceil |C|/7 \rceil}{|C|} \cdot ov(C)$. Let C'' denote the set of the resulting paths together with the cycles of length ≤ 6 in C.

2.2'. Construct a weighted digraph D' from D and C'' as follows. Initially, set $D' = D$. Next, for each arc $e = (j_1, j_2)$ in D' such that j_1 and j_2 appear on different cycles or paths (say C and C') of C'', redefine the weight on e to be $ov(e) + ov(P_{j_1}) + ov(P_{j_2})$, where P_{j_1} (resp., P_{j_2}) is a maximum-weight Hamiltonian path on the vertices of C (resp., C') ending at j_1 (resp., starting at j_2). Finally, contract the vertices on each $C \in C''$.

2.3'. Find a maximum-weight matching M in D'. Let P be the set of paths in D corresponding to the arcs in M. That is, for each arc $(j_1, j_2) \in M$, add (j_1, j_2) and the arcs in P_{j_1}, P_{j_2} to P.

Lemma 20. The second version of Algorithm 4 achieves a compression ratio of $\frac{2}{3} + \frac{4}{15}(b - 2c)$.

Proof. From the analysis in [8], it suffices to note that for an unbalanced 2-cycle C_i, $b_i > 2c_i$ and $13b_i + c_i \geq 9(b_i + c_i) = 9 \cdot ov(C_i)$. ∎

Theorem 21. There is an **RNC** approximation algorithm for SSP achieving a compression ratio of $\frac{38}{63}$. It runs in $O(\log^2 |S|)$ expected time using $|S|^{3.376}$ processors.

Theorem 22. There is an **RNC** approximation algorithm for SSP achieving an approximation ratio of $2\frac{50}{63}$. It runs in $O(\log^2 |S|)$ expected time using $|S|^{3.376}$ processors.

Proof. According to [2], if SSP can be approximated with a compression ratio of $\frac{1}{2} + \epsilon$, it can be approximated with an approximation ratio of $3 - 2\epsilon$. ∎

5   NC-approximation of shortest superstrings

We adopt the notations and definitions in the last section. Recall that a directed path set (DPS) in a digraph D is a set B of arcs in D such that $D[B]$ is an acyclic digraph in which the indegree and outdegree of each vertex are both at most 1. In [12], Turner showed that the following simple sequential algorithm finds a Hamiltonian path with weight at least $\frac{ov(P_{max})}{2}$ in $OG(S)$. (Recall that P_{max} denotes a maximum-weight Hamiltonian path in $OG(S)$.)

Algorithm GREEDY
Input: $OG(S) = (V, A, ov)$.
1. Initialize B to be the empty set.
2. While the digraph (V, B) is not a Hamiltonian path in $OG(S)$, perform the following: Add to B the largest arc e such that $B \cup \{e\}$ is a DPS in $OG(S)$ but $B \cup \{e'\}$ is not a DPS in $OG(S)$ for all arcs e' with $ov(e) < ov(e')$. (Note: We here assume that the arcs of equal weights in $OG(S)$ are linearly ordered.)
3. Output the digraph (V, B).

Fact 2 [12] Even if the input to GREEDY is changed to an arbitrary weighted complete digraph D, GREEDY finds a Hamiltonian path in D whose weight is at least one-third of the maximum weight of a Hamiltonian path in D.

Algorithm 5
Input: $OG(S) = (V, A, ov)$.
1. Let $c = 1 + \frac{\epsilon}{3}$. In parallel, for each arc $e \in A$, set $lev(e) = \lceil \log_c ov(e) \rceil$ if $ov(e) > 1$, and set $lev(e) = 0$ otherwise.
2. Compute $MaxLev = \max\{lev(e) : e \in A\}$.
3. Set $B = \emptyset$ and $CurLev = MaxLev$.
4. While $CurLev \geq 0$, perform the following steps:
   4.1. Construct an unweighted digraph $D = (V, E)$ by setting $E = B \cup \{e \in A : lev(e) = CurLev\}$.
   4.2 Use Algorithm 2 to compute an MDPS F in D with $B \subseteq F$ and then update B to be F.
   4.3 Decrease $CurLev$ by 1.
5. Output the digraph (V, B).

Lemma 23. Algorithm 5 finds a Hamiltonian path in $OG(S)$ with weight at least $\frac{ov(P_{max})}{3+\epsilon}$.

Proof. The proof is very similar to that of Lemma 3 in [3]. Let P_{out} be the output of Algorithm 5. Since $OG(S)$ is a complete (weighted) digraph, P_{out} is certainly a Hamiltonian path in $OG(S)$. We next show that $ov(P_{out}) \geq \frac{ov(P_{max})}{3+\epsilon}$.

Define a weighted digraph $H = (V, A, w_H)$ by setting $w_H(e) = c^{lev(e)}$. Let $\succ$ be a total order satisfying the following condition: If either $w_H(e) > w_H(e')$, or $w_H(e) = w_H(e')$ and e is contained in P_{out} but e' is not contained in P_{out}, then $e \succ e'$. When $e \succ e'$, we say that e is *larger* than e'.

Let W be the maximum weight of a Hamiltonian path in H. Obviously, if the arcs in H are sorted in nondecreasing order using $\succ$, then given H as input to GREEDY (in place of $OG(S)$), GREEDY will output P_{out}. Thus, by

Fact 2, $w_H(P_{out}) \geq \frac{W}{3}$. On the other hand, $w_H(P_{out}) < c \cdot ov(P_{out})$ because $w_H(e) < c \cdot ov(e)$ for all $e \in A$. Therefore, $c \cdot ov(P_{out}) > \frac{W}{3} \geq \frac{ov(P_{max})}{3}$ since $w_H(e) \geq ov(e)$ for all $e \in A$. Recalling that $c = 1 + \frac{\epsilon}{3}$, we now have $ov(P_{out}) > \frac{ov(P_{max})}{3c} = \frac{ov(P_{max})}{3+\epsilon}$. ∎

Theorem 24. There is an **NC** approximation algorithm for SSP with a compression ratio of $\frac{1}{3+\epsilon}$ for any $\epsilon > 0$. It runs in $O(\log^3 n \cdot \log_{1+\epsilon/3} |S|)$ time with $|S|^2$ processors.

Acknowledgments I am indebted to Professor Magnús M. Halldórsson for many helpful comments. I also wish to thank Doctor Artur Czumaj for pointing out an error in an earlier version of this paper. My thanks also to Professor Takumi Kasai for discussion and encouragement.

References

1. C. Armen and C. Stein, A $2\frac{3}{4}$-Approximation Algorithm for the Shortest Superstring Problem, unpublished manuscript, 1994.
2. A. Blum, T. Jiang, M. Li, J. Tromp, and M. Yannakakis, Linear Approximation of Shortest Superstrings, in: *Proc. 23rd ACM Symp. on Theory of Computing* (ACM, 1991) 328-336.
3. A. Czumaj, L. Gasieniec, M. Piotrow, and W. Rytter, Parallel and Sequential Approximation of Shortest Superstrings, in: *Proc. 4th Scandinavian Workshop on Algorithm Theory*, Lecture Notes in Computer Science, Vol. 824 (Springer, Berlin, 1994) 95-106.
4. J. Gallant, D. Maier, and J. Storer, On Finding Minimal Length Superstrings, *Journal of Computer and System Sciences* **20** (1980) 50-58.
5. N. Karmarkar and R.M. Karp, An Efficient Approximation Scheme for the One-Dimensional Bin Packing Problem, in: *Proc. 23rd IEEE Symp. on Foundations of Computer Science* (IEEE, 1982) 312-320.
6. M. Li, Towards a DNA Sequencing Theory (Learning a String), in: *Proc. 31st IEEE Symp. on Foundations of Computer Science* (IEEE, 1990) 125-134.
7. H. Peltola, H. Soderlund, J. Tarhio, and E. Ukkonen, Algorithms for Some String Matching Problems Arising in Molecular Genetics, in: *Proc. 2nd IFIP Congress* (1983) 53-64.
8. S. Rao Kosaraju, J.K. Park, and C. Stein, Long Tours and Short Superstrings, in: *Proc. 35th IEEE Symp. on Foundations of Computer Science* (IEEE, 1994) 166-177.
9. J. Storer, *Data Compression: Methods and Theory* (Computer Science Press, 1988).
10. J. Tarhio and E. Ukkonen, A Greedy Approximation Algorithm for Constructing Shortest Common Superstrings, *Theoretical Computer Science* **57** (1988) 131-145.
11. S.-H. Teng and F. Yao, Approximating Shortest Superstrings, in: *Proc. 34th IEEE Symp. on Foundations of Computer Science* (IEEE, 1993) 158-165.
12. J.-S. Turner, Approximation Algorithms for the Shortest Common Superstring Problem, *Information and Computation* **83** (1989) 1-20.

On the Construction of Classes of Suffix Trees for Square Matrices: Algorithms and Applications

Raffaele Giancarlo[1]* and Roberto Grossi[2]**

[1] Dipartimento di Matematica ed Applicazioni, Università di Palermo, Italy
[2] Dipartimento di Sistemi e Informatica, Università di Firenze, Italy

Abstract. Given an $n \times n$ $TEXT$ matrix with entries defined over an ordered alphabet Σ, we introduce 4^{n-1} classes of index data structures for $TEXT$. Those indices are informally the two-dimensional analog of the suffix tree of a string [15], allowing on-line searches and statistics to be performed on $TEXT$. We provide *one* simple algorithm that efficiently builds any chosen index in those classes in $O(n^2 \log n)$ *worst case* time using $O(n^2)$ space. The algorithm can be modified to require optimal $O(n^2)$ *expected time* for bounded Σ.

1 Introduction

The development of a uniform framework in which to cast the study of a class of data structures seems to be a worthwhile task since it usually leads to a better understanding of the class and to the design of better algorithms for the construction of the data structures in that class. For instance, the dichromatic framework developed by Guibas and Sedgewick [9] for balanced search trees led to new balanced search tree schemes as well as to new interesting implementations of the algorithms for the management of B-trees [3] and of AVL trees [9].

In this abstract, we provide a uniform framework for the study of index data structures for a two-dimensional matrix $TEXT$, i.e., data structures that are the two-dimensional analog of the suffix tree for a string [15]. Informally, given an $n \times n$ matrix $TEXT$ (referred to as the text) with entries defined over an ordered alphabet Σ, an index for the text is a tree compactly representing all square submatrices of $TEXT$. Two kinds of queries must be supported by the index: (a) statistical information about the text, for instance, find the largest repeated square submatrix in $TEXT$; (b) *on-line* pattern matching for an $m \times m$ pattern matrix PAT, in which all of the occurrences of PAT as a submatrix of $TEXT$ should be found in time that is independent of the size n^2 of $TEXT$. The study

---

* The work of this author is partially supported by MURST Grant "Algoritmi, Modelli di Calcolo e Strutture Informative". This work was done while the author was with AT&T Bell Laboratories, Murray Hill, NJ. U.S.A. Email: raffaele@altair.math.unipa.it

** Work supported by MURST of Italy. Part of this work was done while the author was visiting AT&T Bell Laboratories. Email: grossi@di.unipi.it

of indices is motivated by many important applications in low level image processing [16] and compression [18], pattern recognition [11], visual databases [10] and geographic information systems [14]. We obtain:

(1) 4^{n-1} classes of indices for $TEXT$. Each class contains $\Pi_{i=2}^{n}(2i - 1)!$ indices that we show to be isomorphic, i.e., all indices in a class have the same topological structure. Every index is associated with a different representation of an $n \times n$ matrix as a string. We provide a *concise description* for each index, in the form of string of n "instructions" to build it.

(2) *One* simple algorithm that takes in input the matrix $TEXT$ and, as a parameter, the concise description of an index I_{TEXT} from one class in (1). It builds I_{TEXT} in $O(n^2 \log n)$ worst case time using $O(n^2)$ space. Such an algorithm can be seen as a "compiler" for the indices defined in (1).

(3) *One* simple algorithm, a nonobvious variation of the one in (2), which requires $O(n^2)$ *expected time*. The probabilistic assumptions are extensions of the ones in [19] and seem to be very mild and realistic.

(4) Each index in a class defined in (1) supports a wide variety of statistical queries about $TEXT$, many of which can be answered in optimal time. Moreover, we have *one* pattern matching algorithm that answers the on-line query for PAT in $O(m^2 \log |\Sigma| + occ)$ worst case time, where occ is the number of occurrences of PAT in $TEXT$. It can be seen as a general pattern matcher for the indices defined in (1).

We have the following remarks:

• The general problem of building an index representing *all submatrices* of an *arbitrary* $n \times w$ matrix $TEXT$ has been considered in [4, 6, 8]. The index proposed in [8] can be built in $O(\max(n, w)^2 \min(n, w)^3)$ worst case time, while the ones proposed in [4] and [6] can be built in $O(\max(n, w) \min(n, w)^2 \log(nw))$ worst case time. All three indices support a wide variety of queries (see [4, 6, 8]). Moreover, it has been shown in [4] that any index representing compactly *all* submatrices of $TEXT$ must require $\Omega(\max(n, w) \min(n, w)^2)$ space, while an index representing *only the square* submatrices of $TEXT$ can be stored in $O(nw)$ space. It is worth pointing out that many images and maps are represented and queried as square matrices (e.g., see [17]).

• Up to date, the known indices for square matrices are the PAT-*tree* by Gonnet [8], the *Lsuffix tree* by Giancarlo [5] and the *two dimensional suffix trie* by Storer [18]. We show that they are members of two classes defined in (1). Moreover, we provide $4^{n-1} - 2$ new classes of indices. That is important from the applications point of view since much of the on-going research in spatial data structures tries to understand which linear representation of a square matrix best captures a given set of neighborhood properties of the given matrix [14]. Since each index defined in (1) corresponds to a distinct linear representation of a matrix, we provide the freedom to choose from a large number of linear representations for the construction of indices, guaranteeing efficient algorithms (for the worst and expected case) for each index.

• From the algorithmic viewpoint, the framework from which we derive the indices in (1) allows us to blend together, improve and generalize in a nonobvious way many ideas that have been developed over the years for string and array pattern matching (see [2, 5, 6, 12, 13]). For instance, a widely used technique is "naming." It consists of assigning integers to $\Theta(n^2 \log n)$ square submatrices of $TEXT$. So far, it is implemented using the pattern matching techniques of Karp et al. [12] and requires $O(n^2 \log n)$ time and space for an $n \times n$ matrix. We have a new technique to implement "naming" that takes optimal $O(n^2)$ space. In the worst case, the time bound is $O(n^2 \log n)$ and it can be reduced to $O(n^2)$ for the expected case. For instance, we can now use our naming technique (applied to strings) to get a simple algorithm that builds suffix arrays in the same time and space bounds as in [13].

- The effectiveness of the framework and its associated techniques is suggested by the following facts. In the worst case, using the algorithm in (2), we speed up the construction of the *PAT-tree* [8] from $O(n^4)$ to $O(n^2 \log n)$ time and we obtain the same time bound as in [5] for the construction of the *Lsuffix tree*. For the expected case, the algorithm in (3) improves by a $\log n$ factor the algorithms [5, 8].

Due to space limitations, we limit ourselves to define the indices in (1) and to a very sketchy overview of the algorithms. Detailed presentations of all the results claimed here can be found in [7].

2 Templates and Macro Strings

Throughout the paper, let A be a generic $n \times n$ matrix. Intuitively, a linear representation of A is a "flattening" function that represents A as a one-dimensional string. We define templates and macro strings for doing it. We start with the definition of a *shape*, which is a special character in the set $\{\mathcal{IN}, \mathcal{SW}, \mathcal{NW}, \mathcal{SE}, \mathcal{NE}\}$. It is used to denote a way of extending an $s \times s$ matrix S into an $(s+1) \times (s+1)$ matrix S' such that S is submatrix of S' (see Fig. 1):

- $\mathcal{SW}$: Append a row of length s to the bottom (South) of S and a column of length $s+1$ to the left (West) of S. Shape $\mathcal{NW}$ is defined similarly.
- $\mathcal{SE}$: Append a row of length s to the bottom (South) of S and a column of length $s+1$ to the right (East) of S. Shape $\mathcal{NE}$ is defined similarly.

The definition of shapes applies to any length of row and column. The special shape $\mathcal{IN}$ encodes the 1×1 submatrix that is generated from the empty submatrix by creating a square.

A *template* T is a string of shapes, starting with the shape $T[1] = \mathcal{IN}$ (see Figs. 2 and 4b). Its shape $T[i]$ extends an $(i-1) \times (i-1)$ matrix S into an $i \times i$ matrix S' for $2 \leq i \leq |T|$, where $|T|$ is the number of shapes in T. We adopt for templates the same notation that is commonly used for strings. Notice that there are 4^{n-1} distinct templates of length n.

A template $T[1:n]$ can be thought of as a sequence of n "onion peeling" instructions to represent the $n \times n$ matrix A as a string of n^2 characters from Σ. Indeed T describes a partition of A in terms of shapes, obtained by superimposing T on A as shown in Fig. 4. We need a further notion.

Let $\hat{\Sigma} = \cup_{s=1}^{\infty} \Sigma^{2s-1}$ be the set of *macro characters*, which are strings interpreted as being atomic (composed of subatomic parts given by the characters of Σ). Two macro characters m_1 and m_2 are equal if they are equal as strings, and they can be concatenated if $m_1 \in \Sigma^{2s-1}$ and $m_2 \in \Sigma^{2s+1}$, for some $s \geq 1$. A *macro string* b of length n is the concatenation of n macro characters such that the first macro character $b[1] \in \Sigma$ (see Fig. 4). We adopt for macro strings the same terminology as for strings. For example, the *macro prefix* of b of length i is $b[1:i]$. However, a macro substring $b[i:j]$ is called a *chunk* to emphasize the fact that $b[i:j]$ is *not* a macro string, when $i > 1$.

We show how to produce a macro string a as a linear representation of A as follows. For $1 \leq s \leq n$, let $a_1, \ldots, a_{2s-1}$ be the entries of A in the subrow and subcolumn corresponding to $\mathcal{T}[s]$, in clockwise order (see Fig. 4). Choose a permutation π_{2s-1} of the integers $1, 2, \ldots, 2s-1$, and let b_s be the string of permuted entries $a_{\pi_{2s-1}(1)} a_{\pi_{2s-1}(2)} \cdots a_{\pi_{2s-1}(2s-1)}$. Permutation π_{2s-1} is stored in entry $\Pi[s]$ of a *permutation array* $\Pi[1, n]$. The linear representation of A is the macro string $a = b_1 b_2 \ldots b_n$, given the template $\mathcal{T}$ and the permutation array Π. See Fig. 4 for an example. Notice that there is a one-to-one correspondence between A and its linear representation as a macro string. Clearly, there are $4^{n-1} \Pi_{i=1}^{n} (2i-1)!$ distinct representations for A in our framework. The linear representations introduced in [5, 8, 18] are a special case of ours.

3 An Index for a Square Matrix, Given $\mathcal{T}$ and Π

In analogy with the suffix tree of a string [15], we want to define an index for A (actually we will define several classes of indices) built using only n^2 square submatrices of A. Such submatrices will be chosen to be *maximal* with respect to the "submatrix of" relation, so that the index will satisfy the following two constraints:

• *Completeness:* For each square submatrix B of A, there must be a unique path of arcs from the root that represents B as a macro string.

• *Common Submatrix:* Let Π_u and Π_v be the paths of arcs from the root ending respectively in a node u and one of its descendants v. If Π_u represents a submatrix B of A then Π_v must represent a larger submatrix C having B as submatrix.

We first need the notion of trie built on macro strings. Then, we define an index for matrix A with respect to the template $\mathcal{T}$ and the permutation array Π that satisfies the above two constraints.

Consider a set S of macro strings such that no macro string is prefix of another one. The set S corresponds to a collection of matrices, given $\mathcal{T}$ and Π. A trie representing the macro strings in S is defined in a way similar to a trie for strings: **(1)** Each arc is labeled with a macro character. **(2)** All arcs outgoing the same node are labeled with different macro characters, which correspond to the *same type of shape* and are of the same length as strings. **(3)** For each macro string b in S, there is a leaf v such that the concatenation of the labels along the path from the root to v gives b.

We compact tries for macro strings by compressing chains of nodes with outdegree one into a single arc. The label on that new arc is the concatenation of the macro characters along the arcs of the chain, i.e., it is a chunk. So, we define a *compacted trie* for S as the trie representing the macro strings in S, in which there are no nodes of outdegree one and the arcs are labeled with chunks. For an example see Fig. 5.

To define an index I_A for A as a compacted trie, first we take the augmented matrix $A\hat{\$}[-n+1{:}2n, -n+1{:}2n]$ having A in its center, whereas the rest of $A\hat{\$}$'s entries are filled with distinct instances of $\$ \notin \Sigma$ (see Fig. 3). We extend the alphabet $\hat{\Sigma}$ of macro characters accordingly.

Next, the choice of the n^2 submatrices of $A\hat{\$}$ to be stored as macro strings in I_A depends on T. For $1 \le i, j \le n$, let $A_{i,j}$ denote the $n \times n$ submatrix of $A\hat{\$}$ that has *origin* in (i, j) and is induced by T. That is, $T[1]$ is placed in position (i, j) of $A\hat{\$}$ and the adjacent subrows and subcolumns of A are "covered" by $T[2, n]$ (see Fig. 3). Let $a_{i,j}$ be the macro string of length n representing $A_{i,j}$ with respect to T and Π. Since all $\$$'s are distinct, no $a_{i,j}$ is prefix of another. (From now on, when referring to i, j and $A_{i,j}$, we will implicitly assume that $1 \le i, j \le n$.)

The *index* I_A of matrix A with respect to template T and permutation array Π is the compacted trie built on the macro strings in $S = \{a_{i,j} : 1 \le i, j \le n\}$. Therefore I_A satisfies the additional constraint: The concatenation of the chunks labeling the arcs on the path from the root to a leaf gives exactly one macro string $a_{i,j} \in S$. That leaf is labeled with the origin (i, j) since it corresponds to $A_{i,j}$ via T and Π. See Fig. 5 for an example of index. We remark that I_A can be stored in $O(n^2)$ space. Furthermore, it satisfies what we have informally called the completeness and common submatrix constraints:

Lemma 1. *Fix a template T and a permutation array Π. Let A be an $n \times n$ matrix, I_A be its index, and b be the macro string corresponding to a square matrix B, with respect to T, Π. Then B is a submatrix of A if, and only if, there is a unique path from the root in I_A corresponding to b. Moreover, all the origins of the occurrences of B as a submatrix of A are found in the leaves of I_A descending from that path.*

From the definition of index just given, it can be easily established that the indices can be partitioned into 4^{n-1} distinct classes, one for each template. Within each class, there is one index for each permutation array Π, out of $\prod_{i=1}^{n}(2i-1)!$ ones. We prove in the full paper that such indices turn out to be isomorphic [7].

4 Worst Case Construction — Main Components

We describe now our two-phase algorithm for building the index I_A for an $n \times n$ matrix A, given T and Π. It results convenient to introduce an abstract problem, interesting in its own right, which we refer to as *Names on Demand* (*NOD* for short). We state it in terms of a matrix $F[1{:}p, 1{:}p]$, with $p = 3n$, whose entries are integers in $[1, P]$, where $P = p^2$. We use the shorthand $[\![(i, j), k]\!]$ to denote the submatrix $F[i{:}i + 2^k - 1, j{:}j + 2^k - 1]$, for $1 \leq i, j \leq p$ and $0 \leq k \leq \log n$. Notice that some entries of $[\![(i, j), k]\!]$ might be taken out of the boundaries of F. In that case we simply assume that they exist and are unique (i.e., equal to distinct \$'s). NOD consists of the following:

- NOD-*Preprocessing*: Partition the submatrices of F, of side at most n and a power of two, into equivalence classes. Two submatrices of equal side are in the same equivalence class if and only if they are equal.

- NOD-*Query*: We are given, on-line, a set $\mathcal{Q} = \{[\![(i_r, j_r), k_r]\!] : 2^{k_r} \leq n$ for $1 \leq r \leq q\}$ of $q = \Theta(P)$ submatrices of F. A unique integer in $[1, P]$ (the *name*) must be assigned to each submatrix in $\mathcal{Q}$ in such a way that two submatrices of *equal* side get assigned equal names if and only if they are equal. (We say that those names are *consistent*.) Queries may be repeated with different sets $\mathcal{Q}$ of submatrices.

Notice that we are allowing submatrices having different side to get the same name, since their side distinguishes them. Moreover, when a matrix $[\![(i, j), k]\!]$ is involved in two distinct queries, the name assigned to it may change. We also remark that if we use the pattern matching techniques of Karp et al. [12], NOD can be solved in $O(P \log P)$ time and space for the preprocessing and in $O(q)$ time for the query. In Section 5, we briefly discuss our new techniques to reduce the space to $O(P)$ for the preprocessing, while preserving the time performance of the preprocessing and the query. Based upon NOD, the two phases of the algorithm building I_A can be outlined as follows:

- **Phase One of Construction**: It applies the preprocessing step of NOD, with $F = A\hat{\$}$.

- **Phase Two of Construction**: It produces a sequence of refinement trees, denoted $D^{(r)}$, for $r = \log n, \ldots, 0$, each of which is a better and better approximation of I_A. More precisely, there is an initial step in which $D^{(\log n)}$ is built, followed by $\log n$ refinement steps. For $0 < r \leq \log n$, step r transforms tree $D^{(r)}$ in $D^{(r-1)}$. Except for a few minor differences, the high level scheme of the refinement algorithm is the sequential version of the parallel algorithm in [2] for the construction of the suffix tree of a string. Essential to each refinement step is the partition of $O(n^2)$ submatrices of $A\hat{\$}$, each of side at most n, into equivalence classes. However, the choice of the submatrices depends on the refinement step and we can make no assumption on which submatrix will be chosen. Here we use the query procedure of NOD. Indeed, let H be the set of matrices that need to be compared during one refinement step. Divide each matrix $M \in H$ into four maximal (possibly overlapping) submatrices M_i, for $i = 1 \ldots 4$, each having side a power of two and covering a distinct corner of M. Let $\mathcal{Q}$ be the new set of matrices so obtained. Assign names to the matrices in $\mathcal{Q}$ (through the procedure *Query*). Thus two matrices $M, M' \in H$ are equal if and only if (a) they have the same side and (b) the name for M_i is equal to the name for M_i', for $i = 1 \ldots 4$. Therefore, using the names for matrices in $\mathcal{Q}$, we can group together matrices in H being equal in linear time, by a simple lexicographic sort [1].

5 Algorithms for Names on Demand

We compute $\log n + 1$ partitions of the positions (i, j) of F: for $0 \leq k \leq \log n$, the k-th partition Γ_k groups together into the same equivalence class all (i, j)'s such that the submatrices $[(i, j), k]$ are equal. The partitions are suitably "packed" into a set of arrays that require globally $O(P)$ memory cells. Then, given a set $\mathcal{Q}$ of submatrices of F having side a power of two for which we want to compute names, the query procedure consists of "unpacking" part of the partitions.

- **Preprocessing:** A *representation* of the partitions $\Gamma_0, \ldots, \Gamma_{\log n}$ is given by a *ranking* $\beta : [1, p] \times [1, p] \to [1, P]$ along with $\log n + 1$ *binary sequences* $B_0, B_1, \ldots, B_{\log n}$ of $P + 1$ bits each, such that for each $k = 0, 1, \ldots, \log n$:

(S1) B_k is a "characteristic" vector of the partition Γ_k satisfying the following conditions. Matrices in the same equivalence class of Γ_k are assigned to contiguous positions of B_k. Such a ranking is given by β, i.e., $\beta(i, j) = f$ if and only if $B_k[f]$ has been assigned to $[(i, j), k]$. All entries of $B_k[1:P]$ are set to 0, except where a new equivalence class begins, i.e., each class is encoded by a 1 followed by a run of 0's. (Notice that $B_k[1]$ is always set to 1.)

It is possible to pack the first P bits of B_k into an array $\hat{B}_k$ of $2P/\log P$ cells, containing integers into $[0, P^{1/2} - 1]$. (The $P + 1$st bit of B_k is always set to 1, so there is no need to explicitly give it in $\hat{B}_k$.) The ranking β and the arrays $\hat{B}_k$, $0 \leq k \leq \log n$, are called a *succinct representation* of the partitions $\Gamma_0, \ldots, \Gamma_{\log n}$. That representation can be stored in $O(P) = O(n^2)$ space and it can be computed in $O(P \log P) = O(n^2 \log n)$ time and $O(P) = O(n^2)$ space [7].

- **Query:** We are given a set $\mathcal{Q}$ of $O(P)$ submatrices of F, whose side is at most n and a power of two. We use the following "high level strategy" to assign consistent names to the submatrices in $\mathcal{Q}$:

Algorithm Query(Q)

Q1. Lexicographically sort the pairs $(k, \beta(i, j))$, for all $[(i, j), k] \in \mathcal{Q}$, to get a sorted list $\mathcal{L}$. For a fixed k, let $\mathcal{L}_k$ be the contiguous portion of $\mathcal{L}$ that corresponds to the submatrices in $\mathcal{Q}$ having the same side 2^k.

Q2. For $k = 0, 1, \ldots, \log n$, repeat the following on the list $\mathcal{L}_k$. Using the information computed during the preprocessing step about Γ_k, partition $\mathcal{L}_k$ in such a way that equivalent matrices are contiguous in $\mathcal{L}_k$ and the boundaries between adjacent equivalence classes are known. Using the knowledge of the boundaries, assign a distinct integer to each class. That integer is the name of the matrices in the class.

Query($\mathcal{Q}$) can be implemented to take $O(P)$ time and space. In order to do that, we need to define a suitable transducer that "marks" the boundaries of classes in $\mathcal{L}_k$, given B_k. The computation of such a transducer is sped up by means of a RAM simulation program [1]. The details are in [7].

Theorem 2. *Let A be an $n \times n$ matrix and let I_A be the index for A with respect to template T and permutation array Π. Index I_A can be built in $O(n^2 \log n)$ worst case time and $O(n^2)$ space.*

6 Expected Linear Time Construction

We present now an optimal $O(n^2)$ expected time construction of the index I_A for an $n \times n$ matrix A. We require $|\Sigma| = O(1)$. The analysis of our algorithm works under the following hypothesis.

- **Hp 1:** The expected size of the largest repeated square submatrix in A is asymptotically smaller than $(cL)^2$, for some constant $c > 1$, where L is the largest power of two such that $L^2 \leq \lfloor \log_{|\Sigma|} n^2 \rfloor$.

Hp 1 holds under mild probabilistic models, which have been introduced and discussed by Szpankowski for strings (see [19], equations (2.3b) and (2.10) with $b = 1$, and point (iv), p.1185), but they generalize to square matrices as well [20]. The crucial implication of Hp 1 is that, on the average, the internal nodes of I_A store macro strings of length at most cL. Thus assume w.l.o.g. that cL is an integer, and let CT be the compacted trie obtained from I_A by pruning the subtrees rooted at the nodes storing the macro strings of length larger than cL.

- **Main idea:** Build CT in $O(n^2)$ worst case time, and then apply the refinement steps to expand the leaves of CT in lazy fashion.

The linear time construction of CT is the key point of our new refining algorithm, and it does not seem to be solvable with already known techniques. We proceed in order, sketching how to modify the algorithm in Section 4 to implement the above idea.

We start with the *Names on Demand*. *NOD-Preprocessing* is simulated in $O(n^2)$ worst case time, for all submatrices of side at most L, with a RAM program [1]. Within those bounds, we are able to produce the two-dimensional analog of the *sort history* tree SHT for strings [13]. For the remaining submatrices, of side a power of two larger than L, we resort to the *NOD*-Preprocessing in Section 4, obtaining a succinct representation $SR = (\beta, \{B_k\})$, where $k = \log L + 1, \ldots, \log n$. But we stop as soon as we find the first k such that $B_k = \cdots = B_{\log n} = 1^{n^2+1}$. We have $k = \log L + O(\log c) = \log L + O(1)$ by Hp 1, thus taking only $O(n^2)$ extra time on the average.

NOD-Query still takes $O(n^2)$ time in the worst case. It partitions the set $\mathcal{Q}$ of submatrices into $\mathcal{Q}_1 = \{[(i,j), k] \in \mathcal{Q} : k \leq \log L\}$ and $\mathcal{Q}_2 = \mathcal{Q} - \mathcal{Q}_1$. The equivalence classes for $\mathcal{Q}_1$ are obtained with a visit of SHT (details are omitted). The ones for $\mathcal{Q}_2$ are found as in Section 5, using SR.

- **Phase One** of the algorithm applies the new version of *NOD*-Preprocessing, with $F = A\hat{\$}$.

- **Phase Two** of the algorithm builds CT in $O(n^2)$ worst case time, and applies the refinement steps in lazy fashion. Let S be the macro strings that should be stored in the leaves of I_A, and let S' be the set of their distinct macro prefixes of length equal to cL. Notice that CT is isomorphic to the compacted trie built on the macro strings in S'. Thus assume to have correctly built CT. To obtain I_A, first we augment CT through the creation of new leaves: A new leaf w is created as a child of a leaf $v \in CT$ if there is a macro string in S whose prefix is stored in v. Then the subtrees composed of the new leaves and their parents are refined to obtain I_A from CT.

For such a refining, consider the last 2^h refinement steps producing the trees $D^{(2^h)}, \ldots, D^{(0)}$ in Section 4, for $0 \leq h \leq \log \log n$. Let $lazy(h)$ be those refinement steps modified to work for all the subtrees composed of the new leaves and their parents in the augmented CT. Notice that $lazy(\log \log n)$ correctly produces I_A as a refining of CT. However, Hp 1 implies that, on the average, $lazy(\hat{h})$ produces I_A for some $\hat{h} = O(1)$ as well! Thus we execute $lazy(0), lazy(1), \ldots$ until $lazy(\hat{h})$ does not obtain further refining. This way, the number of refinements steps is still $O(\log n)$ in the worst case, but it becomes $O(2^{\hat{h}}) = O(1)$ on the average. Therefore, the overall refinement of CT into I_A can be done in $O(n^2)$ expected time.

It remains to sketch how CT is built in $O(n^2)$ worst case time. Consider the macro strings in S' as strings over Σ of length $(cL)^2 = O(\log_{|\Sigma|} n^2)$.

1. Lexicographically sort the strings in S' producing an ordered list $\mathcal{R}$.
2. Compute the length of the longest common prefix LCP between any two contiguous strings in $\mathcal{R}$ (e.g., see [13]).
3. Build the compacted trie CT' on the ordered list $\mathcal{R}$ of strings, using the LCP information (e.g., see [5]).
4. Compute CT from CT', using the following fact on $x, y \in S'$: $LCP(x,y) = \ell$ when x, y are seen as macro strings if and only if $\ell^2 \leq LCP(x,y) < (\ell+1)^2$ when x, y are seen as ordinary strings.

Steps 1–4 require $O(n^2)$ worst case time each, noting that the size of S', $\mathcal{R}$, CT' and CT is $O(n^2)$. However, implementing Step 1 is a challenging task, because the characters of the strings in S' can be arbitrarily permuted with T and Π. Again, we do it with a RAM simulation program [1]. Due to lack of space, the technical discussion is deferred to the full paper.

Theorem 3. *The index I_A for an $n \times n$ matrix A, with respect to template T and permutation array Π, can be built in $O(n^2)$ expected time for a bounded alphabet Σ.*

Acknowledgements. We are grateful to Dany Breslauer and Paolo Ferragina for having read a preliminary version of this abstract.

References

1. A.V. Aho, J.E. Hopcroft, and J.D. Ullman. *The Design and Analysis of Computer Algorithms.* Addison-Wesley, Reading, MA., 1974.
2. A. Apostolico, C. Iliopoulos, G. Landau, B. Schieber, and U. Vishkin. Parallel construction of a suffix tree with applications. *Algorithmica*, 3:347–365, 1988.

3. R. Bayer and E.M. McCreight. Organization and maintenance of large ordered indices. *Acta Informatica*, 1:173–189, 1972.
4. R. Giancarlo. An index data structure for matrices, with applications to fast two-dimensional pattern matching. In *Proc. of Workshop on Algorithms and Data Structures, LNCS-Springer-Verlag*, pages 337–348, 1993.
5. R. Giancarlo. The suffix tree of a square matrix, with applications. In *Proc. Fourth Symposium on Discrete Algorithms*, pages 402–411. ACM-SIAM, 1993. To appear in *SIAM J. on Computing*, 1995.
6. R. Giancarlo and R. Grossi. Parallel construction and query of suffix trees for two-dimensional matrices. In *Proc. of the 5-th ACM Symposium on Parallel Algorithms and Architectures*, pages 86–97, 1993.
7. R. Giancarlo and R. Grossi. On the construction of classes of index data structure for square matrices: algorithms and applications. AT&T Bell Labs. Technical Memorandum 11272-940110-03, 1994.
8. G.H. Gonnet. Efficient searching of text and pictures- Extended Abstract. Technical report, University Of Waterloo- OED-88-02, 1988.
9. L.J. Guibas and R. Sedgewick. A dichromatic framework for balanced trees. In *Proc. 19th Symposium on Foundations of Computer Science*, pages 8–21. IEEE, 1978.
10. R. Jain. Workshop report on visual information systems. Technical report, National Science Foundation, 1992.
11. P. Johansen. Combinatorial pattern recognition, the method and the program package. Technical report, DIKU 94/10, 1994.
12. R.M. Karp, R. Miller, and A. Rosenberg. Rapid identification of repeated patterns in strings, arrays and trees. In *Proc. 4th Symposium on Theory of Computing*, pages 125–136. ACM, 1972.
13. U. Manber and E. Myers. Suffix arrays: a new method for on-line string searches. *SIAM Journal of Computing 22*, 5 (1993), 935–948.
14. D.M. Mark. Neighbor-based properties of some orderings of two-dimensional space. *Geographical Analysis*, 22:145–157, 1990.
15. E.M. McCreight. A space economical suffix tree construction algorithm. *J. of ACM*, 23:262–272, 1976.
16. A. Rosenfeld and A.C. Kak. *Digital Picture Processing*. Academic Press, 1982.
17. H. Samet. *The Design and Analysis of Spatial Data Structures*. Addison-Wesley, NY, 1990.
18. J.A. Storer. Two dimensional suffix tries and their use in lossless sliding window image compression. In *Proc. 6th Symposium on Combinatorial Pattern Matching*. To appear, LNCS, 1995.
19. W. Szpankowski. A generalized suffix tree and its (un)expected asymptotic behaviour. *SIAM J. on Computing*, 22:1176–1198, 1993.
20. W. Szpankowski. Private communication.

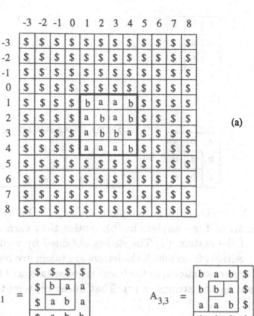

(a)

Fig. 1. The shape $\mathcal{SW}$. The other shapes can be obtained by a 90 degrees rotation of the shaded region.

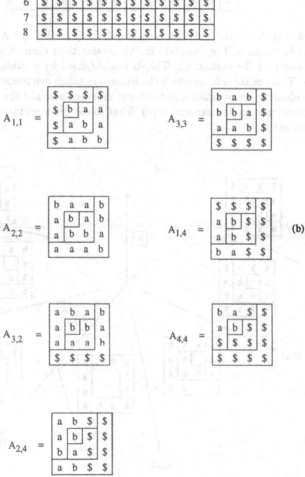

(b)

IN, SE, SE, SE, SE

IN, SW, NE, SE, SW

Fig. 2. Two templates, the top one being used in [5, 18].

Fig. 3. (a) The matrix $A\hat{\$}$, corresponding to the matrix A (in the center). We have omitted the indices for the $'s, which are clear from the entries in which they appear. (b) The $A_{i,j}$'s. We are only showing the ones such that $A[i,j]\hat{\$} = b$. They are partitioned according to the template $\mathcal{IN}, \mathcal{SE}, \mathcal{NW}, \mathcal{SE}$.

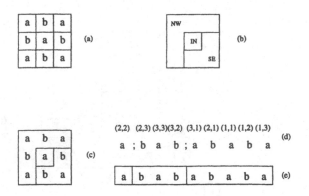

Fig. 4. (a) A matrix. (b) The template $\mathcal{IN}$, $\mathcal{SE}$, $\mathcal{NW}$. (c) A partition of the matrix in (a) in terms of the template in (b). Notice that each shape covers a subrow and subcolumn of the matrix. (d) The strings obtained by visiting the shapes in clockwise order. The positions from which the letters are taken are reported on top. (e) We adopt three identity permutations, so the linear representation of the matrix in (a) is obtained by concatenating the strings in (c). That linear representation is the one introduced by Gonnet [8].

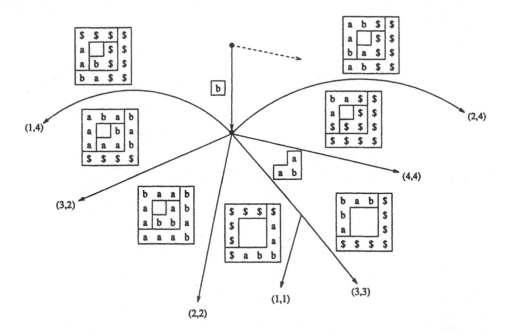

Fig. 5. A compacted trie for the matrices in Fig. 3b. It is also the index for the matrix in Fig. 3a (we are only showing the part of the index that corresponds to the matrices in Fig. 3b. The linear representation of such matrices is obtained from the template in Fig. 3b (we have omitted the permutation array since it only affects the way we "read" the letters in the shapes).

How to Use the Minimal Separators of a Graph for its Chordal Triangulation

Andreas Parra* and Petra Scheffler

TU Berlin, MA 6-1,
D-10623 Berlin, Germany
Email: {parra, scheffle}@math.tu-berlin.de

Abstract. We prove a 1-1 correspondence between maximal sets of pairwise parallel minimal separators of a graph and its minimal chordal triangulations. This yields polynomial-time algorithms to determine the minimum fill-in and the treewidth in several graph classes. We apply the approach to d-trapezoid graphs for which we give the first polynomial-time algorithms that determine the minimum fill-in resp. the treewidth.

1 Introduction

Throughout this paper, we consider simple, finite and undirected graphs. A chordal triangulation of a graph is a supergraph that does not contain any induced cycle of length greater than three. Here, we will *not* consider planar triangulations that have in common with our topic only the name.

The problem of finding a chordal triangulation that optimizes an objective function for a given graph has given rise to much interest in recent years, see [7, 21] for a survey. Usually, two versions are considered. First, in the MINIMUM FILL-IN problem, we look for a triangulation where the number of additional edges is minimum. It has applications in VLSI-design [24] and in sparse matrix factorization [30]. Second, the TREEWIDTH problem corresponds to finding a triangulation with the smallest maximum clique size. The treewidth is widely studied, mainly because many $\mathcal{NP}$-hard problems are solvable in linear time when they are restricted to graphs of bounded treewidth, see e. g. [3]. All known algorithms of that kind rely on a given tree-decomposition of small width.

This has motivated much research on algorithms which find optimal tree-decompositions. Indeed, it is $\mathcal{NP}$-hard to determine the treewidth of a graph [2]. For several graph classes, however, there are polynomial time algorithms. These are cographs, permutation graphs, circle graphs, circular-arc graphs and co-comparability graphs of bounded dimension [5, 4, 16, 29, 20]. For co-bipartite and hence for co-comparability graphs the problem is $\mathcal{NP}$-hard [2]. MINIMUM FILL-IN is also $\mathcal{NP}$-hard for co-bipartite graphs [30]. Yet, there are polynomial time algorithms for this problem on cographs, bipartite permutation graphs and multitolerance graphs [9, 28, 26].

Recently, Bodlaender [6] presented a linear time algorithm to decide whether

---

* The research of this author has been supported by the graduate school "Algorithmic Discrete Mathematics" by the Deutsche Forschungsgemeinschaft, grant We 1265/2-1.

the treewidth of a graph is larger than some fixed constant k. Indeed, the algorithm is rather complicated and the hidden constant is comparatively large. Historically the first algorithms for this problem have time complexity $\mathcal{O}(n^{k+2})$ [2, 27]. They systematically check for all separators S of size k whether it is possible to triangulate the graph in such a way that S remains a separator. The correctness of this approach rests on the fact that a chordal graph with treewidth smaller than $k + 1$ does not have larger separators. These algorithms lead to a triangulation into a k-tree, i.e., a chordal graph in that all maximal cliques have equal size $k+1$. In [17], Kloks et al. tried to generalize the algorithm of [2]. They used minimal separators instead of separators of size k, and they suggested an algorithm that solves TREEWIDTH and MINIMUM FILL-IN for any graph with the only restriction that the number of minimal separators is polynomially bounded. Unfortunately, the algorithm in [17] is not correct [18]. We feel that the difficulties arise mainly from the unsuitable definition of 'minimal triangulations'. Also, we shall not neglect the structure that links the minimal separators with each other.

In our paper, we simply consider inclusion minimal triangulations. By this, we are able to give correct algorithms for MINIMUM FILL-IN and TREEWIDTH that make use of the minimal separators. For a graph G, we define its separator graph $\Sigma(G)$. The vertices of $\Sigma(G)$ are the minimal separators of G and two separators are connected by an edge if one of them does not contain two vertices that are separated by the other. We prove that every inclusion minimal triangulation of a graph corresponds to a maximal clique of its separator graph. This characterization of triangulations can be used for efficient algorithms to solve MINIMUM FILL-IN and TREEWIDTH in classes of graphs where the minimal separators of G and the maximal cliques of $\Sigma(G)$ can be computed in polynomial time. Moreover, we develop algorithms for special classes of intersection graphs where the intersection model yields a transitive orientation of the separator graph.

In the first two sections of our paper, we give the preliminaries about chordal triangulations and prove some simple properties of minimal separators. In Section 4, we prove our main theorem. In Section 5, we define the d-trapezoid graphs and obtain information about their separators. Finally, we describe the algorithms MINIMUM FILL-IN and TREEWIDTH on d-trapezoid graphs, respectively.

2 Chordal Triangulations

For a graph $G = (V, E)$, we denote by $n = |V|$ its size, by $G[W]$ the subgraph induced by $W \subseteq V$, and $G \setminus W = G[V \setminus W]$. We call any complete subgraph of G a *clique*. See [14] for all graph-theoretic notions not defined here.

A graph H is called *chordal* (or *triangulated*) if it does not contain any induced cycle of length greater than 3. Several characterizations of chordal graphs are known (cf. [14]). One in terms of minimal separators is given below in Section 3.

Any chordal supergraph $H \supseteq G$ is called a *triangulation* of G. Always, the complete graph with the same vertex set is a trivial chordal supergraph. We are looking for optimal triangulations. In the literature, two parameters for optimality are considered: the fill-in and the treewidth. A triangulation $H = (V, F)$ of

$G = (V, E)$ is called *minimal* if any intermediate graph $H' = (V, F')$, with $E \subseteq F' \subset F$, is not chordal. The *fill-in* of H with respect to G is $f(H, G) = |F \setminus E|$. The *minimum fill-in* of G is $f(G) = min\{f(H, G) : H$ is a triangulation of $G\}$. The *clique number* of H is $\omega(H) = max\{|C| : C$ is a clique in $H\}$. The *treewidth* of G is $tw(G) = min\{\omega(H) : H$ is a triangulation of $G\} - 1$.

Given a graph $G = (V, E)$, the problem MINIMUM FILL-IN is to find a triangulation H of G such that $f(H, G) = f(G)$, and the problem TREEWIDTH is to find a triangulation H of G such that $\omega(H) = tw(G) + 1$. Analogously, in INTERVAL COMPLETION and PATHWIDTH, only those triangulations of G are considered that are interval graphs. In general, the four problems are $\mathcal{NP}$-hard even for co-bipartite graphs [30, 2]. Notice that for each of these problems the function we have to minimize depends monotonously nondecreasing on the number of additional edges, and we are interested in the minimum. Thus, w.l.o.g., we may confine ourselves to minimal triangulations.

3 Minimal Separators

A vertex set $S \subseteq V$ is an a, b-*separator* for two non adjacent vertices $a, b \in V$ if a and b are contained in different connected components of $G \setminus S$. An a, b-separator S is *minimal* if no proper subset of S separates the same pair of vertices. We denote the set of all minimal a, b-separators by $\Delta_{a,b}(G)$ and the set of all *minimal separators* in G by $\Delta(G)$, i.e., $\Delta(G) = \bigcup_{a,b \in V} \Delta_{a,b}(G)$. Observe that one minimal separator may well be contained in another. We denote by $c(G \setminus S)$ the set of the connected components of $G \setminus S$. A component $C \in c(G \setminus S)$ is called *a full component* if every vertex $s \in S$ is adjacent to at least one $v \in C$. The following crucial property of minimal separators is well known and proved already in [14, 11, 19]. It implies that a separator S is minimal if and only if there exist at least two full components of $G \setminus S$.

Lemma 1. *Let S be a separator of G and $C, D \in c(G \setminus S)$ with $C \neq D$. Then the following statements are equivalent:*
1. *C and D are full components of $G \setminus S$.*
2. *For any $c \in C$ and any $d \in D$, S is a minimal c, d-separator.*
3. *There are some $c \in C$ and $d \in D$ such that $S \in \Delta_{c,d}(G)$.*

Definition 2. Let $S, T \in \Delta(G)$. We say the separator S *crosses* T if there are some components $C, D \in c(G \setminus T)$ with $C \neq D$ such that $S \cap C \neq \emptyset \neq S \cap D$.

Lemma 3. *Let $S, T \in \Delta(G)$. If S crosses T then T crosses S.*

Proof. Let $s, s' \in S$ be in different components of $G \setminus T$, and let $C, D \in c(G \setminus S)$ be two full components of $G \setminus S$. By definition of a full connected component, there are s, s'-paths in $G[C \cup \{s, s'\}]$ and in $G[D \cup \{s, s'\}]$, respectively. Either of these paths intersects the s, s'-separator T. Hence, $T \cap C \neq \emptyset \neq T \cap D$. $\square$

Notice, that the crossing relation is not symmetric in general.

Definition 4. Let $S, T \in \Delta(G)$. We call S and T *parallel* if they do not cross each other.

Equivalently, S is parallel to T if there is a $C \in c(G \setminus T)$ such that $S \subseteq T \cup C$. In the following, we write $S \# T$ if the two minimal separators cross each other and $S \| T$ if they are parallel. Clearly, $S \| T$ if $S \subset T$. By Lemma 3 follows:

Lemma 5. *Let $S, T \in \Delta(G)$ and S is a clique. Then S and T are parallel.*

Dirac characterized chordal graphs in terms of minimal separators:

Theorem 6 [11]. *A graph is chordal iff every minimal separator is a clique.*

Corollary 7. *Let H be a chordal graph and $S, T \in \Delta(H)$. Then $S \| T$.*

If G is not chordal then some of its minimal separators S are no cliques. In order to triangulate G we have to destroy each of them: either by making it complete or by adding edges between its full components so that S is no longer a minimal separator. For $S \in \Delta(G)$, we denote by G_S the graph obtained by adding all missing edges between vertices in S thus making S a clique. For $\{S_1, S_2, \ldots, S_k\} \subseteq \Delta(G)$, we denote by $G_{S_1, S_2, \ldots, S_k}$ the graph obtained by adding the edges $\{uv :$ there exists some S_i with $u, v \in S_i\}$. The next two lemmas describe the connection between minimal separators in G and those in its supergraphs that are obtained by changing some minimal separators into cliques. Let $S_1, S_2, \ldots, S_k \in \Delta(G)$ and $H = G_{S_1, S_2, \ldots, S_k}$.

Lemma 8. *Let $T \in \Delta_{a,b}(G)$ such that $T \| S_i$ in G, for all $i \in \{1, \ldots, k\}$. Then T is a minimal a, b-separator of H' for all H' such that $G \subseteq H' \subseteq H$.*

Proof. Consider the two full components $C_a, C_b \in c(G \setminus T)$ containing a and b, respectively. Since $T \| S_i$, any S_i contains only vertices of one connected component of $G \setminus T$. Hence, C_a and C_b remain full components in H' for all $G \subseteq H' \subseteq H$. □

Lemma 9. *Let $S_i \| S_j$ for all $i, j \in \{1, \ldots, k\}$. Then every minimal a, b-separator T in H is also a minimal a, b-separator in G and $T \| S_i$ in G for all $i \in \{1, \ldots, k\}$.*

Proof. In H, all S_i are cliques, and hence $T \| S_i$, for $i \in \{1, \ldots, k\}$, by Lemma 5. We claim $T \in \Delta_{a,b}(G)$ and $c(G \setminus T) = c(H \setminus T)$. Suppose the opposite. Clearly, T separates a and b in G. Thus, there is some $T' \subseteq T$ such that $T' \in \Delta_{a,b}(G)$ and there is some S_i that crosses T' in G. By Lemma 3, then there are $C \neq D \in c(G \setminus S_i)$ such that $T' \cap C \neq \emptyset \neq T' \cap D$. Since all S_j with $j \neq i$ are parallel to S_i, we have $c(H \setminus S_i) = c(G \setminus S_i)$. This yields $T \# S_i$ in H, a contradiction. The second part of our claim implies directly that $S_i \| T$ in G for $i \in \{1, \ldots, k\}$. □

4 The Separator Graph and Minimal Triangulations

Now we are able to prove our main theorem. It concerns the relation between minimal triangulations and maximal sets of pairwise parallel minimal separators.

Theorem 10.
(i) *Let $\{S_1, S_2, \ldots, S_k\}$ be a maximal (with respect to inclusion) set of pairwise parallel minimal separators of G. Then $H = G_{S_1, S_2, \ldots, S_k}$ is a minimal triangulation of G and $\Delta(H) = \{S_1, S_2, \ldots, S_k\}$.*
(ii) *Let H be a minimal triangulation of G. Then $\Delta(H)$ is a maximal set of pairwise parallel minimal separators of G and $H = G_{\Delta(H)}$.*

Proof. (i) Consider $H = G_{S_1,S_2,...,S_k}$ for a maximal set of pairwise parallel minimal separators of G. By Lemma 9, $\Delta(H) \subseteq \{S_1, S_2, \ldots, S_k\}$. Hence, every minimal separator of H is a clique and thus H is a triangulation of G by Theorem 6. On the other hand, Lemma 8 implies that $\{S_1, S_2, \ldots, S_k\} \subseteq \Delta(H')$ for all H' with $G \subseteq H' \subseteq H$. This together with Theorem 6 yields that H is a minimal triangulation of G. Moreover, $\Delta(H) = \{S_1, S_2, \ldots, S_k\}$.

(ii) First, consider any $S \in \Delta(H)$. We claim $S \in \Delta(G)$ and $c(G \setminus S) = c(H \setminus S)$. Suppose not. Then, since S is a separator also in $G \subseteq H$, there is some $S' \subseteq S$ with $S' \in \Delta(G)$ such that at least one of the triangulating edges connects two different connected components of $G \setminus S'$. Consider the subgraph $H' = H \setminus \{cd \in E(H) : c \in C, d \in D \text{ for } C \neq D \in c(G \setminus S')\}$. We have $G \subseteq H' \subseteq H$ and H is a minimal triangulation of G. Consequently, there is a chordless cycle Y of length at least four in H' which contains vertices of at least two different components of $G \setminus S'$. Hence, Y passes trough S' at least twice. But then, Y is not chordless in H', since $S' \subseteq S$ is a clique in H'.

Second, by Lemma 7, the elements of $\Delta(H)$ are pairwise parallel in H. Since $c(G \setminus S) = c(H \setminus S)$ for all $S \in \Delta(H)$, the same holds in G.

Next, we show that $H = G_{\Delta(H)}$. By Theorem 6, $H \supseteq G_{\Delta(H)}$. Suppose that $H \subseteq G_{\Delta(H)}$ is not true. Then, there is at least one edge $ab \in E(H)$ that does not belong to $G_{\Delta(H)}$. As H is a minimal triangulation of G, there is an induced cycle a, c, b, d in $H \setminus \{ab\}$. Observe that a larger induced cycle cannot be triangulated with only one additional edge. Consequently, a minimal separator $S \in \Delta_{c,d}(H)$ exists which contains a and b, a contradiction to $ab \notin E(G_{\Delta(H)})$.

Finally, we show that $\Delta(H)$ is maximal with respect to inclusion. Suppose that there is some $T \in \Delta(G) \setminus \Delta(H)$ such that $T \| S$ for all $S \in \Delta(H)$. Then, by Lemma 8, $T \in \Delta(G_{\Delta(H)})$ in contradiction to $H = G_{\Delta(H)}$. □

This theorem gives a general characterization for the role of minimal separators in finding minimal triangulations. It may be used as proposed in [17] to solve TREEWIDTH and MINIMUM FILL-IN for graphs of several classes.

We define the *separator graph* $\Sigma(G) = (V_\Sigma, E_\Sigma)$ for a graph G by $V_\Sigma = \Delta(G)$ and $E_\Sigma = \{ST : S \| T\}$. Then, we have another formulation of Theorem 10 and, as a result, a new characterization of chordal graphs.

Theorem 11. *A graph H is a minimal triangulation of G if and only if $H = G_C$, where C is a maximal (with respect to inclusion) clique of $\Sigma(G)$.*

Corollary 12. *A graph G is chordal iff its separator graph $\Sigma(G)$ is complete.*

The first part of an algorithm for triangulating a graph G that makes use of the above theorem is to compute all minimal separators of G. Kloks and Kratsch give an algorithm for this problem that needs polynomial time per separator [19]. It is not difficult to modify this algorithm so that it also reports whether two separators are parallel. The algorithm is efficient for graphs in which the number of minimal separators is polynomially bounded in n. If, moreover, the maximal cliques C of $\Sigma(G)$ can be computed in polynomial time, then we are able to calculate for each C the corresponding minimal triangulation G_C and hence to solve MINIMUM FILL-IN and TREEWIDTH.

In particular, we get efficient algorithms for some classes of intersection graphs, where the minimal separators can be computed efficiently by scanlines and the intersection model yields a transitive orientation of the separator graph. Using these additional properties, we are able to model the number of additional edges and the clique size of a minimal triangulation by weights in the separator graph. Then, finding an optimal triangulation of G corresponds to determining a minimum weighted maximal clique in the comparability graph $\Sigma(G)$. The latter problem can be solved in polynomial time [23].

Our approach generalizes the algorithms for TREEWIDTH in permutation graphs and co-comparability graphs of bounded dimension [4, 20] and those for MINIMUM FILL-IN in bipartite permutation graphs and bounded multitolerance graphs [28, 26]. In the following sections, we demonstrate the method for d-trapezoid graphs.

5 d-Trapezoid Graphs

Let us first recall some order-theoretic notions. A *partial order* is denoted by $P = (V, <_P)$, where $<_P$ is an irreflexive, transitive binary relation on the ground set V. Two elements $u, v \in V$ are *comparable* if $u <_P v$ or $v <_P u$, otherwise they are *incomparable*. A set of pairwise comparable elements is called a *chain*. The *transitive reduction* of $(V, <_P)$ is the binary relation $<'_P$ obtained from $<_P$ by deleting all pairs $u <_P v$ such that there is some $w \in V$ with $u <_P w <_P v$. A partial order $P = (V, <_P)$ is an *interval order* if there is a collection $\{I_v = [l_v, r_v] : v \in V, l_v, r_v \in I\!R\}$ of closed intervals on the line such that, for all $v, w \in V$, $v <_P w$ iff $r_v < l_w$. The *interval dimension* of P, denoted by $idim(P)$, is the smallest number m of interval orders $(V, <_{Q_1}), \ldots, (V, <_{Q_m})$ such that $<_P = <_{Q_1} \cap \ldots \cap <_{Q_m}$.

The *co-comparability graph* $G^c(P) = (V, E)$ of $P = (V, <_P)$ is defined by $uv \in E$ iff u and v are incomparable, for all $u, v \in V$ with $u \neq v$. More generally, a graph G is called a *co-comparability graph* if there exists a partial order P such that $G = G^c(P)$. Clearly, interval graphs are exactly the co-comparability graphs of interval orders. It is known that the interval dimension is a comparability invariant, i.e., $idim(P) = idim(Q)$ for all P and Q with $G^c(P) = G^c(Q)$, see [15]. Therefore, we can define the *interval dimension* of a co-comparability graph G as $idim(P)$, for any P with $G = G^c(P)$.

Trapezoid graphs were introduced by Dagan et al. as a generalization of both the interval graphs and the permutation graphs [10]. They are the intersection graphs of trapezoids, where each trapezoid has its parallel lines on two horizontal lines. In [22], Ma and Spinrad presented an $\mathcal{O}(n^2)$ recognition algorithm, and in [12], Felsner et al. gave $\mathcal{O}(n \log n)$ algorithms for the chromatic number, weighted independent set, clique cover, and maximum weighted clique problems of trapezoid graphs. We generalize the definition of trapezoid graphs in the following way, where $d \in I\!N_0$, $\overline{d} = \{0, \ldots, d\}$ and $D^0, \ldots, D^d$ are parallel lines in $I\!R^2$.

Definition 13. A graph $G = (V, E)$ is a *d-trapezoid graph* if there are families of intervals $\mathcal{I}^i = \{I_v^i = [l_v^i, r_v^i] : v \in V, \; l_v^i, r_v^i \in D^i\}$ on D^i for all $i \in \overline{d}$, satisfying $vw \in E \iff Q_v \cap Q_w \neq \emptyset$, for all $v, w \in V$ with $v \neq w$, where Q_x denotes the closed polygon $l_x^0, \ldots, l_x^d, r_x^d, \ldots, r_x^0$.

This definition appeared first in [13], where Flotow has shown that any power of a d-trapezoid graph is again a d-trapezoid graph. Note that trapezoid graphs are 1-trapezoid graphs, and interval graphs are 0-trapezoid graphs.

We define a *trapezoid order* associated with a d-trapezoid graph as the partial order $P = (V, <_P)$ obtained from the d-trapezoid representation of $G = (V, E)$ by setting $v <_P w$ iff Q_v lies totally to the left of Q_w, i.e., $r_v^i < l_w^i$ for all $i \in \overline{d}$. Then $G = G^c(P)$, and this yields the following fact.

Theorem 14. *The d-trapezoid graphs are exactly the co-comparability graphs with interval dimension at most $d + 1$.*

In the following we assume that $G = (V, E)$ is an arbitrary d-trapezoid graph with a given d-trapezoid representation $\{\mathcal{I}^i = \{I_v^i = [l_v^i, r_v^i] : v \in V\} : i \in \overline{d}\}$.

Next, we define *scanlines*. Given one point s^i on each line D^i, a *scanline* $\overline{S}$ is the polyline $s^0, \ldots, s^d$. Let S denote the set of vertices $v \in V$ with $Q_v \cap \overline{S} \neq \emptyset$, and let S_l resp. S_r denote the set of vertices whose d-trapezoids Q_v lie totally to the left resp. to the right of $\overline{S}$. In the following lemmas, we show that the minimal separators correspond to scanlines with a certain intersection property. The first fact was already shown for permutation graphs in [4]. The proof for d-trapezoid graphs is very similar.

Lemma 15. *Let $S \in \Delta_{a,b}(G)$, and let $C_a, C_b \in c(G \setminus S)$ be the full components containing a resp. b. Then, there is a scanline $\overline{S}$ corresponding to S such that $C_a \subseteq S_l$ and $C_b \subseteq S_r$ or vice versa.*

Lemma 16. *Let $\overline{S}$ be a scanline that satisfies*

$(*)$ *for every $s \in S$, there are some $w_l \in S_l$ and $w_r \in S_r$ such that*
$Q_{w_l} \cap Q_s \neq \emptyset \neq Q_{w_r} \cap Q_s$.

Then there are full components $C, D \in c(G \setminus S)$ such that $C \subseteq S_l$ and $D \subseteq S_r$.

Proof. By symmetry, it is sufficient to show that a full component $C \subseteq S_l$ exists. Clearly, S_l and S_r are separated by S. By $(*)$, there is some $i \in \overline{d}$ such that $W^i = \{w \in S_l :$ there is some $s \in S$ such that $I_w^i \cap I_s^i \neq \emptyset\} \neq \emptyset$. Choose $w \in W^i$ with maximum r_w^i. If w is adjacent to every element of S then, clearly, w belongs to some full component C of $G \setminus S$. Thus, assume that $sw \notin E$ for $s \in S$. Then, Q_w lies totally to the left of Q_s. By $(*)$, there are some $j \in \overline{d}$ and $x \in S_l$ such that $I_x^j \cap I_s^j \neq \emptyset$. This yields $r_w^j < r_x^j$ and $j \neq i$, because otherwise this contradicts $sw \notin E$. Hence, since $r_x^i \leq r_w^i$, x and w are adjacent. We obtain that the set W which consists of w and, for every $s \in S$ with $Q_s \cap Q_w = \emptyset$, of one vertex x as described just above, is a subset of a full component C of $G \setminus S$. □

Theorem 17. *Let $\Delta = \Delta(G)$ be the set of minimal separators of G. Then, $\Delta = \{S \subseteq V :$ there is a scanline $\overline{S}$ that satisfies condition $(*)$ of Lemma 16\}.*

This is implied directly by the two preceding lemmas together with Lemma 1. In the following, we assume that there is fixed such a scanline $\overline{S}$ for every $S \in \Delta$.

Lemma 18. *Let $S, T \in \Delta$. S is parallel to T iff either $(<)$ or $(>)$ holds.*

    $(<)$ $S \subseteq T \cup T_l$ and $T \subseteq S \cup S_r$.

    $(>)$ $T \subseteq S \cup S_l$ and $S \subseteq T \cup T_r$.

Proof. "$\Rightarrow$": By the definition of parallel separators, $S \subseteq T \cup T_l$ or $S \subseteq T \cup T_r$. Hence, it is sufficient to show that if $S \subseteq T \cup T_l$ and $S \not\subseteq T$, then $T \subseteq S \cup S_r$.

This is obvious if $\overline{S} \cap \overline{T} = \emptyset$. Thus, assume $\overline{S} \cap \overline{T} \neq \emptyset$. Let $s \in S \cap T_l$. By Theorem 17, there is some $w \in S_r$ that is adjacent to s. Let $i \in \overline{d}$ be such that $I_s^i \cap I_w^i \neq \emptyset$. Then, $s^i < l_w^i < r_s^i < t^i$. On the other side, because of $\overline{S} \cap \overline{T} \neq \emptyset$, there is some $j \in \overline{d}$ such that $r_s^j < t^j \leq s^j < l_w^j$. Hence, $w \in T$, and $T \subseteq S \cup S_r$.

"$\Leftarrow$": W.l.o.g., let $(<)$ be fulfilled with some $s \in S \cap T_l$. By Lemma 16, there is some full component C of $G \setminus S$ such that $C \subseteq S_l$. Let $w \in C$. Since $T \subseteq S \cup S_r$, $w \notin T$. Further, if $w \in T_r$, then $\overline{T}$ would lie totally to the left of $\overline{S}$, which contradicts $S \cap T_l \neq \emptyset$. Thus, $C \subseteq T_l$. Now, since C is a full component of $G \setminus S$, $G[C \cup S \setminus T]$ is connected. Consequently, $S \| T$. $\square$

Theorem 19. *For any two $S, T \in \Delta$, define $S < T$ if they satisfy condition $(<)$. Then, $(\Delta, <)$ is a partial order.*

Proof. Clearly, only transitivity is to show. Suppose, some $S, T, U \in \Delta$ exist such that $S < T < U$, but $S \not< U$. W.l.o.g., let there be some $s \in S \cap U_r$. Since $T < U$, $s \notin T$. Then $S < T$ implies that $s \in T_l$. Hence, $\overline{U}$ lies totally to the left of $\overline{T}$, which contradicts the fact that there are some $w \in T \cap U_l$ or $w \in U \cap T_r$. $\square$

By Lemma 18, the separator graph $\Sigma(G)$ is the comparability graph of this partial order $(\Delta, <)$. Since the cliques of a comparability graph $G(P)$ correspond to the chains of P, Theorem 10 leads us immediately to the following fact.

Theorem 20. *A graph H is a minimal triangulation of G iff there is a maximal chain $\mathcal{C} = S_0, \ldots, S_m$ in $(\Delta, <)$ such that $H = G_{\mathcal{C}}$.*

As a conclusion, we obtain a new, geometric proof for the next corollary. Observe that this result is already known by [25].

Corollary 21. *Every minimal triangulation H of a d-trapezoid graph G is an interval graph.*

Proof. By Theorem 20, there is a maximal chain $\mathcal{C}$ such that $H = G_{\mathcal{C}}$. We show that $G_{\mathcal{C}}$ is still a d-trapezoid graph and, consequently, a co-comparability graph. This yields the claim, since the interval graphs are just the chordal co-comparability graphs, see [14].

For every $S \in \mathcal{C}$ and $s \in S$, replace r_s^d by s^d if $r_s^d < s^d$, and replace l_s^d by s^d if $l_s^d > s^d$, respectively. This new d-trapezoid graph G' clearly contains $G_{\mathcal{C}}$. Suppose that $G' \supset G_{\mathcal{C}}$. Then, there is some $s \in S$ with $S \in \mathcal{C}$ such that, for some $t \notin S$, we have $st \in E(G') \setminus E(G_{\mathcal{C}})$. W.l.o.g., assume that r_s^d was replaced by s^d. In this case, Q_t lies totally to the left of $\overline{S}$ in the representation of G. On the other side, since $st \notin E$, I_t^d lies totally to the right of I_s^d. As Q_s intersects with $\overline{S}$, this yields a contradiction. Thus, $G' = H$. $\square$

6 Minimum Fill-In of d-Trapezoid Graphs

Lemma 22. *Let $S, T, U \in \Delta$. If $S < T < U$, then $S \cap U \subseteq T$.*

Theorem 23. *Given a d-trapezoid graph G together with a representation, there is an algorithm* MinFill-In(G) *which computes a solution for* MINIMUM FILL-IN *and* INTERVAL COMPLETION *of G in $\mathcal{O}(\max\{n^{2.376(d+1)}, n^{2d+4}\})$ time.*

Procedure MinFill-In(G)

Input: A d-trapezoid intersection model of a d-trapezoid graph $G = (V, E)$.

Output: A set of additional edges F such that $G \cup F$ solves MINIMUM FILL-IN.

1. Construct $\Delta = \Delta(G)$.
2. Compute the partial order $(\Delta, <)$.
3. Construct the transitive reduction $<'$ of $(\Delta, <)$.
4. Assign the number ω_S of missing edges in S to every $S \in \Delta$, i.e., set $\omega_S = |\{vw \notin E : v, w \in S\}|$.
5. Set $(\delta, <_\delta) = (\Delta, <')$.
6. For every $S, T \in \Delta$ with $S <' T$, add a new vertex ST to δ, and replace $S <_\delta T$ by $S <_\delta ST <_\delta T$.
7. Assign the number $\omega_{ST} = -|\{vw \notin E : v, w \in S \cap T\}|$ to every new vertex $ST \in \delta \setminus \Delta$.
8. Compute a minimum weighted maximal chain $\mathcal{C}'$ in $(\delta, <_\delta)$ and restrict it to $\mathcal{C} = \mathcal{C}' \cap \Delta$.
9. Return $F = \{vw \notin E : \text{there is some } S \in \mathcal{C} \text{ such that } v, w \in S\}$.

Proof. By Theorem 20, a solution for MINIMUM FILL-IN or, by Corollary 21, for INTERVAL COMPLETION corresponds to a maximal chain in $(\Delta, <)$ with a minimum number of new edges in $G_\mathcal{C}$. This is achieved by a minimum weighted maximal chain in $(\delta, <_\delta)$ that is restricted to Δ. Because a maximal chain in $(\Delta, <)$ corresponds exactly to a maximal chain in $(\delta, <_\delta)$ and, further, if $\mathcal{C} = S^0, S^1, \ldots, S^m$ is a maximal chain in $(\Delta, <)$, then, by construction and Lemma 22, the weight of the chain $S^0, S^0 S^1, S^1, \ldots, S^{m-1} S^m, S^m$ in δ coincides with the fill-in of $G_\mathcal{C}$.

For each line D^i, there are at most $2n$ scanline points that intersect with mutually different intervals I_v^i, namely l_v^i and $r_v^i + \epsilon$, for every $v \in V$ and a sufficient small number ϵ such that $r_v^i + \epsilon$ is smaller than every interval end point that is right from r_v^i. Let $\overline{\Delta}$ be the set of scanlines that consists of all combinations of those scanline points. Hence, $|\overline{\Delta}| \leq (2n)^{d+1}$. By using a suitable data structure, the construction of $\overline{\Delta}'$ such that all elements of the corresponding Δ' are distinct and satisfy condition (*) of Lemma 16, can be done in $\mathcal{O}(n^{d+3})$ time. Theorem 17 yields $\Delta' = \Delta(G) = \Delta$. Subsequently, by Lemma 18, the partial order $(\Delta, <)$ can be computed in $\mathcal{O}(n^{2d+3})$ time. The transitive reduction $(\Delta, <')$ can be found in $\mathcal{O}(n^{2.376(d+1)})$ time [1, 8]. The steps 4 to 6 clearly take less than this time, while step 7 needs $\mathcal{O}(n^{2d+4})$ time.

Similar to the maximum weighted chain algorithm [23], a minimum weighted chain $\mathcal{C}'$ in $(\delta, <_\delta)$ and its restriction $\mathcal{C}$ to Δ can be computed in $\mathcal{O}(n^{2d+2})$ time:

8.1. While there is some minimal element S of $(\Delta, <')$ with undefined λ_S do:

8.2.   Set $\lambda_S = \omega_S$.

8.3.   Set $\lambda_{ST} = \lambda_S + \omega_{ST}$, for all $T \in \Delta$ such that $S <' T$.

8.4. While there is some $T \in \Delta$ with undefined λ_T such that λ_S is defined for all $S \in \Delta$ with $S <' T$ do:

8.5.   Pick some $S <' T$ such that λ_{ST} is minimum.

8.6.   Set $\lambda_T = \lambda_{ST} + \omega_T$ and assign a pointer from T to S.

8.7.   Set $\lambda_{TU} = \lambda_T + \omega_{TU}$, for all $U \in \Delta$ such that $T <' U$

8.8. Pick some maximal element T of $(\Delta, <')$ such that λ_T is minimum.

8.9. Let $\mathcal{C}$ consist of T and, successively, of all pointed predecessors.   □

7   Treewidth of d-Trapezoid Graphs

Lemma 24. *For $S, T \in \Delta$ with $S < T$, consider the induced subgraph $G(S, T) = G[S \cup T \cup \{v : v \in S_r \text{ and } v \in T_l\}]$. If there is no $U \in \Delta$ with $S < U < T$, then $G(S, T)_{S,T}$ is chordal.*

Proof. Suppose not. Let $\mathcal{C}$ be a maximal chain in $(\Delta, <)$ that contains S and T. By Theorem 20, $G_{\mathcal{C}}$ is chordal. Hence, there is some edge vw in $G_{\mathcal{C}}$ with $vw \notin E(G(S, T)_{S,T})$ such that $v, w \in U$ for some $U \in \mathcal{C}$. W.l.o.g., let $w \notin S$ and $v \notin T$. Therefore, by Lemma 18, $w \in S_r$ and $v \in T_l$. Since $S <' T$, either $U < S, T$ or $S, T < U$. W.l.o.g., let $U < S, T$. As $w \in U \cap S_r$, this contradicts Lemma 18.   □

Procedure Treewidth(G, k)

Input: A d-trapezoid intersection model of a d-trapezoid graph $G = (V, E)$ and a number $k \in \{1, \ldots, n - 2\}$.

Output: YES if $tw(G) \leq k$, otherwise NO.

1. Construct $\Delta = \Delta(G)$.

2. Compute the partial order $(\Delta, <)$.

3. Construct the transitive reduction $<'$ of $(\Delta, <)$.

4. Set $(\delta, <_\delta) = (\Delta, <')$.

5. Add a new element Min to δ, set $Min <_\delta S$ for all minimal elements S of $(\delta, <_\delta)$, and let $\overline{Min}$ be a scanline that lies totally to the left of all d-trapezoids.

6. Add a new element Max to δ, set $S <_\delta Max$ for all maximal elements S of $(\delta, <_\delta)$, and let $\overline{Max}$ be a scanline that lies totally to the right of all d-trapezoids.

7. Assign the clique number $\omega_{S,T} = \omega(G(S, T)_{S,T})$ to every $S <_\delta T$.

8. While there is some $S <_\delta T$ with $\omega_{S,T} > k + 1$ do:

9.   Remove the comparability $S <_\delta T$ from $(\delta, <_\delta)$.

10. Return YES if Min and Max are comparable in the transitive closure of $(\delta, <_\delta)$, otherwise NO.

Theorem 25. *There is an algorithm* Treewidth(G, k) *that solves* TREEWIDTH *and* PATHWIDTH *for a d-trapezoid graph G which is given together with a d-trapezoid representation, in time $\mathcal{O}(\max\{n^{2.376(d+1)}, n^{2d+4}\})$.*

Proof. As shown in Theorem 23, the steps 1 to 4 need $\mathcal{O}(n^{2.376(d+1)})$ time. Clearly, the steps 5 and 6 take less than this time. By Lemma 24, $G(S,T)_{S,T}$ is chordal. Then, by the algorithm of [14] for maximum clique on chordal graphs, step 7 can be performed in $\mathcal{O}(n^{2d+4})$ time.

Clearly, if $G(S,T)_{S,T}$ contains a clique with more than $k+1$ elements, an optimal triangulation cannot contain both S and T if $tw(G) \leq k$. Hence, if *Min* and *Max* finally are incomparable in $(\delta, <_\delta)$, there is no maximal chain $\mathcal{C}$ in $(\Delta, <)$ such that $\omega(G_\mathcal{C}) \leq k+1$. By Theorem 20, $tw(G) > k$.

On the other side, if, after the while loop is executed, there is a maximal chain $\mathcal{C} = S^0, \ldots, S^m$ such that *Min* $<_\delta S^0 <_\delta \ldots <_\delta S^m <_\delta$ *Max*, then $G_\mathcal{C}$ is a minimal triangulation of G and, by Corollary 21, an interval graph. We have $\omega(G_\mathcal{C}) \leq k+1$, because, by Lemma 18, a maximal clique K of $G_\mathcal{C}$ fulfills $K \subseteq S^i \cup S^i_r$ or $K \subseteq S^i \cup S^i_l$, for every $i \in \overline{m}$. Hence, there are some $S <_\delta T$ such that $K \subseteq G(S,T)_{S,T}$.

Note that such a chain $\mathcal{C}$ can be computed within our time bounds. With binary search, $tw(G)$ resp. $pw(G)$ can be calculated in at most $\log n$ executions of the while loop and step 10. $\qquad\Box$

8 Concluding Remarks

We have proved a close relation between minimal chordal triangulations of a graph and the maximal cliques of its separator graph. The quite general theorem can be the starting point for a lot of ongoing research. First of all, one should try to use the characterization in order to find efficient algorithms for TREEWIDTH and MINIMUM FILL-IN in further classes of graphs. Actually, it would be very interesting to investigate the connection between properties of graphs and properties of their separator graphs for several classes. Possibly, one could find a useful structure in the separator graphs of planar graphs, and thus solve the open TREEWIDTH problem for them. Both, a polynomial time algorithm or a $\mathcal{NP}$-completeness proof, could make use of our results. Finally, it would be interesting to find a similar characterization for triangulations into an interval graph. Here, it is not sufficient to consider only minimal triangulations.

References

1. A.V. Aho, M.R. Garey, J.D. Ullman: The transitive reduction of a directed graph. *SIAM J. Comput.* **1** (1972) 131–137
2. S. Arnborg, D.G. Corneil, A. Proskurowski: Complexity of finding embeddings in a k-tree. *SIAM J. Algebraic Discrete Meth.* **8** (1987) 277–284
3. S. Arnborg, J. Lagergren, D. Seese: Easy problems for tree-decomposable graphs. *J. Algorithms* **12** (1991) 308–340
4. H. Bodlaender, T. Kloks, D. Kratsch: Treewidth and pathwidth of permutation graphs. *Proc. 20th ICALP*, LNCS 700. Berlin: Springer 1993, 114–125
5. H. Bodlaender, R.H. Möhring: The pathwidth and treewidth of cographs. *SIAM J. Discrete Math.* **6** (1993) 181–188
6. H. Bodlaender: A linear time algorithm for finding tree-decompositions of small treewidth. *Proc. 25th STOC* 1993, 226–234

7. H. Bodlaender: A tourist guide through treewidth. *Acta Cybernetica* 11 (1993) 1–23
8. D. Coppersmith, S. Winograd: Matrix multiplication via arithmetic progressions. *Proc. 19th STOC* 1987, 1–6
9. D.G. Corneil, Y. Perl, L. Stewart: Cographs: recognition, applications and algorithms. *Congr. Numer.* **43** (1984) 249–258
10. I. Dagan, M.C. Golumbic, I. Pinter: Trapezoid graphs and their coloring. *Discrete Appl. Math.* **21** (1988) 35–46
11. G.A. Dirac: On rigid circuit graphs. *Abh. Math. Sem. Univ. Hamburg* **25**(1961)71–76
12. S. Felsner, R. Müller, L. Wernisch: Optimal algorithms for trapezoid graphs. Technical Report 368/1993, Technische Universität Berlin, 1993
13. C. Flotow: On powers of m-trapezoid graphs. Technical Report, Universität Hamburg, Fachbereich Mathematik, 1993
14. M.C. Golumbic: *Algorithmic Graph Theory and Perfect Graphs*. New York: Academic Press 1980
15. M. Habib, D. Kelly, R.H. Möhring: Interval dimension is a comparability invariant. *Discrete Math.* **88** (1991) 211–229
16. T. Kloks: Treewidth of circle graphs. *Proc. 4th ISAAC*, Lecture Notes in Computer Science 762. Berlin: Springer 1993, 108–117
17. T. Kloks, H. Bodlaender, H. Müller, D. Kratsch: Computing treewidth and minimum fill-in: All you need are the minimal separators. *Proc. 1st ESA*, LNCS 726. Berlin: Springer 1993, 260–271
18. T. Kloks, H. Bodlaender, H. Müller, D. Kratsch: Erratum to the ESA'93 Proceedings. *Proc. 2nd ESA*, LNCS 855. Berlin: Springer 1994, p. 508
19. T. Kloks, D. Kratsch: Finding all minimal separators of a graph. *Proc. 11th STACS*, Lecture Notes in Computer Science 775. Berlin: Springer 1994, 759–768
20. T. Kloks, D. Kratsch, J. Spinrad: Treewidth and pathwidth of cocomparability graphs of bounded dimension. Technical Report 93/46, Eindhoven University, 1993
21. T. Kloks: *Treewidth. Computations and Approximations*. Lecture Notes in Computer Science 842. Berlin: Springer 1994
22. T. Ma, J. Spinrad: An $O(n^2)$ time algorithm for the 2-chain cover problem and related problems. *Proc. 2nd SODA* 1991, 363–372
23. R.H Möhring: Algorithmic aspects of comparability graphs and interval graphs. In: I. Rival (ed.): *Graphs and Order*. Dordrecht: Kluwer Academic Publ. 1985, 41–101
24. R.H. Möhring: Graph problems related to gate matrix layout and PLA folding. In: G. Tinhofer, E. Mayr, H. Noltemeier, M. Sysło (eds.): *Computational Graph Theory*. Wien, New York: Springer 1990, 17–52
25. R.H. Möhring: Triangulating graphs without asteroidal triples. Technical Report 365/1993, Technische Universität Berlin, 1993
26. A. Parra: Triangulating multitolerance graphs. Technical Report 392/1994, Technische Universität Berlin, 1994
27. P. Scheffler: Dynamic programming algorithms for tree-decomposition problems. Technical Report P-MATH-28/86, Weierstraß-Institut f. Mathematik Berlin, 1986
28. J. Spinrad, A. Brandstädt, L. Stewart: Bipartite permutation graphs. *Discrete Appl. Math.* **18** (1987) 279–292
29. R. Sundaram, K.S. Singh, C. Pandu Rangan: Treewidth of circular-arc graphs. *SIAM J. Discrete Math.* **7** (1994) 647–655
30. M. Yannakakis: Computing the minimum fill-in is NP-complete. *SIAM J. Algebraic Discrete Meth.* **2** (1981) 77–79

Fast Gossiping by Short Messages*

J.-C. Bermond,[1] L. Gargano,[2] A. A. Rescigno,[2] and U. Vaccaro[2]

[1] I3S, CNRS, Université de Nice, 06903 Sophia Antipolis Cedex, France
[2] Dipartimento di Informatica, Università di Salerno, 84081 Baronissi (SA), Italy.

Abstract. Gossiping is the process of information diffusion in which each node of a network holds a packet that must be communicated to all other nodes in the network. We consider the problem of gossiping in communication networks under the restriction that communicating nodes can exchange up to a fixed number p of packets at each round. In the first part of the paper we study the extremal case $p = 1$ and we exactly determine the optimal number of communication rounds to perform gossiping for several classes of graphs, including Hamiltonian graphs and complete k-ary trees. For arbitrary graphs we give asymptotically matching upper and lower bounds. We also study the case of arbitrary p and we exactly determine the optimal number of communication rounds to perform gossiping under this hypothesis for complete graphs, hypercubes, rings, and paths.

1 Introduction

Gossiping (also called total exchange or all–to–all communication) in distributed systems is the process of distribution of information known to each processor to every other processor of the system. This process of information dissemination is carried out by means of a sequence of message transmissions between adjacent nodes in the network.

Rossiping is a fundamental primitive in distributed memory multiprocessor system. There are a number of situations in multiprocessor computation, such as global processor synchronization, where gossiping occurs. Moreover, the gossiping problem is implicit in a large class of parallel computation problems, such as linear system solving, Discrete Fourier Transform, and sorting, where both input and output data are required to be distributed across the network [7]. Due to the interesting theoretical questions it poses and its numerous practical applications, gossiping has been widely studied under various communication models. Hedetniemi, Hedetniemi and Liestman [12] provide a survey of the area. Two

* The work of the first author was partially supported by the French GDR/PRC Project PRS. He wants also to thank the Dipartimento di Informatica ed Applicazioni of the Università di Salerno, where part of his research was done, for inviting him. The work of the last three authors was partially supported by Progetto Finalizzato Sistemi Informatici e Calcolo Parallelo of C.N.R. under Grant No. 92.01622.PF69 and by the Italian Ministry of the University and Scientific Research, Project: Algoritmi, Modelli di Calcolo e Strutture Informative.

more recent surveys paper collecting the latest results are [9, 14]. The reader can also profitably see the book [16].

The great majority of the previous work on gossiping has considered the case in which the packets known to a processor at any given time during the execution of the gossiping protocol can be freely concatenated and the resulting (longer) message can be transmitted in a constant amount of time, that is, it has been assumed that the time required to transmit a message is independent from its length. While this assumption is reasonable for short messages, it is clearly unrealistic in case the size of the messages becomes large. Notice that most of the gossiping protocols proposed in the literature require the transmission, in the last rounds of the execution of the protocol, of messages of size $\Theta(n)$, where n is the number of nodes in the network. Therefore, it would be interesting to have gossiping protocols that require only the transmission of bounded length messages between processors. In this paper we consider the problem of gossiping in communication networks under the restriction that communicating nodes can exchange up to a fixed number p of packets at each round.

1.1 The Model

Consider a communication network modeled by a graph $G = (V, E)$ where the node set V represents the set of processors of the network and E represents the set of the communication lines between processors.

Initially each node holds a packet that must be transmitted to any other node in the network by a sequence of *calls* between adjacent processors. During each call, communicating nodes can exchange up to p packets, where p is an *a priori* fixed integer. We assume that each processor can participate in at most one call at time. Therefore, we can see the gossiping process as a sequence of *rounds*: During each round a disjoint set of edges (matching) is selected and the nodes that are end vertices of these edges make a call. This communication model is usually referred to as *telephone model* [12] or *Full–Duplex 1–Port* (F_1) [15]. We denote by $g_{F_1}(p, G)$ the minimum possible number of rounds to complete the gossiping process in the network G subject to the above conditions. Another popular communication model is the *mail model* [12] or *Half–Duplex 1–Port* (H_1) [15], in which in each round any node can either send a message to one of its neighbors or receive a message from it but not simultaneously. The problem of estimating $g_{H_1}(p, G)$ has been considered in [4]. Analogous problems in bus networks have been considered in [10, 13]. Optimal bounds on $g_{H_1}(1, G)$ when the edges of G are subject to random failures are given in [8]. Packet routing in interconnection networks in the F_1 model has been considered in [1].

1.2 Results

We first study the extremal case in which gossiping is to be performed under the restriction that communicating nodes can exchange *exactly* one packet at each round. We provide several lower bounds on the gossiping time $g_{F_1}(1, G)$ and we provide matching upper bounds for Hamiltonian graphs, complete trees, and complete bipartite graphs. For general graphs we provide asymptotically tight upper and lower bounds. Subsequently, we study the case of arbitrary p and we compute exactly $g_{F_1}(p, G)$ for complete graphs, hypercubes, rings and paths. Our result for hypercubes allows us to improve the corresponding result in the H_1 model given in [4]. Due to the space limits, all proofs are omitted. We refer to the full version [5] for all omitted proofs and some additional results.

2 Gossiping by exchanging one packet at time

In this section we study $g_{F_1}(1, G)$, that is the minimum possible number of rounds to complete gossiping in a graph G under the condition that at each round communicating nodes can exchange *exactly one* packet. In order to avoid overburdening the notation, we will simply write $g(G)$ to denote $g_{F_1}(1, G)$.

2.1 Lower bounds on $g(G)$

In this section we give some lower bounds on the time needed to complete the gossiping process.

Lemma 2.1 *For any graph $G = (V, E)$, with $|V| = n$, let $\mu(G)$ be the size of a maximum matching in G, then $g(G) \geq \left\lceil \frac{n(n-1)}{2\mu(G)} \right\rceil$.*

Lemma 2.2 *Let $X \subset V$ be a vertex cutset of the graph $G = (V, E)$ whose removal disconnects G into the connected components $V_1, \ldots, V_d$, then $g(G) \geq \left\lceil \sum_{i=1}^{d} \frac{\max\{|V_i|, n-|V_i|\}}{|M_X|} \right\rceil$, where $|M_X|$ is the size of a maximum matching M_X in G such that any edge in it has an endpoint in X and the other in $V - X$.*

Corollary 2.1 *Let $\alpha(G)$ be the independence number of G, then*

$$g(G) \geq \left\lceil \frac{\alpha(G)(n-1)}{n - \alpha(G)} \right\rceil. \tag{1}$$

Let T be a tree and v one of its nodes, we indicate the connected components into which the node set of T is splitted by the removal of v by $V_1(v), \ldots, V_{\deg(v)}(v)$, ordered so that $|V_1(v)| \geq \ldots \geq |V_{\deg(v)}(v)|$.

Corollary 2.2 *Let T be a tree on n nodes of maximum degree $\Delta = \max_{v \in V} \deg(v)$, then $g(T) \geq \max_{v \,:\, \deg(v) = \Delta} L(v)$, where*

$$L(v) = \begin{cases} (\deg(v) - 1)n + 1 & \text{if } |V_1(v)| \leq n/2; \\ (\deg(v) - 2)n + 1 + 2|V_1(v)| & \text{if } |V_1(v)| > n/2. \end{cases}$$

2.2 Upper bounds

We will determine exactly $g(G)$ for several classes of graphs, including Hamiltonian graphs and complete k-ary trees. We will also provide good upper bounds for general graphs. We first note that in any graph $G = (V, E)$ the size of a maximum matching $\mu(G)$ is at most $\lfloor |V|/2 \rfloor$. Therefore, from Lemma 2.1 we get that the gossiping time $g(G)$ of *any* graph with n nodes is always lower bounded by

$$g(G) \geq \begin{cases} n-1 \text{ if } n \text{ is even;} \\ n \quad \text{ if } n \text{ is odd.} \end{cases} \tag{2}$$

We can prove that this lower bound is attained by Hamiltonian graphs.

Let $C_n = (V, E)$ denote the ring of length n; we assume the vertex set be $V = \{0, \ldots, n-1\}$ and the edge set be $E = \{(v, w) : 1 = |v - w| (\text{mod } n)\}^2$.

Lemma 2.3 $\qquad g(C_n) \leq \begin{cases} n-1 & \text{if } n \text{ is even;} \\ n & \text{if } n \text{ is odd.} \end{cases}$

Proof. We shall discuss only the case n even. The case n odd is similar. For each integer t define the perfect matching in C_n given by

$$M_t = \begin{cases} \{(v, w) : v \text{ is even and } w = v + 1\} & \text{if } t \text{ is even} \\ \{(v, w) : v \text{ is odd and } w = v + 1(\text{mod } n)\} & \text{if } t \text{ is odd;} \end{cases} \tag{3}$$

notice that M_t and M_{t+1} are disjoint for each t. The gossiping algorithm is shown in Figure 1. The easy proof of correctness is left to the reader.

Gossiping-even(C_n)
Round $t = 1$: each node v sends its own packet to the node w such that $(v, w) \in M_1$;
Round $t = 2$: each node v sends its own packet to the node w such that $(v, w) \in M_2$;
Round t, $3 \leq t \leq n - 1$: For each node v let w be the node such that $(v, w) \in M_t$, node v sends a new packet to w, namely v sends the packet it has first got among those v has neither received from w nor sent to w in any previous round.

Figure 1: Gossiping Algorithm in C_n, n even.

$\square$

From Lemma 2.3 and (2) we immediately get

Theorem 2.1 *For any Hamiltonian graph G on n vertices we have*

$$g(G) = \begin{cases} n-1 & \text{if } n \text{ is even;} \\ n & \text{if } n \text{ is odd.} \end{cases}$$

---

[2] Here and in the rest of the paper with $x = a \pmod{b}$ we denote the unique integer $0 \leq x < b$ such that $x = qb + a$.

2.3 Trees

In this section we investigate the gossiping time in trees. We first give an upper bound on the gossiping time in *any* tree and afterwards we exactly compute the gossiping time of *k*-ary trees.

Consider a tree $T = (V, E)$. We recall that for each node v the set $V_1(v)$ denotes the largest of the connected components into which T is splitted by the removal of v. Let $\vartheta = \max |V_1(v)|$, where the maximum is taken over all the internal nodes v having exactly $\deg(v) - 1$ leaves as neighbors; notice that any other internal node u has $|V_1(u)| \leq \vartheta - 1$.

Call *pre–leaf* any node v such that $|V_1(v)| = \vartheta$ and denote by π the maximum degree of a node in the subgraph consisting only of the edges (u, f) where f is either a leaf or a pre–leaf of T.

Finally, let λ be the maximum number of leaves connected to a same node and $\Delta = \max_{v \in V} \deg(v)$.

Theorem 2.2 *For any tree T on n nodes $g(T) \leq (\vartheta - 1)\Delta + \pi + (n - \vartheta - 1)\lambda$.*

The gossiping algorithm in T is given in the Figure 2.

Gossiping-tree(T)
Phase 1
[Color each edge (u, v) of T with color $c(u, v) = c(v, u) \in \{0, \dots, \Delta - 1\}$.]
Round t, for $t = 1, \dots, \Delta(\vartheta - 1)$: For each node u, if there is an edge (u, v) such that $c(u, v) = t - 1 (\bmod \Delta)$ then u sends a new packet to v, namely u sends to v a packet among those that u has neither sent to v nor received from v in a previous round, if such a packet exists, otherwise u sends nothing.
Phase 2
[Give to each edge (u, f), where f is a leaf or a pre–leaf, a color $c'(u, f) \in \{0, \dots, \pi - 1\}$]
Round $\Delta(\vartheta - 1) + t$, for $t = 1, \dots, \pi$: For each leaf or pre–leaf f, if there is an edge (u, f) with $c'(u, f) = t - 1$, then u sends to f a packet among those that u has neither sent to f nor received from f in a previous round, if any.
Phase 3
[Give to each edge (u, f), where f is a leaf of T, a color $c''(u, f) \in \{0, \dots, \lambda - 1\}$.]
Round $\Delta(\vartheta - 1) + \pi + t$, for $t = 1, \dots, (n - \vartheta - 1)\lambda$: For each leaf f, if the edge (u, f) on f has $c''(u, f) = t - 1 (\bmod \lambda)$ then u sends to f any packet f does not know.

Figure 2: Gossiping Algorithm in a tree T.

Let δ denote the minimum degree of an internal node in T. It is easy to see that we can upper bound ϑ by $n - \delta$. Therefore, from Theorem 2.2 we have the following upper bound on $g(T)$ that is expressed only in terms of degree properties of the nodes in T.

Corollary 2.3 *For any tree T on n nodes $g(T) \leq (n - \delta)\Delta + (\delta - 1)\lambda$.*

Given a connected graph $G = (V, E)$, denote by $\mathcal{T}$ the set of all spanning trees of G and for any vertex $v \in V$ denote by $\deg_T(v)$ the degree of v in $T \in \mathcal{T}$. Define $d(G) = \min_{T \in \mathcal{T}} \max_{v \in V} \deg_T(v)$. The following corollary is immediate.

Corollary 2.4 *For any connected graph $G = (V, E)$ with n vertices*

$$g(G) \leq (n - 1)d(G). \tag{4}$$

We point out that, altough the problem of computing $d(G)$ is NP-hard, there exists an efficient algorithm to compute a spanning tree of maximum degree at most $d(G) + 1$ (see [11]). From Corollary 2.2 and Corollary 2.4 we have that for any tree with n nodes and maximum degree Δ it holds $n\Delta - n + 1 \leq g(T) \leq n\Delta - \Delta$. Let us consider now the tree $S_{n,\Delta}$ of Figure 3. If $\Delta = n - 1$ then $S_{n,n-1}$ is the star on n nodes and from Corollary 2.2 and Theorem 2.2 we have $g(S_{n,n-1}) = (n-1)^2$. If $\Delta > 2$ is constant with respect to $n > 2\Delta$ then from Corollary 2.2 and Theorem 2.2 we get $\Delta(n - 1) - (\Delta - 1) \leq g(S_{n,\Delta}) \leq \Delta(n - 1) - 2$. It is not difficult to obtain a specific gossiping algorithm attaining the lower bound. Therefore, we have that for any n and Δ there exists a graph $G_{n,\Delta}$ with n vertices and maximum degree Δ such that $g(G_{n,\Delta}) = \Omega((n - 1)\Delta)$, hence the bound (4) is asymptotically tight. In [8] it is conjectured that for any graph G it holds $g_{H_1}(1, G) = \Omega(nd(G))$. This conjecture, if true, together with Corollary 2.4 would imply the rather interesting fact that for any graph G it holds $g(G) = \Theta(nd(G))$.

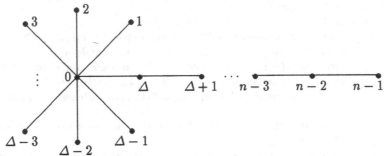

Figure 3: Tree $S_{n,\Delta}$

We shall now exactly compute the gossiping time of k-ary trees, that is, rooted trees in which each internal node has exactly k sons. Let $T_{k,n}$ denote *any* k-ary tree with n nodes.

Let us first notice that for $n = k + 1$ the tree $T_{k,n}$ is the star $S_{k+1,k}$. Consider then a tree $T_{k,n}$ with $n \geq 2k + 1$ nodes. Let u be a node of $T_{k,n}$ whose sons are all leaves, by Corollary 2.2 we get

$$g(T_{k,n}) \geq \max_v L(v) \geq L(u) = \begin{cases} kn + 1 & \text{if } n = 2k + 1 \\ (k + 1)(n - 1) - k & \text{if } n \geq 3k + 1. \end{cases} \tag{5}$$

We show now that (5) holds with equality. Applying Theorem 2.2 to $T_{k,n}$ we get that

$$g(T_{k,n}) \leq (\vartheta - 1)\Delta + \pi + (n - \vartheta - 1)\lambda = (\vartheta - 1)(k + 1) + \pi + (n - \vartheta - 1)k. \tag{6}$$

Unless exactly $k-1$ sons of the root are leaves (cf. the tree in Figure 4) $T_{k,n}$ has $\vartheta = n - k - 1$ and $\pi \leq \Delta = k + 1$, that by (6) and (5) gives

$$g(T_{k,n}) = (n - k - 2)(k + 1) + k + 1 + k^2 = (k + 1)(n - 1) - k.$$

Consider now the remaining case when $T_{k,n}$ is the tree of Figure 4. The only pre-leaf is the root, and $\vartheta = n - k$. If $n \geq 3k + 1$ we have $\pi = k$ and from (6) we get

$$g(T_{k,n}) \leq (n - k - 1)(k + 1) + k + (k - 1)k = (n - 1)(k + 1) - k;$$

if $n = 2k + 1$ we have $\pi = \Delta = k + 1$ and $g(T_{k,2k+1}) \leq kn + 1.$

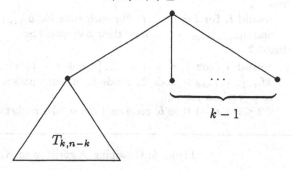

Figure 4

Therefore, we have proved the following result.

Theorem 2.3 *For any k-ary tree on n nodes $T_{k,n}$ it holds that*

$$g(T_{k,n}) = \begin{cases} k^2 & \text{if } n = k + 1 \\ 2k^2 + k + 1 & \text{if } n = 2k + 1 \\ (k + 1)(n - 1) - k & \text{if } n \geq 3k + 1. \end{cases}$$

The particular case $k = 1$ of above theorem deserves to be explicitly stated.

Corollary 2.5 *For the path on n nodes P_n we have* $g(P_n) = \begin{cases} 1 & \text{if } n = 2 \\ 4 & \text{if } n = 3 \\ 2n - 3 & \text{if } n \geq 4. \end{cases}$

2.4 Complete bipartite graphs

Let $K_{r,s} = (V(K_{r,s}), E(K_{r,s}))$ be the complete bipartite graph on the node set $V(K_{r,s}) = \{a_0, \ldots, a_{r-1}\} \cup \{b_0, \ldots, b_{s-1}\}$, with $\{a_1, \ldots, a_{r-1}\} \cap \{b_0, \ldots, b_{s-1}\} = \emptyset$, $r \geq s$, and edge set $E(K_{r,s}) = \{a_0, \ldots, a_{r-1}\} \times \{b_0, \ldots, b_{s-1}\}$. In the next theorem we determine the gossiping time of $K_{r,s}$.

Theorem 2.4 *For each r and s, $r \geq s \geq 1$, it holds* $g(K_{r,s}) = \lceil (r + s - 1)r/s \rceil$.

Proof. The lower bound $g(K_{r,s}) \geq \lceil (r+s-1)r/s \rceil$ is an immediate consequence of Corollary 2.1 since the complete bipartite graph has $\alpha(K_{r,s}) = r$.

In order to give a gossiping algorithm in $K_{r,s}$ requiring $\lceil (r+s-1)r/s \rceil$ communication rounds, we define the matchings

$$M_j = \{(b_i, a_{i+j \, (\text{mod } r)}) \; : \; 0 \leq i \leq s-1\},$$

for $j = 0, \ldots, r-1$. The algorithm is shown in Figure 5.

Gossiping–bipartite($K_{r,s}$)
Phase 1
    round t, for $t = 1, \ldots, r$: For each edge $(b_i, a_{i+t-1 \, (\text{mod } r)}) \in M_{t-1}$ nodes b_i
    and $a_{i+t-1 \, (\text{mod } r)}$ exchange their own packets;
Phase 2
    round t, for $t = r+1, \ldots, \lceil r(r+s-1)/s \rceil$: For each edge $(b_i, a_j) \in$
    $M_{(t-1-r)s \, (\text{mod } r)}$ node b_i sends to a_j any packet that a_j has not received
    in a previous round;
    if $t \leq r+s-1$ then b_i receives from a_j the packet of $b_{i+t-r(\text{mod } s)}$.

Figure 5: Gossiping Algorithm in $K_{r,s}$.

2.5 Generalized Petersen Graphs

In Section 2.2 we have seen that Hamiltonian graphs have the minimum possible gossiping time among all graphs with n nodes. A natural question to ask is to see if there are non–Hamiltonian graphs on n vertices with gossiping time equal to n if n is odd and $n-1$ if n is even. A quick check shows that this is not the case for rectangular grids $G_{t,s}$ with both t and s odd [3]. In fact, we know that $\alpha(G_{t,s}) = \lceil \frac{s \cdot t}{2} \rceil$ and from Corollary 2.1 we get $g(G_{t,s}) \geq s \cdot t + 1$. Moreover, it is also easy to check that the gossiping time of the Petersen graph on 10 vertices is at least 10. Therefore, one could be tempted to conjecture that the gossiping time $g(G)$ of a graph G is equal to the minimum possible only if G is Hamiltonian. This conjecture, although nice sounding, would be wrong as the following classes of graphs, including the Generalized Petersen Graphs, shows.

Let $P_{k,\pi}$ be the graph consisting of two cycles of size k connected by a perfect matching in the following way: given a permutation π of $\{0, \ldots, k-1\}$ the graph $P_{k,\pi} = (V(P_{k,\pi}), E(P_{k,\pi}))$ has vertex set $V(P_{k,\pi}) = \{a_0, \ldots, a_{k-1}\} \cup \{b_0, \ldots, b_{k-1}\}$ and edge set

$$E(P_{k,\pi}) = \{(a_i, a_{i+1(\text{mod } k)}) \; : \; 0 \leq i < k\} \cup \{(b_i, b_{i+1(\text{mod } k)}) \; : \; 0 \leq i < k\}$$
$$\cup \{(a_i, b_{\pi(i)}) \; : \; 0 \leq i < k\}.$$

---

[3] It is well known that all rectangular grids $G_{t,s}$ are Hamiltonian but for values of t and s both odd.

The Petersen Graph has $k = 5$ and $\pi(i) = 3i \pmod 5$, for $i = 0, 1, 2, 3, 4$; Generalized Petersen Graphs (GPG) have k odd and $\pi(s \cdot i \pmod k)) = i$, $i = 0, \ldots, k - 1$, for a fixed integer s. From Lemma 2.1 we know that $g(P_{k,\pi}) \geq |V(P_{k,\pi})| - 1 = 2k - 1$. We can prove the following theorem.

Theorem 2.5 *For any k and π such that $P_{k,\pi}$ is 3-edge-colorable, we have $g(P_{k,\pi}) = 2k - 1$.*

Notice that each cubic GPG, other than the Petersen graph itself, is 3-edge-colorable. Moreover, the class of 3-edge-colorable $P_{k,\pi}$'s includes the family of non Hamiltonian GPGs with $k = 5 \pmod 6$ and $s = 2$ (see [2] and references therein quoted).

3 Gossiping by exchanging more than one packet at time

In this section we shall study the minimum number of time units $g_{F_1}(p, G)$ necessary to perform gossiping in a graph G, under the restriction that at each time instant communicating nodes can exchange up to p packets, p fixed but arbitrary otherwise. We assume that p is smaller than the number of nodes of the graph G, otherwise the problem is equivalent to the classical one. Again, for ease of notation, we shall write $g(p, G)$ to denote $g_{F_1}(p, G)$.

3.1 Lower Bounds

First of all we shall present a simple lower bound on $g(p, G)$ based on elementary counting arguments. Nonetheless, we shall prove in the sequel that the obtained lower bound is tight for complete graphs with an even number of nodes and for hypercubes. In order to derive the lower bound, let us define $I(p, t)$ as the maximum number of packets a vertex can have possibly received after t communication rounds in *any* graph. Since at each round i, with $1 \leq i \leq t$, any vertex can receive at most $\min\{p, 2^{i-1}\}$ packets, it follows that $I(p, t) = 1 + \sum_{i=1}^{t} \min\{p, 2^{i-1}\}$, or, equivalently

$$I(p, t) = 1 + \sum_{i=1}^{\lceil \log p \rceil} 2^{i-1} + p(t - \lceil \log p \rceil) = 2^{\lceil \log p \rceil} + p(t - \lceil \log p \rceil) \qquad (7)$$

for any $t \geq \lceil \log p \rceil$. Therefore, for any graph $G = (V, E)$, the gossiping time $g(p, G)$ is always lower bounded by the smallest integer t^* for which $I(p, t^*) \geq |V|$. Since t^* is obviously greater or equal to $\lceil \log |V| \rceil \geq \lceil \log p \rceil$, we can use (7) and obtain $g(p, G) \geq \lceil \log p \rceil + \left\lceil \frac{1}{p}(|V| - 2^{\lceil \log p \rceil}) \right\rceil$. Moreover, notice that if the number of nodes in the graph is odd then at each round there is a node that does not receive any message. This implies that after any round t there exists a node who can have possibly received at most $I(p, t - 1)$ packets. Therefore, $g(p, G) \geq \lceil \log p \rceil + \left\lceil \frac{1}{p}(|V| - 2^{\lceil \log p \rceil}) \right\rceil + 1$. The above arguments give the following lemma.

Lemma 3.1 *For any graph* $G = (V, E)$, $|V| = n$, *and integer* p *such that* $2^{\lceil \log p \rceil} \leq n$ *we have*

$$g(p, G) \geq \begin{cases} \lceil \log p \rceil + \left\lceil \frac{1}{p}(n - 2^{\lceil \log p \rceil}) \right\rceil & \text{if } n \text{ is even,} \\ \lceil \log p \rceil + \left\lceil \frac{1}{p}(n - 2^{\lceil \log p \rceil}) \right\rceil + 1 & \text{if } n \text{ is odd.} \end{cases}$$

3.2 Rings and Paths

Let $g(\infty, G)$ denote the gossiping time of the graph G in absence of any restriction on the size of the messages. It is obvious that for each p it holds $g(p, G) \geq g(\infty, G)$, it is possible to see that equality holds for any $p \geq 2$ when G is either the ring C_n or the path P_n on n nodes. It is well known that [14]

$$g(\infty, P_n) = 2 \left\lceil \frac{n}{2} \right\rceil - 1 \quad \text{and} \quad g(\infty, C_n) = \begin{cases} n/2 & \text{if } n \text{ is even,} \\ (n+3)/2 & \text{if } n \text{ is odd.} \end{cases}$$

We just point out that it is easy to see that the algorithms attaining $g(\infty, C_n)$ and $g(\infty, P_n)$ do not need to send more than 2 packets at time. Therefore the following results hold.

Theorem 3.1 *For each* $n \geq 3$ *and* $p \geq 2$ *it holds*

$$g(p, C_n) = g(2, C_n) = \begin{cases} n/2 & \text{if } n \text{ is even,} \\ (n+3)/2 & \text{if } n \text{ is odd.} \end{cases}$$

Theorem 3.2 *For each* $n \geq 2$ *and* $p \geq 2$ *it holds* $g(p, P_n) = g(2, P_n) = 2\lceil \frac{n}{2} \rceil - 1$.

3.3 Complete graphs

Let K_n be the complete graph on n nodes. We recall that $g(\infty, K_n)$ is equal to $\lceil \log n \rceil$ if n is even, and $\lceil \log n \rceil + 1$ if n is odd.

Theorem 3.3 *For each even integer* n *and integer* p *such that* $2^{\lceil \log p \rceil} \leq n$ *it holds* $g(p, K_n) = \lceil \log p \rceil + \left\lceil \frac{n - 2^{\lceil \log p \rceil}}{p} \right\rceil$.

Theorem 3.4 *For each odd integer* N *and integer* p *such that* $2^{\lceil \log p \rceil} \leq N + 1$ *it holds* $\lceil \log p \rceil + \left\lceil \frac{N - 2^{\lceil \log p \rceil}}{p} \right\rceil + 1 \leq g(p, K_N) \leq \lceil \log p \rceil + \left\lceil \frac{N+1 - 2^{\lceil \log p \rceil}}{p} \right\rceil + 2$.

For N odd, we believe that the true value of $g(p, K_N)$ is $\lceil \log p \rceil + \left\lceil \frac{N - 2^{\lceil \log p \rceil}}{p} \right\rceil + 1$; we can verify this equality for small values of N and p. In case $p = 2$, Theorem 3.1 and Lemma 3.1 tell us that $g(2, K_N) = (N + 3)/2 = g(2, C_N)$, for each odd $N \geq 2$. Moreover, we can prove that

Theorem 3.5 *If* p *is a multiple of* 4 *then* $g(p, K_N) = \lceil \log p \rceil + \left\lceil \frac{N - 2^{\lceil \log p \rceil}}{p} \right\rceil + 1$.

3.4 Hypercube

Let H_d be the d-dimensional hypercube with 2^d nodes. We can prove

Theorem 3.6 *For each integer* $p < 2^d$ *it holds*

$$g(p, H_d) = \lceil \log p \rceil + \left\lceil \frac{1}{p} \left(2^d - 2^{\lceil \log p \rceil} \right) \right\rceil.$$

Remark 3.1 It is worth pointing out that the obvious inequality $g_{H_1}(p, G) \leq 2g_{F_1}(p, G)$ and above theorem allow us to improve the upper bound on $g_{H_1}(p, H_d)$ given by Theorem 4 of [4] for all values of p not power of two. Indeed, the authors of [4] have $g_{H_1}(p, H_d) \leq 2d + 2^{d+1}/p - 2/p$ while from Theorem 3.6 and above inequality we get

Theorem 3.7 *For each integer* $p < 2^d$ *we have*

$$g_{H_1}(p, H_d) \leq 2\lceil \log p \rceil + 2 \left\lceil \frac{1}{p} \left(2^d - 2^{\lceil \log p \rceil} \right) \right\rceil.$$

4 Concluding Remarks and Open Problems

We have considered the problem of gossiping in communication networks under the restriction that communicating nodes can exchange up to a fixed number p of packets at each round. In the extremal case $p = 1$ we have given optimal algorithms to perform gossiping in several classes of graphs, including Hamiltonian graphs, paths, complete k-ary trees, and complete bipartite graphs. For arbitrary graphs we gave asymptotically matching upper and lower bounds.

In the case of arbitrary p we have determined the optimal number of communication rounds to perform gossiping under this hypothesis for complete graphs, hypercubes, rings, paths and complete bipartite graphs $K_{r,r}$. Several open problems remain in the area. We list the most important of them here.

• It would be interesting to determine the computational complexity of computing $g_{F_1}(1, G)$ $(g_{F_1}(p, G))$ for general graphs, it is very likely that it is NP–hard. (We know that computing $g_{F_1}(\infty, G)$ is NP–hard, see [15]).

• We have left open the problem of determining the gossiping time $g_{F_1}(1, G_{t,s})$, and more generally $g_{F_1}(p, G_{t,s})$, of rectangular grids $G_{t,s}$ with both t and s odd. We know from Corollary 2.1 that $g_{F_1}(1, G_{t,s}) \geq st + 1$. Does equality holds? We can prove that $g_{F_1}(1, G_{3,3}) = 10$. A general upper bound on $g_{F_1}(1, G_{t,s})$ can be obtained by observing that $G_{t,s} = P_t \times P_s$, where P_t and P_s are the paths on t and s nodes, respectively, and $\times$ denotes the cartesian graph product. Now, given two graphs $G = (V, E)$ and $H = (W, F)$ it is easy to see that $g_{F_1}(1, G \times H) \leq \min\{g_{F_1}(1, G) + |V|g_{F_1}(1, H), g_{F_1}(1, H) + |W|g_{F_1}(1, G)\}$ that, together with Corollary 2.5, immediately gives $g_{F_1}(1, G_{t,s}) \leq 2ts - 3 - \max\{t, s\}$.

• We know from (2) that for any graph G with n vertices one has $g_{F_1}(1, G) \geq n$ if n is odd, $g_{F_1}(1, G) \geq n - 1$ if n is even and from Theorem 2.1 we get that the equality holds for Hamiltonian graphs. It would be interesting to characterize

the class of graphs for which this lower bound is tight. We know from the results of Section 2.5 that this class is larger than the class of the Hamiltonian graphs.
• Finally, we mention that in [6] we have analyzed the minimum total number of calls necessary to perform gossiping under the restriction that communicating nodes can exchange up to p packets during each call. In the full version of the present paper [5] we will also present results concerning the construction of sparse interconnection networks with gossiping time equal to that of the complete graph.

References

1. N. Alon, F.R.K. Chung, and R.L. Graham "Routing Permutations on Graphs via Matchings", *Proc. 25th ACM Symposium on the Theory of Computing (STOC '93)*, San Diego, CA (1993), 583–591.
2. B. Alspach, "The Classification of Hamiltonian Generalized Petersen Graphs", *J. Combinatorial Theory, Series B*, 34 (1983), 293 - 312.
3. A. Bagchi, E.F. Schmeichel, and S.L. Hakimi, "Sequential Information Dissemination by Packets", *Networks*, 22 (1992), 317–333.
4. A. Bagchi, E.F. Schmeichel, and S.L. Hakimi, "Parallel Information Dissemination by Packets", *SIAM J. on Computing*, 23 (1994), 355-372.
5. J.-C. Bermond, L. Gargano, A. Rescigno, and U. Vaccaro, "Fast Gossiping by Short Messages", manuscript available from the authors.
6. J.-C. Bermond, L. Gargano, A. Rescigno, and U. Vaccaro, *in preparation*.
7. D. P. Bertsekas, and J. N. Tsitsiklis, *Parallel and Distributed Computation: Numerical Methods*, Prentice-Hall, Englewood Cliffs, NJ, 1989.
8. B.S. Chlebus, K. Diks, A. Pelc, "Optimal Gossiping with Short Unreliable Messages", *Discr. Appl. Math.*, 53 (1994), 15–24.
9. P. Fraignaud, E. Lazard, "Methods and Problems of Communication in Usual Networks", *Discr. Appl. Math.*, 53 (1994), 79–134.
10. S. Fujita, "Gossiping in Mesh–Bus Computers by Packets with Bounded Length", *IPS Japan SIGAL*, 36-6 (1993), 41–48.
11. M. Fürer and B. Raghavachari, "Approximating the Minimum Degree Spanning Tree to within One from the Optimal Degree", *Proc. of SODA '92*, Orlando, FL (1992), 317–324.
12. S. M. Hedetniemi, S. T. Hedetniemi, and A. Liestman, "A Survey of Gossiping and Broadcasting in Communication Networks", *Networks*, 18 (1988), 129–134.
13. A. Hily and D. Sotteau, "Communications in Bus Networks", in: *Parallel and Distributed Computing*, M. Cosnard, A. Ferreira, and J. Peters (Eds.), Lectures Notes in Computer Science, 805, Springer - Verlag, (1994),197–206.
14. J. Hromkovič, R. Klasing, B. Monien, and R. Peine, "Dissemination of Information in Interconnection Networks (Broadcasting and Gossiping)", to appear in: F. Hsu, D.-Z. Du (Eds.) *Combinatorial Network Theory*, Science Press & AMS.
15. D. W. Krumme, K.N. Venkataraman, and G. Cybenko, "Gossiping in Minimal Time", *SIAM J. on Computing*, 21 (1992), 111–139.
16. J. de Rumeur, *Communication dans les Reseaux de Processeur*, MASSON, Paris (1994).

Break Finite Automata Public Key Cryptosystem

Feng Bao and Yoshihide Igarashi

Department of Computer Science,
Gunma University, Kiryu, 376 Japan

Abstract. In this paper we break a 10-year's standing public key cryptosystem, Finite Automata Public Key Cryptosystem(FAPKC for short). The security of FAPKC was mainly based on the difficulty of finding a special common left factor of two given matrix polynomials. We prove a simple but previously unknown property of the input-memory finite automata. By this property, we reduce the basis of the FAPKC's security to the same problem in **module** matrix polynomial rings. The problem turns out to be easily solved. Hence, we can break FAPKC by constructing decryption automata from the encryption automaton(public key). We describe a modification of FAPKC which can resist above attack.

1 Introduction

1.1 About FAPKC

Finite Automaton Public Key Cryptosystem, denoted by FAPKC, is a public key cryptosystem based on the invertibility theory of finite automata. For the invertibility theory of finite automata, the reader may be referred to [1,3,5-9,11,17-19,22]. FAPKC is a 10-year's standing public key cryptosystem that first appeared in [20], and its modified versions were given in the following year [21]. The identity-based finite automaton public key cryptosystem and digital signature schemes were proposed in CRYPTO-CHINA'92 [23]. Since FAPKC is a stream cipher, it possesses an advantageous property that plaintexts need not be divided into blocks. Its speed is very fast(much faster than RSA), and its key size is comparable with RSA's [12] [24]. FAPKC can be easily implemented. since it includes only logic operations. FAPKC was practically used in some local networks in China, and the product of digital signature based on FAPKC was developed for practical use. Although almost all the relevant references about FAPKC were published in China, the copies of [21] were distributed at ICALP'88 when Tao, the inventor of FAPKC, presented [22] at the conference. FAPKC was also mentioned in Section 5.3 of Salomaa's book [16].

1.2 Weakly Invertible Finite Automata

Since the relevant references are not conveniently accessed by the reader, we try to make this paper self-contained. Invertible finite automata were first proposed by Huffman as "information-lossless finite state logical machines" [11]. Since then, they have been much studied [1,3,8,9,13-15,25]. In [8] and [9], "generalized finite automata" were used instead of "sequential machines" and "logic machines". Reference [17] is a book on this subject, in which both invertible finite automata and weakly invertible finite automata have been intensively studied.

Formally speaking, a finite automaton is a five-tuple $M =< X, Y, S, \delta, \lambda >$, where X is a finite input alphabet, Y is a finite output alphabet, S is a finite set of internal states, δ is a next-state function, $\delta : S \times X \to S$ and λ is an output function, $\lambda : S \times X \to Y$.

We can naturally extend δ to a mapping from $S \times X^*$ to S, where X^* denotes the set of all finite strings over X (including the empty string ε): For any $s \in S$, $\alpha \in X^*$ and $x \in X$, define $\delta(s, \varepsilon) = s$ and $\delta(s, \alpha x) = \delta(\delta(s, \alpha), x)$. Similarly, λ can be naturally extended to a mapping from $S \times (X^* \cup X^\omega)$ to $Y^* \cup Y^\omega$, where X^ω denotes the set of all infinite strings over X: For any $s \in S$, $\alpha \in X^* \cup X^\omega$ and $x \in X$, define $\lambda(s, \varepsilon) = \varepsilon$ and $\lambda(s, x\alpha) = \lambda(s, x)\lambda(\delta(s, x), \alpha)$.

Our finite automaton is a transducer converting a string over the input alphabet into a length-preserving string over the output alphabet, rather than just an acceptor.

Definition 1. Let $M =< X, Y, S, \delta, \lambda >$ be a finite automaton. If for any $s \in S$ and any $\alpha, \alpha' \in X^\omega$, $\lambda(s, \alpha) = \lambda(s, \alpha')$ always implies $\alpha = \alpha'$, then M is said to be a *weakly invertible finite automaton* (WIFA for short).

Definition 2. A finite automaton $M =< X, Y, S, \delta, \lambda >$ is said to be a WIFA *with delay* τ if for any $s \in S$, $a, a' \in X$ and $\alpha, \alpha' \in X^\tau$, $\lambda(s, a\alpha) = \lambda(s, a'\alpha')$ always implies $a = a'$, where τ is a non-negative integer and X^τ is the set of all strings of length τ over X.

Proposition 3. *A finite automaton M is a WIFA if and only if M is a WIFA with delay τ for some $\tau \leq n(n-1)/2$, where n is the number of states of M.*

Definition 4. Let $M =< X, Y, S, \delta, \lambda >$ and $M' =< Y, X, S', \delta', \lambda' >$ be a pair of finite automata. If for any $s \in S$ there exists $s' \in S'$ such that for any $\alpha \in X^\omega$, there exists $\alpha_0 \in X^\tau$ satisfying $\lambda'(s', \lambda(s, \alpha)) = \alpha_0\alpha$, then M is called a *weak inverse with delay* τ of M. Such a state s' is called the *match state* of s, and such a string $\alpha_0 = x_{-\tau} \cdots x_{-1}$ the *delay prefix*.

Proposition 5. *A finite automaton M is a WIFA with delay τ if and only if M has a weak inverse with delay τ.*

For the proofs of above two propositions, the reader may be referred to [9] [17].

Definition 6. Let $M_1 =< X, Y, S_1, \delta_1, \lambda_1 >$ and $M_2 =< Y, Z, S_2, \delta_2, \lambda_2 >$ be a pair of finite automata. The *composition* of M_1 and M_2 is a new finite automaton $M_1 \cdot M_2 = < X, Z, S_1 \times S_2, \delta, \lambda >$, where $\delta(< s_1, s_2 >, x) = < \delta_1(s_1, x), \delta_2(s_2, \lambda_1(s_1, x)) >$ and $\lambda(< s_1, s_2 >, x) = \lambda_2(s_2, \lambda_1(s_1, x))$ for any $x \in X$ and $< s_1, s_2 > \in S_1 \times S_2$.

$M_1 \cdot M_2$ is the natural concatenation of M_1 and M_2 (i.e., M_2 takes the output of M_1 as its input). it is not difficult to see that if M_1 is a WIFA with delay τ_1 and M_2 is a WIFA with delay τ_2, then $M_1 \cdot M_2$ is a WIFA with delay $\tau_1 + \tau_2$.

Definition 7. Let $M_1 = < X, Y, S_1, \delta_1, \lambda_1 >$ and $M_2 = < X, Y, S_2, \delta_2, \lambda_2 >$ be a pair of finite automata. States $s_1 \in S_1$ and $s_2 \in S_2$ are said to be *equivalent* if for any $\alpha \in X^*$, $\lambda_1(s_1, \alpha) = \lambda_2(s_2, \alpha)$. Finite automata M_1 and M_2 are said to be *equivalent* if for any state $s_1 \in S_1$, there exists a state $s_2 \in S_2$ equivalent to s_1, and for any $s_2 \in S_2$, there exists $s_1 \in S_1$ equivalent to s_2.

1.3 The Basic Idea of Breaking FAPKC

In FAPKC, the plaintext is encrypted by a nonlinear WIFA with delay τ (typically $\tau > 15$) which is the composition of a nonlinear WIFA with delay 0 and a linear WIFA with delay τ. The security of FAPKC depends on the following facts: (a) It is difficult to construct a weak inverse with delay τ of any given nonlinear WIFA with delay τ. (b) It is easy to construct a weak inverse with delay τ of any given linear WIFA with delay τ. (c) It is easy to construct a weak inverse with delay 0 of any given nonlinear WIFA with delay 0.

The principle of FAPKC is as shown in Fig. 1, where M_0 is a nonlinear WIFA with delay 0, M_1 is a linear WIFA with delay τ, M_0^{-1} is a weak inverse with delay 0 of M_0, M_1^{-1} is a weak inverse with delay τ of M_1, s_0^{-1} is a match state of s_0 and s_1^{-1} is a match state of s_1.

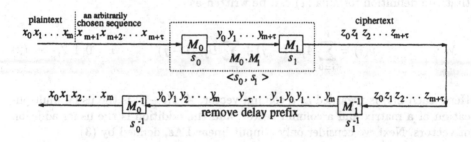

Fig. 1. The principle of FAPKC.

In FAPKC, M_0 and M_1 are kept as the secret key, from which M_0^{-1} and M_1^{-1}, the decryption automata, can be easily derived. The public key is a sub-automaton of the composition of M_0 and M_1 (not exactly the $M_0 \cdot M_1$, see Section 3 for details). We call it the encryption automaton. To separate M_0 and M_1 from the encryption automaton, we needed to find a special left common factor of two given matrix polynomials. However, that is a very difficult problem. So far, no divisibility theory has been established for matrix polynomial rings.

In this paper, we prove a previously unknown property of WIFAs. Then, by this property, we reduce the problem of separating M_0 and M_1 from the encryption automaton to a much easier problem: finding a special left common factor of two given matrix polynomials in module matrix polynomial rings. Hence, we can break FAPKC by successfully finding decryption automata from the public key.

2   Linear WIFAs

Finite automata used in FAPKC are actually sequential machines. In order to keep consistent with the references, we still use the term "finite automaton"(FA for short). Finite automata considered here are of the form $M =< X, Y, S, \delta, \lambda >$, where $X = Y =$ the l-dimensional linear space over $GF(2) = \{0, 1\}$ (typically $l = 8$, so that FAPKC encrypts byte by byte), and δ and λ are specified by the following *definition formula*.

$$M : \quad y(i) = f(x(i), x(i-1), \cdots, x(i-r), y(i-1), \cdots, y(i-t)), \quad i = 0, 1, 2, \cdots \quad (1)$$

In the definition formula (1), $x(i)$ represents the input at time i, $y(i)$ represents the output at time i, and each of $x(i)$ and $y(i)$ is a binary l-dimensional vector. Obviously, S is the $l(r+t)$-dimensional linear space over $GF(2)$. A finite automaton defined by (1) is so-called r-input and t-output memory FA, where $< x(-1), \cdots, x(-r), y(-1), \cdots, y(-t) >$ is the initial state. All finite automata used in FAPKC are of this type.

The M defined by (1) is called a linear FA, if f is a linear function. In this case, there exist $l \times l$ matrices over $GF(2)$, $A_0, A_1, \cdots, A_r, B_1, B_2, \cdots, B_t$ such that the definition formula (1) can be written as

$$M : \qquad y(i) = \sum_{j=0}^{r} A_j x(i-j) + \sum_{j=1}^{t} B_j y(i-j), \qquad i = 0, 1, 2, \cdots \quad (2)$$

Here, each $x(i)$ is regarded as a column vector, $A_j x(i-j)$ is the usual multiplication of a matrix and a column vector, and the addition is the usual addition of vectors. Next we consider only r-input linear FAs, defined by (3).

$$M : \qquad y(i) = A_0 x(i) + A_1 x(i-1) + \cdots + A_r x(i-r), \qquad i = 0, 1, 2, \cdots \quad (3)$$

The linear coefficients $A_0, A_1, A_2, \cdots, A_r$ uniquely determine the finite automaton M. We can decide from $A_0, A_1, A_2, \cdots, A_r$ whether M is a WIFA with delay r. Furthermore, if M is a WIFA with delay r, we can construct a weak inverse with delay r of M.

Let $s_0 =< x(-1), x(-2), \cdots, x(-r) >$ be an arbitrarily given initial state of M. Let the *time variate* i take $0, 1, 2, \cdots, r$ in (3). Then we obtain

$$\begin{pmatrix} y(0) \\ y(1) \\ \vdots \\ y(r) \end{pmatrix} = \begin{pmatrix} A_0 & (0) & \cdots & (0) \\ A_1 & A_0 & \cdots & (0) \\ \vdots & \vdots & \ddots & \vdots \\ A_r & A_{r-1} & \cdots & A_0 \end{pmatrix} \begin{pmatrix} x(0) \\ x(1) \\ \vdots \\ x(r) \end{pmatrix} + \begin{pmatrix} A_1 & \cdots & A_r \\ \vdots & \vdots & \vdots \\ A_r & \cdots & (0) \\ (0) & \cdots & (0) \end{pmatrix} \begin{pmatrix} x(-1) \\ x(-2) \\ \vdots \\ x(-r) \end{pmatrix} \quad (4)$$

where (0) denotes the zero $l \times l$ matrix over $GF(2)$.

Proposition 8. *There exists a fast algorithm for finding an invertible* $(r+1)l \times (r+1)l$ *matrix* $\mathbf{P}$ *from* $A_0, A_1, A_2, \cdots, A_r$, *such that*

$$
\mathbf{P} \begin{pmatrix} A_0 & (0) & \cdots & (0) \\ A_1 & A_0 & \cdots & (0) \\ \vdots & \vdots & \ddots & \vdots \\ A_r & A_{r-1} & \cdots & A_0 \end{pmatrix} = \begin{pmatrix} A_{0,0} & (0) & \cdots & (0) \\ A_{1,1} & A_{1,0} & \cdots & (0) \\ \vdots & \vdots & \ddots & \vdots \\ A_{r,r} & A_{r,r-1} & \cdots & A_{r,0} \end{pmatrix} \tag{5}
$$

where $A_{i,j}$'s *satisfy the following condition: for any* $0 \leq h \leq r$,

$$
\forall x_1, x_2, \cdots, x_h \in X, \exists x_0 \in X, A_{h,1}x_1 + A_{h,2}x_2 + A_{h,h}x_h = A_{h,0}x_0 \tag{6}
$$

We explain how to find $\mathbf{P}$ in Appendix.

Proposition 9. *The finite automaton* M *is a WIFA with delay* r *if and only if* $A_{0,0}$ *is an invertible matrix, i.e., of full rank.*

If M is a WIFA with delay r, we can construct a weak inverse with delay r of M from $\mathbf{P}$ easily. Adding the time variate i into (4), we have

$$
\begin{pmatrix} y(i) \\ y(i+1) \\ \vdots \\ y(i+r) \end{pmatrix} = \begin{pmatrix} A_0 & (0) & \cdots & (0) \\ A_1 & A_0 & \cdots & (0) \\ \vdots & \vdots & \ddots & \vdots \\ A_r & A_{r-1} & \cdots & A_0 \end{pmatrix} \begin{pmatrix} x(i) \\ x(i+1) \\ \vdots \\ x(i+r) \end{pmatrix} + \begin{pmatrix} A_1 & \cdots & A_r \\ \vdots & \vdots & \vdots \\ A_r & \cdots & (0) \\ (0) & \cdots & (0) \end{pmatrix} \begin{pmatrix} x(i-1) \\ x(i-2) \\ \vdots \\ x(i-r) \end{pmatrix},
$$
$$
i = 0, 1, 2, \cdots \tag{7}
$$

Multiply the the matrix $\mathbf{P}$ from left to (7), we obtain

$$
\mathbf{P} \begin{pmatrix} y(i) \\ y(i+1) \\ \vdots \\ y(i+r) \end{pmatrix} = \begin{pmatrix} A_{0,0} & (0) & \cdots & (0) \\ A_{1,1} & A_{1,0} & \cdots & (0) \\ \vdots & \vdots & \ddots & \vdots \\ A_{r,r} & A_{r,r-1} & \cdots & A_{r,0} \end{pmatrix} \begin{pmatrix} x(i) \\ x(i+1) \\ \vdots \\ x(i+r) \end{pmatrix} + \mathbf{P} \begin{pmatrix} A_1 & \cdots & A_r \\ \vdots & \vdots & \vdots \\ A_r & \cdots & (0) \\ (0) & \cdots & (0) \end{pmatrix} \begin{pmatrix} x(i-1) \\ x(i-2) \\ \vdots \\ x(i-r) \end{pmatrix},
$$
$$
i = 0, 1, 2, \cdots \tag{8}
$$

Denote $\mathbf{P} = \begin{pmatrix} P_{0,0} & P_{0,1} & \cdots & P_{0,r} \\ P_{1,0} & P_{1,1} & \cdots & P_{1,r} \\ \vdots & \vdots & \vdots & \vdots \\ P_{r,0} & P_{r,1} & \cdots & P_{r,r} \end{pmatrix}$ and $\mathbf{P} \begin{pmatrix} A_1 & \cdots & A_r \\ \vdots & \vdots & \vdots \\ A_r & \cdots & (0) \\ (0) & \cdots & (0) \end{pmatrix} = \begin{pmatrix} Q_{0,1} & Q_{0,2} & \cdots & Q_{0,r} \\ Q_{1,1} & Q_{1,2} & \cdots & Q_{1,r} \\ \vdots & \vdots & \vdots & \vdots \\ Q_{r,1} & Q_{r,2} & \cdots & Q_{r,r} \end{pmatrix}$,

where $P_{i,j}$ and $Q_{i,j}$ are $l \times l$ matrices. Then from the first line of (8), we have

$$
x(i) = A_{0,0}^{-1} \sum_{j=0}^{r} P_{0,j} y(i+j) + A_{0,0}^{-1} \sum_{j=1}^{r} Q_{0,j} x(i-j), \qquad i = 0, 1, 2, \cdots \tag{9}
$$

Denote $P_j = A_{0,0}^{-1}P_{0,j}$ and $Q_j = A_{0,0}^{-1}Q_{0,j}$ for $j = 0, 1, 2, \cdots, r$. We can rewrite (9) as (10),

$$x(i) = \sum_{j=0}^{r} P_j y(i+j) + \sum_{j=1}^{r} Q_j x(i-j), \qquad i = 0, 1, 2, \cdots \qquad (10)$$

For any initial state $s = < x(-1), x(-2), \cdots, x(-r) >$ of M and any $x(0), x(1), \cdots, x(n+r) \in X$, if $y(0)y(1)\cdots y(n+r) = \lambda_M(s, x(0)x(1)\cdots x(n+r))$ (i.e., (2) holds for $i = 0, 1, \cdots, n+r$). Then from (10) we can calculate $x(0), x(1), \cdots, x(n)$ one by one. Hence, (10) specifies a weak inverse with delay r of M, although (10) itself is not a definition formula.

3 FAPKC and Matrix Polynomials

In this section, we explain FAPKC in detail. As introduced in Section 1.3, the secret key of FAPKC consists of a nonlinear WIFA with delay 0, M_0 and a linear WIFA with delay τ, M_1. The definition formula of M_0 is

$$M_0: \qquad y(i) = \sum_{j=0}^{r} B_j x(i-j) + \sum_{j=1}^{r-1} \bar{B}_j x(i-j) \cdot x(i-j-1), \qquad i = 0, 1, 2, \cdots \quad (11)$$

where $B_0, B_1, B_2, \cdots, B_r, \bar{B}_1, \bar{B}_2, \cdots, \bar{B}_{r-1}$ are $l \times l$ matrices over $GF(2)$, and B_0 is an invertible matrix. The operation $x \cdot y$ is defined to be $(a_1b_1, a_2b_2, \cdots, a_lb_l)^T$ for $x = (a_1, a_2, \cdots, a_l)^T$ and $y = (b_1, b_2, \cdots, b_n)^T$. (It does not matter what the operation '·' is. Actually, we only need '·' to be a nonlinear operation from two vectors to one vector. The operation '·' defined above is adopted in the practical FAPKC.) M_0 is an r-input memory FA. It is easy to see that the r-output memory M_0^{-1} is a weak inverse with delay 0 of M_0.

$$M_0^{-1}: x(i) = B_0^{-1}\left(y(i) + \sum_{j=1}^{r} B_j x(i-j) + \sum_{j=1}^{r-1} \bar{B}_j x(i-j) \cdot x(i-j-1)\right)$$
$$(12)$$

For any initial state $s_0 = < x(-1), x(-2), \cdots, x(-r) >$ of M_0, its match state in M_0^{-1} is also $< x(-1), x(-2), \cdots, x(-r) >$.

The definition formula of M_1 is

$$M_1: \qquad z(i) = A_0 y(i) + A_1 y(i-1) + \cdots + A_\tau y(i-\tau) \qquad i = 0, 1, 2, \cdots \quad (13)$$

The encrypt automaton is the composition of M_0 and M_1, denoted by $M = M_0 \circ M_1$. The definition formula of M is obtained by substituting (11) into (13).

$$M: z(i) = \sum_{t=0}^{\tau} A_t \left(\sum_{j=0}^{r} B_j x(i-j-t) + \sum_{j=1}^{r-1} \bar{B}_j x(i-j-t) \cdot x(i-j-t-1)\right)$$
$$(14)$$

Actually, $M = M_0 \circ M_1$ is not exactly the same as $M_0 \cdot M_1$, but it is equivalent to a sub-automaton of $M_0 \cdot M_1$. That is, for any state $s = <x(-1), x(-2), \cdots, x(-r - \tau)>$ of $M = M_0 \circ M_1$, s is equivalent to the state $<s_0, s_1>$ of $M_0 \cdot M_1$, where $s_0 = <x(-1), x(-2), \cdots, x(-r)>$ and $s_1 = <y(-1), y(-2), \cdots, y(-\tau)>$, and

$$y(-h) = \sum_{j=0}^{r} B_j x(-h-j) + \sum_{j=1}^{r-1} \bar{B}_j x(-h-j) \cdot x(-h-j-1), h = 1, 2, \cdots, \tau \quad (15)$$

Now let us look back at (14). It can be simplified as (16).

$$M: \quad z(i) = \sum_{j=0}^{r+\tau} C_j x(i-j) + \sum_{j=1}^{r+\tau-1} \bar{C}_j x(i-j) \cdot x(i-j-1), \quad i = 0, 1, 2, \cdots \quad (16)$$

$$C_j = \sum_{h+t=j} A_h B_t, j = 0, 1, \cdots, r+\tau; \bar{C}_j = \sum_{h+t=j} A_h \bar{B}_t, j = 1, 2, \cdots, r+\tau-1 \quad (17)$$

Hence, FAPKC works in the following way:

1. Choose M_0 and M_1 as defined above. All the A_j, B_j and $\bar{B}_j$ are kept as secret key. (There are some criterions about the way how we choose A_j, B_j and $\bar{B}_j$ [2] [12] [24].)
2. Calculate C_j and $\bar{C}_j$ from A_j, B_j and $\bar{B}_j$ as defined by (17), and arbitrarily choose $s = <x(-1), x(-2), \cdots, x(-r-\tau)>$ as the initial state. Then C_j, $\bar{C}_j$ and s are made public.
3. From $s = <x(-1), x(-2), \cdots, x(-r-\tau)>$, get s_0 and s_1 as defined in (15).
4. To encrypt plaintext $x(0)x(1)\cdots x(m)$, first arbitrarily choose $x(m+1)x(m+2)\cdots x(m+\tau) \in X^\tau$. Then, input $x(0)x(1)x(2)\cdots x(m+\tau)$ into $M = M_0 \circ M_1$ with initial state s. The output $z(0)z(1)z(2)\cdots z(m+\tau)$ is the ciphertext.
5. To decrypt $z(0)z(1)z(2)\cdots z(m+\tau)$, first use M_1^{-1} (as defined in Section 2) and s_1 to obtain $y(0)y(1)\cdots y(m)$. Then supply $y(0)y(1)\cdots y(m)$ into M_0^{-1} with initial state s_0, and obtain $x(0)x(1)\cdots x(m)$ as the output.

One of possible attacks to FAPKC is to reduce (16) to simultaneous quadratic equations over $GF(2)$, and solve them. However, it is known that solving nonlinear equations over $GF(2)$ is very difficult if the number of its arguments is large. Another way is the ciphertext attack through randomized searching. FAPKC can resist such a kind of attacks by taking $\tau > 15$ and choosing uniformly *increasing ranks* for M_1 [2].

Now we consider the possibility of separating M_0 and M_1 from $M = M_0 \circ M_1$. We denote the $l \times l$ matrix polynomial ring over $GF(2)$ by $\mathbf{MP}(x)$. More precisely, $\mathbf{MP}(x)$ is the set of all the polynomials like $\mathbf{G}(x) = G_0 + G_1 x + \cdots + G_n x^n$, where $G_0, G_1, \cdots, G_n$ are $l \times l$ matrices over $GF(2)$ and x is the formal variable.

Let $\mathbf{A}(x) = \sum_{j=0}^{\tau} A_j x^j$, $\mathbf{B}(x) = \sum_{j=0}^{r} B_j x^j$, $\bar{\mathbf{B}}(x) = \sum_{j=1}^{r-1} \bar{B}_j x^{j-1}$, $\mathbf{C}(x) = \sum_{j=0}^{r+\tau} C_j x^j$ and $\bar{\mathbf{C}}(x) = \sum_{j=1}^{r+\tau-1} \bar{C}_j x^{j-1}$. Then from (17), we have $\mathbf{C}(x) = \mathbf{A}(x)\mathbf{B}(x)$ and $\bar{\mathbf{C}}(x) = \mathbf{A}(x)\bar{\mathbf{B}}(x)$.

Problem 1: For $\mathbf{C}(x)$ and $\bar{\mathbf{C}}(x)$ given above, find $\mathbf{A}'(x) = \sum_{j=0}^{\tau} A_j' x^j$, $\mathbf{B}'(x) = \sum_{j=0}^{r} B_j' x^j$ (where B_0' is invertible) and $\bar{\mathbf{B}}'(x) = \sum_{j=1}^{r-1} \bar{B}_j' x^{j-1}$ in $\mathbf{MP}(x)$, such that $\mathbf{C}(x) = \mathbf{A}'(x)\mathbf{B}'(x)$ and $\bar{\mathbf{C}}(x) = \mathbf{A}'(x)\bar{\mathbf{B}}'(x)$.

If we could solve Problem 1 by a fast algorithm, then we could break FAPKC. This is because in such a case we could obtain a nonlinear WIFA with delay 0, M_0' from $B_0', B_1', \cdots, B_r', \bar{B}_1, \bar{B}_2, \cdots, \bar{B}_{r-1}$, and a linear WIFA with delay τ, M_1' from $A_0', A_1', \cdots, A_\tau'$, while $M_0' \circ M_1' = M_0 \circ M_1$. Hence, we could break FAPKC by constructing $M_0'^{-1}$ and $M_1'^{-1}$. However, it is very difficult to solve Problem 1. Since $\mathbf{MP}(x)$ is noncommunitive and contains divisors of zero, establishing divisibility theory for $\mathbf{MP}(x)$ seems to be difficult [10]. Although polynomial time algorithms for factorization of polynomials over a finite field exist [4], any feasible algorithm for factoring matrix polynomials is not yet known.

In the next section, we prove a simple but previously unknown property for the input memory WIFAs. By this property we can reduce the secure base of FAPKC to Problem 2, which is much easier than Problem 1.

Problem 2: For $\mathbf{C}(x)$ and $\bar{\mathbf{C}}(x)$ given above, find $\mathbf{A}'(x) = \sum_{j=0}^{\tau} A_j' x^j$, $\mathbf{B}'(x) = \sum_{j=0}^{r} B_j' x^j$ (where B_0' is invertible) and $\bar{\mathbf{B}}'(x) = \sum_{j=1}^{r-1} \bar{B}_j' x^{j-1}$ in $\mathbf{MP}(x)$, such that $\mathbf{C}(x) = \mathbf{A}'(x)\mathbf{B}'(x) \pmod{x^{\tau+1}}$ and $\bar{\mathbf{C}}(x) = \mathbf{A}'(x)\bar{\mathbf{B}}'(x) \pmod{x^{\tau-1}}$.

4 A Property of WIFAs

In this section, we prove that only those terms related with $x(i), x(i-1), \cdots, x(i-\tau)$ determine whether an FA is a WIFA with delay τ. Let M_f and M_{f+g} be a pair of $(r+\tau)$-input memory FAs. Their definition formulae are

$$M_f: \quad y(i) = f(x(i), x(i-1), \cdots, x(i-r-\tau)), \quad i = 0, 1, 2, \cdots \quad (18)$$

$$M_{f+g}: y(i) = f(x(i), \cdots, x(i-r-\tau)) + g(x(i-\tau-1), \cdots, x(i-r-\tau)), i = 0, 1, 2, \cdots \quad (19)$$

Theorem 10. *The finite automaton M_f is a WIFA with delay τ if and only if M_{f+g} is a WIFA with delay τ.*

Proof. ($\Longrightarrow$) The proof is simply from the definition of WIFA with delay τ. For any state $s = \langle x(-1), x(-2), \cdots, x(-r-\tau) \rangle$ of M_f and any $x(0), x(1), \ldots, x(\tau)$, $x(0)', x(1)', \ldots, x(\tau)' \in X$, $\lambda_{M_f}(s, x(0)x(1)\cdots x(\tau)) = \lambda_{M_f}(s, x(0)'x(1)'\cdots x(\tau)')$ implies $x(0) = x(0)'$, i.e.,

$$\begin{cases} f(x(0), x(-1), x(-2), \cdots, x(-r-\tau)) = f(x(0)', x(-1), x(-2), \cdots, x(-r-\tau)) \\ f(x(1), x(0), x(-1), \cdots, x(1-r-\tau)) = f(x(1)'x(0)', x(-1), \cdots, x(1-r-\tau)) \\ \qquad\qquad\vdots \\ f(x(\tau), \cdots, x(0), x(-1), \cdots, x(-r)) = f(x(\tau)', \cdots, x(0)', x(-1), \cdots, x(-r)) \end{cases}$$

implies $x(0) = x(0)'$. Adding $g(x(-\tau-1), \cdots, x(-\tau-r))$, $g(x(-\tau), \cdots, x(1-\tau-r))$, $\cdots$, $g(x(-1), x(-2), \cdots, x(-r))$ to both sides of equations above respectively, $\lambda_{M_{f+g}}(s, x(0)x(1)\cdots x(\tau)) = \lambda_{M_{f+g}}(s, x(0)'x(1)'\cdots x(\tau)')$ implies $x(0) = x(0)'$.

From the arbitrariness of s, $x(0), x(1), \ldots, x(\tau), x(0)', x(1)', \ldots,$ and $x(\tau)'$, M_{f+G} is a WIFA with delay τ. ($\Longleftarrow$) The opposit direction can be proved similarly. $\square$

Next we show that we can construct a weak inverse with delay τ of M_{f+g} from a weak inverse with delay τ of M_f.

Theorem 11. *If M_f^{-1} is a weak inverse with delay τ of M_f, then a weak inverse with delay τ of M_{f+g} can be constructed from M_f^{-1}.*

Proof. Let $s = < x(-1), x(-2), \cdots, x(-r - \tau) >$ be an initial state of M_f and M_{f+g}, and let s^{-1} be the match state of s of M_f^{-1}. We replace the first τ output of M_f^{-1} by $x(-\tau), x(1 - \tau), \cdots, x(-1)$. The principle of constructing M_{f+g}^{-1} is given in Fig. 2. $\square$

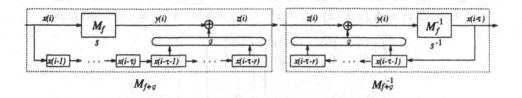

Fig. 2. M_{f+g} and its weak inverse with delay τ M_{f+g}^{-1}.

5 Break FAPKC

Let M be the encryption WIFA with delay τ in FAPKC, as defined by (16). Then $M = M_0 \circ M_1$, where M_0 and M_1 are defined by (11) and (13), respectively.

Theorem 12. *If Problem 2 can be solved by a fast algorithm, then we can break FAPKC.*

Proof. Suppose that we can find $\mathbf{A}'(x) = \sum_{j=0}^{\tau} A'_j x^j$, $\mathbf{B}'(x) = \sum_{j=0}^{r} B'_j x^j$ (where B'_0 is invertible) and $\bar{\mathbf{B}}'(x) = \sum_{j=1}^{r-1} \bar{B}'_j x^{j-1}$ in $\mathbf{MP}(x)$, such that $\mathbf{C}(x) = \mathbf{A}'(x)\mathbf{B}'(x) \pmod{x^{\tau+1}}$ and $\bar{\mathbf{C}}(x) = \mathbf{A}'(x)\bar{\mathbf{B}}'(x) \pmod{x^{\tau-1}}$. We construct a nonlinear WIFA with delay 0, M_0' from $B'_0, B'_1, \cdots, B'_r, \bar{B}_1, \bar{B}_2, \cdots, \bar{B}_{r-1}$, and a linear WIFA with delay τ, M_1' from $A'_0, A'_1, \cdots, A'_\tau$. ($M_1'$ is a linear WIFA with delay τ from Proposition 9 and Theorem 10.)

It is easy to construct $M_0'^{-1}$ and $M_1'^{-1}$ from $B'_0, B'_1, \cdots, B'_r, \bar{B}_1, \bar{B}_2, \cdots, \bar{B}_{r-1}$ and $A'_0, A'_1, \cdots, A'_\tau$ respectively. Hence, we can construct a weak inverse with delay τ of $M' = M'_0 \circ M'_1$, denoted by M'^{-1}.

$$M' : \qquad z(i) = \sum_{j=0}^{r+\tau} C'_j x(i-j) + \sum_{j=1}^{r+\tau-1} \bar{C}'_j x(i-j) \cdot x(i-j-1), \qquad i = 0, 1, 2, \cdots$$

Let $\mathbf{C}'(x) = \mathbf{A}'(x)\mathbf{B}'(x) = \sum_{j=0}^{r+\tau} C'_j x^j$ and $\bar{\mathbf{C}}'(x) = \mathbf{A}'(x)\bar{\mathbf{B}}'(x) = \sum_{j=0}^{r+\tau} \bar{C}'_j x^j$. Since $\mathbf{C}(x) = \mathbf{C}'(x) \pmod{x^{\tau+1}}$ and $\bar{\mathbf{C}}(x) = \bar{\mathbf{C}}'(x) \pmod{x^{\tau-1}}$, from Theorem 11, we can construct the weak inverse with delay τ of M, M^{-1} from M'^{-1}. $\square$

Now we illustrate that Problem 2 can be easily solved. Let $\mathbf{A}(x) = \sum_{j=0}^{r} A_j x^j$, $\mathbf{B}(x) = \sum_{j=0}^{r} B_j x^j$, $\bar{\mathbf{B}}(x) = \sum_{j=1}^{r-1} \bar{B}_j x^{j-1}$, $\mathbf{C}(x) = \sum_{j=0}^{r+\tau} C_j x^j$, $\bar{\mathbf{C}}(x) = \sum_{j=1}^{r+\tau-1} \bar{C}_j x^{j-1}$, $\mathbf{C}(x) = \mathbf{A}(x)\mathbf{B}(x)$ and $\bar{\mathbf{C}}(x) = \mathbf{A}(x)\bar{\mathbf{B}}(x)$, as defined in Section 3. Our aim is to find $\mathbf{A}'(x) = \sum_{j=0}^{r} A'_j x^j$, $\mathbf{B}'(x) = \sum_{j=0}^{r} B'_j x^j$ and $\bar{\mathbf{B}}'(x) = \sum_{j=1}^{r-1} \bar{B}'_j x^{j-1}$ such that $\mathbf{C}(x) = \mathbf{A}'(x)\mathbf{B}'(x) \pmod{x^{\tau+1}}$, $\bar{\mathbf{C}}(x) = \mathbf{A}'(x)\bar{\mathbf{B}}(x)$ $\pmod{x^{\tau-1}}$, and B'_0 is invertible.

Let $\mathbf{B}'(x) = I$ (the $l \times l$ unit matrix), $\mathbf{A}'(x) = \sum_{j=0}^{r} C_j x^j$. We can find $\bar{\mathbf{B}}'(x)$ such that $\bar{\mathbf{C}}(x) = \mathbf{A}'(x)\bar{\mathbf{B}}(x) \pmod{x^{\tau-1}}$, by solving a linear equation group

$$\begin{pmatrix} C_0 & (0) & \cdots & (0) \\ C_1 & C_0 & \cdots & (0) \\ \vdots & \vdots & \ddots & \vdots \\ C_{\tau-2} & C_{\tau-3} & \cdots & C_0 \end{pmatrix} \begin{pmatrix} \bar{B}'_1 \\ \bar{B}'_2 \\ \vdots \\ \bar{B}'_{\tau-1} \end{pmatrix} = \begin{pmatrix} \bar{C}_1 \\ \bar{C}_2 \\ \vdots \\ \bar{C}_{\tau-1} \end{pmatrix} \qquad (20)$$

The linear equation group (20) definitely has solutions. Actually, one simple solution is

$$\begin{pmatrix} \bar{B}'_1 \\ \bar{B}'_2 \\ \vdots \\ \bar{B}'_{\tau-1} \end{pmatrix} = \begin{pmatrix} B_0 & (0) & \cdots & (0) \\ B_1 & B_0 & \cdots & (0) \\ \vdots & \vdots & \ddots & \vdots \\ B_{\tau-2} & B_{\tau-3} & \cdots & B_0 \end{pmatrix}^{-1} \begin{pmatrix} \bar{B}_1 \\ \bar{B}_2 \\ \vdots \\ \bar{B}_{\tau-1} \end{pmatrix} \qquad (21)$$

However, we can only obtain $\bar{B}'_1, \bar{B}'_2, \cdots, \bar{B}'_{\tau-1}$ by solving (20) since $\mathbf{A}(x) = \sum_{j=0}^{r} A_j x^j$, $\mathbf{B}(x) = \sum_{j=0}^{r} B_j x^j$ and $\bar{\mathbf{B}}(x) = \sum_{j=1}^{r-1} \bar{B}_j x^{j-1}$ are kept secret in FAPKC.

6 A Modification

We have shown that FAPKC can be broken by constructing the decryption automata from the encryption automaton. The key point of insecure FAPKC is that M_0 is a WIFA with delay 0, i.e., B_0 is an invertible matrix. A natural modification to FAPKC is to change M_0 into a WIFA with delay τ, based on Theorem 10 and Theorem 11.

$$M_0 : \qquad y(i) = \sum_{j=0}^{\tau} B_j x(i-j) + \sum_{j=1}^{\tau-1} \bar{B}_j x(i-\tau-j) \cdot x(i-\tau-j-1), \qquad i = 0, 1, 2, \cdots$$

Here $B_0, B_1, \cdots, B_\tau$ are taken to be the linear coefficients of a linear WIFA with delay τ. Apparently, M_0 is a nonlinear WIFA with delay τ. M_1 is the same as before. Let the encryption automaton be

$$M : z(i) = \sum_{j=0}^{2\tau} C_j x(i-j) + \sum_{j=1}^{\tau} \bar{C}_j x(i-\tau-j) \cdot x(i-\tau-j-1), i = 0, 1, 2, \cdots$$

Let $\mathbf{A}(x) = \sum_{j=0}^{\tau} A_j x^j$, $\mathbf{B}(x) = \sum_{j=0}^{\tau} B_j x^j$, $\bar{\mathbf{B}}(x) = \sum_{j=1}^{\tau-1} \bar{B}_j x^{j-1}$, $\mathbf{C}(x) = \sum_{j=0}^{2\tau} C_j x^j$ and $\bar{\mathbf{C}}(x) = \sum_{j=1}^{\tau} \bar{C}_j x^{j-1}$. Then $\mathbf{C}(x) = \mathbf{A}(x)\mathbf{B}(x)$ and $\bar{\mathbf{C}}(x) = (\mathbf{A}(x)\bar{\mathbf{B}}(x) \bmod x^\tau)$. This is also a variation of the version in [23].

This modified FAPKC can resist the attack described in this paper. To resist the attack in which $\mathbf{A}'(x)$ is taken to be $\mathbf{C}(x)$, $\mathbf{B}'(x)$ is taken to be I and $\bar{\mathbf{B}}'(x)$ is calculated by solving the linear equation group, we should choose $\mathbf{A}(x)$, $\mathbf{B}(x)$ and $\bar{\mathbf{B}}(x)$ carefully so that they satisfy some tradeoff. However, further analyses are needed before we can give any definite conclusion on its security.

References

1. F. Bao, "Limited error-propagation, self-synchronization and finite input memory FSMs as weak inverses", in Advances in Chinese Computer Science, Vol. 3, World Scientific, Singapore, 1991, pp. 1-24.
2. F. Bao, Y. Igarashi, "A randomized algorithm to finite automata public key cryptosystem", in Proc. of ISAAC'94, LNCS 834, Springer-Verlag, 1994, pp. 678-686.
3. F. Bao, Y. Igarashi, X. Yu, "Some results on the decomposability of weakly invertible finite automata", IPSJ Technical Report, 94-AL, November, 1994, pp. 17-24.
4. E. Berlekamp, "Factoring polynomials over large finite field", Math. Comp. Vol. 24, 1970, pp. 713-735.
5. S. Chen, "On the structure of finite automata of which M' is an (weak) inverse with delay τ", J. of Computer Science and Technology, Vol. 1, No. 2, 1986, pp. 54-59.
6. S. Chen, "On the structures of (weak) inverses of an (weakly) invertible finite automata", J. of Computer Science and Technology, Vol. 1, No. 3, 1986, pp. 92-100.
7. S. Chen, R. Tao, "The structure of weak inverses of a finite automaton with bounded error propagation", Kexue Tongbao, Vol. 32, No. 10, 1987, pp. 713-714.
8. S. Even, "Generalized automata and their information losslessness", in Switching Circuit Theory and Logic Design, 1962, pp. 144-147.
9. S. Even, "On Information lossless automata of finite order", IEEE Trans. on Electric Computer, Vol. 14, No. 4, 1965, pp. 561-569.
10. I. Gohberg, P. Lancaster, L. Rodman, Matrix Polynomials, Academic Press, New York.
11. D. A. Huffman, "Canonical forms for information-lossless finite-state logic machines", IRE Trans. on Circuit Theory, Vol. CT-6, Special Supplements, May, 1959, pp. 41-59.
12. J. Li, X. Gao, "Realization of finite automata public key cryptosystem and digital signature", in Proc. of the Second National Conference on Cryptography, CRYPTO-CHINA'92, pp. 110-115. (in Chinese)
13. J. L. Massey, M. K. Sain, "Inverse of linear sequential circuits", IEEE Trans. on Computers, Vol. 17, No. 4, 1968, pp. 330-337.

14. J. L. Massey, A. Gubser, A. Fisger, et al., "A selfsynchronizing digital scrambler for cryptographic protection of data", in Proc. of International Zurich Seminar on Digital Communications, March, 1984.

15. R. R. Olson, "A note on feedforward inverses for linear sequential circuits", IEEE Trans. on Computers, Vol. 19, No. 12, 1970, pp. 1216-1221.

16. A. Salomaa, Public-Key Cryptography, EATCS Monographs on Theoretical Computer Science, Vol. 23, Springer-Verlag, 1990.

17. R. Tao, Invertibility of Finite Automata, Science Press, 1979, Beijing. (in Chinese)

18. R. Tao, "On the relation between bounded error propagation and feedforward inverse", Kexue Tongbao, Vol. 27, No. 7, 1982, pp. 406-408. (in Chinese)

19. R. Tao, "On the structure of feedforward inverse", Science in China, A, Vol. 26, No. 12, 1983, pp. 1073-1078. (in Chinese)

20. R. Tao, S. Chen, "Finite automata public key cryptosystem and digital signature", Computer Acta, Vol. 8, No. 6, 1985, pp. 401-409. (in Chinese)

21. R. Tao, S. Chen, "Two varieties of finite automata public key cryptosystem and digital signature", J. of Computer Science and Technology, Vol. 1, No. 1, pp. 9-18.

22. R. Tao, "Invertibility of linear finite automata over a ring", in Proc. of ICALP'88, LNCS 317, Springer-Verlag, 1988, pp. 489-501.

23. R. Tao, S. Chen, "An implementation of identity-based cryptosystems and signature schemes by finite automaton public key cryptosystems", in Proc. of the Second National Conference on Cryptography, CRYPTO-CHINA'92, pp. 87-104. (in Chinese)

24. H. Zhang, Z. Qin, et al., "The software implementation of FA public key cryptosystem", in Proc. of the Second National Conference on Cryptography, CRYPTO-CHINA'92, pp. 105-109. (in Chinese)

25. X. Zhu, "On the structure of binary feedforward inverses with delay 2", J. of Computer Science and Technology, Vol. 4, No. 2, 1989, pp. 163-171.

Appendix

The algorithm for finding $\mathbf{P}$ was originally described in [17] as "Ra-Rb algorithm". $\mathbf{P}$ actually is a sequence of linear row transformations. We use the following example with $r = 2$ to show how the algorithm works. First operatea row transformation so that A_0 is changed to a matrix B whose first several rows are linear indepedent and the rest rows are all zero.

$$
\begin{pmatrix} A_0 \ (0) \ (0) \\ A_1 \ A_0 \ (0) \\ A_2 \ A_1 \ A_0 \end{pmatrix} \Rightarrow \begin{pmatrix} B \ (0) \ (0) \\ C \ B \ (0) \\ D \ C \ B \end{pmatrix} = \begin{pmatrix} \begin{pmatrix} upper B \\ all zero \end{pmatrix} & \begin{pmatrix} all zero \\ all zero \end{pmatrix} & \begin{pmatrix} all zero \\ all zero \end{pmatrix} \\ \begin{pmatrix} upper C \\ lower C \end{pmatrix} & \begin{pmatrix} upper B \\ all zero \end{pmatrix} & \begin{pmatrix} all zero \\ all zero \end{pmatrix} \\ \begin{pmatrix} upper D \\ lower D \end{pmatrix} & \begin{pmatrix} upper C \\ lower C \end{pmatrix} & \begin{pmatrix} upper B \\ all zero \end{pmatrix} \end{pmatrix} \begin{matrix} (1) \\ \\ (2) \\ \\ (3) \end{matrix}
$$

Then we operate the following row transformations to the above matrix: shift (3) to (2), (2) to (1) and (1) to (3). Now we get a new matrix $\begin{pmatrix} A_0' \ (0) \ (0) \\ A_1' \ A_0' \ (0) \\ A_{2,2} \ A_{2,1} \ A_{2,0} \end{pmatrix}$, where $A_{2,2}$, $A_{2,1}$ and $A_{2,0}$ satisfy (6) in Section 2. Next, we continue applying the same procedure to the matrix $\begin{pmatrix} A_0' \ (0) \ (0) \\ A_1' \ A_0' \ (0) \end{pmatrix}$. Then we obtain the final matrix we want.

Stochastic Graphs Have Short Memory: Fully Dynamic Connectivity in Poly-Log Expected Time*

S. Nikoletseas[1], J. Reif[2], P. Spirakis[1] and M. Yung[3]

[1] Computer Technology Institute, Patras, Greece
[2] Department of Computer Science, Duke University
[3] IBM Research Division, T. J. Watson Research Center

Abstract. This paper introduces *average case* analysis of *fully dynamic* graph connectivity (when the operations are edge insertions and deletions). To this end we introduce the model of *stochastic graph processes*, i.e. dynamically changing random graphs with random equiprobable edge insertions and deletions, which generalizes Erdös and Renyi's 35 year-old random graph process. As the stochastic graph process continues indefinitely, all potential edge locations (in $V \times V$) may be repeatedly inspected (and learned) by the algorithm. This learning of the structure seems to imply that traditional random graph analysis methods cannot be employed (since an observed edge is not a random event anymore). However, we show that a small (logarithmic) number of dynamic random updates are enough to allow our algorithm to re-examine edges as if they were *random with respect to certain events* (i.e. the graph "forgets" its structure). This *short memory* property of the stochastic graph process enables us to present an algorithm for graph connectivity which admits an *amortized expected* cost of $O(\log^3 n)$ time per update. In contrast, the best known deterministic worst-case algorithms for fully dynamic connectivity require $n^{1/2}$ time per update.

1 Introduction

1.1 Dynamic graph problems and previous work

In dynamic graph problems one would like to answer queries on a graph that is undergoing a sequence of updates. A problem is called *fully dynamic* when the updates include both insertions and deletions of edges; this case is often more complex than the *partially dynamic* (where only one type of update is allowed, e.g. only insertions). The goal of a dynamic graph algorithm is to update the solution after a change (i.e. *on-line*), doing so more efficiently than recomputing

* This research was partially supported by the EEC ESPRIT Basic Research Projects ALCOM II and GEPPCOM. Also supported by DARPA/ISTO Contracts N00014-88-K-0458, DARPA N00014-91-J-1985, N00014-91-C-0114, NASA subcontract 550-63 of prime contract NAS5-30428, US-Israel Binational NSF Grant 88-00282/2, and NSF Grant NSF-IRI-91-00681.

it at that point from scratch. The adaptivity requirements usually make dynamic algorithms and dynamic data structures more difficult to design and analyze than their static counterparts.

Graph connectivity is one of the most basic problems with numerous applications and various algorithms in different settings. The best known result for worst-case fully-dynamic connectivity was (for quite a number of years) the very basic algorithm due to Frederikson ([9]) which takes $O(\sqrt{|E|})$ update time and which initiated a clustering technique. Very recently this was improved by a novel sparsification technique to $O(\sqrt{n} \log(|E|/n))$ by [7].

We remark that all the previous results ([7], [9], [10], [13] and [14]) on efficient fully-dynamic structures for general graphs were based on clustering techniques. This has led to solutions of an inherent time bound of $O(n^{\epsilon})$, for some $\epsilon < 1$, since the key problem encountered by these techniques is that the algorithm must somehow balance: (i) the work investing in maintaining the component of the cluster structure, and (ii) the work on the cluster structure (connecting the components).

In this work we initiate the study of *average case* analysis of fully dynamic algorithms and techniques to achieve much more efficient (i.e., poly-log per update) expected amortized time complexity. Designing algorithms with good average case behavior is directed towards better solving the typical case and better capturing (via probabilistic methods) its complexity– rather than analyzing the worst possible case. Basic good average case graph algorithms were presented in various settings: sequential, parallel, distributed, NP-hard and so on (e.g. [2], [5], [17], [18], [19], [21], [22], [28]). For these investigations, various random graph models have been employed. We are not aware of any average case investigation concerning fully-dynamic graph theoretic problems which has taken place prior to our effort.

The investigation of the average case of fully dynamic graphs suggests random graph updates. We would like to perform any sequence of three kinds of operations:

1. *Property-Query* (parameter): Returns *true* iff the property holds (or returns a sub-graph as a witness to the property). For a connectivity query (u, v), a "true" answer means that the vertices u, v are in the same connected component.
2. *Insert* (x,y): Inserts a new edge joining x to y (assuming $\{x, y\} \notin E$).
3. *Delete* (x,y): Deletes the edge $\{x, y\}$ (assuming $\{x, y\} \in E$).

In this we assume random updates (insert and delete operations) have equal probability, $1/2$. The edge to be deleted (inserted) is randomly chosen in E (E^c, i.e., the set of edges not in E).

1.2 Our results and related recent work

We consider the above randomly changing graphs, which thus give rise to a new probabilistic process on graphs which we call a *stochastic graph process* (which may be of independent interest). We then use this model to analyze our algorithm for connectivity. The time complexity per update is $O(\log^3 n)$, both with high probability and on the average. Our complexity measure is amortized time (amortization over a long enough finite sequence of operations).

We show that the stochastic graph processes have an important property (which we call *short memory*) which allows our analysis to overcome various dependencies.

Recent work. The technical report which presented a preliminary version of this work has already motivated further work on average case dynamic graph algorithms. In particular, D. Alberts and M. Rauch ([1]) presented a nice modification of our model. In our model the adversary chooses a random operation and a random edge location each time (i.e., $\log n + 1$ random bits), while in theirs, the operation is determined arbitrarily whereas the edge location ($\log n$ bits) is still random. They do not attempt to show polylogarithmic time bounds (only sublinear), which makes their analysis (which can not rely on properties of random graphs) much simpler than ours.

Also recently, a novel coin-flipping algorithm for dynamic graph problems that performs in expected poly-logarithmic (Las-Vegas Algorithm) was given in [15]. The models, and thus the results, of the above works are incomparable with the results reported here.

2 Definition of Stochastic Graph Processes

In this section we define a process on the set of vertices of a graph.

Definition 1. A stochastic graph process (sgp) on $V = \{1, 2, \ldots, n\}$ is a Markov Chain $G^* = \{G_t\}_0^\infty$ whose states are graphs on V.

Definition 2. A stochastic graph process on $V = \{1, 2, \ldots, n\}$ is called fair (fsgp) if

1. G_0 is the empty graph.
2. There is a $t_1 > 0$ such that $\forall t \leq t_1$, G_t is obtained by G_{t-1} by an addition of an edge uniformly at random among all edges in E_{t-1}^c. (Up to t_1 an edge is added at each $t \leq t_1$).
3. $\forall t \geq t_1$, G_t is obtained from G_{t-1} by either the addition of one edge, which happens with probability $1/2$ (all the new edges being equiprobable), or by the deletion of one existing edge, which happens with probability $1/2$ (and all existing edges are equiprobable to be selected).

Remark: The stochastic graph process which includes only steps (1) and (2) above and $t_1 \leq \binom{n}{2}$ was used by Erdös and Renyi ([8]) to define $G_{n,M}$, which was called the *random graph process* as the number of inserted edges M progresses; this was used in analyzing various graph theoretic and algorithmic issues in the last 35 years.

Lemma 3. *Let G^* be an fsgp on $V = \{1, 2, \ldots, n\}$. Let $G_t = (V, E_t)$ be the state of the process at time t, then G_t is also a random graph from $G_{n,M}$ (for some M).*

Proof. See [23]. □

The fsgp will be used as our model. The "steps" denoted by the variable t (time) are those which change the graph, namely the insertions and deletions.

We may assume (w.l.o.g.) that we start from a random G_{t_1}, which is approximately a graph in $G_{n,p}$ of $p = c/n$ (for appropriate large enough c) where $t_1 = p \binom{n}{2}$, (see [3] for results on the close approximation between the random graph spaces with respect to various properties). The initial stage (of building by insertions only) can be easily handled (e.g., by a union-find algorithm).

One may investigate the behavior of stochastic graphs when the edge set cardinality is bounded from above (or possibly from below as well):

Definition 4. 1. An fsgp is called truncated from below by L (denoted $tb - fsgp(L)$) if the edge deletion probability becomes zero (rather than $1/2$) for each $t > t_1$ such that $|E_t| \leq L$ (where $0 \leq t_1 \leq L \leq \binom{n}{2}$ is called the minimum number of edges).
  2. An fsgp is called truncated from above by U (denoted $ta - fsgp(U)$) if the edge insertion probability becomes zero (rather than $1/2$) for each $t > t_1$ such that $|E_t| \leq U$ (where $0 \leq t_1 \leq U \leq \binom{n}{2}$ is called the maximum number of edges).
  3. An fsgp is called bounded by L, U (denoted $b - fsgp(L, U)$) if it is both $ta - fsgp(U)$ and $tb - fsgp(L)$ for $L \leq U$.

In general, our analysis raises the question of "holding time" for a property of a stochastic graph, which we analyze for our connectivity properties. This is somewhat analogous to "threshold properties" of the random graph process.

3 General description of the algorithm

3.1 The Data Structures and the Edge Re-Activation Technique

We assume the input to the algorithm to be a stochastic graph process of the form b-fsgp(L,U) with $L = cn$ (for an appropriate constant c greater than the giant component threshold for random graphs) and $U = \binom{n}{2}$. This is w.l.o.g. as we can also handle sparser graphs (applying, a deterministic maintenance algorithm to this range). In fact, the adaptation of the graph algorithm and data structures in use to the global condition (namely, the number of current edges) defines various graph epochs and is a first technique which we employ. For purpose of the analysis we will first assume graphs with $U = \lambda n \log n$ (where λ

is below the connectivity threshold constant for random graphs). Denser graphs results easily follow.

Our algorithm maintains a forest of spanning trees, one per connected component of the graph. The algorithm guarantees that the tree $T(\Gamma)$ of the graph's giant component Γ, has diameter bounded above by $\Delta = \lambda_1 \log^2 n$ most of the time, with high probability ($\lambda_1 > 1$ is an appropriate constant). The trees are *rooted* and the roots are suitably maintained. The tree is directed towards the root; each tree node points to a neighbor in the direction of the root. The pointer directions are maintained within the stated time bounds. (Where the context is clear, we simply denote $T(\Gamma)$ by T).

In addition to the forest of rooted trees, we also maintain the graph adjacency matrix and each edge in the matrix may have a status label which is one of (1) "tree" edge (2) "retired" edge or (3) "pool" edge.

Active edges are pool or tree edges. The pool can be imagined to be a set of unused random edges, useful for our dynamic maintenance operations. The pool is being exhausted by edges selected to be used in insert/delete operations. We look at many edges at one operation and discard them while at most we add one edge to the pool. Thus, we may soon exhaust it and have no "pool", as we have looked at the entire graph and cannot use the fact that it is random. This seems like an inherent problem. However, we develop a technique which allows us to refill the pool dynamically with edges that could be seen as "random" for later operations, in two ways: (1) from superfluous insertions and (2) from the set of retired edges, because (as we shall show later) retired edges are *reactivated* after a certain number of random insertions and deletions counted from the time they became retired.

Edge reactivation allows re-use of seen edges efficiently, because their past use conditions only a certain number of future operations and then this bias is "forgotten" by the stochastic process, due to the effect of the random updates in the time passed.

We keep for each vertex its component name (e.g. the root of the corresponding tree). This component name allows connectivity queries to be answered in $O(1)$ time with certainty. For each tree we also keep its size (in number of vertices).

Our algorithm's main goal is to maintain the correct labeling of the components.

3.2 The Dynamic Algorithm

High level description: Graph Activation and Graph Retirement Periods. The algorithm is initialized by running the linear expected time connected components algorithm of [18] by using a fraction of at least cn edges of the graph (for our analysis we use a larger c than the constant of that algorithm). This construction guarantees the creation of a random spanning tree in each component. With probability at least $1 - n^{-\alpha_1}$ (where $\alpha_1 > 2$ a constant) the algorithm constructs a giant component (call it Γ) and many small ones (the giant component has at least ϵn vertices and the sizes of the other components are at most $O(\log n))$(see e.g. [3]).

With small probability ($\leq n^{-\alpha_1}$) the construction fails to produce a giant component. Then, we enter a *graph retirement epoch*. The giant component tree construction is called a **total reconstruction**. A *successful* total reconstruction produces a giant component and logarithmic ones, with a $\log n$-diameter tree spanning the giant component. The algorithm then enters a *graph activation epoch*.

The graph process proceeds forever through a sequence of graph epochs which alternate between graph activation and graph retirement epochs.

A graph activation epoch is entered by a *successful* total reconstruction. It lasts at most A operations (A is equal to $\lambda_1 n^2 \log n$ where $\lambda_1 > 1$ is a constant); after A operations a reconstruction is attempted. The epoch may (with very small probability) end before all A operations are done. This can happen only when a deletion operation disconnects the giant component tree and the attempted fast reconnection of the tree (by the excess existing edges) fails.

A graph retirement epoch starts when the (previous) graph activation epoch ends. It lasts at least R operations (R is equal to $\lambda_2 n \log^2 n$, where λ_2 is a constant > 1). At the end of a graph retirement's R operations, a total reconstruction is attempted. As we show, it will succeed with high probability. In such a case, a new graph activation epoch is entered. However, with small probability, the total reconstruction may fail. Then, the graph retirement epoch continues for another set of R operations and again a total reconstruction is attempted. This may continue until a successful total reconstruction.

The formal structure of the dynamic algorithm can be found in [23].

The Dynamic Operations in a Graph Activation Epoch. In a graph activation epoch we use the data structures to process updates at $\log^3 n$-time cost per operation with high likelihood, as we will show in the analysis section.

Inductively assume during an activation period that there are $\Lambda_1, \epsilon, \alpha$ ($0 < \epsilon < 1, \Lambda_1 > 0, \alpha > 1$):
(a) The distance between the root and any node in the giant tree T cannot exceed $\Lambda_1 \log^2 n$,
(b) At least a constant fraction $\epsilon|T|$ of the giant tree nodes are at distance up to $\alpha \log n$ from the root (e.g., $\epsilon \geq 1/2$).

The induction holds at the beginning of an activation period (basis), by construction. Assume that it holds before a random edge update. We perform the insertion-deletion operations according to the following rules:

- **Insertion of a random edge:**
    Case b1: The edge joins vertices of the same tree. We just put it in the pool.
    Case b2: The new edge joins two trees. We update the root and the component name by changing the name and pointers at nodes of the smaller tree. (The time needed is linear in the size of the smaller tree).

– **Deletion of a random edge**

Case c1: The edge is either a pool or a retired edge. We just delete it from all data structures.

Case c2: The edge is a tree edge of a small tree. We sequentially update the small component (it will either reconnect by a pool edge or it splits into two small trees, in which case we relabel the smaller of the two). This also takes linear time in the tree size.

Case c3: The edge is a tree edge of the tree T of the giant component. Let the deleted edge be $e = \{u, v\}$. Let $T(u), T(v)$ be the two pieces in which the tree is split by the removal of e, such that $u \in T(u)$ and $v \in T(v)$. From each of the vertices u, v we start a procedure called neighborhood search. Each neighborhood search consists of a sequence of phases. The two neighborhood searches out of u, v are interleaved in the sense that we execute one phase of each in turn. Here we present the neighborhood search out of u: ($k_1 > 1$ is an appropriate constant).

Neighborhood Search (u):

Start a breadth-first-search (BFS) out of u until $k_1 \log n$ nodes are visited. The visit of the ith node designates the start of the ith phase. A phase may return "success" or "failure". The neighborhood search may:

1. Finish with no success, in which case the algorithm undergoes total reconstruction and the current activation epoch is ended (this may be called a "fatal event").
2. The nodes connected to u may be exhausted before the search finishes. In this case the giant component is just disconnected into a still-giant and a $O(\log n)$ piece. In this case, we just rename the midget component.
3. The search may report a number of successes (successful reconnection of the two tree pieces). We then choose the reconnection which is closer to the root and reconnect the two pieces.

Phase i (visit of node w): If w has no pool edges out of it then end this phase. Else, choose one pool edge out of w at random, out of its pool edges. Let this edge be $e' = \{w, w'\}$.

1. Check whether u points to v by its pointer towards the root R of the tree, or vice-versa. If u points to v then R belongs to $T(v)$ else R belongs to $T(u)$. Since w is in the same piece with u, we follow the path from w' to the root R (inductively this takes $O(\log^2 n)$ time). If e is not used in that path and $R \in T(v)$ then e' reconnects the two pieces. In all other cases, e' does not reconnect the two pieces. In such a case, end the phase and return "failure" after putting e' in retirement.
2. Provided that $e' = \{w, w'\}$ reconnects the two pieces, assume without loss of generality that $R \in T(v)$. We follow the path from u to the root and in $O(\log^2 n)$ time we determine the distance $d(u, R)$ of u from the root in the tree before the reconnection. Let $d(u, w)$ be the distance from u to w in $T(u)$. Check whether $d(u, R) \geq (k_1 + \alpha) \log n$ (i.e. whether the piece chopped away

is a distant one). Also check whether $d(w', R) < \alpha \log n$ (test (*)), where $d(w', R)$ is the distance of w' from R in $T(v)$.

In the analysis we show that, for each e' which reconnects the two pieces, test (*) is satisfied with probability at least $1/2$. Since we try $\Theta(\log n)$ such e', test (*) will be found to hold at least once with very high probability. Note that (*) implies that, in the reconnected new tree,

$$d_{new}(u, R) \le (k_1 + \alpha) \log n$$

The search returns "success" in the following two cases:

Case 1: $d(u, R) \le (k_1 + \alpha) \log n$ and test (*) holds (Type-1 reconnections).

Case 2: $d(u, R) > (k_1 + \alpha) \log n$ and test (*) holds (Type-2 reconnections).

Note that Type-2 reconnections never increase the giant tree's diameter. However, Type-1 reconnections may increase the giant tree's diameter by at most $k_1 \log n$. The idea here is that, when a random deletion chops off a "distant" piece from the tree then we succeed whp to reconnect it *closer* to the root (Type-2 reconnections). The reconnections of pieces that were already "close" may increase the diameter at most logarithmically, but we will show that the cumulative effect of such increments *will respect the induction hypothesis* over the whole length of the graph activation period.

(end of phase i)

(end of neighborhood search)

If the neighborhood search ends up with success then the giant tree is reconnected (by the successful e') and the graph activation epoch continues.

Edge Reactivation. Graph activation epochs are partitioned into *edge deletion intervals* of $k_3 \log n$ deletions (k_3 an appropriate constant). All edges put in retirement during an edge deletion interval are returned into the pool after a delay of another edge deletion interval. We do not reactivate edges of small components (midgets). Of course each edge under retirement is reactivated at a different time (so edge retirement periods do not correspond necessarily to edge deletion intervals) but we "bulk" edges to be reactivated in blocks for reasons of analysis (see section 4.2).

Also, after a total reconstruction, all edges not seen by the connected components algorithm are put in the pool.

4 The Analysis of the Algorithm

4.1 Analysis of the Graph Activation Epochs

The giant tree reconnects successfully with high probability. Consider two (coupled) phases (i) of the neighborhood searches out of u and v in a random deletion of a giant tree edge e. Each edge $\{w, w'\}$ tried in a phase, is taken from the pool. Hence, it is a non-tree edge which is random with respect to its other end w' (to prove that, see the analysis of edge reactivation). If A, B are the pieces in which the tree T has been broken, let w.l.o.g. $|A| \geq |B|$ and consider in the two coupled phases the edge $\{w, w'\}$ emanating from B. Condition on its existence. Then

$$\Pr\{\{w, w'\} \text{ is such that } w' \in A | \{w, w'\} \text{ exists in pool}\} = \frac{|A|}{|A| + |B|} \geq \frac{1}{2} \quad (1)$$

Total reconstruction, if successful, provides a *random graph* of at least cn edges (for a proof of this see section 4.3). Here c is selected to be a constant much greater than α' ($\alpha'n$ are the edges used by the Karp-Tarjan algorithm during total reconstruction). Thus, the average number of edges emanating from each vertex and belonging in the pool is initially (i.e. in the beginning of the activation phase) $c - \alpha'$.

The proofs of most Lemmas and Theorems below are given in [23].

Lemma 5. *For each node w visited in Neighborhood Search:*

$$\Pr\{\text{at least one pool edge out of } w \text{ exists}\} \geq \frac{1}{4}$$

Let $E_1 = $ the event that the edge $e = \{w, w'\}$ indeed reconnects the tree T, given $\{w, w'\}$ exists. Let $E_2 = $ the event that $\{w, w'\}$ returns success in the neighborhood search, given E_1. From (1), $\Pr\{E_1\} \geq 1/2$.

Lemma 6. $\Pr\{E_2|E_1\} \geq \frac{1}{4}$

From the above it follows that:

Lemma 7. *Each neighborhood search succeeds with probability at least $1 - n^{-\gamma}$. (γ a constant which can be made as large as we want by increasing k_1).*

Next we estimate the expected (and whp) number of successes in each search.

Lemma 8. *The number of successes in each neighborhood search is $\Theta(\log n)$ with high probability.*

Theorem 9. *The induction hypothesis is preserved for the whole length $A = \lambda_1 n^2 \log n$ of the graph activation period, with probability at least $1 - n^{-\delta}$, $\delta > 0$ a constant.*

Proof. See [23]. □

Corollary 10. *During the whole activation period, the time for an operation is $\Theta(\log^3 n)$ with probability $\geq 1 - n^{-\delta}$.*

Theorem 11. *Conditioning on the fact that the graph undergoes a graph activation epoch and provided that the pool is reactivated successfully, each operation in the epoch takes at most $O(\log^3 n)$ time, with probability $\geq 1 - n^{-\delta}$.*

4.2 Edge Reactivation with High Probability

Definition 12. Let x be an edge chosen from the pool and unsuccessfully used to reconnect the two pieces A, B of the tree T which corresponds to the giant component, created by the deletion D of an edge e. Let $L(D)$ be the shortest possible sequence of insertions/deletions following D, such that the last operation in $L(D)$ is a deletion, breaking the tree into pieces A', B' where, by again using x, the probability of reconnecting A', B' is at least as it was when x was used to reconnect A, B. Then $|L(D)|$ is called the retirement interval of edge x (i.e. we want the information that x failed to certifiably reconnect A, B to be "lost"). Then we say that x has been reactivated and can be put in the pool of "random" edges.

Lemma 13. *A failed edge x (as above in Def. 12) will be reactivated when:*

1. *An insertion or a tried edge from pool reconnects the tree (and A, B) and*
2. *The first of the subsequent deletions happens with the property that it deletes a random edge from the tree T.*

Proof. Let $x = \{u, v\}$. Let I_1 be the event "after the deletion of a random edge e, x will reconnect the two pieces A, B, for the first time of Def. 12" and I_2 be the event "after A, B are again reconnected by a tried pool edge and a random edge e' is deleted for the first time, x will reconnect A, B ". We must show:

$$\Pr\{I_1\} \leq \Pr\{I_2/\bar{I}_1\}$$

Let $path(u, v)$ be the path connecting u, v on the tree T before the deletion of e. Then $\Pr\{I_1\} = \Pr\{e \in path(u, v)$ and either u or v are close to the root of $T\}$.

Now, consider $I_2/\bar{I}_1$. For x to have been *tried* and *failed*, it must have been the case that (1) x was at the neighborhood of e and (2) (a) either both u, v belonged to the same piece of T or (2) (b) both were away from the root of T, even when they belonged to different pieces. Notice that the distance of u, v from the root of T *does not change in each piece* from the time x is tried to the time x is re-tried.

Since x is tried again, it belongs to the neighborhood of e'. Thus the neighborhoods of e and e' *intersect*. This means that e' belongs to the part of tree T *which is unchanged up to the deletion of e'* (new insertions of tree edges are happening at the leaves). Thus, for x to succeed, it again must be the case that $e' \in path(u, v)$ in the new tree, with the possible choices of e' the same as for e or better. Thus $\Pr\{I_1\} \leq \Pr\{I_2/\bar{I}_1\}$. $\qquad\square$

Lemma 14. *For each edge x which was used and failed (except for midget edges) there exist $\lambda_1 > 2$, $\lambda_2 > 2$ such that with probability $\geq 1 - n^{-\lambda_2}$*

$$retirement - interval(x) \leq \lambda_1 \log n$$

where $retirement - interval(x)$ is the length of the retirement interval for edge x.

4.3 Analysis of the Graph Retirement Epochs

The purpose of the graph retirement epoch is to re-activate the set E_1 of edges in the giant tree and the smaller trees (the pool edges are "fresh"). Since there are at most $c_1 n$ such edges, a graph retirement epoch needs to delete the whole set E_1. Each random deletion hits (deletes) an edge in E_1 with probability $\Theta(\frac{1}{n})$. Thus,

Pr{a particular $e \in E_1$ avoids to be deleted for $\lambda_2 n \log n$ operations} $\leq$

$$\leq \left(1 - \frac{1}{n}\right)^{\lambda_2 n \log n} \leq e^{-\lambda_2 \log n} = n^{-\lambda_2}$$

Thus

$$\Pr\{\text{all } E_1 \text{ is deleted in graph retirement}\} \geq 1 - n^{-(\lambda_2 - 1)}$$

A single erasure of all E_1 edges may not be enough, because we must have at least an equal number of insertions and deletions (this happens with probability $1/2$) and the Karp-Tarjan algorithm must succeed (the probability for this is $\geq 1 - n^{-\alpha}$). Thus at most $\Theta(\log n)$ repetitions will suffice with high probability.

Lemma 15. *With probability $\geq 1 - n^{-k_1}$ the number of runs in the graph retirement period is at most $k \log n$ and thus its total length is at most $k \lambda_2 n \log^2 n$.*

4.4 The Amortized Expected Time Analysis of the Algorithm

Important Remark: Each reconnection (after an edge deletion) of the giant tree has expected time $O(\log^3 n)$ with high probability. In order to speak about amortized expected time, this has to be done for a finite sequence of operations (else, the linearity of expectation will not apply to infinite sequences and we may need the bounded convergence theorem). It follows that:

Theorem 16. *Our algorithm has $O(\log^3 n)$ amortized cost per operation, with high probability (or amortized expected) in the sense that a long sequence of M operations takes expected time $O(M \log^3 n)$, for large enough M.*

Proof. See [23]. $\qquad\qquad\qquad\qquad\qquad\qquad\qquad\qquad\qquad\qquad\qquad\qquad\square$

References

1. D. Alberts and M. Rauch, "Average Case Analysis of Dynamic Graph Algorithms", 6th SODA, 1995.
2. D. Angluin and L. Valiant, "Fast Probabilistic Algorithms for Hamiltonian Circuits and Matchings", JCSS, vol. 18, pp. 155–193, 1979.
3. B. Bollobás, "Random Graphs", Academic Press, 1985.
4. J. Cheriyan and R. Thurimella, "Algorithms for parallel k-vertex connectivity and sparse certificates", 23rd STOC , pp. 391–401, 1991.
5. D. Coppersmith, P. Raghavan and M. Tompa, "Parallel graph algorithms that are efficient on the average", 28th FOCS, pp. 260–270, 1987.

6. D. Eppstein, G. Italiano, R. Tamassia, R. Tarjan, J. Westbrook and M. Yung, "Maintenance of a minimum spanning forest in a dynamic plane graph", 1st SODA, pp. 1–11, 1990.

7. D. Eppstein, Z. Galil, G. Italiano and A. Nissenzweig, "Sparsification- A technique for speeding up Dynamic Graph Algorithms", 33rd FOCS, 1992.

8. P. Erdös and A. Renyi, "On the evolution of random graphs", Magyar Tud. Akad. Math. Kut. Int. Kozl. 5, pp. 17–61, 1960.

9. G. Frederikson, "Data structures for on-line updating of minimum spanning trees", SIAM J. Comput. , 14, pp. 781–798, 1985.

10. G. Frederikson, "Ambivalent data structures for dynamic 2-edge-connectivity and k-smallest spanning trees", 32nd FOCS, pp. 632–641, 1991.

11. G. Frederikson, "A data structure for dynamically maintaining rooted trees", 4th SODA, 1993.

12. A. Frieze, "Probabilistic Analysis of Graph Algorithms", Feb. 1989.

13. Z. Galil and G. Italiano, "Fully dynamic algorithms for edge connectivity problems", 23rd STOC, 1991.

14. Z. Galil, G. Italiano and N. Sarnak, "Fully Dynamic Planarity Testing", 24th STOC, 1992.

15. M. Rauch Henzinger and V. King, "Randomized Dynamic Algorithms with Polylogarithmic Time per Operation", 27-th STOC, 1995.

16. R. Karp, "Probabilistic Recurrence Relations", 23rd STOC, pp. 190–197, 1991.

17. R. Karp and M. Sipser, "Maximum matching in sparse random graphs", 22nd FOCS, pp. 364–375, 1981.

18. R. Karp and R. Tarjan, "Linear expected time for connectivity problems", 12th STOC, 1980.

19. R. Motwani, "Expanding graphs and the average-case analysis of algorithms for matching and related problems", 21st STOC, pp. 550–561, 1989.

20. S. Nikoletseas and P. Spirakis, "Expander Properties in Random Regular Graphs with Edge Faults", 12th STACS, pp. 421 – 432, 1995.

21. S. Nikoletseas and P. Spirakis, "Near-Optimal Dominating Sets in Dense Random Graphs in Polynomial Expected Time", 19th International Workshop on Graph-Theoretic Concepts in Computer Science (WG), 1993.

22. S. Nikoletseas, K. Palem, P. Spirakis and M. Yung, "Short Vertex Disjoint Paths and Multiconnectivity in Random Graphs: Reliable Network Computing", 21st ICALP, pp. 508 – 515, 1994.

23. S. Nikoletseas, J. Reif, P. Spirakis and M. Yung, "Stochastic Graphs Have Short Memory: Fully Dynamic Connectivity in Poly-Log Expected Time", Technical Report T.R. 94.04.25, Computer Technology Institute (CTI), Patras, 1994.

24. J. Reif, "A topological approach to dynamic graph connectivity", Inform. Process. Lett. , 25, pp. 65–70, 1987.

25. J. Reif and P. Spirakis, "Expected parallel time analysis and sequential space complexity of graph and digraph problems", Algorithmica, 1992.

26. D. Sleator and R. Tarjan, "A data structure for dynamic trees", J. Comput. System Sci., 24, pp. 362–381, 1983.

27. J. Spencer, "Ten Lectures on the Probabilistic Method", SIAM, 1987.

28. P. Spira and A. Pan, "On finding and updating spanning trees and shortest paths", SIAM J. Comput., 4, pp. 375–380, 1975.

On the Number of Random Bits
in Totally Private Computation [*]

Carlo Blundo, Alfredo De Santis, Giuseppe Persiano, and Ugo Vaccaro

Dipartimento di Informatica ed Applicazioni,
Università di Salerno, 84081 Baronissi SA, Italy.

E-mail {carblu,ads,giuper,uv}@dia.unisa.it

Abstract. We consider the classic problem of n honest but curious players with private inputs $x_1, \ldots, x_n$ who wish to compute the value of a fixed function $f(x_1, \cdots, x_n)$ in such way that at the end of the protocol every player knows the value $f(x_1, \cdots, x_n)$. Each pair of players is connected by a secure point-to-point communication channel. The players have unbounded computational resources and they intend to compute f in a totally private way. That is, after the execution of the protocol no coalition of *arbitrary* size can get any information about the inputs of the remaining players other than what can be deduced by their own inputs and the value of f.

We study the amount of randomness needed in totally private protocols. Our main result is a lower bound on the number of random bits needed to compute a function with sensitivity n. As a corollary we obtain that when the private inputs are uniformly distributed and the players have access to a source of uniformly distributed bits, at least $k(n-1)(n-2)/2$ random bits are needed to compute the sum modulo 2^k of n k-bit integers. This result is tight as there are protocols for this problem that use *exactly* this number of random bits.

1 Introduction

We consider the classic problem of n honest but curious players with private inputs $x_1, \ldots, x_n$ who wish to compute the value of a fixed function $f(x_1, \cdots, x_n)$ in such way that at the end of the protocol every player knows the value $f(x_1, \cdots, x_n)$. Each pair of players is connected by a secure point-to-point communication channel. The players have unbounded computational resources and they intend to compute f in a totally private way. That is, after the execution

---

[*] Work partially supported by CNR and MURST Progetto 40% Algoritmi, Modelli di Calcolo e Strutture Informative. Work partially done while: Carlo Blundo was visiting the Department of Computer Science of the Technion, Haifa, Israel; Alfredo De Santis was visiting the International Computer Science Institute (ICSI), Berkeley, CA, U.S.A.; Giuseppe Persiano was visiting the DIMACS Center, New Brunswick, NJ, U.S.A.; Ugo Vaccaro was visiting the Department of Mathematics of Bielefeld University, Germany.

of the protocol no coalition of arbitrary size can get any information about the inputs of the remaining players other than what can be deduced by their own inputs and the value of f. Private computation in this model has been the subject of several papers [1, 7, 8, 15, 18, 5, 6, 16]. Chor and Kushilevitz [7] characterized the boolean functions that can be computed in a totally private way. More precisely, they proved a boolean function $f(x_1, \cdots, x_n)$ is totally private if and only if it can be represented as the XOR of n one-argument boolean functions. In [16], the concept of universality in totally private computation has been investigated.

It is well known that no non-trivial function can be computed privately by means of a deterministic protocol and therefore randomness is an essential ingredient to all secure computations. The subject of this paper is to quantify the amount of randomness needed in protocols for totally secure computations.

Randomness plays an important role in several other areas of Computer Science, most notably Algorithm Design and Complexity. Since random bits are a natural computational resource, the amount of randomness used in computation is an important issue in many applications. Therefore, considerable effort has been devoted both to reduce the number of random bits used by probabilistic algorithms (see for instance [12]) and to analyze the amount of randomness required in order to achieve a given performance [3, 4, 11, 14, 19]. Motivated by the fact that "truly" random bits are hard to obtain, it has also been recently investigated the possibility of using imperfect sources of randomness in randomized algorithms [20]. Our approach is close in spirit to [2] in that we mainly concentrate on the rigourous quantification of the number of random bits the players need to execute a protocol for secure computation. Since different algorithms might use random bits produced by different sources, we first need a uniform measure for the amount of randomness provided by different sources. To this aim, we use the Shannon entropy of the source generating the random bits, since it represents the most general and natural measure of randomness in settings where no limitation is imposed on the computational power of the parties. We also recall the important result of Knuth and Yao [13] that shows that the Shannon entropy of a random variable (i.e., of a memoryless random source) is closely related to a more algorithmically oriented measure of randomness. More precisely, Knuth and Yao have shown that the entropy of a random variable R is approximatively equal to the average number of tosses of an unbiased coin necessary to simulate the outcomes of R.

Our main result is a *tight* lower bound on the number of bits needed to compute the modular sum of the inputs of the players in a totally private way. The importance of the computation of modular sum when totally privacy is required lies in the result of Chor and Kushilevitz that tells us that, in the boolean case, the modular sum is the *only* building block to construct totally-private functions. We prove a lower bound on the entropy of the source supplying the stream of bits used by the players as source of randomness and we obtain that in the setting most studied in literature, that is when the private inputs are uniformly distributed and the players have access to a source of uniformly distributed bits, at least $k(n-1)(n-2)/2$ random bits are needed to compute the sum modulo 2^k of n k-bit integers. Our lower bound is *tight* as Chor and Kushilevitz have presented in [8] a protocol for computing the sum modulo 2^k of n k-bit integers that uses exactly $k(n-1)(n-2)/2$ random bits. To prove our lower bound we make no assumption on the general structure of the protocol; for example, our lower bound holds also for non oblivious protocols. Oblivious protocols are protocols where the decision whether a player i sends a message to player j at round k depends only on i, j, and k and not on his input or random coin tosses. Finally, we discuss extension of our work to settings in which it is only required that subsets of up to t players, for $t < n$, get no information on the inputs of other players. The randomness efficiency has also been studied in [17]. In particular in [17] it has been proved that at least t random bits are needed in a t-private protocol computing an n arguments function f.

2 The Model for Totally Private Computation

In this section we describe the model and then we give formal definitions for totally private protocols.

There is a set $\mathcal{P} = \{P_1, P_2, \ldots, P_n\}$ of n players, each holding a value $x_i \in \mathbf{Z}_q$, $q \geq 2$. They wish to distributely compute a function $f = f(x_1, x_2, \ldots, x_n)$. Player P_i has also a random input r_i taken from a random source R_i, independent from the values x_i's. They compute the value of the function by exchanging messages over a network of private channels, that is channels in which only the sender and the receiver can read the transmitted message. Each message sent by player P_i is function of its private input x_i, its random input r_i, and the messages previously exchanged with the other players. The players are honest but also curious, so they want to accomplish this task in such a way that any subset of them cannot infer any additional information from the execution of the protocol other than what they can obtain from the value of the function and their inputs. An example of a function that can be distributely computed in a private way is the *Sum* function, i.e., $\sum_{i=1}^{n} x_i \bmod q$.

Now, let us introduce the notation used in the following. We denote random variables by capital letters and any value that a random variable can take by lower letters. Thus, for $i = 1, 2, \ldots, n$, X_i will denote the random variable induced by the i-th player's input x_i, F the random variable induced by the value assumed by the function $f(x_1, x_2, \ldots, x_n)$. With c_i we denote all the communication involving the player P_i during the execution of the protocol, while C_i denotes the corresponding random variable. The communication between players P_i and P_j is represented by the random variable $C_{i,j} = C_{j,i}$. Hence, $C_i = C_{i,1} \ldots C_{i,i-1} C_{i,i+1} \ldots C_{i,n}$.

If A_i (a_i) is a random variable (a value) and $W = \{i_1, \ldots, i_m\}$, where $i_1 < i_2 < \cdots < i_m$, then A_W (a_W) denotes the random variable $A_{i_1} \ldots A_{i_m}$ (the value $a_{i_1} \ldots a_{i_m}$). If $Y = \{i_1, \ldots, i_t\} \subseteq \{1, \ldots, n\}$, where $i_1 < i_2 < \cdots < i_t$, then $x_Y = x_{i_1} \ldots x_{i_t}$, $C_Y = C_{i_1} \ldots C_{i_t}$, and $C_{j,Y} = C_{j,i_1} \ldots C_{j,i_t}$.

In an ideal evaluation of a function there is a trusted party. The players give him their private inputs x_i. The trusted party computes the value of the function on those inputs, and reveals it to all players, without revealing any other information. Thus all the information that a coalition W of players can get on the other players' private inputs is all that can be deduced from the values x_W, r_W, and $f = f(x_1, x_2, \ldots, x_n)$.

In a private protocol for function evaluation there is no longer a trusted party. Players exchange messages over the network. At the end of the computation they learn the value of the function. All the information that a coalition W of players can get on the other players' private inputs is all that can be deduced from the values x_W, r_W, $f = f(x_1, x_2, \ldots, x_n)$, and the set of all messages exchanged c_W.

In order for a protocol to be totally private we require that all information that a set W of players can compute on the other players' private inputs is the same that they can compute in an ideal function evaluation. Next definition formalizes this concept. To formally define the concept of total privacy, we consider the Shannon entropy of the random variables. Given two random variables A and B and a joint probability distribution $\{p(a, b)\}_{a \in A, b \in B}$ on their cartesian product, the *conditional entropy* $H(A|B)$, is defined as $H(A|B) = -\sum_{b \in B} \sum_{a \in A} p(b)p(a|b) \log p(a|b)$ (the log is to base 2). We refer the reader to [10] for information-theoretic background.

Definition 1. Let $\mathcal{P} = \{P_1, P_2, \ldots, P_n\}$ be a set of n players, each holding a value x_i. A protocol computing the value $f = f(x_1, x_2, \ldots, x_n)$ is said to be totally private, if the following two properties are satisfied.

1. *Each player can compute the value f.*
   Formally, for any player P_i, it holds that $H(F|C_i X_i) = 0$.
2. *Any coalition of players gets no additional information.*
   Formally, for all $Y, W \subseteq \{1, \ldots, n\}$, $W \cap Y = \emptyset$, it holds that
   $H(X_W|C_Y R_Y X_Y F) = H(X_W|R_Y X_Y F)$.

Notice that, since the random variables R_i's are independent from the private inputs X_i's, we have that, for all $Y, W \subseteq \{1, \ldots, n\}$, $H(X_W|R_Y X_Y F) = H(X_W|X_Y F)$.

Next lemma gives equivalent formulations of Property 2. In particular, Property *2a* is a simpler but equivalent formulation of Property 2. Property *2b* has been used in the literature and we prove here its equivalence to our definition.

Lemma 2. *Property 2. of Definition 1 is equivalent to each of the following:*

2a. For all $Y \subseteq \{1, \ldots, n\}$, it holds that $H(X_W|C_Y R_Y X_Y F) = H(X_W|R_Y X_Y F)$, where $W = \{1, \ldots, n\} \backslash Y$.

2b. For all $Y \subseteq \{1, \ldots, n\}$, for all x_Y, r_Y, c_Y, x_W, x'_W such that the function evaluated at $x_Y x_W$ and $x_Y x'_W$ has the same value f, it holds that
$$Pr(c_Y|x_W, r_Y, x_Y, f) = Pr(c_Y|x'_W, r_Y, x_Y, f)$$
where $W = \{1, \ldots, n\} \backslash Y$ and where the probability is taken over all random inputs r_W.

Proof : First, we prove that Properties 2. and $2a$. are equivalent. Since Property 2. states that $H(X_W|C_Y R_Y X_Y F) = H(X_W|R_Y X_Y F)$ for any disjoint sets Y, W then the same equality holds also for any pair Y and $W = \{1, \ldots, n\}\backslash Y$ and Property $2a$. is satisfied. On the other hand, suppose that Property $2a$. holds, that is for all $Y \subseteq \{1, \ldots, n\}$, we have that the mutual information $I(X_W; C_Y|R_Y X_Y F) = H(X_W|C_Y R_Y X_Y F) - H(X_W|R_Y X_Y F) = 0$, where $W = \{1, \ldots, n\}\backslash Y$. We have to prove that for all $Y, W' \subseteq \{1, \ldots, n\}$, $W' \cap Y = \emptyset$, it holds that $I(X_{W'}; C_Y|R_Y X_Y F) = 0$. Let $W = W' \cup W''$, where $W' \cap W'' = \emptyset$. We have that $0 = I(X_W; C_Y|R_Y X_Y F) = I(X_{W'}; C_Y|R_Y X_Y F) + I(X_{W''}; C_Y|R_Y X_Y F X_{W'}) \geq I(X_{W'}; C_Y|R_Y X_Y F) \geq 0$. Hence, we get that for all $Y, W' \subseteq \{1, \ldots, n\}$, $W' \cap Y = \emptyset$, it holds that $I(X_{W'}; C_Y|R_Y X_Y F) = 0$. Thus, Property 2. is satisfied.

Now, we prove that Properties $2a$. and $2b$. are equivalent. Property $2a$. says that C_Y and X_W are statistical independent given $R_Y X_Y F$, i.e., $Pr(x_W|c_Y, r_Y, x_Y, f) = Pr(x_W|r_Y, x_Y, f)$. This is equivalent to $Pr(c_Y|x_W, r_Y, x_Y, f) = Pr(c_Y|r_Y, x_Y, f)$ which, in turn, is equivalent to $Pr(c_Y|x_W, r_Y, x_Y, f) = Pr(c_Y|x'_W, r_Y, x_Y, f)$, where x_W, x'_W are two sets of values such that f computed at $x_Y x_W$ and $x_Y x'_W$ has the same value f. $\quad\square$

3 Sensitivity

In this section we define the sensitivity of a function f and we prove useful properties of protocols computing functions f with fixed sensitivity.

We say that the function $f : \mathbf{Z}_q^n \to \mathbf{Z}_q$ is *sensitive* to its i-th variable on assignment $\mathbf{x} = (x_1, x_2, \ldots, x_n)$, if it results that $|\{f(x_1, \ldots, x_{i-1}, z_i, x_{i+1}, \ldots, x_n) : z_i \in \mathbf{Z}_q\}| = q$. Hence, a function f is sensitive to its i-th variable, on a fixed assignment $\mathbf{x}$, if its value changes when the i-th variable's value does. We say that the function f is i-*sensitive* if it is sensitive to its i-th variable for any assignment $\mathbf{x}$. The *sensitivity* of a function f, denoted by $\mathcal{S}(f)$, is the number of indices i to which the function f is i-sensitive. For instance, the sensitivity of the *Sum* function is n. More generally, if $(G, \otimes)$ is a group then the function $f : G^n \to G$ defined by $f(x_1, x_2, \cdots, x_n) = x_1 \otimes \cdots \otimes x_n$ has sensitivity n. Notice that our definition of sensitivity differs from other definitions found in literature.

The following lemma relates the sensitivity of a function f to a property of private protocols computing f. If a function $f = f(\mathbf{x})$ is i-sensitive then the communication C_i uniquely determines the input value x_i of player P_i.

Lemma 3. *In any protocol computing an i-sensitive function $f = f(\mathbf{x})$, it results that*

$$H(X_i|C_i) = 0.$$

Proof : If $H(X_i) = 0$ then, as $0 \leq H(X_i|C_i) \leq H(X_i)$, we have $H(X_i|C_i) = 0$. Assume $H(X_i) \geq 0$ and hence there are two values x'_i x''_i with non zero probability. The identity $H(X_i|C_i) = 0$ means that the communication involving player P_i uniquely determines its input value x_i. Would it be otherwise, the player P_i with two different inputs, x_i' and x_i'', would be involved in the same communication c_i, but this leads to a contradiction. Indeed, the protocol, fixed the input $\mathbf{x}$ of all the player but P_i, will output the same value $f(\mathbf{x})$ for P_i holding either x_i' or x_i''. We know that this is impossible since the function f is i-sensitive. Thus, the lemma holds. □

Next lemma states that if a function $f = f(\mathbf{x})$ is i-sensitive then its value is uniquely determined by the communication C_i involving the player P_i.

Lemma 4. *In any protocol computing an i-sensitive function $f = f(\mathbf{x})$, it results that*

$$H(F|C_i) = 0.$$

Proof : If a function $f = f(\mathbf{x})$ is i-sensitive then from Lemma 3 it turns out that $H(X_i|C_i) = 0$. Hence, $I(X_i; F|C_i) \leq H(X_i|C_i) = 0$. Since $I(X_i; F|C_i) = I(F; X_i|C_i) = 0$ and, by definition of private protocol, $H(F|C_iX_i) = 0$, we get $H(F|C_i) = H(F|C_iX_i) = 0$. □

Since the function *Sum* has sensitivity n, then the following corollary holds.

Corollary 5. *In any protocol computing the function $\sum_{i=1}^{n} x_i$ mod q, for any $i = 1, 2, \ldots, n$, it results that $H(X_i|C_i) = 0$ and $H(F|C_i) = 0$.*

4 Protocol's Randomness

To formally define the protocol's randomness we consider the Shannon entropy of the random variables generating the values held by the players and the messages exchanged over the network. The entropy is strictly related to the measure of randomness introduced by Knuth and Yao [13]. Indeed, Knuth and Yao [13] have shown that the Shannon entropy of a random variable is closely related to a more algorithmically oriented measure of randomness. More precisely, they have shown that the entropy of a random variable X is very close to the average number of tosses of an unbiased coin necessary to simulate the outcomes of X. Let A be an algorithm that generates a random variable R with distribution $(p_1, \ldots, p_n)$, using only independent and unbiased random bits in inputs. Denote by $T(A)$ the average number of random bits used by the algorithm A and let $T(R) = \min_A T(A)$. In [13] the following theorem has been proved.

Theorem 6 (Knuth and Yao [13]).

$$H(R) \leq T(R) < H(R) + 2.$$

Thus, the entropy of a random source is very close to the average number of independent unbiased random bits necessary to simulate the source.

The total randomness present in a totally private protocol for n player is equal to the entropy $H(R_1, \cdots, R_n)$ of the players random inputs. The communication's randomness is the randomness of all the communication in the protocol to compute the function $f = f(x_1, x_2, \ldots, x_n)$ once the value $x_1, x_2, \ldots, x_n$ and the probability distribution Π on the set of the values $(x_1, x_2, \ldots, x_n)$ are known. Therefore, for any totally private protocol Δ the communication's randomness used in the execution of Δ is equal to the entropy $H(C_1 \ldots C_n | X_1 \ldots X_n)$. The total randomness and the communication's randomness are related by the following lemma.

Lemma 7. *Let* $\mathcal{P} = \{P_1, P_2, \ldots, P_n\}$ *be a set of n players each holding a value x_i. For any protocol computing a function $f = f(x_1, x_2, \ldots, x_n)$, it holds that*

$$H(R_1 \ldots R_n) \geq H(C_1 \ldots C_n | X_1 \ldots X_n) \geq H(C_1 \ldots C_n) - H(X_1 \ldots X_n).$$

If the sensitivity of f is $S(f) = n$, then

$$H(C_1 \ldots C_n) = H(C_1 \ldots C_n | X_1 \ldots X_n) + H(X_1 \ldots X_n).$$

Proof : Each message is determined by the private and the random inputs; that is, there is a function h such that for all sequences $x_1 \ldots x_n, r_1 \ldots r_n$, we have $(c_1, \cdots, c_n) = h(x_1 \ldots x_n, r_1 \ldots r_n)$. Therefore,

$$
\begin{aligned}
H(C_1 \ldots C_n | X_1 \ldots X_n) &= H(h(X_1 \ldots X_n R_1 \ldots R_n) | X_1 \ldots X_n) \\
&\leq H(X_1 \ldots X_n R_1 \ldots R_n | X_1 \ldots X_n) \\
&= H(R_1 \ldots R_n | X_1 \ldots X_n) = H(R_1 \ldots R_n).
\end{aligned}
$$

The mutual information $I(C_1 \ldots C_n; X_1 \ldots X_n)$ can be written either as $H(C_1 \ldots C_n) - H(C_1 \ldots C_n | X_1 \ldots X_n)$ or as $H(X_1 \ldots X_n) - H(X_1 \ldots X_n | C_1 \ldots C_n)$. Hence,

$$H(C_1 \ldots C_n) = H(C_1 \ldots C_n | X_1 \ldots X_n) + H(X_1 \ldots X_n) - H(X_1 \ldots X_n | C_1 \ldots C_n).$$

Since $H(X_1 \ldots X_n | C_1 \ldots C_n) \geq 0$, we get

$$H(C_1 \ldots C_n) \leq H(C_1 \ldots C_n | X_1 \ldots X_n) + H(X_1 \ldots X_n).$$

If the sensitivity of f is $S(f) = n$, then from Lemma 3 we get

$$0 \leq H(X_1 \ldots X_n | C_1 \ldots C_n) \leq \sum_{i=1}^{n} H(X_i | C_i) = 0.$$

Hence, it follows that $H(X_1 \ldots X_n | C_1 \ldots C_n) = 0$. Thus,

$$H(C_1 \ldots C_n) = H(C_1 \ldots C_n | X_1 \ldots X_n) + H(X_1 \ldots X_n).$$

$\square$

To analyze the randomness needed by a totally private protocol for n player that computes the function $f = f(x_1, x_2, \ldots, x_n)$, we define the *players' randomness* of a totally private protocol Δ, when the probability distribution on the set of the values $(x_1, x_2, \ldots, x_n)$ is Π, to be

$$\mathcal{R}_f(n, \Pi, \Delta) = H(R_1 \ldots R_n).$$

The value $\mathcal{R}_f(n, \Pi, \Delta)$ represents the amount of randomness required by the totally private protocol Δ to evaluate the function $f = f(x_1, x_2, \ldots, x_n)$ when Π is the probability distribution on the set of values $(x_1, x_2, \ldots, x_n)$. Notice that $\mathcal{R}_f(n, \Pi, \Delta)$ depends also on Δ since the probability that participants receive/send a particular message depends both on Π and Δ. Since we are interested in the minimum amount possible of randomness for computing a function $f = f(x_1, x_2, \ldots, x_n)$, we give the following definition.

Definition 8. Let $\mathcal{P} = \{P_1, P_2, \ldots, P_n\}$ be a set of n players each holding a value x_i. The *players' randomness* of a totally private protocol computing the function $f = f(x_1, x_2, \ldots, x_n)$, when Π is the probability distribution on the set of values $(x_1, x_2, \ldots, x_n)$ is defined to be

$$\mathcal{R}_f(n, \Pi) = \inf_{\Delta \in \mathcal{T}} \mathcal{R}_f(n, \Pi, \Delta),$$

where $\mathcal{T}$ is the set of all totally private protocols computing the function $f = f(x_1, x_2, \ldots, x_n)$.

In the sequel, whenever the function f is clear from the context, we will simply write $\mathcal{R}(n, \Pi)$ instead of $\mathcal{R}_f(n, \Pi)$.

5 An Optimal Lower Bound for Totally Private Protocols

We are now ready to prove a lower bound on the number of random bits needed to compute in a totally private way any function with sensitivity n. We start by proving the following preliminary lemmas.

Lemma 9. Let $f = f(x_1, x_2, \ldots, x_n)$ be a function with sensitivity $S(f) = n$, and let $W, Y \subseteq \{1, 2, \ldots, n\}$, $|Y| \leq n - 2$. In any totally private protocol computing f, if $C_{i,W} C_{i,Y} = C_i$, then

$$H(C_{i,W} | C_{i,Y} F) \geq H(X_i | X_Y F).$$

Proof : The following chain of inequalities proves the lemma.

$$
\begin{aligned}
H(C_{i,W} | C_{i,Y} F) &\geq I(C_{i,W}; X_i | C_{i,Y} F) = I(X_i; C_{i,W} | C_{i,Y} F) \\
&= H(X_i | C_{i,Y} F) - H(X_i | C_{i,W} C_{i,Y} F) \\
&= H(X_i | C_{i,Y} F) - H(X_i | C_i F) \geq H(X_i | C_{i,Y} F) - H(X_i | C_i) \\
&= H(X_i | C_{i,Y} F) \text{ (from Lemma 3)} \\
&\geq H(X_i | C_Y F) \geq H(X_i | C_Y R_Y X_Y F) \\
&= H(X_i | R_Y X_Y F) \text{ (from 2. of Definition 1)} \\
&= H(X_i | X_Y F).
\end{aligned}
$$

$\square$

With the symbol $C_i^{(j)}$ we denote the random variable induced by the messages involving the player P_i with the exception of the ones pertinent to the player P_j, i.e., $C_i^{(j)} = C_{i,Y}$, where $Y = \{1, 2, \ldots, n\}\backslash\{i, j\}$.

Lemma 10. *In any totally private protocol computing a function* $f = f(x_1, \ldots, x_n)$, *with sensitivity* $S(f) = n$, *for any* $i \in \{1, 2, \ldots, n\}$, *it holds that*

$$H(C_i) \geq H(F) + \sum_{j \neq i} H(X_i|X_{\sigma(i,j)}F),$$

where $\sigma(i, j) = \{1, 2, \ldots, n\}\backslash\{i, j\}$.

Proof: For the sake of simplicity we will prove the lemma for the communication C_1. Consider the mutual information between F and the communication C_1 involving the player P_1. We have, $I(C_1; F) = H(C_1) - H(C_1|F)$ and $I(F; C_1) = H(F) - H(F|C_1) = H(F)$ (the last equality derives from Lemma 4). Hence, we obtain

$$H(C_1) = H(F) + H(C_1|F) = H(F) + H(C_{1,2}C_{1,3}\ldots C_{1,n}|F)$$

$$= H(F) + H(C_{1,2}|F) + \sum_{j=3}^{n} H(C_{1,j}|C_{1,2}\ldots C_{1,j-1}F)$$

$$\geq H(F) + \sum_{j=2}^{n} H(C_{1,j}|C_1^{(j)}F)$$

$$\geq H(F) + \sum_{j=2}^{n} H(X_1|X_{\sigma(1,j)}F) \quad \text{(from Lemma 9)}.$$

Thus, the lemma holds. □

Lemma 11. *Let* $f = f(x_1, x_2, \ldots, x_n)$ *be a function with sensitivity* $S(f) = n$. *In any totally private protocol computing* f *on any indipendent input, it holds that*

$$H(C_i|C_1 \ldots C_{i-1}) \geq (n - i)H(X_i).$$

Proof: Recall that $\sigma(i, j) = \{1, 2, \ldots, n\}\backslash\{i, j\}$. Since $H(C_{i,1}\ldots C_{i,i-1}|C_1\ldots C_{i-1}) = 0$ and $H(C_{i,i+1}\ldots C_{i,n}|C_1\ldots C_{i-1}C_{i,1}\ldots C_{i,i-1}) = H(C_{i,i+1}\ldots C_{i,n}|C_1\ldots C_{i-1})$, we have

$$H(C_i|C_1\ldots C_{i-1}) = H(C_{i,1}\ldots C_{i,i-1}|C_1\ldots C_{i-1}) + H(C_{i,i+1}\ldots C_{i,n}|C_1\ldots C_{i-1})$$

$$= H(C_{i,i+1}\ldots C_{i,n}|C_1\ldots C_{i-1})$$

$$\geq \sum_{j=i+1}^{n} H(C_{i,j}|C_1\ldots C_{i-1}C_i^{(j)})$$

$$\geq \sum_{j=i+1}^{n} H(C_{i,j}|C_{\sigma(i,j)})$$

Since the protocol is totally private and f has sensitivity n, we have that $H(X_i|C_{\sigma(i,j)}) = H(X_i)$ and $H(X_i|C_{\sigma(i,j)}C_{i,j}) = 0$. Consider the mutual information between X_i and $C_{i,j}$ given $C_{\sigma(i,j)}$. It is equal to $I(X_i; C_{i,j}|C_{\sigma(i,j)}) = H(X_i|C_{\sigma(i,j)}) - H(X_i|C_{i,j}C_{\sigma(i,j)}) = H(X_i)$. Moreover, $I(X_i; C_{i,j}|C_{\sigma(i,j)}) = I(C_{i,j}; X_i|C_{\sigma(i,j)}) = H(C_{i,j}|C_{\sigma(i,j)}) - H(C_{i,j}|C_{\sigma(i,j)}X_i)$. Hence, $H(C_{i,j}|C_{\sigma(i,j)}) = H(C_{i,j}|C_{\sigma(i,j)}X_i) + H(X_i)$. Since the entropy is non negative, $H(C_{i,j}|C_{\sigma(i,j)}) \geq H(X_i)$. Therefore,

$$H(C_i|C_1 \ldots C_{i-1}) \geq \sum_{j=i+1}^{n} H(X_i) = (n-i)H(X_i).$$

Thus, the lemma holds. □

The following theorem states a lower bound on the players' randomness of any totally private protocol computing a function f whose sensitivity is $\mathcal{S}(f) = n$.

Theorem 12. *In any totally private protocol computing the function $f(x_1, ..., x_n)$, with sensitivity $\mathcal{S}(f) = n$, if the X_i's are independent and $H(X) \overset{\triangle}{=} H(X_1) = \cdots = H(X_n) = H(F)$, then the players' randomness $\mathcal{R}(n, \Pi)$ satisfies*

$$\mathcal{R}(n, \Pi) \geq \frac{(n-1)(n-2)}{2}H(X).$$

Proof : We have

$$\begin{aligned}
H(R_1 \ldots R_n) &\geq H(C_1 \ldots C_n) - H(X_1 \ldots X_n) \quad \text{(from Lemma 7)} \\
&= H(C_1 \ldots C_n) - nH(X) \\
&= H(C_1) + \sum_{i=2}^{n} H(C_i|C_1 \ldots C_{i-1}) - nH(X) \\
&\geq H(F) + \sum_{j=2}^{n} H(X_1|X_{\sigma(1,j)}F) + \sum_{i=2}^{n} H(C_i|C_1 \ldots C_{i-1}) - nH(X)
\end{aligned}$$

$$\text{(from Lemma 10).}$$

Since f has sensitivity n and since the X_i's are independent we have that

$$\begin{aligned}
H(R_1 \ldots R_n) &\geq H(F) + (n-1)H(X_1) + \sum_{i=2}^{n} H(C_i|C_1 \ldots C_{i-1}) - nH(X) \\
&= nH(X) + \sum_{i=2}^{n-1} H(C_i|C_1 \ldots C_{i-1}) - nH(X) \\
&= \sum_{i=2}^{n-1} H(C_i|C_1 \ldots C_{i-1}) \\
&\geq \sum_{i=2}^{n-1} (n-i)H(X) \quad \text{(from Lemma 11)} \\
&= \frac{(n-1)(n-2)}{2}H(X).
\end{aligned}$$

□

Corollary 13. *Let Π be the uniform probability distribution on n-tuple of k-bit integers $(x_1, \cdots, x_n)$ and f any function with sensitivity n. Then,*

$$\mathcal{R}(n, \Pi) \geq k \frac{(n-1)(n-2)}{2}.$$

The bound of previous theorem is tight in that the protocol presented in [8], to compute the sum modulo $q = 2^k$ of n values uniformly and independently distributed in $\mathbf{Z}_q$, makes use of $k(n-1)(n-2)/2$ random bits.

6   Extensions

In this section we consider the case in which we want the protocol to be private with respect to a coalition of at most t users.

In order for a protocol to be private against coalitions of size at most t, we require that all information that a set Y, $|Y| \leq t$, of players can compute on the other players' private inputs is the same that they can compute in an ideal function evaluation. Next definition formalizes this concept.

Definition 14. Let $\mathcal{P} = \{P_1, P_2, \ldots, P_n\}$ be a set of n players, each holding a value x_i. A protocol computing the value $f = f(x_1, x_2, \ldots, x_n)$ is said to be t-private, where $1 \leq t \leq n$, if the following two properties are satisfied.

1. *Each player can compute the value f.*
   For any player P_i, it holds that $H(F|C_iX_i) = 0$.
2. *Any coalition of at most t players gets no additional information.*
   For all $Y, W \subseteq \{1, \ldots, n\}$, such that $|Y| \leq t$ and $W \cap Y = \emptyset$, it holds that $H(X_W|C_Y R_Y X_Y F) = H(X_W|R_Y X_Y F)$.

In the case of t-private protocols computing functions f with sensitivity $S(f) = n$, when the X_i's are independent and $H(X) \triangleq H(X_1) = \cdots = H(X_n) = H(F)$, we obtain that the players' randomness is at least

$$\frac{2n(2t + 3 - n) - (t+2)^2 - t}{2(n - t - 1)} H(X).$$

If $t = \alpha n$, where $0 \leq \alpha < 1$, then the above bound can be written as

$$\frac{(\alpha^2 - 4\alpha + 2)n^2 - (6 - 5\alpha)n + 4}{2n(\alpha - 1) + 2} H(X).$$

If $2 - \sqrt{2} < \alpha < 1$, then we get a linear lower bound, namely $\Omega(n)H(X)$; whereas, if $t = n - c$, for some constant c, then from previous bound one can conclude that the players' randomness is at least $\Omega(n^2)H(X)$.

Acknowledgement

This work benefitted from discussions with Benny Chor. The first author wants to thank Benny Chor, Eyal Kushilevitz, and Amos Beimel for helpful discussions during his visit to The Technion.

References

1. M. Ben-Or, S. Goldwasser, and A. Wigderson, *Completeness Theorems for Non-Cryptographic Fault-Tolerant Distributed Computation*, STOC 1988, pp. 1–10.
2. C. Blundo, A. De Santis, and U. Vaccaro, *Randomness in Distribution Protocols*, ICALP 1994, Vol. **820** of LNCS, 1994, pp. 568–579.
3. R. Canetti and O. Goldreich, *Bounds on Tradeoffs Between Randomness and Communication Complexity*, Computational Complexity **3**, pp. 141–167, 1993.
4. S. Chari, P. Rohatgi, and A. Srinivasan, *Randomness-Optimal Unique Element Isolation, with Application to Perfect Matching and Related Problems*, STOC 1993, pp. 458–467.
5. D. Chaum, C. Crépeau, and I. Damgård, *Multiparty Unconditionally Secure Protocols*, STOC 1988, pp. 11–19.
6. B. Chor, M. Gereb-Graus, and E. Kushilevitz, *On The Structure of the Privacy Hierarchy*, J. of Cryptology **7**, 1994, pp. 53–60.
7. B. Chor and E. Kushilevitz, *A Zero-One Law for Boolean Privacy*, SIAM J. Discrete Math., 4, 1991, pp. 36–47.
8. B. Chor and E. Kushilevitz, *A Communication-Privacy Tradeoff for Modular Addition*, Information Processing Letters, Vol. 45, 1993, pp. 205–210.
9. B. Chor and N. Shani, *The Privacy of Dense Symmetric Functions*, to appear in Computational Complexity.
10. T. M. Cover and J. A. Thomas, *Elements of Information Theory*, John Wiley & Sons, 1991.
11. R. Fleischer, H. Jung, and K. Melhorn, *A Time-Randomness Tradeoff for Communication Complexity*, 4th International Workshop on Distributed Algorithms, Vol. **486** of LNCS, 1991, pp. 390–401.
12. R. Impagliazzo and D. Zuckerman, *How to Recycle Random Bits*, FOCS 1989 pp. 248–255.
13. D.E. Knuth and A.C. Yao, *The Complexity of Nonuniform Random Number Generation*, in "Algorithms and Complexity", Academic Press, 1976, pp. 357–428.
14. D. Krizanc, D. Peleg, and E. Upfal, *A Time-Randomness Tradeoff for Oblivious Routing*, STOC 1988, pp. 93–102.
15. E. Kushilevitz, *Privacy and Communication Complexity*, SIAM J. Discrete Math., 5, pp. 273–284.
16. E. Kushilevitz, S. Micali, and R. Ostrowsky, *Universal Boolean Judges and their Characterization*, FOCS 1994, pp. 478–489.
17. E. Kushilevitz and Y. Mansour, *Small Sample Spaces and Privacy*, manuscript.
18. E. Kushilevitz and A. Rosen, *A Randomness-Rounds Tradeoff in Private Computation*, CRYPTO 94, Vol. **839** of LNCS, 1994, pp. 397–410.
19. P. Raghavan and M. Snir, *Memory Versus Randomization in On-line Algorithms*, ICALP 1989, LNCS, 1989, pp. 687–703.
20. D. Zuckerman, *Simulating BPP Using a General Weak Random Source*, FOCS 1991, pp. 79–89.

Lower Time Bounds for Randomized Computation

Rūsiņš Freivalds[1] and Marek Karpinski[2]

[1] Institute of Mathematics and Computer Science, University of Latvia, Raina bulv. 29, Riga, Latvia***
[2] Department of Computer Science, University of Bonn, 53117, Bonn, Germany†

Abstract. It is a fundamental problem in the randomized computation how to separate different randomized time or randomized space classes (c.f., e.g., [KV87, KV88]). We have separated randomized space classes below log n in [FK94]. Now we have succeeded to separate small randomized time classes for multi-tape 2-way Turing machines. Surprisingly, these "small" bounds are of type $n + f(n)$ with $f(n)$ not exceeding linear functions. This new approach to "sublinear" time complexity is a natural counterpart to sublinear space complexity. The latter was introduced by considering the input tape and the work tape as separate devices and distinguishing between the space used for processing information and the space used merely to read the input word from. Likewise, we distinguish between the time used for processing information and the time used merely to read the input word.

1 Introduction

The advantages of using randomization in the design of algorithms have become increasingly evident in the last couple of years. It appears now that these algorithms are more efficient than the purely deterministic ones in terms of running time, hardware size, circuits depth, etc. The advantages of randomized Turing machines over deterministic machines have been studied early starting with [Fr75] where the sets of palindromes were proved to be computable by Monte Carlo off-line Turing machines much faster than by deterministic machines of the same type. Later similar results were obtained for space and reversal complexity for various types of machines [Fr83, Fr85, KF90]. On the other hand, it is universally conjectured that randomness do not always help. However, these conjectures usually cannot be supported by proofs since proving lower bounds in always hard, and proving lower bounds for complexity of randomized machines has turned out to be much harder than proving lower bounds for complexity of deterministic and nondeterministic machines.

*** Research partially supported by Grant No.93-599 from the Latvian Council of Science
† Research partially supported by the International Computer Science Institute, Berkeley, California, by the DFG grant KA 673/4-1, and by the ESPRIT BR Grants 7079 and ECUS030

In [FK94] we proved the first nontrivial small lower space bounds for various types of ranomized machines. In this paper we have proved the first nontrivial small lower time bounds for randomized multitape 2-way Turing machines.

We distinguish between two types of randomized machines: Monte Carlo and probabilistic machines.

We say that a Monte Carlo machine M recognizes language L in time $T(n)$ if there is a positive constant δ such that:

1. For arbitrary $x \in L$, the probability of event "M accepts x in time not exceeding $T(|x|)$" exceeds $1/2 + \delta$,
2. For arbitrary $x \notin L$, the probability of event "M rejects x in time not exceeding $T(|x|)$" exceeds $1/2 + \delta$.

We say that a probabilistic machine M recognizes language L in time $T(n)$ if:

1. For arbitrary $x \in L$, the probability of event "M accepts x in time not exceeding $T(|x|)$" exceeds $1/2$,
2. For arbitrary $x \notin L$, the probability of event "M rejects x in time not exceeding $T(|x|)$" exceeds $1/2$.

In a similar way one defines space complexity of Monte Carlo and probabilistic machines. Probabilistic machines are interesting theoretical devices but they are rather remote from practical needs. Hence much more effort has been spent to study Monte Carlo machines.

There have already been lower time bounds ($const \cdot n^2$ for Monte Carlo off-line Turing machines to recognize palindromes [Fr75, Fr77]). On the other hand, these lower bounds employ the specific restrictions of the machine model. For multitape 2-way Turing machines, lower time bounds are weak even in the case of deterministic machines. In the case of Monte Carlo machines the situation is much worse. We did not even know whether $MCTIME(n) = MCTIME(n^{\log n})$. Nevertheless we prove in Section 2 the first nontrivial lower time bounds for recognition of specific languages by muultitape 2-way Turing machines. This allows us to prove separation theorem for small time complexity classes of Monte Carlo multitape 2-way Turing machines. The method used in this proof generalizes the methods used in [FI77, JKS84].

2   Separation Theorem

We consider a language

$$A = \{x2y | (\exists m)(x \in \{0,1\}^{2^m} \& y \in \{0,1\}^m \& (\exists i)(y \text{ is the binary notation}$$

$$\text{for the integer } i \ \& \text{ the } i\text{-th digit of the word } x \text{ equals } 1))\}$$

Theorem 2.1 *The language A cannot be recognized by a multitape 2-way Monte Carlo Turing machine in time $n + o(n)$.*

Proof. Assume from the contrary that A is recognized by multitape 2-way Monte Carlo Turing machine M in time $n + f(n)$ where $f(n) = o(n)$.

We fix an arbitrarily large integer m and consider the work of M on all the words

$$x2y \tag{1}$$

where $y \in \{0,1\}^m$ and $x \in \{0,1\}^{2^m}$. We call $x2$ the head of the word (1), and y the tail of the word. The length n of the word (1) equals $2^m + m + 1$.

We consider the first moment when the machine M reads the symbol 2 from the input. Let the heads on the work tapes observe the squares $b_1, \ldots, b_r$ at this moment. There remains no more time than $m + f(n)$ till the moment of output. The contents of a square of the worktape can influence the result only if this square is reachable from the squares $b_1, \ldots, b_r$ in no more than $m + f(n)$ steps. The absolute addresses $b_1, \ldots, b_r$ also do not influence the result. Hence all the needed information about the head of the word (1) is encoded in the configuration of the reachable part of the worktapes. We denote the set of all the a priori possible configurations of this part of the worktapes by

$$\{z_1, z_2, \ldots, z_u\} \tag{2}$$

It is easy to see that there is a constant $c > 0$ such that

$$u \leq c^{m+f(n)} \tag{3}$$

Hence all the needed information about the head of the word (1) is encoded by the probability distribution in the set (2). The contradiction obtained below shows the cardinality u of the set (2) is too small for this task.

We will use methods of Shannon's Information Theory for this proof. All the notions and notation not defined in this paper see in [Gal68].

Let the words

$$w_1, \ldots, w_{2^m}$$

denote the lexicographical ordering of all the words in $\{0,1\}^m$.

We consider a random variable $X = (X_1, \ldots, X_{2^m})$ each component of which takes value

$$\begin{cases} 1 \text{ with probability } 1/2 \\ 0 \text{ with probability } 1/2 \end{cases}$$

statistically independently from the other components. (In alternative way to explain the same thing, we consider the head of the word (1) as taken randomly with X_i being the i-th digit of the head of the word.)

Then all the 2^{2^m} possible values of the random varibale X are equiprobable, and each one of these 2^{2^m} values corresponds to a certain head of the word (1). Hence the entropy

$$H(X) = 2^m \tag{4}$$

Now we define a random variable Z taking values $z_1, \ldots, z_u$ out of (2). The probabilities of these values are defined by taking the random varibale X, considering the head of the word (1) corresponding to the particular value of X, processing it by the machine M and observing the configurations of the reachable part of the worktapes.

We estimate the amount of information about X in Z:

$$I(X|Z) = I(Z|X) \leq H(Z) \leq \log_2 u$$

Taking (3) into consideration, we get

$$I(X|Z) \leq (m + f(n))\log_2 c \tag{5}$$

We introduce new random variables $Y_i'(i = 1, 2, \ldots, 2^m)$ taking the values 1, 0 and "no result". The probabilities of these values are defined as the probabilities for the machine M to output 1, 0 or no definite result in $n \dotdiv f(n)$ steps, correspondingly, provided the tail of the word (1) is w_i. Finally, we introduce random variables $Y_i(i = 1, 2, \ldots, 2^m)$ taking the values 1 and 0 only. The probabilities of these values are defined as follows.

$$p(Y_i = 0) = p(Y_i' = 0) + \frac{1}{2}p(Y_i' = \text{"no result"})$$

$$p(Y_i = 1) = p(Y_i' = 1) + \frac{1}{2}p(Y_i' = \text{"no result"})$$

It is easy to see that

$$p(Y_i' = 0) + p(Y_i' = 1) + p(Y_i' = \text{"no result"}) = 1$$

and

$$p(Y_i = 0) + p(Y_i = 1) = 1$$

These probabilities can be calculated from the probabilities of the various values of z_j of the random variable Z and the conditional probabilities of the results of the machine M starting from a configuration z_j and reading w_i from the tail of the input word. This calculation shows that the random variables Y_i do not depend from X immediately but only through the random variable Z.

We consider a lower bound for $I(X|Z)$. We have

$$I(X|Z) = H(X) - H(X|Z)$$

$$H(X|Z) = \sum_{j=1}^{u} p(z_j)H(X|Z = z_j)$$

$$H(X|Z = z_j) = H(X_1, X_2, \ldots, X_{2^m}|z = z_j) \leq \sum_{i=1}^{2^m} H(X_i|Z = z_j) \tag{6}$$

Hence

$$H(X|Z) \leq \sum_{j=1}^{u} p(z_j) \sum_{j=1}^{2^m} H(X_i|Z = z_j) =$$

$$= \sum_{i=1}^{2^m}\sum_{j=1}^{u} p(z_j H(X_i|Z = z_j) = \sum_{i=1}^{2^m} H(X_i|Z) \tag{7}$$

We proceed to estimate $H(X_i|Z)$. We start by trying to prove the intuitively valid inequivalence

$$H(X_i|Z) \leq H(X_i|Y_i) \tag{8}$$

Indeed,

$$H(Y_i|Z) = H(Y_i|X_i, Z)$$

because the distribution of probabilities for Y_i is determined by the random variable Z and do not depend on the value of X. Hence we get

$$H(Z, Y_i) - H(Z) = H(X_i, Z, Y_i) - H(X_i, Z)$$

$$H(X_i, Z) - H(Z) = H(X_i, Z, Y_i) - H(Z, Y_i)$$

$$H(X_i|Z) = H(X_i|Z, Y_i) \leq H(X_i|Y_i)$$

This completes the proof of (8).

It follows from (7) and (8) that

$$H(X|Z) \leq \sum_{i=1}^{2^m} H(X_i|Y_i) \tag{9}$$

We fix an arbitrary $i (1 \leq i \leq 2^m)$ and the values for the random variables $X_1, \ldots, X_{i-1}, X_{i+1}, \ldots, X_{2^m}$. It follows from the definition of X that

$$p(X_i = 0) = p(X_i = 1) = 1/2 \tag{10}$$

The Monte-Carlo machine M recognizes the language A with probability exceeding $1/2 + \delta$. Hence there are positive real numbers ϵ_0 and ϵ_1 such that

$$p(Y_i = 0|X_i = 0) = 1/2 + \delta + \epsilon_0 \tag{11}$$

$$p(Y_i = 1|X_i = 1) = 1/2 + \delta + \epsilon_1$$

It follows from (10) and (11)

$$p(X_i = 0|Y_i = 0) =$$

$$\frac{p(X_i = 0)p(Y_i = 0|X_i = 0)}{p(X_i = 0)p(Y_i = 0|X_i = 0) + p(X_i = 1)p(Y_i = 0|X_i = 1)} =$$

$$= \frac{1/2(1/2 + \delta + \epsilon_0)}{1/2(1/2 + \delta + \epsilon_0) + 1/2(1/2 - \delta - \epsilon_1)} \geq \frac{1/2 + \delta + \epsilon_0}{1 + \epsilon_0} \geq 1/2 + \delta$$

Hence

$$H(X_i|Y_i = 0) \leq H(1/2 + \delta, 1/2 - \delta)$$

Similarly one can prove

$$H(X_i|Y_i = 1) \leq H(1/2 + \delta, 1/2 - \delta)$$

It follows that

$$H(X_i|Y_i) = p(Y_i = 0)H(X_i|Y_i = 0) + p(Y_i = 1)H(X_i|Y_i = 1) \leq$$

$$\leq H(1/2 + \delta, 1/2 - \delta)$$

Combining this inequality and (9), we get

$$H(X|Z) \leq 2^m H(1/2 + \delta, 1/2 - \delta)$$

Taking into account (4) and (6), we get

$$I(X|Z) \geq D \cdot 2^m \tag{12}$$

where $D = 1 - H(1/2 + \delta, 1/2 - \delta) > 0$.

Comparing (5) and (12), we get

$$D \cdot 2^m \leq (m + f(n)) \cdot \log_2 c$$

This implies

$$f(n) \geq 2^m \cdot const$$

$$f(n) \geq n \cdot const$$

Contradiction. □

Binary notation of integers is natural, and it has nice properties. However, from our point of view, it has a serious deficiency. Namely, two consecutive integers can have binary notations differing in very many symbols.

We call a notation $sbin(n)$ of non-negative integers n *superbinary* if it has two properties:

1. The length $|sbin(n)|$ for arbitrary non-negative integer n is no more and no less than $const \cdot \log n$;
2. There is a deterministic real time "clock", i.e. a multitape deterministic Turing machine maintaining $sbin(t)$ on one of its worktapes at every moment t (it is allowed to have several symbols of $sbin(t)$ in the same square of the worktape).

For instance, the following notation of non-negative integers in the alphabet $\Sigma_1 = \{0, 1, \boxed{1}, *, \pm, \overrightarrow{0}, \overleftarrow{0}, \overrightarrow{1}, \overleftarrow{1}\}$ is superbinary:

n	s	$bin(n)$
0		*
1		$\boxed{1}$*
2		$\overrightarrow{1}$ *
3		1 *
4		$\overrightarrow{0}$ *
5		$\boxed{1}$ 0 *
6		$\overrightarrow{1}$ 0 *
7		1 $\overrightarrow{0}$ *
8		1 0 *
9		1 $\boxed{1}$*
10		1 $\overrightarrow{1}$ *
11		1 1 *
12		1 $\overleftarrow{0}$ *
13		$\overleftarrow{0}$ 0 *
14		$\boxed{1}$ 0 0 *
15		$\overrightarrow{1}$ 0 0 *
16		1 $\overrightarrow{0}$ 0 *
17		1 0 $\overrightarrow{0}$*
18		1 0 0 *
19		1 0 $\boxed{1}$*
20		1 0 $\overrightarrow{1}$ *
21		1 0 1 *
22		1 0 $\overleftarrow{0}$ *
23		1 $\boxed{1}$ 0 *
24		1 $\overrightarrow{1}$ 0 *

...

We consider language

$$A' = \{x2y | (\exists i)(\exists j)(x \in \{0,1\}^j \& y \in \Sigma_1^* \& y = sbin(i) \& j \geq i \&$$
$$\&\text{the } i\text{-th digit of the word } x \text{ equals } 1\}$$

Theorem 2.2 *The language A' cannot be recognized by a multitape 2-way Monte Carlo Turing machine in time $n + o(n)$.*

Proof. Essentially the same as for Theorem 2.1. □

We go on to consider more complicated languages. Let M be a deterministic multitape 2-way Turing machine, Σ be the input alphabet of M, $t_M(x)$ be the running time of M on input x, and $g(n)$ be the maximum of $(t_M(x) - |x|)$ over all the words $x \in \Sigma^*$ of length not exceeding n. Let # be a symbol not in $\Sigma \cup \{0, 1, 2\}$.

$$B_M = \{x\#y | x \in A' \text{ and } |y| = t_M(|x\#y|)\}$$

Theorem 2.3 *For arbitrary deterministic multitape 2-way Turing machine, the language B_M cannot be recognized by a multitape 2-way Monte Carlo Turing machine in time $n + o(g(n))$.*

Proof. Essentially the same as for Theorem 2.1. □

Theorem 2.4 *For arbitrary deterministic multitape 2-way Turing machine, the language B_M can be recognized by a multitape 2-way Monte Carlo Turing machine in time $n + 2g(n)$.*

Proof. The TM recognizing B_g performs several actions in parallel:

1. Simulates M on $x\#y$ and stores the value of $t_M(|x\#y|)$ on an additional worktape (in unary notation). (This is done in time $n + g(n)$.)
2. Finds out whether $|y| = t_M(|x\#y|)$. (This is done after the action 1 in no more time than $g(n)$.)
3. Finds out whether $y \in A$. (This is done after the action 1 in no more time than $g(n)$; can be done in parallel with the action 2.)

□

Using Theorem 3.1 (below in Section 3) we can improve this result to get

Theorem 2.5 *For arbitrary deterministic multitape 2-way Turing machine, and for arbitrary $\epsilon > 0$ the language B_M can be recognized by a multitape 2-way deterministic Turing machine in time $n + \epsilon g(n)$.*

This way, $n + const \cdot g(n)$ is optimal time for recognition of the language B_M, and randomization does not help.

3 Reduced Time Complexity

Theorems 2.3 and 2.4 show that the complexity measure $t(n) = n + g(n)$ is a natural complexity measure, any way, no less natural than sublinear space complexity introduced in [SHL65] by separating input tape from the worktapes. Before [SHL65] it was considered that the lowest nontrivial space complexity class was LINSPACE since reading all the input demands at least linear space. Separating input tape allowed to consider LOGSPACE and other important small space complexity classes. For instance, it was proved in [SHL65] that DLOGLOGSPACE contains nonregular languages but every language recognizable in $o(\log\log n)$ space even by nondeterministic Turing machines is regular. However there are regular languages recognizable by Monte Carlo 2-way Turing machines in constant space[Fr81], and there are nontrivial space classes defined by $\log\log\log n$, $\log \cdots \log n$, $\log^* n$ for Monte Carlo 2-way Turing machines[KV87].

In this section we prove some results motivating the naturalness of our approach to sublinear time complexity.

Theorem 3.1 *If $n + g(n)$ is the running time of a deterministic (nondeterministic, Monte Carlo, alternating) multitape 2-way Turing machine recognizing a language L, then the language L can be recognized by a deterministic (nondeterministic, Monte Carlo, alternating) multitape 2-way Turing machine in time $n + \frac{1}{2}g(n)$.*

Proof. Every worktape of the given machine is simulated by a worktape with a larger alphabet. Content of four squares from the old tape is stored in a single square of the new tape. An additional worktape is used to make a copy of the input tape (again with 4 symbols stored in a single square of the worktape) thus allowing more speedy retrieval. If the time interval between two consequtive moments of reading a new input symbol by the given machine is between 2 and 4, the simulating machine reads the corresponding input symbols with time interval 1. If the time interval exceeds 4, the simulating machine speeds up. □

Corollary 3.1 *If a language L is recognized by a deterministic (nondeterministic, Monte Carlo, alternating) multitape 2-way Turing machine in time $n + const$, then the language L can be recognized by a deterministic (nondeterministic, Monte Carlo, alternating) multitape 2-way Turing machine in real time.*

It is well-known that even deterministic multitape 2-way Turing machines recognize some non-regular languages in real time, e.g. the language

$$\{x2x | x \in \{0,1\}^*\}$$

For the sake of brevity we introduce a new term for the complexity measure "running time minus the length of the input word". We will call it *reduced running time*.

So far we have shown that the reduced running time can be effectively decreased arbitrary constant number of times (Theorem 3.1). Nonregular languages can be recognized in reduced time 0. Now we go on to show that the reduced time complexity can be very slowly growing: $\log n$, $\log \log n$, $\log \log \log n$, ..., $\log^* n$, etc.

Theorem 3.2 *For arbitrary integer $k \geq 1$, there is a language L_k such that:*

1. *For arbitrary $\epsilon > 0$, the language L_k can be recognized in time $n + \epsilon \cdot \underbrace{\log \log \ldots \log n}_{k \ times}$ by a deterministic 2-way Turing machine;*

2. *No Monte Carlo 2-way Turing machine can recognize the language L_k in time $n + o\big(\underbrace{\log \log \ldots \log n}_{k \ times}\big)$.*

Proof. It suffices to prove the existence of a deterministic Turing machine with the reduced running time $\underbrace{\log \log \cdot \log n}_{k \ times}$, and our theorem will be implied by Theorems 2.3, 2.4 and 2.5.

For $k = 1$ we consider a deterministic 2-way Turing machine which recognizes whether the input word is of type

$$sbin(0) * sbin(1) * sbin(2) * sbin(3) * \ldots * sbin(i)$$

After the real time recognition whether the input word is of the needed type the machine additionally reads the last fragment $sbin(i)$ once more. The running time exceeds the length of the input word by $|sbin(i)| = const \cdot \log i = const' \cdot \log n$.

For $k > 1$ the same idea is used iteratively. Let

$$m_1 = |sbin(m_2)|$$

$$m_2 = |sbin(m_3)|$$

$$m_3 = |sbin(m_4)|$$

$$\ldots$$

$$m_k = |sbin(m_{k+1})|$$

The input word is supposed to be of type

$$sbin(0) * sbin(1) * \ldots * sbin(m_{k+1}) * *sbin(0) * sbin(1) * \ldots * sbin(m_k * * \ldots$$

$$\ldots * *sbin(0) * sbin(1) * \ldots * sbin(m_1)$$

□

Theorem 3.3 *There is a language L_* such that:*

1. *For arbitrary $\epsilon > 0$, the language L_k can be recognized in time $n + \epsilon \cdot \log^* n$ by a deterministic 2-way Turing machine;*
2. *No Monte Carlo 2-way Turing machine can recognize L_k in time $n + o(\log^* n)$.*

Like other complexity measures, the reduced time complexity turns out to be sensitive in choosing the level of determinism (deterministic, nondeterministic, probabilistic, Monte Carlo, alternating) of the machine.

Theorem 3.4 *There is a language A' such that:*

1. *A' can be recognized in real time by a nondeterministic 1-way Turing machine,*
2. *$\bar{A}'$ can be recognized in real time by a nondeterministic 1-way Turing machine,*
3. *No deterministic or Monte Carlo 2-way Turing machine can recognize A' in $n + o(n)$ time.*

Proof. The language A' from Theorem 2.2 has the needed properties. □

4 Advantages of Randomization

In spite of the huge literature on complexity advantages of randomized machines over deterministic ones there had not been proved any running time advantages of Monte Carlo multitape 2-way Turing machines over determinstic machines of the same type.

Theorem 4.1 *There is a language E such that:*

1. *For arbitrary $\epsilon > 0$, the language E can be recognized in real time by a Monte Carlo multitape 2-way machine with probability $1 - \epsilon$,*
2. *E cannot be recognized in time $n + o(n)$ by a deterministic multitape 2-way Turing machine.*

Sketch of proof. The language E consists of all the words in the form

$$x_1 2 x_2 2 y_1 2 y_2 3 u(1) 2 u(2) 2 \ldots 2 u(i) 3 v(1) 2 v(2) 2 \ldots 2 v(i)$$

such that

$$(\exists m)(\exists j)(x_1 = 1^m \& x_2 = 1^{2^m} \& y_2 = 1^j \& 2^m \leq i \leq j \leq 8i \&$$

$$\& (\forall k)(1 \leq k \leq i \rightarrow u(k) \in \{0,1\}^m \cdot 2 \cdot 1^{m^3}) \& \text{the string}$$

$$v(1), v(2), \ldots, v(i) \text{ is permutation of } u(1), u(2), \ldots, u(i)$$

1. The prefix $x_1 2 x_2 2 y_1 2 y_2$ is designed to allow the randomized machine time to prepare the random parameters and organize the worktapes.

   Let c denote $\lceil \frac{2}{\epsilon} \rceil$. A random integer r is chosen ($1 \leq r \leq c \cdot 2^m$). A random prime number p execeding $c \cdot 2^m$ is chosen among the first $c \cdot j$ prime numbers (i.e. among the random integers of size not exceeding $\log_2(c \cdot j \cdot \ln j)$ being prime numbers). The random number theorem implies that on average the length of the prefix suffices for the randomized machine to generate and test such a random prime.

   When the blocks $u(1), u(2), \ldots, u(i)$ are read from the input, and $u(k) = bin(z) \cdot 2 \cdot 1^{m^3}$, the number r^z (mod p) is added to a counter. When the blocks $v(1), v(2), \ldots, v(i)$ are read from the input, and $v(k) = bin(z) \cdot 2 \cdot 1^{m^3}$, the number r^z (mod p) is subtracted from the counter. If the input word is in the language E, at the end the counter is empty.

   If the input word is not in E, the properties of Vandermonde determinant show that no more than the fraction $\frac{1}{c}$ of all the random numbers r would make the totals of r^z (not r^z (mod p)) for $u(1), u(2), \ldots, u(i)$ and $v(1), v(2), \ldots, v(i)$ equal.

   If

$$\sum_u r^z \neq \sum_v r^z$$

but the totals are congruent modulo p, then the difference is a multiple of p. If inequal totals are congruent modules $p_1, p_2, \ldots, p_s$ then the difference is a multiple of the product $p_1 \cdot p_2 \cdot \ldots \cdot p_s$. Since the difference of totals cannot be larger than the maximum value of the total, the fraction of "defective" prime modulos is small.

Computing of r^z (mod p) is not real-time but we have added blocks 1^{m^3} when constructing the blocks $u(1), \ldots, u(i), v(1), \ldots, v(i)$. This way, the values

$$r^0 \quad (\bmod \ p)$$
$$r^1 \quad (\bmod \ p)$$
$$r^2 \quad (\bmod \ p)$$
$$r^4 \quad (\bmod \ p)$$
$$\ldots$$
$$r^{2^m} \quad (\bmod \ p)$$

are preprocessed, and added to the counter when the block $u(k)$ (containing binary notation of z) is read from the input.

2. Notion of Kolmogorov complexity is used for this part of the proof.

Assume from the contrary that there is such a deterministic machine. Take a large integer m and a binary string α such that $|\alpha| = 2^m$ and α has nearly maximal Kolmogorov complexity.

We denote

$$\alpha = \alpha_1 \alpha_2 \ldots \alpha_{2^m}$$

We construct $u(1)2u(2)2 \ldots 2u(i)$ taking the strings z in the lexicographical order $(00000, 00001, 00010, 00011, \ldots)$. If the current α_b equals 0, we take one block $u(k)$ corresponding to this z. If the current α_b equals 1, we take two blocks $u(k)$ and $u(k+1)$ corresponding to this z. This way, $2^m \leq 2 \cdot 2^m$. We consider a special graph representing the information transfer in the multitape Turing machine. The vertices of the graph correspond to the time moments and positions of the heads on the tapes. If there are d tapes (including the input tape) then d new vertices correspond to every time moment (one vertice per head). The vertices are connected if:

(a) either the time moments are the same or adjacent,

(b) or the same square of the tape is visited again at some different moment (but there have been no visits to this square of the tape between these moments).

Since the input word corresponds to α which has a high Kolmogorov complexity, the number of sqaures on the worktapes is to be at least linear with respect to $|\alpha|$. This implies the nearly-linear diameter of the graph. Hence there are fragments of $u(1)2u(2)2 \ldots 2u(i)$ and $v(1)2v(2)2 \ldots 2v(i)$ containing the same linear number of z's such that the distance in the graph is of linear size. Next, we consider cuts in this graph separating these fragments. Since the cuts are disjoint sets of vertices, and the diameter is nearly-linear, there is a cut of about-logarithmic size. However the cut is supposed to contain all the information about α, otherwise the machine can be fooled. But one cannot compress α into about-logarithmic size. Contradiction.

$\square$

5 Acknowledgement

The idea to use prime modulos for the randomized algorithm in the proof of Theorem 4.1 was proposed to the authors by Andris Ambainis. We thank also Leonid Levin and Peter Gacs for interesting discussions.

References

[Fr75] Freivalds, R., *Fast computations by probabilistic Turing machines*, Proceedings of Latvian State University, 233(1975), pp. 201-205 (Russian)

[Fr77] Freivalds, R., *Probabilistic machines can use lsess runnning time*, Information Processing'77 (Proc. IFIP Congress'77), North Holland, 1977, pp. 839-842

[Fr79] Freivalds, R., *Speeding up recognition of some sets by usage of random number generators*, Problemi kibernetiki, 36(1979), pp. 209-224 (Russian)

[Fr81] Freivalds, R., *Probabilistic two-way machines*, LNCS, 118(1981), pp.33-45

[Fr83] Freivalds, R., *Space and reversal complexity of probabilistic one-way Turing machines*, LNCS, 158(1983), pp. 159-170

[Fr85] Freivalds, R., *Space and reversal complexity of probabilistic one-way Turing machines*, Annals of Discrete Mathematics, 24(1985), pp. 39-50

[FI77] Freivalds, R., Ikaunieks, E., *On advantages of nondeterministic machines over probabilistic ones*, Izvestiya VUZ. Matematika, No.2(177), 1977, pp.108-123 (Russian)

[FK94] Freivalds, R., Karpinski, M., *Lower space bounds for randomized computation*, Lecture Notes in Computer Science, vol. 820(1994), pp. 580-592

[Gal68] Gallager, R.G., *Information Theory and Reliable Communication.* John Wiley, NY, 1968

[JKS84] Ja'Ja', J., Prasanna Kumar, V.K., Simon, J., *Information transfer under different sets of protocols*, SIAM J. Computation, 13(1984), pp.840-849

[KF90] Kaneps, J. and Freivalds R., *Minimal nontrivial space complexity of probabilistic one-way Turing machines*, LNCS, 452(1990), pp. 355-361

[KV87] Karpinski, M. and Verbeek, R., *On the Monte Carlo space constructible functions and space separation results for probabilistic complexity classes*, Information and Computation, 75(1987), pp. 178-189

[KV88] Karpinski, M. and Verbeek, R., *Randomness, probability, and the separation of Monte Carlo time and space*, LNCS, 270(1988), pp. 189-207

[KV93] Karpinski, M. and Verbeek, R., *On randomized versus deterministic computation*, Proc. ICALP'93, LNCS, 700(1993), pp. 227-240

[LV93] Ming Li, Paul Vitanyi, *An Introduction to Kolmogorov Complexity and Its Applications*, Springer, 1993

[SHL65] Stearns, R.E., Hartmanis, J., Lewis, P.M., *Hiearchies of memory limited computations*, Proc. IEEE Conference on Switch., Circuit Theory and Logical Design, 1965, pp. 179-190

New Collapse Consequences of NP Having Small Circuits

Johannes Köbler[1] and Osamu Watanabe[*2]

[1] Abteilung Theoretische Informatik, Universität Ulm, 89069 Ulm, GERMANY
koebler@informatik.uni-ulm.de
[2] Dept. of Computer Science, Tokyo Institute of Technology, Tokyo 152, JAPAN
watanabe@cs.titech.ac.jp

Abstract. We show that if a self-reducible set has polynomial-size circuits, then it is low for the probabilistic class ZPP(NP). As a consequence we get a deeper collapse of the polynomial-time hierarchy PH to ZPP(NP) under the assumption that NP has polynomial-size circuits. This improves on the well-known result of Karp, Lipton, and Sipser (1980) stating a collapse of PH to its second level Σ_2^P under the same assumption. As a further consequence, we derive new collapse consequences under the assumption that complexity classes like UP, FewP, and $C_=P$ have polynomial-size circuits.

Finally, we investigate the circuit-size complexity of several language classes. In particular, we show that for every fixed polynomial s, there is a set in ZPP(NP) which does not have $O(s(n))$-size circuits.

1 Introduction

The question whether intractable sets can be efficiently decided by non-uniform models of computation has motivated much work in structural complexity theory. In research from the early 1980's to the present, a variety of results has been obtained showing that this is impossible under plausible assumptions (see, e.g., the survey [18]). A typical model for non-uniform computations are circuit families. In the notation of Karp and Lipton [22], sets decidable by polynomial-size circuits are precisely the sets in P/poly, i.e., they are decidable in polynomial time with the help of a polynomial length bounded advice function [31].

Karp and Lipton (together with Sipser) [22] proved that no NP-complete set has polynomial size circuits (in symbols NP $\not\subseteq$ P/poly) unless the polynomial time hierarchy collapses to its second level. The proof given in [22] exploits a certain kind of self-reducibility of the well-known NP complete problem SAT. More generally, it is shown in [7, 8] that every (Turing) self-reducible set in P/poly is low for the second level Σ_2^P of the polynomial time hierarchy. Intuitively speaking, a set is low for a relativizable complexity class if it gives no additional power when used as an oracle for that class.

In this paper, we show that every self-reducible set in P/poly is also low for the probabilistic class ZPP(NP). Since for every oracle A, $\Sigma_2^P(A) =$

---

* Part of this work has been done while visiting Universität Ulm (supported in part by the guest scientific program of Universität Ulm).

$\exists \cdot \text{ZPP}(\text{NP}(A))$, lowness for $\text{ZPP}(\text{NP})$ implies lowness for Σ_2^P. As a consequence of our lowness result we get a deeper collapse of the polynomial-time hierarchy to $\text{ZPP}(\text{NP})$ under the assumption that NP has polynomial-size circuits. At least in some relativized world, the new collapse level is quite close to optimal: there is an oracle relative to which NP is contained in P/poly but PH does not collapse to $\text{P}(\text{NP})$ [17, 35]. Our proof heavily uses the universal hashing technique [13, 33] and builds on ideas from [2, 14, 24]. A central and new notion used for the design of a *zero error* probabilistic algorithm is the concept of half-collisions.

Based on our lowness result, we obtain new collapse consequences under the assumption that complexity classes like NP, UP, FewP, and $\text{C}_{=}\text{P}$ have polynomial-size circuits. We further obtain new *relativizable* collapses for the case that Mod_mP, PSPACE, or EXP have polynomial-size circuits.

Very recently, Bshouty, Cleve, Kannan, and Tamon [10] building on a result from [19] have shown that the class of all circuits is exactly learnable in (randomized) expected polynomial time with equivalence queries and the aid of an NP oracle. This immediately implies that every set A in P/poly has an advice function in $\text{FZPP}(\text{NP}(A))$. More precisely, since the circuit produced by the probabilistic learning algorithm of [10] depends on the outcome of the coin flips, the $\text{FZPP}(\text{NP}(A))$ transducer T computes a *multi-valued* advice function, i.e., on input 0^n, T accepts with probability at least $1/2$, and on every accepting path, T outputs some circuit that correctly decides all instances of length n w.r.t. A. Using the technique in [10] we are able to show that every self-reducible set A in P/poly has an advice function in $\text{FZPP}(\text{NP})$; thus providing an alternative way to deduce the $\text{ZPP}(\text{NP})$ lowness of all self-reducible sets in P/poly. However, since our main interest is in the collapse consequences of intractable problems having polynomial-size circuits we prefer to give a self-contained proof of the lowness result.

As a further application, we derive new circuit-size lower bounds. In particular, we show by relativizing proof techniques that for every fixed polynomial s, there is a set in $\text{ZPP}(\text{NP})$ which does not have $O(s(n))$-size circuits. This improves on the result of Kannan [21] that for every polynomial s, the class $\Sigma_2^P \cap \Pi_2^P$ contains such a set. It further follows that in every relativized world, there exist sets in the class $\text{ZPEXP}(\text{NP})$ that do not have polynomial-size circuits. It should be noted here that there is a non-relativizing proof for a stronger result. As a corollary to the result in [4], which is proved by a non-relativizing technique, it is provable that $\text{MA}_{\text{exp}} \cap \text{co-MA}_{\text{exp}}$ (a subclass of $\text{ZPEXP}(\text{NP})$) contains non P/poly sets [12, 34].

The paper is organized as follows. Section 2 introduces notation and defines the self-reducibility used in the paper. In Section 3, we prove the $\text{ZPP}(\text{NP})$ lowness of all self-reducible sets in P/poly. In Section 4, we state the collapse consequences, and in Section 5, we derive the new circuit-size lower bounds. Because of space limitations some of the proofs have been omitted.

2 Preliminaries and notation

All languages are over the binary alphabet $\Sigma = \{0, 1\}$. The *length* of a string $x \in \Sigma^*$ is denoted by $|x|$. $\Sigma^{\leq n}$ ($\Sigma^{<n}$) is the set of all strings of length at most n (resp., of length smaller than n). For a language A, $A^{=n} = A \cap \Sigma^n$ and $A^{\leq n} = A \cap \Sigma^{\leq n}$. The cardinality of a finite set A is denoted by $|A|$. The *characteristic function* of A is defined as $A(x) = 1$ if $x \in A$, and $A(x) = 0$, otherwise. For a class $\mathcal{C}$ of sets, co-$\mathcal{C}$ denotes the class $\{\Sigma^* - A \mid A \in \mathcal{C}\}$. To encode pairs (or tuples) of strings we use a standard polynomial-time computable pairing function denoted by $\langle \cdot, \cdot \rangle$ whose inverses are also computable in polynomial time. Where intent is clear we write $f(x_1, \ldots, x_k)$ in place of $f(\langle x_1, \ldots, x_k \rangle)$. $\mathcal{N}$ denotes the set of non-negative integers. Throughout the paper, the base of log is 2.

The textbooks [9, 11, 25, 30, 32] can be consulted for the standard notations used in the paper and for basic results in complexity theory. For definitions of probabilistic complexity classes like ZPP see also [15]. Next we define the kind of self-reducibility that we use in this paper.

Definition 1. Let $\succ$ be an irreflexive and transitive order relation on Σ^*. A sequence $x_0, x_1, \ldots, x_k$ of strings is called a $\succ$-chain (of length k) from x_0 to x_k if $x_0 \succ x_1 \succ \cdots \succ x_k$. Relation $\succ$ is called *length checkable* if there is a polynomial q such that

1. for all $x, y \in \Sigma^*$, $x \succ y$ implies $|y| \leq q(|x|)$,
2. the language $\{\langle x, y, k \rangle \mid$ there is a $\succ$-chain of length k from x to $y\}$ is in NP.

Definition 2. A set A is *self-reducible*, if there is a polynomial-time oracle machine M and a length checkable order relation $\succ$ such that $A = L(M, A)$ and on any input x, M queries the oracle only about strings $y \prec x$.

It is straightforward to check that the polynomially related self-reducible sets introduced by Ko [23] as well as the length-decreasing and word-decreasing self-reducible sets of Balcázar [6] are self-reducible in our sense. Furthermore, it is well-known that complexity classes like NP, Σ_k^P, Π_k^P, $k \geq 1$, PP, $C_=P$, $\text{Mod}_m P$, $m \geq 2$, PSPACE, and EXP have many-one complete self-reducible sets (see, for example, [9, 6, 29]).

Karp and Lipton [22] introduced the notion of advice functions in order to characterize non-uniform complexity classes. A function $h : \mathcal{N} \to \Sigma^*$ is called a *polynomial-length function* if for some polynomial p and for all $n \geq 0$, $|h(n)| = p(n)$. For a class $\mathcal{C}$ of sets, let $\mathcal{C}/\text{poly}$ be the class of sets A such that there is a set $I \in \mathcal{C}$ and a polynomial-length function h such that

$$\forall n, \forall x \in \Sigma^{\leq n} \, [\, x \in A \Leftrightarrow \langle x, h(n) \rangle \in I \,].$$

Function h is called an *advice function* for A, and I is the corresponding *interpreter set*.

In this paper, we will heavily make use of the "hashing technique", which has been very fruitful in complexity theory. Here we review some notations and facts about hash families. We also extend the notion of "collision" and introduce the concept of a "half collision" which is central to our proof technique.

Sipser [33] used universal hashing, originally invented by Carter and Wegman [13], to decide (probabilistically) whether a finite set X is large or small. A linear hash function h from Σ^m to Σ^k is given by a Boolean (k, m)-matrix (a_{ij}) and maps any string $x = x_1 \ldots x_m$ to some string $y = y_1 \ldots y_k$, where y_i is the inner product $a_i \cdot x = \sum_{j=1}^m a_{ij} x_j \pmod 2$ of the i-th row a_i and x.

Let $x \in \Sigma^m, Y \subseteq \Sigma^m$, and let h be a linear hash function from Σ^m to Σ^k. We say that x has a *collision on* Y *w.r.t.* h if there exists a string $y \in Y$, different from x, such that $h(x) = h(y)$. In general, for any $X \subseteq \Sigma^m$, and any family $H = \{h_1, \ldots, h_l\}$ of linear hash functions from Σ^m to Σ^k, X *has a collision on* Y *w.r.t.* H (*Collision*(X, Y, H) for short) if there is some $x \in X$ that has a collision on Y w.r.t. any h_i in H. That is,

$$\text{Collision}(X, Y, H) \Leftrightarrow \exists x \in X \, \exists y_1, \ldots, y_l \in Y : x \notin \{y_1, \ldots y_l\}$$
$$\text{and for all } i = 1, \ldots, l : h_i(x) = h_i(y_i).$$

If X has a collision on itself w.r.t. H, we simply say that X *has a collision w.r.t.* H. Next we extend the notion of "collision" in the following way. For any X and $Y \subseteq \Sigma^m$, and any family $H = \{h_1, \ldots, h_l\}$ of linear hash functions, we say that X *has a half-collision on* Y *w.r.t.* H (*Half-Collision*(X, Y, H) for short) if there is some $x \in X$ that has a collision on Y w.r.t. at least $\lceil l/2 \rceil$ many of the hash functions h_i in H. That is,

$$\text{Half-Collision}(X, Y, H) \Leftrightarrow \exists x \in X \, \exists y_1, \ldots, y_l \in Y : x \notin \{y_1, \ldots y_l\}$$
$$\text{and } |\{i \mid 1 \leq i \leq l, h_i(x) = h_i(y_i)\}| \geq \lceil l/2 \rceil.$$

An important relationship between collisions and half-collisions is the following one: If X has a collision w.r.t. H on $Y = Y_1 \cup Y_2$, then X must have a half-collision w.r.t. H either on Y_1 or on Y_2.

We denote the set of all families $H = \{h_1, \ldots, h_l\}$ of l linear hash functions from Σ^m to Σ^k by $\mathcal{H}(l, m, k)$. The following theorem is proved by a pigeon-hole argument. It says that every sufficiently large set must have a collision w.r.t. any hash family.

Theorem 3. [33] *For any hash family $H \in \mathcal{H}(l, m, k)$ and any set $X \subseteq \Sigma^m$ of cardinality $|X| > l \cdot 2^k$, X must have a collision w.r.t. H.*

On the other hand, we get from the next theorem (called Coding Lemma in [33]) an upper bound on the collision probability for sufficiently small sets.

Theorem 4. [33] *Let $X \subseteq \Sigma^m$ be a set of cardinality at most 2^{k-1}. If we choose a hash family H uniformly random from $\mathcal{H}(k, m, k)$, then the probability that X has a collision w.r.t. H is at most $1/2$.*

We will also make use of the following extension of Theorem 4 which can be proved along the same lines.

Theorem 5. *Let $X \subseteq \Sigma^m$ be a set of cardinality at most 2^{k-s}. If we choose a hash family H uniformly random from $\mathcal{H}(l, m, k)$, then the probability that X has a collision w.r.t. H is at most $2^{k-s(l+1)}$.*

By combining Theorem 3 and Theorem 5, a rough estimation for the cardinality of a nonempty set $X \subseteq \Sigma^m$ can be obtained with high probability: choose $n \geq 0$ and for every $k = 1, \ldots, m$, randomly guess a hash family H_k from $\mathcal{H}(n + k, m, k)$; let $k_{max} \geq 0$ be the maximum $k \leq m$ such that for all $i \leq k$, X has a collision w.r.t. H_i; then we have that $|X| \leq (n + k_{max} + 1)2^{k_{max}+1}$, and with probability at least $1 - 2^{-n-1}$, $|X| > 2^{k_{max}-1}$.

Gavaldà [14] extended Sipser's Coding Lemma (Theorem 4) to the case of a collection $\mathcal{C}$ of exponentially many sets. In this paper we make use of a corresponding result for the case of half-collisions.

Theorem 6. *Let $X \subseteq \Sigma^m$ and let $\mathcal{C}$ be a collection of at most 2^n subsets of Σ^m, each of which has cardinality at most 2^{k-s-2}. If we choose a hash family H uniformly random from $\mathcal{H}(l, m, k)$, then the probability that X has a half-collision on some $Y \in \mathcal{C}$ w.r.t. H is at most $|X| \cdot 2^{n-sl/2}$.*

3 Lowness of self-reducible sets in P/poly

In this section, we prove the following theorem.

Theorem 7. *Every self-reducible set A in the class $(NP \cap co\text{-}NP)/poly$ is low for $ZPP(NP)$.*

Let I_A be an interpreter set and h_A be an advice function for A. We construct a probabilistic algorithm T_A and an NP oracle L_A having the following properties:

a) The expected running time of $T_A^{L_A}$ is polynomially bounded.

b) On every computation path on input 0^n, $T_A^{L_A}$ outputs some information that can be used to determine the membership of any x up to length n to A by some strong NP computation (in the sense of [26]).

Using these properties, we can prove the lowness of A for $ZPP(NP)$ as follows: In order to simulate any $NP(A)$ computation, we first precompute the above mentioned information for A (up to some length) by $T_A^{L_A}$, and then by using this information, we can simulate the $NP(A)$ computation by some $NP(NP \cap co\text{-}NP)$ computation. Note that the precomputation (performed by $T_A^{L_A}$) can be done in $ZPP(NP)$, and since $NP(NP \cap co\text{-}NP) = NP$, the remaining computation can be done in NP. Hence, $NP(A) \subseteq ZPP(NP)$, which implies further that $ZPP(NP(A)) \subseteq ZPP(ZPP(NP))$ $(= ZPP(NP)$ [37]).

We will now make the term "information" precise. For this, we need some additional notation. Let M_{self} be a polynomial-time oracle machine, let $\succ$ be a length checkable order relation, and let q be a polynomial witnessing the self-reducibility of A. We assume that $|h_A(q(n))| = p(n)$ for some fixed polynomial $p > 0$. In the following, we fix n and consider instances (to A) of length up to $q(n)$ as well as advice strings of length exactly $p(n)$.

- A *sample* is (the encoding of) a set of pairs of the form $\langle x_i, A(x_i) \rangle$, where the x_i's are instances.
- For any sample $S = \{\langle x_1, b_1 \rangle, \ldots, \langle x_k, b_k \rangle\}$, let *Consistent(S)* be the set $\{w \in \Sigma^{p(n)} \mid \forall i\,(1 \le i \le k) : I_A(x_i, w) = b_i\}$ of all advice strings w that are consistent with S.
- For any sample S and any instance x, let *Accept(x, S)* (resp., *Reject(x, S)*) be the set of all consistent advice strings that *accept* x (resp., *reject* x). That is, $Accept(x, S) = \{w \in Consistent(S) \mid I_A(x, w) = 1\}$ and $Reject(x, S) = \{w \in Consistent(S) \mid I_A(x, w) = 0\}$.
- Let *Correct(x, S)* be the set $\{w \in Consistent(S) \mid I_A(x, w) = A(x)\}$ of consistent advice strings that decide x correctly, and let *Incorrect(x, S)* be the complementary set $\{w \in Consistent(S) \mid I_A(x, w) \ne A(x)\}$.

The above condition b) can now be precisely stated as follows:

b) On every computation path on input 0^n, $T_A^{L_A}$ outputs a pair (S, H) consisting of a sample S and a hash family $H \in \mathcal{H}(q(n) + k, p(n), k)$, for some k, $1 \le k \le p(n)$, such that for all x up to length n, *Consistent(S)* has a half-collision on *Correct(x, S)* w.r.t. H, but not on *Incorrect(x, S)*.

Once we have a pair (S, H) satisfying condition b), we can determine whether an instance x of length up to n is in A by simply checking on which one of *Accept(x, S)* or *Reject(x, S)*, *Consistent(x, S)* has a half-collision w.r.t. H. Since condition b) guarantees that the half-collision can always be found, this checking can be done by a strong NP computation. Let us now prove our main lemma.

Lemma 8. *For any self-reducible set A in* (NP $\cap$ co-NP)/poly, *there exist a probabilistic transducer T_A and an oracle L_A in NP satisfying the above conditions a) and b).*

Proof. We use the notation introduced so far. Further, we denote by $\Sigma^{\preceq n}$ the set $\{y \mid \exists x \in \Sigma^{\le n}, x \succeq y\}$. It is clear that $\Sigma^{\le n} \subseteq \Sigma^{\preceq n} \subseteq \Sigma^{\le q(n)}$ (recall that $q(n)$ is a length bound on the queries occuring in the self-reducing tree produced by M_{self} on any instance of length n). Let c be a fixed constant such that $q(n) + p(n) + 1 \le 2^{\lfloor c \log n \rfloor - 2}$ for all sufficiently large n. (Recall that $p(n)$ is the advice length for the set of all instances of length up to $q(n)$.)

A description of T_A is given in Figure 1. Starting with the empty sample, T_A enters the main loop. During each execution of the loop, T_A first randomly guesses a series of $p(n)$ many hash families $H_k \in \mathcal{H}(q(n) + k, p(n), k), 1 \le k \le p(n)$. Then T_A computes the integer $d = k_{max} - \lfloor c \log n \rfloor$, where k_{max} is the maximum integer $k \in \{0, \ldots, p(n)\}$ such that *Consistent(S)* has a collision w.r.t. all hash families $H_i, 1 \le i \le k$. (W.l.o.g. we can assume that d is positive.)

Note, in particular, that *Consistent(S)* has a collision w.r.t. H_d; thus, for every instance x, *Consistent(S)* has a half-collision w.r.t. H_d on either *Correct(x, S)* or *Incorrect(x, S)*.

If there exists a string $x \in \Sigma^{\preceq n}$ such that *Consistent(S)* has a half-collision on *Incorrect(x, S)* w.r.t. H_d, then this string is added to the sample S, and T_A

input 0^n
$S := \emptyset$
loop
    for $k = 1, \ldots, p(n)$, choose H_k randomly from $\mathcal{H}(q(n) + k, p(n), k)$,
    $k_{max} := \max\{k \mid \forall i \leq k, \; Consistent(S) \text{ has a collision w.r.t. } H_i\}$
    $d := k_{max} - \lfloor c \log n \rfloor$
    if there exists an $x \in \Sigma^{\leq n}$ such that $Consistent(S)$ has a half-collision
    on $Incorrect(x, S)$ w.r.t. H_d
    then
        use oracle L_A to find such a string x and to determine $A(x)$
        $S := S \cup \{\langle x, A(x)\rangle\}$
    else exit(loop) **end**
end loop
output (S, H_d)

Fig. 1. The probabilistic algorithm T_A.

reenters the loop. We will describe below how T_A uses the NP oracle L_A to find such an x (if it exists). Otherwise, the pair (S, H_d) has the desired properties as stated above, and T_A outputs the pair (S, H_d).

The intuition behind the choice of the value for d (depending on k_{max}) is as follows:

- d is still large enough to ensure that for a suitable polynomial t and for a random hash family $H \in \mathcal{H}(q(n) + d, p(n), d)$, the probability is exponentially small that $Consistent(S)$ has a half-collision w.r.t. H on some set $Incorrect(x, S)$ of size smaller than $c(S)/t(n)$.
- On the other hand, with high probability, d is so small that $Consistent(S)$ has a collision w.r.t. every hash family $H \in \mathcal{H}(q(n) + d, p(n), d)$. This is important since in order to estimate the success probability of T_A we have to consider the *conditional* probability for $Consistent(S)$ having a half-collision on X *given that* $Consistent(S)$ has a collision w.r.t. H_d.

A more precise analysis follows. Let S be a sample and let d be the corresponding integer as determined by T_A (i.e., $d = k_{max} - \lfloor c \log n \rfloor$, where $k_{max} = p(n)$ or $Consistent(S)$ does not have a collision w.r.t. some hash family $H_{k_{max}+1} \in \mathcal{H}(q(n) + k_{max} + 1, p(n), k_{max} + 1)$). We first estimate the probability that w.r.t. a uniformly at random chosen hash family $H \in \mathcal{H}(q(n) + d, p(n), d)$, $Consistent(S)$ has a half-collision on some set $Incorrect(x, S)$ of relatively small size. Let $\mathcal{C}$ be the collection of all sets X of the form $Accept(x, S)$ or $Reject(x, S)$ for some $x \in \Sigma^{\leq q(n)}$ such that $|X| \leq c(S)2^{-2\lfloor c \log n \rfloor - 5}$.

Claim. The probability of $Consistent(S)$ having a half-collision on some $X \in \mathcal{C}$ w.r.t. a uniformly at random chosen hash family $H \in \mathcal{H}(q(n) + d, p(n), d)$ is at most $2^{-q(n)-1}$. (Proof omitted.)

Now consider an arbitrary execution of the main loop during which S is expanded by some instance x. Since $c(S \cup \{\langle x, A(x)\rangle\}) = c(S) - |Incorrect(x, S)|$, the expected number of loop iterations is polynomially bounded provided that there is some fixed polynomial t such that $|Incorrect(x, S)| \leq c(S)/t(n)$ holds only with low probability.

Claim. There is a polynomial t such that in each execution of the main loop, with probability at most 2^{-n} an instance x with $|Incorrect(x, S)| \leq c(S)/t(n)$ is selected. (Proof omitted.)

We finally show how T_A can find a string $x \in \Sigma^{\leq n}$ such that $Consistent(S)$ has a half-collision on $Incorrect(x, S)$ w.r.t. H_d (if such an x exists). Define oracle L_A as follows:

$$L_A = \{\langle 0^n, x, k, b, S, H\rangle \mid \text{there is a } \succ\text{-chain of length } k \text{ from some string}$$
$$y \in \Sigma^{\leq n} \text{ to some string } z \leq x \text{ such that there is a computation path}$$
$$\pi \text{ of } M_{self} \text{ on input } z \text{ fulfilling the following properties:}$$

- if a query q is answered 'yes' ('no'), then $Consistent(S)$ has a half-collision on $Accept(q, S)$ (resp., $Reject(q, S)$) w.r.t. H,
- if π is accepting (rejecting), then $Consistent(S)$) has a half-collision on $Reject(z, S)$ (resp., $Accept(z, S)$) w.r.t. H,
- if $b = 1$, then π is accepting $\}$.

If the tuple $\langle 0^n, 1^{q(n)}, 0, 0, S, H_d\rangle$ is not in T_A, then it follows by the definition of T_A that w.r.t. H_d, $Consistent(S)$ does not have a half-collision on any of the sets $Incorrect(x, S)$, $x \in \Sigma^{\leq n}$.

Otherwise, by asking queries of the form $\langle 0^n, 1^{q(n)}, i, 0, S, H_d\rangle$, T_A can compute by binary search i_{max} as the maximum value $i \leq 2^{q(n)+1}$ such that $\langle 0^n, 1^{q(n)}, i, 0, S, H_d\rangle$ is in L_A (a similar idea is used in [27]). Knowing i_{max}, T_A can find the lexicographically smallest string x_{min} such that $\langle 0^n, x_{min}, i_{max}, 0, S, H_d\rangle$ is in L_A. Since for all $q \prec x_{min}$, $Consistent(S)$ does not have a half-collision on $Incorrect(q, S)$ w.r.t. H_d, it is easy to see that $Consistent(S)$ must have a half-collision on $Incorrect(x_{min}, S)$ w.r.t. H_d. Finally, T_A can determine $A(x_{min})$ by asking whether $\langle 0^n, x_{min}, i_{max}, 1, S, H_d\rangle$ is in L_A. □

4 Collapse consequences

As a direct consequence of Theorem 7 we get an improvement of Karp, Lipton, and Sipser's result [22] that NP is not contained in P/poly unless the polynomial-time hierarchy collapses to Σ_2^P.

Corollary 9. *If* NP *is contained in* (NP ∩ co-NP)/poly *then the polynomial-time hierarchy collapses to* ZPP(NP).

The collapse to ZPP(NP) in Corollary 9 is quite close to optimal, at least in some relativized world [17, 35]: there is an oracle relative to which NP is contained in P/poly but the polynomial-time hierarchy does not collapse to P(NP).

In the rest of this section we report some other interesting collapses which can be easily derived using (by now) standard techniques, and which have also been pointed out independently by several researchers to the second author. First, it is straightforward to check that Theorem 7 relativizes: For any oracle B, if A is a self-reducible set in the class $(NP(B) \cap co\text{-}NP(B))/poly$, then $NP(A)$ is contained in $ZPP(NP(B))$. Consequently, Theorem 7 generalizes to the following result.

Theorem 10. *If A is a self-reducible set in the class $(\Sigma_k^P \cap \Pi_k^P)/poly$, then $NP(A) \subseteq ZPP(\Sigma_k^P)$.*

As a direct consequence of Theorem 10 we get an improvement of results in [1, 20] stating (for $k = 1$) that Σ_k^P is not contained in $(\Sigma_k^P \cap \Pi_k^P)/poly$ unless the polynomial-time hierarchy collapses to Σ_{k+1}^P.

Corollary 11. *Let $k \geq 1$. If Σ_k^P is contained in $(\Sigma_k^P \cap \Pi_k^P)/poly$, then the polynomial-time hierarchy collapses to $ZPP(\Sigma_k^P)$.*

Proof. Since Σ_k^P contains complete self-reducible languages, the assumption that Σ_k^P is contained in $(\Sigma_k^P \cap \Pi_k^P)/poly$ implies that $\Sigma_{k+1}^P = NP(\Sigma_k^P) \subseteq ZPP(\Sigma_k^P)$.

A further consequence of Theorem 10 is the following improvement of a result due to Yap [36] stating that Π_k^P is not contained in $\Sigma_k^P/poly$ unless the polynomial-time hierarchy collapses to Σ_{k+2}^P.

Corollary 12. *Let $k \geq 1$. If $\Pi_k^P \subseteq \Sigma_k^P/poly$, then PH collapses to $ZPP(\Sigma_{k+1}^P)$.*

Proof. The assumption that Π_k^P is contained in $\Sigma_k^P/poly$ implies that Σ_{k+1}^P is contained in $\Sigma_k^P/poly \subseteq (\Sigma_{k+1}^P \cap \Pi_{k+1}^P)/poly$. Hence we can apply Corollary 11. □

As corollaries to Theorem 10, we also have similar collapse results for many other complexity classes. What follows are some typical examples. (Harry Buhrman pointed out that Corollary 14 can also be derived from Theorem 15.)

Corollary 13.

i) *For $\mathcal{K} \in \{UP, FewP\}$, if $\mathcal{K} \subseteq (NP \cap co\text{-}NP)/poly$ then $\mathcal{K}$ is low for $ZPP(NP)$.*
ii) *For every $k \geq 1$, if $C_=P \subseteq (\Sigma_k^P \cap \Pi_k^P)/poly$ then $CH = ZPP(\Sigma_k^P)$.*

Corollary 14. *For every $k \geq 1$, and in every relativized world,*

i) *For $m \geq 2$, if $Mod_mP \subseteq (\Sigma_k^P \cap \Pi_k^P)/poly$ then $Mod_mP \subseteq PH = ZPP(\Sigma_k^P)$.*
ii) *If $PSPACE \subseteq (\Sigma_k^P \cap \Pi_k^P)/poly$ then $PSPACE = ZPP(\Sigma_k^P)$.*
iii) *If $EXP \subseteq (\Sigma_k^P \cap \Pi_k^P)/poly$ then $EXP = ZPP(\Sigma_k^P)$.*

Since our proof technique is relativizable, the above results hold for every relativized world. On the other hand, some of the results are further improved by non-relativizable arguments.

Theorem 15. [28, 4, 3] *For $\mathcal{K} \in \{PP, Mod_m P, PSPACE, EXP\}$, if $\mathcal{K} \subseteq P/poly$ then $\mathcal{K} \subseteq MA$.*

5 Circuit complexity

Kannan [21] proved that for every fixed polynomial s, there is a set in $\Sigma_2^P \cap \Pi_2^P$ which does not have $O(s(n))$-size circuits. Using a padding argument, he obtained the existence of sets in $NEXP(NP) \cap co\text{-}NEXP(NP)$ not having polynomial-size circuits.

Theorem 16. [21]

1. *For every polynomial s, there is a set A_s in $\Sigma_2^P \cap \Pi_2^P$ not having $O(s(n))$-size circuits.*
2. *For every increasing time-constructible super-polynomial function $f(n)$, there is a set A_f in $NTIME[f(n)](NP) \cap co\text{-}NTIME[f(n)](NP)$ not having polynomial size circuits.*

Here with our proof technique, we can modify the proof of this theorem to obtain the following corollaries.

Corollary 17. *For every fixed polynomial s, there is a set A_s in $ZPP(NP)$ that does not have $O(s(n))$-size circuits.*

Corollary 18. *For every increasing time-constructible super-polynomial function $f(n)$, there is a set A_f in $ZPTIME[f(n)](NP)$ that does not have polynomial-size circuits.*

Corollary 19. *In every relativized world, $ZPEXP(NP)$ contains sets that do not have polynomial-size circuits.*

We remark again that the above results are provable by relativizable arguments, and thus they hold for every relativized world. On the other hand, these results are strengthened by using some proof technique that is not relativizable. Harry Buhrman [12] and independently Thomas Thierauf [34] pointed out to the authors that Theorem 15 can be used to show that $MA_{exp} \cap co\text{-}MA_{exp}$ contains non P/poly sets, where MA_{exp} denotes the exponential-time version of Babai's class MA [5]. It was further communicated to us by Harry Buhrman [12] that Corollary 18 can be improved in the unrelativized setting from $ZPTIME[f(n)](NP)$ to $MA[f(n)] \cap co\text{-}MA[f(n)]$. Here a question of interest is whether the same result is provable by some relativizable technique. Notice that it is not possible to extend Corollary 19 by relativizing techniques to the class $EXP(NP)$, since there exist recursive oracles relative to which all sets in $EXP(NP)$ have polynomial size circuits [35, 17].

6 Concluding remarks

An interesting question concerning complexity classes that are known to contain non P/poly sets but possibly don't have complete sets is whether an explicit non P/poly set can be *constructed* in that class. For example, by Corollary 19 we know that the class ZPEXP(NP) must contain sets that do not have polynomial-size circuits. But we were not able to give a *constructive* proof of this fact. To our knowledge, it is not even known whether the existence of a non P/poly set can be constructively proved within the class NEXP(NP) ∩ co-NEXP(NP).

Acknowledgments

For helpful discussions and suggestions regarding this work we are very grateful to H. Buhrman, R. Gavaldà, L. Hemaspaandra, M. Ogihara, U. Schöning, R. Schuler, and T. Thierauf. We also thank H. Buhrman, L. Hemaspaandra, and M. Ogihara for permitting us to include their observations in the paper.

References

1. M. ABADI, J. FEIGENBAUM, AND J. KILIAN. On hiding information from an oracle. *Journal of Computer and System Sciences* **39** (1989) 21–30.
2. D. ANGLUIN. Queries and concept learning. *Machine Learning* **2** (1988) 319–342.
3. L. BABAI, L. FORTNOW. Arithmetization: A new method in structural complexity. *Computational Complexity* **1** (1991) 41–66.
4. L. BABAI, L. FORTNOW, C. LUND. Non-deterministic exponential time has two-prover interactive protocols. *Computational Complexity* **1** (1991) 1–40.
5. L. BABAI AND S. MORAN. Arthur-Merlin games: a randomized proof system and a hierarchy of complexity classes. *Journal of Computer and System Sciences* **36** (1988) 254–276.
6. J.L. BALCÁZAR. Self-reducibility. *Journal of Computer and System Sciences* **41** (1990) 367–388.
7. J.L. BALCÁZAR, R. BOOK, AND U. SCHÖNING. Sparse sets, lowness and highness. *SIAM Journal on Computing* **23** (1986) 679–688.
8. J.L. BALCÁZAR, R. BOOK, AND U. SCHÖNING. The polynomial-time hierarchy and sparse oracles. *Journal of the ACM* **33(3)** (1986) 603–617.
9. J.L. BALCÁZAR, J. DÍAZ, J. GABARRÓ. *Structural Complexity Theory.* (Springer, Berlin, 1988 and 1990).
10. N.H. BSHOUTY, R. CLEVE, S. KANNAN, AND C. TAMON. Oracles and queries that are sufficient for exact learning. *Proc. 7th ACM Conference on Computational Learning Theory* (1994) 130–139.
11. D.P. BOVET, P. CRESCENZI. *Introduction to the Theory of Complexity.* Prentice-Hall, 1993.
12. H. BUHRMAN, *personal communication.*
13. J.L. CARTER AND M.N. WEGMAN. Universal classes of hash functions. *Journal of Computer and System Sciences* **18** (1979) 143–154.
14. R. GAVALDÀ. Bounding the complexity of advice functions. *Proc. 7th Structure in Complexity Theory Conference* (IEEE, New York, 1992) 249–254.

15. J. GILL. Computational complexity of probabilistic complexity classes. *SIAM Journal on Computing* **6** (1977) 675–695.

16. J. HARTMANIS AND Y. YESHA. Computation times of NP sets of different densities. *Theoretical Computer Science* **34** (1984) 17–32.

17. H. HELLER. On relativized exponential and probabilistic complexity classes. *Information and Control* **71** (1986) 231–243.

18. L. HEMACHANDRA, M. OGIWARA, AND O. WATANABE. How hard are sparse sets? *Proc. 7th Structure in Complexity Theory Conference* (IEEE, New York, 1992) 222–238.

19. M.R. JERRUM, L.G. VALIANT, V.V. VAZIRANI. Random generation of combinatorial structures from a uniform distribution. *Theoretical Computer Science* **43** (1986) 169–188.

20. J. KÄMPER. Non-uniform proof systems: A new framework to describe non-uniform and probabilistic complexity classes. *Theoretical Computer Science* **85(2)** (1991) 305–331.

21. R. KANNAN. Circuit-size lower bounds and non-reducibility to sparse sets. *Information and Control* **55** (1982) 40–56.

22. R.M. KARP AND R.J. LIPTON. Some connections between nonuniform and uniform complexity classes. *Proc. 12th ACM Symposium Theory of Computing* (1980) 302–309.

23. K. KO. On self-reducibility and weak p-selectivity. *Journal of Computer and System Sciences* **26** (1983) 209–221.

24. J. KÖBLER. Locating P/poly optimally in the extended low hierarchy. *Theoretical Computer Science* **134** (1994) 263–285.

25. J. KÖBLER, U. SCHÖNING, J. TORÁN. *The Graph Isomorphism Problem: Its Structural Complexity*. Birkhäuser, Boston, 1993.

26. T.J. LONG. Strong nondeterministic polynomial-time reducibilities. *Theoretical Computer Science* **21** (1982) 1–25.

27. A. LOZANO AND J. TORÁN. Self-reducible sets of small density. *Mathematical Systems Theory* **24** (1991) 83–100.

28. C. LUND, L. FORTNOW, H. KARLOFF, N. NISAN. Algebraic methods for interactive proof systems. *Journal of the ACM* **39(4)** (1992) 859–868.

29. M. OGIWARA AND A. LOZANO. On sparse hard sets for counting classes. *Theoretical Computer Science* **112** (1993) 255–275.

30. C.H. PAPADIMITRIOU. *Computational Complexity*. Addison-Wesley, 1994.

31. N. PIPPENGER. On simultaneous resource bounds. *Proc. 20th Symposium on Foundations of Computer Science* (IEEE, New York, 1979) 307–311.

32. U. SCHÖNING. *Complexity and Structure*. Lecture Notes in Computer Science, Vol. 211 (Springer, Berlin, 1986).

33. M. SIPSER. A complexity theoretic approach to randomness. *Proc. 15th ACM Symposium Theory of Computing* (1983) 330–335.

34. T. THIERAUF, *personal communication*.

35. C. WILSON. Relativized circuit complexity. *Journal of Computer and System Sciences* **31(2)** (1985) 169–181.

36. C. YAP. Some consequences of non-uniform conditions on uniform classes. *Theoretical Computer Science* **26** (1983) 287–300.

37. S. ZACHOS. Robustness of probabilistic computational complexity classes under definitional perturbations. *Information and Control* **54** (1982) 143–154.

The complexity of searching succinctly represented graphs[†]

José L. Balcázar*

Abstract

The standard complexity classes of Complexity Theory do not allow for direct classification of most of the problems solved by heuristic graph search algorithms. The reason is that, in their standard definition, complexity classes are specifically tailored to explicit, instead of implicit, graphs of state or problem reduction spaces. But the usual practice works to a large extent, in some areas of Computer Science, over implicit graphs. To allow for more precise comparisons with standard complexity classes, we introduce here a model for the analysis of algorithms on graphs given by vertex expansion procedures. It is based on previously studied concepts of "succinct representation" techniques, and allows us to prove PSPACE-completeness or EXPTIME-completeness of specific, natural problems on implicit graphs, such as those solved by A*, AO*, and other best-first search strategies.

1.   The problem and the model

Heuristic search algorithms play a noticeable role among the techniques developed for attacking many relevant, practically important problems currently considered as intractable. Our aim here is to promote the use of a tool taken from Structural Complexity theory, namely "succinct representations", to gain further understanding of some aspects of the computational problems to which these algorithms are applied.

Many Combinatorial Optimization algorithms and problems from Operations Research have been given a lot of attention from the complexity theory community, being actually a main source of examples of completeness for various complexity classes (see [7], [9], [11], [12]). However, for optimization algorithms based on informed search methods from AI, some of which follow analogous intuitions,

[†] Work supported by the EC through the Esprit BRA Program (project 7141, AL-COM II) and through the HCM Program (project CHRX-CT93-0415, COLORET Network).

* Dept. LSI, Univ. Politècnica de Catalunya, Edif. U, Pau Gargallo 5, 08028 Barcelona, Spain: balqui@lsi.upc.es

much less seems to be known. Some information about problems on AND/OR graphs is readily obtained from the rich body of literature about the complexity-theoretic approach to games via alternation (let us mention in particular [28]). But for simpler search problems, all the complexity-theoretic analyses known to this author are based on concepts, tools, and conventions essentially different from those of Structural Complexity. This paper suggests the use of "succinct representations" to bridge this gap.

Inputs to realistic graph searching

From the point of view of a naive complexity-theorist, the problems solved by A* and other best-first strategies are not difficult: essentially, construction and optimization of paths in graphs, captured by nondeterministic logarithmic space complexity classes. In the AI case, the graph is a so-called "state space", characterized by the fact that the exploration of a vertex is done with knowledge of additional information, gathered along the path to it or computed on the spot. The algorithm may decide to expand a vertex, and this operation consists of computing all its successors. In most cases, what is wanted is to find a path from a start vertex to one of a set of goal vertices, often with the additional request of minimizing a cost function.

Actually, starting from Dijkstra's algorithm, and continuing with many other applications of algorithm design techniques, polynomial time algorithms solving that sort of problems exist, offering good practical performance. Most of them are currently available, precompiled in libraries, ready to use very easily, as executable subroutines of C++ programs [19]. Variants of A* can be seen as a refinements of Dijkstra's algorithm through the use of so-called "admissible" heuristic information (see also the detailed discussion in [17]).

But from the complexity-theoretic approach, one would consider that the input to the algorithm is the whole graph, and this would be already unacceptable to practitioners: in most practical cases, the search graph is huge enough that even linear time on its size is infeasible by far. This accounts for the labeling of "hard", often applied to the problems solved by such AI algorithms, and for the fact that these algorithms can be employed to attack well-known NP-complete problems. It is a task for complexity theory (which we address here) to find a framework to analyze these problems in a more satisfactory way, taking into account the astonishing sizes of the searched graphs.

There are several results on the formal analysis of the complexity of these algorithms; see [25] (in particular, chapters 5 and 6). The context is often probabilistic, the analysis is given mostly as a function of the length of the optimal output path, and made in terms of other quantities, such as the number of different expanded vertices, or the total number of vertex expansions (which in A* is exponentially higher since, in the bad context of a nonmonotonic admissible heuristic, the same vertex may be chosen for expansion exponentially many times). Such frameworks do not allow for easy translation into standard complexity classes. In other cases the analysis is restricted to the use of the algorithm to solve some specific, particular problem.

We want to supplement the view obtained from the probabilistic average analysis, which is of course very useful but depends on several simplifying assumptions and on the hypothesis of some specific probability distribution [25]; we will see how to model these algorithms in such a way that the standard complexity classes give informative properties of (the problems solved by) these algorithms, in as much generality as possible, and in terms of worst-case complexity.

Our point of view is that, in order to obtain a consistent approach, informative for workers of other areas, one must consider that the input to such search algorithms is *not the state space graph*, but is instead the (hopefully efficient) *procedure for vertex expansion*. Note that, if simply to expand a vertex is computationally expensive, then the mere feasibility of all such algorithms is at stake. Moreover, this condition that expansion better be efficient implies that the outdegree of the state space graph must be appropriately bounded, lest simply the production of a too long list of successors be unaffordable.

It is not clear a priori how complexity classes could cope with the idea of having a vertex expansion procedure as input. As it turns out, Complexity Theory has the right intuitive tool to put to use: here we suggest to resort to an appropriately refined notion of succinct representations.

Indeed, one could argue that the vertex expansion procedure is, in a sense, a description of the whole graph, allowing for efficient local treatment. One well-studied form of succinct representation is based on encoding the transition matrix: a small, fast enough procedure receives as input (the codes of) two vertices and outputs simply the bit indicating presence or absence of an edge between them (or even, more elaboratedly, its cost). But this is not appropriate here: the search algorithms we want to model frequently assume that more information can be obtained from each vertex, including the complete list of successors. It may be infeasible to extract this list from the transition matrix, for instance when the number of actual successors is not high but the total number of vertices is large.

We propose here to use as succinct representation of the graph the formalization of a procedure that, given a vertex, produces the list of successors, including, when appropriate, the cost corresponding to each edge. It must be feasible, so certainly a mild condition on it is to require it to work in polynomial time, which marks a weak feasibility limit. Therefore, we assume that it can be implemented with a family of boolean circuits: it is well-known that this model exactly characterizes (via completeness) polynomial time computations [1]. Let us set this idea in the perspective of previous research on succinct input representation.

Problems on succinct representations

The study of algorithmic graph problems on succinct representations was introduced independently in [6] and [29], and has been proven useful in other applications of Complexity Theory. Essentially, the idea is as follows: while traditional models of computation assume their input represented, under some reasonable encoding, as a string of letters over some alphabet (frequently binary), it is interesting as well to see what happens to computational problems when their instances are encoded in some other, hopefully more compact, form.

In principle, the intuitive consequences of such input conventions are contradictory: being, in general, shorter than the full description, they allow for less running time (which is a function of the length of the shorter input); but since only very regular instances allow for substantial savings in the encoding, it may be the case that the algorithms operate faster on these instances. So, a careful and detailed analysis of each input convention is necessary.

In the case of decisional problems on graphs, the input is not the graph itself but (the encoding of) a boolean circuit which computes its transition matrix. For this model, [6] and [29] classified as complete in some class of the polynomial time hierarchy or the counting hierarchy (or higher up) many specific problems on succinct instances; many of these are polynomially solvable for standard input representation. Then [24] pointed out that the use of projection reducibility allows one to prove that all known NP-complete problems become NEXP-complete under succinct input representation via circuits (see also Chapter 20 of [21]). This observation is extended in [2], by means of logtime reducibility, to a very general result (the Conversion Lemma) relating the complexity of problems on standard input convention to their respective succinct versions, independently of the complexity classes in which they lie. This last result also abstracts from the previously used graph-theoretic setting, since the Conversion Lemma works on binary encodings of arbitrary data types.

Subsequently, the notion of succinct representation and variants of the results from [2] have been employed in [5] to characterize the computational complexity of a number of problems from Logic Programming and Databases (specifically, disjunctive Datalog queries under various semantics).

This tool could be used as such for our study of graph search algorithms, but would admit a serious criticism: the practical applications are less restricted than this model assumes. The work done so far only assumes the ability of the input circuit to answer presence or absence of an edge, given both endpoints; practical applications assume the ability to expand a vertex, i.e. generate all its successors, and this may be infeasible from the plain "transition matrix" implementation. We propose here to introduce a new model for succinct inputs, better suited to the application we have in mind, which closely parallels the idea of giving, as input to the algorithm, a procedure to expand nodes. We will see that this approach gives more informative characterizations of the hardness of graph search problems.

2. Preliminaries

Most of our notions of Complexity Theory regarding models of computation, complexity classes, and reducibilities, are standard ([1], [21]). We assume decisional problems to be encoded as sets of strings over the standard binary alphabet. Likewise, functional problems are encoded as partial functions from strings to strings. It is a well-known fact (heavily employed when computers are in use) that binary strings can represent numbers and many other combinatorial structures such as graphs. For a string x, we denote by $|x|$ its length. Real-valued functions are assumed to be rounded to the nearest positive integer. Graphs are assumed to be directed. We assume that $\langle .,. \rangle$ is an easily computable pairing function, and that both arguments can be easily computed from the result.

Our model of computation is a variant of the multitape Turing machine; it is well-known that other more realistic models are equivalent to this one modulo polynomial time overheads and constant factor space overheads, so that all the complexity classes we mention are invariant with respect to the machine model employed.

The slight difference between the standard Turing machine model and ours consists in that, since we will need to work with sublinear time computations, we cannot afford sequential access to the input. Instead, we use indexing machines [3], in which there is a specific "indexing tape" which points to the input tape. On an input of length n, when the machine enters a specific "read" state with $i \leq n$ written on the indexing tape, then in one step the ith input symbol is transferred to the head of the first worktape. A special "error" mark appears there if the contents of the indexing tape is $i > n$. Note that this is nothing but a straightforward formalization of the idea of direct access to the input, as if it is stored in RAM.

We will mention the following generic complexity classes of decisional problems: for space or time bounds f, respectively, we have DSPACE(f), corresponding to problems A such that whether $x \in A$ that can be solved within memory space $f(|x|)$, and DTIME(f), corresponding to problems A such that whether $x \in A$ that can be solved within computation time $f(|x|)$. Many classes have more familiar names, and thus LOGSPACE = DSPACE($\log n$), P = $\bigcup_{k \in \mathbb{N}}$ DTIME(n^k), PSPACE = $\bigcup_{k \in \mathbb{N}}$ DSPACE(n^k), and EXPTIME = $\bigcup_{k \in \mathbb{N}}$ DTIME(2^{n^k}). We will briefly mention in the wrap-up discussion the well-known class NP, which is defined by polynomial time nondeterministic machines.

We also need two concepts of reducibility. The first one is standard: a decisional problem A is polynomial-time m-reducible to a problem B, denoted $A \leq_m B$, if there is a polynomial-time computable function f for which $x \in A \iff f(x) \in B$ holds for all x. So f provides a way of solving A in polynomial time given any arbitrary way of solving B in polynomial time.

Our main results are based on polylog-time m-reducibility. A function f is computable in polylogarithmic time if there is an indexing machine that, on input x and j, with $j \leq |f(x)|$ outputs the jth bit of $f(x)$ in time $\log^{O(1)} |x|$. Analogously to the polynomial-time case, a polylog-time reduction between problem

A and problem B is a function f computable in polylogarithmic time such that, for all x, $x \in A \iff f(x) \in B$. We say that A is PL-reducible to B, and denote it $A \leq_m^{PL} B$.

The concept of reducibility gives rise immediately to the concepts of hardness and completeness for a complexity class. A problem A is hard for a class D under a reducibility if all problems in D are reducible to A. A problem A is complete for a class D if it is hard for it and also belongs to it.

From the time and space hierarchy theorems of Complexity Theory, it is well-known that LOGSPACE $\neq$ PSPACE and that P $\neq$ EXPTIME. In particular, this implies that a complete problem for PSPACE, under any of our reducibilities, cannot be in LOGSPACE, and actually cannot be solved in subpolynomial memory space (here subpolynomial means $o(n^{\alpha})$ for all α). Similarly, a problem that is complete for EXPTIME cannot be solved in polynomial, nor even subexponential, running time. As of now, it is possible that P $=$ PSPACE, so that completeness for PSPACE does not rule out the possible existence of a polynomial time algorithm; but such an algorithm would imply immediately polynomial time algorithms for all problems in PSPACE, including all NP-complete problems and many others seemingly harder than them.

Let f be a polylog-time reducibility. We say that f is polylog-time covered (or PL-covered) if the values j such that the bit $\langle i, j \rangle$ of $f(x)$ is 1 are predictable in advance, only knowing i and $|x|$; more formally, if there is an indexing machine that, on inputs i and n (in binary) and in polynomial time (equivalently, polylog time when inputs are in unary), computes a list L such that, whenever bit $\langle i, j \rangle$ of $f(x)$ is 1 for some x of length n, we have $j \in L$.

This notion will be used in the following context: the output of f will be interpreted as an instance of a decisional problem on (directed) graphs, described by a transition matrix. Bit $\langle i, j \rangle$ indicates presence or absence of the edge $\langle i, j \rangle$ in the output graph. Coverability means, therefore, that, when all x with $|x| = n$ are considered, and for each i, there are only $\log^{O(1)} n$ many potential outgoing edges from vertex i that may appear in any of the graphs $f(x)$, and that all their other endpoints are easily computable. This technical condition is necessary in some of our statements, and can be extended easily, if necessary, to the case in which the graphs are weighted.

3. The new model of succinct instance representation

Following the informal description given in the introduction, we now choose to represent graphs as follows. First, we restrict ourselves to graphs without isolated vertices. For a graph of less than n vertices and degree d, its representation is a boolean circuit of $\log n$ inputs and $d \log n$ outputs. Without loss of generality, we assume that n can be read out from the circuit easily. Vertices are numbered, reserving code 0 to be an additional dummy vertex name. Assume that vertex v_k corresponding to number k has as successors $v_{i_1} \ldots v_{i_p}$ with $p \leq d$; then p among the d groups of $\log n$ output gates must each give one of the numbers i_j, written in binary with zeroes to the left up to size $\log n$, and without repetition; the remaining outputs must be all zero, i.e., describe the additional dummy vertex.

Clearly the representation is not unique, but given a circuit c of $\log n$ inputs and $d \log n$ outputs, there is a single graph G_c represented by it: take as vertices the positive integers that can be written with $\log n$ inputs, set edges by feeding each to the circuit to find out the successors of each vertex and discarding vertex zero whenever it shows up, and finally remove any isolated vertices that might be left. When c is a syntactically incorrect circuit in which the number of outputs is not a multiple of the number of inputs, and which therefore does not represent any graph, then it is mapped to some fixed, trivial graph $G_c = G_0$.

Note the following fact: it is not necessary that the vertices are numbered by consecutive numbers. As a consequence of our convention, we may afford as many numbers as convenient that do not correspond to any vertex. It suffices that the circuit is able to detect them, to output an empty list of successors, and to never output them as successors of other vertex numbers. In this way, they are represented as isolated and will be discarded by our interpretation.

Let A be any decisional problem on graphs. We denote by sA the succinct version of A, defined as follows: $sA = \{c \mid G_c \in A\}$. We state now the main property of this representation, from the complexity-theoretic point of view.

1. Theorem. Let A and B be decisional problems defined on graphs. If $A \leq_m^{PL} B$, via a PL-covered reduction, then $sA \leq_m sB$.

We sketch the proof. Let f be the reduction. The machine computing f gets as input an instance of A, i.e. (an encoding of) a graph G on, say, n vertices, and a pair of vertices of $f(G)$, and finds whether this pair is indeed an edge of $f(G)$. Via a now standard transformation (see [1]), a boolean circuit can be constructed which computes any requested bit of $f(x)$. It needs that individual bits of G be provided when necessary. Examination of the circuit simulating M shows that, using the given succinct representation of G, extra circuitry can be added to obtain these bits. In this way, the whole circuit calculating each bit of $f(G)$ is obtained; call this circuit C'.

Now the hypothesis of coverability must be used to obtain from C' a bona fide circuit representation according to our definition. For this, a similar circuit, simulating this time the machine computing the covering, is prepended to an appropriate number of replicas of C', thus computing in parallel the necessary list of succesors of any given vertex of $f(G)$. Further details will be provided in the unabridged version of the present paper. □

If the succinct representation of some graph becomes exponentially smaller, the complexity of a problem A may jump up by about one exponential when passing to sA. It is not difficult to see that it does not jump up more:

2. Lemma. If $A \in \mathrm{DTIME}(f(n))$, for f at least linear, then $sA \in \mathrm{DTIME}(f(2^{O(n)}))$; if $A \in \mathrm{DSPACE}(f(n))$, for f at least logarithmic, then $sA \in \mathrm{DSPACE}(f(2^{O(n)}))$; and similarly for nondeterministic complexity classes.

We end this section with the following technical fact, necessary in later proofs: the "succinct representation" operator has a kind of "inverse", given by what we will call *trivial graph* representation. Let w be a binary string. Then $\mathrm{tg}(w)$ is a "chain" graph of m nodes, each but the last having a single edge to the next, the last to itself, and m being selected so that its binary representation is $1w$. Note that it is exponentially large in $|w|$. Overloading the operator, let $\mathrm{tg}(L) = \{\mathrm{tg}(w) \mid w \in L\}$.

3. Lemma. Every set L of binary strings is m-reducible in polynomial time to the succinct representation of $\mathrm{tg}(L)$.

The proof is omitted and appears in the unabridged version.

4. Application to graph search

The graph accessibility problem (GAP) is: given a directed graph, a source vertex, and a goal vertex (or set of them), decide whether there is a path connecting the source vertex to a goal. In the weighted variant, costs are associated to the edges; an integer k is also given, and the question is whether such a path of total cost at most k exists.

Some broadly employed heuristic search algorithms solve precisely this problem, with the peculiarity that the graph is *not* given explicitly. Instead, a procedure is given to "expand" a node, i.e. to find the list of all successors of a given vertex, together with the costs of the edges in the weighted variant.

Vertices are described, in these cases, by some kind of "configuration" or specification of "how the vertex looks like" (e.g. pieces on a board); essentially, the information that must be present in this description is (at least) the necessary data to compute feasibly all successors. We assume that the number of vertices (or configurations) may reach an exponential on the number of bits used to store a configuration.

The point is now that the internal representation of the configuration in a computer program, whatever means are used for it, can be read as a binary number (a form of Gödel numbering, in a sense) and directly interpreted as *the* number of the vertex. Thus, the vertex number abstracts the description of "how the vertex looks like" and contains information about how to find the successors. Syntactically incorrect descriptions correspond to numbers that do not represent vertices, and will be treated as isolated vertices by the circuit representation, as indicated in the previous section.

We assume that the "expand" procedure is feasible, i.e. takes time polynomial on the size of (the codenumber of) the vertex. Every reasonable application of graph search will need this condition; otherwise the search will be infeasible just because some node expansions, to start with, take far too long. This also implies that both the degree and the number of bits needed to specify costs are also polynomial on the size of the number of the vertex. Note that this bound may be as low as polylog on the size of the whole graph.

Equivalently, under this condition, we can formalize the expansion procedure as a boolean circuit, via the standard simulation of polynomial time computations by boolean circuits. This simulation is feasible, and any other sequential programming language can be assumed instead for the expansion procedure, since all these can be compiled into circuits in polynomial time. This circuit is the input to a search algorithm based on node expansions. The output wanted is given, in general, by an optimality condition, but a (somewhat artificial) decisional problem can be associated to it in a standard way [7]; and the optimization problem is no easier (and possibly harder) than the decisional problem.

In this case, the decisional problem is exactly (the weighted version of) sGAP, since the solution is a path and the input is a circuit describing the vertex expansion procedure to construct the graph. Our aim in this section is to use the setup from the previous section to show:

4. *Theorem.* sGAP is PSPACE-complete.

We omit the proof.

Of course, some particular search problems lie in complexity classes potentially below PSPACE, and particular algorithms for them may be more efficient. To bypass our hardness result, however, the price is to design specific algorithms extracting from their input more information than plainly a way of expanding vertices.

5. Application to AND/OR graphs

Another problem of wide practical interest is the search in AND/OR graphs, also called "problem reduction spaces" or also "alternating graphs".** In these, each nongoal vertex carries a label as "universal" or "existential". A solution is no longer a path, but a subgraph in which each existential vertex has exactly one successor and each universal vertex has all its successors; all the leaves must be goal nodes. A prime set of examples of application is given by games (although the problem appears in many other guises).

The AND/OR graph problem (AGAP, also standing for alternating GAP) is: given an AND/OR directed graph, a start vertex, and a set of goal vertices, decide whether there is a solution subgraph. The obvious weighted variant is also useful.

Again, algorithms (like AO*) exist for this problem, and various relevant algorithms are specifically tailored to games, in which the minimax principle is applied. Let us mention only alpha-beta pruning and SSS* [25]. The complexity of searching AND/OR graphs can be assessed from the abundant studies on the

---

** The reason is that, without loss of generality, one can assume that through each path AND (universal) vertices and OR (existential) vertices alternate. The problem is often modeled equivalently with hypergraphs [26].

complexity of games. In particular, in [28] a number of games are proved to be EXPTIME-complete.

In that reference, a game is defined as given by the set of positions (encoded as words of a given length, and split into existential and universal) and a polynomial time algorithm to check validity of a move. There is a condition that "the board cannot be enlarged during the play", i.e. that valid moves are between positions encoded by words of the same length. Succinct representations as defined here (i.e. easy enough expansion algorithms) do not necessarily exist for the games from [28], and for some of them the (high) degree of each vertex makes them impossible. But, actually, careful examination proves that for some of the EXPTIME-complete games from [28], polynomial time expansion algorithms (i.e. succinct representations as defined here) do exist. In particular, the "game" on propositional formulas labeled G_2 in that paper has easy expansion algorithms, is EXPTIME-complete, and reduces to sAGAP (i.e. can be solved by AO*). A proof that sAGAP is EXPTIME-complete follows immediately from this fact.

5. *Theorem.* sAGAP is EXPTIME-complete.

This implies, by the time hierarchy theorem (see [1]), that no polynomial (nor even subexponential) time solution exists at all for this problem: EXPTIME-complete problems are provably intractable. The heuristic search methods such as AO* *are therefore unavoidable* to obtain solutions to practical cases in feasible computer time, and *will never be proved to provide a guarantee* of feasibility since such feasibility provably does not hold. (Note that no such provability can be stated for A* since it is possible, as of now, that P = PSPACE, or equivalently that polynomial time algorithms for sGAP exist.)

An alternative proof is possible, based on the GAME problem proved P-complete in [14] (essentially, a reformulation of AGAP); the completeness can be proved to hold under polylog time PL-covered reductions, so that an application of theorem 1, analogous to the one in the previous section, yields the result.

6. Discussion

This paper has proposed a new model of succinct representation, and has studied the complexity of search problems for implicit graphs under this model. We have proved completeness theorems, and argued that the relevance of this approach stems from the fact that many practical search algorithms can be seen as exploring an exponentially large graph for which a feasible vertex expansion procedure exists, such as A* or AO*. But there is a different context in which a similar consideration has been made: certain classes of total multivalued non-deterministically computable functions. These include PLS [12] and functions related to parity arguments [22].

Consider the following functional version based on a given local search algorithm: given an input, find the precise output of the local search started on it.

This can be seen as the search for a path in the exponentially large graph defined by the neighborhood structure of the local search scheme; indeed, accordingly, for hard enough local search schemes this problem is PSPACE-hard [12], [20]. However, other functional versions in which any local optimum suffices as output lie probably lower: in PLS [12]. The approach used here is, thus, consistent with the known results, and might be informative as of properties of these local search classes and problems. Similar considerations can be made with respect to the classes introduced in [22] based on "parity" or on "pigeonhole" arguments (PPA, PPAD, and PPP).

A major obstacle seems to be the fact that all these classes of functions below functional NP correspond to total functions, i.e. some sort of the so-called "promise problems" in which only inputs fulfilling certain "promise" are to be solved correctly. It is not clear to the author what could be the appropriate notion for obtaining from this intuition more general results about the classes PLS, PPA, PPAD, and PPP.

Acknowledgements

The following persons have been very helpful, either through personal discussions, proofreading, or both: Núria Castell, Josep Diaz, Ricard Gavaldà, Montse Hermo, Antoni Lozano, Pedro Meseguer, and Carme Torras. Special thanks are due to Georg Gottlob, for indicating that the scope of applications of the succinct representation technique was broader than I thought, and for suggesting the use of the (transitive) polylog time reducibility instead of log time reducibility.

7. References

[1] J. L. Balcázar, J. Diaz, J. Gabarró: *Structural Complexity I*. Springer-Verlag 1988.

[2] J. L. Balcázar, A. Lozano, J. Torán: "The complexity of algorithmic problems on succinct instances". In: *Computer Science: Research and Applications*, R. Baeza-Yates and U. Manber (eds.), Plenum 1992.

[3] A. K. Chandra, D. C. Kozen, L. J. Stockmeyer: "Alternation". *Journal of the ACM* 28 (1981), 114–133.

[4] A. K. Chandra, L. J. Stockmeyer, U. Vishkin: "Constant depth reducibility". *SIAM Journal on Computing* 13 (1984), 423–439.

[5] T. Eiter, G. Gottlob, H. Mannila: "Adding disjunction to DATALOG". In: Proc. ACM Symp. on Principles of Database Systems 1994.

[6] H. Galperin, A. Wigderson: "Succinct representations of graphs". *Information and Control* 56 (1983), 183–198.

[7] M. R. Garey, D. S. Johnson: *Computers and intractability: a guide to the theory of NP-completeness*. Freeman 1979.

[8] B. L. Golden, W. R. Stewart: Empirical analysis of heuristics. In: *The Traveling Salesman Problem: A Guided Tour of Combinatorial Optimization*, E. L. Lawler, J. K. Lenstra, A. H. G. Rinnooy Kan, D. B. Shmoys (eds.), John Wiley and Sons 1985.

[9] R. Greenlaw, H. J. Hoover, W. L. Ruzzo: *A compendium of problems complete for P.* Oxford Univ. Press 1994, in press.

[10] J.-W. Hong: "On some deterministic space complexity problems". *SIAM Journal on Computing* 11 (1982), 591-601.

[11] D. S. Johnson: *A catalog of complexity classes.* In: *Handbook of Theoretical Computer Science*, vol. A: Algorithms and Complexity, J. van Leeuwen (ed.), Elsevier 1990, 67-161.

[12] D. S. Johnson, C. H. Papadimitriou, M. Yannakakis: "How easy is local search?". *Journal of Computer and System Sciences* 37 (1988), 79-100.

[13] N. D. Jones: "Space-bounded reducibility among combinatorial problems". *Journal of Computer and System Sciences* 11 (1975), 68-75.

[14] N. D. Jones, W. T. Laaser: "Complete problems for deterministic polynomial time". *Theoretical Computer Science* 3 (1977), 105-117.

[15] N. D. Jones, Y. E. Lien, W. T. Laaser: "New problems complete for nondeterministic log space". *Math. Systems Theory* 10 (1976), 1-17.

[16] M. Kowaluk, K. W. Wagner: "Vector language: simple description of hard instances". In: Proc. Math. Foundations of Computer Science, Lecture Notes in Computer Science 452, Springer-Verlag 1990, 378-384.

[17] S. Kundu: "A new variant of the A* algorithm which closes a node at most once". *Annals of Math. and Artificial Intelligence* 4 (1991), 157-176.

[18] T. Lengauer, K. W. Wagner: "The correlation between the complexities of the nonhierarchical and hierarchical versions of graph problems". *Journal of Computer and System Sciences* 44 (1992), 63-93.

[19] K. Mehlhorn, S. Näher: "LEDA: a platform for combinatorial and geometric computing. *Communications of the ACM* 38, 1 (1995), 96-101.

[20] C. H. Papadimitriou: "The complexity of the Lin-Kernighan heuristic for the Traveling Salesman Problem". *SIAM Journal on Computing* 21 (1992), 450-465.

[21] C. H. Papadimitriou: *Computational Complexity.* Addison-Wesley 1994.

[22] C. H. Papadimitriou: "On the complexity of the parity argument and other inefficient proofs of existence". *Journal of Computer and System Sciences* 48 (1994), 498-532.

[23] C. H. Papadimitriou, K. Steiglitz: *Combinatorial Optimization: Algorithms and Complexity.* Prentice-Hall 1982.

[24] C. H. Papadimitriou, M. Yannakakis: "A note on succinct representations of graphs". *Information and Control* 71 (1986), 181-185.

[25] J. Pearl: *Heuristics: Intellingent Search Strategies for Computer Problem Solving.* Addison-Wesley 1984.

[26] E. Rich, K. Knight: *Artificial Intelligence.* McGraw-Hill 1991 (2nd edition).

[27] W. J. Savitch: "Relations between nondeterministic and deterministic tape complexities". *Journal of Computer and System Sciences* 4 (1970), 177-192.

[28] L. J. Stockmeyer, A. K. Chandra: "Provably difficult combinatorial games". *SIAM Journal on Computing* 8 (1979), 151-174.

[29] K. W. Wagner: "The complexity of combinatorial problems with succinct input representation". *Acta Informatica* 23 (1986), 325-356.

Optimal shooting: Characterizations and applications[*]

Frank Bauernöppel[678] Evangelos Kranakis[12] Danny Krizanc[12]
Anil Maheshwari[3278] Marc Noy[5] Jörg-Rüdiger Sack[127] Jorge Urrutia[42]

[1] Carleton University, School of Computer Science, Ottawa, ON, Canada
[2] Research supported in part by NSERC grant.
[3] Tata Institute of Fundamental Research, Bombay 400 005, India
[4] University of Ottawa, Department of Computer Science, Ottawa, ON, Canada
[5] Univ. Politecnica de Catalunya, Dept. de Matematicas, Cataluna, Spain
[6] Institut für Informatik, Humboldt-Universität zu Berlin, 10099 Berlin, Germany
[7] Research supported in part by ALMERCO Inc.
[8] Work by the author was carried out during a stay at Carleton University.

1 Introduction

Suppose that an archer is hunting birds flying over hunting grounds described as a bounded region possibly with holes formed by obstacles such as mountains, lakes, dense forests, etc. In an attempt to minimize the number of arrows used, the archer tries to identify pairs of birds that can be pierced by a single arrow; this is possible, if the positions of two birds line up with some point on the hunting grounds. This corresponds to the well known paradigm of "killing two birds with one stone."

The archer problem can be modeled as follows: Let $X = \{p_1, \ldots, p_n\}$ be a collection of points in $\mathbb{R}^3$ (in general position) such that the z-coordinate of each element of X is strictly greater than zero and let S be a compact plane set of $\mathbb{R}^3$, called stage, contained in the hyperplane $H_0 = \{p \in \mathbb{R}^3 : \text{the } z\text{-coordinate of } p \text{ is } 0\}$. Given X and S construct a graph $G(X, S)$ with vertex set X such that two vertices p_i, p_j of $G(X, S)$ are adjacent if the line through p_i and p_j intersects S. $G(X, S)$ will be called the stage graph of X and S.

Applications of stage graphs may arise in several problems such as the positioning of floodlights to illuminate fixed objects in space and the positioning of directional satellite antennae to pick up signals from ground stations, not to mention the traditional problem of "killing two birds with one stone." An important relationship which will be discussed and exploited is to two-processor task scheduling.

In Section 2 we study the planar problem; that is $X = \{p_1, \ldots, p_n\}$ is a planar point-set where each point has a strictly positive y-coordinates and S is a line segment on the x-axes. We prove that in this case, the family of stage graphs generated, is exactly the set of comparability graphs of partial orders of dimension two, i.e., permutation graphs. This immediately yields a polynomial time

---

[*] The full version of this paper is available on the world wide web address http://www/scs.carleton.ca under technical reports.

algorithm for recognizing such graphs. The characterization implies a compact linear space representation for two-dimensional stage graphs. We exploit this fact and the characterization result for the design of several algorithms presented in Section 3.

More specifically, in Section 3.1 we solve the archer's problem. The problem of minimizing the number of arrows the archer needs naturally corresponds to that of finding a maximum matching of stage graphs. Therefore it is possible to solve the problem using the Micali and Vazirani matching algorithm [13]. This results in an $O(\sqrt{n}m)$ algorithm where n and m are the number of vertices and edges of the graph respectively. A more efficient algorithm is obtained when stating the problem as a two-processor task scheduling problem. Efficient algorithms for finding tightest two-processor schedules are known [6, 7]. We follow the approach of [2] that leads to an $O(n+m)$ time algorithm for the scheduling problem [20]. Through vector dominance and using computational geometry techniques we establish that the problem has an $O(n \log^3 n)$ solution. We therefore not only solve the archer's problem efficiently, but also provide a novel and improved algorithm for matching in permutation graphs. Furthermore, if the dependency graph of a scheduling problem is known to be a permutation graph, then we now have an improved two-processor scheduling algorithm (in case that the number of edges is $\Omega(n \log^3 n)$).

The linear-space representation of stage graphs also leads to conceptually simple, new, and improved algorithms for vector dominance and rectangle query problems. These algorithms are discussed in Section 3.2. Let $P = \{p_1, p_2, ..., p_n\}$ be a planar point set of n distinct points $p_i = (x_i, y_i)$, $i = 1, ..., n$. A point p_i is said to *dominate* a point p_j, if $x_i \geq x_j$ and $y_i \geq y_j$ and $i \neq j$. We present new simple and optimal sequential and EREW PRAM algorithms for reporting all dominance pairs. The algorithms also present an improvement in that the previous algorithm [8] for that problem required CREW processors. A problem related to dominances is the *rectangle query problem* for planar point sets P. A query consists of a pair of points (p_i, p_j), where $p_i, p_j \in P$, and we need to answer whether the rectangle formed by the query points is empty or not. The characterization is further exploited to construct an $O(n \log n)$ space data structure which answers rectangle queries in $O(1)$ time. The data structure can be constructed in sequential $O(n \log n)$ time and in $O(\log n)$ time using n EREW PRAM processors. Our parallel rectangle query algorithm improves on previous $(O(n^2)$ space, $O(1)$ query) or $(O(n \log n)$ space, $O(log n)$ query) results.

In Section 4 and 5 we turn to natural generalizations of the planar single stage case by considering stage graphs in three dimensions and stage graphs in the plane with multiple stages respectively. In Section 4 we consider the three-dimensional case where S is a compact convex set. Three-dimensional stage graphs are also shown to correspond to comparability graphs of partial orders. We prove that the set of three-dimensional stage graphs obtained when S is a triangle is exactly the set of comparability graphs of orders of dimension at most three, and thus already for this case, the recognition problem is NP-complete. We provide a characterization of three-dimensional stage graphs in terms of geometric containment orders, that is partial orders arising from containment relations among the elements of families of convex sets on the plane. For example, when S is a circle, $G(S, X)$ is the comparability graph of a circle order [21]. We prove next that not all comparability graphs arise from three-dimensional stage graphs, regardless of the choice and complexity of S.

In Section 5 we consider another generalization of the two-dimensional case to k non-intersecting line segments contained in the x-axis. For fixed n, let $\mathcal{G}_k$ denote the class of graphs (on n points in general position) which can be represented by k line segments as above. We prove upper and lower bounds on the number of line segments needed to represent a graph and establish results on separating the classes $\mathcal{G}_k$. In particular, we prove the existence of graphs which require $\Omega(n^2/\log n)$ line segments for their representation and we show how to construct graphs requiring $\Omega(\sqrt{n})$ line segments for their representation. We also determine the number of line segments required for several common graphs, including lines, cycles, trees and complete bipartite graphs.

2 The Single Stage

Consider a line segment L or "stage" contained in the x-axis of the plane and a set of points $X = \{p_1, \ldots, p_n\}$ in general position with positive y-coordinates. We define a graph $G(X, L)$ with vertex set X in which two vertices are adjacent if the line connecting them intersects L. $G(X, L)$ will be called a $2 - d$-stage graph.

In this section we prove the following theorem which yields an $O(n^2)$ algorithm to recognize $2 - d$-stage graphs.

Theorem 1. *A graph G is a $2 - d$-stage graph if and only if G is a permutation graph.*

Corollary 2. *[22] Recognizing $2 - d$-stage graphs can be done in $O(\min\{n^2, n + m\log n\})$ time.*

3 Applications

In this section, the characterization provided by Theorem 1 is applied to solve the matching problem and to give new simple algorithms for dominance problems on a planar point set.

3.1 Matching/Scheduling Algorithms

In this section we provide an efficient solution to the archer's problem which implies novel and improved solutions to matching in permutation graphs and two-processor task scheduling for permutation graph dependencies. Using the Micali and Vazirani algorithm [13] an $O(\sqrt{n}m)$ algorithm is obtained where n and m are the number of vertices and edges of the graph, respectively.

As pointed out, e.g., in [14], there is a strong relation between maximum matchings in co–comparability graphs and the following scheduling problem: Let $G = (V, E)$ be a directed acyclic graph; let G have n vertices and m edges. Vertex $v \in V$ is a *successor* of a vertex $u \in V$ if there is a directed path from u to v in G. A *two-processor scheduling* for G is an assignment of time units $1, 2, 3, \ldots$ to the vertices $v \in V$ such that

1. each vertex $v \in V$ is assigned exactly one time unit,
2. at most two elements are assigned the same time unit, and
3. if v is a successor of u in G, then u is assigned a lower time unit than v.

The edges of G represent dependencies among the set of n vertices (tasks) to be executed. The largest time unit assigned to a vertex is called the length of the schedule.

It has been shown that the following holds [5]: Given a directed acyclic graph G with n vertices then there is a two-processor scheduling for G of length l iff there is a matching of size $n - l$ in the undirected complement graph G'. Two vertices u and v are connected by an undirected edge in G' iff neither uv nor vu is an edge in graph G. Moreover, a matching in G' can be obtained from a schedule by simply matching all pairs of vertices that are assigned the same time unit. Note that G' is a co–comparability graph.

Efficient algorithms for finding a tightest two-processor schedule are known [6, 7]. We follow the approach of [2] that leads to a linear time algorithm for the scheduling problem [20] when the graph G is transitively closed. This approach uses a vertex numbering assigning numbers $1, 2, \ldots$ to the n vertices of G in increasing order. By $L(u)$ we denote the label of vertex u and by $N(u)$ we denote the list $(L(v_1), L(v_2), \ldots, L(v_k))$ of the labels of the successors v_i of u in G sorted in decreasing order.

Suppose the numbers $1, 2, \ldots, k - 1$ have already been assigned. A vertex u is labeled with value $L(u) = k$, if

1. all successors of u in G are already labeled, and
2. for each other vertex u' fulfilling 1, the sorted list $N(u')$ is lexicographically not smaller than $N(u)$. (Ties are broken arbitrarily.)

Once the vertex labeling is complete, the matching is found in a greedy manner by a list schedule according to decreasing vertex labels: Starting with the highest label n, a vertex is matched with the highest label possible.

The $|V| + |E|$ time algorithm of [20] is optimal when the restriction graph G is given explicitly. This yields a $O(n^2)$ time maximum matching algorithm for the corresponding co–comparability graph G'.

Since permutation graphs can be represented in $O(n)$ space, we are interested in a faster algorithm for this class of graphs. Using the above derived geometric interpretation of permutation graphs we show that simple geometric arguments and data structures suffice to design a matching algorithm whose run-time is sublinear in the number of edges of the graph.

To achieve this, we partition the vertices into *levels*. The level of a vertex $v \in V$ is the length of the longest path from v to a vertex of outdegree zero. It is easy to see that the following holds: If vertex $u \in V$ is at a higher level than vertex $v \in V$, then $L(u) > L(v)$. This can be shown by an inductive argument. Let l_u, l_v be the level of u and v respectively. Then, v has at least one successor at level $l_v - 1 \geq l_u$ whereas u has no successor at this level.

This partitioning into levels corresponds to vertex domination in the geometric representation of a permutation graph. The partition into levels can be done in $O(n \log n)$ time. All that remains is to determine the order in which the vertices within a level are labeled. Instead of determining sorted lists $N(v)$ we use a geometric argument.

Denote by $DomReg(p)$ the upper right quadrant of an axis-aligned coordinate system whose origin is at point p. In this section, a point p dominates a point q if q lies in $DomReg(p)$. Let R be any region of the plane, then $Max(R)$ denotes the maximum label of all labeled points which lie in R, it is set to zero if R contains no labeled point.

Now let u and v be two points on a common level and assume w.l.o.g. that u lies above v, i.e. u's y-coordinate is larger than v's. The intersection of $DomReg(u)$ and $DomReg(v)$ is a quadrant called $SharedQ(u,v)$. Then the region $DomReg(u)$ can be partitioned into $SharedQ(u,v)$ and the remaining half-open rectangle, called $Rec(u)$; similarly, for v. Now, we observe that $L(u) > L(v)$ if and only if $Max(Rec(u))$ is greater than $Max(Rec(v))$.

This reduces the problem of performing a comparison operation of the form "$L(u) > L(v)$" (as needed for sorting each layer) to a comparison between two integers (labels) obtained via Maximum-Labeled-Element-Queries in half-open rectangles. There are different approaches to solving such queries: one is to state the problem as a (dynamic) 3-d range searching problem where the third coordinate is the label, the other, taken here, is to use the traditional range-range priority search trees (see e.g., [16]).

The primary tree stores the points sorted by x-coordinate in its leaves. Located at each internal node of the primary tree is a y-sorted secondary tree containing all points in the subtree; in addition, the secondary tree contains, at each internal node, the maximum label of all elements stored in its subtree.

To perform a Maximum-Labeled-Element-Query, we must find the maximum labeled point in the region bounded by $[x_a, x_b]$ and $[y_a, y_b]$ (where one of the coordinates is infinity). We locate the $O(\log n)$ roots of subtrees spanning the x-range $[x_a, x_b]$ and for each of these we use the $O(\log n)$ secondary trees spanning the y-range to find the maximum labels of all points in the entire half-open rectangle. It is easily observed that the time for a Maximum-Labeled-Element-Query is $O(\log^2 n)$. In total, $O(\log^2 n)$ roots of subtrees are to be located and the Maximum-Labeled-Element is the maximum of the $O(\log^2 n)$ candidate values so obtained in the secondary trees.

Initially we build the primary and secondary trees for all points, setting all labels to zero. The labels for layer i are calculated using (only) the labels of layers 1 to i-1; the locations of all points remain unchanged. Updating the label of a point requires updating the corresponding leaf and the nodes on the path to the root. The total time per label-update is therefore $O(\log n)$. Next we summarize the algorithm.

1. Build a range-range priority search tree on all points, setting the labels of all points equal to zero.
2. Partition the point set into layers, $1, 2, \ldots, k$
3. Label the n_1 points on the first layer $1, 2, \ldots, n_1$
4. For layer $i = 2, 3, \ldots, k$ do
   sort the points on layer i (using the above described comparison operator)
   assign consecutive labels to the points
   update the labels in the search tree structure.
5. perform a greedy matching on the labeled graph.

If an optimal sorting algorithm is used, the total number of queries can be bounded by $\sum_{i=1}^{k} n_i \log n_i$, where n_i denotes the cardinality of layer i. The matching can be done by a sequence of $O(n)$ Maximum-Labeled-Element-Queries using the quadrant $DomReg(p)$ for finding point q to be matched with p. (This requires $O(n)$ deletions as well.) ¿From the above it follows that the total time complexity of the matching algorithm is $O(n \log^3 n)$.

Theorem 3. *A maximum matching in permutational graphs can be computed in $O(n \log^3 n)$ time where n is the number of vertices of G.*

Corollary 4. *The problem of minimizing the total number of arrows to kill all n birds can be solved in $O(n \log^3 n)$ time.*

Since complement graphs of permutation graphs are permutation graphs we get the following result.

Theorem 5. *The two-processor task scheduling for dependency graphs known to be permutation graphs can be solved in $O(n \log^3 n)$ time where n is the number of processes (not dependencies).*

3.2 Dominance Algorithms

Dominance problems arise naturally in a variety of applications and they are directly related to well studied geometric and non-geometric problems. These problems include: range searching, finding maximal elements and minimal layers, computing a largest area empty rectangle in a point set, determining the longest common sequence between two strings, and interval/rectangle intersection problems, etc. Dominance problems are also needed to be solved for our maximum matching algorithm. Here we show how to solve dominance problems by using the compact representation of the stage graphs.

Let $P = \{p_1, p_2, ..., p_n\}$ be a planar point set of n distinct points $p_i = (x_i, y_i)$, $i = 1, ..., n$. A point p_i is said to *dominate* a point p_j, if $x_i \geq x_j$ and $y_i \geq y_j$ and $i \neq j$. The *dominance problem* is to enumerate all dominances of a given point set. Preparata and Shamos [16] presented an optimal sequential algorithm for this problem and it runs in $O(n \log n + k)$ time, where k is the total number of dominance pairs. Using the compact representation of stage graphs we obtain a simple sequential algorithm for this problem. Goodrich [8] presented a parallel algorithm for this problem; it runs in $O(\log n)$ time using $O(n + k/\log n)$ CREW PRAM processors. By adapting the above sequential algorithm we are able to present a simple optimal parallel algorithm for this problem which runs on an EREW PRAM.

Without loss of generality assume that the points of the set $P = \{p_1, p_2, ..., p_n\}$, where $p_i = (x_i, y_i)$, $i = 1, ..., n$, are sorted with respect to increasing x-coordinate. Therefore, we relabel each point p_i by its index i. From now on, we refer to a point p_i by its index i. Let Y be the array consisting of labels of points in P, sorted with respect to increasing y-coordinate; i.e. Y is a permutation of $\{1, ..., n\}$. Let i appear at the position pos_i, where $1 \leq pos_i \leq n$, in Y. From the above definitions it follows that a point i dominates a point $j \in P$ if and only if $i > j$ and $pos_i > pos_j$. Hence the points dominated by i are the elements of the subarray $Y[1, ..., pos_i]$ which are less than i. So the dominance problem reduces to that of reporting all elements of subarray $Y[1, ..., i - 1]$ which are less than $Y[i]$, for all $i \in \{2, ..., n\}$.

The sequential algorithm has $O(\log n)$ merge stages. In order to simplify notation, we present the last merge stage. Assume that we know all dominances for each point within subarrays $Y[1, ..., n/2]$ and $Y[n/2 + 1, ..., n]$. We wish to compute dominances for each point in $Y[1, ..., n]$. Observe that we need only compute the points dominated by $Y[n/2 + 1, ..., n]$ in $Y[1, ..., n/2]$ since no point in $Y[1, ..., n/2]$ dominates any point in $Y[n/2 + 1, ..., n]$. The dominances are computed as follows. First note that the arrays $Y[1, ..., n/2]$ and $Y[n/2 + 1, ..., n]$ have already been sorted in increasing order during the recursion. Now rank each element of $Y[n/2 + 1, ..., n]$ in $Y[1, ..., n/2]$ (see [3]). Suppose an element $Y[i]$, where $n/2 + 1 \leq i \leq n$ is ranked at the position j $(1 \leq j \leq n/2)$ in $Y[1, ..., n/2]$. The points dominated by $Y[i]$ in $Y[1, ..., n/2]$ are $Y[1], Y[2], ..., Y[j]$. After cross ranking, we can report dominances in time proportional to the number of dominance pairs. We summarize the result in the following theorem.

Theorem 6. *All dominance pairs of an n-point planar set can be computed in $O(n \log n + k)$ time using $O(n)$ space, where k is the total number of dominance pairs.*

Now we parallelize the above algorithm by using the results of [3, 10]. The parallel-merge sort algorithm of [3] cross-ranks elements of each subarray during each stage of merging. As observed above, after cross ranking, the problem reduces to that of reporting subarrays $Y[1, ..., j]$ for an appropriate j, where $1 \leq j \leq n/2$, for each $Y[i]$, where $n/2 + 1 \leq i \leq n$. Subarrays can be optimally reported on EREW PRAM by the algorithm of [10]. We summarize the result in the following theorem.

Theorem 7. *All dominances of an n-point planar set can be computed in $O(\log n)$ time using $O(n + k/\log n)$ processors on the EREW PRAM, where k is the total number of dominances.*

The other problem we study in this section is a *rectangle query problem* for a planar point set P. A query consists of a pair of points (p_i, p_j), where $p_i, p_j \in P$, and we need to answer whether the rectangle formed by the query points is empty or not. Given $O(n^2)$ space, we can answer such queries in $O(1)$ time, since we can precompute information about each pair of points in P. The space can be reduced to $O(n \log n)$ using range search data structures [16], but unfortunately the query time increases to $O(\log n)$. We provide an $O(n \log n)$ size data-structure, where the queries can still be answered in $O(1)$ time. Furthermore, the data-structure is very simple and can be computed in sequential $O(n \log n)$ time and in parallel $O(\log n)$ time using $O(n)$ EREW PRAM processors. That is, we can show:

Theorem 8. *A data structure of size $O(n \log n)$ can be computed in $O(n \log n)$ sequential time and in $O(\log n)$ parallel time using $O(n)$ EREW PRAM processors, using which rectangle queries can be answered in $O(1)$ sequential time.*

4   Stage Graphs in Three Dimensions

In this section we study stage graphs in three dimensions. Let $X = \{p_1, \ldots, p_n\}$ be a collection of points in $\mathbb{R}^3$ such that the z-coordinate of each element of X is strictly greater than zero and S a compact plane-convex set of $\mathbb{R}^3$ contained in the hyperplane $z = 0$ in $\mathbb{R}^3$. Given X and S construct the stage graph $G(X, S)$ with vertex set X such that two vertices p_i, p_j of $G(X, S)$ are adjacent if the line through p_i and p_j intersects S. $G(X, S)$ is called a $3 - d$-stage graph.

We prove first that $3 - d$-stage graphs are also comparability graphs. Next, we show that the set of $3 - d$-stage graphs obtained when S is a triangle is exactly the set of comparability graphs of orders of dimension at most three, and thus in this case, the recognition problem is NP-complete [26]. We also provide a characterization of $3 - d$-stage graphs in terms of geometric containment orders, that is ordered sets arising from containment relations among the elements of families of convex sets on the plane. For instance, we prove that when S is a circle, $G(S, X)$ is the comparability graph of a circle order [21]. Finally, we prove that not all comparability graphs are $3 - d$-stage graphs.

Lemma 9. *Let S be a plane convex set of $\mathbb{R}^3$ and X a collection of points in $\mathbb{R}^3$ with z coordinates greater than zero, then $G(S, X)$ is a comparability graph.*

The orientation induced on $G(S, X)$ induces a partial order $P(X, <)$ on X. $P(X, <)$ will be called a stage order. If in addition we want to specify that a stage order $P(X, <)$ arises from a specific convex set S, we will call $P(X, <)$ an S-stage order. Natural questions arises:

> Is it true that for every ordered set $P(Y, <)$ is a stage order? That is, is it true that for every ordered set $P(Y, <)$ there is a convex set S and a point set X such that $P(Y, <)$ is isomorphic to the stage order generated by X and S? Can we characterize stage orders?

We now show that recognizing graphs with triangular stages is equivalent to recognizing orders of dimension three, which is NP-hard [26].

Theorem 10. *An ordered set $P(X, <)$ is a order for a triangular stage if and only if the dimension of $P(X, <)$ is at most three.*

Corollary 11. *The recognition problem for graphs with triangular stages is NP-hard.*

An algorithmic characterization of $3 - d$-stage graphs is unlikely. Nevertheless, we can give a partial characterization of stage orders in terms of "geometric containment" orders. The proof is involved and is based on the theory of geometric containment and ordered sets. Due to the space limitations, the proof has been omitted.

Theorem 12. *Not every comparability graph is a $3 - d$-stage graph.*

The following theorem provides a characterization of 3d-stage graphs in terms of geometric containment orders.

Theorem 13. *An ordered set $P(Y, <)$ is an S-stage order if and only if $P(Y, <)$ is an S-ordered set.*

5 Multiple Stages

In this section we consider a natural generalization of $2 - d$-stage graphs to the case of more than one line segment or "stage." Suppose we have a collection of n points on the plane in general position, as well as k fixed but arbitrary finite, closed, non-intersecting, straight line segments contained in the x-axis of the plane. We define a graph as follows: Vertices are the given points, and $\{u, v\}$ is an edge if and only if the infinite (straight) line segment uv joining the point u to v intersects one of the k stages. We say that the above graph is represented by the above k stages. For fixed n, let $\mathcal{G}_k$ denote the class of graphs (on n points in general position) which can be represented by k stages as above.

It is easy to show that a simple graph with m edges is in the class $\mathcal{G}_m$. In particular, every graph on n vertices can be represented with at most $n(n-1)/2$ stages. It is clear that we have the following containments for graphs on n vertices:

$$\mathcal{G}_1 \subseteq \mathcal{G}_2 \subseteq \cdots \subseteq \mathcal{G}_{n(n-1)/2}. \tag{1}$$

For any graph G let $\text{st}(G)$ be the minimum number of stages needed in order to represent the graph G as above. In the sequel we consider the size $\text{st}(G)$ as a function of the number of vertices of the graph. We consider upper and lower bounds on the number of stages needed to represent a graph and establish results on separating the above classes of graphs.

First we consider the stage number of several simple graphs, including lines, cycles, trees and complete bipartite graphs.

Theorem 14.

1. *The line graph L_n on n vertices can be represented with a single stage. Hence $st(L_n) = 1$.*
2. *The cycle C_n on n vertices can be represented with a single stage if and only if $n \leq 4$. Moreover, $st(C_n) = 2$, for $n \geq 5$.*
3. *The complete bipartite graph $K_{m,n}$ can be represented with a single stage.*
4. *Every tree can be represented with at most two stages. In addition, caterpillars are precisely the trees representable with one stage.*

For arbitrary graphs we can establish the following upper bound on $st(G)$.

Theorem 15. *Every n vertex graph can be represented with at most $n(n-1)/4 - \Theta(n \log n)$ stages.*

While the above theorem shows that all graphs are representable using $O(n^2)$ stages, none of the examples considered require more than two stages to represent. The next theorem establishes the existence of graphs requiring $\Omega(n^2 / \log n)$ stages. Our proof makes use of the following result of H. Warren [25].

Lemma 16. *If $p_1, \ldots, p_m$ are polynomials in r variables with degree $\leq d$ then the number of sign-patterns is $s(p_1, \ldots, p_m) \leq \left(\frac{4edm}{r} \right)^r$.*

Theorem 17. *There exist graphs which require $\Omega(n^2 / \log n)$ stages for their representation.*

Nevertheless, Theorem 17 gives no indication on how to construct graphs requiring a large number of stages. To give such a construction we use the previous observation that every cycle with 5 or more nodes requires at least two stages for its representation. This means that graphs which are representable with a single stage must have girth ≤ 4. We take advantage of this fact in order to prove the following result.

Theorem 18. *Every graph G with minimal degree d and girth ≥ 5 requires at least $\lfloor d/2 \rfloor$ stages for its representation via stages.*

There are a number of constructions in the literature of d-regular graphs with girth 5. For example, see [17, 18, 19, 15, 1, 24, 12]. An interesting construction of a regular bipartite graph of degree $p+1$, $p^2 + p + 1$ nodes and girth 6, p prime, is the projective plane over the Galois Field on p elements, with $p+1$ lines each line containing exactly $p+1$ points [12][10.15]. It is clear from Theorem 18 that this graph requires at least $(p+1)/2$ stages for its representation, i.e., it is an n vertex graph, G, such that $st(G) = \Omega(\sqrt{n})$. It is known [11][Theorem 4.2] that a graph with $n > 2$ vertices, girth ≥ 5 can have at most $\frac{1}{2}n\sqrt{n-1}$ edges. It follows that $\Theta(\sqrt{n})$ is the highest possible stage number for a graph obtained by Theorem 18.

6 Conclusion

We have introduced the archer's problem and shown that its solution leads to the intersting class of stage graphs which we characterized to be permutation graphs. The characterization which leads to the solution for the archer's problem allowed for the development of improved algorithms for matching in permutation graphs, for a class of two-processor scheduling problems, and for several geometric problems. We answer the natural question of how the archer's problem generalizes to multiple stages and to three-dimensions. In two dimensions we establish upper and lower bounds on the number of stages required to represent graphs. In three dimensions we give characterization results and establish the NP-completeness of the recognition problem already for triangular stages.

There are several interesting open problems suggested by our investigations. The notion of stage number as a graph theoretic parameter seems to be interesting in its own right. This suggests searching for tighter (constructive or not) upper and lower bounds on the stage number of an arbitrary graph, as well as determining the complexity of the recognition problem $G \in \mathcal{G}_k$, both for fixed as well as variable k.

References

1. B. Bollobás, Extremal Graph Theory, Academic Press, 1978.
2. Coffman, E. G. Jr.; Graham, R. L.; "Optimal scheduling for two-processor systems" Acta Informatica 1 (1972), 200-213.
3. R. Cole, "Parallel Merge Sort", SIAM J. Comp. 17:4 (1988), 770-785.
4. A. Datta, A. Maheshwari, J.-R. Sack, "Optimal parallel algorithms for direct dominance problems", LNCS 726 (1993), 109-120.
5. Fujii, M.; Kasami, T.; Ninomiya, K. "Optimal sequencing of two equivalent processors." SIAM J. Appl. Math. 17:4 (1969), 784-789.
6. Gabow, H. N. "An almost-linear algorithm for two-processor scheduling" J. ACM 29:3 (1982), 766-780.
7. Gabow, H. N.; Tarjan, R. E. "A linear-time algorithm for a special case of disjoint set union" J. Comp. and System Sci. 30 (1985), 209-221.
8. M. T. Goodrich, "Intersecting line segments in parallel with an output-sensitive number of processors", SIAM J. Comp., 20 (1991), 737-755.
9. J. JáJá, An introduction to parallel algorithms, Addison-Weseley Publishing Company, 1992.
10. A. Lingas and A. Maheshwari, "Simple optimal parallel algorithm for reporting paths in trees", Symposium on Theoretical Aspects of Computer Science, LNCS, 1994.
11. J. H. van Lint and R. M. Wilson, A Course in Combinatorics, Cambridge University Press, 1992.
12. L. Lovász, Combinatorial Problems and Exercises, North Holland Publishing Company, 1979.
13. S. Micali and V. V. Vazirani, "An $O(\sqrt{V}E)$ algorithm for finding maximum matching in general graphs", in: Proc. 21st Ann. IEEE Symp. Foundations of Computer Science, (1980),17-27.

14. Moitra, A.; Johnson, R., C. "A Parallel algorithm for maximum matching on interval graphs." 18th Intl. Conf. on Parallel Processing III (1989), 114-120.

15. V. Neumann-Lara, "k-Hamiltonian graphs with given girth", in: Infinite and Finite Sets, Colloquia Mathematica Societatis János Bolyai, Keszthely, Hungary (1973), 1133-1142.

16. F. P. Preparata and M. I. Shamos, Computational Geometry: An Introduction, Springer-Verlag, New York, 1985.

17. H. Sachs, Einführung in die Theorie der endlichen Graphen, Teil I - II, Teubner, 1970-1972.

18. H. Sachs, "Regular graphs with given girth and restricted circuits", Journal of the London Mathematical Society, 38 (1963), 423-429.

19. H. Sachs, "On regular graphs with given girth", in: Theory of Graphs and its Applications, (M. Fiedler, ed.), Academic Press, New York (1965), 91-97.

20. Sehti, R. "Scheduling graphs on two processors" SIAM J. Comp. 5:1 (1976), 73-82.

21. J.B. Sidney, S.J. Sidney and J. Urrutia, "Circle Orders, n-gon orders and the crossing number", Order 5 (1988), 1-10.

22. J. Spinrad, "On comparability and permutation graphs", SIAM J. Comp. 14 (1985), 658-670.

23. J. Urrutia, "Partial orders and Euclidean geometry". In Algorithms and Order, I. Rival (ed) (1989), 387-434 Kluwer Academic Publishers.

24. H. Walther and H.-J. Voß, Über Kreise in Graphen, VEB Deutscher Verlag der Wissenschaften, 1974.

25. H. Warren, "Lower bounds for approximation by nonlinear manifolds", Transactions of the AMS 133 (1968), 167-178.

26. M. Yannakakis, "The complexity of the partial order dimension Problem", SIAM J. Discr. Meth. 3 (1982), 351-358.

Placing Resources in a Tree:
Dynamic and Static Algorithms *

Vincenzo Auletta, Domenico Parente and Giuseppe Persiano

Dipartimento di Informatica ed Appl., Università di Salerno, 84081 Baronissi, Italy.
E-mail: {auletta, parente, giuper}@dia.unisa.it

Abstract. We study the classical problem of optimally placing resources in a tree. We give dynamic algorithms that recompute the optimal solution after a weight change in polylogarithmic time for the case of one resource in a general tree and for any constant number of resources in a complete tree. Our algorithms are the first dynamic algorithms for this problem. We also give linear-time algorithms for the static version of the problem for two resources. Previously known algorithms run in time quadratic in the number of vertices. We also discuss an on-line amortized constant time algorithm for placing any number of resources on a line.

1 Introduction

We study the dynamic version of the classical problem of optimally placing resources in a tree that can be described as follows.

Let $\mathcal{T}$ be a tree with N vertices and, for each vertex v, let $w(v)$ be the weight of v. We want to place a constant number k of resources in $\mathcal{T}$. Each resource can satisfy any number of requests at the same time and each request is satisfied using the closest resource. We define the cost of a placement as the sum over all vertices v of the product of the weight $w(v)$ of the vertex times the distance from the closest resource. The optimal placement is the placement with least cost among all placements.

This problem arises in several contexts. For example, we can think of the resources as copies of a read-only file. Each node of a communication network has a copy of the file but, as the file is very large, only a fixed number k of copies are on-line (i.e., reside at disks located at the nodes of the communication network) and the remaining are kept off-line (e.g., on magnetic tapes). The weight of a vertex is the measure of how often the file is read by a process residing at that vertex. It is thus desirable to place the copies of the file so to minimize the average delay incurred into by the programs requesting access to the file. This is the *static* version of the problem.

Due to the changed needs of the users located at the nodes of the communication network, it is possible that the access frequencies of the nodes (and thus the weights of the vertices of the tree) vary and therefore a new optimal placement needs to be recomputed. Moreover, some of the workstations might

---

* Partially supported by Progetto MURST 40%, Algoritmi, Modelli di Calcolo e Strutture Informative.

be unreachable at a given moment because of failure of some of the links in the communication network. Nonetheless, we want all the nodes of each connected component of the tree to have access to the file and thus we have to recompute the optimal placement of k copies of the file in each connected component. This is the *dynamic* version of the problem. We remark here that in our setting, insertion and deletion of nodes are not taken into consideration.

The mechanics of a dynamic algorithm can be described as follows. A weighted graph is presented to the algorithm. The algorithm performs some preprocessing during which it builds some data structure; for our algorithms this preprocessing takes linear time and the size of the data structure is linear in the number of vertices of the tree (thus our algorithms are optimal with respect to both these measures). After the preprocessing stage, an intermixed sequence of the following two types of operations is performed: 1) the weight of a vertex is changed; 2) the optimal placement of the resources in a subtree of the current weighted tree is asked for. The algorithm has to perform each operation before it can see the next request.

Summary of the results. In Section 2, we study the dynamic version of the one-resource problem. We give an algorithm that solves the dynamic one-resource problem for trees of N vertices and depth d in time $O(\min(d, \log N \log d))$, using a preprocessing of time $O(N)$. The algorithm is based on a novel data structure to dynamically maintain the weights of the subtrees of a tree under weight-change operations that is of independent interest. As a corollary we obtain a linear time algorithm for the static one-resource problem. In Section 3, we give a linear time algorithm for the static two-resource problem. The best previously known algorithm takes $O(N^2)$ time for a tree with N vertices [3]. In Section 4 we present a dynamic algorithm that recomputes the optimal placement of k resources in a complete tree with N vertices in polylogarithmic time. Our algorithm needs a linear time preprocessing. As a corollary we obtain a linear time algorithm for the static k-resource problem (k constant). We describe the algorithm for the case $k = 2$ and briefly explain how it can be extended to solve the problem for $k > 2$. In Section 5, we just mention (due to the lack of space) an on-line algorithm for placing k-resources on a line. The algorithm does not receive all the vertices of the input at once. Instead, the line grows one vertex at a time from one of the two extremities; each time a new vertex is given, the algorithm has to recompute the optimal solution. Our algorithm handles the insertion of a new vertex in amortized constant time. The proofs of the main results are found in [1].

To ease exposition, we present our algorithms for the case of binary trees. All algorithms can be modified to handle trees of arbitrary degree.

Related works. The static version of our problem is known under the name of the *k-median problem* and has been extensively studied in the past. Kariv and Hakimi [3] proved that the k-median problem is NP-complete even for planar graphs of maximum degree 3. Papadimitriou [4] proved that the k-median problem in the plane is NP-complete. The k-median problem is solvable in polynomial time for fixed k and for trees. Kariv and Hakimi [3] gave an algorithm that solves the k-median problem on trees with N vertices in time $O(N^2)$.

Notation and definitions. Unless otherwise specified, $\mathcal{T}$ is an N-vertex weighted binary tree with root $\mathcal{R}$, where each vertex v of $\mathcal{T}$ has a weight $w(v) > 0$. We assume, without loss of generality, that each vertex of $\mathcal{T}$ is either a leaf or has two children. For each vertex u, we denote by T_u the subtree of $\mathcal{T}$ rooted in u, by T_u^l and by T_u^r the left and right subtrees of T_u and let $T_u^c = \mathcal{T} - T_u$. Let S be a subtree of $\mathcal{T}$ and u a vertex of S. We call the trees $S \cap T_u^l$, $S \cap T_u^r$, and $S \cap T_u^c$ the trees *adjacent* to u in S. Also we let $W(T_u)$ denote the sum of the weights of the nodes belonging to T_u. The cost $Cost(x, S)$ of serving the vertices in a subset S of $\mathcal{T}$ with a resource located at the vertex x is $Cost(x, S) = \sum_{v \in S} w(v) \cdot \text{dist}(x, v)$, where $\text{dist}(x, v)$ is the distance of x from v. The cost of serving a vertex v from a placement $(x_1, x_2, \cdots, x_k)$ is $Cost(x_1, x_2, \cdots, x_k, v) = w(v)\text{dist}(v, x_1, x_2, \cdots, x_k)$, where $\text{dist}(v, x_1, x_2, \cdots, x_k)$ is the distance of v from the closest of $x_1, x_2, \cdots, x_k$. If S is a subset of the vertices of $\mathcal{T}$, then the cost of placement (x_1, x_2) relative to S is simply the sum of the costs of serving each vertex in S. We define the cost of the optimal placement of one resource in a tree T, $Cost_1(T)$, as $Cost_1(T) = \min_{x \in T} Cost(x, T)$ and $Loc_1(T)$ as the set of vertices which obtain the optimal cost. $Cost_2(T)$ and $Loc_2(T)$ are defined similarly.

2 An algorithm for the dynamic one-resource problem

In this section we give a dynamic algorithm that, after a linear-time preprocessing, recomputes in time $O(\min(d, \log N \log d))$ the optimal placement of one resource in a tree with N vertices and depth d after a weight change. Our algorithm is based on the properties of the optimal placement that are summarized in the following lemma.

Lemma 1. *Let $\mathcal{T}$ be a weighted binary tree with N nodes and u a vertex of $\mathcal{T}$. Then,*
1. $u \in Loc_1(\mathcal{T})$ if and only if, for all trees S adjacent to u in $\mathcal{T}$, it holds that $W(S) \leq \frac{W(\mathcal{T})}{2}$;
2. $Loc_1(\mathcal{T})$ belongs to the tree S adjacent to u in $\mathcal{T}$ if and only if $W(S) > \frac{W(\mathcal{T})}{2}$.

Corollary 2. *For all weighted binary trees $\mathcal{T}$, $|Loc_1(\mathcal{T})| \leq 2$. Moreover, if $Loc_1(\mathcal{T}) = \{u, v\}$, then u and v are adjacent.*

From the above lemma a simple dynamic algorithm that recomputes in time $O(d)$ an optimal placement of one resource and its cost after a weight change can be derived. In the rest of the section we present the algorithm 1-Dynamic that recomputes in time $O(\log N \log d)$ an optimal placement (along with its cost) of one resource after a weight change.

2.1 Algorithm 1-Dynamic

Algorithm 1-Dynamic uses the data structure *Tree-Sum*. The data structure *Tree-Sum* consists of an initialization procedure $Init(\mathcal{T})$, that takes as input a weighted binary tree $\mathcal{T}$ and returns in time $O(N)$ an instance, $\mathcal{S}$, of a *Tree-Sum* data structure; of a procedure $Change(\mathcal{S}, u, \delta)$ that increments the weight of the vertex u by δ and returns an updated version of $\mathcal{S}$; and $Opt(\mathcal{S})$ that returns a vertex $v \in Loc_1(T)$. *Change* and *Opt* have each running time $O(\log N \log d)$. The *Tree-Sum* data structure is an extension of the *Path-Sum* data structure discussed in [6] (see also [2]) that consists of three procedures: the procedure $Init\text{-}Path(I)$, that takes a sequence of l integers $I = (x_1, x_2, \cdots, x_l)$ and returns the initialized *Path-Sum* data structure S_I in $O(l)$ time; the procedure $Update(S_I, j, \delta)$ that returns an updated version of S_I with the effect of setting $x_j = x_j + \delta$ and $Min(S_I, W)$ that gives the minimum j such that $\sum_{i=1}^{j} x_i > W$.

A sequence of pairwise distinct vertices of $\mathcal{T}$, $P = (v_1, \ldots, v_k)$, $k > 0$, is a *path* if, for $i = 1, 2, \ldots, k - 1$, v_{i+1} is the parent of v_i. The following lemma is from [5].

Lemma 3. *Let $\mathcal{T}$ be a tree of N vertices. Then $\mathcal{T}$ can be partitioned in $O(N)$ time into a set $\mathcal{P} = (P_1, \ldots, P_k)$ of disjoint paths such that each path of $\mathcal{T}$ is contained in the union of $O(\log N)$ paths of $\mathcal{P}$. We call such a partition in paths an* efficient *partition of $\mathcal{T}$.*

Implementation of Init, Change *and* Opt. Let us now describe how to implement the procedures *Init*, *Change* and *Opt* using the *Path-Sum* data structure and the concept of efficient partition. Let $\mathcal{P} = (P_1, \ldots, P_k)$ be an efficient partition of $\mathcal{T}$ and, for each $1 \le i \le k$, let $P_i = (v_{i,1}, \ldots, v_{i,h_i})$. We define the father index, $Father(P_i)$, of a path P_i not containing the root in the following way. Let u be the parent of the vertex v_{i,h_i} (that is the highest level vertex in P_i) and P_j be the path that contains u. Then, $Father(P_i)$ is defined to be the index of u in P_j. The *Father* of the path containing the root is undefined.

Procedure $Init(\mathcal{T})$ computes an efficient partition $\mathcal{P}$ and, for each path P_i of $\mathcal{P}$, initializes the *Path-Sum* data structures S_{P_i} and returns the *Tree-Sum* data structure $\mathcal{S} = (S_{P_1}, \ldots, S_{P_k})$. S_{P_i} is obtained by running $Init\text{-}Path$ on input the sequence $V_i = (V_{i,1}, \ldots, V_{i,h_i})$, where $V_{i,1} = W(T_{v_{i,1}})$ and $V_{i,j} = W(T_{v_{i,j}}) - W(T_{v_{i,j-1}})$, for $j > 1$.

The procedure $Change(\mathcal{S}, v^*, \delta)$ is implemented as follows. Let $P = v_1 \ldots v_m$ be the path from $v_1 = v^*$ to $v_m = \mathcal{R}$. By Lemma 3, the path P is covered by $l = O(\log N)$ paths of $\mathcal{P}$, say $Q_1, \ldots, Q_l$. As the weights of only the trees rooted in $v_1 \ldots, v_m$ are affected by the change of $w(v^*)$, we only need to update the data structures $S_{Q_1}, \ldots, S_{Q_l}$ relative to $Q_1, \ldots, Q_l$. Let j be the index of v^* in Q_1. The operation $Change(\mathcal{S}, v^*, \delta)$ performs the sequence of operations $Update(S_{Q_1}, j, \delta)$, $Update(S_{Q_2}, Father(Q_1), \delta)$, $\ldots$, $Update(S_{Q_l}, Father(Q_{l-1}), \delta)$. Since a path of $\mathcal{P}$ has length at most d, then each operation $Update$ takes time $O(\log d)$. Hence $Change$ takes time $O(\log N \log d)$.

To show how to implement $Opt(\mathcal{S})$, let us first observe that, by Lemma 3, the path Q from $\mathcal{R}$ to an optimal placement is covered by $O(\log N)$ paths each of length at most d. The problem is to determine the vertices of Q and we already have exhibited a strategy that takes time $O(d)$. We now describe how procedure $Opt(S)$ searches for an optimal placement and computes its optimal cost without scanning the entire path Q. Let $Q_1, \ldots, Q_l$, $l = O(\log N)$, be the (so far unknown) paths of the efficient partition which cover the path Q from $\mathcal{R}$ to a vertex in $Loc_1(\mathcal{T})$ and let Q_1 be the path containing $\mathcal{R}$. To determine Q_{i+1} from Q_i, $1 \leq i < l$, procedure $Opt(S)$ runs the operation $Min(S_{Q_i}, W(\mathcal{T})/2)$. This returns the index of the lowest level vertex v in Q_i such that $W(T_v) > W(\mathcal{T})/2$. Since at most one among $W(T_v^l)$ and $W(T_v^r)$, say $W(T_v^\alpha)$, is greater than $W(\mathcal{T})/2$, then from the second condition of Lemma 1, $Loc_1(T)$ is located in the subtree T_v^α. Hence Q_{i+1} is the path containing the neighbour vertex of v belonging to T_v^α. If such a neighbour does not exist, then $Opt(S)$ returns $v \in Loc_1(\mathcal{T})$. Since we can have at most $O(\log N)$ different paths, each of length at most d, then $Opt(S)$ takes time $O(\log N \log d)$.

We can now state the main result of the section.

Theorem 4. *Algorithm 1-Dynamic solves the dynamic version of the problem of placing 1 resource in a tree of N vertices and depth d in time $O(\log N \log d)$ after a linear-time preprocessing.*

The algorithm we have just presented only returns the location of an optimal placement but not its cost. The cost of the optimal solution can be computed using similar techniques. We omit the technical details.

We also remark that the data structure *Tree-Sum* described above can be modified to handle also the operation $Opt(S, u)$ that computes the optimal placements of one resource in the subtrees T_u and T_u^c in time $O(\log N \log d)$.

3 An algorithm for the static two-resource problem

In this section we present the algorithm 2-Static for computing an optimal placement of two resources in $\mathcal{T}$. We start with an informal discussion of the ideas behind the algorithm. We define the function $PseudoCost(u, \mathcal{T})$ as the sum of the costs of the optimal solutions of the problem with one resource in T_u and T_u^c. $PseudoCost(\mathcal{T})$ is then defined as the minimum of $PseudoCost(u, \mathcal{T})$ over all the vertices u of $\mathcal{T}$. Algorithm 2-Static computes $PseudoCost(u, \mathcal{T})$ for all vertices u of the tree and sets $Cost_2(\mathcal{T}) = PseudoCost(\mathcal{T})$. We shall prove that the optimal cost of placing two resources in $\mathcal{T}$ is equal to $PseudoCost(\mathcal{T})$. An optimal placement is, then, obtained by considering a vertex v such that $PseudoCost(v, \mathcal{T}) = PseudoCost(\mathcal{T})$ and taking two vertices $x_1 \in Loc_1(T_v)$ and $x_2 \in Loc_1(T_v^c)$. Let us proceed more formally.

Lemma 5. *For any weighted binary tree $\mathcal{T}$, the cost $Cost_2(\mathcal{T})$ of an optimal placement of two resources in $\mathcal{T}$ is equal to $PseudoCost(\mathcal{T})$. Moreover, if v is a vertex such that $PseudoCost(v, \mathcal{T}) = Cost_2(\mathcal{T})$, then any pair of vertices $x_1 \in Loc_1(T_v)$ and $x_2 \in Loc_1(T_v^c)$ is an optimal placement of two resources in $\mathcal{T}$.*

Lemma 6. *Let u be a vertex of $\mathcal{T}$, S_1, S_2 be two subtrees of $\mathcal{T}$ adjacent to u with $W(S_1) > W(S_2)$, and let opt_1 be a vertex in $Loc_1(S_1)$. Then an optimal placement of a resource in $S = S_1 \cup S_2 \cup \{u\}$ lies on the path from u to opt_1.*

We have thus reduced the problem of computing $Cost_2(\mathcal{T})$ and $Loc_2(\mathcal{T})$ to the problem of computing $PseudoCost(u, \mathcal{T})$ and $Loc_1(T_u)$ and $Loc_1(T_u^c)$, for all vertices u of $\mathcal{T}$.

Let us now describe the algorithm 2-Static. Suppose we have computed $W(T_u)$ and $W(T_u^c)$ for all vertices u of $\mathcal{T}$. The values $PseudoCost(\cdot, \cdot)$ are computed in two phases: the first phase computes, for each vertex u, the value $Cost_1(T_u)$ and a vertex $opt_u \in Loc_1(T_u)$; the second phase computes, for each vertex u, the value $Cost_1(T_u^c)$ and a vertex $opt_{u^c} \in Loc_1(T_u^c)$.

The first phase starts from the leaves of $\mathcal{T}$ and works *bottom up* as follows. If u is a leaf then $Cost_1(T_u) = 0$ and $opt_u = u$. Otherwise, let v and z be the two children of u and let opt_v and opt_z be the optimal placements of a resource in T_v and T_z, respectively. Then, by Lemma 6, there exists a vertex $opt_u \in Loc_1(T_u)$ on the path from the optimal placement in the heavier of T_v of T_z to u. The vertex opt_u is found by scanning this path. In order to compute also $Cost(opt_u, T_u)$, each time we move along this path from a vertex x to a vertex y we compute $Cost(y, T_u)$ as $Cost(y, T_u) = Cost(y, T_u) + 2W(T_x) - W(T_u)$.

The second phase starts from $\mathcal{R}$ and works *top down*. Let u be a vertex other than $\mathcal{R}$, v, z be its parent and its sibling, respectively and opt_v^c and opt_z be the optimal placements of a resource in T_v^c and T_z. Assume that T_z is heavier that T_v^c (the other case is similar). Then by Lemma 6 there exists a vertex $opt_u^c \in Loc_1(T_u^c)$ on the path from opt_z to u. Working as in the previous phase, the algorithm finds opt_u^c and computes $Cost(opt_u^c, T_u^c)$.

We can now state the main result of the section.

Theorem 7. *The algorithm 2-Static finds the best placement of two resources in a weighted binary tree $\mathcal{T}$ in time $O(N)$.*

Proof. The correctness of the algorithm (described above) follows from Lemmas 5 and 6. The computation of $W(T_u)$ and $W(T_u^c)$ for all vertices u of $\mathcal{T}$ can be easily carried out in $O(N)$ time. Let us now analyze the running time of the bottom-up phase (a similar reasoning works also for the top-down phase). Vertex opt_u is computed by scanning the path from opt_S to u, where S is the heavier of the subtrees T_u^l and T_u^r, searching for a vertex satisfying condition 1 of Lemma 1. Since all the vertices of this path, except for opt_u, will not be scanned in the rest of the bottom up phase, we have that the bottom-up phase stops after $O(N)$ steps.

4 An algorithm for the dynamic problem on complete binary trees

In this section we present a dynamic algorithm 2-Dynamic that updates the optimal placement of two resources in a complete binary tree with N vertices after the weight of a vertex has been changed. Then we discuss how to extend the algorithm to compute the optimal placement of a constant number of resources greater than 2.

Algorithm 2-Dynamic uses a set of functions, that we call *canonical functions*, defined over the vertices of the tree. The canonical functions have the property that if a table of the values for all the vertices is available then we can compute, for any vertex v, the optimal placements of two resources in T_v and T_v^c along with their costs in time $O(\log^3 N)$. Moreover, a table of all values of the canonical functions can be computed in time $O(N)$ and updated in time $O(\log^3 N)$ in case the weight of a vertex changes.

The following theorem presents the main result of this section.

Theorem 8. *Algorithm 2-Dynamic after a linear-time preprocessing solves the dynamic version of the problem of placing two resources in a complete binary tree with N nodes in time $O(\log^3 N)$.*

The rest of this section is organized as follows. In the next subsection we define the canonical functions and state some simple lemmas. In Subsection 4.2, we present algorithm *Pre-Proc* that computes the table of values of the canonical functions. In Subsection 4.3 we present algorithm *Update* that updates the table of values of the canonical functions. Finally, we show how to compute, for any vertex v, the optimal placements of two resources in T_v and T_v^c.

4.1 The canonical functions

For the rest of this section $\mathcal{T}$ is a complete weighted binary tree with N vertices and root $\mathcal{R}$. For $v \in \mathcal{T}$, we denote by height(v) the level of a vertex v defined as the distance of v from the leaves in T_v. Notice that height$(v) < \log(N + 1)$. For a vertex v and integer $l \leq$ height(v), we denote by $T(v, l)$ the tree of depth l rooted at v; thus T_v is simply $T(v, \text{height}(v))$. Also, we define the *group* $G(v, l)$ as the set of all nodes $u \in T_v$ that are at distance l from v.

If R is a subset of the vertices of $\mathcal{T}$ we denote by dist(R, v) the minimal distance between v and the vertices of T. The distance between two subsets of vertices R, S is obviously defined as dist$(R, S) = \min_{v \in S} \text{dist}(R, v)$.

If l is an integer by $Cost(l, T)$ we denote the quantity $Cost(x, T)$ where x is a vertex at distance l from the lowest common ancestor (in short, lca) of T and x is not a descendant of the lca of T. Notice that $Cost(l, T)$ is well defined as $Cost(x, T)$ is the same for all x's not descendant of the lca of T that are at the same distance from the lca of T.

Definition of W, X and M. For a vertex v and an integer $l \leq$ height(v), we define $W(v, l)$ as the sum of the weights of all the vertices of $T(v, l)$. Thus, $W(v)$ is equal to $W(v, \text{height}(v))$.

For a vertex v and an integer $l \leq$ height(v), define $M(v, l)$ as the minimum cost of serving the tree T_v with a resource in $G(v, l)$ and $X(v, l)$ as the vertex in $G(v, l)$ that achieves this minimum. Similarly, for a vertex v and integers

$0 \leq l_1 \leq l_2 \leq \text{height}(v)$, we define $\mathrm{M}(v, l_1, l_2)$ as

$$\mathrm{M}(v, l_1, l_2) = \min_{\substack{x_1 \in G(v, l_1) \\ x_2 \in G(v, l_2)}} Cost(x_1, x_2, T_v)$$

and $\mathrm{X}(v, l_1, l_2)$ as a pair $x_1 \in G(v, l_1)$, $x_2 \in G(v, l_2)$ of vertices such that $Cost(x_1, x_2, T_v) = \mathrm{M}(v, l_1, l_2)$. Thus, $X(v, l_1, l_2)$ is the placement of one resource in $G(v, l_1)$ and the other in $G(v, l_2)$ that minimizes the cost of serving T_v. Obviously, the cost, $Cost_2(T_v)$ of the optimal placement is equal to the minimum over all l_1, l_2 of $\mathrm{M}(v, l_1, l_2)$. Later, we will show how to compute $Cost_2(T_v^c)$.

Definition of N and Y. We now define the functions N and Y that will provide a way to reduce a 2-resource problem to a 1-resource problem (in general, a k-resource problem is reduced to a $(k-1)$-resource problem) and this will be crucial to obtain algorithms *Pre-Proc* and *Update*. Indeed, as we shall see in Section 4.2, it is possible to express the values $\mathrm{X}(\cdot, \cdot, \cdot)$ (that is the solution to a 2-resource problem) in terms of $\mathrm{X}(\cdot, \cdot)$ (that is the solution to a 1-resource problem) and of the function Y. Similarly for the function $\mathrm{M}(\cdot, \cdot, \cdot)$.

Consider a vertex v, and let $l \leq \text{height}(v)$ and $h < l + 2$. The group $G(v, k)$, with $k = \lceil \frac{l-h}{2} \rceil$, is halfway between $H(v, h)$ (the set of vertices at distance h from v that are not in T_v) and $G(v, l)$. More precisely, for each vertex y in $H(v, h)$ and each vertex z in $G(v, l)$ the midpoint of the path from y to z lies in $G(v, k)$. Moreover, let $S(v, k)$ be the set of vertices in T_v at distance not smaller than k from v. Finally, we define $Y(v, l, h)$ as the vertex z in $G(v, l)$ such that by placing a resource in z and the other in $H(v, h)$ the cost of serving the vertices in $S(v, k)$ is minimized (over z ranging in $G(v, l)$) and $N(v, l, h)$ to be such a minimum cost.

As it is easily seen, both $Y(v, l, l)$ and $Y(v, l, l+1)$ are equal to $X(v, l)$ and both $N(v, l, l)$ and $N(v, l, l+1)$ are equal to $M(v, l)$.

Lemma 9. *Let v be a vertex and k an integer $k \leq \text{height}(v)$. Then for all l and h such that $h < l$ we have that*

$$\sum_{z \in G(v,k)} Cost(h + k, T_z) = \mathrm{N}(v, l, h) - \mathrm{M}(u, l - k) + \mathrm{M}(u, 0) + (h + k)\mathrm{W}(T_u),$$

where u is the ancestor of $Y(v, l, h)$ in $G(v, k)$ and $k = \lceil \frac{l-h}{2} \rceil$.

4.2 Preprocessing

In this section we present a linear time algorithm *Pre-Proc* for computing a table of values for the canonical functions to be used by algorithm 2-Dynamic. Algorithm *Pre-Proc* computes the values in a bottom-up manner: it assumes that the values for the children have already been computed and from these it obtains the values for the parent. To this aim we now state a series of lemmas that express the values of the canonical functions at a node as functions of the values at the children of this node. Obviously, if v_0 and v_1 are the children of a vertex v then, for all integers $1 \leq l \leq \text{height}(v)$, we have that $W(v, l) = W(v_0, l-1) + W(v_1, l-1) + w(v)$.

The next lemma gives an expression for X, M in terms of the values computed at the children. Essentially, it says that, for each vertex v, the optimal placement of one resource among the vertices in $G(v, l)$ is the best placement among the best in the left subtree and the best in the right subtree.

Lemma 10. *Let v be a vertex, v_0 and v_1 be its two children and l an integer $1 \leq l \leq \text{height}(v)$. For $b \in \{0, 1\}$ let $A_b(v, l)$ be the quantity $M(v_b, l-1) + M(v_{1-b}, 0) + (l+1)W(v_{1-b}) + l \cdot w(v)$. Then,*

$$M(v, l) = \min\{A_0(v, l), A_1(v, l)\}$$

and

$$X(v, l) = \begin{cases} X(v_0, l-1), & \text{if } A_0(v, l) < A_1(v, l); \\ X(v_1, l-1), & \text{otherwise.} \end{cases}$$

Moreover, for all vertices v it holds that $X(v, 0) = v$ and $M(v, 0) = M(v_0, 0) + M(v_1, 0) + W(v) - w(v)$.

The following lemma gives an expression for Y and N.

Lemma 11. *Let v be a vertex, v_0 and v_1 be its two children, l and h be integers such that $0 \leq l \leq \text{height}(v)$ and $1 \leq h < l$ and let $k = \lceil \frac{l-h}{2} \rceil$. For $b = 0, 1$ denote by $B_b(v, l, h)$ the quantity*

$$N(v_b, l-1, h+1) + N(v_{1-b}, l-1, h+1) - M(u_{1-b}, l-k) + (l-k)W(u_{1-b}) + M(u_{1-b}, 0),$$

where u_b is the ancestor in $G(v_{1-b}, k-1)$ of $Y(v_{1-b}, l-1, h+1)$. Then, $N(v, l, h) = \min\{B_0(v, l, h), B_1(v, l, h)\}$ and

$$Y(v, l, h) = \begin{cases} Y(v_0, l-1, h+1), & \text{if } B_0(v, l, h) < B_1(v, l, h); \\ Y(v_1, l-1, h+1), & \text{otherwise.} \end{cases}$$

In the next three lemmas we give an expression for $X(v, l_1, l_2)$ and $M(v, l_1, l_2)$ in terms of the values of the two children of v. For sake of clarity we divide the discussion in three parts. We analyze the case $l_1 \neq l_2$ and $l_1 > 0$ in Lemma 12; next we study the case $l_1 = l_2$ in Lemma 13 and finally we consider the case $l_1 = 0$ in Lemma 14.

Lemma 12. *Let v be a vertex and v_0 and v_1 be its two children. Then, for all pairs of integers l_1, l_2, $1 \leq l_1 < l_2 \leq \text{height}(v)$, $X(v, l_1, l_2)$ is the pair of minimum cost among the following four:*
1) $P_1 = X(v_0, l_1 - 1, l_2 - 1)$;
2) $P_2 = X(v_1, l_1 - 1, l_2 - 1)$;
3) $P_3 = (X(v_0, l_1 - 1), Y(v_1, l_2 - 1, l_1 + 1))$;
4) $P_4 = (X(v_1, l_1 - 1), Y(v_0, l_2 - 1, l_1 + 1))$.

Moreover, we have that

$$Cost(P_1, T_v) = M(v_1, 0) + (l_1 + 1) \cdot W(v_1) + M(v_0, l_1 - 1, l_2 - 1);$$
$$Cost(P_2, T_v) = M(v_0, 0) + (l_1 + 1) \cdot W(v_0) + M(v_1, l_1 - 1, l_2 - 1);$$
$$Cost(P_3, T_v) = M(v_0, l_1 - 1) + N(v_1, l_2 - 1, l_1 + 1) +$$
$$Cost(l_1 + 1, T(v_1, k - 1)) + l_1 \cdot w(v);$$
$$Cost(P_4, T_v) = M(v_1, l_1 - 1) + N(v_0, l_2 - 1, l_1 + 1) +$$
$$Cost(l_1 + 1, T(v_0, k - 1)) + l_1 \cdot w(v),$$

where $k = \lceil \frac{l_2 - l_1}{2} \rceil$.

Remark. The above Lemma allows to compute in constant time the value $X(v, l_1, l_2)$ (and $M(v, l_1, l_2)$) from the values at the children of v. Indeed observe that

$$Cost(l_1 + 1, T(v_b, k - 1)) = M(v_b, 0) + (l_1 + 1)W(v_b) - \sum_{z \in G(v_b, k)} Cost(l_1 + k + 1, T_z),$$

and using Lemma 9 we can express that sum in terms of the values of the functions N and M.

Lemma 13. *For all vertices v and integers $1 \leq l \leq$ height(v) $X(v, l, l)$ is the pair of minimal cost among the following three: $X(v_0, l - 1, l - 1)$, $X(v_1, l - 1, l - 1)$ and $(X(v_0, l - 1), X(v_1, l - 1))$.*

Lemma 14. *Let v be a vertex, v_0 and v_1 be its two children and l and integer $1 \leq l \leq$ height(v). Then*

$$M(v, 0, l) = \min_{b=0,1} \Big(M(v_b, 0) + W(v_b) + N(v_{1-b}, l - 1, 1) + Cost(1, T(v_{1-b}, l - 1)) \Big).$$

Moreover, the minimum is achieved by the pair $(v, Y(v_b, l - 1, 1))$.

From the lemmas above it follows that a table of values for the functions X, M, Y, N, W for all the vertices of a tree T can be computed in time linear in the number of vertices.

Lemma 15. *There exists an algorithm Pre-Proc that, on input a complete weighted binary tree T with N vertices computes, in time $O(N)$, the table of values of the canonical functions for all vertices in T.*

4.3 Updating the table

In this section we present the algorithm *Update* that we use to update the table of values of the canonical functions. As we shall see, one weight-change only affects $O(\log^3 N)$ values and these values can be quickly recomputed. Then, the new optimal placement is the pair $X(v, l_1, l_2)$ that has the minimum cost among all such pairs. This minimum can be trivially computed in time $O(\log^2 N)$.

Suppose that the weight of the vertex v^* has been incremented by δ and let $u_0, u_1, \cdots, u_{d-1}, u_d$ be the vertices on the path from the root $\mathcal{R} = u_0$ to $v^* = u_d$. Obviously, the vertices u_j are the only vertices for which we need to update the table of values.

In the following we will use the superscript n to denote the updated values of all the functions considered. The updating of $W(u_j, l)$ is trivial. We now show how the values $X(\cdot, \cdot)$ and $M(\cdot, \cdot)$ are updated. The other values are updated in a similar way.

Updating $X(v, l)$ and $M(v, l)$. We start by showing how to update the values relative to v^* and, for all the vertices u_j, the values for $l = 0$. For $0 \le l \le$ height(v^*), we have that $X^n(v^*, l) = X(v^*, l)$ and $M^n(v^*, l) = M(v^*, l) + l \cdot \delta$. Indeed, all vertices of $G(v^*, l)$ are at the same distance l from v^*. Therefore, for all vertices $z \in G(v^*, l)$ the cost of serving $T(v^*, l)$ with a resource in z increases by $l \cdot \delta$. This means that the vertex that achieves the minimum (that is $X^n(v^*, l)$) stays the same and its cost (that is $M^n(v^*, l)$) increases by $l\delta$.

Moreover, for all u_j, $0 \le j < i$, we have $X^n(u_j, 0) = X(u_j, 0) = u_j$ and $M^n(u_j, 0) = M(u_j, 0) + (i - j)\delta$.

To update the remaining values of the functions X and M we proceed from v^* and move upwards along the path from v^* to u_0 and compute the values for $u_{d-1}, u_{d-2}, \cdots, u_0$. At the i-th iteration, the value of u_{d-i} is computed by using Lemma 10 that expresses the values $X(u_j, l)$ and $M(u_j, l)$ in terms of the values of the functions at u_{d-i+1} (these values have been computed in the previous iteration) and the values of the function at the other child of u_{d-i} (notice these values have not been affected by weight-change in v^*).

Lemma 16. *There exists and algorithm Update that updates the table of the values of the canonical functions in time $O(\log^3 N)$ after the weight of a vertex has been changed.*

Computing the solution. We have already observed that, for each vertex u, $Cost_2(T_u)$ is easily computed as the minimum over $0 \le l_1 \le l_2 \le$ height(u) of $M(u, l_1, l_2)$. On the other hand, computing $Cost_2(T_u^c)$ is equivalent to computing the cost of the optimal placement of two resources in a tree $\mathcal{T}'$ where the weights of all vertices in T_u have been set equal to 0. Thus, all we need to do is to recompute the values of the canonical functions with respect to $\mathcal{T}'$ for the vertices on the path from u to $\mathcal{R}$ and this can be done in time $O(\log^3 N)$. Observe that there is no need to recompute the canonical functions for the descendant of u, instead these values are computed each time they are needed. In fact, all the values of the functions M, N are equal to 0; $X(u, l)$ can be set equal to any vertex in $G(u, l)$ and similarly for $X(u, l_1, l_2)$ and $Y(u, l, h)$.

5 An on–line algorithm for the k-resource problem on a line

In this section we briefly describe an on–line algorithm to compute the optimal placement of k resources on a weighted line. The algorithm proceeds in steps and at the m–th step the weight of v_m, $w(v_m)$, is given to the algorithm and the optimal placement is recomputed. We denote by T_m the line consisting of the vertices $v_1, v_2, \cdots, v_m$. For $i = 1, 2, \cdots, k$, we denote by $Cost_i(m)$ the minimum cost of serving the first m vertices in T_m by placing i resources in T_m and by $X_i(m)$ one of such minimum-cost placements. $P_i(m)$, instead, is the minimum cost of serving the vertices in T_m under the constraint that one resource is placed at vertex v_m and $Y_i(m)$ is one of such minimum placements. The next lemma shows the relations between P and $Cost$.

Lemma 17. *For all m and i, with $1 \leq i \leq k$, it holds that*

$$
Cost_i(m) = \min_{1 \leq j \leq m} \left(P_i(j) + \sum_{l=j+1}^{m} w(l) \cdot (l - j) \right),
$$

$$
P_i(m) = \min_{1 \leq j < m} \left(Cost_{i-1}(j) + \sum_{l=j+1}^{m} w(l) \cdot (m - l) \right).
$$

Based on this lemma, we design two data structures that we call Dyn_Line1 and Dyn_Line2 that maintain dynamically $Cost_i$ and P_i when new vertices are inserted in the line. The algorithm constructs k instances of each data structure. The i-th instance of Dyn_Line1 is used to compute $Cost_i(m)$ and $X_i(m)$, assuming that the values $P_i(j), Y_i(j)$ are known, for $j \leq m$. The i-th instance of Dyn_Line2 instead is used to compute the values of $P_i(m)$ and $Y_i(m)$ given the values of $Cost_{i-1}(j)$ and $X_{i-1}(j)$, for $j < m$. We refer the reader to [1] for a detailed description of the algorithm.

Theorem 18. *There exists an algorithm that recomputes the optimal placement of k resources after a new vertex has been appended in amortized constant time.*

References

1. V. Auletta, D. Parente, G. Persiano, *Dynamic and Static Algorithms for Optimal Placement of Resources in a Tree*, submitted for publication. Available on the WWW at *www.unisa.it.*
2. H. Hampapuram, Fredman M.L., *Optimal Bi-Weighted Binary Trees and the Complexity of Maintaining Partial Sums*, in Proceedings of the 34th IEEE Symposium on Foundations of Computer Science, 1993, pp. 480–485.
3. O. Kariv and S.L. Hakimi, *An Algorithmic Approach to Network Location Problems, Part II: p-medians*, SIAM Journal on Applied Math., 37 (1979), pp.539-560.
4. C. Papadimitriou, *Worst-Case and Probabilistic Analysis of a Geometric Location Problem*, SIAM Journal on Computing, 10 (1981), pp. 542–557.
5. D.D. Sleator, R.E. Tarjan, *A data structure for dynamic trees*, in J. Compu. Syst. Sci., 24:362–381, 1983.
6. A. C. Yao, *On the complexity of Maintaining Partial Sums*, SIAM Journal on Computing, 14 (1985), pp. 277–288.

Shortest Path Queries in Digraphs of Small Treewidth *

SHIVA CHAUDHURI and CHRISTOS D. ZAROLIAGIS

Max-Planck-Institut für Informatik, Im Stadtwald, D-66123 Saarbrücken, Germany
E-mail: {shiva, zaro}@mpi-sb.mpg.de

Abstract. We consider the problem of preprocessing an n-vertex digraph with real edge weights so that subsequent queries for the shortest path or distance between any two vertices can be efficiently answered. We give algorithms that depend on the *treewidth* of the input graph. When the treewidth is a constant, our algorithms can answer distance queries in $O(\alpha(n))$ time after $O(n)$ preprocessing. This improves upon previously known results for the same problem. We also give a dynamic algorithm which, after a change in an edge weight, updates the data structure in time $O(n^\beta)$, for any constant $0 < \beta < 1$. The above two algorithms are based on an algorithm of independent interest: computing a shortest path tree, or finding a negative cycle in linear time.

1 Introduction

Finding shortest paths in digraphs is a well-studied and important problem with many applications, especially in network optimization (see e.g. [1]). The problem is to find paths of minimum weight between vertices in an n-vertex, m-edge digraph with real edge weights (Section 2). In the single-source problem we seek such paths from a specific vertex to all other vertices and in the all-pairs shortest paths (apsp) problem we seek such paths between every pair [1].

For general digraphs the best algorithm for the apsp problem takes $O(nm + n^2 \log n)$ time [14]. An apsp algorithm must output paths between $\Omega(n^2)$ vertex pairs and thus requires this much time and space. A more efficient approach is to preprocess the digraph so that subsequently, *queries* can be efficiently answered. A query specifies two vertices and a *shortest path query* asks for a minimum weight path between them, while a *distance query* only asks for the weight of such a path. This approach is particularly promising when the digraph is sparse i.e. $m = O(n)$. An interesting subclass of sparse digraphs, namely *outerplanar* digraphs, has been intensively studied. In [12] it was shown that after $O(n)$ preprocessing, a shortest path or distance query is answered in $O(L+\log n)$ time (where L is the number of edges of the reported path). In [8], a different approach reduces the distance query time to $O(\log n)$ (with the same preprocessing time). Recently, in [13], the distance query time is improved to $O(\alpha(n))$, where $\alpha(n)$

* This work was partially supported by the EU ESPRIT Basic Research Action No. 7141 (ALCOM II).

is the inverse of (the well-known) Ackermann's function and is a very slowly growing function.

Another important subclass of sparse graphs is the class of graphs with bounded treewidth. The study of graphs using the *treewidth* as a parameter was pioneered by Robertson and Seymour [17] and continued by many others (see e.g. [3, 4, 5]). Roughly speaking, the treewidth of a graph G is a parameter which measures how close is the structure of G to a tree. (A formal definition is given in Section 2.) Graphs of treewidth t are also known as partial t-trees and have at most tn edges. In [6], the same bounds as in [8] are achieved for the above problem on digraphs with treewidth at most 2. Classifying graphs based on treewidth is useful because diverse properties of graphs can be captured by a single parameter. For instance, the class of graphs of bounded treewidth includes series-parallel graphs, outerplanar graphs, graphs with bounded bandwidth and graphs with bounded cutwidth [3, 5]. Thus giving efficient algorithms parameterized by treewidth is an important step in the development of better algorithms for sparse graphs.

In this paper we consider the above problem for digraphs of small treewidth. Our main result is an algorithm that, for digraphs of constant treewidth, after $O(n)$ preprocessing answers a distance query in $O(\alpha(n))$ time and a shortest path query in $O(L\alpha(n))$ time. This improves the results in [6, 8, 12, 13] in two ways: it improves the distance query time and applies to a larger class of graphs. The data structures in [12, 13] are not dynamic, while those in [6, 8] are dynamic. After a change in the weight of an edge, these data structures can be updated in $O(\log n)$ time. We also give a dynamic data structure that does not achieve this update bound, but does achieve a sublinear one. In particular, we can perform updates in $O(n^\beta)$ time, for any constant $0 < \beta < 1$, maintaining the previous query times.

We actually show a trade-off between the preprocessing and query times which is parameterized by the treewidth of the graph and an integer $1 \le k \le \alpha(n)$. Specifically, for a digraph of treewidth t and any integer $1 \le k \le \alpha(n)$, we give an algorithm that achieves distance (resp. shortest path) query time $O(t^4 k)$ (resp. $O(t^4 kL)$). The preprocessing bound required is $O(t^4 n \log n)$, when $k = 1$, $O(t^4 n \log^* n)$, when $k = 2$, and decreases rapidly to $O(t^4 n)$ when $k = \alpha(n)$ (Section 4). We note that graphs of treewidth t may have $\Omega(tn)$ edges.

Concerning the single-source problem, most algorithms either construct a *shortest path tree* rooted at a given vertex, or find a negative weight cycle. Constructing a shortest path tree is often easier when the digraph has non-negative edge weights. For general digraphs with non-negative real edge weights the best algorithm takes $O(m + n \log n)$ time [14] to construct the shortest path tree. If the digraph contains negative real edge weights, then one needs $O(nm)$ time to either construct a shortest path tree, or find a negative weight cycle [1]. For outerplanar digraphs, in $O(n)$ time, a shortest path tree can be constructed [8, 11], or a negative cycle can be found [15]. For *planar* digraphs with positive real edge weights, an $O(n)$ time algorithm is given in [16]. With negative but integer weights, the same paper gives an $O(n^{4/3} \log n)$ time algorithm which constructs a shortest path tree, or finds a negative cycle. In the case of negative real edge weights, the results for planar digraphs in [9, 10], imply an algorithm that in $O(n\sqrt{\log\log n})$ time either computes a shortest path tree, or decides that the graph contains a negative cycle. (We note that this algorithm does not find the cycle.) The best algorithm to construct a shortest path tree, or find a negative cycle in a planar digraph takes (in the worst case) $O(n^{1.5}\log n)$ time [15].

We also give here an $O(n)$ time algorithm that, for digraphs of constant treewidth, either constructs a shortest path tree or finds a negative cycle (Section 3). This generalizes the results in [15] for outerplanar digraphs. To the best of our knowledge, this is the most general class of graphs for which the complexity of computing a shortest path tree matches that of finding a negative cycle.

All of our algorithms start by computing a *tree-decomposition* of the input digraph G. The tree decomposition of a graph with constant treewidth can be computed in $O(n)$ time [4]. The main idea behind our algorithms is the following. We define a certain value for each node of the tree-decomposition of G, and an associative operator on these values. We then show that the shortest path problem reduces to computing the product of these values along paths in the tree-decomposition. Algorithms to compute the product of node values along paths in a tree are given in [2, 7]. Our preprocessing vs. query time trade-off arises from a similar trade-off in [2, 7]. The dynamization of our data structures is based on the above ideas and on a graph partitioning result which is of independent interest. Due to space limitations some proofs have been shortened or omitted.

2 Preliminaries

In this paper, we will be concerned with finding shortest paths or distances between vertices of a directed graph. Thus, we assume that we are given an *n-vertex weighted digraph* G, i.e. a digraph $G = (V(G), E(G))$ and a weight function $wt : E(G) \longrightarrow \mathbb{R}$. We call $wt(u, v)$ the *weight* of edge $\langle u, v \rangle$. The weight of a path in G is the sum of the weights of the edges on the path. For $u, v \in V(G)$, a *shortest path* in G from u to v is a path whose weight is minimum among all paths from u to v. The *distance* from u to v, written as $\delta(u, v)$ or $\delta_G(u, v)$, is the weight of a shortest path from u to v in G. A cycle in G is a (simple) path starting and ending at the same vertex. If the weight of a cycle in G is less than zero, then we will say that G contains a *negative cycle*. It is well-known [1] that shortest paths exist in G, iff G does not contain a negative cycle.

For a subgraph H of G, and vertices $x, y \in V(H)$, we shall denote by $\delta_H(x, y)$ the distance of a shortest path from x to y in H. A *shortest path tree* rooted at $v \in V(G)$, is a tree such that $\forall w \in V(G)$, the tree path from v to w is a shortest path in G from v to w.

Let G be a (directed or undirected) graph and let $W \subseteq V(G)$. Then by $G[W]$ we shall denote the subgraph of G induced on W. Let V_1, V_2 and S be disjoint subsets of $V(G)$. We say that S is a *separator for V_1 and V_2*, or that S *separates V_1 from V_2*, iff every path from a vertex in V_1 (resp. V_2) to a vertex in V_2 (resp. V_1) passes through a vertex in S. Let H be a subgraph of G. A *cut-set* *for H* is a set of vertices $C(H) \subseteq V(H)$, whose removal separates H from the rest of the graph.

Definition 1. Let H be a digraph, with V_1, V_2 and U a partition of $V(H)$ such that U is a separator for V_1 and V_2. Let H_1 and H_2 be subgraphs of H such that $V(H_1) = V_1 \cup U$, $V(H_2) = V_2 \cup U$ and $E(H_1) \cup E(H_2) = E(H)$. We say that H_1' is a graph obtained by *absorbing* H_2 *into* H_1, if H_1' is obtained from H_1 by adding edges $\langle u, v \rangle$, with weight $\delta_{H_2}(u, v)$ or $\delta_H(u, v)$, for each pair $u, v \in U$. (In case of multiple edges, retain the one with minimum weight.)

Absorbing preserves distances in a digraph, as the following lemma shows. This allows us to absorb the subgraph on one side of the separator and restrict our attention to the remaining subgraph, which maybe is smaller.

Lemma 2. Let H, H_1, H_2 and H_1' be as in Definition 1. Then, for all $x, y \in V(H_1')$, $\delta_{H_1'}(x, y) = \delta_H(x, y)$.

Proof. It is enough to show that $\delta_H(x, y) \leq \delta_{H_1'}(x, y)$ and $\delta_{H_1'}(x, y) \leq \delta_H(x, y)$. Call an edge $\langle u, v \rangle$ in H_1' an H_2-edge if it has weight $\delta_{H_2}(u, v)$ and an H-edge if its weight is $\delta_H(u, v)$.

Case 1: $\delta_H(x, y) \leq \delta_{H_1'}(x, y)$. Consider a shortest path from x to y in H_1'. Construct a walk from x to y in H by replacing, in the above path from H_1', all H_2-edges by a path in H_2 of the same weight and all H-edges by a path in H of the same weight (both of which exist, by construction). Now this walk has weight $\delta_{H_1'}(x, y)$ and a shortest path in H_1' from x to y cannot weight more.

Case 2: $\delta_{H_1'}(x, y) \leq \delta_H(x, y)$. Consider a shortest path from x to y in H. Find all maximal (w.r.t. the number of edges) subpaths that are contained in H_2. These paths must start and end in vertices in U. Let W be the weight of one such path (in H_2) from u to v, $u, v \in U$. Then H_1' has an edge $\langle u, v \rangle$ with weight either $\delta_{H_2}(u, v)$ or $\delta_H(u, v)$, both of which are at most W. Construct a path from x to y in H_1' by replacing each such subpath by the corresponding H_2-edge or H-edge in H_1'. The resulting path has weight at most $\delta_H(x, y)$. $\square$

A *tree-decomposition* of a (directed or undirected) graph G is a pair (X, T) where $T = (V(T), E(T))$ is a tree and X is a family $\{X_i | i \in V(T)\}$ of subsets of $V(G)$, such that $\cup_{i \in V(T)} X_i = V(G)$ and also the following conditions hold:

- *(edge mapping)* $\forall (v, w) \in E(G)$, there exists an $i \in V(T)$ with $v \in X_i$ and $w \in X_i$.
- *(continuity)* $\forall i, j, k \in V(T)$, if j lies on the path from i to k in T, then $X_i \cap X_k \subseteq X_j$, or equivalently: $\forall v \in V(G)$, the nodes $\{i \in V(T) | v \in X_i\}$ induce a connected subtree of T.

The *treewidth* of a tree-decomposition is $\max_{i \in V(T)} |X_i| - 1$. The treewidth of G is the minimum treewidth over all possible tree-decompositions of G.

Fact 3. *[4] (a) For all constant $t \in \mathbb{N}$, there exists an $O(n)$ time algorithm which tests whether a given n-vertex graph G has treewidth $\leq t$ and if so, outputs a tree-decomposition (X, T) of G with treewidth $\leq t$, where $|V(T)| = n - t$. (b) We can, in $O(n)$ time, convert (X, T) into another tree-decomposition (X_b, T_b) of G with treewidth t, where T_b is a binary tree and $|V(T_b)| \leq 2(n - t)$.*

Part (b) of the above fact follows by the usual binarization of an arbitrary tree. We will use this in Section 5. Given a tree-decomposition of G, we can quickly find separators in G, as the following proposition shows.

Proposition 4. *[17] Let G be a graph and let (X, T) be its tree-decomposition. Also let $e = (i, j) \in E(T)$ and let T_1 and T_2 be the two subtrees obtained by removing e from T. Then $X_i \cap X_j$ separates $\cup_{m \in V(T_1)} X_m$ from $\cup_{m \in V(T_2)} X_m$.*

3 Constructing a shortest path tree

Call a tuple (a, b, c) a *distance tuple* if a, b are arbitrary symbols and $c \in \mathbb{R}$. For two distance tuples, $(a_1, b_1, c_1), (a_2, b_2, c_2)$, define their product $(a_1, b_1, c_1) \otimes (a_2, b_2, c_2) = (a_1, b_2, c_1 + c_2)$ if $b_1 = a_2$ and as nonexistent otherwise.

For a set of distance tuples, M, define $\mathrm{minmap}(M)$ to be the set $\{(a, b, c) : (a, b, c) \in M \text{ and } \forall (a', b', c') \in M \text{ if } a' = a, b' = b, \text{ then } c \leq c'\}$, i.e. among all tuples with the same first and second components, minmap retains only the tuples with the smallest third component.

Let M_1 and M_2 be sets of distance tuples. Define the operator $\circ$ by $M_1 \circ M_2 = \mathrm{minmap}(M)$, where $M = \{x \otimes y : x \in M_1, y \in M_2\}$. It is not difficult to show that $\circ$ is an associative operator.

Let G be a digraph with real edge weights. Note that in the above definition, if M_1 and M_2 have tuples of the form (a, b, x) where $a, b \in V(G)$ and x is the weight of a path from a to b, then $M_1 \circ M_2$ computes tuples (a, b, y) where y is the (shortest) distance from a to b using only the paths represented in M_1 and M_2.

For $X, Y \subseteq V(G)$, not necessarily distinct, define $P(X, Y) = \{(a, b, \delta_G(a, b)) : a \in X, b \in Y\}$. We will write $S(X)$ for $P(X, X)$. (By definition, $S(X)$ contains tuples $(x, x, 0)$, $\forall x \in X$.)

Definition 5. Let G be an n-vertex weighted digraph without negative cycles and let (X, T) be a tree decomposition of G, with treewidth t. Then, for $i \in V(T)$, we define $\gamma(i) = S(X_i)$.

The following lemma shows that we can compute $\delta(a, b)$ by computing the product of the γ values on the path in T between nodes i and j such that $a \in X_i$ and $b \in X_j$.

Lemma 6. *Let G, (X, T) and $\gamma(i)$, for $i \in V(T)$, be as in Definition 5. Let $v_1, \ldots, v_p$ be a path in T. Then $\gamma(v_1) \circ \ldots \circ \gamma(v_p) = P(X_{v_1}, X_{v_p})$.*

Proof. It is not hard to show, from the definitions of $P(X, Y)$ and of $\circ$, that $P(X,Y) \circ P(Z,W) = \{(x, w, d) : x \in X, w \in W, d$ is the weight of the shortest x to w path that includes a vertex in $Y \cap Z$ (this vertex may be x or w)$\}$.

We prove the lemma by induction on p. If $p = 1$, then the lemma holds by the definition of $\gamma(v_1)$. If $p > 1$, then by the inductive hypothesis, $\gamma(v_1) \circ \cdots \circ \gamma(v_{p-1}) = P(X_{v_1}, X_{v_{p-1}})$. By definition, $\gamma(v_p) = S(X_{v_p})$. By Proposition 4, *all* paths from a vertex in X_{v_1} to a vertex in X_{v_p} include a vertex from $X_{v_{p-1}} \cap X_{v_p}$. Hence, by the characterization above, $P(X_{v_1}, X_{v_{p-1}}) \circ S(X_{v_p}) = \{(x, y, \delta_G(x, y)) : x \in X_{v_1}, y \in X_{v_p}\} = P(X_{v_1}, X_{v_p})$. $\square$

The following lemma shows that we can efficiently compute the γ values for each node of a tree-decomposition. The algorithm repeatedly shrinks the tree, by absorbing the subgraphs corresponding to leaves. When the tree is reduced to a single node, the algorithm computes γ using brute force, for this node, since the distances are preserved during absorption. Then, it reverses the shrinking process and expands the tree, using the γ values already computed to compute γ values for the newly expanded nodes.

Lemma 7. *Let G be an n-vertex weighted digraph and let (X, T) be the tree decomposition of G, of treewidth t. For each pair u, v such that $u, v \in X_i$ for some $i \in V(T)$, let $Dist(u, v) = \delta(u, v)$ and $Int(u, v) = x$ where x is some intermediate vertex on a shortest path from u to v. (If $wt(u, v) = \delta(u, v)$, then $Int(u, v) = null$.) Then in $O(t^4 n)$ time, we can either find a negative cycle in G, or compute the values $Dist(u, v)$ and $Int(u, v)$ for each such pair u, v.*

Proof. Initially all the values $Dist(u, v)$ are set to ∞ and $Int(u, v)$ to null. We give an inductive algorithm.

We use induction on $|V(T)|$. Choose a leaf, l, of T. Run the Bellman-Ford algorithm on $G[X_l]$ in time $O(t^4)$. If $G[X_l]$ contains a negative cycle, it will be found, so henceforth assume that $G[X_l]$ does not contain a negative cycle. Update the values for pairs $u, v \in X_l$ as follows: if the weight of the shortest path found is less than the current value of $Dist(u, v)$, then set $Dist(u, v)$ to the new value and $Int(u, v)$ to any intermediate vertex on the shortest path found. If $wt(u, v)$ is equal to the weight of the shortest path found, then set $Int(u, v) = null$.

If $|V(T)| = 1$, we are done. Otherwise remove l from T and call the resulting tree T'. Let $V' = \cup_{i \in V(T')} X_i$ and construct G' by absorbing $G[X_l]$ into $G[V']$, where the weight of each added edge $\langle u, v \rangle$ is $\delta_{G[X_l]}(u, v)$. Then, for any vertices $u, v \in V'$, $\delta_{G'}(u, v) = \delta_G(u, v)$, by Lemma 2. In particular, if G contains a negative cycle, so does G'. Note that $(X - X_l, T')$ is a tree-decomposition for G'. Inductively run the algorithm on G'. If a negative cycle is found in G', then a negative cycle in G can be found by replacing any edges added during the absorption by their corresponding paths in $G[X_l]$. Hence, we may assume that G' does not contain a negative cycle.

For $a, b \in V'$, $Dist(a, b) = \delta_{G'}(a, b) = \delta_G(a, b)$, as desired. If $Int(a, b) = x \neq$ null, then x is an intermediate vertex on a shortest a to b path in G' and hence also in G, as desired. If $Int(a, b) = $ null, then $\langle a, b \rangle$ is a shortest path in G'. If $wt(a, b) > Dist(a, b)$, then this edge must have been added during the absorption. Correct the value $Int(a, b)$ by setting it to some intermediate vertex on the corresponding a to b shortest path found in $G[X_l]$. After this, all Int values are correct for $a, b \in V'$.

Construct a digraph G'' by absorbing $G[V']$ into $G[X_l]$, with each added edge $\langle u, v \rangle$ having weight $\delta_G(u, v)$. By Lemma 2, $\delta_{G''}(x, y) = \delta_G(x, y)$, $\forall x, y \in X_l$. Run the Bellman-Ford algorithm on G'' to recompute all pairs shortest paths. Update the values $Dist(a, b)$ and $Int(a, b)$ for $a, b \in X_l$ as before.

For $a, b \in X_l$, $Dist(a, b) = \delta_{G''}(a, b) = \delta_G(a, b)$ as desired. For $a, b \in V' \cap X_l$, $Int(a, b)$ is not changed since $Dist(a, b)$ is already $\delta_G(a, b)$. If either a or b does not belong to $V' \cap X_l$, $Int(a, b) = $ an intermediate vertex on a shortest path in G'' and hence in G, or $Int(a, b) = $ null in which case $wt(a, b) = \delta_G(a, b)$. Thus, the values computed are correct for all pairs a, b which completes the induction. The time analysis follows easily. $\square$

Therefore, we can assume in the following that G does not contain a negative cycle. We will now briefly describe how a shortest path tree, rooted at a given vertex s, is computed. Perform a DFS of T starting at vertex i, where $s \in X_i$, storing at each vertex $j \in V(T)$ the product of the γ values on the path from i to j. Let $y \in V(G)$ and let $j \in V(T)$ such that $y \in X_j$. By Lemma 6, the value stored at vertex j during the DFS, is $P(X_i, X_j)$ which contains the tuple $(s, y, \delta(s, y))$. This implies that for each $y \in V(G)$, we have the distance $\delta(s, y)$. Having the distances, we construct the actual tree by performing a kind of BFS (starting at s) in G based on these distances. Hence, we conclude:

Theorem 8. *Let G and (X, T) be as in Definition 5. Let $s \in V(G)$. In $O(t^4 n)$ time we can compute a shortest path tree rooted at s.*

4 Shortest path and distance queries

For a function f let $f^{(1)}(n) = f(n)$; $f^{(i)}(n) = f(f^{(i-1)}(n))$, $i > 1$. Define $I_0(n) = \lceil \frac{n}{2} \rceil$ and $I_k(n) = \min\{j \mid I_{k-1}^{(j)}(n) \leq 1\}$, $k \geq 1$. The functions $I_k(n)$ decrease rapidly as k increases; note, for example, that $I_1(n) = \lceil \log n \rceil$ and $I_2(n) = \log^* n$. Finally, define $\alpha(n) = \min\{j \mid I_j(n) \leq j\}$. The following theorem was proved in [2, 7].

Theorem 9. *Let $\bullet$ be an associative operator defined on a set S, such that for $q, r \in S$, $q \bullet r$ can be computed in $O(m)$ time. Let T be a tree with n nodes such that each node is labelled with an element from S. Then: (i) for each $k \geq 1$, after $O(mnI_k(n))$ preprocessing, the composition of labels along any path in the tree can be computed in $O(mk)$ time; and (ii) after $O(mn)$ preprocessing, the composition of labels along any path in the tree can be computed in $O(m\alpha(n))$ time.*

We use this in the proof of the following.

Theorem 10. *For any integer t and any $k \geq 1$, let G be an n-vertex weighted digraph of treewidth at most t, whose tree-decomposition can be found in $T(n, t)$ time. Then, the following hold: (i) After $O(t^4 n I_k(n) + T(n, t))$ time and space preprocessing, distance queries in G can be answered in time $O(t^4 k)$. (ii) After $O(t^4 n + T(n, t))$ time and space preprocessing, distance queries in G can be answered in time $O(t^4 \alpha(n))$.*

Proof. First, we compute the tree-decomposition (X, T) of G. By Lemma 7, we compute values $Dist(u, v)$ for u, v such that $u, v \in X_i$ for some $i \in V(T)$. From these values, we can easily compute $\gamma(i)$, $\forall i \in V(T)$. By Theorem 9 we preprocess T so that product queries on γ can be answered. Given a query, $u, v \in V(G)$, let i, j be vertices of T such that $u \in X_i$ and $v \in X_j$. We ask for the product of the γ values on the path between i and j. By Lemma 6, the answer to this query contains the information about $\delta(u, v)$. The bounds follow easily by the ones given in Theorem 9 and by the fact that the composition of any two γ values can be computed in $O(t^4)$ time. $\square$

Theorem 11. *For any integer t and any $k \geq 1$, let G be an n-vertex weighted digraph of treewidth at most t, whose tree-decomposition can be found in $T(n, t)$ time. Then, the following hold: (i) After $O(t^4 n I_k(n) + T(n, t))$ preprocessing, we can answer shortest path queries in G in time $O(t^5 k L)$, where L is the length of the reported path. (ii) After $O(t^4 n + T(n, t))$ preprocessing, we can answer shortest path queries in G in time $O(t^5 \alpha(n) L)$, where L is the length of the reported path.*

Proof. We first compute a tree decomposition (X, T) of G. In the preprocessing phase, we compute the following data structures. Using Lemma 7, we compute the values $Dist(u, v)$ and $Int(u, v)$, for all pairs $u, v \in X_i$, for some $i \in V(T)$. From the $Dist$ values, we compute $\gamma(i)$, $\forall i \in V(T)$. We use Theorem 10 to compute a data structure in $O(t^4 n I_k(n))$ (or in $O(t^4 n)$) time so that distance queries can be answered in time $O(t^4 k)$ (or $O(t^4 \alpha(n))$). Root the tree T arbitrarily. Define, for each vertex $v \in V(G)$, $h(v)$ to be the tree node i such that $v \in X_i$ and i is the closest such node to the root of the tree. Preprocess T so that $h(v)$ can be found in constant time. Such a preprocessing can easily be done with, say, a DFS of T. Further, preprocess T so that lowest common ancestor (LCA) queries can be answered in constant time. Clearly, the time for the preprocessing is dominated by the time required by Theorem 10.

Let the query be for the shortest path between u and v. We first show that it is sufficient to consider the case when $h(u)$ is a descendant of $h(v)$ in T, or vice versa. Suppose $h(u)$ and $h(v)$ are not descendants of each other. Then let i be the LCA of the two. By Proposition 4, a shortest path from u to v passes through some vertex $z \neq u, v$ in X_i, and $\delta(u, v) = \delta(u, z) + \delta(z, v)$. By $O(t)$ queries, we can find this vertex z and then find the shortest paths from u to z and from z to v, and $h(u)$ and $h(v)$ are both descendants of $h(z)$.

Therefore, assume $h(u)$ is a descendant of $h(v)$. (A similar argument holds when $h(v)$ is a descendant of $h(u)$.) The query algorithm first checks if u and

v both belong to X_i, for some $i \in V(T)$. In particular, if there exists such an X_i, then u and v appear together in $X_{h(u)}$. If they do, then, if $Int(u,v) =$ null, the algorithm returns the edge $\langle u, v \rangle$. If $Int(u,v) = x \neq$ null, the algorithm recursively queries for the shortest paths from u to x and from x to v, and returns the concatenation of these two paths. Therefore, assume that u and v do not appear together in any X_i. Let p be the parent of $h(u)$ in T. Then, by Proposition 4, there exists a vertex $z \in X_p$ such that a shortest path from u to v passes through z, hence, $\delta(u,v) = \delta(u,z) + \delta(z,v)$. (Note that z may be v.) This vertex can be found with $O(t)$ distance queries. The algorithm recursively queries for the shortest paths from u to z and from z to v, and returns the concatenation of these two paths.

A simple induction shows that the query algorithm returns a path in $O(t^5 kL)$ (or $O(t^5 \alpha(n)L)$) time, where L is the number of edges of the reported path. □

Hence, the results claimed in the Introduction, for digraphs of constant treewidth, are now immediate from Fact 3 and Theorems 10 and 11.

5 Dynamization

In this section we shall give our dynamic data structures and algorithms. The following lemma about graph partitions plays a key role. (The proof is omitted for lack of space.)

Lemma 12. *Given an n-vertex digraph G, a binary tree-decomposition of G of treewidth t and a positive integer $1 \leq m \leq n$, we can, in $O(t^2 n)$ time, divide G into $q \leq 16n/m$ subgraphs $H_1, \ldots, H_q$, and construct another subgraph H' such that: (i) H_i has at most tm vertices and a cut-set $C(H_i)$ of size at most $3t$; (ii) H' is the induced subgraph on vertices $\cup_{i=1}^{q} C(H_i)$, augmented with edges $\langle x, y \rangle$, $x, y \in C(H_i)$ for each $1 \leq i \leq q$; and (iii) we have a binary tree decomposition of treewidth t for each H_i and a binary tree decomposition for H' of treewidth $3t$.*

Our dynamic algorithm works as follows. Using the above Lemma, it divides the digraph into subgraphs with disjoint edge sets and small cut-sets, and constructs another (smaller) digraph – the reduced digraph – by absorbing each subgraph. The sizes of the subgraphs are chosen so that the subgraphs and the reduced digraph both have size roughly $\sqrt{n}$. The algorithm then constructs a query data structure for each subgraph and for the reduced digraph. Queries can be efficiently answered by querying these data structures. Since the edge sets are disjoint, a change in the weight of an edge affects the data structure for only one subgraph. Then the data structure of this subgraph is updated. This may result in new distances between vertices in its cut-set, which appear in the reduced digraph as changes in the weights of edges between these cut-set vertices. Since the cut-set is small, the weights of only a few edges in the reduced digraph change. The data structure for the reduced digraph is updated to reflect these changes. Thus an update in the original digraph is accomplished by a small number of updates in subgraphs of size $\sqrt{n}$. This idea is recursively applied below to further reduce the update time.

Let $Dyn(G, P, U, Q)$ be a dynamic data structure for a digraph G, where $O(P)$ is the preprocessing time and space to be set up, $O(Q)$ is the time to answer a distance query and $O(U)$ is the time to update it after the modification of an edge-weight.

Theorem 13. *For all positive integers t, r, given an n-vertex weighted digraph G, and a binary tree-decomposition of G of treewidth t, we can construct the following dynamic data structures: (i) $Dyn(G, c^r t^3 n, c^{2r} t^2 n^{(1/2)^{r-1}}, c^{2r} t^2 \alpha(n))$; and (ii) $Dyn(G, c^r t^3 n I_k(n), c^{2r} t^2 n^{(1/2)^{r-1}}, c^{2r} t^2 k)$, where $c = 3^{r+2} t$.*

Proof. We shall prove part (i). Part (ii) can be proved similarly. We use induction on r. For $r = 1$, the basis is given by the static data structure of Theorem 10, with updates implemented by simply recomputing the data structure.

We use the notation $D(G, n, r, t)$ for $Dyn(G, c^r t^3 n, c^{2r} t^2 n^{(1/2)^{r-1}}, c^{2r} t^2 \alpha(n))$. Assume that the theorem holds for any $r' < r$. We show how to construct $D(G, n, r, t)$.

Use Lemma 12 (with parameter $m = 8\sqrt{n}$) to divide G into subgraphs $H_1, \ldots, H_q$, $q \leq 2\sqrt{n}$, each with at most $8t\sqrt{n}$ vertices and construct H' which has at most $(3t)2\sqrt{n}$ vertices. Define G_i to be H_i with all edges joining pairs of vertices in its cut-set deleted. Define G' to be H' with edges $\langle x, y \rangle$ weighted $\delta_{G_i}(x, y)$ for each pair $x, y \in C(G_i)$, $1 \leq i \leq q$. Replace multiple edges by the edge of minimum weight. Note that G' is exactly the graph obtained by absorbing $G_1, G_2, \ldots, G_q$ into the rest of the graph. By Lemma 2, it follows that $\delta_{G'}(x, y) = \delta_G(x, y)$, $\forall x, y \in V(G')$.

Let $u \in V(G_i), v \in V(G_j) - V(G_i)$. Then, any path from u to v must pass through a vertex in each of the cut-sets of G_i and G_j. Then we have $\delta_G(u, v) = \min\{\delta_{G_i}(u, x) + \delta_{G'}(x, y) + \delta_{G_j}(y, v) : x \in C(G_i), y \in C(G_j)\}$. Similarly, for $u, v \in V(G_i)$, we have $\delta_G(u, v) = \min\{\delta_{G_i}(u, v), \min\{\delta_{G_i}(u, x) + \delta_{G'}(x, y) + \delta_{G_i}(y, v) : x, y \in C(G_i)\}\}$. If we are able to make queries of the form $\delta_{G_i}(x, y)$ and $\delta_{G'}(x, y)$, the above directly yields a query algorithm for any pair of vertices x, y.

Write n_i for $|V(G_i)|$ and n' for $|V(G')|$. Note that Lemma 12 also gives us a tree-decomposition of treewidth t for each subgraph G_i, and a tree-decomposition of treewidth $3t$ for G'. Thus we can inductively construct $D(G_i, n_i, r - 1, t)$ for each $1 \leq i \leq q$, which enables us to answer queries of the form $\delta_{G_i}(x, y)$, and $D(G', n', r - 1, 3t)$ which enables us to answer queries of the form $\delta_{G'}(x, y)$. The data structure $D(G, n, r, t)$ is the union of the above data structures.

The update procedure is the following: note that $E(G_i) \cap E(G_j) = \emptyset$, $i \neq j$ and $E(G_i) \cap E(G') = \emptyset$, i.e. each edge of G belongs to exactly one of the G_i's or to G'. Suppose the cost of an edge belonging to G_i is changed. Then, we update the data structure for G_i. This may result in new values for $\delta_{G_i}(x, y)$, $x, y \in C(G_i)$. We query the updated data structure for $\delta_{G_i}(x, y)$, $x, y \in C(G_i)$ and change the weights of the corresponding edges of G', updating the data structure for G' after each change. That the procedure is correct follows from the fact that changing the cost of an edge in G_i does not change $\delta_{G_j}(x, y)$, $x, y \in C(G_j)$ when $j \neq i$. Thus, after we change, in G', the cost of edges $\langle x, y \rangle$, $x, y \in C(G_i)$, we have $\delta_{G'}(u, v) = \delta_G(u, v)$, $u, v \in V(G')$, again, by repeated applications of Lemma 2. After the last update, the data structure for G' yields correct distances in G, between vertices in $V(G')$. Now suppose we change the cost of an edge belonging to G'. Then the distances $\delta_{G_i}(x, y)$ do not change. Thus, in this case, we simply update the data structure for G'. This completes the description of the preprocessing and update algorithms.

Let the time taken for preprocessing, querying and updating $D(G, n, r, t)$ be $P(r, t)n$, $Q(r, t)\alpha(n)$ and $U(r, t)n^{(1/2)^{r-1}}$, respectively. Writing $N = \max\{n_i : 1 \leq i \leq q\}$, we have the following recurrences:

$$P(r, t)n \leq t^4 n + \sum_{i=1}^{q} P(r - 1, t)N + P(r - 1, 3t)n'$$

$$Q(r, t)\alpha(n) \leq (3t)^2[2Q(r - 1, t)\alpha(N) + Q(r - 1, 3t)\alpha(n')]$$

$$U(r, t)n^{(1/2)^{r-1}} \leq U(r - 1, t)N^{(1/2)^{r-2}} +$$
$$(3t)^2[Q(r - 1, t)\alpha(N) + U(r - 1, 3t)(n')^{(1/2)^{r-2}}]$$

The terms in the recurrence for $P(r, t)n$ are for constructing the G_i's and G using Lemma 12, for constructing $D(G_i, n_i, r - 1, t)$ for each G_i and for constructing $D(G', n', r - 1, 3t)$. The terms in the recurrence for $Q(r, t)\alpha(n)$ are for the two queries in G_i and G_j and for the query in G', which have to be made for each pair of vertices, one in the cut-set of G_i and one of G_j. The terms in the update recurrence are for updating G_i, and then updating the edges in G' between vertices in the cut-set of G_i.

By construction, $n', N \leq 8t\sqrt{n}$. The sum of the number of vertices in each G_i cannot exceed the number of vertices in the initial tree decomposition, so $\sum_{i=1}^{q} n_i \leq 2tn$. Making these substitutions in the above recurrences and estimating gives:

$$P(r, t)n \leq t^4 n + 2tnP(r - 1, t) + 8t\sqrt{n}P(r - 1, 3t) \leq 9tP(r - 1, 3t)n$$
$$Q(r, t)\alpha(n) \leq (3t)^2[2Q(r - 1, t)\alpha(8t\sqrt{n}) + Q(r - 1, 3t)\alpha(8t\sqrt{n})]$$
$$\leq 3(3t)^2 Q(r - 1, 3t)\alpha(n)$$
$$U(r, t)n^{(1/2)^{r-1}} \leq U(r - 1, t)(8t\sqrt{n})^{(1/2)^{r-2}} + (3t)^2[Q(r - 1, t)\alpha(8t\sqrt{n})$$
$$+ U(r - 1, 3t)(8t\sqrt{n})^{(1/2)^{r-2}}] \leq (3t)^2 16tU(r - 1, 3t)n^{(1/2)^{r-1}}$$

It is easily verified that the claimed bounds satisfy the recurrences above. Thus we can construct $D(G, n, r, t)$, completing the induction. $\square$

The next theorem follows directly from Fact 3 and Theorem 13 with $r = 1 - \log \beta$.

Theorem 14. *Let $k \geq 1$ be any constant integer and let $0 < \beta < 1$ be any constant. Given an n-vertex weighted digraph G of constant treewidth, we can construct: (i) $Dyn(G, n, n^\beta, \alpha(n))$; and (ii) $Dyn(G, nI_k(n), n^\beta, k)$.*

The algorithms described above give answers to distance queries only. They can be modified to answer path queries as well, in time $O(kL)$ (or $O(L\alpha(n))$). Also, before running our update procedure after a change in the weight of an edge, we have to ensure that this change does not create a negative cycle in G. This can be easily tested as follows. Let $\langle u, v \rangle$ be an edge with weight $wt(u, v)$ and let $wt'(u, v)$ be its new weight. Clearly, the new weight $wt'(u, v)$ creates a negative cycle in G iff $\delta_G(v, u) + wt'(u, v) < 0$. This test takes time proportional to that of finding $\delta_G(v, u)$ and hence does not affect our update bound.

Acknowledgement. We would like to thank Hans Bodlaender for many interesting discussions concerning the treewidth of graphs.

References

1. R. Ahuja, T. Magnanti and J. Orlin, "Network Flows", Prentice-Hall, 1993.
2. N. Alon and B. Schieber, "Optimal Preprocessing for Answering On-line Product Queries", Tech. Rep. No. 71/87, Tel-Aviv University, 1987.
3. S. Arnborg, "Efficient Algorithms for Combinatorial Problems on Graphs with Bounded Decomposability - A Survey", *BIT*, 25, pp.2-23, 1985.
4. H. Bodlaender, "A Linear Time Algorithm for Finding Tree-decompositions of Small Treewidth", *Proc. 25th ACM STOC*, pp.226-234, 1993.
5. H. Bodlaender, "A Tourist Guide through Treewidth", *Acta Cybernetica*, Vol.11, No.1-2, pp.1-21, 1993.
6. H. Bodlaender, "Dynamic Algorithms for Graphs with Treewidth 2", *Proc. 19th WG'93*, LNCS 790, pp.112-124, Springer-Verlag, 1994.
7. B. Chazelle, "Computing on a Free Tree via Complexity-Preserving Mappings", *Algorithmica*, 2, pp.337-361, 1987.
8. H. Djidjev, G. Pantziou and C. Zaroliagis, "On-line and Dynamic Algorithms for Shortest Path Problems", *Proc. 12th STACS*, 1995, LNCS 900, pp.193-204, Springer-Verlag.
9. E. Feuerstein and A.M. Spaccamela, "Dynamic Algorithms for Shortest Paths in Planar Graphs", *Theor. Computer Science*, 116 (1993), pp.359-371.
10. G.N. Frederickson, "Fast algorithms for shortest paths in planar graphs, with applications", *SIAM J. on Computing*, 16 (1987), pp.1004-1022.
11. G.N. Frederickson, "Planar Graph Decomposition and All Pairs Shortest Paths", *J. ACM*, 38(1991), pp.162-204.
12. G.N. Frederickson, "Searching among Intervals and Compact Routing Tables", *Proc. 20th ICALP*, 1993, LNCS 700, pp.28-39, Springer-Verlag.
13. G.N. Frederickson, "Using Cellular Graph Embeddings in Solving All Pairs Shortest Path Problems", accepted in *J. of Algorithms*, 1994.
14. M. Fredman and R. Tarjan, "Fibonacci heaps and their uses in improved network optimization algorithms", *J. ACM*, 34(1987), pp. 596-615.
15. D. Kavvadias, G. Pantziou, P. Spirakis and C. Zaroliagis, "Efficient Sequential and Parallel Algorithms for the Negative Cycle Problem", *Proc. 5th ISAAC*, 1994, LNCS 834, pp.270-278, Springer-Verlag.
16. P. Klein, S. Rao, M. Rauch and S. Subramanian, "Faster shortest-path algorithms for planar graphs", *Proc. 26th ACM STOC*, 1994, pp.27-37.
17. N. Robertson and P. Seymour, "Graph Minors II: Algorithmic Aspects of Treewidth", *J. Algorithms*, 7(1986), pp.309-322.

A Dynamic Programming Algorithm for Constructing Optimal Prefix-Free Codes for Unequal Letter Costs

M. J. Golin[1] and Günter Rote[2]

[1] Hong Kong UST, Clear Water Bay, Kowloon, Hong Kong. E-mail: *golin@cs.ust.hk*
[2] Technische Universität Graz, Institut für Mathematik, Steyrergasse 30, A-8010 Graz, Austria. E-mail: *rote@ftug.dnet.tu-graz.ac.at*

Abstract. We consider the problem of constructing prefix-free codes of minimum cost when the encoding alphabet contains letters of unequal length. The complexity of this problem has been unclear for thirty years with the only algorithm known for its solution involving a reduction to integer linear programming. In this paper we introduce a new dynamic programming solution to the problem. It optimally encodes n words in $O\left(n^{C+2}\right)$ time, if the costs of the letters are integers between 1 and C. While still leaving open the question of whether the general problem is solvable in polynomial time our algorithm seems to be the first one that runs in polynomial time for fixed letter costs.

1 Introduction

In this paper we will present an algorithm for constructing optimal-cost prefix-free codes when the letters of the alphabet from which the codewords are constructed have different lengths (costs).

To define the problem we assume that we have messages which consist of sequences from an alphabet of n source symbols. Suppose we wish to transmit them over a channel admitting an *encoding alphabet* $\Sigma = \{\alpha_1, \ldots, \alpha_r\}$ containing r characters. The length (cost, transmission time) of letter α_i is $c_i = length(\alpha_i)$, where c_i is a positive rational number. A codeword w is a string of characters in Σ, i.e., $w \in \Sigma^+$. The *length* of $w = \alpha_{i_1}\alpha_{i_1}\ldots\alpha_{i_k}$ is the sum of the lengths of its component letters, $length(w) = \sum_{j=1}^{k} c_{i_j}$. As an example consider the Morse-code alphabet $\{\cdot, -\}$. If $c_1 = 1$, $c_2 = 2$ then $length(\cdot - \cdot) = 4$.

Codeword $w = \alpha_{i_1}\alpha_{i_2}\ldots\alpha_{i_k}$ is a *prefix* of codeword $w' = \alpha'_{i_1}\alpha'_{i_2}\ldots\alpha'_{i_{k'}}$ if $k \leq k'$ and $\alpha_j = \alpha'_j$ for all $j = 1, \ldots, k$. A set of codewords $W = \{w_1, \ldots, w_n\}$ is *prefix-free* if no codeword is a prefix of another codeword.

A prefix-free *code* assigns a codeword w_i to each of the n source symbols, such that the set $W = \{w_1, \ldots, w_n\}$ is prefix-free. Let $p_1, \ldots, p_n$ be the probabilities with which the source symbols occur; these are also the probabilities with which the respective codewords will be used. The number p_i is also called the *weight* or *frequency* of codeword w_i. The cost of the code W is $C(W) = \sum_{i \leq n} length(w_i) \cdot p_i$. This is the expected length of the string needed to transmit one source symbol.

Given $c_1, \ldots, c_r$, and $p_1, \ldots, p_n$, the *Optimal Coding Problem* is to find a code W of n prefix-free codewords with minimum cost $C(W)$.

If $r = 2$ and $c_1 = c_2$ then the problem can be solved by a very well known $O(n \log n)$ greedy-type algorithm due to Huffman, see [11]. The resulting optimal binary code is called a Huffman code. This greedy algorithm cannot be adapted to solve the general optimal coding problem (in the next section we will see why). In fact the only solution known for the general problem seems to be the one proposed by Karp when he formulated it in 1961 [8]. He recast the problem as an integer linear program, which he solved by cutting plane methods. Thus his algorithm does not run in polynomial time.

Since then, different authors have studied various aspects of the problem such as finding bounds on the cost of the solution [1, 7] or solving the special case in which the all codewords are equally likely to occur ($p_i = 1/n$ for all i) [3, 4, 5, 10, 12], but not much is known about the general problem, not even if it is NP-hard or solvable in polynomial time.

In this paper we describe a dynamic programming approach to the general problem. In fact, our algorithm may be viewed as a dynamic programming solution of the integer programming formulation of Karp [8, Theorem 3].

We assume that the letter costs c_i are integers with $c_1 \leq c_2 \leq \cdots \leq c_r$, and we denote by $C = c_r$ the largest cost of a single letter. Our algorithm will run in $O\left(n^{C+2}\right)$ time. If $gcd(c_1, c_2, \ldots, c_r) > 1$, we can of course reduce C by scaling the costs. While this algorithm does not settle the long standing question of the problem's complexity it appears to be the first solution that runs in polynomial time for fixed letter costs.

In the next section we describe the problem in detail, showing how it can be transformed into a problem on trees. We explain why the standard Huffman encoding algorithm fails. In section 3 we will introduce the procedure of truncating a tree at a given level, which will allow us to define the subproblems that we will solve in the dynamic programming algorithm. The details of this algorithm are given in section 4.

2 Some Facts Concerning Trees

We will assume throughout that the numbers $c_1, \ldots, c_r$ are fixed positive integers and $1 > p_1 \geq p_2 \geq \cdots \geq p_n > 0$ are positive reals. We will also assume $n \geq r$ because otherwise we can reduce the problem to one using only letters with length $c_1, \ldots, c_n$.

Let $W = \{w_1, \ldots, w_n\}$ be a prefix-free set of codewords. We will find it convenient to follow the standard practice and represent W by a tree T. Each node in T can have up to r children, carrying distinct labels between 1 and r. To build T perform the following for each $w = \alpha_{i_1} \alpha_{i_1} \ldots \alpha_{i_k} \in W$: Start from the root and draw the path consisting of an edge labeled i_1 followed by an edge labeled i_2 followed by an edge labeled i_3 an so on until all k characters have been processed. The i-th edge leaving a node corresponds to writing character α_i in the codeword. See Figure 1.

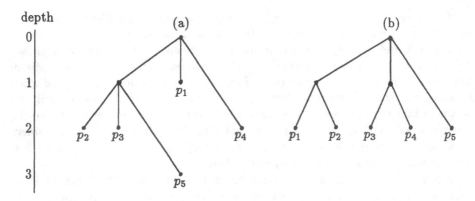

Fig. 1. Two codetrees for 5 codewords using the three letter alphabet $\Sigma = \{\alpha_1, \alpha_2, \alpha_3\}$ with respective letter costs $c_1 = c_2 = 1$ and $c_3 = 2$. Figure (a) encodes the set of words $W = \{\alpha_1\alpha_1, \alpha_1\alpha_2, \alpha_1\alpha_3, \alpha_2, \alpha_3\}$, and Figure (b) encodes the set $W = \{\alpha_1\alpha_1, \alpha_1\alpha_2, \alpha_2\alpha_1, \alpha_2\alpha_2, \alpha_3\}$.

This process will construct a tree corresponding to W; the tree will have n leaves, each of which corresponds to a codeword in W. Furthermore every tree with n leaves will correspond to a different prefix-free set of n codewords.

We construct our trees so that the *length* of an edge connecting a node to its i-th child is c_i. The *depth* of a node $v \in T$, denoted by $depth(v)$, is the sum of the lengths of the edges on the path connecting the root to the node. The root has depth 0. Note that if T represents a code W and a leaf v represents $w \in W$ then our definitions ensure that $depth(v) = length(w)$. As usual, the *height* of a tree is the maximum depth of its leaves.

For example, in Figure 1(a) the codeword $\alpha_1\alpha_3$ is mapped to the leaf v associated with p_5 and $depth(v) = 3 = length(\alpha_1\alpha_3)$.

Suppose now that T is a tree with n leaves. Number the leaves in the sequence $v = (v_1, \ldots, v_n)$; v_i is the codeword assigned to the i-th input symbol, having probability p_i. Define the cost of the tree under the labeling to be its weighted external path length $cost(T, v) = \sum_{i=1}^{n} depth(v_i) \cdot p_i$.

The following lemma is obvious:

Lemma 1 *Let T be a fixed tree with n leaves. If v is a labeling of the nodes such that*

$$depth(v_1) \le depth(v_2) \le \cdots \le depth(v_n) \tag{1}$$

then $cost(T, v)$ is minimum over all labelings of the tree.

Proof. We want to find a permutation π which minimizes the inner product of two vectors $(p_i)_{i=1,\ldots,n}$ and $(depth(v_{\pi(i)}))_{i=1,\ldots,n}$, where the entries of the second vector may be arbitrarily permuted. It is well-known that the minimum is achieved by permuting one vector into increasing order and the other one into decreasing order, see [6, p. 261]. $\square$

This lemma implies that in the optimal cost labeling the deepest node is assigned the smallest probability, the next deepest node the second smallest probability and so on, up to the shallowest node which is assigned the highest probability (see Figure 1). Such a code is called a *monotone code*, cf. Karp [8, Section IV].

Since we are interested in minimum cost trees we will restrict our attention to monotone codes. Thus $cost(T)$ can be defined without specifying a labeling of the leaves because the labeling is implied by T.

The Optimal Coding Problem is now seen to be equivalent to the following tree problem: Given $(c_1, \ldots, c_r)$ and $p_1 \geq p_2 \geq \cdots \geq p_n$ find the tree T^* with n leaves with minimum cost over all trees with n leaves, i.e.,

$$cost(T^*) = \min\{ cost(T) : T \text{ has } n \text{ leaves} \}.$$

From here on we will restrict ourselves to discussing the tree version of the problem in place of the original coding formulation.

For example, in Figure 1, $(c_1, c_2, c_3) = (1, 1, 2)$. The tree in (a) has minimum cost for the probabilities $(p_1, p_2, p_3, p_4, p_5) = (\frac{9}{10}, \frac{1}{40}, \frac{1}{40}, \frac{1}{40}, \frac{1}{40})$; and the tree in (b) has minimum cost for the probabilities $(p_1, p_2, p_3, p_4, p_5) = (\frac{1}{5}, \frac{1}{5}, \frac{1}{5}, \frac{1}{5}, \frac{1}{5})$.

Let us see why the standard Huffman encoding algorithm cannot be adapted to work in the general case. The Huffman encoding algorithm works by constructing the tree from the leaves up; it takes the two leaves with lowest probability p_n, p_{n-1} and combines them to form a new node with probability $p_n + p_{n-1}$ and then recurses.

Lemma 1 tells us that p_n and p_{n-1} will be assigned to the two deepest nodes in the tree. In the standard Huffman case of $r = 2$, $c_1 = c_2 = 1$, it is always true that the deepest node's sibling is also a deepest node. Therefore there is a minimum cost labeling of the tree in which the leaves assigned weights p_n and p_{n-1} are siblings of each other. Algorithmically, this implies that these two leaves can be combined together like in the Huffman algorithm.

In the general case of $r = 2$, $c_1 < c_2$, however, the deepest and second deepest node are not necessarily siblings, and thus p_n and p_{n-1} cannot be combined.

3 Truncated Subtrees of the Optimal Tree

Instead of building the trees from the leaves up we will construct them from the root down. To do so we introduce the following notation: let T be a tree with n leaves. We define its *i-th-level truncation* to be the tree T_i containing the root of T along with all other nodes in T whose parents have depth at most i (see Figure 3):

$$T_i = root(T) \cup \{u \in T : depth(parent(u)) \leq i\}$$

In other words, we remove the children of all nodes with depth bigger than i.

We will successively build the tree T from the root down, starting with T_0 and computing T_{i+1} from T_i until the tree is complete. When we build T_{i+1} from

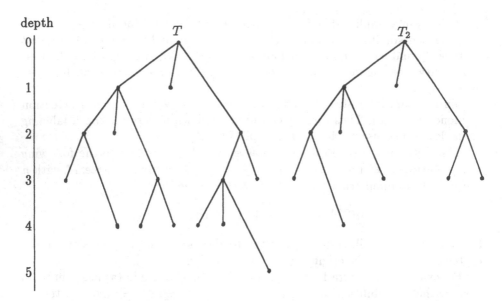

Fig. 2. In this example $r = 3$, $c_1 = c_2 = 1$, $c_3 = 2$. Tree T_2 is the 2-level-truncation of T; $sig_2(T_2) = (2; 4, 1)$.

T_i we only need to know the *total* number of leaves on the first $i + 1$ levels 0 up to i in T_i, and the number of leaves of T_i on each level $i + 1$, $i + 2$, ... (There are no leaves beyond depth $i + C$.) We therefore define the *signature* of T_i to be the $(C + 1)$-tuple

$$sig_i(T_i) = (m; l_1, l_2, \ldots, l_C)$$

in which $m = |\{\, v \in T_i \,:\, v \text{ is a leaf}, depth(v) \le i \,\}|$, is the number of leaves in T_i with depth at most i and

$$l_k = |\{\, u \in T_i \,:\, depth(u) = i + k \,\}|, \quad k = 1, \ldots, C,$$

is the number of nodes in T_i of depth $i + k$.

The definition of the signature $sig_i(T)$ makes sense for any tree T whose height is at most $i + C$. Note that the truncation level i is not part of the signature, although the way in which $sig_i(T_i)$ is computed depends on i. Clearly, the truncation operation cannot increase the total number of leaves, and hence we must have $m + l_1 + l_2 + \cdots + l_C \le n$ if T_i is a truncation of a code tree for n symbols.

Let $u \in T_i$ be a node with $depth(u) \le i$. Then u is a leaf in T_i if and only if the corresponding node is a leaf in T. Therefore m is just the number of leaves in T with depth at most i. A node in T_i with depth greater than i must be a leaf, but it can correspond to a leaf or an internal node in T. This means that by Lemma 1 we can optimally assign the m largest weights to the m highest leaves in T_i. The smaller weights must wait to be allocated to the descendents of the remaining leaves of T_i. Accordingly, we define the cost of T_i as

$$cost_i(T_i) = \sum_{t=1}^{m} depth(v_t) \cdot p_t + i \cdot \sum_{t=m+1}^{n} p_t$$

where, as usual, $v_1, \ldots, v_m$ are the m highest leaves of T_i ordered by depth. The first term in this sum reflects the cost of the paths to the m leaves which have already been assigned, whereas the second term reflects only part of the cost for reaching the remaining leaves, namely only the part until level i.

Note that when T is a tree with n leaves and i is equal to the height of T or larger, then $T_i = T$, $cost_i(T_i)$ agrees with $cost(T)$ as defined in the previous section, and we have $sig_i(T_i) = (n; 0, \ldots, 0)$.

By $opt[m; l_1, \ldots, l_C]$ we denote the minimum cost of a tree with signature $(m; l_1, \ldots, l_C)$. More precisely,

$$opt[m; l_1, \ldots, l_C] = \min\{ \ cost_i(T) : i \geq 0, \ T \text{ is a tree with depth at most } i + C,$$
$$\text{and } sig_i(T) = (m; l_1, \ldots, l_C) \ \}.$$

The crucial fact concerning truncations is that every truncation of an optimal tree must have minimum cost among all trees with the same signature.

Lemma 2 *Let T_i be a tree which realizes the minimum in $opt[m; l_1, \ldots, l_C]$, i.e., T_i is has depth at most $i + C$, $sig_i(T_i) = (m; l_1, \ldots, l_C)$, and $cost_i(T_i) = opt[m; l_1, \ldots, l_C]$. Let T_{i-1} be the $(i-1)$-st level truncation of T_i. Then T_{i-1} has also the minimum cost among the trees with the same signature. In other words, if $sig_{i-1}(T_{i-1}) = (m'; l'_1, \ldots, l'_C)$ then $cost_{i-1}(T_{i-1}) = opt[m'; l'_1, \ldots, l'_C]$.*

Proof. The difference between T_{i-1} and T_i is only that T_i has a certain number q of internal nodes $u_1, \ldots, u_q$ of depth i, and the children of these nodes (which must be leaves) are deleted in T_{i-1} so that $u_1, \ldots, u_q$ become leaves. Let $T^1, \ldots, T^q$ denote the subtrees rooted at the nodes $u_1, \ldots, u_q$ in T_i.

Now consider another tree T'_{i-1} with the same signature sig_{i-1} as T_{i-1} and let $u'_1, \ldots, u'_q$ be an arbitrary set of q leaves of depth i in T'_{i-1}. If we replace these leaves by the subtrees $T^1, \ldots, T^q$ we get a tree T'_i having the same sig_i-signature as T_i. Furthermore, we have

$$cost_i(T_i) = cost_{i-1}(T_{i-1}) + \sum_{t=m'+1}^{n} p_t,$$

$$cost_i(T'_i) = cost_{i-1}(T'_{i-1}) + \sum_{t=m'+1}^{n} p_t,$$

and thus

$$cost_{i-1}(T'_{i-1}) - cost_{i-1}(T_{i-1}) = cost_i(T'_i) - cost_i(T_i) \geq 0,$$

since T_i is optimal among all trees with signature $sig_i(T_i)$. This means that T_{i-1} must also be optimal among all trees with $sig_{i-1}(T_{i-1})$. $\square$

Corollary 3 *Let T^* be a minimum cost tree, and let T_i be the i-th level truncation of T^*, for some i. Then T_i has the minimum cost among the trees with the same signature as T_i.*

Proof. By applying the previous lemma inductively, we can conclude from the optimality of T_i among the trees with the same signature to the optimality of its j-level truncation T_j for any j. (The j-level truncation of the i-level truncation of a tree T is equal to the j-level truncation of T, for $j \leq i$.) Since $cost(T^*) = opt(n, 0, \ldots, 0)$ the statement follows. $\square$

A certain complication will arise in our algorithm from the fact that, for $r \geq 3$, some internal nodes of an optimal tree may have fewer than r children. We can evade this difficulty by allowing trees with more than n leaves and padding the sequence of probabilities with sufficiently many zeros: $(p_1, p_2, \ldots, p_n, 0, 0, 0, \ldots)$. In this way we may restrict ourselves to full r-ary trees, where every node has either 0 or r children, and $opt[m; l_1, \ldots, l_C]$ is well-defined even when $m + l_1 + l_2 + \cdots + l_C > n$. We will describe below how we can avoid dealing with such signatures.

4 The Dynamic Programming Algorithm

As mentioned before, our algorithm will compute an optimal tree T_{i+1} with a given signature, having height at most $i + 1 + C$, from an optimal tree T_i which is the i-th level truncation of T_{i+1}. Let $sig_i(T_i) = (m; l_1, \ldots, l_C)$. The difference between T_i and T_{i+1} is that a certain number of the l_1 leaves of T_i at level $i + 1$ become internal nodes in T_{i+1}. To compute the signature of T_{i+1} from the signature of T_i, it is useful to encode the properties of the encoding alphabet Σ in the vector

$$(d_1, d_2, \ldots, d_C),$$

where d_j is the number of code letters α_i of length $c_i = j$. For example, when $r = 3$, $c_1 = c_2 = 2$, $c_3 = 5$, we have $(d_1, \ldots, d_C) = (0, 2, 0, 0, 1)$. If q is the number of internal nodes on level $i + 1$ in T_{i+1}, $(0 \leq q \leq l_1)$, we have

$$sig_{i+1}(T_{i+1}) = (m + l_1; l_2, \ldots, l_C, 0) + q \cdot (-1; d_1, d_2, \ldots, d_C). \tag{2}$$

First, the vector $(l_1, \ldots, l_C)$ is shifted to the left because we change from sig_i to sig_{i+1}, and then q leaves of level $i + 1$ are expanded, i.e., they are replaced by $qd_1, \ldots, qd_C$ leaves at depth $i + 2, \ldots, i + 1 + C$, respectively. We call the signature in (2) *the q-th successor* of the signature $(m; l_1, \ldots, l_C)$.

The cost of T_{i+1} is simply given by

$$cost_{i+1}(T_{i+1}) = cost_i(T_i) + \sum_{t=m+1}^{n} p_t,$$

independent of q: some leaves $v \in T_i$ of depth $i + 1$ can become "permanent" leaves, in which case the corresponding probability p_t is multiplied by $depth(v) =$

$i + 1$ in the expression for $cost_{i+1}(T_{i+1})$. The remaining probabilities p_t for $m < t \leq n$ are also multiplied by $i + 1$, and thus the multiplier of $\sum_{t=m+1}^{n} p_t$ just increases from i to $i + 1$. The remaining terms are unchanged.

To see that we indeed make some progress when going from $sig_i(T_i)$ to $sig_{i+1}(T_{i+1})$ and don't run into a cycle, note that when $q > 0$ then $m+l_1+\cdots+l_C$ increases by $(-1 + d_1 + \cdots + d_C)q = (r - 1)q > 0$. When $q = 0$ we look at the vector

$$(m + l_1 + \cdots + l_C, \; m + l_1 + \cdots + l_{C-1}, \; \ldots, \; m + l_1 + l_2, \; m + l_1, \; m) \qquad (3)$$

which is derived from the signature $(m; l_1, \ldots, l_C)$. It increases elementwise, and at least one element increases strictly unless $l_1 = l_2 = \cdots = l_C = 0$. Thus, by processing all signatures in lexicographically increasing order of the derived vector (3), we can ensure that we have an order which is consistent with the successor relation defined in (2).

The above process may lead to signatures with more than n leaves, i.e., $m+l_1+l_2+\cdots+l_C > n$. As described above, we could handle this by introducing *phantom* leaves of weight $p_t = 0$.

However we will show that we can simply "reduce" such a signature to a signature with $m + l_1 + l_2 + \cdots + l_C \leq n$. Suppose we have a truncated tree T_i whose signature indicates that it has more than n leaves. If we assigned the weights $(p_1, p_2, \ldots, p_n, 0, 0, \ldots)$ to the leaves of this tree, only the n highest leaves would receive positive weights, whereas the remaining leaves could be omitted from the tree without changing the result. The modified tree would receive a signature fulfilling $m + l_1 + l_2 + \cdots + l_C = n$. However, it is conceivable that in going from T_i to $T_{i+1}, T_{i+2}, \ldots, T_{i+j}, \ldots$ some leaves at level $i + 1$, $i + 2$, $\ldots, i + j, \ldots$ are expanded, and a node which was previously not among the n highest nodes of T_i is now among the n highest nodes of T_{i+j}. In such a case it would be an error to omit this node. The next lemma shows that this does not happen for an optimal tree.

Lemma 4 *Consider an optimal tree T which is augmented by "phantom" leaves of weight $p_t = 0$ so that all internal nodes have r children. Let T_i be the i-th level truncation of T. Some leaves of T_i are "real" in that they correspond to "real" nodes of T (internal nodes or leaves with positive weight p_t), whereas other nodes correspond to the phantom leaves of T.*

Then the real leaves of T_i are among the n highest leaves of T_i.

Proof. Before proving this we must clarify the meaning of the phrase "*the n highest leaves*" in the last statement because it does not refer to a unique set. It means that there is a set H of n leaves of T_i, which includes all real leaves, and we have $depth(v) \leq depth(w)$ for any $v \in H$ and any leaf $w \notin H$. This remark applies similarly to other places in the paper, whenever we speak of the highest leaves, the deepest nodes, etc.

In the tree T, we have $depth(v) \leq depth(w)$ when v is a real leaf and w is a phantom leaf, by Lemma 1. The same inequality holds when v is an internal node, because every internal node has at least one real leaf as a descendent. The real nodes of T_i form a subtree of the original optimal tree, which has n leaves, and therefore there are at most n real leaves in T_i. By the previous remark, they are at least as high as any phantom leaf. $\qquad\square$

Assume that $sig_i(T_i) = (m; l_1, \ldots, l_C)$ with $m+l_1+l_2+\cdots+l_j > n$ and $l_j > 0$, for some $1 \le j \le C$. The above lemma says that we may ignore a leaf of level $i + j$, replacing the signature $(m; l_1, \ldots, l_j, \ldots, l_C)$ by $(m; l_1, \ldots, l_j - 1, \ldots, l_C)$, without sacrificing the possibility of extending T_i to an optimal tree. (If $m > n$ then, analogously, we may change $(m; l_1, \ldots, l_C)$ to $(m - 1; l_1, \ldots, l_C)$.)

By repeated application of this process we get the following reduction procedure $reduce(m, l_1, \ldots, l_C)$:

Replace m by $m' := \min\{m, n\}$. For $j = 1, \ldots, C$, successively replace l_j by $l'_j := \min\{l_j, n - (m' + \sum_{k=1}^{j-1} l'_k)\}$.
Use $(m', l'_1, \ldots, l'_C)$ instead of $(m, l_1, \ldots, l_C)$.

To put it differently, we can also start at the right and try to fulfill the condition $m + l_1 + l_2 + \cdots + l_C \le n$ by successively reducing l_C, l_{C-1}, ... to 0 until the condition can be achieved by reducing some l_j to a positive value.

4.1 Statement of the Algorithm

We have a $(C+1)$-dimensional array $\mathtt{OPT}[m; l_1, \ldots, l_C]$ with entries for the index range $1 \le m + l_1 + \cdots + l_C \le n$, $m, l_1, \ldots, l_C \ge 0$. The entries are initialized to ∞, and their final value will be equal to the correct value of $opt[m; l_1, \ldots, l_C]$ as defined in the previous section.

Rather than following the usual practice of setting up a dynamic programming recursion which expresses $\mathtt{OPT}[m; l_1, \ldots, l_C]$ as the minimum of certain terms involving other array elements which have been computed earlier, we will look at every element and see what influence it has on the successive elements. In this way the algorithm is more convenient to formulate.

1. **Initialization.** Initialize all entries of the array $\mathtt{OPT}$ to ∞.
   Set $\mathtt{OPT}[0; d_1, d_2, \ldots, d_C] := 0$;
2. **Loop.** Process the array entries $\mathtt{OPT}[m; l_1, \ldots, l_C]$ with
   $(l_1, \ldots, l_C) \ne (0, \ldots, 0)$ in lexicographically increasing order of
   $(m + l_1 + \cdots + l_C, \; m + l_1 + \cdots + l_{C-1}, \; \ldots, \; m + l_1, \; m)$.
   For each entry $\mathtt{OPT}[m; l_1, \ldots, l_C]$ do the following:

2a. $\qquad new\_cost := \mathtt{OPT}[m; l_1, \ldots, l_C] + \sum_{t=m+1}^{n} p_t$;

$\qquad$ **for** $q := 0$ **to** l_1 **do**

2b. $\qquad\qquad$ Let $(m', l'_1, \ldots, l'_C) := (m + l_1, l_2, \ldots, l_C, 0) + q \cdot (-1, d_1, \ldots, d_C)$;

2c. $\qquad\qquad$ If $m' + l'_1 + \cdots + l'_C > n$ then $reduce(m', l'_1, \ldots, l'_C)$ as described above in the text;

2d. $\qquad\qquad$ $\mathtt{OPT}[m'; l'_1, \ldots, l'_C] := \min\{\mathtt{OPT}[m'; l'_1, \ldots, l'_C], new\_cost\}$;

$\qquad$ **end for**;

3. **Termination.** $\mathtt{OPT}[n; 0, \ldots, 0]$ denotes the cost of the optimal code.

To actually construct the optimal tree, we have to augment the algorithm by storing a pointer with every array entry, and whenever $\mathtt{OPT}[m'; l'_1, \ldots, l'_C]$ is improved we update the pointer to remember where the current optimal value came from. At the end we backtrack from $\mathtt{OPT}[n; 0, \ldots, 0]$ to recover the optimal solution. This is standard dynamic programming practice and we omit the details. (If our computational model does not allow us to store a pointer to an array element in one word we must multiply the space complexity by C.)

Theorem 1 *The algorithm described above finds an optimal prefix-free code over an alphabet with integer letter costs between 1 and C for n symbols with given probabilities $p_1, \ldots, p_n$ in $O\Big(C \cdot \binom{n+C+2}{C+2}\Big) = O(n^{C+2})$ time using $O\Big(\binom{n+C+1}{C+1}\Big) = O(n^{C+1})$ space.*

Proof. Let us first prove correctness. If we modify the algorithm by working with an array $\mathtt{OPT}$ without bounds and omitting the reduction step 2c, correctness of the resulting (infinite) algorithm follows from the preceding discussion, in particular from Lemma 2 and Corollary 3. We only have to modify Step 3 by computing the minimum of $\{\ \mathtt{OPT}[m; l_1, \ldots, l_C]\ :\ m \geq n$ and $l_1, \ldots, l_C \geq 0\ \}$. (We can make the algorithm finite by specifying some a-priori bound on the number of nodes that a subtree of an optimal tree can have, even when it is augmented with phantom leaves to a full tree. For example, every internal node must clearly have at least two "real" children. Thus the number of internal nodes is at most $n - 1$, and hence the total number of phantom leaves can be at most $(n - 1)(r - 2)$.)

To show that the algorithm with Step 2c is correct, we have to show two things: (a) The value computed by the algorithm is the cost of some actual code. (b) The algorithm does not make a mistake by missing the optimal code.

For (a), note that if the algorithm computes a value x for $\mathtt{OPT}[m; l_1, \ldots, l_C]$ this means there is a tree T_i of the respective signature with $cost_i(T_i) = x$. Now, if the algorithm chooses to store this value x in some other entry $\mathtt{OPT}[m'; l'_1, \ldots, l'_C]$ with $m' = m$, $l'_1 \leq l_1, \ldots, l'_C \leq l_C$, this means just that we have to throw away some leaves at depth $i + 1$, $i + 2$, $\ldots$, $i + C$, but it is still true that there is a reduced tree T'_i with signature $(m'; l'_1, \ldots, l'_C)$ having cost x. The same holds true for the case $m > n = m'$, where we throw away the deepest leaves among the leaves of depth at most i. Thus, what the algorithm computes is the cost of a valid solution.

For showing (b), we invoke Lemma 4. It says that a leaf which is "ignored" by reducing the signature would be a phantom leaf in the optimal solution. Hence we do not lose anything by ignoring this leaf. A detailed formal proof will be given in the full paper.

Finally, let us show that the reduction of the signature does not lead to a conflict with the lexicographic order in which the elements are processed. If we check what effect the procedure *reduce* has on the vector $(m' + l'_1 + \cdots + l'_C,\ m' + l'_1 + \cdots + l'_{C-1},\ \ldots,\ m' + l'_1,\ m')$ we see that it only reduces components which are bigger than n to n. It can therefore not lead to a signature which comes *before* the currently processed array element $\mathtt{OPT}[m; l_1, \ldots, l_C]$ in the lexicographic order.

The worst thing that can happen is that the reduced signature $(m'; l'_1, \ldots, l'_C)$ is equal to $(m; l_1, \ldots, l_C)$ and the array element $\texttt{OPT}[m; l_1, \ldots, l_C]$ tries to improve itself in Step 2d. Since *new_cost* is bigger than $\texttt{OPT}[m; l_1, \ldots, l_C]$ this has no effect, and the possibility $(m'; l'_1, \ldots, l'_C) = (m; l_1, \ldots, l_C)$ can be ignored without harm.

The number of entries of the array $\texttt{OPT}$ is $\binom{n+C+1}{C+1} - 1$, which gives the space bound. For the time bound note that one loop iteration in Step 2 takes $O(C(l_1 + 1))$ time. By summing this over the range of array indices we get the claimed time bound. □

In a practical implementation of the algorithm many improvements are possible. Suppose that we know an upper bound for the cost of an optimal tree, for example the cost of a tree that was constructed by some heuristic method. Then we can skip the processing of any array element $\texttt{OPT}[m; l_1, \ldots, l_C]$ whose value exceeds this bound. Of course, if we use a sharper lower bound than $\texttt{OPT}[m; l_1, \ldots, l_C]$ for the value of a tree that can be built from a subtree with signature $(m; l_1, \ldots, l_C)$ and cost $\texttt{OPT}[m; l_1, \ldots, l_C]$, we may be able to eliminate more array elements.

The dynamic programming algorithm can be regarded as a shortest path problem on an acyclic directed graph whose nodes correspond to the array entries $\texttt{OPT}[m; l_1, \ldots, l_C]$ and which has an arc connecting $\texttt{OPT}[m; l_1, \ldots, l_C]$ with $\texttt{OPT}[m'; l'_1, \ldots, l'_C]$ whenever $\texttt{OPT}[m'; l'_1, \ldots, l'_C]$ is updated in the loop for the entry $\texttt{OPT}[m; l_1, \ldots, l_C]$. The weight of such an arc is $\sum_{t=m+1}^{n} p_t$. We are looking for the shortest path from $\texttt{OPT}[0; d_1, d_2, \ldots, d_C]$ to $\texttt{OPT}[n; 0, \ldots, 0]$ in this network. Since the weights are non-negative, we could use Dijkstra's shortest path algorithm and always process the array entry whose entry is minimal among the entries which have not been processed. Also the whole range of heuristic search algorithms is open to be considered, cf. [9].

We may also start from the terminal node $\texttt{OPT}[n; 0, \ldots, 0]$ and carry out the whole process backwards, effectively building up the optimal tree bottom-up from the leaves to the root.

4.2 Numeric Results

We have tested the algorithm for computing optimal codes for the Roman alphabet plus "space", using the probabilities which are given in [2, p. 52] and reproduced in Karp [8]. The results are shown in the following table.

Luckily, the optimal code turned out to have the same cost with both algorithms, although, for the first example, we got a different code than Karp. If the running times have any meaning at all, they show that even for $C = 3$, when the running time is $O(n^5)$, our algorithm is still quite competitive, even without any of the improvements indicated after Theorem 1.

Acknowledgements: The authors would like to thank Dr. Jacob Ecco for introducing them to the Morse Code puzzle which sparked this investigation. The first author will also like to acknowledge that this work was partially supported by HK RGC/CRG grant number 181/93E.

Encoding alphabet with 2 letters, $c_1 = 1$, $c_2 = 2$		
algorithm	optimal cost	running time
Karp [8]	5.8599	1 minute[a]
ours	5.8599	13 seconds[b]

Alphabet with 3 letters, $(c_1, c_2, c_3) = (2, 3, 3)$		
algorithm	optimal cost	running time
Karp [8]	6.7324	5 minutes[a]
ours	6.7324	38 seconds[b]

[a] experimental program on an IBM 704, probably written in FORTRAN.

[b] straightforward experimental implementation in MAPLE on a PC with Intel 80486 processor, 33 MHz, and 16 MByte of RAM.

Table 1. Coding the letters of the English alphabet. Comparison of our algorithm with the results of Karp (1960).

References

1. Doris Altenkamp and Kurt Mehlhorn, "Codes: Unequal Probabilities, Unequal Letter Costs," *J. Assoc. Comput. Mach.* **27** (3) (July 1980), 412–427.
2. L. Brouillon, *Science and Information Theory*, Academic Press, New York 1956.
3. N. Cot, "A linear-time ordering procedure with applications to variable length encoding," *Proc. 8th Annual Princeton Conference on Information Sciences and Systems*, (1974), pp. 460–463.
4. N. Cot, "Complexity of the Variable-length Encoding Problem," *Proceedings of the 6th Southeast Conference on Combinatorics, Graph Theory and Computing*, (1975), pp. 211–224.
5. Mordecai Golin and Neal Young, "Prefix Codes: Equiprobable Words, Unequal Letter Costs," *Proceedings of the 21st International Colloquium on Automata, Languages and Programming (ICALP '94)*, (July 1994), Lecture Notes in Computer Science **820**, pp. 605–617.
6. G. H. Hardy, J. E. Littlewood, and G. Pólya, *Inequalities*, Cambridge University Press, Cambridge 1967.
7. Sanjiv Kapoor and Edward Reingold, "Optimum Lopsided Binary Trees," *Journal of the Association for Computing Machinery* **36** (3) (July 1989), 573–590.
8. R. Karp, "Minimum-Redundancy Coding for the Discrete Noiseless Channel," *IRE Transactions on Information Theory* **IT-7** (1961), 27–39.
9. Nils J. Nilsson, *Principles of Artificial Intelligence*, Tioga, Palo Alto 1980.
10. Y. Perl, M. R. Garey, and S. Even, "Efficient generation of optimal prefix code: Equiprobable words using unequal cost letters," *Journal of the Association for Computing Machinery* **22** (2) (April 1975), 202–214,
11. Robert Sedgewick, *Algorithms*, 2nd ed., Addison-Wesley, Reading, Mass. (1988).
12. L. E. Stanfel, "Tree Structures for Optimal Searching," *Journal of the Association for Computing Machinery* **17** (3) (July 1970), 508–517.

Parallel Algorithms with Optimal Speedup for Bounded Treewidth*

Hans L. Bodlaender[1] and Torben Hagerup[2]

[1] Department of Computer Science, Utrecht University, P.O. Box 80.089, 3508 TB Utrecht, the Netherlands. Email: hansb@cs.ruu.nl
[2] Max-Planck-Institut für Informatik, Im Stadtwald, D–66123 Saarbrücken, Germany. Email: torben@mpi-sb.mpg.de

Abstract. We describe the first parallel algorithm with optimal speedup for constructing minimum-width tree decompositions of graphs of bounded treewidth. On n-vertex input graphs, the algorithm works in $O((\log n)^2)$ time using $O(n)$ operations on the EREW PRAM. We also give faster parallel algorithms with optimal speedup for the problem of deciding whether the treewidth of an input graph is bounded by a given constant and for a variety of problems on graphs of bounded treewidth, including all decision problems expressible in monadic second-order logic. On n-vertex input graphs, the algorithms use $O(n)$ operations together with $O(\log n \log^* n)$ time on the EREW PRAM, or $O(\log n)$ time on the CRCW PRAM.

1 Introduction

The concept of treewidth has proved to be a useful tool in the design of graph algorithms: Many important classes of graphs have bounded treewidth, and many important graph problems that are otherwise quite hard can be solved efficiently on graphs of bounded treewidth. A *tree decomposition* of an undirected graph $G = (V, E)$ is a pair $(T, \mathcal{U})$, where $T = (X, F)$ is a tree and $\mathcal{U} = \{U_x \mid x \in X\}$ is a family of subsets of V called *bags*, one for each node in T, such that

- $\bigcup_{x \in X} U_x = V$ (every vertex in G occurs in some bag);
- for all $(v, w) \in E$, there exists an $x \in X$ such that $\{v, w\} \subseteq U_x$ (every edge in G is "internal" to some bag);
- for all $x, y, z \in X$, if y is on the path from x to z in T, then $U_x \cap U_z \subseteq U_y$ (every vertex in G occurs in the bags in a connected part of T, i.e., in a subtree).

The *width* of a tree decomposition $(T, \{U_x \mid x \in X\})$ is $\max_{x \in X} |U_x| - 1$. The *treewidth* of a graph G, denoted $tw(G)$, is the smallest treewidth of any tree decomposition of G. *Path decompositions* and *pathwidth* are defined analogously, with the tree T restricted to be a path.

* This research was partially supported by the ESPRIT Basic Research Actions Program of the EU under contract No. 7141 (project ALCOM II).

The majority of efficient algorithms for graphs of bounded treewidth depend not only on a guarantee that the treewidth of an input graph is small, but in fact on the availability of a minimum-width tree decomposition of the input graph, so that the construction of minimum-width tree decompositions for graphs of bounded treewidth is a key problem. A quest for the fastest possible algorithm for this problem [2, 20, 16, 19, 5] led to the linear-time algorithm of [5], which eliminated the bottleneck in a large number of algorithms for bounded treewidth.

In the setting of parallel computation, the situation is similar. Many otherwise difficult graph problems can be solved in NC (i.e., in polylogarithmic time with a polynomial amount of hardware) on graphs of bounded treewidth, and again the need for a minimum-width tree decomposition is a serious bottleneck. The best parallel algorithms for computing tree decompositions of width k of graphs of treewidth k, for fixed k, run on the CRCW PRAM using $O((\log n)^2)$ time and $O(n^{2k+6})$ processors [9], or $O(\log n)$ time and $O(n^{3k+4})$ processors [4]. Although these algorithms are fast, they are extremely wasteful in terms of processors, in view of the linear-time sequential algorithm. A related result was obtained by Wanke [22], who showed that the problem of deciding whether the treewidth of an input graph is bounded by a constant k belongs to the complexity class LOGCFL; this result also does not seem to lead to parallel algorithms that are efficient from the point of view of processor utilization. If we relax the requirements by allowing tree decompositions of width $O(k)$, rather than exactly k, more algorithms come into play: Lagergren [16] finds a decomposition of width $\leq 6k + 5$ in $O((\log n)^3)$ time using n processors, and Reed's sequential $O(n \log n)$-time algorithm [19] for obtaining a decomposition of width $\leq 4k + 3$ can be parallelized to yield an algorithm that works in $O((\log n)^2)$ time using $O(n/\log n)$ processors. The parallel version of Reed's algorithm uses $O(n \log n)$ *operations*, i.e., has a time-processor product of $O(n \log n)$, and is the most efficient of the parallel algorithms discussed above. Still, since the problem can be solved in linear sequential time, it does not have optimal speedup, which requires a time-processor product of $O(n)$.

We describe an EREW PRAM algorithm for constructing minimum-width tree decompositions for graphs of bounded treewidth in $O((\log n)^2)$ time using $O(n)$ operations. The algorithm achieves optimal speedup and is the first parallel algorithm to do so. Moreover, the new algorithm is second in speed only to Bodlaender's algorithm [4], but uses a weaker model of computation (the EREW PRAM versus the CRCW PRAM), on which Bodlaender's algorithm can be simulated only in the same time of $O((\log n)^2)$. The new result immediately implies that a large number of problems on graphs of bounded treewidth can now be solved by parallel algorithms with optimal speedup.

A subroutine used in the construction algorithm but of independent interest is a parallel version of an algorithm due to Bodlaender and Kloks [7]. The algorithm takes as input a tree decomposition of bounded width of a graph G and outputs a minimum-width tree decomposition of G, thus blurring the distinction between the "exact" and the "approximate" construction algorithms discussed above. The algorithm runs in $O(\log n)$ time using $O(n)$ operations on the EREW PRAM.

While we cannot compute tree decompositions faster than in $O((\log n)^2)$ time, it turns out that we can give faster algorithms for the related problem of deciding whether the treewidth of an input graph is bounded by a given constant k. The algorithms have optimal speedup (i.e., use $O(n)$ operations) and run in $O(\log n \log^* n)$ time on the EREW PRAM, or in $O(\log n)$ time on the CRCW PRAM. We achieve the same resource bounds for a number of problems on graphs of bounded treewidth, including all problems expressible in monadic second-order logic. These algorithms operate without an explicit tree decomposition and so bypass the (time) bottleneck of our construction algorithm. Furthermore, we achieve the same results for path decompositions and pathwidth as for tree decompositions and treewidth.

The paper combines graph reduction, derandomization and the bounded adjacency-list search technique. The latter tries to circumvent the difficulties caused by high-degree vertices in parallel algorithms for sparse graphs by letting each neighbor of a high-degree vertex v inspect only a piece of constant size of the adjacency list of v near its own entry, rather than the whole adjacency list. The bounded adjacency-list search technique was used previously (although not named) in [13]; there, as here, it serves to eliminate the need both for concurrent reading and writing and for superlinear space.

All graphs in this paper are undirected, loopless and without multiple edges.

2 Minimizing Decomposition Width

In this section we show how to obtain a minimum-width tree decomposition of a graph G from any tree decomposition of G of bounded width. We begin with a simple observation, also made in [15], that allows us to assume that tree decompositions are rooted, binary and of logarithmic depth whenever this is convenient. In representation terms, every nonroot node in a rooted tree knows its parent, and a rooted tree is *binary* if no node has more than two children. In the lemma below, the input parameter k is qualified as being a constant, the meaning of which is that k can be any positive integer, but that the $O(\cdots)$ of the lemma may (and will) hide factors that depend on k.

Lemma 1. *For all constants $k \geq 1$ and all integers $n \geq 2$, the following problem can be solved on an EREW PRAM using $O(\log n)$ time, $O(n)$ operations and $O(n)$ space: Given a tree decomposition of width k of an n-vertex graph G, compute a rooted, binary tree decomposition of G of depth $O(\log n)$ and width at most $3k + 2$.*

Proof. One can simply use an algorithm of [4], except that the nonoptimal tree-contraction algorithm of Miller and Reif [18] should be replaced by an optimal tree-contraction algorithm, e.g. [1]. □

We will use the phrase "balancing a tree decomposition" to describe an application of the algorithm implicit in the preceding lemma. The remaining goal in the present section is to prove the result below.

Theorem 2. *For all constants $k \geq 1$ and all integers $n \geq 2$, the following problem can be solved on an EREW PRAM using $O(\log n)$ time, $O(n)$ operations and $O(n)$ space: Given a tree decomposition of width k of an n-vertex graph G, construct a minimum-width tree decomposition of G.*

The corresponding decision problem ("Is $tw(G) \leq l$?", for some given l) can be solved by a straightforward parallelization of the decision algorithm of [7]. The latter algorithm essentially consists of a constant number of leaves-to-root passes over a tree decomposition of the input graph which, in light of Lemma 1, can be taken to be binary and of logarithmic depth. The processing of each node takes constant time, and all nodes on the same level in the tree can be processed in parallel. If the nodes in the tree decomposition are first sorted by their levels, which can be done in $O(\log n)$ time using $O(n)$ operations [10, 21], it is easy to process the whole tree in $O(\log n)$ time using $O(n)$ operations. The sequential construction algorithm shares the same overall structure, but the processing of a node no longer necessarily takes constant time, so that an amortization argument is needed to bound the total running time by $O(n)$. Since this appears to stand in the way of a direct parallelization, we choose a somewhat different approach.

Suppose that the input graph is $G = (V, E)$. Close inspection of the algorithm of [7] (we omit the details, some of which were hinted at above) reveals that $O(\log n)$ time and $O(n)$ operations suffice to compute a certain *implicit representation* of the desired tree decomposition $(T = (X, F), \{U_x \mid x \in X\})$ consisting of the binary tree T (without the bags U_x) together with, for each $v \in V$, a collection $\mathcal{P}_v$ of disjoint simple paths in T whose union contains a node $x \in X$ if and only if $v \in U_x$. Rather than directly specifying the set of vertices contained in each bag, the implicit representation thus presents the set of bags containing each vertex in the form of a collection of disjoint paths. For each $v \in V$, a path in $\mathcal{P}_v$ with end nodes x and y is represented by marking both x and y with the triple (x, y, v); a node may be marked with several triples but, of course, at most with $k + 1$.

By the preceding discussion (in particular, note that $|X| = O(n)$), proving Theorem 2 boils down to showing the following.

Lemma 3. *For all constants $k \geq 1$ and all integers $n \geq 2$, the following problem can be solved on an EREW PRAM using $O(\log n)$ time, $O(n)$ operations and $O(n)$ space: Given a rooted binary n-node tree T and a collection $\mathcal{P}$ of paths in T, each of which is labeled by an integer and represented, at each of its endpoints, by a triple specifying its endpoints and label, such that no node in T belongs to more than $k + 1$ paths in $\mathcal{P}$, mark each node x in T with the set of all labels of paths in $\mathcal{P}$ containing x.*

Proof sketch. We formulate the problem as an instance of tree contraction and then appeal to the existence of generic tree-contraction schemes with the stated resource bounds. During the contraction we keep track of the set of paths with exactly one endpoint in a supernode, and during the subsequent expansion we keep track of the set of paths with a part inside and a part outside a supernode. □

With a similar (but, in fact, easier) argument one can also show the following result.

Theorem 4. *For all constants $k, l \geq 1$ and all integers $n \geq 2$, the following problem can be solved on an EREW PRAM using $O(\log n)$ time, $O(n)$ operations and $O(n)$ space: Given a tree decomposition of width k of an n-vertex graph G, decide whether the pathwidth of G is at most l and, if so, construct a minimum-width path decomposition of G.*

3   A Structural Lemma

In this section we provide the basis for showing that any sufficiently large connected graph of bounded treewidth admits a large number of reductions of certain types. Moreover, given any adjacency-list representation of the graph, a large fraction of these reductions can be identified efficiently.

The *boundary* of a subgraph H of a graph G is the set of those vertices in G that have at least one neighbor in H, but do not themselves belong to H. Let d, k, $n_{\min}$ and $n_{\max}$ be positive integers, to be characterized more closely in the following. A vertex will be called *small* if its degree is bounded by d, and *large* otherwise. Given a graph G of treewidth at most k, we are essentially looking for connected subgraphs of G consisting of between $n_{\min}$ and $n_{\max}$ small vertices and with a boundary of size at most $2(k+1)$. It turns out that such subgraphs may not occur in G at all, for which reason we have to replace the connectedness condition by a weaker, more complicated condition described below after the introduction of additional terminology.

Two vertices are said to be *twins* if they have the same set of neighbors. By analogy, we call two subgraphs of a common graph twins if they have the same boundary. A *weakly connected component* of a subgraph H of a graph G is a connected component of the graph obtained from H by the introduction of an edge between each pair of nonneighbors in H with a common small neighbor in G; a weakly connected component of H may thus comprise several (usual) connected components of H, linked indirectly via small common neighbors in the boundary of H. A subgraph that consists of a single weakly connected component is *weakly connected*. Given an adjacency-list representation of a graph G, two disjoint subgraphs H_1 and H_2 of G are said to be *acquainted* if the intersection of their boundaries contains a vertex in whose adjacency list some entry of a vertex in H_1 is separated from some entry of a vertex in H_2 by a distance of at most d. This definition, which embodies the bounded adjacency-list search technique, reflects the fact that H_1 can "discover" H_2 by searching through a piece of length at most $2d + 1$ of the adjacency list of each of its boundary vertices.

A *valley* in a graph G is a weakly connected subgraph of G induced by a set of at most $n_{\max}$ small vertices and with a boundary of size at most $2(k+1)$. A *plain* (or $(d, k, n_{\min}, n_{\max})$-plain, for emphasis) in G (relative to a particular adjacency-list representation of G) is an induced subgraph of G of at least $n_{\min}$

and at most $n_{\max}$ vertices, whose weakly connected components are pairwise acquainted twin valleys. We now state the main result of this section; the proof is omitted from this extended abstract.

Lemma 5. *For all constants* $k, n_{\min} \geq 1$, *there are constants* $d, n_{\max} \geq 1$ *and* $c > 0$ *such that every connected graph with* $n > n_{\max}$ *vertices and treewidth at most* k *contains at least* cn *disjoint* $(d, k, n_{\min}, n_{\max})$-*plains (relative to any adjacency-list representation).*

4 Constructing Tree Decompositions

In this section we show that minimum-width tree decompositions of n-vertex graphs of bounded treewidth can be constructed on an EREW PRAM using $O((\log n)^2)$ time and $O(n)$ operations. More precisely, given an n-vertex graph G and a constant k, our algorithm outputs either a tree decomposition of G of treewidth $tw(G)$ or an indication of the fact that $tw(G) > k$.

The algorithm is based on the graph-reduction technique: A connected input graph of treewidth $\leq k$ is successively replaced by smaller and smaller graphs in a series of *reductions* until a constant-size graph results. Starting from a minimum-width tree decomposition of the final constant-size graph, the reductions are then undone one by one in the reverse order of their application, where, in undoing a reduction that originally replaced a graph G' by a smaller graph G'', a minimum-width tree decomposition of G' is derived from one of G''. At the end of this process we obtain a minimum-width tree decomposition of the input graph.

Suppose that u and v are vertices in a graph G' that are either adjacent or twins and let G'' be the graph obtained from G' by removing u and its incident edges after first inserting an edge between v and each neighbor of u that was not previously a neighbor of v; we will call u and v *reduction partners* and say that G'' is obtained from G' by reduction *on* the pair $\{u, v\}$. A tree decomposition of G'' can be obtained from any tree decomposition of G' by replacing each occurrence of u in a bag by v if u and v are adjacent in G', and by removing all occurrences of u if u and v are twins in G'; hence $tw(G'') \leq tw(G')$. On the other hand, $tw(G') \leq tw(G'') + 1$, since a tree decomposition of G' can be obtained from any tree decomposition of G'' by replacing each occurrence of v in a bag by occurrences of both u and v — we will say that v is *expanded*. If G' is of bounded treewidth, we can therefore undo the reduction transforming G' into G'' by applying the width-minimizing procedure of Theorem 2 to derive a minimum-width tree decomposition of G' from one of G''.

For a fast parallel algorithm it clearly does not suffice to remove vertices one by one. It is easy to see, however, that the scheme described in the preceding paragraph remains valid if, rather than reducing on a single pair of vertices, we reduce simultaneously on an arbitrary collection of pairs that are sufficiently far apart in the graph not to interfere with each other. The only difference is that the treewidth of G' may now be as much as twice that of G'', plus one (each vertex in a bag may need to be expanded into two vertices), which is still fine for the width-minimizing procedure.

Theorem 6. *For all constants $k \geq 1$ and all integers $n \geq 2$, the following prob-lem can be solved on an EREW PRAM using $O((\log n)^2)$ time, $O(n)$ operations and $O(n)$ space: Given an n-vertex graph G, construct a minimum-width tree decomposition of G or decide (correctly) that $tw(G) > k$.*

Proof. For the time being assume that G is connected and of treewidth at most k. We will apply Lemma 5 to G with $n_{\min} = 2$. Hence let the constants c and d be as in the lemma and define the concepts *small* and *acquainted* accordingly. The lemma implies that G contains at least cn distinct pairs $\{u,v\}$ of small vertices such that u and v are either adjacent or acquainted twins; to see this, note that each of the cn disjoint plains whose existence is guaranteed by the lemma contains either a valley of at least two vertices, hence a small vertex with a small neighbor, or two or more acquainted twin valleys of one vertex each. Furthermore, the set R of all such pairs can be computed in constant time using $O(n)$ operations, since it suffices to let each small vertex inspect all its neighbors and all vertices with which it is acquainted; with some care, this can be done without concurrent reading.

We cannot necessarily execute all reductions corresponding to pairs in R, since vertices in distinct pairs may coincide, be adjacent or have adjacent entries in some adjacency list, which hinders the simultaneous execution of the associated reductions. In order to deal with this complication, we construct a *conflict graph* with a vertex for each pair in R and an edge between two vertices if the corresponding reductions exclude each other for one of the reasons mentioned above. It is easy to see that the conflict graph is of bounded degree and can be constructed in constant time using $O(n)$ operations. We compute an independent set of size $\Omega(n)$ in the conflict graph, which can be done in $O(\log n)$ time using $O(n)$ operations [13, Lemma 7(b)]. Finally we execute the reductions on the pairs in the independent set, which takes constant time using $O(n)$ operations.

The reductions described above change G into a smaller graph G'. Let us now see that we can undo the reductions in the sense of deriving a minimum-width tree decomposition of G from one of G'. We already observed that all that is involved is to expand certain vertices into the corresponding pair of reduction partners, after which we can finish using the width-minimizing procedure of Theorem 2. Allowing concurrent reading, the task would be trivial — processors collectively inspecting the whole tree decomposition could simply expand each such node after looking up its partner in a table. In order to avoid concurrent reading from the table, we begin by balancing the given tree decomposition of G' (Lemma 1); this may increase its width, but only by a constant factor. We then process the resulting balanced tree decomposition $(T = (X,F), \{U_x \mid x \in X\})$ in topological order, i.e., each node is processed before all of its children. The processing of a node x in T expands all vertices in U_x that need to be expanded. If x is the root of T, this is easy. If not, the identity of the reduction partner of each relevant vertex $v \in U_x$ can be passed to x from its parent y, except if v occurs in U_x for the first time (i.e., $v \notin U_y$). For each vertex v the latter happens only at a single tree node x, however, so that in this case we can use table lookup to find the reduction partner of v without any risk of concurrent reading.

The balanced tree decomposition can be processed as described in $O(\log n)$ time using $O(n)$ operations.

The graph G' derived from G is connected and of treewidth at most k, so that a new batch of reductions can be applied to G'. Since G' is smaller than G by a constant factor, as measured by the number of vertices, $O(\log n)$ successive stages of simultaneous reductions suffice to reduce the input graph to a graph of constant size. Provided that the representation of the graph at hand is compacted after each stage by means of prefix summation, the number of operations and the space needed decrease geometrically over the stages, so that the whole process uses $O((\log n)^2)$ time, $O(n)$ operations and $O(n)$ space. Undoing the reductions is no more expensive. This proves Theorem 6 for connected input graphs of treewidth at most k.

Suppose now that the input graph G is of treewidth at most k, but not connected. Our approach will be to apply the algorithm developed above not to G, but to an auxiliary connected graph H obtained from G by introducing a new vertex r and an edge between r and a single vertex in each connected component of G. Except in the trivial case in which G has no edges, G and H have the same treewidth, so that a minimum-width tree decomposition of G can be obtained from a minimum-width tree decomposition of H by removing the occurrences of r from all bags. In order to select a vertex from each connected component of G, we can apply the first part of the reduction algorithm to G in a preprocessing phase: Each connected component of G, being of treewidth at most k, is reduced to constant size, at which point the selection is easy, and the component can be removed (to prevent subsequent stages operating on larger components from being too expensive).

If the treewidth of the input graph G is larger than k, one or more of its connected components may fail to be reduced to constant size within the time bound established for graphs of treewidth at most k, or one of the intermediate graphs encountered while undoing reductions may have treewidth larger than k. In either case, the algorithm can stop and announce that $tw(G) > k$. Finally, it is easy to see from the description of the algorithm that even if $tw(G) > k$, the algorithm never performs an illegal action such as concurrent reading, and any output produced by the algorithm is a correct minimum-width tree decomposition of G. □

By applying first the algorithm of Theorem 6 and then that of Theorem 4, we get the following result.

Corollary 7. *For all constants $k \geq 1$ and all integers $n \geq 2$, the following problem can be solved on an EREW PRAM using $O((\log n)^2)$ time, $O(n)$ operations and $O(n)$ space: Given an n-vertex graph G, construct a minimum-width path decomposition of G or decide (correctly) that the pathwidth of G is larger than k.*

5 Deciding Treewidth

An important bottleneck for the running time of the algorithm in the previous section is the repeated application of the algorithm of Theorem 2 while undoing the reductions. When we aim for a decision algorithm only, we can follow a different approach: We will not undo reductions, but instead make sure that all reductions preserve treewidth. We actually describe a generic algorithm, whose instantiations solve various decision problems on graphs of bounded treewidth; in the more general setting, reductions must not affect membership in the class of graphs to be recognized.

Our algorithm can be viewed as a parallelization of a linear-time sequential algorithm due to Arnborg et al. [3]. A first parallel version of this algorithm was given in [6]. The algorithm described there is randomized, works only for graphs of bounded degree and uses $O(\log n)$ expected time and $O(n \log n)$ expected operations on n-vertex input graphs. The algorithm given in this section works for arbitrary graphs, uses $O(n)$ operations and is deterministic, but at a cost of an extra factor of $O(\log^* n)$ in the running time. The algorithm of [3] uses an amount of space bounded by a polynomial, but a polynomial whose degree is large and unspecified. We reduce this to $O(n)$ by means of the bounded adjacency-list search technique.

A *terminal graph* is a triple $G = (V, E, Z)$, where (V, E) is a graph and $Z \subseteq V$ is an ordered set of distinguished vertices in G. The vertices in Z and those in $V \backslash Z$ are called the *terminals* and the *internal vertices* of G, respectively. A terminal graph is *open* if there are no edges between terminals. For $l \geq 0$, an *l-terminal graph* is a terminal graph with exactly l terminals. Let $\mathcal{H}_l$ denote the class of l-terminal graphs, for $l \geq 0$.

Given two l-terminal graphs G_1 and G_2, for some $l \geq 0$, we define $G_1 \oplus G_2$ as the graph obtained by taking the disjoint union of G_1 and G_2 and then identifying the ith terminals in G_1 and G_2, for $i = 1, \ldots, l$, dropping possible multiple edges.

Let $\mathcal{G}$ be a class of graphs. We define an equivalence relation $\sim_\mathcal{G}$ on the set of terminal graphs as follows: $G_1 \sim_\mathcal{G} G_2$ if and only if for some l, G_1 and G_2 both have l terminals, and for all $H \in \mathcal{H}_l$, we have $G_1 \oplus H \in \mathcal{G}$ if and only if $G_2 \oplus H \in \mathcal{G}$. Informally, G_1 and G_2 are equivalent under $\sim_\mathcal{G}$ if any occurrence of G_1 in a bigger graph can be replaced by an occurrence of G_2 without affecting membership of the bigger graph in $\mathcal{G}$. We say that a class $\mathcal{G}$ or its defining property P (i.e., $G \in \mathcal{G}$ if and only if $P(G)$) is of *finite index* if, for every $l \geq 0$, $\mathcal{H}_l$ is split into a finite number of equivalence classes under $\sim_\mathcal{G}$. (Instead of finite index, such graph classes are also known as *regular* or *finite state*.) Many important properties are known to be of finite index.

Theorem 8. *For every graph property P of finite index and for all constants $k \geq 1$ and all integers $n \geq 2$, the problem of deciding whether $P(G) \wedge (tw(G) \leq k)$ for an n-vertex input graph G can be solved on an EREW PRAM using $O(\log n \log^* n)$ time, $O(n)$ operations and $O(n)$ space.*

Proof. We will assume that $P(G)$ implies that G is connected. This is not really a restriction. We can process each connected component separately and subsequently combine the results for the connected components, but we omit the details.

It was shown in [17] that the class of graphs of treewidth at most k is of finite index, and one easily observes that finite index is closed under intersection (see, e.g., [8]). Hence $\mathcal{G} = \{G \mid P(G) \wedge (tw(G) \leq k)\}$ is of finite index. Let $\mathcal{R}$ be a finite set of open terminal graphs that contains at least one element of each equivalence class of $\sim_{\mathcal{G}}$ comprising one or more open terminal graphs with at most $2(k+1)$ terminals, and take $n_{\min}$ as the largest number of vertices of any graph in $\mathcal{R}$. By Lemma 5, we can choose integers $d, n_{\max} \geq 1$ and $c > 0$ such that G contains at least cn disjoint $(d, k, n_{\min}, n_{\max})$-plains (for any adjacency-list representation of G).

The significance of $n_{\min}$ is that any open terminal graph with at least $n_{\min}$ vertices and at most $2(k+1)$ terminals has a smaller equivalent terminal graph in $\mathcal{R}$. In particular, each plain H together with its boundary B and all edges joining a vertex in H and a vertex in B, with the vertices in B considered as terminals (call this an *extended plain*), is such an open terminal graph, so that it can be replaced by a smaller terminal graph in $\mathcal{R}$. Considering isomorphic graphs as identical, there is only a finite number of different extended plains, all of which can therefore be mapped to equivalent smaller open terminal graphs by means of a finite table T. Each entry in T corresponds to a reduction in a natural way.

The algorithm proceeds in a number of phases. In each phase, each vertex determines whether it belongs to a plain and, if so, looks up a corresponding reduction in T. This can be done in constant time: It suffices to let each vertex u inspect those vertices and edges that lie on a path of length at most $n_{\max}$ from u such that the entries of any two consecutive edges (v, w) and (w, x) on the path are separated by a distance of at most d in the adjacency list of w; this can be done without concurrent reading. The reductions found by two distinct vertices may not be simultaneously executable: The plain containing one vertex may intersect the plain containing the other vertex or its boundary, or two vertices, one from each plain, may have adjacent entries in some common boundary vertex. Because we only replace open terminal graphs by other open terminal graphs, however, these are the only ways in which two reductions can interfere with each other. As in Section 4, we construct a conflict graph of bounded degree on the vertices belonging to plains, compute an independent set of size $\Omega(n)$ in the conflict graph and execute the corresponding reductions, which reduces the size of the graph by at least a constant factor. After $O(\log n)$ stages, either we are left with a graph of constant size, whose membership in $\mathcal{G}$ can be determined directly, or the input graph did not belong to $\mathcal{G}$.

The only part of a stage that takes more than constant time with a linear number of processors is the computation of an independent set in the conflict graph. For this, we employ in the first $O(\log^* n)$ stages the algorithm of [13, Lemma 7(b)], which uses $O(\log m)$ time and $O(m)$ operations, where m is the number of vertices in the conflict graph. In the remaining phases, we use the algorithm of [12, Theorem 3], which needs $O(\log^* m)$ time and $O(m \log^* m)$ operations. The total time is $O(\log n \log^* n)$, and a simple simulation argument that schedules compactions of the representation conveniently (see [14, Section 4]) shows that the algorithm can be carried out using $O(n)$ operations. □

The theorem implies, in particular, that the problem of deciding whether the treewidth of a given graph is at most k, for constant k, can be solved in $O(\log n \log^* n)$ time with $O(n)$ operations. Moreover, the same result can be shown to hold for pathwidth. Many well-known graph properties are of finite index. For instance, this is true of all problems that can be expressed in monadic second-order logic, such as Hamiltonicity and l-colorability. This was first shown by Courcelle [11]; see [8] for a possibly more accessible proof.

Theorem 8 is nonconstructive: An algorithm with the stated properties is merely shown to exist. To actually exhibit the algorithm, we must be able to compute the number $n_{\min}$ and to construct the table T. If we have a terminating algorithm that decides whether two given terminal graphs are equivalent under $\sim_\mathcal{G}$ or under any refinement (subdivision) of $\sim_\mathcal{G}$ that still has a finite number of equivalence classes, this can be done by a method described in [3] (in a general algebraic setting). For the case in which k is a constant and $\mathcal{G}$ is the class of graphs of treewidth at most k, such an explicit decision algorithm was exhibited in [17]. If k is a constant and $\mathcal{G}$ is the set of those graphs of treewidth at most k that satisfy a property P expressed in monadic second-order logic, then an algorithm that decides a subdivision of $\sim_\mathcal{G}$ with a finite number of equivalence classes can be obtained by combining results implicit in [8, 11, 17].

It is also possible to apply the parallel reduction techniques to problems that are of *finite integer index*, in the sense of [6]. This allows deciding on the size of a maximum independent set, minimum vertex cover, minimum dominating set and others on graphs of bounded treewidth in $O(\log n \log^* n)$ time using $O(n)$ operations on an EREW PRAM.

On the CRCW PRAM we can give a faster solution, as expressed in the theorem below. We omit the proof, noting only that it is based on derandomization of a related randomized algorithm.

Theorem 9. *For every graph property P of finite index and for all constants $k \geq 1$ and all integers $n \geq 2$, the problem of deciding whether $P(G) \wedge (tw(G) \leq k)$ for an n-vertex input graph G can be solved on a CRCW PRAM using $O(\log n)$ time, $O(n)$ operations and $O(n)$ space.*

References

1. K. Abrahamson, N. Dadoun, D. G. Kirkpatrick, and T. Przytycka. A simple parallel tree contraction algorithm. *J. Algorithms* **10** (1989) 287–302.
2. S. Arnborg, D. G. Corneil, and A. Proskurowski. Complexity of finding embeddings in a k-tree. *SIAM J. Alg. Disc. Meth.* **8** (1987) 277–284.
3. S. Arnborg, B. Courcelle, A. Proskurowski, and D. Seese. An algebraic theory of graph reduction. *J. ACM* **40** (1993) 1134–1164.
4. H. L. Bodlaender. NC-algorithms for graphs with small treewidth. *Proc. 14th International Workshop on Graph-Theoretic Concepts in Computer Science (WG 1988), Springer-Verlag, LNCS vol. 344, pages 1–10.*

5. H. L. Bodlaender. A linear time algorithm for finding tree-decompositions of small treewidth. In *Proc. of the 25th Annual Symposium on Theory of Computing (STOC 1993)*, pages 226–234.

6. H. L. Bodlaender. On reduction algorithms for graphs with small treewidth. In *Proc. 19th International Workshop on Graph-Theoretic Concepts in Computer Science (WG 1993), Springer-Verlag, LNCS vol. 790*, pages 45–56.

7. H. L. Bodlaender and T. Kloks. Efficient and constructive algorithms for the pathwidth and treewidth of graphs. Techn. Rep. RUU-CS-93-27, Dept. of Computer Science, Utrecht University, Utrecht, the Netherlands, 1993. A preliminary version appeared in *Proc. 18th International Colloquium on Automata, Languages and Programming (ICALP 1991), Springer-Verlag, LNCS vol. 510*, pages 544–555.

8. R. B. Borie, R. G. Parker, and C. A. Tovey. Automatic generation of linear-time algorithms from predicate calculus descriptions of problems on recursively constructed graph families. *Algorithmica* **7** (1992) 555–581.

9. N. Chandrasekharan and S. T. Hedetniemi. Fast parallel algorithms for tree decomposing and parsing partial k-trees. In *Proc. 26th Annual Allerton Conference on Communication, Control, and Computing*, Urbana-Champaign, Illinois, 1988.

10. R. Cole and U. Vishkin. Deterministic coin tossing with applications to optimal parallel list ranking. *Inform. and Control* **70** (1986) 32–53.

11. B. Courcelle. The monadic second-order logic of graphs. I. Recognizable sets of finite graphs. *Inform. and Comput.* **85** (1990) 12–75.

12. A. V. Goldberg, S. A. Plotkin, and G. E. Shannon. Parallel symmetry-breaking in sparse graphs. *SIAM J. Disc. Math.* **1** (1988) 434–446.

13. T. Hagerup. Optimal parallel algorithms on planar graphs. *Inform. and Comput.* **84** (1990) 71–96.

14. T. Hagerup, M. Chrobak, and K. Diks. Optimal parallel 5-colouring of planar graphs. *SIAM J. Comput.* **18** (1989) 288–300.

15. T. Hagerup, J. Katajainen, N. Nishimura, and P. Ragde. Characterizations of k-terminal flow networks and computing network flows in partial k-trees. In *Proc. 6th Annual ACM-SIAM Symposium on Discrete Algorithms (SODA 1995)*, pages 641–649.

16. J. Lagergren. Efficient parallel algorithms for tree-decomposition and related problems. In *Proc. 31st Annual Symposium on Foundations of Computer Science (FOCS 1990)*, pages 173–182.

17. J. Lagergren and S. Arnborg. Finding minimal forbidden minors using a finite congruence. In *Proc. 18th International Colloquium on Automata, Languages and Programming, (ICALP 1991), Springer-Verlag, LNCS vol. 510*, pages 532–543.

18. G. L. Miller and J. H. Reif. Parallel tree contraction and its application. In *Proc. 26th Annual Symposium on Foundations of Computer Science (FOCS 1985)*, pages 478–489.

19. B. A. Reed. Finding approximate separators and computing tree width quickly. In *Proc. 24th Annual Symposium on Theory of Computing, (STOC 1992)*, pages 221–228.

20. N. Robertson and P. D. Seymour. Graph minors. XIII. The disjoint paths problem. Manuscript, 1986.

21. R. A. Wagner and Y. Han. Parallel algorithms for bucket sorting and the data dependent prefix problem. In *Proc. 1986 International Conference on Parallel Processing*, pages 924–930.

22. E. Wanke. Bounded tree-width and LOGCFL. *J. Algorithms* **16** (1994) 470–491.

Approximating Minimum Cuts under Insertions

Monika Rauch Henzinger*

Department of Computer Science, Cornell University,
Ithaca, NY 14853, USA

Abstract. This paper presents insertions-only algorithms for maintaining the exact and approximate size of the minimum edge cut and the minimum vertex cut of a graph. The algorithms output the approximate or exact size k in time $O(1)$ or $O(\log n)$ and a cut of size k in time linear in its size. The amortized time per insertion is $O(1/\epsilon^2)$ for a $(2 + \epsilon)$-approximation, $O((\log \lambda)((\log n)/\epsilon)^2)$ for a $(1 + \epsilon)$-approximation, and $O(\lambda \log n)$ for the exact size of the minimum edge cut, where n is the number of nodes in the graph, λ is the size of the minimum cut and $\epsilon > 0$. The $(2 + \epsilon)$-approximation algorithm and the exact algorithm are deterministic, the $(1 + \epsilon)$-approximation algorithm is randomized. The algorithms are optimal in the sense that the time needed for m insertions matches the time needed by the best static algorithm on a m-edge graph. We also present a static 2-approximation algorithm for the size κ of the minimum vertex cut in a graph, which takes time $O(n^2 \min(\sqrt{n}, \kappa))$. This is a factor of κ faster than the best algorithm for computing the exact size, which takes time $O(\kappa^2 n^2 + \kappa^3 n^{1.5})$. We give an insertions-only algorithm for maintaining a $(2 + \epsilon)$-approximation of the minimum vertex cut with amortized insertion time $O(n(\log \kappa)/\epsilon)$.

1 Introduction

Computing the connectivity of a graph is a fundamental problem with applications to chip design, system reliability, and communications networks. Since in many of these applications the underlying graph can change incrementally, it is important to efficiently maintain the connectivity of the graph during changes.

Two vertices of an unweighted, undirected graph G are *k-edge (k-vertex) connected* if there are k edge-disjoint (vertex-disjoint) paths between them. The graph G is *k-edge (k-vertex) connected* if all its vertices are k-edge (k-vertex) connected. A *minimum edge (vertex) cut* of a graph G is a minimum cardinality set of edges (vertices) of G whose removal disconnects G. By Menger's theorem [13] determining the size of the minimum cut of an (unweighted) graph is equivalent to computing the edge connectivity of G. Computing the cardinality of the minimum vertex cut κ is equivalent to computing the vertex connectivity of G. Note that k-vertex connectivity implies k-edge connectivity, but not vice versa.

* Maiden Name: Monika H. Rauch. This research was supported by an NSF CAREER Award.

Given an initial graph G with n nodes and m_0 edges an *incremental* algorithm maintains G during an arbitrary sequence of the following operations.

- *Insert(u, v)*: insert the edge (u, v) into G.
- *Query-Size*: return the exact (approximate) size of a minimum cut in G.
- *Query-Cut*: return an exact (approximate) minimum cut of G.

We say that k is a *c-approximation* of λ (κ) if $\lambda \le k \le c\lambda$ ($\kappa \le k \le c\kappa$). This paper presents simple incremental algorithms for maintaining the exact and approximate size of the minimum edge cut and the approximate size of the minimum vertex cut. The basic idea is to use the static algorithm for computing the solution and to build a data structure that quickly tests if the solution computed last by the static algorithm is still the correct solution for the current graph. If yes, the data structure is updated, otherwise, a new solution is computed using the static algorithm. The difficulty lies in finding the appropriate data structure and to amortize the cost for the static algorithm over previous insertions. The algorithms can output the exact or approximate size k of the minimum cut in time $O(1)$ or $O(\log n)$ and a cut of size k in time proportional of its size.

We denote by m the total number of edges in the graph, consisting of m_0 initial edges and m_1 inserted edges, by δ the minimum degree in the final graph, by λ the size of the minimum edge cut in the final graph, and by κ the size of the minimum vertex cut in the final graph. Note that $\kappa \le \lambda$ and $\lambda n = O(m_0 + m_1)$.

Minimum Edge Cuts. Given a static exact, $(1+\epsilon)$-, or $(2+\epsilon)$-approximate minimum edge cut algorithm with running time $T(n, m)$ we convert it into an incremental algorithm such that the total time for m insertions is $O(T(n, m))$, resp. $O(T(n, m)/\epsilon)$. Thus, the algorithms are optimal in the sense that they match the performance of the best static algorithm (up to a factor of $O(1/\epsilon)$). For constant ϵ the overhead of making the algorithm incremental contributes only a constant factor to the running time.

The first algorithm maintains a $(2 + \epsilon)$-approximation of λ in total time $O((m_0 + \lambda n)/\epsilon + m_1/\epsilon^2)$ for any $1 \ge \epsilon > 0$. It is an incremental version of Matula's static $(2 + \epsilon)$-approximation algorithm, which takes time $O(m/\epsilon)$ [17]. If the initial graph is empty, our amortized time per operation is $O(1/\epsilon^2)$.

The second algorithm is an incremental version of Gabow's (exact) minimum edge cut algorithm [8], which computes the size of the minimum edge cut in time $O(m + \lambda^2 n \log(n^2/m))$. Our incremental algorithm takes time $O(m_0 + m_1\alpha(n, n) + \lambda^2 n \log(n/(m_0 + 1)))$. Thus, if the initial graph is empty, the amortized time per insertion is $O(\lambda \log n)$. Apart from answering *Query-Size* or *Query-Cut* operations, our algorithm can answer queries that ask if two given nodes are separated by a cut of size λ in amortized time $O(\alpha(n, n))$. Gabow's algorithm implies an algorithm which allows insertions and deletions of edges in worst-case time $O(m \log(n^2/m))$. Our algorithm can be easily modified to improve this bound to $O(\lambda \log n)$ amortized time per insertion and $O(m \log(n^2/m))$ worst-case time per deletion. If λ is a constant, our running time is close to the running time of the best "special purpose" algorithms: Determining if two nodes are connected, 2-edge-connected, or 3-edge-connected takes amortized time $O(\alpha(m, n))$ per insertion[20, 11, 16].

Finally we combine the previous two (deterministic) algorithms with random sampling to achieve an incremental Monte Carlo algorithm that maintains a $(1 + \epsilon)$-approximation of the minimum edge cut with high probability. The total expected time for m_1 insertions is $O((m_0 + m_1)(\log \lambda)((\log n)/\epsilon)^2)$. Thus, if the initial graph is empty, the amortized expected time per insertion is $O((\log \lambda)(\log n/\epsilon)^2)$. This technique was introduced by Karger [15] in his static algorithm, which needs time $O(m + n((\log n)/\epsilon)^3)$ to compute a $(1 + \epsilon)$-approximation of λ.

Related results: (1) For any fixed k, Dinitz [3] gives an algorithm that supports *Same-k-Component?* queries and edge insertions in a $(k-1)$-edge-connected graph. It takes time $O(m_0 + k^2 n \log(n/k) + (q + n)\alpha(q, n))$, where q is the total number of operations. (2) For any fixed k, Eppstein et. al [5] give a (fully dynamic) algorithm that tests if the whole graph is k-edge-connected in $O(1)$ time. The times per edge insertion or deletion is $O(k^2 n \log(n/k))$. (3) Karger [14] gives a randomized algorithm, which maintains a $\sqrt{1 + 2/\epsilon}$-approximation of the minimum edge cut in expected time $\tilde{O}(n^\epsilon)$ per insertion. Thus, for a $(2 + \epsilon)$-approximation it takes time $\tilde{O}(n^{2/(3+4\epsilon+\epsilon^2)})$ per insertion and for a $(1 + \epsilon)$-approximation it takes time $\tilde{O}(n^{1/\epsilon})$. Our algorithms are an exponential improvement. He also gives a (fully dynamic) $\sqrt{1 + 2/\epsilon}$-approximation algorithm with $\tilde{O}(n^{1/2+\epsilon})$ time per edge insertions and deletions. (4) Recently, Dinitz and Nutov [4] gave an algorithm that maintains all cuts of size λ or $\lambda + 1$. The total time for m_1 edge insertions and q *Same-$(\lambda + 1)$-Component?* queries is $O((n + m_1)\alpha(m_1, n) + q\alpha(q, n))$ for odd λ and $O(m_1 + n \log n + q\alpha(q, n))$ for even λ.

Minimum Vertex Cuts. We present a static algorithm that computes a 2-approximation of the size κ of the minimum vertex cut, i.e. it computes the exact size of κ if $\kappa \leq \delta/2$ and it returns $\delta/2$ if $\kappa > \delta/2$. It takes time $O(n^2 \min(\sqrt{n}, \kappa))$ which is a speed-up of a factor of at least κ over the fastest exact algorithm for computing κ, which takes time $O(\kappa^2 n^2 + \kappa^3 n^{1.5})$ [1, 10, 18]. Using this 2-approximation algorithm as subroutine gives an incremental algorithm with total time $O(m_0 + \kappa n^2 (\log \kappa)/\epsilon + m_1 n)$. If the initial graph is empty, the amortized time per insertion is $O(n(\log \kappa)/\epsilon)$.

Section 2 presents the basic structure that is common to all incremental algorithms in this paper. In Section 3 we give some basic definitions. Section 4 presents the incremental algorithms for the minimum edge cut, Section 5 gives the results for the minimum vertex cut.

2 A Generic Incremental Algorithm

To maintain the exact or approximate minimum edge or vertex cut in a graph we use the following generic algorithm.
1. Compute the solution in the initial graph using the static algorithm.
2. **while** the current solution is correct **do**
    if the new operation is a query **then** output the current solution
    else add the new edge to the graph.
    endwhile

3. Compute a new solution using previous solutions.

Goto 2.

The algorithm decomposes the sequence of insertions into subsequences, between which a new solution is computed in Step 3. The difficulty lies in deciding (1) how to quickly test if the current solution is still correct and (2) how to efficiently compute a new solution using previous solutions. To analyze the running time we amortize the cost of computing a new solution over the sequence of insertions after the last new computation of a solution. We restrict our description to answering *Query-Size* operations. However, it is straightforward to extend the algorithms to answer *Query-Cut* operations.

3 Basic Definitions

Let $G = (V, E)$ be an undirected graph. A *maximal spanning forest decomposition (msfd) of order* k is a decomposition of a graph G into k edge-disjoint spanning forests F_i, $1 \leq i \leq k$, such that F_i is a maximal spanning forest of $G \setminus (F_1 \cup F_2 \cup \ldots \cup F_{i-1})$. If two nodes are connected in F_i, they are i-edge connected. Nagamochi and Ibaraki [18] give a linear time algorithm (referred to as *decomposition algorithm (DA)*) that computes a special msfd, called *DA-msfd*, of order m in time $O(m + n)$. A DA-msfd fulfills two additional conditions: (1) For all $1 \leq i \leq k$ if x and y are connected in F_i, then x and y are i-vertex connected in G. (2) The graph G is k-vertex connected iff $F_1 \cup \ldots \cup F_\delta$ is k-vertex connected. The decomposition algorithm also determines a linear order on the nodes, called the *maximum cardinality search order (mcs-order)*.

An edge (x, y) is *contracted* if x is identified with y and all self-loops (but not parallel edges) are discarded. A contraction reduces the number of nodes in G, but does not reduce the size of the minimum edge cut. We *contract* a forest F if we contract all edges of G that are in F.

If a graph is k-edge connected, it contains $\Omega(kn)$ edges:

Lemma 1. [13] *If a n-node graph is k-edge connected, then it contains at least $kn/2$ edges.*

4 Incremental Algorithms for the Minimum Edge Cut

4.1 An Incremental $(2 + \epsilon)$-Approximation

Using the generic algorithm of Section 2 we create an incremental algorithm that maintains a $(2 + \epsilon)$-approximation of the minimum cut. We describe below (1) how to test the correctness of the current solution and (2) how to efficiently compute a new solution. (1) Let k be the current solution, i.e. $k/(2 + \epsilon) \leq \lambda \leq k$. At the start of each subsequence of insertions we contract G in Procedure *Contract* until the number m' of edges in the contracted graph is at most $kn'/(2 + \epsilon'/2)$. Lemma 1 shows that $\lambda \leq k$ as long as $m' < (k + 1)n'/2$. Thus to test the correctness of the current solution k we simply check if $m' < (k+1)n'/2$. (2) Since $m' \le kn'/(2+\epsilon'/2)$ at the start and $m' \ge (k+1)n'/2$ at the end of a subsequence, we know that the subsequence consists of u insertions with $u \ge (k+1)n'/2 - kn'/(2+\epsilon'/2) = \Omega(kn'\epsilon)$. We use this fact to amortize the cost of computing the next solution, which is $O(u+kn'/\epsilon)$, over each insertion in the subsequence. To compute a new solution we repeatedly increment k until we find the first k such that there exists a node with degree $< k$ in the graph. The latter is tested by computing a DA-msfd, contracting F_p with $p = \lceil k/(2+\epsilon) \rceil$, and checking if in the resulting n'-node graph (a) $n' > 1$ and (b) the number of edges $m' \le kn'/(2+\epsilon'/2)$. The algorithm is given below. We denote by $G' = (V', E')$ with $n' = |V'|$ and $m' = |E'|$ the graph resulting from the contractions.

An Incremental $(2+\epsilon)$-Approximation Algorithm
1. $\epsilon' = \epsilon/4$, $j = 0$.
   Compute a $(2+\epsilon')$-approximation k of λ using the static algorithm and a DA-msfd $F_1^{(0)}, \ldots, F_m^{(0)}$ of $G' = (V', E')$ of order m.
   Contract(G', k, p, j).
2. $N = \emptyset$.
   while $m' < (k+1)n'/2$ **do**
       if the new operation is a query **then** return k
       else add the inserted edge to N.
   endwhile
3. Contract(G', k, p, j).
   while $n' = 1$ **do**
       $k = k(1+\epsilon')$, $p = \lceil k/(2+\epsilon') \rceil$, $V' = V$, $E' = N \cup \cup_{1 \le j, q \le p} F_q^{(l)}$.
       Contract all edges in $\cup_{l \le j} F_p^{(l)}$.
       Contract(G', k, p, j).
   endwhile
   Goto 2.

Contract(G', k, p, j)
    repeat $j = j + 1$.
        Compute a DA-msfd $F_1^{(j)}, \ldots, F_{m'}^{(j)}$ of G' of order m'.
        Contract all edges in the forest $F_p^{(j)}$ and update m' and n'.
    until $m' \le kn'/(2+\epsilon'/2)$

Lemma 2. *The algorithm returns a $(2+\epsilon)$-approximation of λ for $0 < \epsilon \le 4$.*

Proof. We show by induction that $k/(2+\epsilon) \le \lambda \le k$. Initially k is a $(2+\epsilon)$-approximation and, thus, $k/(2+\epsilon) \le \lambda \le k$. Assume inductively that the claim holds and consider the next insertion. Note that insertions only increase λ. We distinguish two cases:

(1) If $m' < (k+1)n'/2$ after the insertion, then Lemma 1 shows that $\lambda \le k$ also after the insertion. By induction $k/(2+\epsilon) \le \lambda$ holds.

(2) If $m' = (k+1)n'/2$ after the insertion, then the algorithm repeatedly contracts G' and multiplies k by $(1+\epsilon)$ until it finds the smallest k such that the contractions stop with $n' > 1$ and $m' \le kn'/(2+\epsilon/2)$. This implies that there exists a node in G' with degree at most $2(k/(2+\epsilon/2)) \le k$, implying that $\lambda \le k$. If k is unchanged, it follows by induction that $k/(2+\epsilon) \le \lambda$. Otherwise, k is the smallest value for which the contractions terminate with $n' > 1$. Thus, contracting $F_{p'}$ for $p' = \lceil k/((2+\epsilon')(1+\epsilon')) \rceil$ contracted G' to one node. This implies that all nodes in G' are connected in $F_{p'}$ and thus p'-edge connected. Thus $\lambda \ge p' = \lceil k/((2+\epsilon')(1+\epsilon')) \rceil \ge k/(2+4\epsilon') = k/(2+\epsilon)$ for $\epsilon' \le 1$.

To analyze the running time we show that contracting all edges in $\cup_{l \le j} F_p^{(l)}$ requires $O(n)$ contractions.

Lemma 3. *Let $k^{(j)}$ be the value of k after the jth and before the $j+1$st msfd is computed. For $p \ge k^{(j)}/(2+\epsilon')$ the graph $\cup_{l \le j} F_p^{(l)}$ is a forest.*

Let k_0 be the initial value of k returned by the static algorithm and let $k_i = k_0(1+\epsilon')^i$. We denote by *Phase i* all steps that are executed while $k = k_i$. Let u_i be the number of insertions during Phase i.

Lemma 4. *Phase i takes time $O(k_i n + u_i/\epsilon^2)$.*

Since $\sum_i k_i = O(\lambda/\epsilon)$ and $\sum_i u_i = m_1$, the theorem follows.

Theorem 5. *Given a graph with n nodes and m_0 edges the total time for inserting m_1 edges and maintaining a $(2+\epsilon)$-approximation of the minimum cut is*

$$O((m_0 + \lambda n)/\epsilon + m_1/\epsilon^2),$$

where λ is the size of the minimum cut in the final graph and $0 < \epsilon \le 4$. The space is linear in the time.

4.2 An Incremental Exact Algorithm

In this section we present a deterministic incremental algorithm that maintains λ. The basic idea for testing efficiently if the current solution is still correct is to compute and store *all* minimum edge cuts when λ is increased by 1. If an insertion increments the size of one of these cuts, it is no longer minimum. Thus, the current solution is correct as long as there exists still a minimum cut whose size has not been increased.

To store all minimum edge cuts we use the *cactus tree* representation [2]. A cactus tree of a graph $G = (V, E)$ is a graph $G_c = (V_c, E_c)$ with a weight function w defined as follows: There is a mapping $\phi : V \to V_c$ such that

(1) every node in V maps to exactly one node in V_c and every node in V_c corresponds to a (possibly empty) subset of V,

(2) $\phi(u) = \phi(v)$ iff u and v are at least $\lambda(G) + 1$- edge connected,

(3) each minimum cut in G_c corresponds to a minimum cut in G, each minimum cut in G corresponds to at least one minimum cut in G_c.

(4) If λ is odd, every edge of E_c has weight λ and G_c is a tree. If λ is even, two simple cycles of G_c have at most one common node, every edge that does not belong to a cycle has weight λ, and every edge that belongs to a cycle has weight $\lambda/2$.

As observed by Dinitz [3], given a cactus-tree the data structure of [11, 16] can maintain the cactus tree for a fixed value of λ such that the total time for u insertions is $O((u+n)\alpha(u+n,n))$. Determining if there exists a minimum cut whose size has not been increased takes constant time.

To quickly compute the cactus tree representation we use an algorithm by Gabow [9]. It takes time $O(m_0 + \lambda^2 n \log(n^2/(m_0 + 1)))$ to compute the cactus tree in the initial graph. Additionally given a cactus tree and a sequence of insertions that increase the minimum cut size by 1, the new cactus tree can be computed in time $O(m' \log(n^2/m'))$, where m' is the current number of edges.

This leads to the following algorithm.

An Incremental Minimum Edge Cut Algorithm

1. Compute the size λ of the minimum cut, a DA-msfd $F_1^{(0)}, \ldots, F_m^{(0)}$ of order m, and a cactus-tree of $\cup_{i \le \lambda} F_i^{(0)}$.
   $j = 0$.
2. $N = \emptyset$.
   while there is ≥ 1 minimum cut of size λ **do**
     if the next operation is a query **then return** λ
     else insert the new edge into the cactus tree and into N.
   endwhile
3. $j = j + 1$, $\lambda = \lambda + 1$.
   Compute a DA-msfd $F_1^{(j)}, \ldots, F_m^{(j)}$ of order m of $\cup_{i \le \lambda} F_i^{(j-1)} \cup N$ such that $\cup_{i \le \lambda} F_i^{(j-1)} \subset \cup_{i \le \lambda} F_i^{(j)}$ and compute the cactus tree of $\cup_{i \le \lambda} F_i^{(j)}$.
   Goto 2.

Theorem 6. *Given a graph G with n nodes and m_0 edges the total time for inserting m_1 edges and maintaining a minimum edge cut of G is*

$$O(m_0 + m_1 \alpha(n,n) + \lambda^2 n \log(n/(m_0 + 1))),$$

where λ is the size of the minimum cut in the final graph. The size of the minimum cut can be output in constant time, a query if two given nodes are separated by a minimum cut can be answered in amortized time $O(\alpha(n,n))$.

4.3   A Randomized $(1 + \epsilon)$-Approximation for the Minimum Cut

In this section we present an incremental Monte Carlo algorithm that maintains a $(1 + \epsilon)$-approximation of the minimum cut.

Karger [14] pointed out that dynamically approximating connectivity can be reduced to dynamically maintaining exact connectivity in $O(\log n)$-connected graph using randomized sparsification. We use this idea to maintain a $(1 + \epsilon)$-approximation of the minimum cut as follows: We put each inserted edge with probability p into a subgraph $G(p)$ of G and maintain $G(p)$ incrementally. The following lemma shows that the resulting incremental algorithm maintains a $(1 + \epsilon)$-approximation of the minimum cut with high probability.

Lemma 7. [14] *Let G be any graph with minimum cut λ and let $p = 2(d + 2)(\ln n)/(\epsilon^2\lambda)$ for any $\epsilon \le 1$. (1) With probability $1 - O(1/n^d)$ the size of the minimum cut in $G(p)$ is $\Theta((\log n)/\epsilon^2)$. (2) With probability $1 - O(1/n^d)$ the minimum cut in $G(p)$ corresponds to a $(1 + \epsilon)$-minimal cut of G.*

To quickly find for each minimum cut in $G(p)$ the size of the corresponding cut in G, we keep the data structure of [11] also in a dynamic tree data structure [19], which we use to "label" each cut represented in the data structure by the size of its corresponding cut in G (we omit the details). This increases the cost for maintaining the data structure during u insertions to $O((m_0 + u)\log n)$. Additionally, we choose p large enough so that the graph $G(p)$ has to be rebuilt only when the size of the minimum cut in G has increased. The latter is tested incrementally by maintaining a 3-approximation of $\lambda(G)$. This leads to the following algorithm:

An Incremental $(1 + \epsilon)$-Approximation Algorithm
1. Compute a 3-approximation k of the size of the minimum cut in G.
   While $k \le 16\ln n$, maintain the exact minimum cut of G incrementally.
   Initialize the incremental 3-approximation algorithm.
   $p = 8(d + 4)(\ln n)/(k\epsilon^2)$.
   Construct $G(p)$ by sampling every edge with probability p. Compute the cactus tree of $G(p)$ and store it in an incremental exact data structure.
2. $N = \emptyset$.
   while $\lambda(G(p))$ is not increased **do**
       if the next operation is a query **then**
           return the size in G of a minimum cut in $G(p)$
       else add the new edge to N and sample it. If sampling is successful,
           insert it into the incremental exact data structure as a new edge
           of $G(p)$, otherwise insert it as a new edge of $G \setminus G(p)$.
   endwhile
3. Add all edges of N to the incremental $(2 + \epsilon)$-approximation algorithm to
   determine a new 3-approximation k'.
   if $k' > 3k$ **then** $k = k'$, $p = 8(d + 4)(\ln n)/(k\epsilon^2)$.
   Construct $G(p)$ by sampling every edge with probability p.
   Compute a new cactus-tree of $G(p)$ and store it in an incremental exact
   data structure.
   Goto 2.

The probability that the algorithm does not maintain a $(1 + \epsilon)$-approximation is bounded by the sum of the probabilities that at any step the algorithm does not compute a $(1 + \epsilon)$-approximation, which is $O(1/n^{d+2})$. Since there are $O(n^2)$ insertions, the probability of failure is $O(1/n^d)$.

Theorem 8. *Let G be a graph with n nodes and m_0 initial edges. For any $\epsilon \leq 1$ and any $d > 0$ the given algorithm maintains a $(1 + \epsilon)$-approximation of the minimum edge cut of G with probability $1 - O(1/n^d)$. The expected time for inserting m_1 edges is*

$$O((m_0 + m_1)(\log n)^2 (\log \lambda)/\epsilon^2)$$

where λ is the size of the minimum cut in the final graph. Each query can be answered in time $O(\log n)$.

5 A 2-Approximation of the Minimum Vertex Cut Algorithm

5.1 A Static Algorithm

Let δ' be $\delta/2$. We give an algorithm that outputs the minimum vertex connectivity κ, if $\kappa \leq \delta/2$ and it outputs $\delta/2$ otherwise. It takes time $O(n^2 \min(\sqrt{n}, \kappa))$. The best algorithm to compute the exact solution takes time $O(\kappa^2 n^2 + \kappa^3 n^{3/2})$([10] with [1, 18]). Our algorithms follows ideas similar to [17].

Even and Tarjan [7] give an algorithm to compute the minimum vertex cut $\kappa(a, b)$ between the nodes a and b in time $O(\min(m\sqrt{n}, \kappa))$, which we call the *pair algorithm (PA)*. The exact minimum vertex cut algorithm [10] makes $O(k^2 + n)$ calls to PA. We reduce the number of calls to $\lceil n/\delta' \rceil$ using the following two observations:

1. The last δ' nodes in the mcs-order computed by DA are pairwise δ'-connected.
2. Given δ' pairwise δ'-connected nodes $L = \{v_1, \ldots, v'_\delta\}$ and a node $a \notin L$, if a node b is added to G and connected to every v_j, then a is δ'-connected to each v_j in G iff a is δ'-connected to b.

We fix a node a of G, and repeatedly find δ' pairwise δ'-connected nodes (using DA), and test if all of them are δ'-connected to a (using PA). Since each call to DA has to select δ' not yet selected δ'-connected nodes, the last δ' nodes in the mcs-order have to be different from all previously selected nodes. We guarantee that the mcs-order determined by DA starts with a followed by all previously connected nodes: Since DA (started at a) puts nodes highly connected to a first in mcs-order, we connect all previously selected nodes with n parallel edges to a. This leads to the following algorithm.

A 2-Approximation Algorithm for the Minimum Vertex Cut
Compute a DA-msfd $F_1, \ldots F_m$ of G and replace G by $G_0 = \cup_{i \leq \delta} F_i$.
Create a graph G_1 by adding a node a to G_0 and connecting a by n parallel edges to each of the last $\delta/2$ nodes in mcs-order.
$k = \delta/2, i = 1$.
while there exists a node in G_i not incident to a **do**
    Use DA to compute a msfd of G_i starting at a.
    Create a new node b and connect b to each of the last $\delta/2$ nodes in

mcs-order by an edge. Call the new graph G_i'.
Use PA to compute the minimum vertex cut $\kappa(a, b)$ in G_i'.
if $\kappa(a, b) < k$ then $k = \kappa(a, b)$.
Create a new graph G_{i+1} by connecting in G_i the last $\delta/2$ nodes in
mcs-order by n parallel edges to a.
$i = i + 1$.
endwhile

The following lemmata prove that instead of computing the $\kappa(s, t)$ for every
pair (s, t) of vertices of G, it suffices to compute $\kappa(a, b)$ for every node b. They
are extensions of lemmata in [6]. We use $\Gamma(a)$ to denote the set of neighbors of
a.

Lemma 9. *For any i, let $L_i = \{v_1, \ldots, v_{\delta'}\}$ be a set of pairwise δ'-connected
nodes in G_{i-1} that are not incident to a.*
1. *Let G_1 be the graph created from G_0 by creating a new node a and adding n
   edges (a, v_j) for every $v_j \in L_0$.*
   (a) If $\kappa(G_0) \leq \delta'$, then $\min_{u \notin \Gamma(a)} \kappa(G_1, a, u) = \kappa(G_0)$.
   (b) If $\kappa(G_0) > \delta'$, then $\kappa(G_1, a, u) = \delta'$ for every node $u \notin \Gamma(a)$.
2. *For any $i > 1$, let G_i be the graph constructed from G_{i-1} by adding n edges
   (a, v_j) for every $v_j \in L_i$. If for all nodes v_j of L_i, $\kappa(G_{i-1}, a, v_j) \geq l$ for some
   $l > 0$, then $\min(l, \min_{u \notin \Gamma(a)} \kappa(G_{i-1}, a, u)) = \min(l, \min_{u \notin \Gamma(a)} \kappa(G_i, a, u))$.*

Lemma 10. *Let $L = \{v_1, \ldots, v_{\delta'}\}$ be a set of nodes that are pairwise δ'-connected
in G and let a be a node of G. Let G' be the graph created from G by adding a
vertex b and edges (b, v_j) for every j. Let κ_{min} be $\min_{1 \leq j \leq \delta'} \kappa(G, a, v_j)$. (1) If
$\kappa_{min} < \delta'$ then $\kappa(G', a, b) = \kappa_{min}$. (2) If $\kappa_{min} \geq \delta'$, then $\kappa(G', a, b) = \delta'$.*

Lemma 11. *If $\kappa(G) < \delta'$ then $k = \kappa(G)$. If $\kappa(G) \geq \delta'$, then $k = \delta'$.*

To complete the proof of correctness we have to show that the last $\delta/2$ nodes
in mcs-order are indeed $\delta/2$-connected. By the properties of a DA-msfd, it suffices
to show that they are connected in $F_{\delta/2}$.

Lemma 12. *The last $\delta/2$ nodes in mcs-order are connected in $F_{\delta/2}$.*

Theorem 13. *A 2-approximation of the minimum vertex cut can be computed
in time $O(n^2 \min(\sqrt{n}, \kappa))$, where κ is the size of the minimum vertex cut.*

5.2 An incremental $(2 + \epsilon)$-approximation of the minimum vertex cut

Using the generic incremental algorithm, we construct an incremental $(2 + \epsilon)$-
approximation algorithm. We compute a 2-approximation of κ in the initial
graph. Let $b_1, \ldots b_{n/\delta'}$ be all the nodes b created by the static 2-approximation
algorithm. Note that PA computes a maximum (a, b_i) flow in a directed graph.
To test if the current solution is still correct, we maintain (1) the minimum

degree in G, (2) $\kappa(a, b_i)$, (3) the residual graphs of all maximum (a, b_i)-flows, and (4) a breadth-first search tree in each residual graph rooted at a. For each b_i such that $\kappa(a, b_i)$ is minimum we try to incrementally maintain the maximum flow from a to b_i by augmenting the breadth-first search tree. Thus, for any i the total time spent for incrementing the maximum flow from a to b_i by 1 is $O(m + n)$ (see also [12]).

As shown in Lemma 11, $k := \min_i \kappa(a, b_i)$ cannot be larger than $\delta/2$ and as long as $k < \delta/2$, $k = \kappa(G)$. If $k = \delta/2$ and the minimum degree has no increased by a factor of $(1 + \epsilon/2)$, it also shows that $k \leq \kappa \leq \delta(1 + \epsilon/2) = (2 + \epsilon)k$. Thus, in either case, k is a $(2 + \epsilon)$-approximation of κ. If these conditions are no longer fulfilled, we rebuild from scratch. Note that this happens $O((\log \kappa)/\epsilon)$ times.

An Incremental $(2 + \epsilon)$-Approximation Algorithm
1. Determine the minimum degree δ, compute a DA-msfd $F_1, \ldots F_m$ of G and replace G by $\cup_{i \leq \delta} F_i$.
   Compute a 2-approximation k of G using the static algorithm storing all the maximum flows from a to b_i and their residual graphs.
   For each such residual graph keep a breadth-first search tree rooted at a.
2. **while** the minimum degree in G is $\leq \delta(1 + \epsilon)$ or $k = \delta/2$ **do**
   if the next operation is a query **then return** k
   else try to augment the maximum flow $\kappa(a, b_i)$ by adding the new edge to its residual graph.
   Maintain k as the minimum of all $\kappa(a, b_i)$.
   endwhile
   Goto 1.

Theorem 14. *Let G be a graph with n nodes and m_0 initial edges. The total time for maintaining a $(2 + \epsilon)$-approximation of the minimum vertex cut κ during m_1 insertions is*

$$O(m_0 + \kappa n^2 (\log \kappa)/\epsilon + m_1 n).$$

Each query can be answered in constant time.

Acknowledgements

We want to thank David Karger for useful discussions. We are also thankful to Yefim Dinitz for helpful comments on the paper.

References

1. J. Cheriyan, M. Y. Kao, and R. Thurimella, "Scan-first search and sparse certificates—an improved parallel algorithm for k-vertex connectivity", *SIAM Journal on Computing*, 22, 1993, 157–174.
2. E. A. Dinitz, A. V. Karzanov, and M.V. Lomonosov, "On the structure of the system of minimal edge cuts in a graph", *Studies in Discrete Optimization*, 1990, 290–306.

3. Ye. Dinitz, "Maintaining the 4-edge-connected components of a graph on-line", *Proc. 2nd Israel Symp. on Theory of Computing and Systems (ISTCS'93)*, IEEE Computing Society press, 1993, 88–97.

4. Ye. Dinitz, Z. Nutov, "A 2-level cactus model for the system of minimum and minimum+1 edge-cuts in a graph and its incremental maintenance", to appear in *Proc. 27nd Symp. on Theory of Computing*, 1995.

5. D. Eppstein, Z. Galil, G. F. Italiano, A. Nissenzweig, "Sparsification - A Technique for Speeding up Dynamic Graph Algorithms" *Proc. 33rd Symp. on Foundations of Computer Science*, 1992, 60–69.

6. S. Even, "An algorithm for determining whether the connectivity of a graph is at least k" *SIAM Journal on Computing*, 4, 1975, 393–396.

7. S. Even and R. E. Tarjan, "Network flow and testing graph connectivity", *SIAM Journal on Computing*, 4, 1975, 507–518.

8. H. N. Gabow, "A matroid approach to finding edge connectivity and packing arborescences" *Proc. 23rd Symp. on Theory of Computing*, 1991, 112–122.

9. H. N. Gabow, "Applications of a poset representation to edge connectivity and graph rigidity" *Proc. 32nd Symp. on Foundations of Computer Science*, 1991, 812–821.

10. Z. Galil, "Finding the vertex connectivity of graphs", *SIAM Journal on Computing*, 1980, 197–199.

11. Z. Galil and G. P. Italiano, "Maintaining the 3-edge-connected components of a graph on-line", *SIAM Journal on Computing*, 1993, 11–28.

12. A. Ya. Gordon, "One algorithm for the solution of the minimax assignment problem", *Studies in Discrete Optimization*, A. A. Fridman (Ed.), Nauka, Moscow, 1976, 327–333 (in Russian).

13. F. Harary, "Graph Theory", *Addison-Wesley, Reading, MA*, 1969.

14. D. Karger, "Using randomized sparsification to approximate minimum cuts" *Proc. 5th Symp. on Discrete Algorithms*, 1994, 424–432.

15. D. Karger, "Random sampling in cut, flow, and network design problems", *Proc. 26rd Symp. on Theory of Computing*, 1994, 648–657.

16. H. La Poutré, "Maintenance of 2- and 3-connected components of graphs, Part II: 2- and 3-edge-connected components and 2-vertex-connected components", Tech.Rep. RUU-CS-90-27, Utrecht University, 1990.

17. D. W. Matula, "A linear time $2+\epsilon$ approximation algorithm for edge connectivity" *Proc. 4th Symp. on Discrete Algorithms*, 1993, 500–504.

18. H. Nagamochi and T. Ibaraki, "Linear time algorithms for finding a sparse k-connected spanning subgraph of a k-connected graph", *Algorithmica* 7, 1992, 583–596.

19. D. D. Sleator, R. E. Tarjan, "A data structure for dynamic trees" *J. Comput. System Sci.* 24, 1983, 362–381.

20. J. Westbrook and R. E. Tarjan, "Maintaining bridge-connected and biconnected components on-line", *Algorithmica* (7) 5/6, 1992, 433–464.

Linear Time Algorithms for Dominating Pairs in Asteroidal Triple-free Graphs

(Extended Abstract)

Derek G. Corneil[1], Stephan Olariu[2], Lorna Stewart[3]

[1] Department of Computer Science, University of Toronto
Toronto, Ontario, Canada M5S 1A4
[2] Department of Computer Science, Old Dominion University
Norfolk VA 23529-0162, USA
[3] Department of Computing Science, University of Alberta
Edmonton, Alberta, Canada T6G 2H1

Abstract. An independent set of three of vertices is called an *asteroidal triple* if between each pair in the triple there exists a path that avoids the neighbourhood of the third. A graph is asteroidal triple-free (AT-free, for short) if it contains no asteroidal triple. The motivation for this investigation is provided, in part, by the fact that AT-free graphs offer a common generalization of interval, permutation, trapezoid, and cocomparability graphs. Previously, the authors have given an existential proof of the fact that every connected AT-free graph contains a dominating pair, that is, a pair of vertices such that every path joining them is a dominating set in the graph. The main contribution of this paper is a constructive proof of the existence of dominating pairs in connected AT-free graphs. The resulting simple algorithm, based on the well-known Lexicographic Breadth-First Search, can be implemented to run in time linear in the size of the input, whereas the best algorithm previously known for this problem has complexity $O(|V|^3)$ for input graph $G = (V, E)$. In addition, we indicate how our algorithm can be extended to find, in time linear in the size of the input, all dominating pairs in a connected AT-free graph with diameter greater than three. A remarkable feature of the extended algorithm is that, even though there may be $O(|V|^2)$ dominating pairs, the algorithm can compute and represent them in linear time.

1 Introduction

Considerable attention has been paid to exploiting algorithmically different aspects of the linear structure exhibited by various families of graphs. Examples of such families include interval graphs [16], permutation graphs [12], trapezoid graphs [6, 11], and cocomparability graphs [14].

The linearity of these four classes is usually described in terms of ad-hoc properties of each of these classes of graphs. For example, in the case of interval graphs, the linearity property is traditionally expressed in terms of a linear order on the set of maximal cliques [4, 5]. For permutation graphs the linear behaviour is explained in terms of the underlying partial order of dimension two [1], for cocomparability graphs the linear behaviour is expressed in terms of the well-known linear structure of comparability graphs [15], and so on.

As it turns out, the classes mentioned above are all subfamilies of a class of graphs called the asteroidal triple-free graphs (AT-free graphs, for short). An independent set of three vertices is called an *asteroidal triple* if between any pair in the triple there exists a path that avoids the neighbourhood of the third. AT-free graphs were introduced over three decades ago by Lekkerkerker and Boland [16] who showed that a graph is an interval graph if and only if it is chordal and AT-free. Thus, Lekkerkerker and Boland's result may be viewed as showing that the absence of asteroidal triples imposes the linear structure on chordal graphs that results in interval graphs. Cocomparability graphs were shown to be AT-free in [14]. Recently, the authors [7] have studied AT-free graphs with the stated goal of identifying the "agent" responsible for the linear behaviour observed in the four subfamilies. Specifically, in [7] we presented evidence that the property of being asteroidal triple-free is what is enforcing the linear behaviour of these classes.

One strong "certificate" of linearity is the existence of a *dominating pair* of vertices, that is, a pair of vertices with the property that every path connecting them is a dominating set. In [7], the authors gave an existential proof of the fact that every connected AT-free graph contains a dominating pair.

The main contribution of this paper is a constructive proof of the existence of dominating pairs in connected AT-free graphs. A remarkable feature of our approach is that the resulting simple algorithm, based on the well-known Lexicographic Breadth-First Search of [17], can easily be implemented to run in time $O(|V| + |E|)$ for input graph $G = (V, E)$. In addition, our algorithm can be extended to find, in time linear in the size of the input, all dominating pairs in a connected AT-free graph with diameter greater than three.

To put our result in perspective, we note that previously, the most efficient algorithm for finding a dominating pair in a graph $G = (V, E)$ was the straightforward $O(|V|^3)$ algorithm described in [2].

For each of the four families mentioned above, vertices that occupy the extreme positions in the corresponding intersection model [13] constitute a dominating pair. It is interesting to note, however, that a linear time algorithm for finding a dominating pair was not previously known, even for cocomparability graphs, a strict subclass of AT-free graphs.

The remainder of this paper is organized as follows: Sect. 2 contains some relevant terminology and background; Sect. 3 is a description of the Lexicographic Breadth-First Search algorithm of [17] along with some properties of that algorithm; in Sect. 4 we present an algorithm which finds a dominating pair in a connected AT-free graph; in Sect. 5, we show how the dominating pair algorithm can be extended to find all dominating pairs in a connected AT-free graph with sufficiently large diameter; and Sect. 6 contains our conclusions.

2 Background

All the graphs in this work are finite with no loops or multiple edges. In addition to standard graph theoretic terminology compatible with [3], we shall define some new terms. We let $d(v)$ denote the *degree* of vertex v; $d(u, v)$ denotes the *distance* between vertices u and v in a graph, that is, the number of edges on a shortest path joining u and v. In addition, we let $\text{diam}(G)$ denote the *diameter* of the graph G, that is, $\max_{u,v \in G} d(u, v)$. Two vertices u and v with $d(u, v) = \text{diam}(G)$ are said to *achieve the diameter*. Given a graph $G = (V, E)$ and a vertex x, we let $N(x)$ denote the set of neighbours of x; $N'(x)$ denotes the set of neighbours of x in the complement $\overline{G}$ of G.

All the paths in this work are assumed to be induced, unless stated otherwise. We refer to a path joining vertices x and y as an x, y-path. We say that a vertex u *intercepts a path* π if u is adjacent to at least one vertex on π; otherwise, u is said to *miss* π. Let $G = (V, E)$ be a graph, π a path in G, x a vertex of G, and X a subset of V. Let $V(\pi)$ be the vertices of G that are on the path π. We shall use the following notation: $\pi - x$ refers to the subgraph of G induced by the vertices $V(\pi) - \{x\}$, $\pi + x$ refers to the subgraph of G induced by the vertices $V(\pi) \cup \{x\}$, and $\pi \cup X$ refers to the subgraph of G induced by the vertices $V(\pi) \cup X$.

For a connected AT-free graph with a pair of vertices x, y we let $D(x, y)$ denote the set of vertices that intercept all x, y-paths. Note that (x, y) is a dominating pair if and only if $D(x, y) = V$. We say that vertices u and v are *unrelated with respect to* x if $u \notin D(v, x)$ and $v \notin D(u, x)$. A vertex x of an AT-free graph G is called *pokable* if the graph obtained from G by adding a pendant vertex adjacent to x is AT-free. It is not hard to see that if an AT-free graph G contains no unrelated vertices with respect to x, then x is pokable. A *pokable dominating pair* is a dominating pair such that both vertices are pokable. A vertex x is a *pokable dominating pair vertex* if x is pokable and there exists y such that (x, y) is a dominating pair. To illustrate these definitions, consider the graph $G = (V, E)$ of Fig. 1.

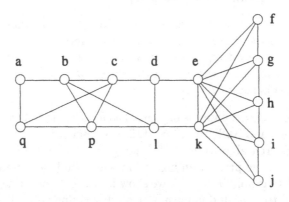

Fig. 1. A connected AT-free graph G

In this graph, $D(c, l) = \{b, c, d, k, l, p, q\}$, $D(c, e) = V \backslash \{a\}$, and $D(a, e) = D(q, i) = V$. Any pair consisting of one vertex from $\{a, q\}$ and one vertex from $\{e, f, g, h, i, j, k\}$ is a dominating pair; a is pokable and h is not pokable (adding a pendant vertex h' adjacent to h would create the AT $\{f, j, h'\}$).

3 Lexicographic Breadth-First Search

Our dominating pair algorithm invokes Procedure LBFS (short for Lexicographic Breadth-First Search) which, when given a connected graph G and a vertex x of G, returns a numbering of the vertices of G. We reproduce below the details of LBFS from [17].

Procedure LBFS(G, x);
{Input: a connected graph $G = (V, E)$ and a distinguished vertex x of G;
Output: a numbering σ of the vertices of G}
begin
  label$(x) \leftarrow |V|$;
  for each vertex v in $V - \{x\}$ **do**
  label$(v) \leftarrow \Lambda$;
  for $i \leftarrow |V|$ **downto** 1 **do begin**
    pick an unnumbered vertex v with (lexicographically) the largest label;
    $\sigma(v) \leftarrow i$; {assign to v number i}
    for each unnumbered vertex u in $N(v)$ **do**
      append i to label(u)
  end
end; {LBFS}

Notice that the numbering returned by LBFS is not unique. One numbering that could result from LBFS(G, q), where G is the graph of Fig. 1, is: $\sigma(q) = 14$, $\sigma(p) = 13$, $\sigma(c) = 12$, $\sigma(a) = 11$, $\sigma(b) = 10$, $\sigma(l) = 9$, $\sigma(d) = 8$, $\sigma(k) = 7$, $\sigma(e) = 6$, $\sigma(i) = 5$, $\sigma(h) = 4$, $\sigma(j) = 3$, $\sigma(g) = 2$, $\sigma(f) = 1$.

A few definitions relating to LBFS are in order at this point. Let x be an arbitrary vertex of a connected graph G and consider running LBFS(G, x). For vertices a, b of G we write $a \prec b$ whenever $\sigma(a) < \sigma(b)$ and we shall say that b is *larger* than a. To make the notation more manageable we shall write $v_1 \prec v_2 \prec \ldots \prec v_k$ as a shorthand for $v_1 \prec v_2$, $v_2 \prec v_3$, $\ldots$, $v_{k-1} \prec v_k$. We shall denote by $\lhd$ the lexicographic total order of the set of LBFS labels. We let $\lambda(a, b)$ denote the label of a when b was about to be numbered. Given vertices a, b, c with $a \prec c$ and $b \prec c$, we shall say that a and b *are tied at* c if $\lambda(a, c) = \lambda(b, c)$. Given a vertex y, an a, b-path is said to be y-*majorizing* if all the vertices on the path are larger than y.

We assume that $G = (V, E)$ is an arbitrary connected graph and LBFS(G, x) has been invoked, where x is an arbitrary vertex of G. We shall make use of the following properties of LBFS, which follow from the lexicographic ordering of labels.

Proposition 1. *Let a, b, c be vertices of G satisfying: $a \prec b$, $b \prec c$, $ac \in E$ and $bc \notin E$. Then, there exists a vertex d in G, adjacent to b but not to a and such that $c \prec d$.* □

Proposition 2. (Monotonicity Property) *Let a, b, c, d be vertices of G such that $a \prec c$ or $a = c$, $b \prec c$ or $b = c$, and $c \prec d$. If $\lambda(a,d) \lhd \lambda(b,d)$, then $\lambda(a,c) \lhd \lambda(b,c)$.* □

Lemma 3. *Let a, b, b', c be vertices of G such that $a \prec b \prec c \prec b'$, $bb' \in E$, and $b'c \notin E$. Then, a and c cannot be tied at b'.*

Proof. By Proposition 1 applied to vertices b, c, b' we find a vertex c' adjacent to c, but not to b, and such that $b' \prec c'$. Write $C = \{t \mid tc \in E, tb \notin E, b' \prec t\}$. Clearly, $c' \in C$. In fact, we select c' to be the *largest* vertex in C.

If the statement is false, then a and c are tied at b'. Since $b' \prec c'$ and since $cc' \in E$ we must have $ac' \in E$. Now, Proposition 1 applied to vertices a, b, c' yields a vertex b'' adjacent to b, but not to a, such that $c' \prec b''$.

Since $b' \prec b''$, the assumption that a and c are tied at b' guarantees that b'' is not adjacent to c. Therefore, Proposition 1 can be applied to vertices b, c, b'' yielding a vertex c'' adjacent to c, but not to b, and such that $b'' \prec c''$. Since $b' \prec c' \prec b'' \prec c''$, it must be that $c'' \in C$, contradicting that c' is the largest vertex in C. □

Lemma 4. *Let y, a, b be pairwise non-adjacent vertices of G such that $y \prec a$ and $a \prec b$. If a and y are not tied at b, then y misses a y-majorizing a,b-path.*

Proof. Assume that a and y are not tied at b. We must exhibit a y-majorizing a,b-path missed by y.

Since $y \prec a$, $\lambda(y,a) \lhd \lambda(a,a)$ or $\lambda(y,a) = \lambda(a,a)$. Therefore, by the Monotonicity Property (Proposition 2), $\lambda(y,b) \lhd \lambda(a,b)$ or $\lambda(y,b) = \lambda(a,b)$. Now, since a and y are not tied at b, $\lambda(y,b) \lhd \lambda(a,b)$. Consequently, we find a vertex a_1 adjacent to a, but not to y, and such that $b \prec a_1$. (Vertex a_1 is chosen to be the largest satisfying these conditions.) We may assume that a_1 is not adjacent to b for otherwise the path a, a_1, b is the desired y-majorizing path.

Now, Lemma 3 guarantees that b and y cannot be tied at a_1. Thus, we find a vertex b_1 adjacent to b, but not to y, and such that $a_1 \prec b_1$. (As before, we select as b_1 the largest vertex with this property.) Trivially, we may assume that b_1 is adjacent to neither a nor a_1, else we have the desired y-majorizing path. Again, Lemma 3 tells us that a_1 and y cannot be tied at b_1 and so we find a vertex a_2 adjacent to a_1, but not to y, and such $b_1 \prec a_2$. (As before, we select as a_2 the largest vertex with this property.) It is easy to verify that a_2 is not adjacent to a (by the choice of a_1), b, or b_1.

Continuing as above, we obtain two chordless y-majorizing paths $a = a_0, a_1, a_2, \ldots$ and $b = b_0, b_1, b_2, \ldots$, both missed by y. If no vertex on the first path is adjacent to a vertex on the second one, then the paths are infinite, contradicting that G is finite. Therefore, such an adjacency must exist, yielding the desired a,b-path. □

4 The Dominating Pair Algorithm

Our dominating pair algorithm takes as input a connected AT-free graph G and returns a pokable dominating pair of G. The algorithm provides a constructive proof of the existence of pokable dominating pairs in connected AT-free graphs (an existential proof of this fact was given in [7]).

The four properties of LBFS specified in the previous section hold for every connected graph G. The proof of correctness of the dominating pair algorithm relies on two additional properties of LBFS which hold when the input graph is a connected AT-free graph. We present these properties, with sketches of their proofs, next. The complete proofs can be found in [10].

Theorem 5. Let $G = (V, E)$ be a connected AT-free graph and let x and y be arbitrary vertices of G. Let $\prec$ be the vertex ordering corresponding to a numbering produced by LBFS(G, x). The subgraph of G induced by y and all vertices z with $y \prec z$, contains no unrelated vertices with respect to y.

Proof sketch. If the statement is false, we find a vertex y and vertices u, v with $y \prec u \prec v$, such that u and v are unrelated with respect to y. This implies the existence of chordless paths $\pi(y, u) : y = u_1, u_2, \ldots, u_p = u$ missed by v, and $\pi(y, v) : y = v_1, v_2, \ldots, v_q = v$ missed by u, with the vertices on both paths, except for y, numbered by LBFS before y.

Using Lemmas 3 and 4, we can prove that $u \prec v_2 \prec v_3$ and that u and y are tied at v_3.

Proposition 1 applied to vertices $y \prec u$ and $u \prec v_2$, guarantees the existence of a vertex u' adjacent to u, but not to y, and such that $v_2 \prec u'$. Since u and y are tied at v_3, it must be the case that $u' \prec v_3$. If u' is adjacent to v_3 then we have a u, v-path missed by y, contradicting that the graph is AT-free. Thus u' is not adjacent to v_3. But now, Lemma 3 guarantees that y and u' cannot be tied at v_3. Further, Lemma 4 tells us that there must exist a y-majorizing u', v_3-path missed by y. This path extends in the obvious way to a y-majorizing u, v-path missed by y, contradicting that the graph is AT-free. This completes the proof of Theorem 5. □

We observe that, if G contains no unrelated vertices with respect to vertex v then v is pokable. This observation and Theorem 5 combined imply that each vertex y of G is pokable in the subgraph of G induced by y and all vertices z with $y \prec z$. In particular, the last vertex numbered by LBFS(G, x) is pokable in G.

One additional theorem about LBFS, specialized to connected AT-free graphs, will lead to the dominating pair algorithm.

Theorem 6. Let $G = (V, E)$ be a connected AT-free graph and suppose that G contains no vertices unrelated with respect to vertex x of G. Let $\prec$ be a vertex ordering corresponding to a numbering produced by LBFS(G, x). Then, for all vertices u, v in V with $u \prec v$, $v \in D(u, x)$.

Proof sketch. Suppose the theorem is false. Let v be the largest vertex in V for which there exists a vertex u with $u \prec v$ and $v \notin D(u, x)$. We choose a vertex u with $u \prec v$, and a minimum length u, x-path $\pi : u = u_1, u_2, \ldots, u_k = x$ such that π is missed by v, $u_1 \prec u_2 \prec \ldots \prec u_k$, and, for all $1 < i \leq k$, u_i is as large as possible.

Observe that $u_1 \prec v \prec u_2$, for otherwise we contradict the choice of π. Now Proposition 1 guarantees the existence of a vertex v_2 adjacent to $v = v_1$, but not to u_1, and such that $u_2 \prec v_2$.

It is easily seen that v_2 is non-adjacent to u_i for all $i > 2$, for otherwise u and v are unrelated with respect to x. This immediately implies that $v_2 \prec u_3$ (else we contradict the choice of v) and $u_2 v_2 \in E$ (else we contradict the choice of both u and v).

Now apply Proposition 1 to vertices u_2, v_2, u_3; we find a vertex v_3 adjacent to v_2 but not to u_2 and such that $u_3 \prec v_3$.

Furthermore, it can be proved that, for any $i \geq 3$, Proposition 1 applied to vertices u_i, v_i, and u_{i+1} guarantees the existence of a vertex v_{i+1} adjacent to v_i but not to u_i and such that $u_{i+1} \prec v_{i+1}$. But this leads to a contradiction: v_k must exist and it must be that $x = u_k \prec v_k$, which is absurd. □

Theorem 6 implies that if G contains no vertices unrelated with respect to x, then (x, y) is a dominating pair in the subgraph of G induced by y and all vertices z with $y \prec z$. In particular, x and the last vertex numbered by LBFS(G, x) constitute a dominating pair of G.

We are now in a position to spell out the details of the dominating pair algorithm.

Procedure DP(G);
{Input: a connected AT-free graph G;
Output: (y, z) a pokable dominating pair of G}
begin
  choose an arbitrary vertex x of G;
  if $N'(x) = \emptyset$ **then** return (x, x);
  LBFS(G, x);
  let y be the vertex numbered last by LBFS(G, x);
  LBFS(G, y);
  let z be the vertex numbered last by LBFS(G, y);
  return(y, z)
end; {DP}

As an example, we refer again to the graph G of Fig. 1. We saw earlier that a possible numbering resulting from LBFS(G, q) corresponds to the ordering:

$$f \prec g \prec j \prec h \prec i \prec e \prec k \prec d \prec l \prec b \prec a \prec c \prec p \prec q.$$

LBFS(G, f) may produce the ordering:

$$a \prec q \prec b \prec p \prec c \prec l \prec d \prec j \prec i \prec h \prec g \prec k \prec e \prec f.$$

Thus, $DP(G)$ may output the pokable dominating pair (a, f).

Finally, we state the following result.

Theorem 7. *Procedure DP finds a pokable dominating pair in a connected AT-free graph, $G = (V, E)$, in $O(|V| + |E|)$ time.*

Proof. Clearly, (x, x) is a pokable dominating pair of G if $N'(x) = \emptyset$. Otherwise, by Theorem 5, G contains no unrelated vertices with respect to y and, hence, by Theorem 6, (y, z) is a dominating pair of G. In addition, Theorem 5 implies that both y and z are pokable in G. It is clear that a linear time implementation is possible (see [17] for details of a linear time implementation of LBFS). $\quad\square$

5 Computing All Dominating Pairs

Since dominating pairs play an important role in the study of AT-free graphs and, intuitively, correspond to the extreme endpoints of the linear structure of the graph, it is interesting to ask whether the above algorithm can be the basis of an efficient algorithm to find all of the dominating pairs in a connected AT-free graph. It turns out that we can indeed extend the algorithm to find all dominating pairs in a connected AT-free graph provided that the graph has diameter greater than three. We now give a brief overview of the extended algorithm. The proofs of the theorems and the details of the algorithm can be found in [10].

The algorithm relies on an extended variant of LBFS that, given a connected AT-free graph G and a pokable dominating pair vertex x of G, computes the sets $D(v, x)$ for all vertices v of G. (Recall that $D(v, x)$ denotes the set of vertices that intercept all v, x-paths.)

Theorem 8. *Let $G = (V, E)$ be a connected AT-free graph and let x be a pokable dominating pair vertex of G. Implicit representations for the sets $D(v, x)$, for every vertex v of G, can be computed in $O(|V| + |E|)$ time.* $\quad\square$

The next theorem indicates how the sets $D(v, x)$, for all vertices v of G, can be used to compute all the dominating pairs in a connected AT-free graph, G, with diameter greater than three.

Theorem 9. *Let G be a connected AT-free graph with $diam(G) > 3$. There exist nonempty, disjoint sets X and Y of vertices of G such that (x, y) is a dominating pair if and only if $x \in X$ and $y \in Y$ (or $x \in Y$ and $y \in X$).* $\quad\square$

We note that Theorem 9 is best possible in the sense that for AT-free graphs of diameter less than 4, the sets X and Y are not guaranteed to exist. To wit, C_5, and the graph of Fig. 2, provide counterexamples of diameter 2 and 3, respectively.

Now, the algorithm works as follows. The first step is to find a pokable dominating pair (x, z), which is done by LBFS in linear time. Then, $D(v, x)$ is computed, for all vertices v, in linear time. Now, Y is the set of all vertices y

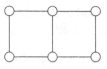

Fig. 2. An AT-free graph of diameter 3 for which the sets X and Y do not exist

with $D(y, x) = V$. (Clearly, $z \in Y$.) Finally, we compute $D(v, z)$, for all vertices v, and let X be the set of all vertices w with $D(w, z) = V$. (Clearly, $x \in X$.) We state the correctness theorem without proof.

Theorem 10. *For a connected AT-free graph $G = (V, E)$ with diam$(G) > 3$, sets X and Y, such that (x, y) is a dominating pair of G if and only if $x \in X$ and $y \in Y$ (or $x \in Y$ and $y \in X$), can be computed in $O(|V| + |E|)$ time.* □

For example, with the graph G of Fig. 1 as input, the extended algorithm returns $X = \{a, q\}$, $Y = \{e, f, g, h, i, j, k\}$ or $X = \{e, f, g, h, i, j, k\}$, $Y = \{a, q\}$ (depending upon the initial choice of pokable dominating pair (x, z)).

Let $G = (V, E)$ be a connected AT-free graph with diam$(G) > 3$. Notice that, even though there may be $O(|V|^2)$ dominating pairs in G, the algorithm of [10] can compute and represent them in linear time, by virtue of Theorem 9. A similar comment applies to the sets $D(v, x)$ for all $v \in V$; even though the sum of the cardinalities of the sets may be $O(|V|^2)$, the algorithm of [10] can compute an implicit representation of them in linear time.

We conclude with a corollary, which follows from the fact that some minimum cardinality connected dominating set must be a shortest path between the vertices of a dominating pair (proved in [9]). Once X and Y have been found, a minimum distance dominating pair can be found in linear time by performing a breadth-first search starting at X until a vertex of Y is encountered. In [8], the authors presented a linear time algorithm to compute a dominating path in an arbitrary connected AT-free graph, but that algorithm does not guarantee a minimum cardinality dominating path. The method of the present paper does guarantee a minimum cardinality dominating path for connected AT-free graphs with diameter greater than three.

Corollary 11. *Let $G = (V, E)$ be a connected AT-free graph with diameter greater than three. A minimum cardinality connected dominating set of G can be computed in $O(|V| + |E|)$ time.* □

6 Conclusions

We have presented a linear time algorithm, based on the well-known Lexicographic Breadth-First Search of [17], for finding a pokable dominating pair in a connected AT-free graph, $G = (V, E)$. The algorithm provides a constructive proof of the existence of pokable dominating pairs in connected AT-free graphs (an existential proof of this fact was given in [7]). It is an improvement over the previously known $O(|V|^3)$ algorithm of [2]. In addition, we extended the dominating pair algorithm to find all dominating pairs in a connected AT-free graph, $G = (V, E)$, with diameter greater than three. Even though there may be $O(|V|^2)$ dominating pairs, the extended algorithm can compute and implicitly represent them in $O(|V| + |E|)$ time. We remark that the simpler maximum cardinality search (MCS) of Tarjan and Yannakakis [18] cannot take the place of LBFS in these algorithms.

Acknowledgements

D. Corneil and L. Stewart wish to thank the Natural Sciences and Engineering Research Council of Canada for financial assistance. S. Olariu was supported, in part, by NSF grant CCR-9407180.

References

1. K. A. Baker, P. C. Fishburn, and F. S. Roberts, Partial orders of dimension two, *Networks*, 2 (1971), 11–28.
2. H. Balakrishnan, A. Rajaraman and C. Pandu Rangan, Connected domination and Steiner set on asteroidal triple-free graphs, *Proc. Workshop on Algorithms and Data Structures, WADS'93*, Montreal, Canada, August 1993, LNCS, Vol. 709, F. Dehne, J.-R. Sack, N. Santoro, S. Whitesides (Eds.), Springer-Verlag, Berlin, 1993, 131–141.
3. J. A. Bondy, U. S. R. Murty, *Graph Theory with Applications*, North-Holland, Amsterdam, 1976.
4. K. S. Booth and G. S. Lueker, Testing for the consecutive ones property, interval graphs and graph planarity using PQ-tree algorithms, *Journal of Comput. Syst. Sci.*, 13 (1976), 335–379.
5. K. S. Booth and G. S. Lueker, A linear time algorithm for deciding interval graph isomorphism, *Journal of the ACM*, 26 (1979), 183–195.
6. D.G. Corneil and P.A. Kamula, Extensions of permutation and interval graphs, *Proceedings 18th Southeastern Conference on Combinatorics, Graph Theory and Computing*, 1987, 267–276.
7. D.G. Corneil, S. Olariu and L. Stewart, Asteroidal triple-free graphs, *Proc. 19th International Workshop on Graph Theoretic Concepts in Computer Science, WG'93*, Utrecht, The Netherlands, June 1993, LNCS, Vol. 790, J. van Leeuwen (Ed.), Springer-Verlag, Berlin, 1994, 211–224.
8. D.G. Corneil, S. Olariu and L. Stewart, A linear time algorithm to compute a dominating path in an AT-free graph, *Information Processing Letters*, to appear.
9. D.G. Corneil, S. Olariu and L. Stewart, Asteroidal triple-free graphs, Technical Report TR-94-31, Department of Computer Science, Old Dominion University, November 1994, submitted for publication.
10. D.G. Corneil, S. Olariu and L. Stewart, Linear time algorithms for dominating pairs in asteroidal triple-free graphs, Technical Report 294/95, Department of Computer Science, University of Toronto, January 1995, submitted for publication.
11. I. Dagan, M.C. Golumbic and R.Y. Pinter, Trapezoid graphs and their coloring, *Discrete Applied Mathematics*, 21 (1988), 35–46.

12. S. Even, A. Pnueli and A. Lempel, Permutation graphs and transitive graphs, *Journal of the ACM*, 19 (1972), 400-410.
13. M.C. Golumbic. *Algorithmic Graph Theory and Perfect Graphs*. Academic Press, New York, 1980.
14. M.C. Golumbic, C.L. Monma and W.T. Trotter Jr., Tolerance graphs, *Discrete Applied Mathematics*, 9 (1984), 157–170.
15. D. Kratsch and L. Stewart, Domination on cocomparability graphs, *SIAM Journal on Discrete Mathematics*, 6 (1993), 400–417.
16. C.G. Lekkerkerker and J.C. Boland, Representation of a finite graph by a set of intervals on the real line, *Fundamenta Mathematicae*, 51 (1962), 45-64.
17. D.J. Rose, R.E. Tarjan, G.S. Lueker, Algorithmic aspects of vertex elimination on graphs, *SIAM Journal on Computing*, 5 (1976), 266-283.
18. R.E. Tarjan and M. Yannakakis, Simple linear-time algorithms to test chordality of graphs, test acyclicity of hypergraphs and selectively reduce acyclic hypergraphs, *SIAM Journal on Computing*, 13 (1984), 566-579.

On-line Resource Management with Applications to Routing and Scheduling *

Stefano Leonardi, Alberto Marchetti-Spaccamela

Dipartimento di Informatica e Sistemistica
Università di Roma "La Sapienza",
via Salaria 113, 00198-Roma, Italia.
e-mail: {leonardi,marchetti}@dis.uniroma1.it.

Abstract. We propose a general framework to model on-line resource management problems as linear programming problems. We consider both min cost problems and max benefit problems and propose logarithmic competitive algorithms that are optimal up to a constant factor.

The proposed framework provides a general methodology that applies to a wide class of on-line problems. Some of these on-line problems are studied for the first time in this work such as shop scheduling problems, packet routing and real and {0, 1} knapsack problems. It also allows to model some already studied on-line problems such as on-line multiprocessor scheduling and virtual circuit routing; in some cases the known results are improved.

1 Introduction

In on-line problems the input instance is presented to the algorithm in an on-line fashion: every time a new piece of the input instance is available to the algorithm the partial solution must be incrementally updated. Thus, the algorithm makes decisions based only on partial information about the input instance, the part that has been disclosed. On-line problems have received considerable attention in the last few years both for their theoretical appeal and their interest in practical applications [1,2,4,9, 12,15,16].

An *On-line resource management problem* is an on-line optimization problem defined by a sequence of jobs that are presented one at a time, each of which can be scheduled with several alternatives, i.e. using different kinds of resources. We consider two different classes of on-line resource management problems depending on the objective function.

1. *Cost problems*. The objective function requires to minimize the amount of resources necessary to handle all the jobs in a sequence.
2. *Benefit problems*. The objective function requires to maximize the benefit gained over the sequence of jobs that are processed without exceeding the available resources. Eventually, the request of processing a job could be rejected.

In this paper we propose a general methodology based on linear programming to deal with several on-line resource management problems both for the cost and the benefit version.

---
* This work was partly supported by ESPRIT BRA Alcom II under contract No.7141, by Italian Ministry of Scientific Research Project 40% "Algoritmi, Modelli di Calcolo e Strutture Informative" and by CNR project "Trasporti II".

Competitive analysis. The performance of an algorithm for an on-line problem is traditionally measured using *competitive analysis* [12,16]. Competitive analysis compares the solution obtained by an on-line algorithm with the optimal solution found by the *optimal off-line algorithm* that has complete knowledge of the problem instance.

Following the literature, an algorithm for an on-line cost problem is c-competitive if for all sequences of jobs the cost associated with the algorithm is no more than c times the cost associated with the optimal off-line algorithm, up to an additive constant factor. Similarly, we say that an algorithm for an on-line benefit problem is c-competitive if for all sequences of jobs the benefit obtained by the algorithm is no less than $1/c$ the benefit obtained by the optimal off-line algorithm, up to an additive constant factor. In the case of on-line randomized algorithms [7] it is considered the expected cost or benefit.

Previous work. Previous work on this kind of on-line problems regards multiprocessor scheduling, routing and financial problems. Two on-line cost problems have been studied in multiprocessor scheduling: minimization of the makespan of a scheduling of a set of jobs of possibly unknown duration time that can be also preempted [15] be preempted and on-line load balancing of the scheduling of jobs on a set of parallel machines [4,6].

In virtual circuit routing two nodes of a computer network that wish to exchange a large number of data packets at a certain transmission rate establish a path between themselves, called a *virtual circuit*, and route each of the packets along that path.

The cost version consists essentially in balancing the load (bandwidth requirement to support all the scheduled transmissions) on the links of the network [1]. Moreover, transmissions can have a limited duration or an unlimited duration, and preemption with rerouting [11] can be eventually allowed.

The on-line benefit problem in virtual circuit routing in a general network has been studied in [2] under the assumption that a communication requires only a small fraction of the maximum bandwidth available on any link. All these results can be presented as special cases of the class of problems studied in this paper. If a communication can ask a big fraction of the bandwidth the problem becomes much harder and it has been solved only for particular network topologies such as trees [3], meshes and hypercubes [5].

Finally we mention financial problems. In [9] the authors have considered the currency exchange problem; in this problem the goal is to determine an on-line strategy that allows to maximize the profit obtained by changing a given amount from one currency to another. It has been proved that in the case of a unidirectional problem (in which the only possible exchange is from one currency to another one) there is an optimal strategy that guarantees a logarithmic competitive ratio; if we consider the bidirectional case then it has been shown an exponential lower bound on the competitiveness of any strategy.

Results of the paper. The results of the paper can be summarized as follows.

1. In section 2 we model the benefit version of an on-line resource management problem as a "on-line linear programming problem". We consider both the case in which variables can assume real values and the 0-1 case. For all these problems we

present logarithmic competitive algorithms. We will also show that the bounds are tight up to a constant factor.

2. In section 3 we consider the cost version of a resource management problem and we propose a logarithmic-competitive algorithm that is optimal up to a constant factor.

3. The class of on-line benefit problems directly includes financial problems such as the unidirectional currency exchange problem [9] but also problems for which no competitive on-line algorithms were known such as the knapsack problem, the multidimensional knapsack, the multiple choice knapsack (we consider both the case in which variables are in [0, 1] and the 0-1 case).

4. In section 4.1 we consider a benefit version of the job shop scheduling problem. In this case a job is specified by a profit, a deadline and a set of operations that must be processed in a given order. Each operation can be performed only on a subset of machines and once started cannot be preempted. We obtain logarithmic upper bounds through a formulation of the problem as an integer linear programming problem. The running time of the algorithm that decides how to process the job does not depend on the total number of jobs but only on the current job. We also show a logarithmic lower bound on the competitiveness of any on-line algorithm.

5. In section 4.2 we propose on-line algorithms for on-line packet routing. We are interested in the cost problem of minimizing the maximum bandwidth requirements over any link that allow to route all the packets within a given deadline.

The techniques used in the paper resemble the ones previously used in on-line load balancing and virtual circuit routing [1,2,4,6]. In particular we use the idea of associating to the constraints a cost function that is exponential in its current occupancy. We remark that the obtained bounds are very close to the known bounds for specific problems even though they are very general and can be applied to a large class of problems. Namely in all cases they are far from the best known solution for a constant factor and in some cases they allow to improve the known bounds. As an example in section 4.3 we improve the bound obtained in [2] for the virtual circuit routing problem by a constant factor. The lower bounds that have been presented for specific problems can be extended to show that our algorithms are optimal [2,9].

In this extended abstract we will present the main ideas and a few proofs. The remaining proofs are in the full version of the paper.

2 An on-line methodology for benefit problems

An on-line resource management benefit problem is characterized by a set of distinct resources $r_i, i = 1, \ldots, m$, each of which is available for a maximum amount b_i. A sequence of jobs $\sigma = w_1, \ldots, w_n$, is presented to the algorithm. Each job w_j has benefit per unit p_j and it can be scheduled with k_j different alternatives. The r-th alternative for job w_j is defined by m values $a_{ij}^r, i = 1, \ldots, m$, where a_{ij}^r is a non negative real number indicating the amount of resource i necessary to schedule job w_j with alternative r.

A variable x_j^r is associated with the r-th alternative for job w_j. The on-line resource management problem requires to assign values to the variables x_j^r before the successive jobs in the sequence and their alternatives are known. The goal is to maximize the profit without violating the constraints on the resources. We can formalize the off-line version of the problem as follows:

$$max \sum_{j=1}^{n} p_j \sum_{r=1}^{k_j} x_j^r$$

$$s.t. \sum_{j=1}^{n} \sum_{r=1}^{k_j} a_{ij}^r x_j^r \le b_i, i = 1, \ldots, m$$

$$0 \le \sum_{r=1}^{k_j} x_j^r \le 1, j = 1, \ldots, n$$

We first study the problem in which the variables x_j^r assume real values in $[0, 1]$ and later the $\{0, 1\}$ case.

In the case of real variables, the above problem generalizes several financial problems. For instance we can model the on-line problem in which we need to define the size of a sequence of investments in such a way to not exceed the use of limited resources with the goal of obtaining the maximum profit from an economic activity. Furthermore, it is possible to model the on-line unidirectional currency exchange problem that has already been studied in [9] and other problems such as the on-line non-integer knapsack problem also with multiple choice, the on-line multidimensional knapsack problem, as well as general fractional packing problems. The case with integer variables models on-line scheduling and routing problems that we will present in section 4.

2.1 The real variables case

In this subsection we study the on-line methodology with real variables, namely $x_j^r \in [0, 1]$. We assume that each job can be globally scheduled for at most 1 unit summing all the associated variables. x_j^r indicates the percentage of job w_j scheduled with the r-th alternative.

The competitive ratio that we will achieve will depend on the p_j, a_{ij}^r and m values. For this reason we assume that the benefit of a single job has value in the range $[1, P]$, with $P \ge 1$, while for each non-zero coefficient a_{ij}^r the ratio a_{ij}^r/b_i has value in the range $[1/B, 1]$, with $B \ge 1$. It is straightforward to extend our algorithm to the case in which P and B are ratios between the maximum and the minimum value. In the following we assume that values B and P are known to the algorithm.

Let $\lambda_i(j) = \frac{1}{b_i} \sum_{k=1}^{j-1} \sum_{r=1}^{k_j} a_{ij}^r x_j^r$ be the fraction of the capacity of constraint i that has already been assigned when job w_j is presented. The algorithm associate to each constraint a cost using an exponential function of the current occupancy; namely, let $\mu = 2mBP$ and let $c_i(j) = \frac{1}{m} \mu^{\lambda_i(j)}$ be the "cost" of constraint i when the job w_j is presented. We say that the global cost of an alternative r is the weighted sum $C_r(j) = \sum_{i=1}^{m} \frac{a_{ij}^r}{b_i} c_i(j)$.

Informally, when job w_j is presented the algorithm initially assigns $x_j^r := 0$, $r = 1, \ldots, k_j$ and computes the global cost associated to each alternative. The basic step of the algorithm is as follows: 1) a fraction of the job at most equal to $\frac{1}{\log \mu}$ is added to one of the alternatives with minimum global cost if this cost is less or equal than the benefit p_j and this assignment does not violate the constraint $\sum_{r=1}^{k_j} x_j^r \leq 1$; 2) the cost of each constraint is updated. The basic step is repeated until all the alternatives have a global cost greater than p_j or the job is globally assigned for a value equal to 1. Hence, at the end of the iteration a job w_j will be scheduled with a possibly empty sequence of alternatives $r_j^1, \ldots, r_j^{s_j}$ with fractions $\overline{x}_j^{r_j^1}, \ldots, \overline{x}_j^{r_j^{s_j}}$.

More formally, let $\lambda_i^s(j) = \lambda_i(j) + \frac{1}{b_i} \sum_{l=1}^{s-1} a_{ij}^{r_j^l} \overline{x}_j^{r_j^l}$ be the fraction of resource r_i that has been assigned after that job w_j has been scheduled with alternative r_j^{s-1}, and let $c_i^s(j) = \frac{1}{m} \mu^{\lambda_i^s(j)}$. We denote with r_j^l one of the alternative with minimum global cost $C_{r_j^l}(j) = \sum_{i=1}^{m} \frac{a_{ij}^{r_j^l}}{b_i} c_i^l(j)$.

The values of variables x_j^r associated to job w_j are assigned with the following algorithm.

$x_j^r := 0, \ r = 1, \ldots, k_j$
$l := 1$
$\lambda_i^l(j) := \lambda_i(j)$
Let r_j^l be one of the alternatives with minimum global cost $C_{r_j^l}$
while $C_{r_j^l}(j) \leq p_j$ and $\sum_{r=1}^{k_j} x_j^r < 1$ do
    $\overline{x}_j^{r_j^l} = min(1 - \sum_{r=1}^{k_j} x_j^r, \frac{1}{\log \mu})$
    $x_j^{r_j^l} := x_j^{r_j^l} + \overline{x}_j^{r_j^l}$
    $\lambda_i^{l+1}(j) := \lambda_i^l(j) + \frac{a_{ij}^{r_j^l}}{b_i} \overline{x}_j^{r_j^l}$
    $l := l + 1$
    Let r_j^l be one of the alternatives with minimum global cost $C_{r_j^l}$
endwhile

Lemma 1. *The solution given by the algorithm is feasible.*

Proof. We need to check that constraints $\sum_{j=1}^{n} \sum_{r=1}^{k_j} a_{ij}^r x_j^r \leq b_i, i = 1, \ldots, m$, are satisfied by the algorithm. By contradiction, assume that for a job w_j the on-line algorithm has increased the fraction scheduled with alternative $r_j^l, l \in \{1, \ldots, k_j\}$ for at most $\frac{1}{\log \mu}$ and that the solution is not feasible. A solution is not feasible if for at least a constraint i the following condition holds before the variable associated to the alternative r_j^l is increased:

$$\lambda_i^l(j) > 1 - \frac{a_{ij}^{r_j^l}}{b_i} \overline{x}_j^{r_j^l} \geq 1 - \frac{1}{\log \mu}.$$

Then the global cost of the alternative r_j^l is at least:

$$\frac{a_{ij}^{r_j^l}}{b_i} c_i^l(j) = \frac{a_{ij}^{r_j^l}}{mb_i} \mu^{\lambda_i^l(j)} > \frac{1}{mB} \mu^{1 - \frac{1}{\log \mu}} = \frac{1}{mB} \frac{\mu}{2} = P \geq p_j.$$

A contradiction follows by noting that in this case the alternative r_j^l does not satisfy the condition of the while loop.

Denote by x_j^{rH} and $x_j^{r*}, r = 1, \ldots, k_j$, the values assigned to the variables associated with job w_j respectively by the on-line algorithm and by the optimal off-line algorithm. Let Q be the set of jobs for which the on-line algorithm has set all the variables to 0. Let $\mathcal{A} = \mathcal{A}' \cup \mathcal{A}''$ be the set of jobs for which the on-line algorithm has set at least one variable to a non-zero value. In particular, for each $j \in \mathcal{A}'$ it holds $\sum_{r=1}^{k_j} x_j^{rH} < 1$, while for each $j \in \mathcal{A}''$ it holds $\sum_{r=1}^{k_j} x_j^{rH} = 1$. Finally, let $c_i(n + 1)$ be the cost of the i-th constraint at the end of the sequence of n jobs.

Next lemma gives an upper bound on the benefit obtained from the optimal off-line algorithm.

Lemma 2. $\sum_{j \in Q \cup \mathcal{A}'} p_j \sum_{r=1}^{k_j} x_j^{r*} \leq 1 + \sum_{j \in \mathcal{A}} p_j$.

Proof. For any job w_j with $j \in Q \cup \mathcal{A}'$ and for any alternative r it holds $p_j < \sum_{i=1}^{m} \frac{a_{ij}^r}{b_i} c_i(j+1) \leq \sum_{i=1}^{m} \frac{a_{ij}^r}{b_i} c_i(n+1)$. The benefit obtained by the off-line algorithm on the set $Q \cup \mathcal{A}'$ of jobs is:

$$\sum_{j \in Q \cup \mathcal{A}'} p_j \sum_{r=1}^{k_j} x_j^{r*} < \sum_{j \in Q \cup \mathcal{A}'} \sum_{r=1}^{k_j} x_j^{r*} \sum_{i=1}^{m} \frac{a_{ij}^r}{b_i} c_i(n+1)$$

$$= \sum_{i=1}^{m} c_i(n+1) \sum_{j \in Q \cup \mathcal{A}'} \sum_{r=1}^{k_j} \frac{a_{ij}^r}{b_i} x_j^{r*} \leq \sum_{i=1}^{m} c_i(n+1).$$

We now prove that $\sum_{i=1}^{m} c_i(n+1) \leq 1 + \sum_{j \in \mathcal{A}} p_j$. Let $r_j^1, \ldots, r_j^{s_j}$, be the sequence of alternatives chosen for job w_j. We show that it is sufficient to prove for each $j \in \mathcal{A}$ and for each l, $l = 1, \ldots, s_j$, that $\sum_{i=1}^{m} [c_i^{l+1}(j) - c_i^l(j)] \leq p_j$. In fact, summing for each time a fraction of a job is scheduled, we get:

$$\sum_{j \in \mathcal{A}} \sum_{l=1}^{s_j} \sum_{i=1}^{m} [c_i^{l+1}(j) - c_i^l(j)] = \sum_{i=1}^{m} \sum_{j \in \mathcal{A}} \sum_{l=1}^{s_j} [c_i^{l+1}(j) - c_i^l(j)] =$$

$$\sum_{i=1}^{m} [c_i(n+1) - c_i(1)] = \sum_{i=1}^{m} [c_i(n+1) - \frac{1}{m}],$$

that implies $\sum_{i=1}^{m} c_i(n+1) \leq 1 + \sum_{j \in \mathcal{A}} p_j$. To prove the lemma observe that

$$c_i^{l+1}(j) - c_i^l(j) \leq \frac{1}{m} (\mu^{\lambda_i^l(j) + \frac{a_{ij}^{r_j^l}}{b_i} \frac{1}{\log \mu}} - \mu^{\lambda_i^l(j)}) = \frac{\mu^{\lambda_i^l(j)}}{m} (2^{\frac{a_{ij}^{r_j^l}}{b_i}} - 1).$$

Since $2^x - 1 \leq x$, if $0 \leq x \leq 1$, we obtain

$$c_i^{l+1}(j) - c_i^l(j) \leq \frac{a_{ij}^{r_j^l}}{b_i m} \mu^{\lambda_i^l(j)} = \frac{a_{ij}^{r_j^l}}{b_i} c_i^l(j).$$

Summing for each constraint, we get:

$$\sum_{i=1}^{m} [c_i^{l+1}(j) - c_i^l(j)] \leq \sum_{i=1}^{m} \frac{a_{ij}^{r_j^l}}{b_i} c_i^l(j) \leq p_j.$$

Theorem 2.1 *The algorithm for on-line resource management benefit problems with real variables is $(1 + log\ \mu)$-competitive.*

Proof. Denote by P and P^* respectively the benefit obtained by the on-line algorithm and by the optimal off-line algorithm. The on-line algorithm assigns $\sum_{r=1}^{k_j} x_j^{rH} \geq \frac{1}{log\mu}$ for each $j \in A$, since $\mu = 2mPB$ implies $\frac{1}{log\ \mu} \leq 1$. Then $P \geq \frac{1}{log\mu} \sum_{j \in A} p_j$. By lemma 2 the benefit obtained by the optimal off-line algorithm is

$$P^* = \sum_{j \in Q \cup A'} p_j \sum_{r=1}^{k_j} x_j^{r*} + \sum_{j \in A''} p_j \sum_{r=1}^{k_j} x_j^{r*} < 1 + \sum_{j \in A} p_j + \sum_{j \in A''} p_j$$

$$= 1 + P\ log\mu + P = 1 + (1 + log\ \mu)P,$$

since the on-line algorithm assigns $\sum_{r=1}^{k_j} x_j^{rH} = 1$ for each $j \in A''$.

The following theorem proves that the algorithm given above is optimal up to a constant factor.

Theorem 2.2 *Any deterministic or randomized algorithm for on-line resource management benefit problems with real variables has competitive ratio $\Omega(log(mPB))$.*

If the coefficients a_{ij}^r can assume negative values, then the competitive ratio depends exponentially on the number of variables with negative coefficients.

Theorem 2.3 *If k variables assume negative coefficients then the competitive ratio of any on-line algorithm is $\Omega((\ln P)^k)$ where P is the ratio between the maximum and the minimum value of the profit p_j.*

2.2 The integer variables case

In this section we propose an algorithm for on-line resource management benefit problems with integer variables, namely $x_j^r \in \{0, 1\}$ for each alternative associated to job w_j. Therefore for each job at most one variable can be set to 1. Notice that the off-line optimization problem is NP-hard.

In the following we pose the following restriction on the maximum amount of a single resource that can be asked from any alternative to schedule a job:

$$\frac{a_{ij}^r}{b_i} \leq \frac{1}{log\mu} \tag{1}$$

Without this restriction as we will show in the following that it is not possible to give deterministic on-line algorithms with polylogarithmic competitive ratio.

In the previous section we have studied the underlying relaxation linear program. The use of techniques introduced in [13] allow to prove probabilistically that there exists an on-line solution for the integer linear program close to the relaxation solution, and that such probabilistic existence can be converted in a deterministic algorithm. Under restriction (1) the solution given by the constructed deterministic algorithm is far from the relaxation solution for at most a constant factor.

A better result is obtained using the following simple algorithm. We use the terminology of the previous subsection to define $C_r(j)$ as the global cost of the alternative r for the job w_j. Let r_j be one of the alternatives for which the global cost $C_{r_j}(j)$ is minimum. The algorithm for job w_j is as follows:

$x_j^r := 0, r = 1, \ldots, k_j$

$If\ C_{r_j}(j) \le p_j$

$then\ x_j^{r_j} = 1$

Theorem 2.4 *The algorithm for on-line resource management benefit problems with integer variables is $(1 + log\mu)$-competitive and $(2 + log\mu)$-strictly competitive.*

In the following we consider the case in which the on-line algorithm has more resources than the off-line algorithm by a logarithmic factor, i.e. the size of the i-th constraint for the on-line algorithm is $b_i log\mu$, while it is b_i for the off-line algorithm.

Theorem 2.5 *If the on-line algorithm has more resources than the off-line algorithm by a logarithmic factor then the competitive ratio is 2 and the strict competitive ratio is 3.*

The on-line virtual circuit routing problem in which the goal is that of maximizing the benefit obtained from scheduling a sequence of communication requests can be formulated within the above framework. Then, lower bounds given in [2] can be extended to the on-line methodology for benefit problems with integer variables. The following theorem shows that, if restriction (1) holds, then the algorithm is optimal up to a constant factor; it also gives an improved lower bounds when restriction (1) does not hold:

Theorem 2.6 *Any deterministic or randomized algorithm for the on-line resource management benefit problem with integer variables under restriction (1) has competitive ratio $\Omega(log\,(mPB))$.*

If a single job can ask a fraction $1/k$ of a resource then the competitive ratio of any algorithm is $\Omega(m^{1/k} + P^{1/k} + B^{1/k})$.

Then, in the general case it is not possible to give competitive algorithms with polylogarithmic competitive ratio.

An important open problem is to determine whether randomized on-line algorithms can perform better. If we assume $P = 1$ and $B = 1$, the lower bound on the competitive ratio of randomized algorithms is $log\,m$ [3] while there is a $\sqrt{m}$-competitive randomized algorithm [10].

Moreover, observe that maximum independent set can be represented within the proposed framework for benefit problems assuming that the number of constraints m is the number of edges in the graph. Since $m = O(n^2)$, with n number of nodes in the graph, the achievement of a $(\sqrt{m})^{1-\epsilon}$ polynomial time competitive randomized algorithm improves the performance of approximate algorithms for maximum independent set (current best algorithm has $n/logn$ approximation ratio [8]).

3 An on-line methodology for cost problems

The previous section introduced an on-line methodology based on linear programming for resource management benefit problems. In this section we study cost problems, in which all the jobs in the sequence have to be scheduled with one among the possible alternatives in order to minimize the maximum amount of each single resource at the end of the sequence.

Let $\{a_{ij}\}, i = 1, \ldots, m$, be the resources associated with the alternative chosen by the on-line algorithm to schedule the job w_j and let $\{a_{ij}^*\}$ be the resources associated with the alternative chosen by the off-line algorithm. Let $b_i(j) = \sum_{k=1}^{j} a_{ik}$ and $b_i^*(j) = \sum_{k=1}^{j} a_{ik}^*$ be the amount of resource r_i needed to the on-line and to the optimal off-line algorithm to schedule the first j jobs. Let $B = max_{i \in 1, \ldots, m} b_i(n)$ and $B^* = max_{i \in 1, \ldots, m} b_i^*(n)$ be the maximum amount of a single resource needed at the end of the sequence respectively by the on-line and by the optimal off-line algorithm.

The algorithm is based on a technique given in [1] for on-line load balancing in virtual circuit routing. In that case the bandwidth on any link is fixed and the goal is that of minimizing the maximum fraction of assigned bandwidth on any link. In our case the goal is instead that of minimizing the maximum amount of a single resource necessary to schedule all the jobs in the sequence. Then the capacity of a single resource is not fixed but it is the function to minimize.

Assume that the on-line algorithm using the doubling technique of [1] knows Λ such that $B^* \leq \Lambda \leq 2B^*$. Moreover, let $\bar{b}_i = \frac{b_i}{\Lambda}$ and $\tilde{a}_{i,j} = \frac{a_{i,j}}{\Lambda}$.

When the j-th job is presented, the on-line algorithm chooses that alternative with coefficients $\{a_{ij}\}$ that minimize

$$\sum_{i=1}^{m} (\alpha^{\bar{b}_i(j) + \tilde{a}_{ij}} - \alpha^{\bar{b}_i(j)}),$$

where α is a suitable constant.

Theorem 3.1 *The algorithm for on-line resource management cost problems is log m-competitive.*

We notice with no proof that a $\Omega(\log m)$ lower bound can be derived from [6]. Therefore, the algorithm proposed in this section is optimal up to a constant factor.

4 On-line scheduling and routing problems

In this section we apply the methodologies developed in the previous sections to model several scheduling and routing problems such as job-shop scheduling, packet routing, multiprocessor scheduling, virtual circuit routing.

Some on-line problems have already been studied such as virtual circuit routing and multiprocessor scheduling, and our methodology allow to found again known results with slight improvements (see subsection 4.3).

4.1 On-line job shop scheduling

In the *job shop scheduling* problem we are given m types of machines $M_1, \ldots, M_m$, and a sequence of n jobs $w_1, \ldots, w_n$. Job w_j consists of a set of λ_j operations $O_{j1}, \ldots, O_{j\lambda_j}$, that have to be processed in the order specified by the sequence. Operation O_{jk} must be processed on a machine of kind M_{jk} requiring processing time t_{jk}. For the sake of simplicity we study a discrete version of the problem in which the processing times are integer values and each operation is scheduled only starting at an integer instant of time. Then, for any integer value t, any machine can process only one job from time t to time $t + 1$. Moreover, once an operation is started it cannot be interrupted.

In the benefit problem we assume that there are b_i machines of kind M_i. A job w_j must be completed within a deadline $d_j \in [1, D]$ that clearly is not less than the sum of the single operation times required to process the job. Moreover, assume that any operation O_{jk} requires processing time $t_{jk} \in [1, T]$. We associate a unit of benefit to each job and the goal of the on-line benefit algorithm is that of maximizing the number of scheduled jobs in the sequence without assigning more than b_i jobs at same time on the same type of machine M_i. In the corresponding on-line cost problem the goal is that of minimizing the number of machines of the same type necessary to schedule all the jobs in the sequence.

Assume that t is the current instant of time. Any job can be scheduled starting from time t and ending at most at time $t + D$. We model the on-line job shop scheduling as an on-line resource management problem as follows:

- The number of constraints is mD. Constraint $c_{ik}, k = 0, \ldots, D-1$, represents the use of machines of type M_i for one unit of time. At time $t + 1$, the m constraints that represent the use of machines between t and $t + 1$ are turned to represent the use of machines between $t + D$ and $t + D + 1$.
- Each job can be scheduled with one of k_j distinct alternatives. The r-th alternative for job w_j with deadline d_j has $a^r_{c_{ik}j} = 1$ if the job w_j is scheduled on a machine of kind M_i for the unit of time represented by constraint c_{ik}, $a^r_{c_{ik}j} = 0$ otherwise.

From the formulation as a on-line resource management problem follows the following theorem that claims the existence of competitive algorithms for on-line job-shop scheduling.

Theorem 4.1 *There is a $O(\log mD)$ competitive algorithm for the benefit version under restriction (1) and the cost version of on-line job shop scheduling.*

Any algorithm for the benefit version of the on-line job shop scheduling problem is $\Omega(\log mT)$-competitive.

Other shop scheduling problems can be modeled with our methodology. For instance on-line *flow shop scheduling* problems in which the operations that form a job must be processed only in a given partial order rather than in a linear order.

4.2 On-line packet routing

A special interesting case of job shop scheduling is *packet routing*. Assume that a packet has to be delivered from a source node to a destination node and that it is required one unit of time to cross each link. A sequence of operations is a sequence of links that form a path from the source node to the destination node. Each alternative for a packet is a possible path from the source node to the destination node together with the schedule at which each link in the path is crossed. The on-line cost version of packet routing we consider consists of minimizing the maximum bandwidth on each link that is necessary to deliver all the packets within a maximum delay. The following result is proved in the full version of the paper:

Theorem 4.2 *Any algorithm for the on-line packet routing problem is $\Omega(\log m)$-competitive.*

Analogously to the case of job shop scheduling problems if we assume that the maximum deadline is of the order of m, then the algorithm proposed for on-line packet routing is optimal up to a constant factor.

4.3 Virtual circuit routing

We have already mentioned that on-line virtual circuit routing for general networks with limited durations can be modeled within the proposed framework. A request asks for the establishment of a communication between two nodes in the network at a certain transmission rate r_j for a limited time t_j. Each link has a bandwidth b_i. If the communication w_j is scheduled with a certain path connecting the two nodes, then the occupied bandwidth is increased by r_j over each link in the path, and a benefit p_j is gained by the algorithm. Assume that the benefit p_j ranges in the interval $[1, P]$, the ratio $\frac{r_i}{b_i}$ in the range $[1/B, 1]$ and the duration t_j is an integer value in the range $[1, T]$.

The reduction for benefit problems with integer variables consider mT constraints, where m is the number of links in the network. Each constraint represents the use of a link for one unit of time in a way similar to that shown for job shop scheduling. Then the algorithm given in section 2.2 is $(2 + log\mu)$-competitive, with $\mu = 2mPB$. This is slightly better than the $2(1 + \log\mu)$ bound of [2].

Moreover, we can also model the on-line load balancing problem in multiprocessor scheduling with unrelated machines and virtual circuit routing, that have already been studied in [1]. Similar results are found again.

5 Conclusions and open problems

In this paper we have presented a framework for on-line resource management problems that includes many problems. We have proposed logarithmic upper bounds optimal up to a constant factor. Many problems are still open. The first open problem is to show whether randomization can help in finding polylogarithmic upper bounds for benefit problems in the general case (i.e. when a job can require also a big fraction of a single resource, namely restriction (1) of section 2.2 is removed).

A related problem is to investigate under which structure of constraints it is possible to provide polylogarithmic randomized algorithms. For instance, particular topologies for routing problems as it has been done in [3] and [5].

A third question is to relax some of the constraints that we assume. Namely, it would be interesting to consider the case in which the parameters are unknown to the on-line algorithm and/or preemption and rescheduling are allowed.

Acknowledgments. A special thank is due to Yossi Azar and Amos Fiat. We would like also to thank Yair Bartal, Anna Karlin and Adi Rosén for useful discussions.

References

1. J. Aspnes, Y. Azar, A. Fiat, S. Plotkin, O. Waarts, "On-line load balancing with applications to machine scheduling and virtual circuit routing", *Proceedings of the 23rd Annual ACM Symposium on Theory of Computing*, 1993.
2. B. Awerbuch, Y. Azar, and S. Plotkin, "Throughput competitive on-line routing", In *Proc. of the 34th Ann. Symp. on Foundations of Computer Science*, 1993.
3. B. Awerbuch, Y. Bartal, A. Fiat, and A. Rosén, " Competitive non-preemptive call control", In *Proc. of the 5th ACM-SIAM Symp. on Discrete Algorithms*, 1994.
4. Y. Azar, A.Z. Broder, A.R. Karlin, "On-line load-balancing", *Proceedings of the 33rd Annual Symposium on Foundations of Computer Science*, pp. 218-225, 1992.
5. B. Awerbuch, R. Gawlick, F.T. Leighton, and Y. Rabani, "On-line admission control and circuit routing for high performance computing and communication", *Proceedings of the 35th Ann. Symp. on Foundations of Computer Science*, 1994.
6. Y. Azar, J. Naor, R. Rom, "The competitiveness of on-line assignments", *Proc. of the 3rd ACM-SIAM Symposium on Discrete Algorithms*, pp. 203-210, 1992.
7. S. Ben-David, A. Borodin, R.M. Karp, G. Tardos and A. Widgerson, "On the power of randomization in on-line algorithms", *Proc. of the 22nd Annual ACM Symposium on Theory of Computing*, pp. 379-386, 1990.
8. R.B. Boppana, M.M. Halldorson, "Approximating maximum independent set by excluding subgraphs", *BIT* 32(2), pp.180-196, 1992.
9. R. El-Yaniv, A. Fiat, R.M. Karp and G. Turpin, "Competitive analysis of financial games", Proc. of the 33rd Ann. Symp. on Foundations of Computer Science, 1992.
10. A. Fiat, personal communication, 1994.
11. J. Garay, I. Gopal, S. Kutten, Y. Mansour, M. Yung, "Efficient on-line call-control algorithms", *Proc. of the 2nd Annual Israel Conference on Theory of Computing and Systems*, 1993.
12. M.S. Manasse, L.A. McGeoch and D. Sleator, Competitive algorithms for on-line problems, *Proc. 20th ACM Symp. on Theory of Computing*, pp. 322-333, 1988.
13. P. Raghavan, "Probabilistic construction of deterministic algorithms: approximating packing integer programs", *Journal of Computer and System Science* 37, pp. 130-143, 1988.
14. D.B. Shmoys, C. Stein, J. Wein, "Improved approximation algorithms for shop scheduling problems", *Proceedings of the 2nd ACM-SIAM Symposium on Discrete Algorithms*, pp. 148-157, 1991.
15. D.B. Shmoys, J. Wein, D.P. Williamson, "Scheduling parallel machines on-line", *Proc. of the 32nd Annual Symposium on Foundations of Computer Science*, 1991.
16. D. Sleator and R.E. Tarjan, Amortized efficiency of list update and paging algorithms, *Communications of ACM* 28, pp. 202-208 (1985).

Alternation in Simple Devices *

H. Petersen [†]

Institut für Informatik der Universität Stuttgart

Breitwiesenstraße 20–22, D-70565 Stuttgart

e-mail: petersen@informatik.uni-stuttgart.de

Abstract

We show that emptiness is undecidable for alternating one-way two-head finite automata operating on unary input. This solves an open problem posed by Geidmanis. Further we show that a conjecture by King concerning the hierarchy of languages accepted by alternating one-way multihead finite automata does not hold. We also consider closure properties of the languages accepted by these devices and obtain as consequences that the Boolean closures of linear and general context-free languages are contained in the lower levels of the hierarchy. Some other simulation techniques are outlined.

1 Introduction

In this paper we will study alternation in a very restricted class of machines, alternating one-way multihead finite automata which were introduced and investigated by King [5]. For the formal definitions we refer to [5] and only recall that such automata move a finite number of heads on a read-only input tape which is bordered at the right by an endmarker. The finite control may contain universal states. If the automaton reaches such a state it accepts only if all successor configurations are accepting. We will denote the class of languages accepted by alternating one-way k-head finite automata by **1AFA(k)**. If we consider decision problems for these classes we tacitly assume that the languages are given by automata of the corresponding class and will denote the classes of automata by the abbreviation 1AFA(k).

One of the questions arising in the investigation of these devices is whether simple automata benefit from the ability to use alternation. We can show that the emptiness problem even for 1AFA(2) with unary input alphabets is undecidable. This problem is decidable for nondeterministic one-way multihead automata with unary alphabets and undecidable for deterministic two-head one-way finite automata with a binary input alphabet.

---

*This work was done at the University of Hamburg

[†]Supported in part by ESPRIT Basic Research Action WG 6317: Algebraic and Syntactic Methods in Computer Science (ASMICS 2)

Then we will show that a conjecture by King fails stating that a certain language over a unary alphabet is not contained in **1AFA(2)**. In the development we will exhibit some techniques for "programming" a 1AFA(2) that seem interesting in their own right.

An important topic concerning classes of formal languages are closure properties. We observe here that the classes **1AFA(k)** are closed under complementation. Combining this property with inclusions previously shown we obtain rich classes of languages already contained in the lower levels of the hierarchy of languages accepted by these automata.

In the last section we describe a translational technique for these automata showing that a large number of heads can be simulated on a suitably modified input by a fixed and rather small number of heads.

A few remarks on notation: The automata receive an input w with a right endmarker. Therefore the input tape has $|w| + 1$ squares and h_i, the position of head i, is a number from 0 (initial position) to $|w|$. Often we will not use h_i but $d_i = |w| - h_i$, the *distance* of head i to the endmarker. Note that the initial distance of all heads is $|w|$.

2 Results

A useful property of 1AFA(k) has been shown in [5] and will be used throughout the paper.

Lemma 1 *A 1AFA(k) which can detect coincidence of heads (the heads are sensing) can be simulated by a 1AFA(k) lacking this feature.*

It is known that languages accepted by 1AFA(1) and 2AFA(1) are regular. King has shown that **1AFA(2)** contains the nonregular unary language $L_1 = \{0^{2^n} \mid n \geq 2\}$ while all unary languages accepted even by nondeterministic multihead pushdown automata are regular [4]. In those cases where the languages accepted are regular effective transformations of the devices into finite automata are given in the proofs and emptiness therefore is decidable. It is also known that emptiness is undecidable for deterministic one-way two-head automata over alphabets with at least two symbols, a result due to Greibach [7].

Geidmanis considered the emptiness problem of 1AFA(k) over unary alphabets and gave a proof that this problem is undecidable for $k \geq 3$ [2, 3]. He posed as an open problem whether or not the emptiness problem for 1AFA(2) operating on unary input is decidable. Here we present the negative solution to this question which has a surprisingly elementary proof.

Theorem 1 *The emptiness problem for 1AFA(2) operating on unary input is undecidable.*

Proof: We describe a reduction of the halting problem for two-counter automata which are started with the input on one of their counters (the other counter is empty). By the universality of such machines [6] this problem is undecidable and undecidability of the given problem follows.

Let a two-counter automaton M and an input i be given. A 1AFA(2) A simulating M on i first universally checks that its input-length is divisible by 2^i, not divisible by 2^{i+1} and 3. The input length n can therefore be written as

$$n = 2^i 3^0 5^s r$$

where $r \geq 1$ is not divisible by 2, 3 or 5. Now A starts a simulation of M. The factorization of the distance of the first head from the endmarker encodes both counters, where the exponent of 2 represents the first counter and the exponent of 3 the second. If M tests a counter for zero A universally branches to verify that the head distance is not divisible by 2 or 3. If M decrements one of the counters A uses its second head to divide d_1 by 2 resp. 3. Both heads can be easily aligned afterwards by Lemma 1. If finally M increments a counter A multiplies the number stored by its first head by $\frac{2}{5}$ resp. $\frac{3}{5}$. The distances of A's heads are d_1 and d_2. Then $d_2 = \frac{2}{5}d_1$ is equivalent to $\frac{d_2}{2} = \frac{d_1}{5}$. This condition can be universally verified by moving the second head two squares while moving the first head five times. The operation transforms $d_1 = 2^a 3^b 5^c r'$ into $d_1' = 2^{a+1} 3^b 5^{c-1} r'$. The second counter of M is simulated similarly.

If A is able to complete the simulation of M it accepts its input, otherwise it rejects. If M halts with input i then A accepts every input of sufficient length and satisfying the conditions on divisibility, hence the language accepted is not empty (and even infinite due to the factor r). Conversely if M does not halt there will be no input which A accepts. The initial exponent s of 5 gives an upper bound on the number of increment operations that can be performed in the course of the simulation. Since it can be arbitrarily large a halting computation can be completed for sufficiently large s. □

On unary input a 1AFA(2) can be simulated by an alternating one-counter automaton. The latter automaton guesses the input length on its counter and universally verifies it. Then it simulates the 1AFA(2) in an obvious way. This shows

Corollary 1 *The emptiness problem for alternating one-counter automata operating on unary input is undecidable even if the counter makes at most one reversal on every path of the computation.*

By the usual arguments and the effective closure of 1AFA(2) under complementation (see below) we get

Corollary 2 *Equivalence, finiteness, inclusion, disjointness, and universe problem are undecidable for 1AFA(2) as well as for alternating one-counter automata operating on unary input.*

Similar results hold for alternating one-way two-tape automata.

One of the main open problems in the theory of alternating finite automata is whether $k + 1$ heads are better than k for one-way devices, i.e. the question whether **1AFA(k)** $\neq$ **1AFA(k+1)** for $k \geq 2$ (for $k = 1$ the inequality is true because of L_1). For deterministic and nondeterministic automata this problem has been solved in [9], even for automata with sensing heads that can detect each others.

King [5, p. 161] defined the following family of languages:

$$L_k = \left\{ 0^{2^{2^{\cdot^{\cdot^{\cdot^{2^n}}}}}} \mid n \geq 1 \right\}$$

where k is the number of 2's in the exponent. King conjectured that L_2 is not in **1AFA(2)**. This would separate the second and third level of the hierarchy since $L_k \in \mathbf{1AFA(3)}$ for all k. We can however show that $L_2 \in \mathbf{1AFA(2)}$.

In the following we will be interested in distances of head-positions and introduce two values ℓ and r. The distance of the trailing head to the leading head of a 1AFA(2) will be denoted by $\ell = h_1 - h_2 = d_2 - d_1$. The distance d_1 of the leading head to the (right) endmarker will be denoted by r. We write bin(n) for the binary representation of n without leading zeros.

Lemma 2 *A 1AFA(2) can compare ℓ and r.*

Proof: The 1AFA(2) moves the trailing head twice as fast as the leading head towards the endmarker. Depending on the head that reaches the endmarker first it determines the relation between ℓ and r. ☐

By adapting the "speed" of the heads it is also possible to compare multiples of ℓ and r.

Lemma 3 *A 1AFA(2) can check bin(r) for membership in a fixed regular set.*

Proof: The 1AFA(2) moves the trailing head onto the position of the leading head (Lemma 1), guesses if r is even or odd, if it guessed odd advances both heads, moves one of the heads to the central square (which is universally verified by Lemma 1) and repeats this process with the new r until all relevant digits have been processed. All guesses are fed into a finite automaton simulated in the finite control of the 1AFA(2). ☐

Lemma 4 *If $r \geq \ell$ then a 1AFA(2) can check bin(ℓ) for membership in a fixed regular set.*

Proof: Move both heads in parallel until $\ell = r$ (Lemma 2), then check r (Lemma 3). ☐

Theorem 2 *The language L_2 is in $\mathbf{1AFA(2)}$.*

Proof: The 1AFA(2) accepting L_2 first verifies that the input is of the form 0^{2^m} for some $m \geq 2$, if $m = 2$ it immediately accepts. Then it moves one head to such a position that $\ell \leq r$ and bin$(r) \in 100^*1^*$ and advances its second head to the same position. Then it repeats the following steps:

- If $\text{bin}(r) = 1011$ then accept.

- If $\text{bin}(r) \notin 10\{1,0\}^* \setminus 10^* = 10\{1,0\}^*1\{1,0\}^*$ then reject.

- Move one of the heads to a position such that the new r' and ℓ' satisfy $\text{bin}(\ell') \in 10^*$ (ℓ' is a power of 2) and $\ell' \leq r' \leq 2\ell'$.

- Move the trailing to the position of the leading head and then both one step.

All of the operations on the heads are possible by the preceding lemmas.

The above algorithm transforms a distance $2^p + q$ (where $q < 2^{p-1}$) into $2^{p-1} + (q-1)$. Since it starts with $2^{m-1} + 2^r - 1$ and terminates with $2^3 + 3$ it correctly verifies that $m = 2^r$ for some $r \geq 1$. $\qquad\square$

By using the algorithm recursively it is possible to show that $L_k \in \mathbf{1AFA(2)}$ for every k.

Next we will consider closure properties of the classes $\mathbf{1AFA(k)}$. It has been remarked in [1] that the complements of languages accepted by nondeterministic one-way k-head finite automata are in $\mathbf{1AFA(k)}$ (in fact the 1AFA(k) possess universal states only). Here we generalize this statement.

Lemma 5 *For every 1AFA(k) A there is an equivalent automaton A' that has no infinite computation tree. Further all leaves of computation trees are configurations with all heads on the endmarker.*

Proof: Since the movement of heads is unidirectional the only way a path in the computation tree extends to infinity is by repeating a state in the finite control while the heads remain stationary. We will show how to rule out such behaviour.

Let A have the set of states S and introduce as states for A' all pairs from $S \times \{1, \ldots |S|\}$. Whenever there is a transition in A from state s_1 to s_2 moving a head we let all transitions from (s_1, i) to $(s_2, 1)$ for $1 \leq i \leq |S|$ be transitions of A'. If there is a transition from s_1 to s_2 which does not move a head there are transitions from (s_1, i) to $(s_2, i+1)$ for $1 \leq i \leq |S| - 1$ in A'. We will indicate why A and A' are equivalent. Acceptance of 1AFA(k) is defined using a finite accepting computation tree and such a tree can be reduced such that no configuration repeats on a path [5]. In the case of A this means that after at most $|S|$ steps a head moves. Such a reduced tree can clearly be transformed into an accepting tree for A' by counting the stationary moves in the second component of its states. Conversely an accepting computation tree of A' is simply transformed into a tree of A by projection on the first component of each state.

Now we outline how to modify the automaton such that it has as leaves only configurations with all heads on the endmarker. If there is an accepting state which allows further transitions we introduce a state which existentially chooses between two copies of the old state, one of which accepts and allows no further transition, and another which is identical to the old one but non-accepting. Then we make all blocking states (for certain input symbols other than the endmarker no transition is possible) non-accepting and add transitions to new states which simply advance all heads and either accept or reject according to the original decision.   □

Theorem 3 *The classes 1AFA(k) are closed under complement and consequently they are closed under Boolean operations.*

Proof: By Lemma 5 we can assume that all computation trees of the automata are finite and in normalized form. To obtain an automaton accepting the complement we exchange the role of accepting/nonaccepting blocking states and the role of universal/existential states. By induction on the depth of the (partial) computation trees we have that every accepting configuration becomes rejecting and vice versa.   □

In [5] it has been shown that **1AFA(3)** contains the context-free languages and **1AFA(2)** contains the linear context-free languages. We therefore get

Corollary 3 *1AFA(3) and 1AFA(2) contain the Boolean closures of context-free and linear context-free languages resp.*

Since 1NFA(k) are restricted 1AFA(k) we also have

Corollary 4 *The Boolean closure of 1NFA(k) is contained in 1AFA(k). The containment is proper for $k \geq 2$.*

Properness follows from the fact that languages accepted by 1NFA(k) with a unary input alphabet are regular [4] while $L_1 \in$ **1AFA(2)**.

It is interesting to compare this "efficiency" in terms of heads required for Boolean operations with automata lacking alternation. Consider two languages L_i and L_j recognized by one-way finite automata with i and j heads. If the automata are deterministic then $L_i \cap L_j$ and $L_i \cup L_j$ can be accepted by automata with $i + j - 1$ heads [7, Remark 4, p. 392]. Further the complements can be recognized by devices of the same type.

If the automata are nondeterministic then the parallel simulation from [7] gives $i + j - 1$ heads for $L_i \cap L_j$ while $\max(i, j)$ clearly suffices for $L_i \cup L_j$. No increase in the number of heads will in general give an automaton recognizing the complement of L_i or L_j as follows from [9].

By the preceding result alternation allows us to perform all these operations on languages from **1AFA(k)** without leaving the class.

As a last point we will briefly sketch how a one-way finite automaton can use alternation for simulating certain aspects of two-way automata. By leaving one head stationary on the first symbol such an automaton can simulate heads that move from right to left by guessing the symbol under a head and universally verifying it with the help of the head on the first symbol. By generalizing the simulation of linear context-free grammars it is also possible to simulate pairs of heads starting at both ends of the input and meeting somewhere inside the word. As a further example of the power of alternation we can show a translational result that in a similar way holds for various two-way automata. Let T_k for $k \geq 1$ be the following operation on words [8, p. 226]:

$$T_k(w) = w^{|w|^{k-1}}$$

Then we have

Theorem 4 *If $L \in 1AFA(k)$ then $T_k(L) \in 1AFA(4)$.*

Proof: At several places in this proof we have to divide the distance d_j of some head j to the endmarker by a number n stored as the distance of head 1 to the endmarker. This can be achieved with the help of two other auxiliary heads by placing one of them nondeterministically at a distance of d_j/n and then branching universally to align this head with head j and continue the computation or check the guess. The latter is done by placing the second auxiliary head n positions to the right of head j and then align them while moving the first auxiliary head one position. This is repeated until both heads reach the endmarker.

Let L be accepted by 1AFA(k) M. We will describe a 1AFA(4) M' that on input $T_k(w)$ simulates M on w. We note that $|T_k(w)| = |w|^k$ and M' first guesses $|w|$ as the distance d_1 of its head 1 to the endmarker. M' can check it with the help of the three other heads by dividing the input length by this distance k times as described above. After guessing $|w|$ M' with heads 2 and 3 verifies that the input is a repetition of w. In another universal branch the simulation is initiated. The position h_i of head i of M is stored as $h_i \cdot |w|^{i-1}$ in head position h'_2 of M' that holds the sum of all these encodings (here $0 \leq h_i \leq |w| - 1$, if a head reaches the endmarker it is stored in the finite control). If head i of M moves right M' moves its second head $|w|^{i-1}$ squares. For this purpose M' guesses $\lfloor |w|^{i-1}/2 \rfloor$ as the distance of heads 3 and 4 where head 3 is aligned with head 2 and head 4 is placed to the right of heads 2 and 3. Then M' universally either adds this distance twice to the position of head 2 (as well as the remainder of the division) or verifies that the distance is $\lfloor |w|^{i-1}/2 \rfloor$. This can be done by the division method and the observation that there is sufficient space to the right of head 4.

The symbol under head i of M can be determined by dividing $d'_2 - 1$ (the distance to the endmarker minus 1) by $|w|^{i-1}$ which can be done in a universal branch. By repeated subtraction of $|w|$ M' can also observe whether head i is on the last symbol and thus keep $h_i \leq |w| - 1$. For these last operations M' temporarily ignores the endmarker and thus transforms d'_2 into $d'_2 - 1$ (without moving head 2). We have

$$d'_2 - 1 = |w|^k - 1 - \sum_{i=1}^{k} h_i \cdot |w|^{(i-1)}$$

$$= (|w| - 1) \cdot \sum_{i=0}^{k-1} |w|^i - \sum_{i=1}^{k} h_i \cdot |w|^{(i-1)}$$

$$= \sum_{i=1}^{k} (|w| - h_i - 1) \cdot |w|^{(i-1)}$$

$$= \sum_{i=1}^{k}(d_i - 1) \cdot |w|^{(i-1)}$$

which shows that the correct distance is computed. □

It is remarkable that the previous translation is independent of the size of the alphabet and even works for unary languages.

3 Discussion

We have demonstrated the power of alternation in a very simple model of computation. Some open problems still remain:

- Do additional heads increase the power of one-way alternating finite automata?

- What is the relation of these automata to the two-way variety?

- Are there natural examples of languages not accepted by one-way alternating finite automata?

Acknowledgement: The author is indebted to K.N.King for help and encouragement.

References

[1] M.Ya.Alberts: *Possibilities of various types of alternating automata*, Cybernetics 6 (1986) (translated from Russian) 714–720.

[2] D.Geidmanis: *On possibilities of one-way synchronized and alternating automata*, Proc. MFCS 1990 (B.Rovan ed.) LNCS 452, 292–299.

[3] D.Geidmanis: *Unsolvability of the emptiness problem for alternating 1-way multi-head and multi-tape finite automata over single-letter alphabet*, Computers and Art. Int. 10 (1991) 133–141.

[4] M.A.Harrison, O.H.Ibarra: *Multitape and multihead pushdown automata*, Inf. Contr. 13 (1968) 433–470.

[5] K.N.King: *Alternating multihead finite automata*, Theor. Comput. Sci. 61 (1988) 149–174.

[6] M.L.Minsky: *Recursive unsolvability of Post's problem of "tag" and other topics in theory of Turing machines*, Annals of Mathematics, 74, 3 (1961), 437–455.

[7] A.L.Rosenberg: *On multi-head finite automata*, IBM Journal Res. and Dev. 10 (1966) 388–394.

[8] K.Wagner, G.Wechsung: *Computational Complexity*, Reidel, Dordrecht (1986).

[9] A.C.Yao, R.L.Rivest: $k+1$ *heads are better than* k, JACM 25 (1978) 337–340.

Hybrid Automata with Finite Bisimulations*

Thomas A. Henzinger

Department of Computer Science, Cornell University, Ithaca, NY 14850

Abstract. The analysis, verification, and control of hybrid automata with finite bisimulations can be reduced to finite-state problems. We advocate a time-abstract, phase-based methodology for checking if a given hybrid automaton has a finite bisimulation. First, we factor the automaton into two components, a boolean automaton with a discrete dynamics on the finite state space $\mathbb{B}^m$ and a euclidean automaton with a continuous dynamics on the infinite state space $\mathbb{R}^n$. Second, we investigate the phase portrait of the euclidean component. In this fashion, we obtain new decidability results for hybrid systems as well as new, uniform proofs of known decidability results.

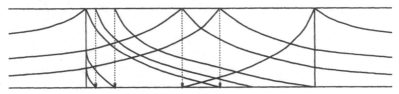

1 Introduction

A *hybrid automaton* [2] is a mathematical model for a digital program that interacts with an analog environment. Hybrid automata are useful for analyzing embedded systems [4, 10, 14, 21, 22]. We advocate the view of hybrid automata as infinite-state transition systems. We call an infinite-state transition system *finitary* and *effective* if it has a finite bisimulation that can be constructed effectively, and we examine the question of which hybrid automata are finitary and effective. This question is important for the analysis of embedded systems, because the problems of language emptiness, model checking, and controller synthesis are decidable for hybrid automata that are finitary and effective [1, 13, 19].

There have been several results of the type "the emptiness problem is undecidable for hybrid automata with clocks and a stopwatch" [2, 12, 15]. While these results leave little hope for finding a general class of finitary hybrid automata (those with finite bisimulations), they must be brought into accord with our experience with the symbolic model checker HyTech for hybrid automata [4]. Although HyTech is guaranteed to terminate only for finitary hybrid automata, the procedure does terminate on many automata with clocks and a stopwatch, including the well-known example of a leaking gas burner [9]. It is this apparent discrepancy between verification theory and practice we set out to explain in this paper.

Consider, for example, the two water-level controllers shown in Figure 1. The real-valued variable x_2 represents the water level in a tank. The water level x_2 increases at the rate of 1 m s^{-1} if the valve at the bottom of the tank is shut, and x_2 decreases at the rate of 2 m s^{-1} if the valve is open. The controller W_1 uses a clock x_1 to open the valve 2 s after the water level hits 9 m, and it closes the valve 2 s after the water level falls to 6 m. The controller W_2 opens the

* This research was supported in part by the NSF grant CCR-9200794, by the AFOSR contract F49620-93-1-0056, and by the DARPA grant NAG2-892.

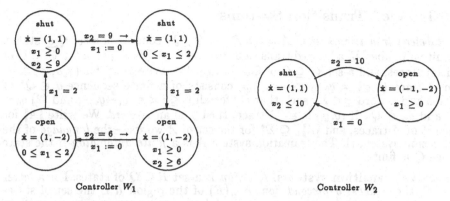

Fig. 1. Two water-level controllers

valve when the water level hits 10 m, and keeps the valve open for precisely the amount of time that it was closed. Neither controller falls into a known decidable subclass of hybrid automata. If we run HyTech to check if the water level stays within a given range, the verification succeeds only for controller W_1. We explain this phenomenon, and we provide a technique that allows us to see, a priori, that controller W_1 is finitary, while controller W_2 is not.

Hybrid automata are time-invariant (autonomous) in the sense that a transition depends only on the current state of the automaton, and not on the current time. All timing aspects are explicitly encoded in the state space, say, through the values of clock variables. Time-invariance allows us to take a phase view of hybrid automata, which abstracts time and gives geometric insight into the state-transition structure of hybrid automata. The phase domain was first exploited by Alur and Dill for proving that the so-called region quotient is a finite bisimulation of a timed automaton [3]. This paper can be seen as an extension of and a tribute to their work.

Consider, for example, a finitary hybrid automaton H with four control locations, two clocks, and one stopwatch. There are two extreme approaches for checking that H is finitary. The *specific* approach looks at the phase portrait of H in the 5D space $\mathbb{B}^2 \times \mathbb{R}^3$. The *generic* approach studies the possible phase portraits of hybrid automata with clocks and a stopwatch. While the specific approach may be unnecessarily concrete and expensive, the general approach is too abstract to yield the desired result (there are hybrid automata with clocks and a stopwatch that are not finitary). Inspired by the theory of dynamical systems, we advocate a natural intermediate approach. We factor the hybrid automaton $H = (B, E)$ into a boolean component B, with a discrete dynamics on the state space $\mathbb{B}^2$, and a euclidean component E, with a continuous dynamics on the state space $\mathbb{R}^3$. A sufficient condition for the finitariness of H, then, is the finitariness of E. We check that the euclidean automaton E is finitary by looking at the phase portrait of E, which does contain clocks and a stopwatch, but also takes into account the concrete set of constraints that H puts on these variables.

This approach leads, first, to a finer distinction between the decidability and undecidability of hybrid automata than previous results indicate and, second, to a uniform explanation of many previous decidability results [2, 6, 20, 21]. While also geometric in its intuition, the hybrid-automaton model differs from the related approach of [18]. First, their dynamical systems are deterministic and our euclidean automata are nondeterministic, both as far as discrete and continuous progress is concerned. Second, we consider the product of euclidean automata with boolean automata, which results in multiple copies of euclidean state spaces.

2 Labeled Transition Systems

A *(labeled) transition system* $A = (Q, \Sigma, \rightarrow, \overleftarrow{Q}, \overrightarrow{Q})$ consists of a set Q of states, a finite alphabet Σ, a labeled transition relation $\rightarrow \subseteq Q \times \Sigma \times Q$, a set $\overleftarrow{Q} \subseteq Q$ of initial states, and a set $\overrightarrow{Q} \subseteq Q$ of final states. We write $q \xrightarrow{\sigma} q'$ for $(q, \sigma, q') \in \rightarrow$. An *A-trace* $(\underline{q}, \underline{\sigma}) = q_0 \sigma_0 q_1 \cdots \sigma_{n-1} q_n$ consists of a finite sequence $\underline{q} \in Q^*$ of states and a word $\underline{\sigma} \in \Sigma^*$ such that (1) for all $0 \leq i < n$, $q_i \xrightarrow{\sigma_i} q_{i+1}$, and (2) $q_0 \in \overleftarrow{Q}$ and $q_n \in \overrightarrow{Q}$. If $(\underline{q}, \underline{\sigma})$ is an *A*-trace, then $\underline{\sigma}$ is an *A-word*. We write $[\![A]\!]$ for the set of *A*-traces, and $[\![A]\!]_L \subseteq \Sigma^*$ for the set of *A*-words —the *language* of the transition system *A*. The transition system *A* is a *finite automaton* if the state space *Q* is finite.

Effective transition systems. A *region* is a set $R \subseteq Q$ of states. For a letter $\sigma \in \Sigma$, the *weakest σ-precondition* $pre_\sigma(R)$ of the region *R* is the set of states from which a state in *R* can be reached by a σ-transition: $q \in pre_\sigma(R)$ iff $(\exists q' \in R \mid q \xrightarrow{\sigma} q')$. We write $dom(pre_\sigma)$ for the region $\{q \in Q \mid pre_\sigma(\{q\}) \neq \emptyset\}$, and $pre(R)$ for the region $\bigcup_{\sigma \in \Sigma} pre_\sigma(R)$. The transition system *A* is *effective* if there is a class $\mathcal{R}$ of effectively representable regions such that (1) $\overleftarrow{Q}, \overrightarrow{Q} \in \mathcal{R}$, (2) $\mathcal{R}$ is effectively closed under all boolean operations and all pre_σ-operations, for $\sigma \in \Sigma$, and (3) the emptiness problem is decidable for the regions in $\mathcal{R}$. For example, every Turing machine is effective.

Finitary transition systems. Let $\sim \subseteq Q^2$ be an equivalence relation of states, and let $Q/_\sim$ be the corresponding partition of the state space *Q*. A $\sim$-*block* is a union of $\sim$-equivalence classes. For a region $R \subseteq Q$, let $R/_\sim$ be the smallest $\sim$-block that contains *R*. For two regions $R, R' \subseteq Q$, let $R \xrightarrow{\sigma}_\exists R'$ if there are two states $q \in R$ and $q' \in R'$ with $q \xrightarrow{\sigma} q'$. The *quotient system* $A/_\sim$ is the transition system with the state space $Q/_\sim$, the alphabet Σ, the transition relation $\rightarrow_\exists$, the initial region $\overleftarrow{Q}/_\sim$, and the final region $\overrightarrow{Q}/_\sim$.

The equivalence $\sim$ is *finite* if it has a finite number of equivalence classes. The equivalence $\sim$ is a *bisimulation* for the transition system *A* if $\overleftarrow{Q}$ is a $\sim$-block and for all letters $\sigma \in \Sigma$ and all $\sim$-equivalence classes *R*, the region $pre_\sigma(R)$ is a $\sim$-block. The two states $p, q \in Q$ are *bisimilar*, written $p \approx q$, if there is a bisimulation $\sim$ such that $p \sim q$. If *p* and *q* are bisimilar, then (1) $p \in \overleftarrow{Q}$ iff $q \in \overleftarrow{Q}$; (2) if $p \xrightarrow{\sigma} p'$, then there is a state $q' \in Q$ such that $q \xrightarrow{\sigma} q'$ and $p' \approx q'$; and (3) if $q \xrightarrow{\sigma} q'$, then there is a state $p' \in Q$ such that $p \xrightarrow{\sigma} p'$ and $p' \approx q'$.

Proposition 1. *For every transition system A, if $\sim$ is a bisimulation for A, then* $[\![A]\!]_L = [\![A/_\sim]\!]_L$.

The transition system *A* is *finitary* if there is a finite bisimulation for *A*. In particular, every finite automaton is finitary. Examples of finitary transition systems with infinite state spaces are affine transition systems [17], finite automata with unbounded but unreliable communication channels [5], and finite automata with real-valued clocks [3]. By Proposition 1, the languages of the finitary transition systems are the regular languages. The bisimilarity partition $Q/_\approx$ for the transition system *A* can be computed by successive approximation:

```
procedure BisimApprox:
    Q/~ := {Q, Q - Q};
    while {assert ≈ ⊆ ~} there are two regions R, R' ∈ Q/~ and a letter
        σ ∈ Σ such that ∅ ⊂ R ∩ pre_σ(R') ⊂ R do
        Q/~ := (Q/~ - {R}) ∪ {R ∩ pre_σ(R'), R - pre_σ(R')}
        od
    {assert ≈ = ~}.
```

Each step of the procedure BisimApprox is effectively computable if A is effective. The successive approximation converges —i.e., the procedure BisimApprox terminates— iff A is finitary. Implementations of the procedure BisimApprox are discussed in [16, 23] for finite automata and in [7, 17] for arbitrary transition systems.

Questions about transition systems.[2] The *emptiness problem* for transition systems asks, given a transition system A, is the language $[A]_L$ empty. If A is effective and finitary, then the emptiness problem can be solved by first computing the bisimilarity quotient $A/_{\approx}$, and then checking the emptiness of the finite quotient system $A/_{\approx}$.

Theorem 2. *The emptiness problem is decidable for transition systems that are effective and finitary.*

Instead of computing the quotient system $A/_{\approx}$, it typically is more efficient to compute, by successive approximation, the region $pre^*(\overrightarrow{Q})$ of states from which a final state can be reached by any finite number of transitions:

> procedure ReachApprox:
> $R := \overrightarrow{Q}$;
>   while {assert $R \subseteq pre^*(\overrightarrow{Q})$ and R is a $\approx$-block} $pre(R) \not\subseteq R$ do
> $R := R \cup pre(R)$;
>   if $R \cap \overrightarrow{Q} \neq \emptyset$ then return "$[A]_L \neq \emptyset$" fi
>   od;
>   {assert $R = pre^*(R)$} return "$[A]_L = \emptyset$".

Again, each step of the procedure ReachApprox is effectively computable if A is effective, and the procedure terminates if A is finitary. While in the worst case both ReachApprox and BisimApprox proceed identically, the successive approximation of $pre^*(\overrightarrow{Q})$ may converge in fewer iterations than the successive approximation of $Q/_{\approx}$; indeed, the successive approximation of $pre^*(\overrightarrow{Q})$ may converge even if the successive approximation of $Q/_{\approx}$ does not.

Operations on transition systems. Let $A_1 = (Q_1, \Sigma_1, \to_1, \overleftarrow{Q}_1, \overrightarrow{Q}_1)$ and $A_2 = (Q_2, \Sigma_2, \to_2, \overleftarrow{Q}_2, \overrightarrow{Q}_2)$ be two transition systems. An *alphabet combinator* γ is a partial function on $\Sigma_1 \times \Sigma_2$. We write $\gamma(\Sigma_1, \Sigma_2)$ for the range of γ. Let $(q_1, q_2) \overset{\sigma}{\Rightarrow} (q_1', q_2')$ if $\gamma(\sigma_1, \sigma_2) = \sigma$, $q_1 \overset{\sigma_1}{\to} q_1'$, and $q_2 \overset{\sigma_2}{\to} q_2'$. The *product* $A_1 \times_\gamma A_2$ is the transition system with the state space $Q_1 \times Q_2$, the alphabet $\gamma(\Sigma_1, \Sigma_2)$, the transition relation $\Rightarrow$, the initial region $\overleftarrow{Q}_1 \times \overleftarrow{Q}_2$, and the final region $\overrightarrow{Q}_1 \times \overrightarrow{Q}_2$. It follows that the product of two effective transition systems is again effective.

Proposition 3. *For all transition systems A_1 and A_2, if $\sim_1$ is a bisimulation for A_1, and $\sim_2$ is a bisimulation for A_2, then $\sim_1 \times \sim_2$ is a bisimulation for $A_1 \times_\gamma A_2$ (where $(p_1, p_2) \sim_1 \times \sim_2 (q_1, q_2)$ iff $p_1 \sim_1 q_1$ and $p_2 \sim_2 q_2$).*

Corollary 4. *The product of two finitary transition systems is finitary.*

The finitariness of transition systems, however, is not preserved by the reversal of the transition relation. Let $A = (Q, \Sigma, \to, \overleftarrow{Q}, \overrightarrow{Q})$ be a transition system. The *reverse system* A^{-1} is the transition system with the state space Q, the alphabet Σ, the transition relation $\leftarrow$ (where $q \overset{\sigma}{\leftarrow} q'$ iff $q' \overset{\sigma}{\to} q$), the initial region $\overrightarrow{Q}$, and the final region $\overleftarrow{Q}$. For a language $L \subseteq \Sigma^*$, the *reverse language* $L^{-1} \subseteq \Sigma^*$ contains all words from L read backwards.

---

[2] In the full paper we discuss, in addition to the emptiness problem, also the model-checking problem and the control problem for transition systems. The full paper should be requested from tah@cs.cornell.edu.

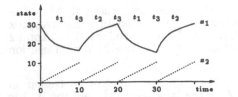

$$t_1: 0 \leq x_2 \leq 10 \wedge 10\dot{x}_1 = -x_1 \wedge \dot{x}_2 = 1$$
$$t_2: 0 \leq x_2 \leq 10 \wedge 10\dot{x}_1 = -x_1 + 50 \wedge \dot{x}_2 = 1$$
$$t_3: x_2 = 10 \wedge x_1' = x_1 \wedge x_2' = 0$$

Fig. 2. A trace of the temperature controller T

Proposition 5. *For every transition system A, $[A^{-1}]_L = [A]_L^{-1}$.*

Corollary 6. *The emptiness problem is decidable for transition systems whose reversal is effective and finitary.*

The reversal operation is interesting, as A^{-1} may be effective or finitary if A is not.[3] Consider, for example, the transition system $\hat{A}$ with the state space N, a unary alphabet, the transition relation $(n + 1) \rightarrow n$, for all $n \in$ N, a finite initial region $\hat{Q} \subset$ N, and a nonempty finite final region $\hat{Q} \subset$ N. Then $\hat{A}$ is not finitary, and $\hat{A}^{-1}$ is finitary.

3 Euclidean Automata

Linear-time semantics: curves, jumps, and trajectories. The *dimension* of a euclidean automaton is a nonnegative integer n. The *state space* of a euclidean automaton of dimension n is the euclidean space $\mathbb{R}^n$. A *state* is a point $\mathbf{x} \in \mathbb{R}^n$ of the state space. A *region*, then, is a set $R \subseteq \mathbb{R}^n$ of points.

A *curve* (δ, f) consists of a positive real $\delta \in \mathbb{R}_{>0}$ —the *duration* of the curve— and a differentiable function $f: [0, \delta] \rightarrow \mathbb{R}^n$. The states $f(0)$ and $f(\delta)$ are called the *source* and the *target* of the curve. Given $t \in [0, \delta]$, we write $f'(t)$ for the first derivative $df(t)/dt$ of f at t. A (nondeterministic) *activity* is a set of curves. The activity F can be *slowed down* if $(\delta, f) \in F$ implies that $(\varepsilon, g) \in F$ for all positive durations $\varepsilon < \delta$ and all functions g with $g(t) = f(t \cdot \delta/\varepsilon)$. The activity F can be *sped up* if $(\delta, f) \in F$ implies that $(\varepsilon, g) \in F$ for all durations $\varepsilon > \delta$ and all functions g with $g(t) = f(t \cdot \delta/\varepsilon)$. The activity F is *time-abstract* if F can be sped up and slowed down.

A *jump* is a pair $(\mathbf{x}, \mathbf{x}') \in \mathbb{R}^n \times \mathbb{R}^n$ of states —the *source* $\mathbf{x}$ and the *target* $\mathbf{x}'$. The *duration* of a jump is 0. A (nondeterministic) *action* is a set of jumps.

A *(trajectory) segment* is a curve or a jump. Given a segment τ, we write $\overleftarrow{\tau} \in \mathbb{R}^n$ for the source, $\overrightarrow{\tau} \in \mathbb{R}^n$ for the target, and $|\tau| \in \mathbb{R}_{\geq 0}$ for the duration of τ. The two segments τ_1 and τ_2 are *adjacent* if $\overrightarrow{\tau_1} = \overleftarrow{\tau_2}$. A *trajectory* $\underline{\tau} = \tau_0 \cdots \tau_n$ is a finite sequence of segments such that for all $0 \leq i < n$, the segments τ_i and τ_{i+1} are adjacent. The trajectory $\underline{\tau}$ has the *source* $\overleftarrow{\underline{\tau}} = \overleftarrow{\tau_0}$, the *target* $\overrightarrow{\underline{\tau}} = \overrightarrow{\tau_n}$, and the *duration* $|\underline{\tau}| = \Sigma_{0 \leq i \leq n} |\tau_i|$.

Syntax: region, activity, and action formulas. Let $X = \{x_1, \ldots, x_n\}$, $dX = \{\dot{x}_1, \ldots, \dot{x}_n\}$, and $X' = \{x_1', \ldots, x_n'\}$ be three disjoint ordered sets of real-valued variables. A *region formula* is a formula whose free variables are in X. Every region formula defines a region: the state $\mathbf{x} \in \mathbb{R}^n$ *satisfies* the region formula $\rho(X)$, written $\mathbf{x} \models \rho$, if $\rho(\mathbf{x})$ holds. An *activity formula* is a formula whose free variables are in $X \cup dX$. Every activity formula defines an activity: the curve (δ, f) *satisfies* the activity formula $\varphi(X, dX)$ if $\varphi(f(t), f'(t))$ holds for all $t \in [0, \delta]$. We write $\mathbf{x} \xrightarrow{\varphi} \mathbf{x}'$ if there is a curve (δ, f) with source $\mathbf{x}$ and target $\mathbf{x}'$ that satisfies φ. An *action formula* is a formula whose free variables are in

---

[3] Looking at the reversal of a transition system is equivalent to considering postconditions instead of preconditions.

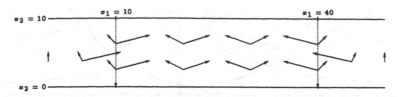

Fig. 3. Phase portrait of the temperature controller T

$X \cup X'$. Every action formula defines an action: the jump $(\mathbf{x}, \mathbf{x}')$ *satisfies* the action formula $\psi(X, X')$, written $\mathbf{x} \xrightarrow{\psi} \mathbf{x}'$, if $\psi(\mathbf{x}, \mathbf{x}')$ holds. By $[\rho]$ ($[\varphi]$; $[\psi]$) we denote the set of states (curves; jumps) that satisfy the corresponding formula.

The *alphabet* $\Sigma = \Phi \uplus \Psi$ of a euclidean automaton is the disjoint union of a finite set Φ of activity formulas —the *activity alphabet*— and a finite set Ψ of action formulas —the *action alphabet*. A Σ-*trace* $(\underline{\sigma}, \underline{\tau}) = (\sigma_0, \tau_0) \cdots (\sigma_n, \tau_n)$ consists of a word $\underline{\sigma} \in \Sigma^*$ and a trajectory $\underline{\tau}$ such that for all $0 \le i \le n$, the segment τ_i satisfies the formula σ_i.

Example 1. Consider the 2D euclidean component T of a temperature-control system. The variable x_1 represents the temperature, and the variable x_2 represents a clock of the controller. If the heater is turned off, then x_1 decreases according to the linear differential equation $\dot{x}_1 = -1/10x_1$, and if the heater is turned on, then x_1 follows $\dot{x}_1 = -1/10x_1 + 5$. We assume that the controller resets the clock x_2 to 0 every 10 s. Thus the alphabet $\Sigma_T = \{t_1, t_2, t_3\}$ consists of the two activity formulas t_1 and t_2, and the action formula t_3 of Figure 2. The figure also shows a sample Σ_T-trace.[4] $\square$

A *euclidean automaton* $E = (n, \Sigma, \overleftarrow{\rho}, \overrightarrow{\rho})$ consists of a dimension n, an alphabet Σ, and two region formulas $\overleftarrow{\rho}$ and $\overrightarrow{\rho}$. The region formula $\overleftarrow{\rho}$ defines the *initial region* of E; the region formula $\overrightarrow{\rho}$ defines the *final region* of E. The Σ-trace $(\underline{\sigma}, \underline{\tau})$ is an E-*trace* if it leads from the initial to the final region —that is, $\overleftarrow{\tau} \models \overleftarrow{\rho}$ and $\overrightarrow{\tau} \models \overrightarrow{\rho}$. If $(\underline{\sigma}, \underline{\tau})$ is an E-trace, then $\underline{\sigma}$ is an E-*word*. We write $[E]$ for the set of E-traces, and $[E]_L \subseteq \Sigma^*$ for the set of E-words —the *language* of the euclidean automaton E.

Phase semantics: flow and jump fields. We represent a vector with the origin $\mathbf{x} \in \mathbb{R}^n$ and the offset $\mathbf{y} \in \mathbb{R}^n$ by the pair $(\mathbf{x}, \mathbf{y})$ of states. We consider two types of vectors. The origin of a *flow vector* represents the source of a curve, and the offset represents the initial tangent of the curve. The origin of a *jump vector* represents the source of a jump, and the offset represents the relative effect of the jump. A (nondeterministic) vector field is a set of vectors. A field of flow vectors is called a *flow field*; a field of jump vectors, a *jump field*.

The phase portrait of an activity is a flow field. The flow vector $(\mathbf{x}, \mathbf{y})$ satisfies the activity formula $\varphi(X, dX)$ if $\varphi(\mathbf{x}, \mathbf{y})$ holds. The activity formula φ, then, defines the flow field $[\varphi]_F$ of all flow vectors that satisfy φ. The activity alphabet Φ defines the flow field $[\Phi]_F = \bigcup_{\varphi \in \Phi} [\varphi]_F$.

The phase portrait of an action is a jump field. The jump vector $(\mathbf{x}, \mathbf{y})$ satisfies the action formula $\psi(X, X')$ if $\psi(\mathbf{x}, \mathbf{x} + \mathbf{y})$ holds. The action formula ψ, then, defines the jump field $[\psi]_J$ of all jump vectors that satisfy ψ. The action alphabet Ψ defines the jump field $[\Psi]_J = \bigcup_{\psi \in \Psi} [\psi]_J$.

The phase portrait $[\Sigma]$ of the alphabet $\Sigma = \Phi \uplus \Psi$ consists of the flow field $[\Phi]_F$ and the jump field $[\Psi]_J$. The *phase portrait* $[E]$ of the euclidean automaton $E = (n, \Sigma, \overleftarrow{\rho}, \overrightarrow{\rho})$ consists of the flow field $[\Phi]_F$, the jump field $[\Psi]_J$, the initial region $[\overleftarrow{\rho}]$, and the final region $[\overrightarrow{\rho}]$.

---

[4] More details about examples and proofs are given in the full paper.

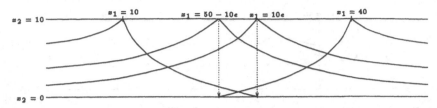

Fig. 4. BisimApprox for the temperature controller T

Example 2. The phase portrait of the temperature controller T is shown in Figure 3. The solid vectors represent sample flow vectors, and the dotted vectors represent sample jump vectors. □

Branching-time semantics: transition systems. With the euclidean automaton $E = (n, \Sigma, \overleftarrow{\rho}, \overrightarrow{\rho})$ we associate the transition relation $\overset{E}{\to} = \bigcup_{\sigma \in \Sigma} \overset{\sigma}{\to}$. The automaton E, then, defines the infinite-state transition system $\langle E \rangle$ with the state space $\mathbb{R}^n$, the alphabet Σ, the transition relation $\overset{E}{\to}$, the initial region $[\overleftarrow{\rho}]$, and the final region $[\overrightarrow{\rho}]$. We say that the euclidean automaton E is effective (finitary) if the corresponding transition system $\langle E \rangle$ is effective (finitary). Since $[E]_L = [\langle E \rangle]_L$, the emptiness problem is decidable for euclidean automata that are both effective and finitary.

3.1 Effective Euclidean Automata

A *linear inequality* on the set V of variables compares a linear integer combination of variables from V with an integer constant. A *linear constraint* on V is a (finite) conjunction of linear inequalities on V. A *linear region formula* is a linear constraint on X. Every linear region formula ρ, then, defines a (convex) rational polyhedron $[\rho] \subseteq \mathbb{R}^n$. A *linear activity formula* $\varphi = \varphi_P \wedge \varphi_S$ consists of a linear region formula φ_P —the *point invariant*— and a linear constraint φ_S on dX —the *slope invariant*. A *linear action formula* ψ is a linear constraint on $X \cup X'$. The euclidean automaton E is *linear* if all activity, action, and region formulas of E are linear. The linear euclidean automaton E is *bounded* if all point invariants of E define bounded regions (i.e., polytopes), and E is *positive* if all point invariants define subsets of $\mathbb{R}_{>0}^n$. The *slope set* of the linear euclidean automaton E is the set of vectors $\mathbf{x} \in \mathbb{R}^n$ such that $\varphi_S(\mathbf{x})$ holds for some slope invariant $\varphi_S(dX)$ of E.

Theorem 7. *Every linear euclidean automaton is effective.*

Proof. The region $R \subseteq \mathbb{R}^n$ is *linear* if R is a finite union of rational polyhedra in $\mathbb{R}^n$. We choose $\mathcal{R}$ to be the class of linear regions. The linear regions can be represented effectively by formulas of the first-order theory $(\mathbb{R}, \leq, +)$ of the reals with order and addition, which is closed under all boolean operations, admits quantifier elimination, and has a decidable satisfiability problem. The weakest precondition of a linear region can be computed using quantifier elimination. □

The procedures BisimApprox and ReachApprox are therefore semi-decision procedure for checking the emptiness of linear euclidean automata. Linearity, on the other hand, does not ensure the existence of a finite bisimulation. To see this, observe that every n-counter machine can be viewed as a linear euclidean automaton of dimension $n + 1$, where the $(n + 1)$-st dimension encodes the control of the machine.

3.2 Finitary Euclidean Automata

To check if a euclidean automaton is finitary, it is convenient to use a phase interpretation of the procedure BisimApprox. Let E be a euclidean automaton

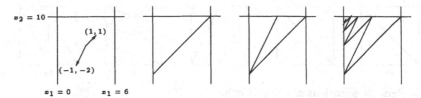

Fig. 5. Phase portrait and BisimApprox for the water-level controller W_2

of dimension n. The partition $Q/_\sim$ of $\mathbb{R}^n$ *is split along* the region $R \subseteq \mathbb{R}^n$ by splitting each region $R' \in Q/_\sim$ into the two parts $R' \cap R$ and $R' - R$ (one of the parts may be empty). We approximate the bisimilarity partition $Q/_\approx$ of E as follows:

1. Initialize $Q/_\sim$ to the result of splitting the state space $\mathbb{R}^n$ along the final region of E;
2. For every activity formula φ of E, split $Q/_\sim$ along $dom(pre_\varphi)$;
3. For every action formula ψ of E, split $Q/_\sim$ along $dom(pre_\psi)$;
4. Repeat until further splitting does not refine $Q/_\sim$:
    4.1. For every activity formula φ of E and every region $R \in Q/_\sim$, split $Q/_\sim$ along $pre_\varphi(R)$;
    4.2. For every action formula ψ of E and every region $R \in Q/_\sim$, split $Q/_\sim$ along $pre_\psi(R)$;
5. Return $Q/_\sim$.

This procedure depends solely on the phase portrait of E, and not on the trajectories of E. Indeed, in two dimensions, a quick visual inspection of the phase portrait suffices to determine if E is finitary.

Example 3. Let t_4: $10 \leq x_1 \leq 40 \wedge x_2 = 10$ define the final region of the temperature controller T. Then the procedure BisimApprox yields the finite partition of the state space that is shown in Figure 4. The result illustrates that the final region $[\![t_4]\!]$ is reachable from all states that satisfy t_5: $0 \leq x_2 \leq 10$. It follows that $[\![t_5]\!]$ is the characteristic region of the temporal formula $\exists \Diamond t_4$, and a finite-state controller can be built to bring the system from any initial state in $[\![t_5]\!]$ to a final state in $[\![t_4]\!]$ (see full paper). If we choose, instead, the final region $[\![10 \leq x_1 \leq 40]\!]$, then the procedure BisimApprox does not converge (see title page). □

Grid automata. In order to study the activities of euclidean automata, we begin by limiting the power of actions. A linear constraint is *rectangular* if it is a conjunction of atomic constraints of the forms $x_i \geq a_i$ and $x_i \leq b_i$, for rational constants a_i and b_i. The activity formula $\varphi = \varphi_P \wedge \varphi_S$ is *rectangular* if the point invariant φ_P is rectangular. A *rectangular action formula* $\psi = \psi_G \wedge \psi_U \wedge \psi_F$ consists of a rectangular constraint ψ_G over X —the *guard*— a conjunction ψ_U of atomic constraints of the form $x_i' = x_i$ —the *update*— and a rectangular constraint ψ_F over X' —the *filter*. The linear euclidean automaton E is *rectangular* if all activity, action, and region formulas of E are rectangular.

The rectangular automaton E is a *grid automaton* if for each action formula ψ of E, the update ψ_U is either *true* or $\bigwedge_{i=1}^{n}(x_i' = x_i)$. The n-dimensional *finite grid* of unit $c \in \mathbb{Q}_{>0}$ and size $d \in \mathbb{N}$ consists of all $(n-1)$-dimensional regions that are defined by formulas of the form $x_i = k \cdot c$, for $k \in \mathbb{N}$ with $|k| \leq d$. During the procedure BisimApprox, the actions of the grid automaton E contribute, independent of its activities, a subset of the finite grid whose unit is determined by the g.c.d. of the constants in E, and whose size is determined by the size of the constants in E.

S_1 S_2 S_3 S_4 S_5

Fig. 6. Periodic partitions of the unit cube

Proposition 8. *Every 1D rectangular automaton is finitary, and there is a bounded 2D grid automaton that is not finitary.*

Example 4. Recall the water-level controller W_2 from the introduction. The euclidean component of W_2 is a 2D grid automaton with the two activity formulas $x_2 \leq 10 \wedge \dot{x}_1 = 1 \wedge \dot{x}_2 = 1$ and $x_1 \geq 0 \wedge \dot{x}_1 = -1 \wedge \dot{x}_2 = -2$, and the two action formulas $x_2 = 10 \wedge x_1' = x_1 \wedge x_2' = x_2$ and $x_1 = 0 \wedge x_1' = x_1 \wedge x_2' = x_2$. If we assume the final region $[x_1 \leq 6]$, then Figure 5 illustrates that there is no finite bisimulation. $\square$

Periodic automata. For periodic activities, it suffices to look at a single grid cell. The n-dimensional *unit cube* U^n is the region $[0, 1]^n$ that is defined by the formula $\bigwedge_{i=1}^{n} (0 \leq x_i \leq 1)$. The unit cube has $2n$ facets of dimension $n-1$, each of which results from intersecting U^n with a region of the form $[x_i = 0]$ or $[x_i = 1]$. By closing the facets of U^n under intersection, we obtain the faces of U^n. In particular, the unit cube has 2^n faces of dimension 0 (i.e., corner points). The corner points of U^n and their negations are called the *boolean vectors* of dimension n. Given a region $R \subseteq \mathbb{R}^n$ and a vector $\mathbf{x} \in \mathbb{R}^n$, by $R \triangleleft \mathbf{x}$ we denote the region $\{\mathbf{y} - k \cdot \mathbf{x} \mid \mathbf{y} \in R \wedge k \in \mathbb{R}_{\geq 0}\}$. The slope set $S \subseteq \mathbb{R}^n$ is *periodic* if S is finite and there is a finite equivalence $\sim \subseteq U^n \times U^n$ on the unit cube such that (1) all faces of U^n are $\sim$-blocks and (2) for all $\sim$-equivalence classes $R \subseteq U^n$, all boolean vectors $\mathbf{u} \in \mathbb{R}^n$, and all slope vectors $\mathbf{x} \in S$, the region $((R + \mathbf{u}) \triangleleft \mathbf{x}) \cap U^n$ is a $\sim$-block. For example, Figure 6 shows that the 2D slope sets

$$S_1 = \{(1,1),(1,0),(0,1)\},$$
$$S_2 = \{(1,1),(1,-1),(-1,1),(-1,-1)\},$$
$$S_3 = \{(1,2),(1,-2),(-1,2),(-1,-2)\},$$
$$S_4 = \{(3,1),(-3,1),(3,-1),(-3,-1)\},$$
$$S_5 = \{(1,2),(-1,-2),(0,1),(0,-1)\}$$

are periodic. The periodic slope sets are closed under subsets and product (see full paper), but not under union. The rectangular automaton E is *periodic* if the slope set of E is periodic.

Theorem 9. *Every bounded periodic grid automaton is finitary.*

The left side of Figure 7 shows the result of applying the procedure BisimApprox to a bounded grid automaton with the slope set S_2 (the grid size is determined by the constants of the automaton). It follows that bounded grid automata with the slope set S_2 can be generalized to admit, in addition to grid actions, also rectangular action formulas with atomic filter constraints of the form $x_i' = c_i$, for $c_i \in \mathbb{Q}$. The result of the procedure BisimApprox is stable under these actions. Similarly, action formulas such as $x_2 = c \wedge x_1' = x_1 \wedge x_2' = d$ are admissible with the slope set S_3. In the full paper, we study which actions are safe for periodic slope sets. In particular, with the slope set S_1 all rectangular action formulas are safe, and we can relax the condition of boundedness to the condition of positiveness. The resulting phase portrait, shown on the right side of Figure 7, is the region quotient of timed automata [3].

Clock and stopwatch automata. We next look at the euclidean components of several generalizations of timed automata. Let E be a positive rectangular

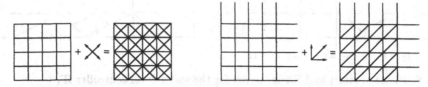

Fig. 7. Phase portrait and BisimApprox for a bounded grid automaton with slope set S_2, and for a positive rectangular automaton with slope set S_1

automaton of dimension n. The variable x_i is a *drifting clock* if there are two slopes $a_i, b_i \in \mathbb{Q}_{>0}$ such that each slope invariant of E contains the conjunct $a_i \le \dot{x}_i \le b_i$, and no other occurrences of $\dot{x}_i$. If $a_i = b_i$, then x_i is a *clock*; if $a_i = b_i = 1$, then x_i is a *precise clock*. The variable x_i is a *stopwatch* if there are two slopes $a_i, b_i \in \mathbb{Q}_{>0}$ such that each slope invariant of E contains either the conjunct $a_i \le \dot{x}_i \le b_i$ or the conjunct $\dot{x}_i = 0$, and no other occurrences of $\dot{x}_i$. The positive rectangular automaton E is a *(drifting) clock automaton* if all variables of E are (drifting) clocks. The positive rectangular automaton E is a *stopwatch automaton* if all variables of E are stopwatches and each slope invariant of E implies $\dot{x}_i = 0$ for all but one $1 \le i \le n$. Stopwatch automata are useful for modeling real-time multi-tasking [20].

Theorem 10. *Every clock automaton and every stopwatch automaton is periodic and finitary, and there is a bounded 2D drifting-clock automaton that is not finitary.*

The right side of Figure 7, which shows the bisimilarity quotient in the case of two precise clocks, indicates that automata with two precise clocks can be generalized to admit, in addition to clock activities, also slope invariants of the forms $\dot{x}_1 = \dot{x}_2$, $0 \le \dot{x}_1 \le \dot{x}_2 \le 1$, etc. (see full paper). The result of the procedure BisimApprox is stable under activities with these slope invariants. If, however, two clocks with different slopes are compared by a guard, filter, or point invariant, or both a clock and a stopwatch advance according to a slope invariant, then emptiness is undecidable [2, 12], and it is not difficult to check that the procedure BisimApprox does not terminate. Finitariness, on the other hand, is only sufficient but not necessary for decidability; indeed, the emptiness problem is decidable for drifting-clock automata [12, 24].

Example 5. While not periodic, the euclidean component of the water-level controller W_1 from the introduction is finitary. Figure 8 illustrates the finite bisimilarity relation of W_1 for the final region $[1 \le x_2 \le 12]$. From that we can construct a finite-state controller that keeps the water level between 1 m and 12 m. In the full paper, we give a somewhat general class of finitary hybrid automata that includes W_1. □

Time-abstract products of finitary automata. Let $E_1 = (m, \Phi_1 \uplus \Psi_1, \overleftarrow{\rho_1}, \overrightarrow{\rho_1})$ and $E_2 = (n, \Phi_2 \uplus \Psi_2, \overleftarrow{\rho_2}, \overrightarrow{\rho_2})$ be two euclidean automata. Given two region formulas $\rho_1(x_1, \ldots, x_m)$ and $\rho_2(x_1, \ldots, x_n)$, let $\gamma(\rho_1, \rho_2)$ stand for the region formula $\rho_1(x_1, \ldots, x_m) \wedge \rho_2(x_{m+1}, \ldots, x_{m+n})$, and adopt analogous conventions for activity and action formulas. The *product* $E_1 \times_\gamma E_2$ is the euclidean automaton with the dimension $m + n$, the alphabet $\gamma(\Phi_1, \Phi_2) \uplus \gamma(\Psi_1, \Psi_2)$, the initial region formula $\gamma(\overleftarrow{\rho_1}, \overleftarrow{\rho_2})$, and the final region formula $\gamma(\overrightarrow{\rho_1}, \overrightarrow{\rho_2})$.

As the transition relation of a euclidean automaton abstracts the duration of transitions, the transition relation of the product $E_1 \times_\gamma E_2$ typically is more restrictive than the product of the individual transition relations. This observation motivates the following subclasses of euclidean automata. The euclidean automaton E can be *slowed down* (*sped up*) if for all activity formulas $\varphi \in \Phi$, the activity $[\varphi]$ can be slowed down (sped up). The euclidean automaton E is

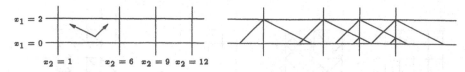

Fig. 8. Phase portrait and BisimApprox for the water-level controller W_1

time-abstract if it can be both slowed down and sped up —i.e., all activities of E are time-abstract. It is useful to model system components in a time-abstract way if the time scales of individual system components are not known a priori. For instance, unlike $\dot{x}_1 = 1 \wedge \dot{x}_2 = 1$, the slope invariant $\dot{x}_1 = \dot{x}_2$ defines a time-abstract activity with the same phase portrait [12]. Similarly, the slope invariants $\dot{x}_1 + 2\dot{x}_2 \geq 5$ and $\dot{x}_1 \leq 5 \wedge \dot{x}_2 = \dot{x}_1$ define activities that can be sped up and slowed down, respectively.

Theorem 11. *Let E_1 and E_2 be two euclidean automata such that one of E_1 and E_2 is time-abstract, or both E_1 and E_2 can be slowed down, or both E_1 and E_2 can be sped up. Then $\langle E_1 \times_\gamma E_2 \rangle = \langle E_1 \rangle \times_\gamma \langle E_2 \rangle$.*

Corollary 12. *Under the conditions of Theorem 11, if both E_1 and E_2 are effective and finitary, then so is the product $E_1 \times_\gamma E_2$.*

4 Hybrid Automata

A *hybrid automaton* $H = (B, E)$ consists of a finite (or boolean[5]) automaton B and a euclidean automaton E over a common alphabet Σ. The branching-time semantics of H is the transition system $\langle H \rangle = B \times_\gamma \langle E \rangle$, where $\gamma(\sigma, \sigma) = \sigma$, and $\gamma(\sigma_1, \sigma_2)$ is undefined for $\sigma_1 \neq \sigma_2$. The linear-time semantics of H is the language $[\![H]\!]_L = [\![B]\!]_L \cap [\![E]\!]_L$ over the alphabet Σ.

Proposition 3 implies that if the euclidean automaton E is effective or finitary, then so is the hybrid automaton (B, E). It follows that the procedures BisimApprox and ReachApprox are semi-decision procedures for the emptiness problem of hybrid automata with effective euclidean components, and both are decision procedures in the finitary case. The procedure ReachApprox has been implemented, in HYTECH, for hybrid automata with linear euclidean components [4].

If E is a clock automaton with precise clocks, then (B, E) is a *timed automaton* [3]. It is not difficult to check that, in addition to timed automata, also the multirate automata of [2, 21], the suspension automata of [20], and the 1-integrator automata of [6] have finitary euclidean components. We have thus provided an alternative, uniform decidability proof for these classes of hybrid automata. Unlike the original proofs, which are based on clock-translation or digitization techniques, our proof via finitariness provides several advantages: (1) it is not restricted to closed regions; (2) it provides direct insight into the state spaces of hybrid automata by identifying bisimilar states; (3) it guarantees the termination of successive-approximation procedures, such as those implemented in HYTECH; and (4) it reduces problems on hybrid automata, such as model checking and control, to the corresponding problems on finite automata.

We close with two remarks. First, as pointed out before, finitariness is sufficient but not necessary for decidability. In particular, hybrid automata with infinite 1-counter encodable bisimulations are decidable, because emptiness can be reduced to the emptiness problem for pushdown automata [8]. Examples of such automata include 2D grid automata with the slope set S_2 and only one

---

[5] A finite automaton with 2^m locations can be viewed as *boolean automaton* of dimension m, which has the state space $\mathbb{B}^m$.

bounded variable (see full paper). Second, we expect that our phase view of hybrid systems will also lead to a theory of conservative approximations for nonlinear hybrid automata with the property that, unlike in the time domain [11], approximation errors do not accumulate.

Acknowledgments. Our view of hybrid systems has been shaped in collaboration with many people, including Rajeev Alur, Costas Courcoubetis, Pei-Hsin Ho, Peter Kopke, Amir Pnueli, Joseph Sifakis, Anuj Puri, and Pravin Varaiya. We also thank Oded Maler and Howard Wong-Toi for valuable comments.

[1] R. Alur, C. Courcoubetis, N. Halbwachs, T.A. Henzinger, P.-H. Ho, X. Nicollin, A. Olivero, J. Sifakis, S. Yovine. The algorithmic analysis of hybrid systems. *Theoretical Computer Science*, 138:3–34, 1995.
[2] R. Alur, C. Courcoubetis, T.A. Henzinger, P.-H. Ho. Hybrid automata: an algorithmic approach to the specification and verification of hybrid systems. *Hybrid Systems*, Springer LNCS 736, pp. 209–229, 1993.
[3] R. Alur, D.L. Dill. A theory of timed automata. *Theoretical Computer Science*, 126:183–235, 1994.
[4] R. Alur, T.A. Henzinger, P.-H. Ho. Automatic symbolic verification of embedded systems. *IEEE Real-time Systems Symp.*, pp. 2–11, 1993.
[5] P. Abdulla, B. Jonsson. Verifying programs with unreliable channels. *IEEE Symp. Logic in Computer Science*, pp. 160–170, 1993.
[6] A. Bouajjani, R. Echahed, R. Robbana. Verifying invariance properties of timed systems with duration variables. *Formal Techniques in Real-time and Fault-tolerant Systems*, Springer LNCS 863, pp. 193–210, 1994.
[7] A. Bouajjani, J.-C. Fernandez, N. Halbwachs. Minimal model generation. *Computer-aided Verification*, Springer LNCS 531, pp. 197–203, 1990.
[8] A. Bouajjani, R. Robbana. Verifying ω-regular properties for subclasses of linear hybrid systems. *Computer-aided Verification*, Springer LNCS, 1995.
[9] Z. Chaochen, C.A.R. Hoare, A.P. Ravn. A calculus of durations. *Information Processing Letters*, 40:269–276, 1991.
[10] T.A. Henzinger, P.-H. Ho. Model-checking strategies for linear hybrid systems. Workshop on Hybrid Systems and Autonomous Control (Ithaca, NY), 1994.
[11] T.A. Henzinger, P.-H. Ho. Algorithmic analysis of nonlinear hybrid systems. *Computer-aided Verification*, Springer LNCS, 1995.
[12] T.A. Henzinger, P. Kopke, A. Puri, P. Varaiya. What's decidable about hybrid automata? *ACM Symp. Theory of Computing*, 1995.
[13] T.A. Henzinger, X. Nicollin, J. Sifakis, S. Yovine. Symbolic model checking for real-time systems. *Information and Computation*, 111:193–244, 1994.
[14] N. Halbwachs, P. Raymond, and Y.-E. Proy. Verification of linear hybrid systems by means of convex approximation. *Static Analysis Symp.*, Springer LNCS 864, 1994.
[15] Y. Kesten, A. Pnueli, J. Sifakis, S. Yovine. Integration graphs: a class of decidable hybrid systems. *Hybrid Systems*, Springer LNCS 736, pp. 179–208, 1993.
[16] P.C. Kanellakis, S.A. Smolka. CCS expressions, finite-state processes, and three problems of equivalence. *Information and Computation*, 86:43–68, 1990.
[17] D. Lee, M. Yannakakis. Online minimization of transition systems. *ACM Symp. Theory of Computing*, pp. 264–274, 1992.
[18] O. Maler, A. Pnueli. Reachability analysis of planar multi-linear systems. *Computer-aided Verification*, Springer LNCS 697, pp. 194–209, 1993.
[19] O. Maler, A. Pnueli, J. Sifakis. On the synthesis of discrete controllers for timed systems. *Theoretical Aspects of Computer Science*, Springer LNCS, 1995.
[20] J. McManis, P. Varaiya. Suspension automata: a decidable class of hybrid automata. *Computer-aided Verification*, Springer LNCS 818, pp. 105–117, 1994.
[21] X. Nicollin, A. Olivero, J. Sifakis, S. Yovine. An approach to the description and analysis of hybrid systems. *Hybrid Systems*, Springer LNCS 736, pp. 149–178, 1993.
[22] A. Olivero, J. Sifakis, S. Yovine. Using abstractions for the verification of linear hybrid systems. *Computer-aided Verification*, Springer LNCS 818, pp. 81–94, 1994.
[23] R. Paige, R.E. Tarjan. Three partition-refinement algorithms. *SIAM J. Computing*, 16:973–989, 1987.
[24] A. Puri, P. Varaiya. Decidability of hybrid systems with rectangular differential inclusions. *Computer-aided Verification*, Springer LNCS 818, pp. 95–104, 1994.

Generalized Sturmian Languages

Luis-Miguel Lopez[1] and Philippe Narbel[2]

(1) Université Marne-La-Vallée, (2) ENS Lyon et Université Paris 7
e-mails: lopez@univ-mlv.fr, pnarbel@ens.ens-lyon.fr, narbel@litp.ibp.fr

Abstract. The purpose of this paper is to generalize some recent results about Sturmian words and morphisms. This is done by introducing a new mathematical interpretation of them, using the modern topological framework of laminations and train tracks. The main result is a description of a class of languages invariant by the action of a finitely generated monoid of morphisms (where classical Sturmian objects represent the case over an alphabet of two letters). We also discuss how these languages can be effectively constructed, and we show how topological results about Pseudo-Anosov homeomorphisms can be proved in terms of $D0L$-systems theory.

1 Introduction

Sturmian words seem to have been on use during at least two centuries, starting by J. Bernoulli, Christoffel, and Markov in number theory [Mar82, Ven70]. These words were actually named "Sturmian" when Morse and Hedlund were founding *symbolic dynamics* [MH38, MH40], by studying geodesics of a surface represented as infinite words. Later, these Sturmian words became an usual example of a minimal dynamical system [Hed44, GH55, CH73, Cov74, Pau74] and this invigorated the search for interesting combinatorial properties. Factor complexity, geometrical interpretations, generation by infinitely iterated homomorphisms have been some of the subjects for which some results are very recent (see for instance [Mig89, DGB90, DLM94]).

The purpose of this paper is to go back to topology in order to interpret and generalize other results about Sturmian words due to Berstel, Kòsa, Mignosi, Séébold and Wen & Wen in [Kòs87, Séé91, Séé92, MS93, BS94, WW94], mainly saying that the language of all Sturmian words over an alphabet of two letters is the invariant set by the action of a finitely presented monoid of morphisms. Here we get the same kind of results for languages over more than two letters.

We shall see how Sturmian words can be interpreted within Thurston's theory of surfaces. This heavily relies on studying systems of geodesics called *laminations* [Thu, Cas88, Hat88, Wei89], which are sets of closed complete and pairwise disjoint geodesics running on a surface, and which can be finitely represented by *train tracks* [PH93]. Laminations are stable by the action of the *mapping class group* (the group of the homeomorphisms of a surface), which is a finitely generated group whose generators are *Dehn's twists*. On the other hand, geodesics can be interpreted as infinite words, laminations as languages of these words and Dehn's twists as morphisms, so that we get a full translation into the symbolical space: a language of infinite words which is stable by the application of a monoid of morphisms. More formally:

Consider the set of all *projective measured laminations* PM$\mathcal{L}$ on a closed surface Σ. This set has a natural topology. Consider also the set $\mathcal{C}$ of all the different simple curves making the laminations of PM$\mathcal{L}$. On the other hand, the set of all finite and infinite pointed words over a finite alphabet A, denoted by $^\infty A^\infty$, is endowed with the product topology. Well-behaved sets of languages can accordingly be endowed with the topology coming from the corresponding Hausdorff function. Now a geodesic on Σ can be coded such that it corresponds to a word in $^\infty A^\infty$. Thus a lamination can be represented into a language of these words. It can be proved that, first, there is a map γ such that γ^{-1} is injective from a language L of words in $^\infty A^\infty$ to $\mathcal{C}$, and second, γ^{-1} is a homeomorphism between a subset of the power set of L (the set of languages of L) and PM$\mathcal{L}$. The mapping class group G of Σ naturally acts on $\mathcal{C}$ and on PM$\mathcal{L}$. Within the symbolic space, the set of all the morphisms on $^\infty A^\infty$ leaving invariant L forms a monoid for the ordinary composition of maps acting on $^\infty A^\infty$. The next claim is that there exists a representation ϕ of one of its submonoid, denoted by **St**, into G such that every θ in **St** acts on the language L as $\phi(\theta)$ acts on $\gamma^{-1}(L)$. Namely, we have the following theorem:

Theorem 1.1 *Let Σ be a closed orientable surface with a negative curvature. Then, for the corresponding set of projective measured laminations, mapping class group, coding language and morphisms monoid, the following diagram is commutative:*

$$
\begin{array}{ccc}
St \times L & \longrightarrow & L \\
{\scriptstyle \phi \times \gamma^{-1}} \downarrow & & \downarrow {\scriptstyle \gamma^{-1}} \\
G \times \mathcal{C} & \longrightarrow & \mathcal{C}
\end{array}
$$

Replacing $\mathcal{C}$ by PM$\mathcal{L}$, L by $\gamma(PM\mathcal{L})$, gives a stronger commutative diagram.

Here, for the sake of simplicity, we shall mainly restrict the discussion about surfaces with no boundary components. The only such case that we shall treat is the punctured torus because of the following result:

Proposition 1.2 *For the punctured torus case, the invariant language L of the theorem and the morphisms monoid St are the ones found for the classical Sturmian words over two letters in [Kòs87, Séé91, MS93, WW94].*

This result justifies why we call **Sturmian languages**, languages like L. Note that we do not speak about the flat torus because there is no non-trivial finite representation of PM$\mathcal{L}$ on a torus, and that surfaces are assumed to have negative curvature. Note also that since the following construction makes use of laminations, we get a generalization of the way of coding geodesics [Ser86] without being disturbed by fundamental polygon's vertices or by punctures.

We shall also show that there is a strong relationship between Sturmian languages L and boundaries $\partial\theta_i$ of the languages obtained by iterating morphisms θ_i, for all $\theta_i \in$ **St**. For the classical Sturmian case over two letters, $L = \text{Closure}(\bigcup_{\theta_i \in \mathbf{St}} \partial\theta_i)$ is proved by using a number theory theorem due to Crisp, Moran, Pollington, Shiue [CMPS93, BS94]). The consequence of this equality is that L can be effectively constructed and regularly coded [Nar95]. Finally, we shall indicate how to prove a weaker version of a theorem about *pseudo-Anosov homeomorphisms* due to Penner [Pen88], and this by making use of results obtained in $D0L$-systems theory by Ehrenfeucht and Rozenberg [ER83].

2 Sturmian Words

A **pointed word** over a finite alphabet A is a map $w : I \to A$, where I is any interval of $\mathbb{Z}$ which includes zero. The set of all pointed words over A is denoted by $^\infty A^\infty$. When $I = \mathbb{Z}$, then w is **bi-infinite**, and when I properly contains $\mathbb{N}$ or $-\mathbb{N}$, then w is **one-way infinite**. The **origin** of a pointed word w is $w(0)$, and any subset of $^\infty A^\infty$ is a **language**. To homogenize $^\infty A^\infty$, all the words which are not yet bi-infinite are padded to both infinities with some dummy symbol not already in A. The **shift operator** σ on $^\infty A^\infty$ is defined when $w(-1) \in A$ by $\sigma(w(n)) = w(n+1)$, for all $n \in \mathbb{Z}$. We say that v and w are **shift-equivalent** iff there exists $n \in \mathbb{Z}$ such that $v = \sigma^n(w)$. Let us now recall the definition of the classical **Sturmian words**: A bi-infinite word w over two letters u and v is associated as follows to a straight line $y = \alpha x + \beta$ in the Euclidean plane with a slope in $(0,1)$. First, define the sequence $\{s_i\}_{i \in \mathbb{Z}}$ over $\{0,1\}$ by $s_i = [\alpha(i+1) + \beta] - [\alpha i + \beta]$, where $[.]$ is the bottom function; second, map s_i to u if $s_i = 0$, and to v if $s_i = 1$. Note that the words corresponding to slopes in $(1, \infty)$ are just words with slopes in $(0,1)$ and with letters u and v inverted. Also, if β and β' are in the $\mathbb{Z}$-submodule of $\mathbb{R}$ generated by 1 and α (i.e. $\beta - \beta' = p + q\alpha$, $p, q \in \mathbb{Z}$) then the associated words are **shift equivalent**.

There is another way of obtaining the same Sturmian word from the same straight line: start at coordinates $(0, \beta)$ and follow the line in the direction of increasing x-coordinates. Write u each time the line meets a line of equation $x = p$ and v each time the line meets a line of equation $y = q$ $(p, q \in \mathbb{N})$ (this can be done except when the line has a rational slope). A bi-infinite word is obtained by doing the same thing towards the negative x-coordinates. By replacing each occurrence of uv by u, one obtains the same word as before. Geometrically, this can be observed in the following picture:

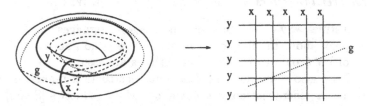

If α is a real number, then we define its **associated α-Sturmian language** as all the Sturmian words obtained by interpreting the lines $y = \alpha x + \beta$, for all $\beta \in [0,1)$. The full **associated Sturmian language** consists of the union of the α-Sturmian languages for all $\alpha \in [0,1] \setminus \mathbb{Q}$.

The next section embeds this framework into topological notions that will lead us to the generalization (notions extensively introduced in [Hat88, Wei89]).

3 Geodesics, Measured Laminations, and Train Tracks

We first assume that the considered surfaces are smooth, oriented, of genus $g \geq 1$ endowed with a metric of constant negative curvature -1. These are the surfaces obtained from closed ones by removing a finite number of open disks, i.e. the **boundary components**, and points, i.e. the **punctures**. In this case, the universal covering of such a surface is the hyperbolic plane $\mathbb{H}^2$ endowed with the Poincaré metric. Thus geodesics running on a surface Σ are quotients of geodesics running on $\mathbb{H}^2$. A **geodesic lamination** $\mathcal{L}$ on Σ is a foliation of a closed subset of Σ by complete simple geodesics (the set of all the lines with the same slope in $\mathbb{R}^2$ is an example). To the left hand, the next figure shows a part of a lamination on the punctured torus:

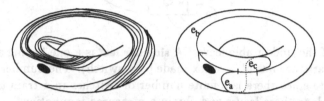

Describing sets of laminations can be quite simplified thanks to the following combinatorial notion: A **train track** τ on a closed surface Σ is an embedding of a *locally oriented trivalent* finite graph, such that at each vertex there is at least one sink edge and one source edge. A trivalent vertex is called a **switch**. For surfaces with boundaries or punctures, graphs may have some vertices with valency 1, at the condition that each such vertex is sent either to the boundary or to a puncture. There are other definitions of train tracks but they are essentially equivalent from the laminations viewpoint [PH93]. A train track of the punctured torus is shown to the right hand of the precedent figure. A train track τ is **admissible** if there is no connected component in $\Sigma \setminus \tau$ with the following shapes (the first three exclude the n-gons with $n = 0, 1, 2$, the last two interact with the border of the surface):

A lamination $\mathcal{L}$ is **carried** by the train track τ if each leaf can be deformed into τ in such a way that the leaf's orientation is either coherent with, or opposite to the local orientation at each visited switch (leaves are like running paths of real trains). A fruitful way of formalizing that is by introducing a measure: A **measured train track** is a pair (τ, w) where τ is a train track with an assignment of non-negative values $w(e)$ to each edge e of τ with a **switching condition**: at each vertex v where e_i is the sink (respec. source) edge and e_j, e_k are the source (respec. sink) edges, we have $w(e_i) = w(e_j) + w(e_k)$. A measured admissible train track (τ, w) is canonically associated to a **measured lamination** $(\mathcal{L}, \mu)$, whose set is denoted by $\mathcal{ML}$, according to the following fundamental construction of Thurston [Hat88, Wei89]:

1) Replace each edge e of the train track by a foliated box $[0, 1] \times [0, w(e)]$, where the set $\{0, 1\} \times [0, w(e)]$ is embedded as a pair of geodesics. The leaves of the foliation are the sets of the form $[0, 1] \times \{x\}$, where $x \in [0, w(e)]$. Thus the measure $w(e)$ gives the widths of the boxes and because of the switching condition, the different boxes can be pasted together: the leaves match together in order to form leaves on the whole surface. The punctured torus train track as shown precedently becomes:

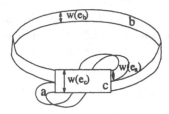

2) Get rid of the undesirable leaves: A **singular leaf** is a one having at least one endpoint (starts or ends at a vertex made by a switch). The number of singular leaves is finite since there is a finite number of switches in a train track. Thus we can cut along these leaves and obtain a measured lamination.

3) Deform the lamination into a geodesic one (always possible and uniquely).

4) Define the measure of this lamination as follows: take an arc cutting transversally the lamination and running from one edge to the opposite one in a box associated to the edge e. The intersection of this arc with the lamination is a finite disjoint union of Cantor sets, and the measure of this intersection is set equal to $w(e)$.

There is a theorem due also to Thurston [Hat88] which says that there is an explicit finite set of train tracks for a surface Σ such that every measured lamination in $M\mathcal{L}$ of Σ is carried by one of these train tracks with some appropriate measure. At this point, measure is the only information which is not finite to describe $M\mathcal{L}$: a measured lamination carried by (τ, w) is represented by the $w(e)$'s for each edge e of τ. We refer to this vector as the **coordinates** of the lamination. A **projective measured lamination** on Σ is an equivalence class in $M\mathcal{L}$ where $(\mathcal{L}_1, \mu_1) \sim (\mathcal{L}_2, \mu_2)$ if they are carried by the same train track and if their coordinates are homothetic. The set of all projective measured laminations of a surface is denoted $PM\mathcal{L}$. The set $M\mathcal{L}$ is endowed with the topology given by the Hausdorff distance between open sets of geodesic leaves, and $PM\mathcal{L}$ with the quotient topology.

A **language** over a finite alphabet can be assigned to any measured train track and therefore to any measured lamination. Indeed, let τ be a train-track and associate a distinct letter a_i to each edge e_i of τ (a letter to each foliated box): this is a **labeled train-track**. Then each leaf of the corresponding carried lamination can be coded by a word which reflects the order it visits the foliated boxes. The **associated language** of the carried lamination over the finite alphabet A consists of all the coding words of all its leaves. Thus, according to the Thurston's theorem, there is a finite set of alphabets such that each element in $M\mathcal{L}$ is represented as a language over one of these. Let us make three remarks:

1) If two measured laminations have their coordinates homothetic, then the associated languages are equal. This can be seen by comparing the laminations leaf by leaf: let (τ, w) and $(\tau, \lambda w)$ $(\lambda > 0)$ be two laminations with homothetic coordinates w and λw. Let ℓ be a leaf of τ; then ℓ has coordinates $t(e)$, $0 \leq t(e) \leq w(e)$, which tell us at which height it crosses the box corresponding to τ's edge e. Then to ℓ corresponds the leaf with coordinates $\lambda t(e)$ in $(\tau, \lambda w)$. These leaves have of course the same associated word. The above process can be reversed (from $(\tau, \lambda w)$ towards (τ, w)), so that we obtain the equality between the two languages.

2) The associated language is a subset of $^\infty A^\infty$, i.e. this language is made of pointed words. This is coherent since geodesics are essentially maps from $\mathbb{R}$ to Σ. In our treated cases, all the words are bi-infinite. However, in the general case, words may be finite or one-way infinite if they meet punctures or boundaries.

3) A **simplified associated language** is obtained by labeling edges such that any pair of letters uniquely determines the path between the two edges they represent. For instance, in the considered train track of the punctured torus, two labels are sufficient.

4 Revisiting Sturmian Words

In this section we obtain the classical Sturmian words in terms of the last topological framework: the train track running on the punctured torus as shown in the precedent figure leads to the Sturmian language over two letters.

Proposition 4.1 *For every irrational number α, there is a measured train track (τ, w) such that the associated language of τ is equal to the associated α-Sturmian language of α (up to countably many of them).*

Proof. Consider the simplified associated language of τ over two letters, say u and v. On the other hand, using the same labels as in the figure, the switching condition says that $w(c) = w(a) + w(b)$. Since two letters, say a and b, are sufficient for the train track, the corresponding lamination coordinate vector is $(w(a), w(b))$. According to the *ergodicity theorem* (see for instance [Kea91]) applied to the geodesic flow on the punctured torus, we know that every infinite leaf must go through the box corresponding to the edge e_a and the box corresponding to e_b with a ratio $w(a)/w(b)$. As a consequence, the measure of the train track defines a "slope" of the corresponding projective class of the lamination.

We claim that the number α is $w(a)/w(b)$, and we assume that this quotient is irrational (if not we get an infinite set of parallel closed curves). Let us fix what is called a **meridian-longitude pair** on the torus, i.e. a pair of non-separating curves on the torus meeting at one point. We call these curves γ_u and γ_v, and we assume that each one meets τ at one point, different from $\gamma_u \cap \gamma_v$. In the universal covering of the torus, γ_u and γ_v are lifted to two lines that can be considered as orthogonal coordinate axes (this was pictured for the geometrical interpretation of the Sturmian words). The result of lifting the measured train track is depicted as follows:

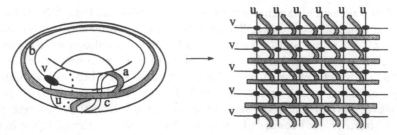

A leaf of a lamination as above can be coded by following it towards some direction and attaching a letter u or v each time it crosses γ_u or γ_v respectively. Doing that towards the opposite direction gives the other half of the word associated to the leaf. Now every line with slope $w(a)/w(b)$ not meeting a point with integral coordinates can be deformed in order to become a curve included in the corresponding geodesic lamination. $\Diamond$

In the last proof, if one wishes to obtain right or left infinite words, one must take a train-track with two more arcs joining the puncture. In that case, a line can meet only once a point with integral coordinates since otherwise $w(a)/w(b)$ would be rational.

5 Generalization for surfaces of genus > 1

We have described the usual Sturmian words in terms of laminations and train tracks. Let us generalize them by applying the same techniques to surfaces of genus larger than the torus. As already said, for all genus, there is a finite set of generic train tracks which describes all the different laminations [Hat88]. For the sake of simplicity, we restrict the presentation to a specific kind of these train tracks, so that one train track is selected for each genus (and so that we get one Sturmian language for each genus). These train tracks on surfaces of genus > 1 are basically obtained by gluing copies of the following two basic train tracks:

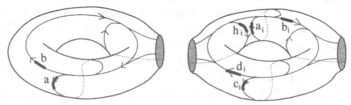

For instance, here is the result for a surface of genus 4:

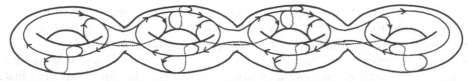

Remark 5.1 *The simplified associated language of the last described train tracks can be lapelled by 2 letters if $g = 1$, and $5(g-1)$ letters if $g > 1$.*

Proof. In the picture of the two basic train tracks, edges which are taken in account are indicated by letters. Whenever a train track has been built by pasting them together, these edges just take distinct letters and any pair of these letters uniquely determines the path between them. ◊

We can now go for the main point of this paper, that is the language made of all the associated languages to a train track (when one makes the measure vary) is invariant by the action of a monoid of morphisms. The **mapping class group** G of a surface is the group of all its homeomorphisms modulo isotopies to the identity map (an excellent overview about G is in [Bir88]). This group acts naturally on the set of all measured laminations PM$\mathcal{L}$. The mapping class group of a closed surface of genus g is generated by $2g + 1$ elements for all the homeomorphisms which let the orientation stable. These generators can be represented by **Dehn twists**, which are homeomorphisms given by cutting the surface along a curve, twisting one extremity of the obtained surface(s) and pasting back the two extremities. Let us denote them by $t_1, s_1, ..., t_g, s_g$, plus one that we denote by u_2 when $g > 1$. For reasons of symmetry, we add $u_3, ..., u_g$, so that we get $3g - 1$ generators:

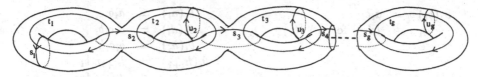

To include the homeomorphisms which do not preserve orientation, we add an element that inverts the longitudinal twists with the meridional ones.

Now recall that $^{\infty}A^{\infty}$ is the set of all the pointed words (finite and infinite) over the alphabet A. This space can be given a metric: let w_1, w_2 be two pointed words, then $d(w_1, w_2) = 0$ iff $w_1 \equiv w_2$, and $d(w_1, w_2) = 2^{-n}$ otherwise, where n is the largest nonnegative integer such that $w_1(k) = w_2(k)$, $|k| \leq n$. Once we have this metric we can define a topology on the subsets of the considered associated languages included in $^{\infty}A^{\infty}$ by using the Hausdorff function which appears to be a distance in this case. If A^+ denotes the free semi-group generated by the letters in A, then a morphism θ on A^+ (also called a **substitution** or a **D0L-system** in formal language theory [RS80]) is uniquely described by the images on A. Since it is not well-defined on pointed words, θ is applied to $^{\infty}A^{\infty}$ quotiented by the shift-equivalence. Let us now sketch a proof for Theorem 1.1:

Proof. Let us consider a closed surface of genus $g \geq 1$ and its mapping class group G. The main idea is to associate a substitution to each Dehn twist's class corresponding to the generators of the mapping class group. We have $3g - 1$ such classes, and the considered train tracks are as precedently described. We denote the classes of the longitudinal twists by t_i, $1 \leq i \leq g$, the meridional ones by s_i, $1 \leq i \leq g$ and the additional meridional ones by u_i, $1 < i < g$ (exactly as shown in the last picture).

Let τ_g be the selected train track for a surface Σ of genus g. By Remark 5.1, its associated language alphabet A consists of either 2 letters if $g = 1$, or $5(g-1)$

if $g > 1$. According to the figure presenting the basic train tracks, we denote them by a_1, b_1 for the first piece of Σ, by a_i, b_i, c_i, d_i, h_i for the ith piece where $1 < i < g$, and by a_g, b_g, h_g for the last piece. Next, we define the following substitutions on A (Id denotes the identity map):

For $\boxed{g = 1}$ then

$$\sigma_1(b_1) = a_1 b_1 \qquad\qquad \theta_1(a_1) = a_1 b_1$$
$$\sigma_1 \equiv Id \quad \text{for } A \setminus \{b_1\} \qquad\qquad \theta_1 \equiv Id \quad \text{for } A \setminus \{a_1\}$$

For $\boxed{g > 1}$ then first,

$$\sigma_1(b_1) = a_1 b_1 \qquad\qquad \theta_1(a_1) = a_1 b_1$$
$$\sigma_1 \equiv Id, \quad \text{for } A \setminus \{b_1\} \qquad\qquad \theta_1(h_2) = h_2 b_1$$
$$\theta_1 \equiv Id, \quad \text{for } A \setminus \{a_1, h_2\}$$

Second, for $1 < i < g$:

$$\sigma_i(b_i) = a_i b_i \qquad\qquad\qquad \theta_i(a_i) = a_i b_i c_i$$
$$\sigma_i \equiv Id, \quad \text{for } A \setminus \{b_i\} \qquad\qquad \theta_i(d_i) = d_i c_i b_i$$
$$\theta_i(h_i) = b_i c_i h_i$$
$$\theta_i(h_{i+1}) = h_{i+1} c_i b_i$$
$$\theta_i \equiv Id, \quad \text{for}$$
$$A \setminus \{a_i, d_i, h_i, h_{i+1}\}$$

$$v_i(b_{i-1}) = b_{i-1} h_i$$
$$v_i(a_i) = h_i a_i$$
$$v_i(b_i) = h_i b_i$$
$$v_i \equiv Id, \quad \text{for } A \setminus \{b_{i-1}, a_i, b_i\}$$

And finally:

$$\sigma_g(b_g) = a_g b_g \qquad\qquad \theta_g(a_g) = a_g b_g$$
$$\sigma_g \equiv Id, \quad \text{for } A \setminus \{b_g\} \qquad\qquad \theta_g(h_g) = h_g b_g$$
$$\theta_g \equiv Id, \quad \text{for } A \setminus \{a_g, h_g\}$$

$$v_g(a_g) = h_g a_g$$
$$v_g(b_g) = h_g b_g$$
$$v_g \equiv Id, \quad \text{for } A \setminus \{a_g, b_g\}$$

We add the substitution corresponding to the generator of the non-preserving orientation homeomorphisms:

$$\iota(a_i) = b_i, \quad \iota(b_i) = a_i, \quad \iota(c_i) = d_i, \quad \iota(d_i) = c_i, \quad \iota(h_i) = h_i.$$

We extend these maps on $^\infty A^\infty$: the monoid generated by them is the searched monoid, that is $\mathbf{St}$. The mapping $\phi : \mathbf{St} \to G$ defined by

$$\phi(\sigma_i) = s_i, \quad \phi(\theta_i) = t_i, \quad \phi(v_i) = u_i \quad \text{for } 1 < i \leq g,$$
$$\phi(\sigma_i) = s_i, \quad \phi(\theta_i) = t_i \quad \text{for } i = 1,$$

is such that the commutation of the first version of the diagram is satisfied for each twist (in order to check that, perform each twist on a piece of curve carried by τ which is affected by this operation, and code the resulting leaf).

Now the map γ between $PM\mathcal{L}$ and L is just the one which associates a language to each class of measured laminations carried by τ_g by coding the way each leaf crosses a foliated box. Deform the measured lamination into a geodesic one: the ergodicity of the geodesic flow tells us that each leaf visits the box corresponding to the edge e with probability $p(e)$. The set of $p(e)$'s form a vector of measures (with sum equal to 1) giving the measure on the train-track (the coordinates of the train track). This proves that γ is a bijection. Since both the source and the target of γ are equipped with the "same type" of metric, that is the Hausdorff distance between closed sets in a compact space, γ is a homeomorphism. Thus we have the second version of the diagram. $\Diamond$

We can now easily check Proposition 1.2: indeed, for the case $g = 1$ of the theorem, the substitutions $\{\iota, \theta_1, \iota \circ \theta_1 \circ \iota\}$ are the same three morphisms as in [Kòs87, Séé91, MS93, WW94].

Let us see now that one can explicitly generate the associated languages: the n-fold composition of a morphism θ on A is denoted by θ^n and its corresponding **substitution language** is the set: $L_\theta = \{w \in A^+ \mid \theta^n(s) = w, \ n \in \mathbb{N}, \ s \in A\}$. The boundary of θ, denoted by $\partial\theta$, is the boundary of the pointed counterpart of L_θ in $(^\infty A^\infty, d)$, obtained by pointing the words of L_θ in every possible way. Using the invariance of the train track's associated language L, we get:

Lemma 5.2 *For an associated language L, then $Closure_d(\bigcup_{\theta \in \mathbf{St}} \partial\theta) \subseteq L$.*

It is an open question whether the converse inclusion is always true. However we can prove it for the classical Sturmian words by using a theorem by Crisp, Moran, Pollington, Shiue [CMPS93, BS94]. That result can be reinterpreted to say that Sturmian words are fixed points under some morphism in **St** iff they represent laminations carried by a train track whose coordinates have a ratio whose continuous fractions have one of these forms (overlines denote periods):

$$\langle 0, r_0, \overline{r_1, ..., r_n} \rangle, \qquad r_n \geq r_0 \geq 1$$
$$\langle 0, 1 + r_0, \overline{r_1, ..., r_n} \rangle, \qquad r_n = r_0 \geq 1$$
$$\langle 0, 1, r_0, \overline{r_1, ..., r_n} \rangle, \qquad r_n > r_0 \geq 1$$

On the other hand, substitutions and homeomorphisms of the torus are tied together as follows: Dehn's twists have a representation into $GL_2(\mathbb{Z})$ (see for instance [Cas88]) which is exactly the matrix representation of a substitution. Indeed, one can prove that the element (i, j) of the matrix of a specific Dehn twist is equal to the number of letters i appearing in the image of the letter j of the corresponding substitution (as given in Theorem 1.1). The slope of the fixed lamination is then given as the largest real eigenvalue > 1 which always exists in this case (see the Perron-Frobenius theorem).

Proposition 5.3 *For the punctured torus case, then $Closure_d(\bigcup_{\theta \in \mathbf{St}} \partial\theta) = L$.*

Proof. (sketch). According to the last remark, each of the fixed points can be generated by iterating the morphisms having an eigenvalue whose continuous fractions have the forms given by the Crisp & al.'s Theorem. These fixed points are thus in the boundaries of these morphisms. Now according to the forms taken by the continuous fractions and since the periods can be made longer and longer, the fixed points have slopes which are dense in $\mathbb{R}$. Therefore, we have that $L \subseteq \mathrm{Closure}_d(\bigcup_{\theta \in \mathbf{St}} \partial\theta)$. Using Lemma 5.2, the result follows. $\Diamond$

Thus according to [Nar95], L may be mapped to a rational language. On the other hand, another result due to Thurston can be reinterpreted. It says that there is a full class of homeomorphisms, the so-called **pseudo-Anosov** homeomorphisms which fix two singular transverse foliations on which they act repulsively for one, and attractively for the other. In the punctured torus case, pseudo-Anosov maps correspond to matrices having two distinct real eigenvalues. Thus Crisp & al.'s Theorem gives an explicit information about their values.

For surfaces of genus $g > 1$, there is a topological result due to Penner [Pen88] which can be proved by formal language theory arguments:

Theorem 5.4 *(simplified version). Let* $\mathbf{St} = \{s_1, ..., s_n\}^*$ *be the morphism monoid associated to a surface of genus g. Then any composition of the s_i's containing at least one s_i for each i corresponds to a pseudo-Anosov homeomorphism.*

Proof. (sketch). According to Thurston, homeomorphisms fall into three types: periodic, reducible (that fixes some simple closed curve up to isotopy on the surface), and pseudo-Anosov. There is no periodic homeomorphism among those considered in Theorem 1.1. For the other cases, we use the following definition: let π be a cyclic permutation of the alphabet A of the associated language, then a morphism is said (A, π)-cyclic if for each $s \in A$, $\theta(s) = s_1...s_m$, then $s_{i+1} = \pi(s_i)$, with $1 \le i \le m-1$, and for each pair $s, t \in A$ such that $\pi(s) = t$, then $\pi(last(\theta(s))) = first(\theta(t))$, where $first(w)$ is the first letter of w, and $last(w)$ its last one. By using the results in [ER83], one may conclude that there is a periodic word in $\partial\theta$ iff θ is (A, π)-cyclic. Compositions of substitutions in **St** as required by the theorem can be checked to be non-cyclic. Thus the only remaining possibility is the pseudo-Anosov case. $\Diamond$

Of course there is more to do. Here are some directions: 1) A combinatorial analysis on the factors generated by iterating the substitutions obtained in Theorem 1.1; 2) A careful analysis of the induced symbolic dynamics; 3) The use of the entire set of generic train tracks as defined in Hatcher (i.e. train tracks on surfaces with punctures with genus > 1); 4) An extensive study of pseudo-Anosov homeomorphisms in the general case in order to generalize the Crisp & al.'s Theorem; 5) The search for the presentations of the monoids **St**.

References

[Bir88]   Birman J.S. – Mapping class groups of surfaces. *In :* *Braids.* – AMS, 1988.

[BS94]   Berstel J. and Séébold P. – Morphismes de Sturm. *Bull. of the Belg. Math . Soc.*, vol. 1, number 2, 1994, pp. 175–190.

[Cas88]   Casson A.J. – *Automorphisms of Surfaces after Nielsen and Thurston.* – London Mathematical Society, 1988.

[CH73]   Coven E.M. and Headlund G. – Sequences with minimal block growth. *Math. Sys. Th.*, vol. 7, 1973, pp. 138–073.

[CMPS93] Crisp D., Moran W., Pollington A. and Shiue P. – Substitution invariant cutting sequences. *J. de Th. des Nombres de Bordeaux*, 5, number 1, 1993, pp. 073–138.

[Cov74]   Coven E.M. – Sequences with minimal block growth II. *Math. Sys. Th.*, vol. 8, 1974, pp. 376–382.

[DGB90]   Dulucq S. and Gouyou-Beauchamps. – Sur les facteurs des suites de Sturm. *Theoretical Computer Science*, vol. 71, 1990, pp. 381–400.

[DLM94]   De Luca A. and Mignosi F. – On some combinatorial properties of Sturmian words. *Theoretical Computer Science*, 1994. – to appear.

[ER83]   Ehrenfeucht A. and Rozenberg G. – Repetitions of subwords in D0L languages. *Information and Control*, vol. 59, 1983, pp. 13–35.

[GH55]   Gottschalk W.H. and Hedlund G.A. – Topological dynamics. *AMS Colloq. Pub.*, vol. 36, 1955.

[Hat88]   Hatcher A.E. – Measured laminations spaces for surfaces from the topological viewpoint. *Topology and its Applications*, vol. 30, 1988, pp. 63–88.

[Hed44]   Hedlund G. – Sturmian minimal sets. *Amer. Journal of Math.*, vol. 66, 1944, pp. 605–620.

[Kea91]   Keane M.S. – *Ergodic theory and subshifts of finite type*, chap. 2. – Oxford Univ. Press, 1991. Bedford, T. and Keane, M. and Series, C., Editors.

[Kòs87]   Kòsa M. – Problem. *Bull. EATCS*, vol. 32, 1987, pp. 331–333.

[Mar82]   Markov A. – Sur une question de Jean Bernoulli. *Math. Ann.*, vol. 19, 1882, pp. 27–36.

[MH38]   Morse M. and Hedlund G.A. – Symbolic dynamics I. *American Journal of Mathematics*, vol. 60, 1938, pp. 807–866.

[MH40]   Morse M. and Hedlund G.A. – Symbolic dynamics II. Sturmian trajectories. *American Journal of Mathematics*, vol. 62, 1940, pp. 1–42.

[Mig89]   Mignosi F. – Infinite words with linear subword complexity. *Theoretical Computer Science*, vol. 65, 1989, pp. 221–242.

[MS93]   Mignosi F. and Séébold P. – Morphismes Sturmiens et règles de Rauzy. *J. de Th. des Nombres de Bordeaux*, vol. 5, number 2, 1993, pp. 221–234.

[Nar95]   Narbel Ph. – The boundary of iterated morphisms on free semi-groups. – 1995. To appear in *International Journal of Algebra and Computation*.

[Pau74]   Paul M. – Minimal symbolic flows having minimal block growth. *Math. Sys. Th.*, vol. 8, 1974, pp. 307–307.

[Pen88]   Penner R.C. – A construction of pseudo-Anosov homeomorphisms. *Transc. of the Amer. Math. Soc.*, vol. 310, number 1, 1988, pp. 179–197.

[PH93]   Penner R.C. and Harer J.L. – *Combinatorics of train tracks.* – Princeton University Press, 1993, *Annals of Math. Studies*. Study 075.

[RS80]   Rozenberg G. and Salomaa A. – *The mathematical theory of L systems.* – Academic press, 1980.

[Séé91]   Séébold P. – Fibonacci mophisms and Sturmian words. *Theoretical Computer Science*, vol. 88, 1991, pp. 367–384.

[Séé92]   Séébold P. – *Problèmes combinatoires liés à la génération de mots infinis ayant des facteurs prescrits.* – Tech. Rep. 92-07, LITP, Paris 7, 1992. Habilitation Thesis.

[Ser86]   Series C. – Geometrical Markov coding of geodesics on surfaces of constant negative curvature. *Ergod. Th. and Dynam. Sys.*, vol. 6, 1986, pp. 601–625.

[Thu]   Thurston W. – *The geometry and topology of 3-manifolds.* – Princeton University Lecture Notes.

[Ven70]   Venkov B.A.– *Elementary Number Theory*– Wolters-Noordhoff, Groningen,1970.

[Wei89]   Weiss H. – The geometry of measured geodesic laminations and measured train tracks. *Ergod. Th. and Dynam. Sys.*, vol. 9, 1989, pp. 587–604.

[WW94]   Wen Z-X. and Wen Z-Y. – Local isomorphisms of invertible substitutions. *C.R. Acad. Sci. Paris*, vol. 318, 1994, pp. 299–304.

Polynomial Closure and Unambiguous Product

Jean-Eric Pin and Pascal Weil [*]

1. Introduction

This paper is a contribution to the algebraic theory of recognizable languages. The main topic of this paper is the *polynomial closure*, an operation that mixes together the operations of union and concatenation. Formally, the *polynomial closure* of a class of languages $\mathcal{L}$ of A^* is the set of languages that are finite unions of marked products of the form $L_0 a_1 L_1 \cdots a_n L_n$, where the a_i's are letters and the L_i's are elements of $\mathcal{L}$. The *unambiguous polynomial closure* is the closure under disjoint union and unambiguous marked product. One can also define, with a slight modification (see section 4) similar operators for languages of A^+.

Our main result is an algebraic characterization of the polynomial closure of a variety of languages. There are several technical difficulties to achieve this result. First, even if $\mathcal{V}$ is a variety of languages, its polynomial closure is not, in general, a variety of languages. The solution to this problem was given in a recent paper by the first author [18]. If the definition of a variety of languages is slightly modified (instead of all boolean operations, only closure under intersection and union are required in the definition), one still has an Eilenberg type theorem. The new classes of languages are called *positive varieties*, but of course, the algebraic counterpart has to be modified too: they are the varieties of finite ordered semigroups or finite ordered monoids. It turns out that the polynomial closure of a variety of languages is always a positive variety. Now, the next question can be asked: given a variety of monoids **V** corresponding to a variety of languages $\mathcal{V}$, describe the variety of ordered monoids corresponding to the polynomial closure of $\mathcal{V}$. Our answer (statement (1)) fits surprisingly well with two other important results on varieties (statements (2) and (3)):

(1) The algebraic operation corresponding to the polynomial closure is the Mal'cev product $\mathbf{W} \circledM \mathbf{V}$, where $\mathbf{W}$ is the variety of finite ordered semigroups $(S, \leq)$ in which $ese \leq e$, for each idempotent e and each element s in S.

(2) The algebraic operation corresponding to the unambiguous polynomial closure is the Mal'cev product $\mathbf{LI} \circledM \mathbf{V}$, where $\mathbf{LI}$ is the variety of semigroups S in which $ese = e$, for each idempotent e and each s in S [20].

(3) The algebraic operation corresponding to the closure under boolean operations and concatenation is the Mal'cev product $\mathbf{A} \circledM \mathbf{V}$, where $\mathbf{A}$ is the variety of aperiodic semigroups (Straubing [30]).

[*] LITP/IBP, Université Paris VI et CNRS, Tour 55-65, 4 Place Jussieu, 75252 Paris Cedex 05, FRANCE. E-mail: `pin@litp.ibp.fr`, `weil@litp.ibp.fr`

The proof of our main result is non-trivial and relies on a deep theorem of Simon [27] on factorization forests. Its importance can probably be better understood on its far-reaching consequences. Due to the lack of place, we indicate some of these consequences. Others can be found in the extended version of this article. First, the polynomial closure leads to natural hierarchies among recognizable languages. Define a boolean algebra as a set of languages of A^* (resp. A^+) closed under finite union and complement. Now, start with the trivial boolean algebra of recognizable languages, and call it level 0. Thus the languages of level 0 are the empty language and A^* (resp. A^+). Then define recursively the higher levels as follows: level $n + 1/2$ is the polynomial closure of level n and level $n + 1$ is the boolean closure of level $n + 1/2$. This defines the Straubing (resp. dot-depth) hierarchy. The main open problem is to know whether each level of these hierarchies is decidable.

Levels 0, 1/2 and 1 of the Straubing hierarchy were known to be decidable. Level 3/2 was also known to be decidable but the proof was quite involved and no practical algorithm was known. We give a simple proof of this last result and show that, given a deterministic n-state automaton $\mathcal{A}$ on the alphabet A, one can decide in time polynomial in $2^{|A|}n$ whether the language accepted by $\mathcal{A}$ is of level 3/2 in the Straubing hierarchy. Decidability of level 2 is still an open question, but we make some progress on this problem. First our main result gives a short proof of a result of Cowan [8] characterizing the languages of level 2 whose syntactic monoid is inverse. Second, we formulate a conjecture for the identities of the variety of monoids corresponding to languages of level 2. More generally, we conjecture that the variety of ordered monoids corresponding to the boolean closure of the polynomial closure of $\mathcal{V}$ is the Mal'cev product $\mathbf{B_1} \textcircled{m} \mathbf{V}$, where $\mathbf{B_1}$ is the variety of finite semigroups corresponding to languages of dot-depth one.

For the dot-depth hierarchy, only levels 0 and 1 were known to be decidable. We show that level 1/2 is also decidable. There is some evidence that level 3/2 is also decidable, but the proof of this result would require some auxiliary algebraic results that will be studied in a future paper.

Another important consequence of our result is the fact that a language L belongs to the unambiguous polynomial closure of a variety of languages $\mathcal{V}$ if and only if both L and its complement belong to the polynomial closure of $\mathbf{V}$. This result has an interesting consequence in logic. Indeed, Thomas [34] (see also [14,17]) showed that Straubing's hierarchy is in one-to-one correspondence with a well known hierarchy of first order logic, the Σ_n hierarchy, obtained by counting the alternative use of existential and universal quantifiers in formulas in prenex normal form. We present analogous results for the Δ_n hierarchy of first order logic. We first show that each level of this logical hierarchy defines a variety of languages. Next we give an effective description of the first levels. For the levels 0 and 1, the corresponding variety is trivial. The variety corresponding to level 2 is the smallest variety of languages closed under non-ambiguous product, introduced by Schützenberger [25].

2. Varieties

All semigroups and monoids considered in this paper are finite or free.

2.1. Varieties of semigroups and ordered semigroups

An *ordered semigroup* $(S, \leq)$ is a semigroup S equipped with an order relation $\leq$ on S such that, for every $u, v, x \in S$, $u \leq v$ implies $ux \leq vx$ and $xu \leq xv$. An *order ideal* of $(S, \leq)$ is a subset I of S such that, if $x \leq y$ and $y \in I$, then $x \in I$. A *morphism of ordered semigroups* $\varphi : (S, \leq) \rightarrow (T, \leq)$ is a semigroup morphism from S into T such that, for every $x, y \in S$, $x \leq y$ implies $x\varphi \leq y\varphi$. A semigroup S can be considered as an ordered semigroup by taking the equality as order relation. Ordered subsemigroups, quotients and products are defined in the natural way.

A *variety of semigroups* (resp. ordered semigroups) is a class of (ordered) semigroups closed under the taking of (ordered) subsemigroups, quotients and finite products. *Varieties of (ordered) monoids* are defined in the same way.

2.2. Identities

Let A be a finite alphabet and let u, v be two words of A^*. A monoid M *separates* u and v if there exists a monoid morphism $\varphi : A^* \rightarrow M$ such that $u\varphi \neq v\varphi$. One defines a distance on A^* as follows: if u and v are elements of A^*, let $r(u, v) = \min\{ |M| \mid M$ separates u and $v \}$ and $d(u, v) = 2^{-r(u,v)}$. By convention, $\min \emptyset = \infty$ and $2^{-\infty} = 0$. Thus $r(u, v)$ measures the size of the smallest monoid which separates u and v. It is not difficult to verify that d is an ultrametric distance function. For this metric, multiplication in A^* is uniformly continuous, so that A^* is a topological monoid. The completion of the metric space (A^*, d) is a monoid, denoted $\hat{A}^*$.

If we consider each finite monoid M as being equipped with the discrete distance, every monoid morphism from A^* onto M is uniformly continuous and can be extended in a unique way into a continuous morphism from $\hat{A}^*$ onto M. Since $\hat{A}^*$ is a completion of A^*, its elements are limits of sequences of words. An important such limit is the ω-power, which traditionally designates the idempotent power of an element of a finite monoid [9,16].

Proposition 2.1. *Let $x \in \hat{A}^*$. The sequence $(x^{n!})_{n \geq 0}$ converges in $\hat{A}^*$ to an idempotent denoted x^ω. Furthermore, if $\mu : \hat{A}^* \rightarrow M$ is a continuous morphism into a finite monoid, then $x^\omega \mu$ is the unique idempotent power of $x\mu$.*

Another useful example is the following. The set 2^A of subsets of A is a semigroup under union and the function $c : A^* \rightarrow 2^A$ defined by $c(a) = \{a\}$ is a semigroup morphism. Thus $c(u)$ is the set of letters occurring in u. Now c extends into a continuous morphism from $\hat{A}^*$ onto 2^A, also denoted c and called the *content* mapping.

Let x, y be elements of $\hat{A}^*$. A monoid (resp. ordered monoid) M *satisfies the identity* $x = y$ (resp. $x \leq y$) if, for every continuous morphism $\varphi : \hat{A}^* \rightarrow M$, $x\varphi = y\varphi$ (resp. $x\varphi \leq y\varphi$). Given a set E of identities of the form $x = y$ (resp.

$x \leq y$), we denote by $[\![E]\!]$ the class of all monoids (resp. ordered monoids) which satisfy all the identities of E. The following result [22] extends two results of Reiterman [23] and Bloom [6].

Theorem 2.2. *Let E be a set of identities of the form $u = v$ (resp. $u \leq v$). Then the class $[\![E]\!]$ forms a variety of monoids (resp. ordered monoids). Conversely, for each variety of monoids (resp. ordered monoids), there exists a set E of identities such that $\mathbf{V} = [\![E]\!]$.*

A similar theory can be developed for varieties of semigroups, using a distance on the free semigroup A^+ instead of the free monoid A^*. Of particular importance for us is the variety **LI** of locally trivial semigroups, defined by the identity $[\![x^\omega y x^\omega = x^\omega]\!]$. Thus a semigroup S is locally trivial if, for all idempotent e of S and for every $s \in S$, $ese = e$. Similarly, the variety $\mathbf{A} = [\![x^\omega = x^{\omega+1}]\!]$ is the variety of aperiodic monoids [1,16].

2.3. Relational morphisms and Mal'cev products

A *relational morphism* between semigroups S and T is a relation $\tau : S \to T$ such that:

(1) $(s\tau)(t\tau) \subseteq (st)\tau$ for all $s, t \in S$,

(2) $(s\tau)$ is non-empty for all $s \in S$,

If S and T are monoids, a third condition is required

(3) $1 \in 1\tau$

Let $\mathbf{V}$ be a variety of monoids (resp. semigroups) and let $\mathbf{W}$ be a variety of semigroups. The *Mal'cev product* $\mathbf{W} \,\textcircled{M}\, \mathbf{V}$ is the class of all monoids (resp. semigroups) M such that there exists a relational morphism $\tau : M \to V$ with $V \in \mathbf{V}$ and $e\tau^{-1} \in \mathbf{W}$ for each idempotent e of V. It is easily verified that $\mathbf{W} \,\textcircled{M}\, \mathbf{V}$ is a variety of monoids (resp. semigroups). The following theorem, obtained by the authors [21], describes a set of identities defining $\mathbf{LI} \,\textcircled{M}\, \mathbf{V}$.

Theorem 2.3. *Let $\mathbf{V}$ be a variety of monoids. Then $\mathbf{LI} \,\textcircled{M}\, \mathbf{V}$ is defined by the identities of the form $x^\omega y x^\omega = x^\omega$, where $x, y \in \hat{A}^*$ for some finite set A and $\mathbf{V}$ satisfies $x = y = x^2$.*

These results can be extended to varieties of ordered monoids as follows. Let $\mathbf{V}$ be a variety of monoids and let $\mathbf{W}$ be a variety of ordered semigroups. The *Mal'cev product* $\mathbf{W} \,\textcircled{M}\, \mathbf{V}$ is the class of all ordered monoids $(M, \leq)$ such that there exists a relational morphism $\tau : M \to V$ with $V \in \mathbf{V}$ and $e\tau^{-1} \in \mathbf{W}$ for each idempotent e of V. It is easily verified that $\mathbf{W} \,\textcircled{M}\, \mathbf{V}$ is a variety of ordered monoids. A defining set of identities for $[\![x^\omega y x^\omega \leq x^\omega]\!] \,\textcircled{M}\, \mathbf{V}$ is given in [21].

Theorem 2.4. *Let $\mathbf{V}$ be a variety of monoids. Then $[\![x^\omega y x^\omega \leq x^\omega]\!] \,\textcircled{M}\, \mathbf{V}$ is defined by the identities of the form $x^\omega y x^\omega \leq x^\omega$, where $x, y \in \hat{A}^*$ for some finite set A and $\mathbf{V}$ satisfies $x = y = x^2$.*

3. Recognizable languages

Recall that a *variety of languages* is a class of recognizable languages closed under finite union, finite intersection, complement, left and right quotients and inverse morphisms between free semigroups. A *positive variety of languages* is a class of recognizable languages closed under finite union, finite intersection, left and right quotients and inverse morphisms between free semigroups.

Eilenberg's variety theorem can be extended to positive varieties if one replaces varieties of semigroups by varieties of ordered semigroups. Let $(S, \leq)$ be an ordered semigroup and let η be a surjective semigroup morphism from A^+ onto S. A language L of A^+ is said to be *recognized* by η if $L = P\eta^{-1}$ for some order ideal P of S. By extension, L is said to be *recognized* by $(S, \leq)$ if there exists a surjective morphism from A^+ onto S that recognizes L. One defines a stable quasiorder $\preceq_L$ and a congruence relation $\sim_L$ on A^+ by setting $u \preceq_L v$ if and only if, for every $x, y \in A^*$, $xvy \in L$ implies $xuy \in L$ and $u \sim_L v$ if and only if $u \preceq_L v$ and $v \preceq_L u$. The congruence $\sim_L$ is called the *syntactic congruence* of L and the quasiorder $\preceq_L$ induces a stable order $\leq_L$ on $S(L) = A^+/\sim_L$. The ordered semigroup $(S(L), \leq_L)$ is called the *syntactic ordered semigroup* of L, the relation $\leq_L$ is called the *syntactic order* of L and the canonical morphism η_L from A^+ onto $S(L)$ is called the *syntactic morphism* of L. The syntactic ordered semigroup is the smallest ordered semigroup that recognizes L. More precisely, an ordered semigroup $(S, \leq)$ recognizes L if and only if $(S(L), \leq_L)$ is a quotient of $(S, \leq)$.

If $\mathbf{V}$ is variety of ordered semigroups and A is a finite alphabet, we denote by $A^+\mathcal{V}$ the set of recognizable languages of A^+ which are recognized by an ordered semigroup of $\mathbf{V}$. Equivalently, $A^+\mathcal{V}$ is the set of recognizable languages of A^+ whose ordered syntactic semigroup belongs to $\mathbf{V}$. It is shown in [26] that the correspondence $\mathbf{V} \to \mathcal{V}$ is a bijective correspondence between varieties of ordered semigroups and positive varieties of languages. A similar result holds if languages are considered as subsets of the free monoid A^*. Then one should consider monoids and varieties of ordered monoids instead of semigroups and varieties of semigroups.

4. Polynomial closure and unambiguous polynomial closure

There are in fact two slightly different notions of polynomial closure, one for +-classes and one for *-classes.

The *polynomial closure* of a class of languages $\mathcal{L}$ of A^+ is the set of languages of A^+ that are finite unions of languages of the form $u_0 L_1 u_1 \cdots L_n u_n$, where $n \geq 0$, the u_i's are words of A^* and the L_i's are elements of $\mathcal{L}$. If $n = 0$, one requires of course that u_0 is not the empty word.

The *polynomial closure* of a class of languages $\mathcal{L}$ of A^* is the set of languages that are finite unions of languages of the form $L_0 a_1 L_1 \cdots a_n L_n$, where the a_i's are letters and the L_i's are elements of $\mathcal{L}$.

By extension, if $\mathcal{V}$ is a +-variety (resp. *-variety), we denote by Pol $\mathcal{V}$ the class of languages such that, for every alphabet A, $A^+\text{Pol } \mathcal{V}$ (resp. $A^*\text{Pol } \mathcal{V}$) is the polynomial closure of $A^+\mathcal{V}$ (resp. $A^*\mathcal{V}$). We also denote by Co-Pol $\mathcal{V}$ the class of languages whose complement is in Pol $\mathcal{V}$ and by BPol $\mathcal{V}$ the boolean closure of Pol $\mathcal{V}$.

Our main result describes the counterpart, on varieties of ordered semigroups, of the operation of polynomial closure on varieties of languages.

Theorem 4.1. *Let* **V** *be a variety of semigroups (resp. monoids) and let* $\mathcal{V}$ *be the corresponding +-variety (resp ∗-variety). Then Pol* $\mathcal{V}$ *is a positive +-variety (resp ∗-variety) and the corresponding variety of semigroups (resp. monoids) is the Mal'cev product* $[x^\omega y x^\omega \leq x^\omega]$ Ⓜ **V**.

In particular, for each alphabet A, A^+Pol $\mathcal{V}$ and A^+Co-Pol $\mathcal{V}$ are closed under finite union and intersection, a result due to Arfi [2,3].

The marked product $L = u_0 L_1 u_1 \cdots L_n u_n$ (resp. $L_0 a_1 L_1 \cdots a_n L_n$) of n languages $L_1, \ldots, L_n$ of A^+ (resp. A^*) is *unambiguous* if every word u of L admits a unique factorization of the form $u_0 v_1 u_1 \cdots v_n u_n$ (resp. $u_0 a_1 u_1 \cdots a_n u_n$) with $v_1 \in L_1, \ldots, v_n \in L_n$. The *unambiguous polynomial closure* of a class of languages $\mathcal{L}$ of A^+ (resp. A^*) is the set of languages that are finite disjoint unions of (marked) unambiguous products of languages of $\mathcal{L}$.

By extension, if $\mathcal{V}$ is a variety of languages, we denote by UPol $\mathcal{V}$ the class of languages such that, for every alphabet A, A^+UPol $\mathcal{V}$ (resp. A^*UPol $\mathcal{V}$) is the unambiguous polynomial closure of $A^+\mathcal{V}$ (resp. $A^*\mathcal{V}$). The following result was established in [15,20] as a generalization of a result of Schützenberger [25].

Theorem 4.2. *Let* **V** *be a variety of monoids (resp. semigroups) and let* $\mathcal{V}$ *be the corresponding ∗-variety (resp. +-variety). Then UPol* $\mathcal{V}$ *is a variety of languages, and the associated variety of monoids (resp. semigroups) is* **LI** Ⓜ **V**.

We give a new characterization of UPol $\mathcal{V}$.

Theorem 4.3. *Let* $\mathcal{V}$ *be a variety of languages. Then Pol* $\mathcal{V} \cap$ *Co-Pol* $\mathcal{V} =$ *UPol* $\mathcal{V}$. *In particular, Pol* $\mathcal{V} \cap$ *Co-Pol* $\mathcal{V}$ *is a variety of languages.*

5. Concatenation hierarchies

Let $\mathcal{V}$ be a variety of languages. The concatenation hierarchy of basis $\mathcal{V}$ is the sequence of varieties $\mathcal{V}_n$ and of positive varieties $\mathcal{V}_{n+1/2}$ defined as follows:
 (1) $\mathcal{V}_0 = \mathcal{V}$
 (2) for every integer $n \geq 0$, $\mathcal{V}_{n+1/2} = $ Pol $\mathcal{V}_n$,
 (3) for every integer $n \geq 0$, $\mathcal{V}_{n+1} = $ BPol $\mathcal{V}_n$.
The corresponding varieties of semigroups and ordered semigroups (resp. monoids and ordered monoids) are denoted $\mathbf{V}_n$ and $\mathbf{V}_{n+1/2}$. Theorem 4.1 gives an explicit relation between $\mathbf{V}_n$ and $\mathbf{V}_{n+1/2}$.

Proposition 5.1. *For every* $n \geq 0$, $\mathbf{V}_{n+1/2} = [x^\omega y x^\omega \leq x^\omega]$ Ⓜ $\mathbf{V}_n$.

5.1. Straubing's hierarchy

This is the hierarchy of positive *-varieties whose level 0 is the trivial variety. The sets of level 1/2 are the finite unions of sets of the form $A^*a_1A^*a_2\cdots a_kA^*$, where the a_i's are letters. It is easy to see directly that level 1/2 is decidable (see Arfi [2,3]). One can also derive this result from our syntactic characterization : a language is of level 1/2 if and only if its ordered syntactic monoid satisfies the identity $x^\omega y x^\omega \le x^\omega$.

This leads to a polynomial algorithm to decide whether the language accepted by a complete deterministic n-state automaton is of level 1/2. This algorithm relies on the notion of *configuration*. Recall that a *subgraph* of a graph $G = (E, V)$ (where $V \subseteq E \times E$) is a graph $G' = (E', V')$ whose set of edges E' is a subset of E. A graph $G' = (E', V')$ is a *quotient* of G if there exists a map π from G onto G' such that $E' = E\pi$. Finally, a *configuration* of G is a quotient of a subgraph of G.

Theorem 5.2. Let $\mathcal{A} = (Q, A, E, i, F)$ be a complete deterministic automaton recognizing a language L and let G be the reflexive and transitive closure of the graph of $\mathcal{A} \times \mathcal{A}$. Then L is of level 1/2 if for every configuration of G of the form

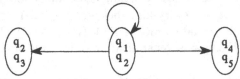

where the q_i's are states of $\mathcal{A}$, the condition $q_4 \in F$ implies $q_5 \in F$. If $\mathcal{A}$ is minimal, this condition is also sufficient. Therefore, one can decide in polynomial time whether the language accepted by a deterministic n-state automaton is of level 1/2.

The sets of level 1 are the finite boolean combinations of languages of level 1/2, which were characterized by Simon [26] : a language of A^* is of level 1 if and only if its syntactic monoid satisfies the identities $x^\omega = x^{\omega+1}$ and $(xy)^\omega = (yx)^\omega$. Simon's result yields an algorithm to decide whether a given recognizable set is of level 1. Actually, it was shown by Stern [29] that one can decide in polynomial time whether the language accepted by a deterministic n-state automaton is of level 1.

The sets of level 3/2 also have a simple description, although this is not a direct consequence of the definition. Indeed, Arfi [2,3] proved that the sets of level 3/2 of A^* are the finite unions of sets of the form $A_0^*a_1A_1^*a_2\cdots a_kA_k^*$, where the a_i's are letters and the A_i's are subsets of A. We derive from our main result the following syntactic characterization.

Theorem 5.3. A language is of level 3/2 if and only if its ordered syntactic monoid satisfies the identity $x^\omega y x^\omega \le x^\omega$ for every x, y such that $c(x) = c(y)$.

Arfi [2,3] proved that level 3/2 is also decidable. But this result relies on a deep result of Hashiguchi, and the corresponding algorithm reduces to a finiteness problem on semigroups of matrices, for which only exponential upper bounds are known. We give below a much more reasonable algorithm. Let $\mathcal{A} = (Q, A, \cdot, i, F)$ be a complete deterministic n-state automaton. Let $\mathcal{B}$ be the automaton that computes the content of a word. Formally, $\mathcal{B} = (2^A, A, \cdot, \emptyset, 2^A)$ where the transition function is defined, for every subset B of A and every letter $a \in A$, by $B \cdot a = B \cup \{a\}$. Consider the product automaton $\mathcal{C} = \mathcal{B} \times \mathcal{A} \times \mathcal{A}$ and let G' be the reflexive and transitive closure of its transition graph.

Theorem 5.4. *Let $\mathcal{A} = (Q, A, E, i, F)$ be a complete automaton recognizing a language L. Then L is of level 3/2 if, for every configuration of G' of the form*

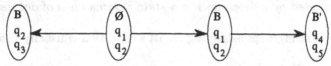

where B and B' are subsets of A and the q_i's are states of $\mathcal{A}$, the condition $q_4 \in F$ implies $q_5 \in F$. This condition is also necessary if $\mathcal{A}$ is minimal. Consequently, there is an algorithm, in time polynomial in $2^{|A|}n$, for testing whether the language of A^* accepted by a deterministic n-state automaton is of level 3/2.

The decidability of level 2 is a challenging open problem, although much progress has been made in recent years [4,5,8,19,31,33,35,36]. In the case of languages whose syntactic monoid is an inverse monoid, a complete characterization was given by Cowan [8], completing partial results of Straubing and the second author [33,35,36]. Our main result gives a much shorter proof of Cowan's result and it is proved in [35,36] that Cowan's result yields the following important corollary.

Corollary 5.5. *It is decidable whether an inverse monoid belongs to $\mathbf{V_2}$.*

5.2. Dot-depth hierarchy

In this hierarchy, introduced by Brzozowski [7], the level 0 is the trivial $+$-variety. The languages of level 1/2 are by definition finite unions of languages of the form $u_0 A^+ u_1 A^+ \cdots u_{k-1} A^+ u_k$, where $k \geq 0$ and $u_0, \ldots, u_k \in A^*$. These languages can also be expressed as finite unions of languages of the form $u_0 A^* u_1 A^* \cdots u_{k-1} A^* u_k$. The syntactic characterization is a simple application of our main result. It also yields to a polynomial algorithm

Proposition 5.6. *A language of A^+ is of dot-depth 1/2 if and only if its ordered syntactic semigroup satisfies the identity $x^\omega y x^\omega \leq x^\omega$.*

Theorem 5.7. *One can decide in polynomial time whether the language accepted by a deterministic n-state automaton is of dot-depth 1/2.*

The sets of dot-depth 1 are the finite boolean combinations of languages of dot-depth 1/2. The syntactic characterization of these languages was settled by Knast and relies on the notion of *graph of a semigroup*. Given a semigroup S, form a graph $G(S)$ as follows: the vertices are the idempotents of S and the edges from e to f are the elements of the form esf.

Theorem 5.8. (Knast [10,11]) *A language of A^+ is of dot-depth 1 if and only if the graph of its syntactic semigroup satisfies the following condition : if e and f are two vertices, p and r edges from e to f, and q and s edges from f to e, then $(pq)^\omega ps(rs)^\omega = (pq)^\omega (rs)^\omega$.*

Theorem 5.9. (Stern [29]) *One can decide in polynomial time whether the language accepted by a deterministic n-state automaton is of dot-depth 1.*

The variety of semigroups satisfying Knast's condition is usually denoted $\mathbf{B_1}$.

6. The sequential calculus

Büchi's sequential calculus is built up from a binary relation symbol $<$ and, for each letter $a \in A$, a unary predicate R_a. To each word u is associated a finite structure $\mathfrak{M}_u = (\{1, \ldots, |u|\}, (R_a)_{a \in A}, <)$ where $R_a = \{i \in \{1, \ldots, |u|\} \mid u(i) = a\}$ is the set of positions of the letter a in u and $<$ is the usual order on $\{1, \ldots, |u|\}$. For instance, if $u = abbaab$, then $R_a = \{1, 4, 5\}$ and $R_b = \{2, 3, 6\}$. Terms, atomic formulæ and first order formulæ are defined in the usual way. A word u satisfies a sentence φ if φ is true when interpreted on the structure $\mathfrak{M}_u$. There is a special convention for the empty word: it satisfies all universal sentences (sentences of the form $\forall x \varphi(x)$) and no existential sentences. To each sentence φ, one associates the sets of words $L(\varphi)$ that satisfy φ. For instance, if $\varphi \equiv \exists i\, R_a i$, then $L(\varphi) = A^* a A^*$. The reader is referred to the survey article [17] for more detail on this logic. The first order definable languages were first characterized by McNaughton and Papert [13] : a recognizable subset of A^* is first-order definable if and only if it is star-free. The correspondence between star-free languages and first order logic is even tighter. Indeed, Thomas has shown that the Straubing hierarchy coincides with the quantifier alternation hierarchy of first order formulæ, defined as follows.

A formula φ is said to be a Σ_n-formula if it is equivalent to a formula of the form $\varphi = Q(x_1, \ldots, x_k)\psi$ where ψ is quantifier free and $Q(x_1, \ldots, x_k)$ is a sequence of n blocks of quantifiers such that the first block contains only existential quantifiers (note that this first block may be empty), the second block universal quantifiers, etc.. Similarly, if $Q(x_1, \ldots, x_k)$ is formed of n alternating blocks of quantifiers beginning with a block of universal quantifiers (which again might be empty), we say that φ is a Π_n-formula.

Denote by Σ_n (resp. Π_n) the class of languages which can be defined by a Σ_n-formula (resp. a Π_n-formula) and by $B\Sigma_n$ the set of boolean combinations of Σ_n-formulæ. Finally, set, for every $n \geq 0$, $\Delta_n = \Sigma_n \cap \Pi_n$. The connection with Straubing's hierarchy can be stated as follows. Denote by $\mathcal{V}_n$ the class of languages of level n. In particular, $\mathcal{V}_{n+1/2}$ is equal to Pol $\mathcal{V}_n$.

Theorem 6.1. (Thomas [34], Perrin and Pin [14])
 (1) *A language is in* $B\Sigma_n$ *if and only if it is in* $\mathcal{V}_n$
 (2) *A language is in* Σ_{n+1} *if and only if it is in Pol* $\mathcal{V}_n$
 (3) *A language is in* Π_{n+1} *if and only if it is in Co-Pol* $\mathcal{V}_n$

We now complete this result by giving a characterization of the Δ_n classes, which follows immediately from Theorems 4.3 and 6.1.

Theorem 6.2. *A language of* A^* *is in* Δ_{n+1} *if and only if it is in UPol* $\mathcal{V}_n$.

Finally, our results on logic can be summarized in the following diagrams

Logical hierarchy

$$\Delta_0 = \Sigma_0 = \Pi_0 = \Delta_1 = B\Sigma_0 \quad \overset{\Sigma_1}{\underset{\Pi_1}{\diamondsuit}} \quad B\Sigma_1 \text{---} \Delta_2 \quad \overset{\Sigma_2}{\underset{\Pi_2}{\diamondsuit}} \quad B\Sigma_2 \text{---} \Delta_3 \quad \overset{\Sigma_3}{\underset{\Pi_3}{\diagup}}$$

Straubing hierarchy

$$\mathcal{V}_0 \quad \overset{\text{Pol } \mathcal{V}_1}{\underset{\text{Co-Pol } \mathcal{V}_1}{\diamondsuit}} \mathcal{V}_1 \text{--- UPol } \mathcal{V}_2 \quad \overset{\text{Pol } \mathcal{V}_2}{\underset{\text{Co-Pol } \mathcal{V}_2}{\diamondsuit}} \mathcal{V}_2 \text{--- UPol } \mathcal{V}_3 \quad \overset{\text{Pol } \mathcal{V}_3 \cdots}{\underset{\text{Co-Pol } \mathcal{V}_3 \cdots}{\diagup}}$$

7. Conclusion and open problems

Let $\mathbf{V}$ be a variety of semigroups and let $\mathcal{V}$ be the corresponding $+$-variety. We have shown that the algebraic counterpart of the operation $\mathcal{V} \to \text{Pol } \mathcal{V}$ on varieties of languages is the operation $\mathbf{V} \to [\![x^\omega y x^\omega \leq x^\omega]\!] \circledM \mathbf{V}$. Similarly, the algebraic counterpart of the operation $\mathcal{V} \to \text{Co-Pol } \mathcal{V}$ is the operation $\mathbf{V} \to [\![x^\omega \leq x^\omega y x^\omega]\!] \circledM \mathbf{V}$. We conjecture that the variety of semigroups (resp. monoids) corresponding to BPol $\mathcal{V}$ is $\mathbf{B_1} \circledM \mathbf{V}$. The conjecture was proved to be true if $\mathbf{V}$ is the trivial variety of monoids, the trivial variety of semigroups or the variety of monoids consisting of all groups [32,10,12]. Note also that every language of BPol $\mathcal{V}$ is recognized by a semigroup of $\mathbf{B_1} \circledM \mathbf{V}$. Finally, it is proved in [21] that the identities of $\mathbf{B_1} \circledM \mathbf{V}$ are

$$(x^\omega p y^\omega q x^\omega)^\omega x^\omega p y^\omega s x^\omega (x^\omega r y^\omega s x^\omega)^\omega = (x^\omega p y^\omega q x^\omega)^\omega (x^\omega r y^\omega s x^\omega)^\omega$$

for all $x, y, p, q, r, s \in \hat{A}^*$ for some finite alphabet A such that $\mathbf{V}$ satisfies $x^2 = x = y = p = q = r = s$.

Acknowledgements

The authors would like to thank Marc Zeitoun for a careful reading of a first version of this article and Jean Goubault for asking a question about the Δ_n hierarchy that led to section 6 of this article.

8. References

[1] J. Almeida, *Finite Semigroups and Universal Algebra*, World Scientific (Series in Algebra, Volume 3), Singapore, 1995, 511 pp.

[2] M. Arfi, Polynomial operations and rational languages, 4th STACS, *Lecture Notes in Computer Science* **247**, (1987) 198–206.

[3] M. Arfi, Opérations polynomiales et hiérarchies de concaténation, *Theoret. Comput. Sci.* **91**, (1991) 71–84.

[4] F. Blanchet-Sadri, On dot-depth two, *Informatique Théorique et Applications* **24**, (1990) 521–529.

[5] F. Blanchet-Sadri, On a complete set of generators for dot-depth two, *Discrete Appl. Math.*, **50**, (1994) 1–25.

[6] S. L. Bloom, Varieties of ordered algebras, *J. Comput. System Sci.* **13**, (1976) 200–212.

[7] J. A. Brzozowski, Hierarchies of aperiodic languages, *RAIRO Inform. Théor.* **10**, (1976) 33–49.

[8] D. Cowan, Inverse monoids of dot-depth 2, *Int. Jour. Alg. and Comp.* **3**, (1993) 411-424.

[9] S. Eilenberg, *Automata, languages and machines*, Vol. B, Academic Press, New York, 1976.

[10] R. Knast, A semigroup characterization of dot-depth one languages, *RAIRO Inform. Théor.* **17**, (1983) 321–330.

[11] R. Knast, Some theorems on graph congruences, *RAIRO Inform. Théor.* **17**, (1983) 331–342.

[12] S. W. Margolis and J.E. Pin, Product of group languages, *FCT Conference, Lecture Notes in Computer Science* **199**, (1985) 285–299.

[13] R. McNaughton and S. Pappert, *Counter-free Automata*, MIT Press, 1971.

[14] D. Perrin and J.E. Pin, First order logic and star-free sets, *J. Comput. System Sci.* **32**, (1986) 393–406.

[15] J.-E. Pin, Propriétés syntactiques du produit non ambigu. *7th ICALP, Lecture Notes in Computer Science* **85**, (1980) 483–499.

[16] J.-E. Pin, *Variétés de langages formels*, Masson, Paris, 1984. English translation: *Varieties of formal languages*, Plenum, New-York, 1986.

[17] J.-E. Pin, Logic, Semigroups and Automata on Words, *Annals of Math. and Artificial Intelligence*, to appear.

[18] J.-E. Pin, A variety theorem without complementation, *Izvestija vuzov. Matematika*, to appear.

[19] J.-E. Pin and H. Straubing, Monoids of upper triangular matrices, *Colloquia Math. Soc. Janos Bolyai 39, Semigroups, Szeged*, (1981) 259–272.

[20] J.-E. Pin, H. Straubing and D. Thérien, Locally trivial categories and unambiguous concatenation, *Journal of Pure and Applied Algebra* **52**, (1988) 297–311.

[21] J.-E. Pin and P. Weil, Free profinite semigroups, Mal'cev products and identities, to appear.

[22] J.-E. Pin and P. Weil, A Reiterman theorem for pseudovarieties of finite first-order structures, to appear.

[23] J. Reiterman, The Birkhoff theorem for finite algebras, *Algebra Universalis* **14**, (1982) 1–10.

[24] M.P. Schützenberger, On finite monoids having only trivial subgroups, *Information and Control* **8**, (1965) 190–194.

[25] M.P. Schützenberger, Sur le produit de concaténation non ambigu, *Semigroup Forum* **13**, (1976) 47–75.

[26] I. Simon, Piecewise testable events, *Proc. 2nd GI Conf., Lecture Notes in Computer Science* **33**, (1975) 214–222.

[27] I. Simon, Factorization forests of finite height, *Theoret. Comput. Sci.* **72**, (1990) 65–94.

[28] I. Simon, The product of rational languages, *Proceedings of ICALP 1993, Lecture Notes in Computer Science* **700**, (1993), 430–444.

[29] J. Stern, Characterization of some classes of regular events, *Theoret. Comput. Sci.* **35**, (1985) 17–42.

[30] H. Straubing, Aperiodic homomorphisms and the concatenation product of recognizable sets, *J. Pure Appl. Algebra* **15** (1979) 319–327.

[31] H. Straubing, Semigroups and languages of dot-depth two, *Theoret. Comput. Sci.* **58** (1988) 361–378.

[32] H. Straubing and D. Thérien, Partially ordered finite monoids and a theorem of I. Simon, *J. of Algebra* **119**, (1985) 393–399.

[33] H. Straubing and P. Weil, On a conjecture concerning dot-depth two languages, *Theoret. Comput. Sci.* **104**, (1992) 161–183.

[34] W. Thomas, Classifying regular events in symbolic logic, *J. Comput. System Sci.* **25**, (1982) 360–375.

[35] P. Weil, Inverse monoids of dot-depth two, *Theoret. Comput. Sci.* **66**, (1989), 233–245.

[36] P. Weil, Some results on the dot-depth hierarchy, *Semigroup Forum* **46** (1993), 352–370.

Lower Bounds on Algebraic Random Access Machines

(extended abstract)

Amir M. Ben-Amram*   and   Zvi Galil**
DIKU, University of Copenhagen   Tel-Aviv University
and Columbia University

We prove general lower bounds for set recognition on random access machines (RAMs) that operate on real numbers with algebraic operations $\{+, -, \times, /\}$, as well as RAMs that use the operations $\{+, -, \times, \lfloor \ \rfloor\}$. We do it by extending a technique formerly used with respect to algebraic computation trees. In the case of algebraic computation trees, the complexity was related to the number of connected components of the set W to be recognized. For RAMs, two similar results apply to the number of connected components of W°, the topological interior of W. Other results use $(\overline{W})^\circ$, the interior of the topological closure of W.

We present theorems that can be applied to a variety of problems and obtain lower bounds, many of them tight, for the following models:

1. A RAM which operates on real numbers, using integers to address memory, and either the operations $\{+, -, \times, /\}$ or $\{+, -, \times, \lfloor \ \rfloor\}$;

2. A RAM of each of the above instruction sets, extended by allowing arbitrary real numbers to be used as memory addresses, and adding a test-for-integer instruction.

3. A RAM of each of the above instruction sets which can compute with arbitrary real numbers, as well as use them for memory addressing, while having the provision that the input is always integer (for one result on this model, we require that all program constants be rational).

1. Introduction

Consider a set $W \subset \Re^n$ and the problem of recognizing W. Ben-Or proved that an algebraic computation tree recognizing W requires height that is at least a logarithm of the number of path-connected components of W [1]. For the exact statement see Theorem 1 in Section 3. Yao proved that under certain conditions on W, a similar lower bound applies to an algebraic computation tree that is only required to give the correct answer for *integer* input [16]. For the exact

---

* Part of this research was performed while this author was a PhD student at Tel-Aviv University.

** Partially supported by NSF grant CCR-93-16209 and CISE Institutional Infrastructure Grant CDA-90-24735.

statement see Theorem 2 in Section 3. The question we ask and answer in this paper is: can we prove similar results for random access machine models with powerful instruction sets?

A well known example, the element distinctness problem, is defined by $W = \{x \in \Re^n | \ x_i \neq x_j \text{ for } 1 \leq i \neq j \leq n\}$. Both Ben-or and Yao used their theorems to obtain an $\Omega(n \log n)$ lower bound on the height of a computation tree that solves this problem. However on a RAM, the problem can be easily solved in time $O(n)$ by storing a flag in address x_i for $i = 1, ..., n$. This indicates that perhaps the answer to our question above is negative. In the case of input from $\Re^n$, this solution may arise an objection: can we expect to be able to use real numbers as memory addresses? This objection does not arise in the case of integer input, so in the integer case it is clear that the application of computation-tree lower bounds to the RAM is indeed problematic. In the case of real-number input too, it is not obvious that they apply.

Our first new results are lower-bound theorems for the real-number Algebraic RAM, i.e., a RAM with operations $\{+, -, \times, /\}$. We consider two main variants of this model. In the standard variant, only integers may be used to address memory locations. For this variant, we prove that recognizing W requires time that is at least a logarithm of the number of connected components of W°.

Referring again to the element distinctness problem, the proof of the lower bound in [1] uses a set W such that both W and W° have $n!$ connected components. Thus the lower bound of $\Omega(n \log n)$ extends to the above model.

We also consider a "non-standard" model which allows arbitrary real numbers to be used as addresses. For this model we prove that the time to recognize W is at least a logarithm of the number of connected components of $(\overline{W})^\circ$, the interior of the closure of W. Moreover, under conditions similar to those required by Yao, the same lower bound applies to programs that are only required to recognize integer elements of W.

In the case of Element Distinctness, it turns out that $(\overline{W})^\circ$ consists of a single component. This explains why the lower bound for Element Distinctness cannot be obtained under these models. However, in some other examples we find that the same lower bound that has been proved for computation trees also holds for the RAM.

We proceed to consider the truncation operation, also known as *floor*. We show that results very similar to the above three propositions also hold for the corresponding models with operations $\{+, -, \times, \lfloor \ \rfloor\}$.

The six lower bound theorems are given in Section 3.

In Section 4 we consider a dozen of examples of set-recognition problems. In some cases the lower bounds proved for algebraic computation trees also apply to random access machines, while in some other they don't. In all of the latter cases but two we show that like in the case of element distinctness, a linear time algorithm exists on the RAM.

Appendix A sketches the proofs of the first two theorems. The next four follow the same outline, but are more complicated and have been omitted for lack of space. An important tool in the proofs is a technique that was introduced by Paul and Simon [14]. Informally, if a RAM recognizes a set W in time bounded by t, we are able to build an algebraic computation tree of height $O(t)$ that recognizes a set V, such that V approximates W in a certain sense. We develop this basic idea into six lemmas that are used in the proofs of the six theorems. In each case, we show that the complexity of the approximate set can be related to that of W via a transformation such as obtaining the interior or closure-interior of the set.

We now review some related results. The original application of the Paul and Simon technique was to prove that algebraic RAMs do not perform better than decision trees for certain problems where decision-tree lower bounds are known (e.g., sorting). Klein and Meyer auf der Heide [8,12] extended lower bounds obtained by the component-counting method to a RAM which can only add and subtract (but not multiply or divide).

The power of *floor* in computation trees has been investigated in numerous papers [2,3,4,7,9,10,11]; Mansour, Schieber and Tiwari [10] also extend their lower bound technique to random access machines. Their result, and most of the others, relate to problems where the time of computation depends on the size of the input value or values. In contrast, in this paper we consider *genuinely time-bounded computations* [13], namely programs whose running time is bounded by a function of n, the number of input values.

Most of the above papers consider models that combine floor with division (sometimes in the form of truncated division of integers); however, we were not able to allow both in our theorems. In fact, it is known that for problems of the kind considered here, this combination adds significant power to the model. A well-known example is Gonzalez' algorithm for the *max gap* problem [15] which runs in linear time, contrasting the $\Omega(n \log n)$ lower bounds given by our theorems. Another example is a linear-time algorithm for the decision problem Min Gap, which also has an $\Omega(n \log n)$ lower bound in the models considered here; this algorithm, which is quite simple, is omitted from this extended abstract.

All of the above results do not allow the RAM to use an arbitrarily-initialized memory. We allow this extension in our lower bounds for problems with real input. When the input is integer, free memory initialization allows every problem to be solved in linear time. We sketch the way this can be done: first encode the input n-tuple in a single number, then use this number to index a table that contains all the right answers. Note that this is an infinite table. All our lower bounds allow the program to make use of finite tables, whose size and contents may depend on n. They thus apply to non-uniform programs in the usual sense.

2. Preliminaries

In this section we define our models of computation and list some mathematical notation used in expressing the results as well as in the proofs.

2.1 Models of Computation

An *algebraic computation tree* or ACT models a program in which each step has one of the following types. (i) An assignment $z \leftarrow u$ where u is either a constant number or a variable. (ii) An arithmetic operation $z \leftarrow u \circ w$ where $\circ \in \{+, -, \times, /\}$. (iii) A comparison $u : 0$. The variables u, w may be either input variables (denoted $x_1, \ldots, x_n$) or results of prior computation. Each arithmetic node has one child; each comparison node has three children associated with the possibilities $>$, $<$ and $=$. Each leaf is labeled with a constant or a variable that represents the output of the computation. An algebraic computation tree is said to recognize a set if each output is a 0 or a 1. A leaf labeled 1 is called *accepting*. The tree is said to recognize $W \subseteq \Re^n$ when an accepting leaf is reached if and only if $(x_1, \ldots, x_n) \in W$. The tree is said to recognize W for a restricted set of inputs (e.g., the integers) if the last condition is fulfilled for inputs of this set; it does not matter which leaf is reached for other inputs.

We denote by $h(T)$ the height of the tree T, which represent the worst-case time complexity of the computation described by T.

Random access machines come in many flavours. In general, A RAM has a program which includes computational primitives, typically arithmetic operations; branches (unconditional or dependent on a comparison), and an output instruction (which may be assumed to stop the program in our case). The variables involved are located in a random access memory and instructions of *indirect addressing* allow the contents of a variable to be used as a memory address.

To specify a RAM model completely, we define three parameters:

The *domain* $\mathcal{D}$ is the set of values that may be manipulated by the machine as "units of data." These are the integers or natural numbers in classical RAM models, and the real numbers or rational numbers in the models considered in this paper.

The *address space* $\mathcal{A}$ is the set of values that may be used as memory addresses. Thus the size of memory is $|\mathcal{A}|$, and the standard RAM model uses $\mathcal{A} = \mathbb{N}$. In case that $\mathcal{A}$ is strictly contained in $\mathcal{D}$, a program may *fault* by attempting to use a value in $\mathcal{D} \setminus \mathcal{A}$ as an address. In this case the program may be considered invalid, or give an undefined results. An instruction that tests whether a value is a valid address may be useful for programming such models.

The *set of primitives* $\mathcal{F}$ defines the capabilities of computational instructions (which are defined to have unit cost). A RAM where $\mathcal{F}$ consists of the field operations $\{+, -, \times, /\}$ is called *algebraic*. We will also consider the operations $\lfloor\ \rfloor$ (truncation to an integer) and $\chi_{\mathbb{Z}}$ (the characteristic function of $\mathbb{Z}$; can be used to test whether a given value is an integer). The last operation can be easily simulated by *floor*, but not vice versa.

For every RAM program, the initial contents of memory cells is assumed to be zero, except for those that hold the input, and possibly another set of cells, whose contents and size may depend on n but not on the input values. Finally, set recognition by a RAM is defined as for a computation tree.

We use the notation $(\mathcal{D}, \mathcal{A}, \mathcal{F})$-RAM for a model characterized by the respective sets.

A *real-number RAM* is a RAM where $\mathcal{D} = \Re$. Its main justification is as a convenient model for studying problems on real numbers, e.g. in computational geometry; but it is also possible that a problem on integer inputs will be solved faster by a program which uses the power of the real-number RAM. Thus, we consider the real-number RAM also in conjunction with integer-constrained problems (in fact, geometrical problems are often encountered in practice with integer-constrained input, because of the use of raster devices and digitized data).

We consider two variants of the real-number RAM. The "standard" version has $\mathcal{A} = \mathbb{N}$, while a stronger, non-standard version has $\mathcal{A} = \Re$. The last choice seems at first to be a far-fetched idealization; however, it may not be so far-fetched in practice. One may think of the real-number RAM as an abstraction of an actual computer which works with a finite representation of real values, say in floating point. In this case, these representations may be used to address an ordinary random-access memory. Another justification for using an idealized model is that it is elucidates problems; and in fact, it is via our analysis of this model that some of our best results for integer-constrained input (even on the integer RAM) have been obtained.

2.2 Various Notations and Terms

For a set $W \subseteq \Re^n$, let $\beta(W)$ be the number of connected components in W. Recall that W° denotes the interior of W, and $\overline{W}$, the topological closure of W. A *boundary set* is a set of empty interior. An *algebraic variety* in $\Re^n$ is the set of common zeros of a set of polynomials; unless trivial (equal to $\Re^n$), a variety is a closed boundary set.

A set $W \subseteq \Re^n$ is called *scale-invariant* if $\mathbf{x} \in W$ implies $\lambda \mathbf{x} \in W$ for all real $\lambda > 0$. A set $W \subseteq \Re^n$ is said to be *rationally dispersed* if, for every $\mathbf{x} \in \Re^n$ and $\varepsilon > 0$, there is a rational point $\mathbf{z}$ such that $\|\mathbf{z} - \mathbf{x}\| < \varepsilon$ and $\mathbf{z} \in W \iff \mathbf{x} \in W$.

We denote by $\hat{\beta}(W)$ is the number of connected components of W which have non-empty interior (called *primary*).

For $W \subseteq \Re^n$, and $\mathbf{x} \in \Re^n$, let $\tilde{\beta}(W) \overset{\text{def}}{=} \lim_{\lambda \to 0^+} \beta(W \cap B(\lambda, \mathbf{0}))$, where $B(\lambda, \mathbf{0})$ is the open ball of radius λ centered at the origin.

Fact 1. *If W is scale invariant then $\beta(W) = \tilde{\beta}(W)$.*

The following constants appear in our theorems: $c_1 = \frac{1}{1 + 2\log_2 3}$ and $c_2 = 1 + \frac{\log_2 3}{1 + 2\log_2 3}$.

3. Lower-Bound Theorems

For algebraic computation trees, the following results were obtained in [1] and [17], respectively (adapted to our notational conventions):

Theorem 1 (Ben-Or). *Let T be an ACT that recognizes $W \subseteq \Re^n$. Then*

$$h(T) \geq \log_9 \beta(W) - \tfrac{1}{2}n.$$

Remark. With a simple change to Ben-Or's proof, we can replace $\beta(W)$ with $\beta(W^\circ)$. For some sets, $\beta(W^\circ)$ is larger.

Theorem 2 (Yao). *Let $W \subseteq \Re^n$ be scale-invariant and rationally dispersed. If T is an ACT that recognizes W for integer input, then*

$$h(T) \geq c_1(\log_2(\hat{\beta}(W) - 1) - c_2 n.$$

Remark. With no essential change in its proof, this theorem can be strengthened by replacing $\hat{\beta}(W)$ with $\beta(W^\circ)$. It is easy to see that always $\hat{\beta}(W) \leq \beta(W^\circ)$.

In this paper we prove:

Theorem 3. *Let $W \subseteq \Re^n$. If W is recognized in time $t(n)$ on $(\Re, \mathbb{N}, \{+, -, \times, /\})$-RAM, then*

$$t(n) \geq \log_9 \beta(W^\circ) - \tfrac{1}{2}n.$$

The program may specify arbitrary initial values for all memory cells.

Theorem 4. *Let $W \subseteq \Re^n$. If W is recognized in time $t(n)$ on $(\Re, \Re, \{+, -, \times, /, \chi_{\mathbb{Z}}\})$-RAM, then*

$$t(n) \geq \log_9 \beta((\overline{W})^\circ) - \tfrac{1}{2}n.$$

The program may specify arbitrary initial values for all memory cells of integer addresses.

Theorem 5. *Let $W \subseteq \Re^n$ be scale invariant and rationally dispersed. If W is recognized for integer input in time $t(n)$ on $(\Re, \Re, \{+, -, \times, /, \chi_{\mathbb{Z}}\})$-RAM, then*

$$t(n) \geq c_1(\log_2(\beta((\overline{W})^\circ) - 1) - c_2 n.$$

The program may use a finite initialized table, which may depend on n.

Theorem 6. *Let $W \subseteq \Re^n$. If W is recognized in time $t(n)$ on $(\Re, \mathbb{N}, \{+, -, \times, \lfloor \rfloor\})$-RAM, then*

$$t(n) \geq \tfrac{1}{3} \log_9 \widetilde{\beta}(W^\circ) - \tfrac{1}{3} n.$$

The program may specify arbitrary initial values for all memory cells.

Theorem 7. *Let $W \subseteq \Re^n$. If W is recognized in time $t(n)$ on $(\Re, \Re, \{+, -, \times, \lfloor \rfloor\})$-RAM, then*

$$t(n) \geq \tfrac{1}{3} \log_9 \widetilde{\beta}((\overline{W})^\circ) - \tfrac{1}{3} n.$$

The program may specify arbitrary initial values for all memory cells.

Theorem 8. *Let $W \subseteq \Re^n$ be scale invariant and rationally dispersed. If W is recognized for integer input in time $t(n)$ on $(\mathbb{Q}, \mathbb{Q}, \{+, -, \times, \lfloor \rfloor\})$-RAM, then*

$$t(n) \geq \tfrac{1}{2} c_1 (\log_2 (\widetilde{\beta}((\overline{W})^\circ) - 1) - \tfrac{1}{2} c_2 n.$$

The program may use a finite initialized table, which may depend on n.

4. Applications

We give some examples of problems for which lower bounds can be proved as corollaries of our theorems. For lack of space, we only quote results without the technical details.

Example 1. *Set Disjointness.* Given two sets $A = \{x_1, \ldots, x_n\}$ and $B = \{y_1, \ldots, y_n\}$ decide whether A and B are disjoint.

$\Omega(n \log n)$ bounds for computation trees follows from Ben-Or's theorem and similarly from Yao's for the case of integer inputs. The same lower bounds follow from Theorems 3 and 6 for the case $\mathcal{D} = \Re$ and $\mathcal{A} = \mathbb{N}$. On the contrary, both models where $\mathcal{D} = \mathcal{A}$ (either real numbers or integers) solve it in linear time with a straight-forward use of indirect addressing.

Example 2. *Sign of Resultant.* Given $x_1, \ldots, x_n; y_1, \ldots, y_n$ decide whether $\mathrm{RES}(\mathbf{x}, \mathbf{y}) = \prod_{i,j} (x_i - y_j) > 0$.

Ben-Or showed that this problem requires $\Omega(n \log n)$ time in an ACT, and the same lower bound holds for all the models considered in Theorems 3–8.

Example 3. *Element Distinctness.* Given $x_1, x_2, \ldots, x_n$ decide whether these n numbers are all distinct.

Example 4. *Min Gap.* Given $n + 1$ numbers $x_1, \ldots, x_n, t$, decide if the minimum difference between a pair of these numbers is bounded above by t.

Frequently this problem is presented in a computational form, where the minimum gap has to be reported; the above decision problem is obviously not harder, and for it we give a lower bound that matches an upper bound known for the computational problem.

Yao's theorem gives an $\Omega(n \log n)$ lower bound for Element Distinctness on computation trees, even when restricted to integer inputs. Obviously, the lower bound also applies to Min Gap. As there is a matching upper bound (by sorting the numbers), the complexity of both problems for the algebraic computation tree is $\Theta(n \log n)$. The lower bound for Min Gap holds for all RAM models as well. However, models where $\mathcal{D} = \mathcal{A}$ can solve Element Distinctness in linear time, with a straight-forward use of indirect addressing. Thus these models separate the complexities of the two problems. On the other hand, Theorems 3 and 6 do yield $\Omega(n \log n)$ lower bounds for Element Distinctness.

Example 5. *Uniform Gap.* Given $n + 1$ numbers $x_1, x_2, \ldots, x_n, \varepsilon$ decide whether the x_i can be permuted into a non-decreasing sequence where the gap between each pair of consecutive elements is ε.

Example 6. *Max Gap.* Given $n + 1$ numbers $x_1, \ldots, x_n, \varepsilon$, decide whether the x_i can be permuted into a non-decreasing sequence such that the gap between each pair of consecutive elements is bounded above by ε.

Ben-Or's theorem gives an $\Omega(n \log n)$ lower bound for Uniform Gap [15]. This also applies to the Max Gap problem because the answer to Uniform Gap is "yes" if and only if the answer to Max Gap on $x_1, \ldots, x_n$ is ε, and the difference between the minimum and the maximum of this set is $(n-1)\varepsilon$. Clearly, Max Gap can be solved in $O(n \log n)$ time by sorting the numbers. Thus the complexity of both problems for the algebraic computation tree is $\Theta(n \log n)$. As in the previous examples, models with $\mathcal{D} = \mathcal{A}$ separate them. Uniform Gap can be solved on these models in linear time by computing the minimum of the $\{x_i\}$, m, and by means of indirect addressing checking whether the rest of the set equals $\{m + \varepsilon, m + 2\varepsilon, \ldots, m + (n-1)\varepsilon\}$. On the contrary, for Max Gap we obtain an $\Omega(n \log n)$ time under all the models considered in Theorems 3–8.

Example 7. *Sign of Permutation.* Given $x_1, \ldots, x_n$ decide whether there exists an even permutation σ such that $x_{\sigma(1)} < x_{\sigma(2)} < \cdots < x_{\sigma(n)}$.

Ben-Or and Yao showed that this problem requires $\Omega(n \log n)$ in an ACT. The same lower bound holds under all the models considered in Theorems 3–8.

Example 8. *Convex Hull.* Given n points in the plane, decide whether they all lie on the convex hull of the set.

Yao considered a slightly different version of this problem, where we have to decide whether the convex hull has n vertices (which means that all the points lie on the hull, and are also in general position). Yao gave an $\Omega(n \log n)$ lower bound for a computation tree that solves this problem, even for integer inputs. The same lower bound for our version of the problem follows by the remark following Theorem 2. The same lower bound is obtained for RAM models by theorems 3–8.

Example 9. *Knapsack.* Given numbers $x_1, \ldots, x_n$ and $M > 0$ decide if there exists some subset $S \subseteq \{1, 2, \ldots, n\}$ such that

$$\sum_{i \in S} x_i = M .$$

Example 10. *Generalized Knapsack.* Given numbers $x_1, \ldots, x_n$ and $M > 0$ decide if there exists a vector $v \in \{0...k\}^n$ such that the inner product $x \cdot v = M$. (For $k = 1$ this is the original problem.)

Example 11. *Approximate Generalized Knapsack.* Given numbers $x_1, \ldots, x_n$ and $M > 0$ decide if there exists a vector $v \in \{0...k\}^n$ such that

$$|x \cdot v - M| \leq \varepsilon M$$

where $k \in \mathbb{N}$, $\varepsilon > 0$ are fixed constants.

For the approximate problem, we apply a result of Meyer auf der Heide [12] to obtain a lower bound of $\Omega(n^2 \log(k + 1))$ on the time to solve this problem under all the models considered in Theorems 3–8.

The exact problems differ from the approximate ones by setting $\varepsilon = 0$. As in previous cases, this change invalidates the results using $(\overline{W})^\circ$. The results that refer to W° still apply.

5. Conclusion

In this paper we showed how topological arguments, previously used to achieve lower bounds for algebraic computation trees, can be applied to random access machines. We considered machines with standard memory access (using integer addresses) and non-standard models that may use arbitrary real addresses. We considered the case of real-number input as well as that of integer input, and instruction sets $\{+, -, \times, /\}$ and $\{+, -, \times, \lfloor \ \rfloor\}$. Our first lower bounds for these models (Theorems 3 and 6, respectively) show that when problems with real input are considered, it is hard to make effective use of integer addresses within these models. This is one motivation for studying the non-standard model.

If the model is extended with the integrality predicate (which is not an algebraic function), some problems may be solved faster without resorting to non-integer addrsses. E.g., Uniform Gap can be solved in linear time. Theorem 4 gives a class of problems that cannot benefit this way. An even stronger extension is adding the truncation ("floor") operation. Together with division, it allows a linear-time solution to max-gap and to min-gap. Obtaining a better characterization of the power of this instruction set is a challenging open problem. Another open problem is to tighten the bounds for the Knapsack and Generalized Knapsack problems (note that here we could not present a faster algorithm in the stronger models. Also, in contrast with other problems we considered, the computation-tree lower bound is not known to be tight).

Finally, a natural extension of this work is to use other characterizations of the set besides the number of connected components. Yao [17] obtained new lower bounds for algebraic computation trees using higher Betti numbers of W (the number of connected components is the 0-th Betti number.) In particular he obtained an $\Omega(n \log(n/k) - cn)$ lower bound for the following

Example 12. *Generalized Element Distinctness.* Given $x_1, x_2, \ldots, x_n$, decide whether some k of these n numbers are equal.

This result has not been extended so far to the RAM.

REFERENCES

[1] M. Ben-Or, "Lower bounds on algebraic computation trees," *Proc. Fifteenth Annual ACM Symp. on Theory of Computing* (1983), 80–86.

[2] N. H. Bshouty, "Lower bounds for the complexity of functions in a realistic RAM model," *Proc. Israel Symposium on the Theory of Computing and Systems*, Haifa, 1992.

[3] N. H. Bshouty, Y. Mansour, B. Schieber and P. Tiwari, "Fast exponentiation with the floor operation," *Computational Complexity* 2 (1992), 244–255.

[4] E. Dittert and M. J. O'Donnell, "Lower bounds for sorting with realistic instruction sets," *IEEE Trans. on Computers* C-34:4 (1985) 311–317.

[5] D. Dobkin and R. J. Lipton, "A lower bound of $\frac{1}{2}n^2$ on linear search programs for the knapsack problem," *J. of Computer and System Sciences* 16 (1978) 413–417.

[6] D. Dobkin and R. J. Lipton, "On the complexity of computations under varying sets of primitives," in *Automata Theory and Formal Languages* (edited by H. Bradhage), Lecture Notes in Computer Science 33, Springer-Verlag (1975), 110–117.

[7] B. Just, F. Meyer auf der Heide and A. Wigderson, "On computations with integer division," *RAIRO Informatique Théorique* 23 (1989), 101–111.

[8] P. Klein and F. Meyer auf der Heide, "A lower time bound for the knapsack problem on random access machines," *Acta Informatica* 19 (1983), 385–395.

[9] K. Lürwer-Brüggemeier and F. Meyer auf der Heide, "Capabilities and complexity of computations with integer division," *Proc. 10th Symp. on Theoretical Aspects of Computer Science*, Lecture Notes in Computer Science 665, Springer-Verlag (1993), 463–472.

[10] Y. Mansour, B. Schieber and P. Tiwari, "Lower bounds for computations with the floor operation," *SIAM J. Comput.* 20:2 (1991), 315–327.

[11] Y. Mansour, B. Schieber and P. Tiwari, "A lower bound for integer greatest common divisor computations," *J. of the ACM* 38 (1991) 453–471.

[12] F. Meyer auf der Heide, "Lower bounds for solving linear diophantine equations on random access machines," *J. of the ACM* 32:4 (1985), 929–937.

[13] F. Meyer auf der Heide, "On genuinely time bounded computations," *Proc. 6th Symp. on Theoretical Aspects of Computer Science*, Lecture Notes in Computer Science 349, Springer-Verlag (1989), 1–16.

[14] W. Paul and J. Simon, "Decision trees and random access machines," in *Logic and Algorithmic*, monographie n°30 de l'enseignement mathématique, Université de Genève 1982.

[15] F. P. Preparata, and M. I. Shamos, *Computational Geometry: An Introduction*, 2nd printing, Springer-Verlag, 1985/1988.

[16] A. C. Yao, "Lower bounds for algebraic computation trees with integer inputs," *SIAM J. Comput.* 20:4 (1991), 655–668.

[17] A. C. Yao, "Decision tree complexity and Betti numbers," *Proc. 26th Annual ACM Symp. on Theory of Computing* (1994).

Appendix A. Proofs of Lower Bound Theorems

We first give some topological facts which are used in the proofs.

Lemma 1. *Let W, A be subsets of some metric space such that A is closed and has empty interior. Then $(\overline{W \setminus A})^\circ = (\overline{W})^\circ = (\overline{W \cup A})^\circ$.*

Lemma 2. *Let W be an open set. Then $\beta((\overline{W})^\circ) \leq \beta(W)$.*

Lemma 3. *Let W be open and A a boundary set. Then $\beta(W \setminus A) \geq \beta(W)$.*

A.1 Proof of Theorem 3

To prove this theorem, we translate the RAM program into a computation tree, which involves determining the effect of indirect addressing operations. This turns out to be especially easy due to the following property of rational functions (functions which are quotients of two polynomials).

Fact 1. *Let $f : \Re^n \to \Re$ be rational. If there is an open non-empty subset of $\Re^n$ on which f only assumes integer values, then f is constant throughout $\Re^n$.*

Using this fact we prove:

Lemma 4. *Let P be a $(\Re, \mathbb{N}, \{+, -, \times, /\})$-RAM program that recognizes a set $W \subseteq \Re^n$ in time bounded by t. The initial contents of the random access memory may be arbitrarily specified by P. Then P can be transformed into an algebraic computation tree, recognizing a set V, such that the following holds:*

(1) $h(T) \leq t$;

(2) V is an open set;

(3) There exists a non-trivial algebraic variety H such that $V = W \setminus H$.

Note that (2) and (3) together yield $V = W^\circ \setminus H$. We next apply the following lemma, essentially from [1].

Lemma 5. *Let T be an ACT recognizing $V \subseteq \Re^n$. Then $\beta(V) \leq 9^{h(T)} 3^n$.*

Consider now a RAM program P recognizing W in time t. Let T, V and H be given by Lemma 4. By lemmas 5 and 3,

$$\beta(W^\circ) \leq \beta(W^\circ \setminus H) = \beta(V) \leq 9^t 3^n.$$

Theorem 3 follows.

A.2 Proof of Theorem 4

We begin with a definition.

Definition 1. *A set $H \subseteq \Re^n$ is a bale if there is a finite set of polynomials $\{P_i, Q_i, \ i \in I\}$ such that*

$$H = \bigcup_{i \in I} \{\, \mathbf{x} \in \Re^n : \frac{P_i(\mathbf{x})}{Q_i(\mathbf{x})} \in \mathbb{Z} \vee Q_i(\mathbf{x}) = 0 \,\}.$$

The above bale is denoted by $\mathcal{B}(P_i, Q_i, i \in I)$. A bale is *non-trivial* if it is different from $\Re^n$. Since a bale is a countable union of algebraic varieties, Baire's category theorem asserts that every non-trivial bale is a closed boundary set.

The following lemma is similar in spirit to Lemma 4, except that the difference between the sets V and W becomes two-sided: we cannot guarantee that we only accept input that is in W.

Lemma 6. *Let P be a $(\Re, \Re, \{+, -, \times, /\})$-RAM program that recognizes a set $W \subseteq \Re^n$ in time bounded by t. The initial contents of the random access memory may be arbitrarily specified by P. Then P can be transformed into an algebraic computation tree, recognizing a set V, such that the following holds:*

(1) $h(T) \leq t$;

(2) V is an open set;

(3) There exists a non-trivial bale H such that $V \cup H = W \cup H$.

Consider now a RAM program P recognizing W in time t. Let T, V and H be given by Lemma 6. Using Lemmas 1 and 6,

$$(\overline{W})^\circ = (\overline{W \cup H})^\circ = (\overline{V \cup H})^\circ = (\overline{V})^\circ$$

by Lemmas 2 and 5,

$$\beta((\overline{W})^\circ) = \beta((\overline{V})^\circ) \leq \beta(V) \leq 9^t 3^n.$$

Theorem 4 follows.

Improved Deterministic PRAM Simulation on the Mesh[*]

(Extended Abstract)

Andrea Pietracaprina[1] and Geppino Pucci[2]

[1] Dipartimento di Matematica Pura e Applicata, Università di Padova,
Via Belzoni 7, I35131 Padova, Italy
[2] Dipartimento di Elettronica e Informatica, Università di Padova,
Via Gradenigo 6/a, I35131 Padova, Italy

Abstract. This paper describes an improved scheme for PRAM simulation on the mesh. The simulation algorithm achieves nearly optimal slowdown by means of a hierarchical distribution technique, which provides a powerful mechanism to control network congestion. The results in this paper improve upon previous works in many directions. Specifically, the scheme requires less powerful expanding graphs and can be made fully constructive for a wide range of memory sizes, with better slowdown than previous constructive schemes.

1 Introduction

An (n, m)-PRAM consists of n RAM processors that have direct access to a shared memory of m *variables*. In a PRAM *step*, executed in unit time, any set of n variables can be read or written in parallel by the processors. Although the PRAM provides a useful and general model for the design of parallel algorithms, its practical realization for large values of m and n is not feasible in any current (or foreseeable) technology, as the number of communication links that can be attached to a single system component is (and will probably stay) bounded by a small constant. In order to fill the gap between ideal and feasible models of computation, we need *simulation schemes* to map PRAM computations onto *Bounded Degree Networks* (BDNs), where processors have local memory modules and are interconnected via point-to-point links in a topology of bounded degree. The crucial components of a simulation scheme are a distribution strategy, to store the PRAM memory among the processors of the BDN, and a simulation algorithm, to serve the memory requests generated by the processors at each step. We say that an (n, m)-PRAM can be simulated by an n-processor BDN with *slowdown* s, if any $T \geq 1$ PRAM steps can be simulated in $O(Ts)$ BDN steps. Efficient simulations give practical significance to the large body of algorithms that have been developed for the PRAM model.

[*] This research was supported, in part, by MURST of Italy and by the ESPRIT III Basic Research Programme of the EC under contract No. 9072 (project GEPPCOM).

1.1 Related Work

In the last decade, a large number of randomized and deterministic PRAM simulations on BDNs appeared in the literature. All randomized schemes are based on the use of universal classes of hash functions to allocate the variables to the local memory modules. The distribution properties of these functions yield very efficient simulations in the probabilistic sense. For instance, logarithmic slowdown can be achieved with high probability on a butterfly network [12]. An adaptation of this result gives a probabilistic scheme with $O(\sqrt{n})$ slowdown for the mesh [4].

In contrast, the development of fast deterministic PRAM simulations appears to be much harder. A simple argument shows that, in order to avoid trivial worst cases, the BDN modules must store several copies of each variable (the number of copies per variable is referred to as the *redundancy* of the scheme), so that accesses are performed on subsets of the copies carefully selected to reduce congestion. Such multicopy approach was pioneered by Mehlhorn and Vishkin in [6], where a scheme to speed-up read operations was devised. Later, Upfal and Wigderson proposed the first general protocol requiring that, in order to read or write a variable, only a majority of its copies be accessed [14]. They also described the allocation of the copies to the memory modules of the simulating machine (*Memory Organization Scheme* or, for short, MOS) in terms of a bipartite graph $G = (V, U)$, where V is the set of PRAM variables, U is the set of modules, and r edges connect each variable to the modules storing its r copies. For m polynomial in n and $r = \Theta(\log n)$, the authors showed that there exist suitably expanding graphs that guarantee a worst-case $O\left(\log n \, (\log \log n)^2\right)$ slowdown on a complete interconnection. The bound was later improved to $O(\log n)$ in [1].

All the above schemes suffer from two major drawbacks. First, the simulating machine is a completely interconnected network of processors (and therefore not a BDN); second, the schemes rely on expanding graphs whose existence is proved through counting arguments but for which no efficient construction is known (indeed, the explicit construction of such graphs is a challenging open problem).

The first problem was tackled in [3, 5], where $O(\log n \log m / \log \log n)$ slowdown simulations were devised for high-bandwidth BDNs. In addition to these simulations, tailored to specific interconnections, one can obtain PRAM simulation schemes for any BDN by using the ones devised for the complete interconnection and simulating, in turn, the complete interconnection on the BDN. Such double simulation introduces an extra factor in the slowdown proportional to the time needed to sort n items on the BDN.

For the second problem, in the last two years, a new line of research has targeted the harder issue of constructivity. The first constructive deterministic schemes with sublinear slowdown were given in [7, 8] for the complete interconnection, for memory sizes ranging from $\Theta(n^{1.5})$ up to $\Theta(n^3)$. More recently, Pietracaprina, Pucci and Sibeyn [11] devised constructive schemes for the mesh to simulate any step of an (n, m)-PRAM, for any $m = O(n^2)$, with slowdown $O(n^{1/2+\eta})$ for constant η, using constant redundancy, and for any $m = O(n^{1.5})$ with slowdown $O(\sqrt{n} \log^{1.59} n)$, using $O(\log^{1.59} n)$ redundancy.

1.2 New Results

In this paper, we provide an improved deterministic PRAM simulation for the mesh, which applies to any memory size polynomial in the number of processors, and attains the best slowdown to date for this interconnection. We follow the lines of [11], where the PRAM variables are replicated among the nodes of the mesh in a hierarchical fashion through the so called Hierarchical Memory Organization Scheme (HMOS), and the access requests are routed across the network using a multistage protocol. Our scheme, however, introduces a novel copy-selection strategy and makes use of a simpler and sharper analysis. The main results of this paper are summarized in the following two theorems:

Theorem 1. *There exists a scheme to simulate one step of an (n, m)-PRAM, with $m = n^{1+\tau}$ variables, for any constant $\tau > 0$, on a square mesh of n nodes, with worst-case slowdown $O(\sqrt{n} \log n)$, using $O(\log^{1.59} n)$ redundancy, and with worst-case slowdown $O(n^{1/2+\eta})$, for any constant $\eta > 0$, using constant redundancy.*

Theorem 2. *The scheme of Theorem 1 can be made fully constructive while maintaining the same slowdowns for $m = O(n^{1.5})$.*

Our simulation improves upon [11] in the following directions:

1. It applies to any memory size m polynomial in n (versus $m = O(n^2)$);
2. It attains an $O(\sqrt{n} \log n)$ slowdown (versus $O(\sqrt{n} \log^{1.59} n)$) using the same redundancy.

It has to be remarked that, in order to extend the scheme to any memory size $m = \Omega(n^2)$, we make use of an expanding graph of constant degree for which no efficient construction is known yet. Note also that an $O(\sqrt{n} \log n)$ slowdown could be attained by simulating the algorithm for the complete interconnection of [1] on the mesh. However, the simulation scheme of [1] is overly complex and the underlying MOS is a nonconstructive expanding graph of logarithmic degree. If such graph were to be implemented by resorting to a random graph, as customary when explicit constructions are not available, it would require larger space than ours for its internal representation. Therefore, the reduction in the degree of the expander operated by our scheme constitutes a step forward towards full constructivity for arbitrary memory sizes.

The rest of this extended abstract is organized as follows. Section 2 describes the HMOS. Section 3 presents the algorithm for simulating a PRAM step. This section is subdivided into two subsections that describe the selection of the copies and the routing protocol, respectively. Finally, Sect. 4 discusses the implementation of the HMOS and how it can be made fully constructive when $m = O(n^2)$.

2 Hierarchical Memory Organization Scheme

The *Hierarchical Memory Organization Scheme* (HMOS), originally introduced in [11], represents the mechanism through which the copies of the m PRAM variables are distributed among the n memory modules of the mesh. Here, we generalize its definition to cover any memory size polynomial in the number of processors. For the sake of clarity, rather than highlighting the differences with the original one, we briefly describe the entire structure of the new HMOS.

The HMOS consists of the set of PRAM variables V plus $k + 1$ sets of logical modules U_i, $0 \leq i \leq k$: the modules in U_i are called *i-modules*. We have $|V| = m = n^{1+\tau}$, for any fixed $\tau > 0$, and $|U_i| = m_i = \Theta\left(n^{1/2^i}\right)$, $0 \leq i \leq k$. First, each variable is replicated into r copies, for some positive constant r, which are stored into distinct 0-modules. Each 0-module, viewed as an indivisible block, is in turn replicated into three copies, which are stored into distinct 1-modules. This process is iterated k times. In general, each $(i - 1)$-module, viewed as an indivisible block, is replicated into three copies, which are stored into distinct *i*-modules, $0 < i \leq k$. Therefore, the components of each *i*-module are copies of $(i - 1)$-modules. It is easy to see that the above process eventually creates 3^{k-i} replicas of each *i*-module, which we call *i-blocks*, and $r3^k$ copies of each variable. Finally, note that k-modules are not replicated, therefore the terms k-module and k-block will be used indistinguishably.

The HMOS is represented using $k + 1$ bipartite graphs, namely (V, U_0) and (U_{i-1}, U_i), $1 \leq i \leq k$ (see Fig. 1). In (V, U_0), each $v \in V$ is adjacent to the r 0-modules that store its copies. Analogously, in (U_{i-1}, U_i), $1 \leq i \leq k$, each $u \in U_{i-1}$ is adjacent to the three *i*-modules storing its copies. The performance of the simulation scheme crucially depends on the expansion properties of such graphs.

Definition 3. Let $G = (V, U)$ be a bipartite graph with input degree r. For $\alpha > 0$, $0 < \epsilon < 1$ and $\mu \leq r$, G is said to have (α, ϵ, μ)-*expansion* if for any subset $S \subseteq V$, $|S| \leq |U|$, and any choice of μ outgoing edges for each node in S, the set $\Gamma^\mu(S) \subseteq U$ reached by the chosen edges has size

$$|\Gamma^\mu(S)| > \alpha |S|^{1-\epsilon} .$$

The existence of suitably expanding graphs was proved in [10].

Theorem 4 [10]. *Let $m = n^{1+\tau}$, with constant $\tau > 0$. For any constant $r \geq 2\tau$, there exists a bipartite graph $G = (V, U_0)$ with $|V| = m$, $|U_0| = n$ and input degree r, which has (α, ϵ, μ)-expansion with $\alpha = \Theta(1)$, $\mu = \lfloor r/2 \rfloor + 1$ and $\epsilon = \tau/\mu$.*

We let (V, U_0) be the bipartite graph given by the above theorem, where r is chosen to make ϵ a constant less than or equal to $1/3$ ($r = \max\{3, \lceil 6\tau \rceil\}$ suffices).

For the other graphs (U_{i-1}, U_i), $1 \leq i \leq k$, we use, as in [11], subgraphs of a well-known combinatorial structure, whose definition is recalled below.

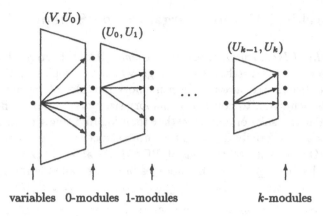

Fig. 1. Structure of the HMOS

Definition 5 [2]. A *Balanced Incomplete Block Design* with parameters m and q, or (m, q)-*BIBD*, is a bipartite graph $G = (V, U)$ with $|V| = \frac{m(m-1)}{q(q-1)}$, $|U| = m$ and input degree q, such that for any two nodes $u_1, u_2 \in U$ there is exactly one node $v \in V$ connected to both.

One basic property of the BIBD, which we will heavily exploit in our simulation, is stated in the following lemma.

Lemma 6 [11]. *Let* $G = (V, U)$ *be an* (m, q)-*BIBD and let* $u \in U$. *Consider a subset* $S \subseteq V$ *such that any node in* S *is adjacent to* u. *For each* v *in* S *fix* $\mu \leq q$ *outgoing edges and let* $\Gamma^\mu(S)$ *denote the set of nodes of* U *reached by these edges. Then*

$$|\Gamma^\mu(S)| \geq (\mu - 1)|S| + 1.$$

Using the above lemma, one can easily show that an (m, q)-BIBD has $((\mu - 1), 1/2, \mu)$-expansion, for any $\mu \leq q$. For $0 < i \leq k$, the graph (U_{i-1}, U_i) is chosen as a subgraph of an $(m_i, 3)$-BIBD, with $m_i = \Theta\left(n^{1/2^i}\right)$, where some inputs are removed so that each node of U_i results adjacent to at most $p_i = \lceil 3m_{i-1}/m_i \rceil = \Theta\left(n^{1/2^i}\right)$ nodes of U_{i-1}. Such subgraphs can be explicitly constructed (see [11]).

The HMOS is physically mapped onto the mesh by storing the i-blocks into distinct submeshes of appropriate size. Recall that there are $3^{k-i}m_i = \Theta\left(3^{k-i}n^{1/2^i}\right)$ i-blocks, $0 \leq i \leq k$, and each i-block contains at most p_i $(i - 1)$-blocks, $1 \leq i \leq k$. We define k nested tessellations of the mesh into submeshes as follows. The outermost tessellation is a subdivision of the mesh into m_k k-*submeshes*, each storing a distinct k-block. Each such submesh is in turn tessellated into p_k $(k - 1)$-*submeshes* storing the component $(k - 1)$-blocks of the

k-block. In general, for $2 \leq i \leq k$, each i-*submesh*, storing a given i-block, is tessellated into p_i $(i-1)$-*submeshes* storing its component $(i-1)$-blocks. Thus, for $1 \leq i \leq k$, we have a total of $\Theta\left(3^{k-i}n^{1/2^i}\right)$ i-submeshes, each of size $t_i = \Theta\left(3^{i-k}n^{1-1/2^i}\right)$. Note that a 1-block is stored into a 1-submesh of t_1 processors, but contains $p_1 > t_1$ 0-blocks. Therefore, we cannot further subdivide the 1-submesh but instead assign $p_1/t_1 = \Theta(3^k)$ 0-blocks to each processor. It is immediate to see that the above strategy evenly distributes the $r3^k m$ copies among the n mesh processors.

3   The Simulation Algorithm

In this section, we describe the simulation of a PRAM step on the mesh. For simplicity, we assume that each of the n processors wants to read or write a distinct variable. (The case of concurrent accesses to the same variable can be handled with minor modifications, without increasing the overall complexity of the simulation.) Let S denote the set of variables requested in the step, and assume that each processor of the mesh emulates a PRAM processor and is in charge of a single variable in S. The algorithm consists of two phases: (1) *Copy Selection* and (2) *Routing*.

3.1   Copy Selection Phase

In the copy selection phase, a suitable set of copies for the variables in S is chosen so that, by accessing these copies, both data consistency and low network congestion are ensured. In order to guarantee consistency, the copies of each variable $v \in S$ are selected according to the rule introduced in [11], which extends the majority protocol of [14] to fit the structure of the HMOS. For any variable v, its $r3^k$ copies are organized in a labeled tree $\mathcal{T}_v$ of $k+2$ levels, numbered, for convenience, from -1 to k. The root, which is at level -1, is labeled with v and has r children, labeled with the names of the 0-modules adjacent to v in (V, U_0). Similarly, for $0 \leq i < k$, an internal node at level i of label u (u is the name of an i-module), has three children, labeled with the names of the $(i+1)$-modules adjacent to u in (U_i, U_{i+1}). Thus, the actual location of the copies of v can be identified by following the $r3^k$ distinct paths from the root to the leaves of $\mathcal{T}_v$. As customary in every multicopy approach, we will provide each copy with a time-stamp, which is updated every time the copy is written.

Let $v \in S$ and $\mu = \lfloor r/2 \rfloor + 1$. Consider a subset C_v of copies of v, and mark the nodes of $\mathcal{T}_v$ on the paths corresponding to these copies.

Definition 7. C_v is a *target set* for v if a majority (μ) of the nodes at level 0 of $\mathcal{T}_v$ is marked, and for each marked node at level i a majority (2) of its children is marked, for $0 \leq i < k$.

A target set for $r = 5$ and $k = 1$ is shown in Fig. 2. Note that a target set contains only $\mu 2^k$ copies out of the $r3^k$ total copies of a variable. A read or write operation on v is performed by accessing a target set of its copies. It can be easily shown that any two target sets for the same variable have nonempty intersection. This ensures that a read operation will always reach at least one most updated copy, which can be retrieved by looking for the copy carrying the most recent time-stamp.

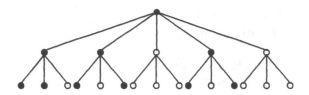

Fig. 2. A target set for $r = 5$ and $k = 1$

The goal of copy selection is to determine a target set C_v for each $v \in S$, with the intent to reduce network congestion in the subsequent routing phase, where the selected copies are physically accessed. The hierarchical structure of the HMOS provides a geographical distribution of the copies into nested regions of the network. By carefully limiting the number of copies that have to be accessed in any block at any level, we reduce the number of packets that will ever be routed to any such region, thereby obtaining small congestion.

The C_v's are determined in $k + 1$ iterations, numbered from 0 to k, in which, for each $v \in S$, the nodes of $\mathcal{T}_v$ are marked in a top-down fashion from the root to the leaves. At the end of the last iteration, the marked nodes in each $\mathcal{T}_v$ will form $\mu 2^k$ root-leaf paths, which identify the copies in C_v, and satisfy the conditions of Definition 7. For each $\mathcal{T}_v$, Iteration 0 marks the root and μ of its children at level 0; Iteration i, $1 \leq i \leq k$, marks two children of each marked node at level $i - 1$. The marking of the nodes is done in parallel on all trees $\mathcal{T}_v$, $v \in S$. Observe that a total of $\mu 2^i n$ nodes are marked in Iteration i, $\mu 2^i$ nodes for each $\mathcal{T}_v$. Such nodes are carefully selected to enforce low congestion in the i-blocks, namely, to control the maximum number of selected copies which reside in any single i-block. In this way, we provide an upper bound on the number of packets sent to any i-submesh during the routing phase.

Consider a node at level i of $\mathcal{T}_v$ with label u and suppose that this node is marked in Iteration i. Eventually, there will be 2^{k-i} distinct root-leaf paths of marked nodes in $\mathcal{T}_v$ which go through that node, corresponding to 2^{k-i} copies in C_v residing in 2^{k-i} distinct blocks of u. Such blocks, which are a subset of all the 3^{k-i} blocks of u, are identified by the paths of marked nodes in the subtree of $\mathcal{T}_v$ rooted at u. The marking of the nodes is done in such a way so that the following property is enforced:

Property 8. *For any T_v and $T_{v'}$ (v and v' need not be distinct) having two marked nodes at level i with the same label u, the two subtrees rooted at these nodes have exactly the same marked nodes.*

This implies that all the physical accesses related to an i-module u will take place in the same 2^{k-i} blocks of u. Moreover, each such block will receive the same pattern of requests. The above property was not enforced by the the selection strategy of [11], and allows us to attain a faster running time for the selection stage and a reduced module congestion.

For $0 \leq i \leq k$, let $A_i \subseteq U_i$ denote the set of i-modules corresponding to nodes in the T_v's marked during Iteration i. The modules in A_i are called *active*. For each $u \in A_i$, its *weight* $w(u)$ is defined as the number of marked nodes in the T_v's labeled with u. From Property 8, we have that for each $u \in A_i$, exactly $w(u)$ copies in $\bigcup_{v \in S} C_v$ will reside in *each* of the selected 2^{k-i} blocks of u, while the other blocks of u will not contain any copy in $\bigcup_{v \in S} C_v$. Exploiting the expansion properties of the HMOS, copy selection ensures that

$$w(u) = O\left(2^i n^{1-\frac{1-\epsilon}{2^i}}\right) ,$$

for any $u \in A_i$, $0 \leq i \leq k$.

We now describe the copy selection algorithm in more detail and sketch its analysis. A complete presentation can be found in [9]. Since Iteration 0 is different from the others, it is described separately.

Iteration 0: For each $v \in S$, r packets are created, each carrying the label of a distinct node at level 0 of T_v. A sequence of *selection steps* are then executed. In the j-th step, $j \geq 1$, for each 0-module u, at most X_j packets carrying label u are selected, and the corresponding nodes in the T_v's are marked. When μ nodes are marked in a T_v, the remaining packets for v are discarded. Iteration 0 ends when exactly μ nodes are marked in each T_v.

By exploiting the (α, ϵ, μ)-expansion of (V, U_0), we can prove the following lemma.

Lemma 9. *Let $a = (1 - \epsilon/2)/(1 - \epsilon) > 1$, and define*

$$c_j = \frac{r}{\alpha} \frac{a^\epsilon}{\left(a^{a^{j-1}}\right)^{\epsilon/2}}, \quad j \geq 1 .$$

Then, choosing $X_j = c_j n^\epsilon$, we have that $O(\log \log n)$ selection steps are sufficient to complete Iteration 0. Moreover, for each active 0-module $u \in A_0$, $w(u) \leq c\mu n^\epsilon$, for some constant $c > 0$.

Proof (Sketch). Let S_j be the number of variables for which less than μ packets have been selected by the end of the j-th selection step, and set $S_0 = n$. By induction on j, we show that $S_j \leq na^{1-a^j}$. The basis is trivial. Suppose now that $S_j > na^{1-a^j}$. Using the expansion of (V, U_0), we conclude that in the j-th selection step at least $X_j \alpha S_j^{1-\epsilon} > rS_{j-1}$ packets have been selected, a contradiction. Since a is a constant greater than 1, we have that $S_T < 1$, for $T = O(\log \log n)$. The proof is completed by observing that $w(u) \leq \sum_j X_j \leq c\mu n^\epsilon$.

Iteration i $(1 \leq i \leq k)$: For each active module $u \in A_{i-1}$, three packets are created, each carrying the name of a distinct i-module adjacent to u in (U_{i-1}, U_i), and the weight $w(u)$. Then, for each i-module in U_i, a maximal subset of packets carrying its name is initially chosen, so that the sum of the weights carried by these packets is at most

$$c\mu 2^{i-1} \left(n^{1-\frac{1-\epsilon}{2^i}} + n^{1-\frac{1-\epsilon}{2^{i-1}}} \right) .$$

For each $u \in A_{i-1}$, if $h < 2$ packets have been selected, then another $2 - h$ arbitrary packets are added to the selected ones. If instead 3 packets were selected, one is discarded. Thus, exactly two packets are selected for each $u \in A_{i-1}$. Finally, for each marked node at level $i-1$ of the $\mathcal{T}_v$'s labeled with u, the two children corresponding to the chosen packets are marked.

By using the expansion property of (U_{i-1}, U_i) stated in Lemma 6, we can prove

Lemma 10. *There is a suitable constant $c \geq 3$ such that, at the end of Iteration i, for each $u \in A_i$, $0 \leq i \leq k$,*

$$w(u) \leq c\mu 2^i n^{1-\frac{1-\epsilon}{2^i}} .$$

Proof (Sketch). The proof proceeds by induction on i. The basis is established by Lemma 9. Suppose that the lemma holds for $i-1$, and let u be an arbitrary i-module. At the end of Iteration i, the weight $w(u)$ is the sum of two terms: $w_1(u)$, coming from the initial selection, and $w_2(u)$, the total weight of the additional packets addressing u which are subsequently chosen to make sure that two packets per $(i-1)$-module are selected. By construction,

$$w_1(u) \leq c\mu 2^{i-1} \left(n^{1-\frac{1-\epsilon}{2^i}} + n^{1-\frac{1-\epsilon}{2^{i-1}}} \right) ,$$

therefore we must only show that the contribution of $w_2(u)$ is not too heavy. For the sake of contradiction, suppose that $w(u) > c\mu 2^i n^{1-\frac{1-\epsilon}{2^i}}$. Then it must be

$$w_2(u) > c\mu 2^{i-1} \left(n^{1-\frac{1-\epsilon}{2^i}} - n^{1-\frac{1-\epsilon}{2^{i-1}}} \right) .$$

Define S_u as the set of modules in A_{i-1} that had a packet contributing to $w_2(u)$. By the inductive hypothesis, the weight of each such module is at most $c\mu 2^{i-1} n^{1-\frac{1-\epsilon}{2^{i-1}}}$, therefore

$$|S_u| \geq \frac{w_2(u)}{c\mu 2^{i-1} n^{1-\frac{1-\epsilon}{2^{i-1}}}} > n^{\frac{1-\epsilon}{2^i}} - 1 .$$

Note that for each module in S_u, at least two packets were not included in the first selection performed in Iteration i. Consider the subgraph of (U_{i-1}, U_i) induced by S_u. Since (U_{i-1}, U_i) is a BIBD, we can apply Lemma 6 and conclude that the edges corresponding to the unselected packets reach at least $|S_u| + 1 > n^{\frac{1-\epsilon}{2^i}}$ modules in U_i. It can be seen that each such i-module accounts for a weight of at least $c\mu 2^{i-1} n^{1-\frac{1-\epsilon}{2^i}}$, for a total weight greater than $c\mu 2^{i-1} n$, which, for $c \geq 3$ is more than the weight of all the packets, a contradiction.

When Iteration k terminates, each processor in charge of a $v \in S$ determines the set C_v of copies to be accessed as those corresponding to the $\mu 2^k$ root-leaf paths of marked nodes in $\mathcal{T}_v$. It can be easily seen that C_v is indeed a target set for v.

In order to determine the running time of copy selection, we recall that ℓn items can be sorted on an n-node mesh in time $O(\ell\sqrt{n})$ [13]. Each selection step of Iteration 0 can be implemented as a constant number of sorting and prefix operations on $O(n)$ items, therefore requiring $O(\sqrt{n})$ time. Since, by Lemma 9, $O(\log\log n)$ executions of such steps are sufficient, Iteration 0 requires $O(\sqrt{n}\log\log n)$ time altogether. Iteration i requires a constant number of sorting and prefix operations on a set of at most $\mu 2^{i-1} n$ items, yielding a running time of $O(\sqrt{n}2^{i-1})$. Therefore, copy selection is completed in time

$$T_{cs} = O\left(\sqrt{n}\log\log n + \sqrt{n}\sum_{i=1}^{k} 2^{i-1}\right) = O(\sqrt{n}\max\{\log\log n, 2^k\}) \ . \tag{1}$$

3.2 Routing Phase

After copy selection is completed, the copies in $\bigcup_{v \in S} C_v$ have to be accessed. A request for each copy is encapsulated in a distinct packet, routed from the requesting processor (*origin*) to the processor storing the copy (*destination*), and back to the origin. We use the same routing protocol of [11], namely, in a sequence of stages, the packets are moved gradually closer to their destinations going through smaller and smaller submeshes.

Consider the origin-destination part of the journey of a packet. It consists of $k + 1$ routing stages, numbered, for convenience, from $k + 1$ down to 1. Stage i, $k+1 \geq i \geq 1$, is executed in parallel and independently in every i-submesh (here, the whole mesh is regarded as a $(k + 1)$-submesh). In Stage i, for $k + 1 \geq i \geq 2$, the packets are routed to arbitrary positions in the $(i - 1)$-submeshes containing their destination, in such a way that the processors of each submesh receive, approximately, the same number of packets. In Stage 1, each packet is sent to its final destination, where the access takes place. In order to estimate the time complexity of the protocol, we need to determine the maximum number of packets sent and received by any processor in each stage. Using Lemmas 9 and 10 we can prove:

Lemma 11. *Let δ_i denote the (maximum) number of packets held by any processor at the beginning of Stage i, $k + 1 \geq i \geq 1$, and let δ_0 be the (maximum) number of packets received by any processor at the end of Stage 1. We have:*

$$\begin{cases} \delta_{k+1} = \mu 2^k \ , \\ \delta_i \ = O\left(\frac{2^i}{3^{i-k}} n^{\epsilon/2^i}\right), \text{ for } k \geq i \geq 1 \ , \\ \delta_0 \ = O(3^k n^\epsilon) \ . \end{cases}$$

The following lemma bounds the running time of each stage, and can be proved using the fact that (δ_i, δ_{i-1})-*routing* on a t_i-node submesh, where each processor sends at most δ_i and receives at most δ_{i-1} packets, can be performed in time $O(\sqrt{\delta_i \delta_{i-1} t_i})$ [13]. Let T_i be the time complexity of Stage i.

Lemma 12. *For $k \geq 1$ we have:*

$$
\begin{cases}
T_{k+1} = O\left(2^k n^{\frac{1}{2}+\frac{\epsilon}{2^{k+1}}}\right), \\
T_i = O\left(2^i 3^{\frac{k-i}{2}} n^{\frac{1}{2}+\frac{3\epsilon-1}{2^{i+1}}}\right), \text{ for } k \geq i \geq 1.
\end{cases}
$$

The return journey of the packets to their origins can be accomplished, within the same time bounds, by running the above protocol backwards. Therefore, the entire routing phase is completed in time

$$
T_{\text{rout}} \leq 2\left(\sum_{i=k+1}^{1} T_i\right) = O\left(2^k n^{\frac{1}{2}+\frac{\epsilon}{2^{k+1}}} + \sum_{i=k}^{1} 2^i 3^{\frac{k-i}{2}} n^{\frac{1}{2}+\frac{3\epsilon-1}{2^{i+1}}}\right). \tag{2}
$$

Let $T_{\text{sim}} = T_{\text{cs}} + T_{\text{rout}}$ denote the overall simulation time for a single PRAM step, and recall that $\epsilon \leq 1/3$. Tedious but straightforward manipulations yield

$$
T_{\text{sim}} = \max\left\{\log\log n, 2^k n^{\frac{\epsilon}{2^{k+1}}}\right\}.
$$

Theorem 1, which was stated in the introduction, follows by choosing suitable values for k.

4 Constructivity Issues

We must point out that the HMOS underlying our simulation algorithm is fully constructive except for the first graph (V, U_0), whose existence is proved by means of a counting argument [10]. A suitable (V, U_0) can be obtained by resorting to a random graph which, as the counting shows, will exhibit the required expansion property with high probability. However, such graph has constant degree and much weaker properties than the ones needed in previous works. For instance, both the schemes in [14] and [1] require nonconstructive graphs of logarithmic degree with $(\Theta(\log n), 0, \Theta(\log n))$-expansion.

The nonconstructiveness of (V, U_0), however, can be avoided when the shared memory size m is within certain ranges. For example, [8] shows how to construct a bipartite graph with $m = O(n^{1.5})$ inputs, n outputs and input degree $r = 3$, which has $(\alpha, 1/3, 2)$-expansion, for some constant α. This graph can be efficiently represented using constant storage, and the address computation requires only logarithmic time. Thus, using this graph as (V, U_0) when $m = O(n^{1.5})$, the HMOS becomes fully constructive and the result of Theorem 2 follows.

Another way to make the construction of (V, U_0) explicit is to use a subgraph of a BIBD, as was done in [11]. Note that the analysis developed in the previous section cannot be applied in this case, since such graph has (α, ϵ, μ) expansion with $\epsilon = 1/2 > 1/3$. However, using the same HMOS as the one in [11] and applying our new copy selection strategy, the results stated in Theorem 2 still hold. More details about these constructive schemes can be found in [9].

References

1. H. Alt, T. Hagerup, K. Mehlhorn, and F.P. Preparata. Deterministic simulation of idealized parallel computers on more realistic ones. *SIAM Journal on Computing*, 16(5):808–835, 1987.

2. M. Hall Jr. *Combinatorial Theory*. John Wiley & Sons, New York NY, second edition, 1986.

3. K. Herley and G. Bilardi. Deterministic simulations of PRAMs on bounded-degree networks. *SIAM Journal on Computing*, 23(2):276–292, April 1994.

4. F.T. Leighton, B. Maggs, A. Ranade, and S. Rao. Randomized routing and sorting on fixed-connection networks. *Journal of Algorithms*, April 1994.

5. F. Luccio, A. Pietracaprina, and G. Pucci. A new scheme for the deterministic simulation of PRAMs in VLSI. *Algorithmica*, 5(4):529–544, 1990.

6. K. Mehlhorn and U. Vishkin. Randomized and deterministic simulations of PRAMs by parallel machines with restricted granularity of parallel memories. *Acta Informatica*, 21:339–374, 1984.

7. A. Pietracaprina and F.P. Preparata. An $O(\sqrt{n})$-worst-case-time solution to the granularity problem. In K.W. Wagner P. Enjalbert, A. Finkel, editor, *Proc. 10th Symp. on Theoretical Aspects of Computer Science*, LNCS 665, pages 110–119, Würzburg, Germany, February 1993. Springer-Verlag.

8. A. Pietracaprina and F.P. Preparata. A practical constructive scheme for deterministic shared-memory access. In *Proc. 5th ACM Symp. on Parallel Algorithms and Architectures*, pages 100–109, 1993.

9. A. Pietracaprina and G. Pucci. Improved deterministic PRAM simulation on the mesh. Technical Report GEPPCOM-16, DEI Università di Padova, Padova, Italy, September 1994.

10. A. Pietracaprina and G. Pucci. Tight bounds on deterministic PRAM emulations with constant redundancy. In J.V. Leeuwen, editor, *Proc. 2nd European Symposium on Algorithms*, LNCS 855, pages 319–400, Utrecht, NL, September 1994. Springer-Verlag.

11. A. Pietracaprina, G. Pucci, and J. Sibeyn. Constructive deterministic PRAM simulation on a mesh-connected computer. In *Proc. 6th ACM Symp. on Parallel Algorithms and Architectures*, pages 248–256, Cape May, NJ, June 1994.

12. A.G. Ranade. How to emulate shared memory. *Journal of Computer and System Sciences*, 42:307–326, 1991. See also *28th IEEE FOCS* (1987).

13. J.F. Sibeyn and M. Kaufmann. Deterministic 1-k routing on meshes, with application to hot-potato worm-hole routing. In *Proc. of the 11th Symp. on Theoretical Aspects of Computer Science*, LNCS 775, pages 237–248, 1994.

14. E. Upfal and A. Widgerson. How to share memory in a distributed system. *Journal of the ACM*, 34(1):116–127, 1987.

On Optimal Polynomial Time Approximations:
P-Levelability vs. Δ-Levelability[1]
(Extended Abstract)

Klaus Ambos-Spies

Mathematisches Institut, Universität Heidelberg
Im Neuenheimer Feld 294, D-69120 Heidelberg, Germany

Abstract. Safe and unsafe polynomial time approximations were introduced by Meyer and Paterson [4] and Yesha [8], respectively. The question of which sets have optimal safe approximations was investigated by several authors (see e.g. [3,6,7]). Orponen et al. [5] showed that a problem has an optimal polynomial time approximation if and only if neither it nor its complement is p-levelable. Recently Duris and Rolim [2] considered the unsafe case and compared the existence of optimal polynomial time approximations for both cases. They left open the question, however, whether there are intractable sets with optimal unsafe approximations and whether there are sets with optimal unsafe approximations but without optimal safe approximations. Here we answer these questions affirmatively. Moreover, we consider a variant of Duris and Rolim's Δ-levelability concept related to the nonexistence of optimal unsafe approximations.

1 Introduction

The concept of safe polynomial time approximations was introduced by Meyer and Paterson in [4]: A polynomial time approximation of a set $A \subseteq \{0,1\}^*$ is a deterministic polynomial time bounded Turing machine M which on each input x outputs either 1 (accept) or 0 (reject) or ? (undetermined) such that all inputs accepted by M are members of A and no member of A is rejected by M. Hence on the determined part of M, dom(M)={x: M(x)=0 or M(x)=1}, M and A agree, so that the approximation M is optimal if there is no other polynominal time approximation M' of A which infinitely extends the determined part of M. Ko and Moore [3] showed that many natural intractable problems – e.g. all E–complete sets and (assuming P≠NP) many NP–complete sets – do not have optimal safe polynomial approximations.

---
[1]This work was supported in part by the HCM program of the European Community under grant CHRX-CT93-0415 (COLORET Network).

As shown by Orponen et al. in [5], the existence of optimal approximations can be phrased in terms of levelability. A set A is p–levelable if for any deterministic Turing machine M accepting A and for any polynomial p there is another machine M' and a polynomial p' such that for infinitely many elements x of A, M does not accept x within p(|x|) steps whereas M' accepts x in p'(|x|) steps. Then it is easy to show that A has an optimal safe polynomial approximation if and only if A and $\overline{A}$ are not p–levelable.

Duris and Rolim [2] compared the above concepts with an unsafe approximation concept introduced by Yesha [8]. An unsafe approximation of a set A is just a standard polynomial time bounded deterministic Turing machine M (with outputs 0 and 1). Then optimizing the approximation amounts to minimizing the set of errors which M makes. Duris and Rolim introduced a levelability concept, Δ–levelability, which implies the nonexistence of optimal unsafe polynomial approximations. Moreover, they introduced a density notion for p–levelability and Δ–levelability to measure the number of strings up to a given length on which every given approximation can be improved. Using these concepts they extended Ko and Moore's results to levelability by any polynomial factor for safe and unsafe approximations.

Comparing safe and unsafe approximations Duris and Rolim showed that there are intractable sets which have an optimal safe approximation but no optimal unsafe approximation. They also constructed a set which -if Δ-levelable at all- is Δ–levelable only with very small density. They did not succeed, however, to produce an intractable set with optimal unsafe approximations and they raised the question whether there is a set which is p–levelable (with high density) but not Δ–levelable. Here we close these gaps by showing that the question of the existence of optimal approximations is independent for the safe and the unsafe case. In particular we construct exponential time sets A, B and C all of which are not Δ-levelable but A and $\overline{A}$ are p-levelable (with high density), B is p-levelable and $\overline{B}$ is not p-levelable, and C and $\overline{C}$ are not p-levelable (in fact C is bi-p-immune). Moreover we prove these results for a weaker notion of Δ-levelability which we call weak Δ-levelability and of which we show that it is equivalent to not having optimal unsafe polynomial time approximations. Finally we show that there are sets which are weakly Δ-levelable but not Δ-levelable.

In the next section we formally introduce the approximation notions and the corresponding levelability concepts and we review some fundamental facts. Then in Section 3 we prove our new results. We conclude this section by introducing some notation.

N denotes the set of natural numbers. $\Sigma=\{0,1\}$ is the binary alphabet and Σ^* is the set of (finite) binary strings. The length of a string x is denoted by |x|. < is the length-lexicographical ordering on Σ^* and z_n is the n–th string under this ordering. Sometimes we identify the nth string z_n with the natural number n.

A subset of Σ^* is called a *language*, a *problem* or simply a *set*. In our notation we do not distinguish between a set and its characteristic function, i.e., $x \in A$ iff $A(x)=1$ and $x \notin A$ iff $A(x)=0$. $|A|$ denotes the cardinality of A; $\overline{A} = \Sigma^*-A$ is the complement of A; $A \Delta B = (A-B) \cup (B-A)$ is the symmetric difference of A and B, and we write $A =^* B$ ($A \subseteq^* B$) to denote that $A \Delta B$ (B-A) is finite. For any number n and any set A we let $A_{=n} = \{x \in A: |x|=n\}$ and $A_{\leq n} = \{x \in A: |x| \leq n\}$.

We will also consider *partial* sets defined by partial (characteristic) functions f: $\Sigma^* \to \{0,1\}$ and again we identify a partial set with its characteristic function. For a partial set Q, $dom(Q) = \{x: Q(x)=0 \text{ or } Q(x)=1\}$ is the *domain* or *determined part* of Q. We say that a partial set Q is *consistent* with a set A if, for all $x \in dom(Q)$, $A(x)=Q(x)$. The union $Q \cup Q'$ of two partial sets Q and Q' is defined by $dom(Q \cup Q') = dom(Q) \cup dom(Q')$ and $(Q \cup Q')(x)=1$ iff $Q(x)=1$ or $Q'(x)=1$.

We will consider the following complexity classes: **P** is the class of polynomial time computable languages, $\mathbf{P_{part}}$ is the class of polynomial time computable partial sets, and **E** $= \cup_{c \geq 1} \mathbf{DTIME}(2^{cn})$ is the class of linear exponential time sets.

We fix a standard polynomial time computable and invertible pairing function $\lambda x,y.<x,y>$ on Σ^* such that, for any string x there is a real $\alpha(x)>0$ such that

$$(1.0) \qquad |\Sigma^{[x]} \cap \Sigma^*_{=n}| \geq \alpha(x) \cdot 2^n \text{ for almost all n}$$

where $\Sigma^{[x]} = \{<x,y>: y \in \Sigma^*\}$ and $\Sigma^{[\leq x]} = \{<x',y>: x'<x \text{ \& } y \in \Sigma^*\}$. Finally we fix a recursive enumeration $\{P_e: e \geq 0\}$ of P such that, for $|x| \geq e$, $P_e(x)$ can be computed in $O(2^{|x|})$ steps (uniformly in e and x).

2 Polynomial Time Approximations and Levelable Sets

We now formally introduce the basic concepts we will deal with. Our definitions are machine independent. The equivalence to the original definitions in terms of Turing machines can be easily shown.

2.1 Definition (Meyer and Paterson [4]). A *safe polynomial time approximation* of a set A is a partial set $Q \in P_{part}$ which is consistent with A, i.e., for every string $x \in dom(Q)$, $A(x)=Q(x)$. The approximation Q of A is *optimal* if for every safe polynomial time appoximation Q' of A, $dom(Q')-dom(Q)$ is finite.

Note that any two optimal approximations of a set A (if there are any) will differ only finitely. Also note that an optimal approximation of A is also an optimal approximation of the complement $\overline{A}$ of A. Another simple observation is that any two safe polynomial time approximations Q and Q' of a set A can be merged to an approximation Q" defined on the union of the domains of Q and Q':

2.2 Proposition. Let Q and Q' be safe polynomial time approximations of A. Then Q∪Q' is also a safe polynomial time approximation of A.

Proof. Straightforward.

By applying Proposition 2.2 we obtain the following characterization of optimality for safe approximations. We omit the simple proof.

2.3 Proposition. For any partial set $Q \in P_{part}$ and for any set A, Q is an optimal safe polynomial time approximation of A if and only if, for every safe polynomial time approximation Q' of A there is a number $c \in N$ such that

$$(2.1) \quad \forall n \in N \, (|(\Sigma^*\text{-dom}(Q))_{\leq n}| \leq |(\Sigma^*\text{-dom}(Q'))_{\leq n}| + c).$$

Orponen, Russo and Schöning [5] observed that the question of the existence of optimal safe approximations is related to p-levelability.

2.4 Definition (Orponen et al. [5]). A set A is *p–levelable* if for any subset $B \in P$ of A there is another subset $B' \in P$ of A such that $|B'-B| = \infty$.

In contrast to the approximation concepts, p-levelability is not symmetric. As one can easily show, there are p-levelable sets with non-p-levelable complement. Intuitively p-levelability says that any safe approximation of A has an infinite extension capturing infinitely many additional *elements* of A.

2.5 Proposition (Orponen et al. [5]). A set A possesses an optimal safe polynomial time approximation if and only if neither A nor $\overline{A}$ is p–levelable.

As observed by Orponen and Schöning, p–levelability is closely related to p–immunity.

2.6 Definition. An infinite set A is *p–immune* if no infinite subset of A is in **P**. A is *bi–p–immune* if A and $\overline{A}$ are p–immune. A is *almost p–immune* if A is the union of a set in **P** and a p–immune set.

2.7 Proposition (Orponen and Schöning [6]). A set A is p–levelable if and only if A is not almost p–immune.

Duris and Rolim [2] refined the concept of p-levelability by adding information on the density of the extended parts.

2.8 Definition (Duris and Rolim [2]). Let f be any unbounded function on the natural numbers. A set A is *p-levelable with density* f if for any subset $B \in P$ of A there is another subset $B' \in P$ of A such that $|B'_{\leq n} - B_{\leq n}| \geq f(n)$ for almost all numbers n. The set A is p-levelable with *exponential density* if A is p-levelable with density $2^{\alpha n}$ for all reals α with $0 < \alpha < 1$.

2.9 Proposition (Duris and Rolim [2]). A set A is p-levelable if and only if A is p-levelable with density f for some unbounded function f on N.

In [8] Yesha considered nonsafe approximations to intractable sets. Duris and Rolim [2] introduced a corresponding levelability concept implying the nonexistence of optimal approximations.

2.10 Definition ([8,2]). (i) An *unsafe polynomial time approximation* of a set A is a set $B \in P$. The set $A \Delta B$ is called the *error set* of the approximation.
(ii) Let f be any unbounded function on the natural numbers. A set A is Δ-*levelable with density* f if for any set $B \in P$ there is another set $B' \in P$ such that
$$(2.2) \qquad |(A \Delta B)_{\leq n}| - |(A \Delta B')_{\leq n}| \geq f(n)$$
for almost all numbers n. A is Δ-*levelable* if A is Δ-levelable with density f for some unbounded function f on N.

Note that, in contrast to p-levelability, Δ-levelability is symmetric, i.e., a set A is Δ-levelable iff its complement $\overline{A}$ is Δ-levelable. Duris and Rolim do not explicitly define when an unsafe approximation is optimal. An direct adaption of the definition of optimality for safe approximations to the unsafe case would yield a trivial notion of optimality: If we would define that a set $B \in P$ is an optimal unsafe polynomial time approximation of a set A if, for any other approximation $B' \in P$, $(A \Delta B) - (A \Delta B')$ is finite, i.e. the error set of the approximation is minimal, then only the polynomial time computable sets had optimal approximations: Namely given any polynomial approximation B of an intractable set A, $\overline{B}$ is a polynomial approximation of A too, $(A \Delta B) \cap (A \Delta \overline{B}) = \emptyset$ and, by intractability of A, $A \Delta B$ is infinite.

So we propose to take the analog of the characterization of optimality given in Proposition 2.3 as a definition of optimality in the unsafe case.

2.11 Definition. (i) An unsafe polynomial time approximation $B \in P$ of a set A is *optimal* if, for any approximation $B' \in P$,
$$(2.3) \qquad \exists k \in N \, \forall n \in N \, (\, |(A \Delta B)_{\leq n}| < |(A \Delta B')_{\leq n}| + k)$$

holds.

(ii) A set A is *weakly Δ-levelable* if no unsafe polynomial time approximation B∈ P of A is optimal.

As the following proposition shows, any Δ-levelable set is weakly Δ-levelable.

2.12 Proposition. Let A and B∈ P be given such that, for any set B'∈ P, (2.3) above or

$$(2.4) \qquad \exists^{\infty}n \ (\ |(A\Delta B)_{\leq n}| < |(A\Delta B')_{\leq n}| \ )$$

holds. Then A is not Δ-levelable. Hence every Δ-levelable set is weakly Δ-levelable.

Proof. For a contradiction assume that A is Δ-levelable. Then there is an unbounded function f on N and a set B'∈ P such that (2.2) holds for almost all numbers n. Since f(n)≥0 for all n and since f is unbounded it follows that (2.4) and (2.3), respectively, fail contrary to assumption. ♦

Below we will show that there are weakly Δ-levelable sets which are not Δ-levelable. By Proposition 2.12 this implies that a definition of the optimality of unsafe approximations based on Δ-levelability instead of weak Δ-levelability will give a very weak optimality concept. The reason for this is clause (2.4) in Proposition 2.12: an approximation B of A could be optimal in this sense and yet there might be another polynomial approximation B' of A such that B' improves B on infinitely many initial segments with unbounded density, i.e., such that lim sup_n |(A\Delta B)_{\leq n}| - |(A\Delta B')_{\leq n}| = ∞. So for defining optimality one should use weak-Δ-levelabiliy as we did above, not Δ-levelability.

3 p-Levelabiliy vs. Δ-Levelability

Ko and Moore [3] and Duris and Rolim [2] have shown that all E–complete sets are p–levelable and Δ–levelable. Since the class of E–complete sets is closed under complementation this implies that there is an intractable set A such that A and $\overline{A}$ are p–levelable and A is Δ–levelable. Duris and Rolim also constructed an exponential time set which is Δ–levelable but not p–levelable. This result can be easily extended to show that there is a set A such that A is Δ–levelable but neither A nor $\overline{A}$ is p–levelable, so that A possesses an optimal safe polynomial time approximation but no optimal unsafe polynomial approximation. Duris and Rolim did not show the existence of non–Δ–levelable sets, however, and they raised the question whether there is a p–levelable set which is not Δ–levelable. Next we give an affirmative answer to this question. In fact we construct a set A which is p-levelable but not weakly Δ–levelable, whence A has an optimal unsafe polynomial time approximation but no optimal safe polynomial time approximation. So the existence of optimal approximations for the safe and unsafe case is independent.

3.1 Theorem. There is a set $A \in E$ such that

    (i) A and $\overline{A}$ are p–levelable (with exponential density)

    (ii) A is not weakly Δ–levelable (hence not Δ–levelable).

Proof (idea). By an effective finite injury argument we inductively define a set A with the required properties in stages. At stage s we will determine the value of $A(x)$ for all strings x of length s. In the construction we will ensure that

$$(3.1) \qquad A \cap \Sigma^{[2e]} =^* \Sigma^{[2e]} \text{ and } A \cap \Sigma^{[2e+1]} =^* \emptyset \quad (e \geq 0)$$

So, for any $e \geq 0$, $A \cap \Sigma^{[\leq e]} \in P$ and $A \cap \Sigma^{[e]} \in P$. Hence, to make A and $\overline{A}$ p–levelable (with exponential density), it suffices to meet the following requirements

$$L_{2e}: \qquad P_e \subseteq A \Rightarrow P_e \subseteq^* \Sigma^{[\leq 2e]}$$

$$L_{2e+1}: \qquad P_e \subseteq \overline{A} \Rightarrow P_e \subseteq^* \Sigma^{[\leq 2e+1]}$$

for all numbers $e \geq 0$. To show that the requirements L_{2e} ($e \geq 0$) together with (3.1) ensure that A is p–levelable with exponential density, fix a subset $C \in P$ of A. It suffices to show that there is another subset $C' \in P$ of A such that $C \cap C' = \emptyset$ and $|C'_{=n}| \geq \alpha 2^n$ for some real $\alpha \in (0,1)$ and almost all numbers n. Now, for an index e of C, the requirement L_{2e} ensures that $C \subseteq^* \Sigma^{[\leq 2e]}$. So, for some number $m > e$, $C \cap \Sigma^{[2m]} = \emptyset$. Let $C' = A \cap \Sigma^{[2m]}$. Then $C \cap C' = \emptyset$ and, by (3.1), $C' \in P$ and $C' =^* \Sigma^{[2m]}$ whence, by (1.0), for almost all n $C'_{=n}$ is of exponential size. Similarly, the requirements L_{2e+1} ($e \geq 0$) ensure that $\overline{A}$ is p–levelable with exponential density.

The strategy for meeting a requirement L_{2e} is as follows: If a string $x \notin \Sigma^{[\leq 2e]}$ shows up in P_e then we let $A(x)=0$ thereby violating the hypothesis of the requirement (so that L_{2e} is trivially met). Note that this action is finitary and does not affect $A \cap \Sigma^{[\leq 2e]}$, whence this strategy is compatible with ensuring (3.1).

To guarantee that A is not weakly Δ–levelable, we will ensure that the polynomial time set

$$B = \bigcup_{e \geq 0} \Sigma^{[2e]}$$

will be an optimal unsafe approximation of A. To do so we will guarantee that

$$(3.2) \qquad \forall e \in N \; \forall n \in N \; (|(A \Delta B)_{\leq n}| \leq |(A \Delta P_e)_{\leq n}| + e + 1).$$

This condition will require a refinement of the strategy for meeting the levelability requirements: the requirement L_{2e+i} is only allowed to diagonalize at x if this action will not interfer with satisfying (3.2) for the numbers $m < 2e+i$. The critical situation is that, for some $m < 2e+i$, $P_m(x) \neq B(x)$ and L_{2e+i} wants to set $A(x) = P_m(x)$ (thereby increasing the difference of $A \Delta B$ and $A \Delta P_m$). Now if $P_m \Delta B$ is finite then for all sufficiently large strings x this conflict can not arise. So in this case satisfaction of L_{2e+i} is only delayed by finitely many stages. On the other hand, if $P_m \Delta B$ is infinite then by letting $A(y) = B(y)$ on sufficiently many strings y for which P_m and B differ, we can make $|(A \Delta B)_{\leq n}|$ much smaller than $|(A \Delta P_m)_{\leq n}|$ so that the action of L_{2e+i} will not injure (3.2). To achieve the latter we meet the following requirements (for all $e, k \geq 0$)

$$NL_{<e,k>}: \quad P_e \Delta B \text{ infinite} \Rightarrow \exists n \, (\, |(A \Delta P_e)_{\leq n}| - |(A \Delta B)_{\leq n}| \geq k \,).$$

The strategy for meeting a requirement $NL_{<e,k>}$ is as follows: At any stage s such that $P_e(x) \neq B(x)$ for some string x of length s and such that

$$\forall n < s \, (\, |(A \Delta P_e)_{\leq n}| - |(A \Delta B)_{\leq n}| < k \,),$$

we ensure that $A(y) = B(y)$ for all strings y of length s. If $P_e \neq^* B$ this can be repeated over and over again whence $|A \Delta P_e|$ is growing more quickly than $|A \Delta B|$ so that eventually the requirement $NL_{<e,k>}$ will be met at some sufficiently large stage. Note that this strategy is compatible with (3.1) and finitary. The latter ensures that the conflicts beween the L and NL requirements can be resolved by a standard finite injury priority construction. The formal construction and its verification can be found in the full version of this paper. ♦

The proof of Theorem 3.1 can be easily modified to show that there are sets B and C which are not weakly Δ-levelable and such that B is p–levelable but $\overline{B}$ is p–immune (hence not p–levelable) and C is bi–p–immune (hence C and $\overline{C}$ are not p–levelable).

3.2 Theorem. There is a set $A \in E$ such that

    (i)   A is p–levelable (with exponential density)

    (ii) $\overline{A}$ is p–immune hence not p–levelable

    (iii)   A is not weakly Δ–levelable.

3.3 Theorem. There is set $A \in E$ such that

    (i)   A is bi–p–immune (hence A and $\overline{A}$ are not p–levelable)

    (ii)   A is not weakly Δ–levelable.

The above theorems together with the results of Duris and Rolim [2] show that the levelability concepts for safe and unsafe polynomial approximations, namely p–levelability and (weak) Δ–levelability, are completely independent. We conclude our comparison of the levelability concepts by showing that weak-Δ-levelability and Δ-levelability differ.

3.4 Theorem. There is a weakly Δ–levelable set $A \in E$ which is not Δ–levelable.

Proof (idea). As in the previous proofs a set A with the required properties is constructed in stages, $A(x)$ for x of length s being defined at stage s. To ensure that A is weakly Δ–levelable it suffices to meet the requirements

$$WL_{<e,k>}: \quad \exists n \in N \, (\, |(A \Delta P_e)_{\leq n}| > |(A \Delta \overline{P}_e)_{\leq n}| + k \,)$$

for $e, k \in N$. Note that the requirements $WL_{<e,k>}$ $(k \geq 0)$ imply that $\overline{P}_e$ witnesses that the unsafe approximation P_e of A is not optimal.

To ensure that A is not Δ–levelable fix any set $B \in P$. Then the requirements

$NL_{<e,k>}$: $P_e \Delta B$ infinite $\Rightarrow$ $\exists n \, (|(A \Delta P_e)_{\leq n}| - |(A \Delta B)_{\leq n}| \geq k)$

from the proof of Theorem 3.1 will ensure that B witnesses the failure of Δ-levelability for A: For any set $B' \in P$, if $B' \Delta B$ is finite then obviously (2.3) holds, and if $B' \Delta B$ is infinite then, for an index e of B', the requirements $NL_{<e,k>}$ ($k \geq 0$) imply (2.4). So A is not Δ-levelable by Proposition 2.12.

To satisfy the requirements $NL_{<e,k>}$ we use the strategy for these requirements in the proof of Theorem 3.1. The strategy for meeting the requirements $WL_{<e,k>}$ is a variant of this strategy where $\overline{P}_e$ plays the role of B. ♦

References

1. J.L. Balcázar and U. Schöning: Bi–immune sets for complexity classes. Math. Systems Theory, 18 (1985) 1–10.
2. P. Duris and J.D.P. Rolim: E–complete sets do not have optimal polynomial time approximations. In: Proc. MFCS '94, Lect. Notes Comput. Sci. 841 (1994) 38–51, Springer Verlag.
3. K. Ko and D. Moore: Completeness, approximation and density. SIAM J. Comput. 10 (1981) 787–796.
4. A.R. Meyer and M.S. Paterson: With what frequency are apparently intractable problems difficult? In: Tech. Rep. TM–126, Laboratory for Computer Science, MIT, 1979.
5. P. Orponen, A. Russo and U. Schöning: Optimal approximations and polynomially levelable sets. SIAM J. Comput. 15 (1986) 399–408.
6. P. Orponen and U. Schöning: The structure of polynomial complexity cores. In: Proc. MFCS '84, Lect. Notes Comput. Sci. 176 (1984) 452–458, Springer Verlag.
7. D.A. Russo: Optimal approximations of complete sets. In: Proc. 1st Annual Conference on Structure in Complexity Theory, Lect. Notes Comput. Sci. 38 (1986) 311–324, Springer Verlag.
8. Y. Yesha: On certain polynomial–time truth–table reducibilities of complete sets to sparse sets. SIAM J. Comput. 12 (1983) 411–425.

Weakly Useful Sequences

Stephen A. Fenner[1]* and Jack H. Lutz[2]** and Elvira Mayordomo[3]***

[1] University of Southern Maine, Portland, Maine, USA. E-mail:
fenner@usm.maine.edu.
[2] Iowa State University, Ames, Iowa, USA. E-mail: lutz@iastate.edu.
[3] University of Zaragoza, Zaragoza, SPAIN. E-mail: emayordomo@mcps.unizar.es.

Abstract. An infinite binary sequence x is defined to be

1. *strongly useful* if there is a recursive time bound within which *every* recursive sequence is Turing reducible to x; and
2. *weakly useful* if there is a recursive time bound within which all the sequences in a non-measure 0 subset of the set of recursive sequences are Turing reducible to x.

Juedes, Lathrop, and Lutz (1994) proved that every weakly useful sequence is strongly deep in the sense of Bennett (1988) and asked whether there are sequences that are weakly useful but not strongly useful.

The present paper answers this question affirmatively. The proof is a direct construction that combines the recent *martingale diagonalization* technique of Lutz (1994) with a new technique, namely, the construction of a sequence that is "recursively deep" with respect to an arbitrary, given uniform reducibility. The *abundance* of such recursively deep sequences is also proven and used to show that every weakly useful sequence is recursively deep with respect to every uniform reducibility.

1 Introduction

It is a truism that the usefulness of a data object does not vary directly with its information content. For example, consider two infinite binary strings, χ_K, the characteristic sequence of the halting problem (whose nth bit is 1 if and only if the nth Turing machine halts on input n), and z, a sequence that is algorithmically random in the sense of Martin-Löf [10]. The following facts are well-known.

1. The first n bits of χ_K can be specified using only $O(\log n)$ bits of information, namely, the *number* of 1's in the first n bits of χ_K [2].

* This research was supported in part by National Science Foundation Grant CCR-9209833.
** This research was supported in part by National Science Foundation Grant CCR-9157382, with matching funds from Rockwell International, Microware Systems Corporation, and Amoco.
*** Work supported by the EC through the Esprit BRA Program (project 7141, AL-COM II) and through the HCM Program (project CHRX-CT93-0415, COLORET Network).

2. The first n bits of z cannot be specified using significantly fewer than n bits of information [10].
3. Oracle access to χ_K would enable one to decide *any* recursive sequence in polynomial time (i.e., decide the nth bit of the sequence in time polynomial in the length of the binary representation of n) [11].
4. Even with oracle access to z, most recursive sequences cannot be computed in polynomial time. (This appears to be folklore, known at least since [3].)

Facts (i) and (ii) tell us that χ_K contains far less information than z. In contrast, facts (iii) and (iv) tell us that χ_K is computationally much more useful than z. That is, the information in χ_K is "more usefully organized" than that in z.

Bennett [3] introduced the notion of *computational depth* (also called "logical depth") in order to quantify the degree to which the information in an object has been organized. In particular, for infinite binary sequences, Bennett defined two "levels" of depth, *strong depth* and *weak depth*, and argued that the above situation arises from the fact that χ_K is strongly deep, while z is not even weakly deep. (The present paper is motivated by the study of computational depth, but does not directly use strong or weak depth, so definitions are omitted here. The interested reader is referred to [3], [7], or [6] for details, and for related aspects of algorithmic information theory.)

Investigating this matter further, Juedes, Lathrop, and Lutz [6] considered two "levels of usefulness" for infinite binary sequences. Specifically, let $\{0,1\}^\infty$ be the set of all infinite binary sequences, let REC be the set of all recursive elements of $\{0,1\}^\infty$, and, for $x \in \{0,1\}^\infty$ and $t: \mathbf{N} \to \mathbf{N}$, let $\mathrm{DTIME}^x(t)$ be the set of all $y \in \{0,1\}^\infty$ for which there exists an oracle Turing machine M that, on input $n \in \mathbf{N}$ with oracle x, computes $y[n]$, the nth bit of y, in at most $t(\ell)$ steps, where ℓ is the number of bits in the binary representation of n. Then a sequence $x \in \{0,1\}^\infty$ is defined to be *strongly useful* if there is a recursive time bound $t: \mathbf{N} \to \mathbf{N}$ such that $\mathrm{DTIME}^x(t)$ contains all of REC. A sequence $x \in \{0,1\}^\infty$ is defined to be *weakly useful* if there is a recursive time bound $t: \mathbf{N} \to \mathbf{N}$ such that $\mathrm{DTIME}^x(t)$ contains a non-measure 0 subset of REC, in the sense of resource-bounded measure [9]. That is, x is weakly useful if access to x enables one to decide a *nonnegligible set* of recursive sequences within some fixed recursive time bound. No recursive or algorithmically random sequence can be weakly useful. It is evident that χ_K is strongly useful, and that every strongly useful sequence is weakly useful.

Juedes, Lathrop, and Lutz [6] generalized Bennett's result that χ_K is strongly deep by proving that *every* weakly useful sequence is strongly deep. This confirmed Bennett's intuitive arguments by establishing a definite relationship between computational depth and computational usefulness. It also substantially extended Bennett's result on χ_K by implying (in combination with known results of recursion theory [10, 13, 4, 5]) that *all* high Turing degrees and *some* low Turing degrees contain strongly deep sequences.

Notwithstanding this progress, Juedes, Lathrop, and Lutz [6] left a critical question open: Do there exist weakly useful sequences that are not strongly useful? The main result of the present paper, proved in Section 4, answers this question affirmatively. This establishes the existence of strongly deep sequences that are not strongly useful. More importantly, it indicates a need for further investigation of the class of weakly useful sequences.

The proof of our main result is a direct construction that combines the *martingale diagonalization* technique recently introduced by Lutz [8] with a new technique, namely, the construction of a sequence that is *recursively F-deep*, where F is an arbitrary uniform reducibility. This notion of uniform recursive depth, defined and investigated in Section 3, is closely related to Bennett's notion of weak depth.

In addition to using specific constructions of recursively F-deep sequences, we prove that, for each uniform reducibility F, *almost every* sequence in REC is recursively F-deep. This implies that every weakly useful sequence is, for every uniform reducibility F, recursively F-deep.

2 Preliminaries

We use N to denote the set of natural numbers (including 0), and we use Q to denote the set of rational numbers. We write $[\![\varphi]\!]$ for the Boolean value of a condition φ, i.e.,

$$[\![\varphi]\!] = \textbf{if } \varphi \textbf{ then } 1 \textbf{ else } 0.$$

Throughout this paper, we identify each set $A \subseteq N$ with its characteristic sequence $\chi_A \in \{0,1\}^\infty$, whose nth bit is $\chi_A[n] = [\![n \in A]\!]$. For any $x, y \in \{0,1\}^* \cup \{0,1\}^\infty$, we write $x \sqsubseteq y$ to mean that x is a prefix of y, and if in addition, $x \neq y$, we may write $x \sqsubset y$.

We fix a recursive bijection $\langle \cdot, \cdot \rangle : N^2 \to N$, monotone in both arguments, such that $i \leq \langle i, j \rangle$ and $j \leq \langle i, j \rangle$ for all $i, j \in N$.

In the proof of Theorem 12, we will deal extensively with *partial characteristic functions*, i.e., functions with domain a subset of N and with range $\{0,1\}$. We will identify binary strings with characteristic functions whose domains are finite initial segments of N. If σ and τ are partial characteristic functions, we let $\mathrm{dom}(\sigma)$ denote the domain of σ, and say that σ and τ are *compatible* if they agree on all elements in $\mathrm{dom}(\sigma) \cap \mathrm{dom}(\tau)$. We say that σ *is extended by* τ ($\sigma \sqsubseteq \tau$) if σ and τ are compatible and $\mathrm{dom}(\sigma) \subseteq \mathrm{dom}(\tau)$ (if in addition $\sigma \neq \tau$, we write $\sigma \sqsubset \tau$). If σ and τ are compatible, we let $\sigma \cup \tau$ be their smallest common extension.

We will often think of N being split up into *columns* $0, 1, 2, \ldots$ where the ith column is $\{\langle i, j \rangle \mid j \in N\}$. If $A \subseteq N$, then the ith *strand of A* is defined as $A_i = \{x \mid \langle i, x \rangle \in A\}$. If σ is a partial characteristic function and $n \in N$, then $\sigma[< n]$ denotes σ restricted to the domain $\{0, \ldots, n-1\}$, and $\sigma[i, < n]$ denotes the unique partial characteristic function τ such that for all x,

$$\tau(x) = \begin{cases} \sigma(\langle i, x \rangle) & \text{if } x < n \text{ and } \sigma(\langle i, x \rangle) \text{ is defined,} \\ \text{undefined} & \text{otherwise.} \end{cases}$$

That is, $\sigma[i, < n]$ results from "excising" the first n bits of σ from the ith column. Inversely, if w is a binary string, then $\{i\} \times w$ denotes the unique partial characteristic function τ such that $\tau(\langle i, x \rangle) = w(x)$ for all $x < |w|$, and is undefined on all other arguments. That is, $\{i\} \times w$ is w "translated" over to the ith column. Of particular importance will be the finite characteristic function defined for an arbitrary $C \subseteq N$ and $k, y \in N$ as

$$\xi^C(k, y) = \bigcup_{k' < k} \{k'\} \times C[k', < y].$$

In other words, $\xi^C(k, y)$ is χ_C restricted to the "rectangle" of width k and height y, with a corner at the origin.

Weakly useful sequences are defined (in Section 1) in terms of *recursive measure*, a special case of the resource-bounded measure developed by Lutz [9]. We very briefly sketch the elements of this theory, referring the reader to [9, 8] for motivation, details, and intuition.

Definition 1. A *martingale* is a function $d: \{0,1\}^* \to [0, \infty)$ such that, for all $w \in \{0,1\}^*$, $d(w) = \frac{d(w0) + d(w1)}{2}$.

Definition 2. A martingale d is *recursive* if there is a total recursive function $\hat{d}: N \times \{0,1\}^* \to Q$ such that, for all $r \in N$ and $w \in \{0,1\}^*$,

$$\left| \hat{d}(r, w) - d(w) \right| \leq 2^{-r}.$$

Definition 3. A martingale d *succeeds* on a sequence $x \in \{0,1\}^\infty$ if

$$\limsup_{n \to \infty} d(x[0 \ldots n-1]) = \infty,$$

where $x[0 \ldots n-1]$ is the n-bit prefix of x. The *success set* of a martingale d is

$$S^\infty[d] = \{x \in \{0,1\}^\infty \mid d \text{ succeeds on } x\}.$$

Definition 4. Let $X \subseteq \{0,1\}^\infty$.

1. X has *recursive measure 0*, and we write $\mu_{\text{rec}}(X) = 0$, if there is a recursive martingale d such that $X \subseteq S^\infty[d]$.
2. X has *recursive measure 1*, and we write $\mu_{\text{rec}}(X) = 1$, if $\mu_{\text{rec}}(X^c) = 0$, where $X^c = \{0,1\}^\infty - X$ is the complement of X.
3. X has *measure 0 in* REC, and we write $\mu(X \mid \text{REC}) = 0$, if $\mu_{\text{rec}}(X \cap \text{REC}) = 0$.
4. X has *measure 1 in* REC, and we write $\mu(X \mid \text{REC}) = 1$, if $\mu(X^c \mid \text{REC}) = 0$. In this case, we say that X contains *almost every* element of REC.

3   Uniform Recursive Depth

In this section we prove that, for every uniform reducibility F, almost every recursive subset of $\mathbf{N}$ has a certain "depth" property with respect to F. This depth property is used in the proof of our main result in Section 4. It is also of independent interest because it is closely related to Bennett's notion of weak depth [3].

We first make our terminology precise. As in [12], we define a *truth-table condition* (briefly, a *tt-condition*) to be an ordered pair $\tau = ((n_1, \ldots, n_k), g)$, where $k, n_1, \ldots, n_k \in \mathbf{N}$ and $g: \{0, 1\}^k \to \{0, 1\}$. We write TTC for the class of all tt-conditions. The *tt-value* of a set $B \subseteq \mathbf{N}$ under a tt-condition $\tau = ((n_1, \ldots, n_k), g)$ is the bit

$$\tau^B = g([\![n_1 \in B]\!] \cdots [\![n_k \in B]\!]).$$

A *truth-table reduction* (briefly, a *tt-reduction*) is a total recursive function $F: \mathbf{N} \to \text{TTC}$. If F is a tt-reduction and $F(x) = ((n_1, \ldots, n_k), g)$, then we call $n_1, \ldots, n_k$ the *queries* made by F on input x. A truth-table reduction F naturally *induces* a function $\widehat{F}: \mathcal{P}(\mathbf{N}) \to \mathcal{P}(\mathbf{N})$ defined by

$$\widehat{F}(B) = \{n \in \mathbf{N} \mid F(n)^B = 1\}.$$

In general, we identify a truth-table reduction F with the induced function $\widehat{F}$, writing F for either function and relying on context to avoid confusion.

The following terminology is convenient for our purposes.

Definition 5. A *uniform reducibility* is a total recursive function $F: \mathbf{N} \times \mathbf{N} \to$ TTC.

If F is a uniform reducibility, then we use the notation $F_k(n) = F(k, n)$ for all $k, n \in \mathbf{N}$. We thus regard a uniform reducibility as a recursive sequence $F_0, F_1, F_2, \ldots$ of tt-reductions.

Definition 6. If F is a uniform reducibility and $A, B \subseteq \mathbf{N}$, then we say that A is *F-reducible* to B, and we write $A \leq_F B$, if there exists $k \in \mathbf{N}$ such that $A = F_k(B)$.

Example 1. Fix a recursive time bound, i.e., a total recursive function $t: \mathbf{N} \to \mathbf{N}$. It is routine to construct a uniform reducibility F such that, for all $A, B \subseteq \mathbf{N}$,

$$A \leq_F B \Longleftrightarrow A \in \text{DTIME}^B(t).$$

Definition 7. Let F be a uniform reducibility. The *upper F-span* of a set $A \subseteq \mathbf{N}$ is the set

$$F^{-1}(A) = \{B \subseteq \mathbf{N} \mid A \leq_F B\}.$$

Definition 8. Let F be a uniform reducibility. A set $A \subseteq \mathbf{N}$ is *recursively F-deep* if $\mu_{\text{rec}}(F^{-1}(A)) = 0$.

Bennett [3] defines a set $A \subseteq \mathbf{N}$ to be *weakly deep* if A is not tt-reducible to any algorithmically random set B. The above definition is similar in spirit, but it (i) replaces "tt-reducible" with "F-reducible," and (ii) replaces "any algorithmically random set B" with "any set B outside a set of recursive measure 0."

Definition 9. A set $A \subseteq \mathbf{N}$ is *recursively weakly deep* if, for every uniform reducibility F, A is recursively F-deep.

It is easy to see that every recursively weakly deep set is weakly deep.

We now prove the main result of this section. Recalling our identification of subsets of $\mathbf{N}$ with their characteristic sequences, we state this result in terms of sequences but, for convenience, prove it in terms of sets.

Theorem 10. *If F is a uniform reducibility, then almost every sequence in* REC *is recursively F-deep.*

Proof sketch Assume the hypothesis. For each $k, n \in \mathbf{N}$ and $A \subseteq \mathbf{N}$, define the set

$$\mathcal{E}_{k,n}^A = \{B \subseteq \mathbf{N} \mid (\forall 0 \leq m < n)[\![m \in A]\!] = [\![m \in F_k(B)]\!]\}.$$

This is the set of all B such that $F_k(B)$ agrees with A on $0, 1, \ldots, n-1$. In particular,

$$F^{-1}(A) = \bigcup_{k=0}^{\infty} \bigcap_{n=0}^{\infty} \mathcal{E}_{k,n}^A.$$

We regard $\mathcal{E}_{k,n}^A$ as an event in the sample space $\mathcal{P}(\mathbf{N})$ with the uniform distribution. Thus we write $\Pr(\mathcal{E}_{k,n}^A)$ for the probability that $B \in \mathcal{E}_{k,n}^A$, where the set $B \subseteq \mathbf{N}$ is chosen probabilistically according to a random experiment in which an independent toss of a fair coin is used to decide membership of each natural number in B.

For each $A \subseteq \mathbf{N}$, define a function $d^A : \{0,1\}^* \to [0, \infty)$ by

$$d^A(w) = \sum_{k=0}^{\infty} \sum_{n=0}^{\infty} 2^{-\frac{k+n}{4}} d_{k,n}^A(w),$$

where, for all $k, n \in \mathbf{N}$ and $w \in \{0,1\}^*$,

$$d_{k,n}^A(w) = \begin{cases} 2^{|w|} \Pr(\mathbf{C}_w \mid \mathcal{E}_{k,n}^A) & \text{if } \Pr(\mathcal{E}_{k,n}^A) > 0, \\ 1 & \text{if } \Pr(\mathcal{E}_{k,n}^A) = 0, \end{cases}$$

where $\mathbf{C}_w = \{A \subset \mathbf{N} \mid w \sqsubseteq \chi_A\}$. It is routine to check that each d^A is a martingale that is recursive in A.

For each $k, n \in \mathbf{N}$ and $A \subseteq \mathbf{N}$, let

$$N_A(k,n) = \left| \left\{ m \mid 0 \leq m < n \text{ and } \Pr(\mathcal{E}_{k,m+1}^A) \leq \tfrac{1}{2} \Pr(\mathcal{E}_{k,m}^A) \right\} \right|,$$

and let

$$X = \left\{ A \subseteq \mathbf{N} \mid (\forall k \in \mathbf{N})(\forall^{\infty} n \in \mathbf{N}) N_A(k,n) > \frac{n}{4} \right\},$$

where the quantifier $(\forall^{\infty} n \in \mathbf{N})$ means "for all but finitely many $n \in \mathbf{N}$."

We use the following four claims (proofs are omitted).

Claim 1 *For all $k, n \in \mathbf{N}$ and $A \subseteq \mathbf{N}$,*

$$\Pr(\mathcal{E}_{k,n}^A) \leq 2^{-N_A(k,n)}.$$

Claim 2 *For all $k, n \in \mathbf{N}$ and $A, B \subseteq \mathbf{N}$ satisfying $A = F_k(B)$,*

$$\liminf_{\ell \to \infty} d_{k,n}^A(\chi_B[0 \ldots \ell - 1]) \geq 2^{N_A(k,n)}.$$

Claim 3 *For all $A \in X$, $F^{-1}(A) \subseteq S^{\infty}[d^A]$.*

Claim 4 $\mu_{\mathrm{rec}}(X) = 1.$

Let

$$\mathcal{D} = \{ A \subseteq \mathbf{N} \mid A \text{ is recursively } F\text{-deep} \}.$$

By Claim 3 and the fact that d^A is recursive in A, we must have $X \cap \mathrm{REC} \subseteq \mathcal{D}$. It follows that $\mathcal{D}^c \cap \mathrm{REC} \subseteq X^c$. Claim 4 tells us that $\mu_{\mathrm{rec}}(X^c) = 0$, and hence

$$\mu(\mathcal{D}^c \mid \mathrm{REC}) = \mu_{\mathrm{rec}}(\mathcal{D}^c \cap \mathrm{REC}) = 0,$$

since any subset of a rec-measure 0 set has rec-measure 0. We thus get $\mu(\mathcal{D} \mid \mathrm{REC}) = 1$, which proves the theorem. $\qquad \qquad \square$ **Theorem 10**

Theorem 11. *Every weakly useful sequence is recursively weakly deep.*

Proof. Assume that A is weakly useful and fix a uniform reducibility F. It suffices to show that $\mu_{\mathrm{rec}}(F^{-1}(A)) = 0$.

Fix a recursive time bound $t : \mathbf{N} \to \mathbf{N}$ such that $\mu(\mathrm{DTIME}^A(t) \mid \mathrm{REC}) \neq 0$. Then there is a uniform reducibility $\widetilde{F}$ such that, for all $B, C, D \subseteq \mathbf{N}$,

$$[\, B \in \mathrm{DTIME}^C(t) \text{ and } C \leq_F D \,] \Longrightarrow B \leq_{\widetilde{F}} D.$$

Let X be the set of all sets that are recursively $\widetilde{F}$-deep. By Theorem 10, $\mu(X \mid \mathrm{REC}) = 1$, so there is a set $B \in X \cap \mathrm{DTIME}^A(t) \cap \mathrm{REC}$. Now $\mu_{\mathrm{rec}}(\widetilde{F}^{-1}(B)) = 0$ (because $B \in X$) and $F^{-1}(A) \subseteq \widetilde{F}^{-1}(B)$ (because $B \in \mathrm{DTIME}^A(t)$), so $\mu_{\mathrm{rec}}(F^{-1}(A)) = 0.$ $\qquad \qquad \square$

4 Main Result

In this section, we prove the existence of weakly useful sequences that are not strongly useful. Our construction uses recursively F-deep sets (for an infinite, nonuniform collection of uniform reducibilities F), but those sets are constructed in a canonical way.

Theorem 12. *There is a sequence that is weakly useful but not strongly useful.*

We include a sketch of the proof of Theorem 12. The proof uses the next proposition, which is of independent interest.

Proposition 13. *If F is a uniform reducibility, then there is a canonical recursive, recursively F-deep set, i.e., a set A such that*

$$\mu_{\text{rec}}(\{B \mid (\exists i) A = F_i(B)\}) = 0,$$

and such that for all $x, i \in \mathbf{N}$, $\Pr_C[F_i(C)[i, < x] = A[i, < x]] \leq 2^{-x}$.

We call A above the *canonical recursively F-deep set.*

Proof sketch of Theorem 12 Our proof is an adaptation of the martingale diagonalization method introduced by Lutz in [8]. We will define H one strand at a time to satisfy the following conditions, where $H_0, H_1, H_2, \ldots$ are the strands of H:

1. Each strand H_k is recursive (although H itself cannot be recursive).
2. If d is any recursive martingale, then there is some k such that d fails on H_k.
3. For every recursive time bound t, there is a recursive set A such that $A \notin \text{DTIME}^H(t)$.

These three conditions suffice for our purposes. By Condition 1, the set $J = \{H_0, H_1, H_2, \ldots\} \subseteq \text{REC}$, and by Condition 2, no recursive martingale can succeed on all its elements. Thus $\mu_{\text{rec}}(J) \neq 0$, which makes H weakly useful, since $J \subseteq \text{DTIME}^H(\text{linear})$. Condition 3 ensures that H is not strongly useful.

Fix an arbitrary enumeration $\{t_k\}_{k \in \mathbf{N}}$ of all recursive time bounds, and an enumeration $\{\tilde{d}_k\}_{k \in \mathbf{N}}$ of all recursive martingales. These enumerations need not be uniform in any sense, since at present we are not trying to control the complexity of H. We will define (in order) a number of different objects for each k:

- a uniform reducibility F^k corresponding to t_k.
- a recursive A^k such that $A^k \notin \text{DTIME}^H(t_k)$ (A^k will be the canonical recursively F^k-deep set (cf. Proposition 13,
- a partial characteristic function α_k of finite domain, compatible with all the previous strands of H (ultimately, $\alpha_k \sqsubseteq H$ for all k),
- martingales $d_{k;q}^{i,j}$ (uniformly recursive over j and q) for all $i, j, q \in \mathbf{N}$ with $i \leq k$, which, taken together, witness that each A^i is recursively F^i-deep, and

– the strand H_k itself, which is designed to make the martingale

$$d'_k = \tilde{d}_k + \sum_{i=0}^{k} \sum_{j=0}^{\infty} \sum_{q=0}^{\infty} d^{i,j}_{k;q} \cdot 2^{-q-j}$$

fail on H_k, thus satisfying Condition 2 above. H_k will also participate in a fixed finite number of diagonalizations against tt-reductions from the A^i to H for $i \leq k$.

Fix $k \in \mathbf{N}$, and assume that all the above objects have been defined for all $k' < k$ (define $\alpha_{-1} = \lambda$). Also assume that for each $k' < k$ we have at our disposal programs to compute (uniformly over j and q) $F_j^{k'}$, $A^{k'}$, $H_{k'}$, and $d^{i,j}_{k';q}$ for all $i \leq k'$. Let $\{M_{j,k}\}_{j \in \mathbf{N}}$ be a recursive enumeration of all oracle Turing machines running in time t_k, and for all j let $M'_{j,k}$ be the same as $M_{j,k}$ except that when $M_{j,k}$ makes a query of the form $\langle x, y \rangle$ for $x < k$, $M'_{j,k}$ instead simulates the answer by computing $H_x(y)$ directly. We let F_j^k be the tt-reduction corresponding to $M'_{j,k}$. Note that on any input, F_j^k only makes queries of the form $\langle x, y \rangle$ for $x \geq k$.

We define A^k to be the canonical recursively F^k-deep set constructed in the proof of Proposition 13, therefore,

Fact 1 *For all* $j, k, x \in \mathbf{N}$, $\Pr_C \left[F_j^k(C)[j, < x] = A^k[j, < x] \right] \leq 2^{-x}$.

Let $H_{<k}$ denote the partial characteristic function that agrees with H on all $\langle x, y \rangle$ with $x < k$, and is undefined otherwise. Given α_{k-1}, which is compatible with $H_{<k}$, we define α_k as follows: let $\langle i, j \rangle = k$. If there is a set $C \sqsupseteq H_{<k} \cup \alpha_{k-1}$ such that $A^i \neq F_j^i(C)$, then we *diagonalize* against F_j^i by letting α_k be the least finite characteristic function extending α_{k-1} that preserves such a miscomputation, i.e., for some C and x such that $A^i(x) \neq F_j^i(x)^C$, α_k will agree with C on all queries made by F_j^i on input x. If no such C exists, let $\alpha_k = \alpha_{k-1}$.

Now fix any i and j with $i \leq k$. We would like to define a martingale that succeeds on all B such that $A^i = F_j^i(B)$. We cannot do this directly, because any given tt-reduction F_j^i from A^i to H might make queries on many different columns at once, and our martingales can only act on one column at a time. Instead, for any $q \in \mathbf{N}$ large enough, the martingales $d^{i,j}_{k';q}$ for all $k' \geq i$ will act together to "succeed as a group" on all sets to which A^i reduces via F_j^i.

The martingale $d^{i,j}_{k;q}$ will be split up into infinitely many martingales

$$d^{i,j}_{k;q} = \sum_{\ell=1}^{\infty} d^{i,j}_{k;q;\ell},$$

where each martingale $d^{i,j}_{k;q;\ell}$ bets a finite number of times. Fix i and j. For any $m \in \mathbf{N}$, let y_m be least such that $v < y_m$ for all queries $\langle u, v \rangle$ made by F_j^i on inputs $\langle j, x \rangle$ for all $x < m$. For any $C \subseteq \mathbf{N}$, let $\mathbf{E}^C(m)$ be the event that

$F_j^i(C)[j, < m] = A^i[j, < m]$, i.e., that $F_j^i(C)$ and A^i agree on the first m elements of the jth column. For all $w \in \{0,1\}^*$, we define

$$d_{k;q;\ell}^{i,j}(w) = 2^{|w|-\ell}.$$

$$\Pr_C \left[\xi^H(k, y_{q\ell}) \cup (\{k\} \times w) \sqsubset C \mid \xi^H(k, y_{q\ell}) \sqsubset C \ \& \ \mathbf{E}^C(q\ell) \right]$$

if $\Pr_C \left[\xi^H(k, y_{q\ell}) \sqsubset C \ \& \ \mathbf{E}^C(q\ell) \right] > 0$. Otherwise, for all w define $d_{k;q;\ell}^{i,j}(w) = 2^{-\ell}$.

We now define H_k. For any y, we assume that $H_k[< y]$ has already been defined, and we set $w = H_k[< y]$. Let

$$H_k(y) = \begin{cases} \alpha_k(\langle k, y \rangle) & \text{if } \alpha_k(\langle k, y \rangle) \text{ is defined,} \\ 0 & \text{if } \alpha_k(\langle k, y \rangle) \text{ is undefined and } d_k'(w0) \le d_k'(w1), \\ 1 & \text{if } \alpha_k(\langle k, y \rangle) \text{ is undefined and } d_k'(w0) > d_k'(w1). \end{cases}$$

Remark. Actually, we cannot do this exactly as stated. A recursive martingale such as d_k' cannot in general be computed exactly, but is only approximated. What we are really comparing are not $d_k'(w0)$ and $d_k'(w1)$, but rather their yth approximations, which *are* computable. Since these approximations are guaranteed to be within 2^{-y} of the actual values, and our sole aim is to make d_k' fail on H_k, it suffices for our purposes to consider only the approximations when doing the comparisons above. The same trick is used in [8].

H_k is evidently recursive (given the last remark), and for cofinitely many y, $H_k(y)$ is chosen so that $d_k'(H_k[< (y+1)]) \le d_k'(H_k[< y]) + 2^{-y}$, the 2^{-y} owing to the error in the approximation of d_k'. Thus d_k' fails on H_k, from which we obtain

Fact 2 *The martingales $\tilde{d}_k$ and $d_{k;q}^{i,j}$ for all and $i \le k$, j, and q all fail on H_k.*

Thus Conditions 1 and 2 are satisfied. Each H_k also preserves the diagonalization commitments made by the $\alpha_{k'}$ for all $k' \le k$, so the following is easily checked:

Fact 3 $\alpha_0 \sqsubseteq \alpha_1 \sqsubseteq \alpha_2 \sqsubseteq \cdots \sqsubset H$.

To verify Condition 3, we show that $A^i \ne F_j^i(H)$ for all i and j. Suppose $A^i = F_j^i(H)$ for some i and j. Let $k_0 = \langle i, j \rangle$, and let $\sigma = H_{<k_0} \cup \alpha_{k_0-1}$. By the definition of α_{k_0}, it must be the case that $A^i = F_j^i(C)$ for all $C \sqsupset \sigma$, otherwise F_j^i would have been diagonalized against by α_{k_0} and would thus fail to reduce A^i to H. Let q_0 be smallest such that $q_0 > i$ and $\sigma(\langle q', y \rangle)$ is undefined for all y and $q' \ge q_0$. We will show that $d_{n;q_0}^{i,j}$ succeeds on H_n for some $n < q_0$, contradicting Fact 2 above.

For any $C \subseteq \mathbf{N}$ and $m \in \mathbf{N}$, we let y_m and $\mathbf{E}^C(m)$ be as before. For any ℓ and $y \ge y_{q_0\ell}$ we have

$$\Pr_C \left[\mathbf{E}^C(q_0\ell) \mid \xi^H(q_0, y) \sqsubset C \right] = 1$$

by the definition of q_0 and $y_{q_0\ell}$, and thus

$$\frac{\mathrm{Pr}_C\left[\mathbf{E}^C(q_0\ell)\mid \xi^H(q_0,y)\sqsubseteq C\right]}{\mathrm{Pr}_C\left[\mathbf{E}^C(q_0\ell)\mid \xi^H(i,y)\sqsubseteq C\right]}$$

$$=\frac{1}{\mathrm{Pr}_C\left[\mathbf{E}^C(q_0\ell)\mid \xi^H(i,y)\sqsubseteq C\right]}$$

$$=\frac{1}{\mathrm{Pr}_C\left[\mathbf{E}^C(q_0\ell)\right]}$$

$$\geq 2^{q_0\ell},$$

the last inequality following from Fact 1. From the definition of $d^{i,j}_{k;q_0;\ell}$, the following inequation can be shown for any ℓ and $y \geq y_{q_0\ell}$ (details are omitted)

$$\prod_{k=i}^{q_0-1} d^{i,j}_{k;q_0;\ell}(H_k[< y]) \geq 2^{-q_0\ell} \cdot \frac{\mathrm{Pr}_C\left[\mathbf{E}^C(q_0\ell)\mid \xi^H(q_0,y)\sqsubseteq C\right]}{\mathrm{Pr}_C\left[\mathbf{E}^C(q_0\ell)\mid \xi^H(i,y)\sqsubseteq C\right]}$$

Therefore,

$$\prod_{k=i}^{q_0-1} d^{i,j}_{k;q_0;\ell}(H_k[< y]) \geq 1$$

for all $y \geq y_{q_0\ell}$, which implies that $d^{i,j}_{k;q_0;\ell}(H_k[< y]) \geq 1$ for at least one k between i and $q_0 - 1$. Since q_0 is fixed and ℓ was chosen arbitrarily, by the Pigeon-Hole Principle there must be some n_0 with $i \leq n_0 < q_0$ such that for infinitely many ℓ, $d^{i,j}_{n_0;q_0;\ell}(H_{n_0}[< y]) \geq 1$ for all $y \geq y_{q_0\ell}$. This in turn implies that the martingale $d^{i,j}_{n_0;q_0}$ succeeds on H_{n_0}, contradicting Fact 2.

Thus $A^i \neq F^i_j(H)$ for all i and j, and Condition 3 is satisfied.

$$\square \ \mathbf{Theorem\ 12}$$

Corollary 14. *There is a sequence that is strongly deep but not strongly useful.*

Proof. This follows immediately from Theorem 12 and the fact [6] that every weakly useful sequence is strongly deep. $\square$

It is easy to verify that weak and strong usefulness are both invariant under tt-equivalence. Thus, Theorem 12 shows that there are weakly useful tt-degrees that are not strongly useful. Our results do not say anything regarding the *Turing* degrees of weakly useful sets, however. In particular, we leave open the question of whether there is a weakly useful Turing degree that is not strongly useful (i.e., whether there is a weakly useful set not Turing equivalent to any strongly useful set). Some facts are known about these degrees. Jockusch [4] neatly characterized the *strongly* useful Turing degrees (under a different name), for example, as being either high or containing complete extensions of first-order Peano arithmetic. This includes some low degrees, but no non-high r.e. degrees. Recently, Stephan [14] has partially strengthened these results, showing that no non-high r.e. Turing degree can be *weakly* useful, either. Therefore, among the r.e. degrees, the strongly useful, weakly useful, and high degrees all coincide.

Acknowledgments

The authors thank Martin Kummer and Frank Stephan for helpful discussions.

References

1. Miklos Ajtai and Ronald Fagin, Reachability is harder for directed than for undirected graphs, *Journal of Symbolic Logic* **55** (1990), pp. 113–150.
2. Y. M. Barzdin', Complexity of programs to determine whether natural numbers not greater than n belong to a recursively enumerable set, *Soviet Mathematics Doklady* **9** (1968), pp. 1251–1254.
3. C. H. Bennett, Logical depth and physical complexity, In R. Herken, editor, *The Universal Turing Machine: A Half-Century Survey*, pp. 227–257. Oxford University Press, 1988.
4. Jr. C. G. Jockusch, Degrees in which the recursive sets are uniformly recursive, *Canadian Journal of Mathematics* **24** (1972), pp. 1092–1099.
5. Jr. C. J. Jockusch and R. I. Soare, Degrees of members of Π_1^0 classes, *Pacific Journal of Mathematics* **40** (1972), pp. 605–616.
6. David W. Juedes, James I. Lathrop, and Jack H. Lutz, Computational depth and reducibility, *Theoretical Computer Science* **132** (1994), pp. 37–70, also appeared in *Proceedings of the Twentieth International Colloquium on Automata, Languages, and Programming*, 1993, pp. 278–288.
7. M. Li and P. M. B. Vitányi, *An Introduction to Kolmogorov Complexity and its Applications*, Springer, 1993.
8. J. H. Lutz, Weakly hard problems, *SIAM Journal on Computing* to appear, also appeared in *Proceedings of the Ninth Structure in Complexity Theory Conference*, 1993, pp. 146–161.
9. J. H. Lutz, Almost everywhere high nonuniform complexity, *Journal of Computer and System Sciences* **44** (1992), pp. 220–258.
10. P. Martin-Löf, On the definition of random sequences, *Information and Control* **9** (1966), pp. 602–619.
11. E. L. Post, Recursively enumerable sets of positive integers and their decision problems, *Bulletin of the American Mathematical Society* **50** (1944), pp. 284–316.
12. H. Rogers, Jr, *Theory of Recursive Functions and Effective Computability*, McGraw - Hill, 1967.
13. G. E. Sacks, *Degrees of Unsolvability*, Princeton University Press, 1966.
14. F. Stephan, 1994, private communication.

Graph Connectivity, Monadic NP and Built-in Relations of Moderate Degree

Thomas Schwentick

Institut für Informatik
Universität Mainz

Abstract. It has been conjectured [FSV93] that an existential second-oder formula, in which the second-order quantification is restricted to unary relations (i.e. a Monadic NP formula), cannot express Graph Connectivity even in the presence of arbitrary built-in relations.

In this paper it is shown that Graph Connectivity cannot be expressed by Monadic NP formulas in the presence of arbitrary built-in relations of degree $n^{o(1)}$. The result is obtained by using a simplified version of a method introduced in [Sch94] that allows the extension of a local winning strategy for Duplicator, one of the two players in Ehrenfeucht games, to a global winning strategy.

1 Introduction

Since the result of Fagin [Fag74] that the complexity class NP coincides with the class of all sets of finite structures that can be characterized by existential second-order formulas (Σ_1^1), there has been a lot of interest in the expressive power of such formulas. Most inexpressibility results are connected to the restriction of Σ_1^1-formulas where the second-order relation variables are unary (MonNP).

Ajtai, [Ajt83], already showed that, in a very strong sense, MonCoNP is not captured by MonNP even in the presence of arbitrary built-in relations.[1] Here MonCoNP is the class of sets of finite structures that can be characterized by monadic universal second-order formulas, i.e. the complement of MonNP.

For a formal definition of built-in relations see Section 2.

Though Ajtai gave a concrete property, derived from "Even Cardinality", that was expressible in MonCoNP but not in MonNP, people were interested in more natural properties. Most of the results in this line of research show that Graph Connectivity can not be expressed with monadic existential second-order formulas even in the presence of built-in relations of several kinds.

It was shown that Graph Connectivity can neither be expressed in MonNP ([Fag75]), nor in the presence of a built-in successor relation ([dR87]), nor in the presence of arbitrary built-in relations of degree $(\log n)^{o(1)}$ ([FSV93]), nor in the presence of a built-in linear order ([Sch94]).

Fagin, Stockmeyer and Vardi [FSV93] conjectured that Graph Connectivity is not expressible in MonNP even in the presence of arbitrary built-in relations.

---

[1] In fact, Ajtai's result is far more general and not restricted to the monadic case.

In this paper we show that Graph Connectivity cannot be expressed in MonNP even in the presence of arbitrary built-in relations of degree $n^{o(1)}$.

For the proof of this result, we introduce a simplified version of a method that was first used in [Sch94]. It is based on the extension, under certain circumstances, of a winning strategy for Duplicator in an Ehrenfeucht game on structures H and H', to a winning strategy on structures G and G' that contain H and H' as substructures. We refer to this method as the Extension Theorem.

In their proof of the above mentioned result, Fagin, Stockmeyer and Vardi [FSV93] make use of a probabilistic construction to obtain graphs with a lot of edges with some desirable property. We show how graphs with enough edges of this kind can be constructed with a straightforward deterministic algorithm.

In Section 2 we give some basic definitions and notations and recall the basic facts about Ehrenfeucht games. In Section 3 we describe the simplified version of the Extension Theorem. In Section 4 we show our main result.

2 Definitions and Notations

2.1 Finite Structures

A *signature* is a finite set of relation symbols $R_1, \ldots, R_s$, each with a fixed arity a_i and constant symbols $c_1, \ldots, c_r$. We do not make use of function symbols.

A *finite S-structure* $\mathcal{A} = (U^{\mathcal{A}}, R_1^{\mathcal{A}}, \ldots, R_s^{\mathcal{A}}, c_1^{\mathcal{A}}, \ldots, c_r^{\mathcal{A}})$ consists of a finite universe $U^{\mathcal{A}}$, relations $R_i^{\mathcal{A}}$ over $U^{\mathcal{A}}$ of arity a_i, and elements $c_i^{\mathcal{A}}$ of $U^{\mathcal{A}}$. Whenever possible we will omit the superscript $\mathcal{A}$.

If S is the disjoint union of S_1 and S_2, the S_1-*reduct* $\mathcal{B}$ of the S-structure $\mathcal{A}$ has universe $U^{\mathcal{A}}$ and those relations and constants of $\mathcal{A}$ that correspond to symbols in S_1. In this case we also call $\mathcal{A}$ an S_2-*expansion* of the S_1-structure $\mathcal{B}$.

Sometimes we want to distinguish certain elements in $U^{\mathcal{A}}$. Then we view $\mathcal{A}$ with the distinguished elements $x_1, \ldots, x_m$ as a $(S \cup \{c_1, \ldots, c_m\})$-structure, where $U^{\mathcal{A}} = U^{\mathcal{A}'}$ and denote the resulting structure with $<\mathcal{A}, x_1, \ldots, x_m>$.

If δ is a function from $U^{\mathcal{A}}$ to $\{0, \ldots, k\}$, we write $(\mathcal{A}, \delta)$ for the $S \cup \{U_0, \ldots, U_k\}$-structure[2] $<\mathcal{A}, U_0^{\mathcal{A}}, \ldots, U_k^{\mathcal{A}}>$, where $x \in U_i^{\mathcal{A}}$ if and only if $\delta(x) = i$.

For a subset $W \subseteq U^{\mathcal{A}}$ that contains all constants $c_i^{\mathcal{A}}$ the *induced substructure* $\mathcal{A} \downarrow W$ has universe W, relations $R_i \cap W^{a_i}$ and constants $c_i^{\mathcal{A}}$.

As an abbreviation we write $\mathcal{A} \downarrow [x_1, \ldots, x_m]$ for the structure $<\mathcal{A} \downarrow \{x_1, \ldots, x_m\}, x_1, \ldots, x_m>$, the substructure of $\mathcal{A}$ which is induced by the x_i and in which all the x_i are distinguished elements.

We say that $x, y \in U^{\mathcal{A}}$, $x \neq y$, are *adjacent* in R_i, if there are $x_1, \ldots, x_{a_i}$ such that $(x_1, \ldots, x_{a_i}) \in R_i^{\mathcal{A}}$ and $x = x_j$ and $y = x_k$ for some $j, k \leq a_i$.

If S' is a subset of S, the S'-degree of a vertex x is the number of vertices that are adjacent with x in some relation of S'. A sequence $x_0, \ldots, x_m$ is called an S'-*path of length* m, if, for every $j < m$, x_j and x_{j+1} are adjacent in some relation of S'. The S'-*distance*, $\delta_{S'}(x, y)$, between x and y is the length of a shortest S'-path from x to y. For $d \geq 0$ the S'-d-neighbourhood of x, $N_d^{S'}(x)$,

---

[2] Here we assume that the unary relation symbols U_i do not occur in S.

consists of all y with $\delta_{S'}(x,y) \le d$. The neighbourhood of an edge or of a set of vertices is defined, accordingly, as the union of the neighbourhoods of the respective vertices.

2.2 Formulas and Expressibility

First-order formulas over a signature S are built of relation and constant symbols from S, the relation symbol $=$, the logical connectives $\wedge, \vee, \neg$, variables $x, y, x_1, \ldots$ and quantifiers $\exists, \forall$ in the usual way.

Second-order formulas may additionally contain relational variables $X, Y, \ldots$. Second-order formulas in prenex normal form, where the second order quantifiers have to be in front of the first order quantifiers, and relational variables are quantified only existentially, are called Σ_1^1-formulas.

We say that a formula ϕ expresses (or characterises) a property P (a set C) of finite structures over some signature S, if for every S-structure $\mathcal{A}$ it holds that $\mathcal{A}$ has property P ($\mathcal{A}$ is in C), if and only if $\mathcal{A} \models \phi$. MonNP is the class of sets of finite structures that can be characterised by Σ_1^1-formulas in which all quantified relation symbols are unary.

We are particularly interested in a modified form of expressibility, where the structures have additional relations, so-called built-in relations. Informally, built-in relations may help a formula to express a property, but the property must not depend on the actual content of the relations.

More formally, let S_1, S_2 be disjoint signatures, and P a property of S_1-structures. Let ϕ be a formula over $S_1 \cup S_2$. We say that ϕ expresses P on the S_2-structure $\mathcal{A}$, if ϕ holds exactly in those S_1-expansions of $\mathcal{A}$ whose S_1-reducts have property P. We say that ϕ expresses P in the presence of S_2-built-in relations, if, for every universe U, there is an S_2-structure with universe U, on which ϕ expresses P. For a function f of natural numbers, we say ϕ expresses P in the presence of built-in relations of degree at most f, if there is a signature S_2 such that, for every universe U, there is an S_2-structure with universe U, in which every vertex has degree at most $f(n)$, and on which ϕ expresses P.

2.3 Ehrenfeucht Games

Ehrenfeucht games were introduced in [Ehr61] and proved to be a useful tool in obtaining inexpressibility results. The rules of a k-round first-order (FO) Ehrenfeucht game are as follows.

There are two players, Spoiler and Duplicator. They play a fixed number, k, of rounds on two finite structures $\mathcal{A}, \mathcal{A}'$ over some signature S. In every round, Spoiler chooses one element of one of the two structures. Then Duplicator chooses an element of the other structure. At the end of the game there are elements $x_1, \ldots, x_k$ of $\mathcal{A}$, and $x_1', \ldots, x_k'$ of $\mathcal{A}'$, chosen, where x_i and x_i' are those chosen in round i. Duplicator wins, if $\mathcal{A} \downarrow [x_1, \ldots, x_m] \cong \mathcal{A}' \downarrow [x_1', \ldots, x_m']$. The importance of Ehrenfeucht games results from the following

Theorem 1. *[Ehr61, Fra54] Let C be a set of finite structures over some signature S. C is first-order definable, iff there is a fixed k, such that, whenever $A \in C$ and $A' \notin C$, then Spoiler has a winning strategy in the k-round FO Ehrenfeucht game on A and A'.*

We write $A \approx_k A'$ if Duplicator has a winning strategy in the k-round FO Ehrenfeucht game on A and A'. We write $<A, x_1, \ldots, x_m> \approx_{k-m} <A', x'_1, \ldots, x'_m>$ if Duplicator still has a winning strategy after m rounds have been played, in which $x_1, \ldots, x_m, x'_1, \ldots, x'_m$ were chosen. If $W \subseteq U^A$ and $W' \subseteq U^{A'}$, we write $<A \downarrow W, x_1, \ldots, x_m> \approx_{k-m} <A' \downarrow W', x'_1, \ldots, x'_m>$, if Duplicator has a $k - m$ rounds winning strategy on the substructures induced by W and W' and those vertices x_i and x'_j in W and W' resp.

Next we define the k-*type*, $\tau_k^A(x)$, of an element x of A and show two useful properties of k-types (cf. [EFT92]), namely

- Duplicator has a winning strategy in the k-round FO game on A and A', iff A and A' have the same set of $(k-1)$-types;
- For a fixed signature the number of different k-types depends only on k.

We define the k-type of a sequence $x_1, \ldots, x_l$ of elements of A inductively.

- $\tau_0^A(x_1, \ldots, x_l)$ is the S-isomorphism class of $A \downarrow [x_1, \ldots, x_l]$.
- $\tau_{k+1}^A(x_1, \ldots, x_l) := \{\tau_k^A(x_1, \ldots, x_l, x) \mid x \in A\}$.

We will sometimes call these types *game types* to distinguish them from isomorphism types. We will make use of the following lemmas, which can easily be proved by induction.

Lemma 2. *Let S be a fixed signature. For every k there is a constant N such that in every S-structure A there are at most N different k-types $\tau_k^A(x)$.*

Lemma 3. *Duplicator has a winning strategy in the k-round FO Ehrenfeucht game on A and A' if and only if $\{\tau_{k-1}^A(x) \mid x \in A\} = \{\tau_{k-1}^{A'}(x) \mid x \in A'\}$.*

For showing that a set C of finite structures is not in MonNP, Ajtai and Fagin [AF90] introduced a more advanced game. We are going to use this game only in the context of graphs instead of general finite structures. Therefore we will describe the game here only for graphs. As usual we view a graph with additional unary relations as a colored graph, where m relations define 2^m colour classes in a natural way. The Ajtai-Fagin (c, k)-game over a set of graphs, C, consists of the following steps.

(1) Duplicator selects a graph $G \in C$.
(2) Spoiler colors G with c colors.
(3) Duplicator selects a graph $G' \notin C$ and colors it with c colors.
(4) Spoiler and Duplicator play a k-round FO Ehrenfeucht game on the colored graphs G and G'.

Theorem 4. *[AF90] A set of graphs, C, is in MonNP, iff there are c and k such that Spoiler has a winning strategy in the Ajtai-Fagin (c, k)-game over C.*

3   How to extend a local winning strategy of Duplicator

3.1   The Basic Ideas of the Extension Theorem

Before we formally state the version of the Extension Theorem we are going to use, we first discuss its main features in a simplified context. Let H and H' be two undirected graphs such that Duplicator has a winning strategy in the k-round Ehrenfeucht game on H and H'. Our question is:

If we extend H and H' in the same way to graphs G and G' respectively, can we also extend Duplicator's winning strategy to the graphs G and G' ?

We will argue in the following that the answer is yes, if the extensions of the graphs fulfil conditions (1) and (2) below. We write δ for the E-distance functions of G and G'. By $N_e(H)$ we denote the E-e-neighbourhood of H, i.e., the subgraph induced by all vertices x with $\delta(x, H) \leq e$. The two conditions are

(1) Duplicator has a winning strategy on $N_{2^k}(H)$ and $N_{2^k}(H')$. Furthermore he can play in such a way that in every round the two chosen vertices have the same distance from H and H' respectively.

(2) There is an isomorphism π from $G - H$ to $G' - H'$ which on $N_{2^k}(H)$ respects the distance from H and H', i.e., $\delta(x, H) = \delta(\pi(x), H')$ for every $x \in N_{2^k}(H) - H$.

The idea for the winning strategy of Duplicator is as follows:

At the beginning of the game we view the vertices in H and H' as *inner vertices* and the vertices outside $N_{2^k}(H)$ and $N_{2^k}(H')$ as *outer vertices*. The status of the vertices in the *buffer area* between inner and outer vertices remains open. By definition, at the beginning, the distance from every outer vertex to every inner vertex is more than 2^k.

During the game, Duplicator's strategy depends on the distance from H of vertex x chosen by Spoiler (where wlog $x \in G$). If x is

an inner vertex, Duplicator chooses a vertex according to his winning strategy on the inner vertices;

an outer vertex, Duplicator plays according to π;

a vertex of the buffer area, then there are two cases:

   − if x is closer to the inner vertices, the area of the inner vertices will be extended to $N_{\delta(x,H)}(H)$ and $N_{\delta(x,H)}(H')$, and Duplicator chooses a vertex according to his winning strategy on the inner vertices,

   − if x is not closer to the inner vertices, the area of the outer vertices will be extended to all vertices outside $N_{\delta(x,H)-1}(H)$ and $N_{\delta(x,H)-1}(H')$, and Duplicator plays according to π.

It is easily shown, by induction on i, that after round i the distance from every outer vertex to every inner vertex is more than 2^{k-i}. **In particular, at the end of the game, no inner vertex is adjacent to any outer vertex.**

Therefore for the induced subgraphs $I \subseteq G$ and $I' \subseteq G'$, consisting only of inner vertices and $O \subseteq G, O' \subseteq G'$, consisting only of outer vertices it holds:

- I and I' are isomorphic because inside Duplicator played according to a winning strategy;
- O and O' are isomorphic, because outside Duplicator played according to π;
- Because there is neither an edge between I and O nor between I' and O', we conclude that $I \cup O$ is isomorphic to $I' \cup O'$, hence the induced subgraphs of all selected vertices are isomorphic.

Thus, Duplicator has indeed a winning strategy on G and G'.

3.2 The Weak Extension Theorem

Now we are going to transform the ideas of the previous discussion into a slightly more general situation. Before, our structures consisted of only the edge relation. Now we have to deal with arbitrary built-in relations too. Accordingly, instead of the distance function induced by the edge relation the distance functions are induced by all built-in relations as described in Section 2.

Theorem 5. *Let $k > 0$. Let S be a signature with relational symbols $R_1 = E, R_2, \ldots, R_s$. Let $\mathcal{A}, \mathcal{A}'$ be finite structures over S. Let H and H' be subsets of $U^{\mathcal{A}}$ and $U^{\mathcal{A}'}$ respectively.*

Duplicator has a winning strategy in the k-round FO Ehrenfeucht game on $\mathcal{A}$ and $\mathcal{A}'$, if the following conditions are fulfilled.

(i) Duplicator has a winning strategy in the k-round Ehrenfeucht game on[3] $(\mathcal{A} \downarrow N_{2^k}(H), \delta)$ and $(\mathcal{A}' \downarrow N_{2^k}(H'), \delta')$, where $\delta(x) := \min_{y \in H} \delta(x, y)$.

(ii) There is an S-isomorphism π from $\mathcal{A} \downarrow (U^{\mathcal{A}} - H)$ to $\mathcal{A}' \downarrow (U^{\mathcal{A}'} - H')$, such that for every $x \in U^{\mathcal{A}} - H$

$$\delta(x) = \delta'(\pi(x)) \qquad or \qquad (\delta(x) > 2^k \ and \ \delta'(\pi(x)) > 2^k).$$

Proof. Duplicator will follow the strategy described above, maintaining a buffer zone of distance 2^{k-q}, after q rounds. We set

$$d_q := \max\{\delta(x_i) \mid i \le q, x_i \text{ inner vertex}\}.$$

Which vertices are considered inner vertices will be defined during the game. We will conclude the statement of the theorem from the following lemma.

Lemma 6. *Duplicator can play in such a way that, for every $q \le k$, after q rounds, it holds that*

(1) for every $i \le q$

(a) either $\delta(x_i) \le d_q$, then $\delta(x_i) = \delta'(x_i')$, (we will call x_i and x_i' in this case inner vertices),

(b) or $\delta(x_i) > d_q + 2^{k-q}$, $\delta'(x_i') > d_q + 2^{k-q}$ and $\pi(x_i) = x_i'$, (we will call x_i and x_i' in this case outer vertices),

---

[3] We write $N_e(H)$ for the set of all vertices with $\delta_S(x, y) \le e$ for some $y \in H$.

(2)

$$<(A \downarrow N_{d_q+2^{k-q}}(H), \delta), x_1, \ldots, x_q > \approx_{k-q} <(A' \downarrow N_{d_q+2^{k-q}}(H'), \delta'), x'_1, \ldots, x'_q >,$$

(3) $d_q \leq 2^k - 2^{k-q}$.

Setting $q := k$, the statement of the theorem follows immediately from this lemma, since from (1)(b) it follows that O and O' are isomorphic and from (2) it follows that I and I' are isomorphic and neither a vertex of I is adjacent to a vertex of O nor a vertex of I' is adjacent to a vertex of O'. □

Proof of Lemma 6. We show the lemma by induction on q.

For $q = 0$ (1) and (3) follow immediately, and (2) follows from (i).

Let Duplicator have played according to (1)-(3) in the first q rounds. Let wlog $x_{q+1} \in U^A$ be the vertex chosen by spoiler in round $q + 1$. We show how duplicator can select x'_{q+1} such that (1)-(3) hold again.

We distinguish two cases.

1. $\underline{\delta(x_{q+1}) \leq d_q + 2^{k-(q+1)}}$

This means that x_{q+1} is closer to the inner vertices, hence, if necessary (i.e. if $\delta(x_{q+1}) > d_q$), the area of the inner vertices will be extended to contain x_{q+1}. By the induction hypothesis we know that

$$<(A \downarrow N_{d_q+2^{k-q}}(H), \delta), x_1, \ldots, x_q > \approx_{k-q} <(A' \downarrow N_{d_q+2^{k-q}}(H'), \delta'), x'_1, \ldots, x'_q > .$$

Therefore there is an $x'_{q+1} \in N_{d_q+2^{k-q}}(H')$ such that

$$<(A \downarrow N_{d_q+2^{k-q}}(H), \delta), x_1, \ldots, x_{q+1} >$$
$$\approx_{k-(q+1)} <(A' \downarrow N_{d_q+2^{k-q}}(H'), \delta'), x'_1, \ldots, x'_{q+1} > .$$

In particular $\delta(x_{q+1}) = \delta'(x'_{q+1})$. Hence (1) follows.

On the other hand, because $d_{q+1} + 2^{k-(q+1)} \leq d_q + 2^{k-q}$, it follows

$$<(A \downarrow N_{d_{q+1}+2^{k-(q+1)}}(H), \delta), x_1, \ldots, x_{q+1} > \approx_{k-(q+1)}$$
$$<(A' \downarrow N_{d_{q+1}+2^{k-(q+1)}}(H'), \delta'), x'_1, \ldots, x'_{q+1} > .$$

So (2) is also fulfilled. Finally

$$d_{q+1} \leq d_q + 2^{k-(q+1)} \leq 2^k - 2^{k-q} + 2^{k-(q+1)} = 2^k - 2^{k-(q+1)},$$

so (3) is fulfilled too.

2. $\underline{\delta(x_{q+1}) > d_q + 2^{k-(q+1)}}$

In this case x_{q+1} will be viewed as outer vertex. Duplicator chooses $x'_{q+1} := \pi(x_{q+1})$. (1) - (3) follow immediately.

□

Remark. Theorem 5 is a simplified version of a theorem shown in [Sch94]. The main difference is, that in the strong Extension Theorem the distance function need not be the one which is induced by the relations of the structure, but has to obey certain other conditions. This allows to deal with structures in which the vertices have a large degree. In particular, the strong Extension Theorem was used to show that Connectivity is not in MonNP even in the presence of a built-in linear order.

4 Connectivity and Built-in Relations of Degree $n^{o(1)}$

In this section we are going to show that Graph Connectivity is not in MonNP even in the presence of built-in relations of degree $n^{o(1)}$.

4.1 The Proof Idea

Let us first review the proof of Fagin, Stockmeyer and Vardi [FSV93] that Graph Connectivity cannot be expressed in MonNP in the presence of built-in relations of degree $(\log n)^{o(1)}$. In their proof Duplicator chooses a cycle G on n vertices for some large n. After Spoiler has coloured G Duplicator chooses two edges $e_1 = (x_1, y_1)$ and $e_2 = (x_2, y_2)$ and obtains G' by switching these edges, i.e., G' contains, instead of e_1 and e_2, the edges (x_1, y_2) and (x_2, y_1), and thus consists of two cycles (cf. Figure 1).

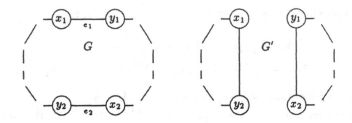

Fig. 1. The graphs G and G' from the proof of Fagin, Stockmeyer and Vardi.

They argue that Spoiler cannot detect the difference if for some d, depending on the number of rounds of the Ehrenfeucht game, e_1 and e_2

- are *good*, i.e. the d-neighbourhoods of x_i and y_i in $G - e_i$ are disjoint,
- have the same local isomorphism type, i.e. their d-neighbourhoods (including the built-in relations) are isomorphic, and
- have disjoint d-neighbourhoods.

Then they show that, with high probability, in a randomly chosen cycle most edges are d-good. As the built-in relations have degree $(\log n)^{o(1)}$ the d-neighbourhoods have size $(\log n)^{o(1)}$. Hence there are $2^{(\log n)^{o(1)}} = n^{o(1)}$ local isomorphism types. So G has enough vertices to assure that, no matter how Spoiler colours G, there will be a pair of edges with the desired properties.

Our proof differs from the original one in that we will construct G explicitly without any probabilistic construction and instead of local isomorphism types we will make use of local game types in the sense of Section 2.3.

Dealing with game types instead of isomorphism types has the advantage that the number of possible game types is bounded by a constant that doesn't depend on the size of the graph. This is because the number of possible game types depends only on the number of rounds of the Ehrenfeucht game and the arity of the given relations, whereas the number of isomorphism types grows exponentially with the size of the neighbourhoods under consideration.

4.2 Handling Built-in Relations

In the following we are going to describe, how, given built-in relations of degree $n^{o(1)}$, a cycle G can be constructed such that it contains many good edges.

To explain the idea we will first look at the case where there is only a single built-in successor relation $\prec$. We call an edge $e = (x, y)$ of a graph G with a built-in successor relation $\prec$ d-good, if in $G - e$ the $\{E, \prec\}$-distance between x and y is more than $2d$, i.e. x and y have disjoint $\{E, \prec\}$-d-neigbourhoods in $G - e$.

Lemma 7. *Let $d > 0$ be an integer, let $\{1, \ldots n\}$ be a set of vertices and $\prec$ a successor relation on $\{1, \ldots n\}$. Then there is an edge relation E on $\{1, \ldots n\}$ such that $(\{1, \ldots n\}, E)$ is a cycle and E contains at least $\frac{n}{4d+2} - 2$ edges which are d-good in*
$$G = (\{1, \ldots n\}, E, \prec) \text{ and have pairwise disjoint } \{E, \prec\}\text{-}d\text{-neighbourhoods.}$$

Proof. Wlog we may assume that $\prec$ is such that $m \prec m + 1$ for every $m < n$.

For every $i \leq \frac{n}{2d+1}$ we define the path p_i to be the sequence of vertices $i(2d+1), i(2d+1)+1, i(2d+1)-1, i(2d+1)+2, \ldots, i(2d+1)+d, i(2d+1)-d$ in that order (cf. Fig.2).

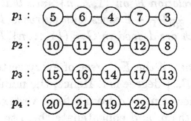

p_1 : ⑤—⑥—④—⑦—③

p_2 : ⑩—⑪—⑨—⑫—⑧

p_3 : ⑮—⑯—⑭—⑰—⑬

p_4 : ⑳—㉑—⑲—㉒—⑱

Fig. 2. The four paths in the case $n = 24, d = 2$.

We construct E by concatenating $\overline{p_1}, p_2, \overline{p_3}, p_4, \ldots$, where $\overline{p}$ stands for p reversed, adding the remaining $(< 3d+1)$ vertices at the end of the resulting path, and connecting the first with the last vertex to close the cycle (cf. Fig. 3).

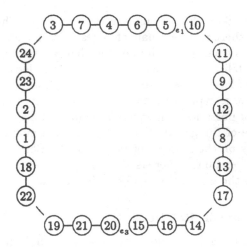

Fig. 3. The resulting graph in the case $n = 24, d = 2$. e_1 and e_3 are 2-good edges.

It is easy to see that,

- for every i, p_i contains all vertices that have $\{\prec\}$-distance at most d from $i(2d+1)$;
- if i is odd, there is an edge e_i between $i(2d+1)$ and $(i+1)(2d+1)$. In $G - e_i$ the vertices $i(2d+1)$ and $(i+1)(2d+1)$ have $\{E, \prec\}$-distance $2d+1$;

This means that, for every odd number i, e_i is a d-good edge. Furthermore, these edges have pairwise disjoint $(E, \prec)$-d-neighbourhoods. □

Now we turn to the case of arbitrary built-in relations (of moderate degree).

Lemma 8. *Let $d, l > 0$, let $\{1, \ldots n\}$ be a set of vertices and let $B_1, \ldots, B_m$ be built-in relations such that the $\{B_1, \ldots, B_m\}$-degree of any vertex is at most $l - 2$. Then there is an edge relation E on $\{1, \ldots n\}$ such that $(\{1, \ldots n\}, E)$ is a cycle and E contains at least $\frac{n}{2l^2(d+1)} - 1$ edges with pairwise disjoint $\{E, B_1, \ldots, B_m\}$-d-neighbourhoods, which are d-good in $G = (\{1, \ldots n\}, E, B_1, \ldots, B_m)$.*

Proof. First, observe that, if the E-degree of every vertex is at most 2, then the maximum $\{E, B_1, \ldots, B_m\}$-degree is l. Hence, the number of vertices that have $(E, B_1, \ldots, B_m)$-distance at most d from a fixed vertex v is at most l^{d+1}.

Again we define paths p_i and concatenate them to a cycle. Let $p_1, \ldots, p_{i-1}$ be already constructed. Let v_i be a vertex that has $\{B_1, \ldots, B_m\}$-distance more than d from all vertices contained in $p_1, \ldots, p_{i-1}$. Such a v_i exists as long as the paths $p_1, \ldots, p_{i-1}$ contain together less than $\frac{n}{l^{d+1}}$ vertices.

Let p_i consist of v_i followed by all vertices with $\{B_1, \ldots, B_m\}$-distance 1 from v_i, followed by all vertices with $\{B_1, \ldots, B_m\}$-distance 2 from v_i and so on up to all vertices with $\{B_1, \ldots, B_m\}$-distance d from v_i. Hence p_i contains at most

l^{d+1} vertices and all vertices which have $\{B_1, \ldots, B_m\}$-distance at most d from v_i are included in p_i.

We note that, in contrast to the case of a single successor relation, here the paths p_i are, in general, of different size.

Again we construct E by concatenating $\overline{p_1}, p_2, \overline{p_3}, p_4, \ldots$, adding all remaining vertices (i.e. those that are not contained in any p_i) at the end of the path and closing the cycle by connecting the first and the last vertex.

For every odd number i we set $e_i := (v_i, v_{i+1})$. It is important to note that in $G - e_i$, all vertices with $\{E, B_1, \ldots, B_m\}$-distance at most d from v_i (v_{i+1} resp.) are contained in p_i (p_{i+1} resp.). Therefore, as in Lemma 7, for every odd i, v_i has $\{E, B_1, \ldots, B_m\}$-distance more than $2d$ from v_{i+1} in $G - e_i$. Furthermore, by construction the $\{E, B_1, \ldots, B_m\}$-d-neighbourhoods of v_i and v_j are disjoint for different odd numbers i and j.

As each single p_i contains at most l^{d+1} vertices, the above construction will yield at least $\frac{n}{2l^2(d+1)} - 1$ good edges. $\qquad\square$

4.3 The Theorem

Theorem 9. *Let $f : \mathbb{N} \to \mathbb{N}$ with $f(n) = n^{o(1)}$. Connectivity is not expressible in MonNP even in the presence of built-in relations of degree at most $f(n)$.*

Proof. We make use of the Ajtai-Fagin game. Let k be a fixed number of rounds and c the number of colours of the Ajtai-Fagin game. Let n be chosen large enough with respect to k, c, and the number and the arities of the built-in relations. Let $B_1, \ldots, B_m$ be relations of degree at most $f(n)$. Let G be constructed as in Lemma 8 with $d := 2^k$ and $l := f(n) + 2$. As G is a cycle it is connected.

Let G be coloured arbitrarily with colours $A_1, \ldots, A_c$. In this proof, distance always means the $\{E, B_1, \ldots, B_m\}$-distance. The outline of the proof is as follows:

(a) We show that there are distinct odd numbers i and j such that Duplicator has a distance respecting winning strategy on p_i and p_j and on p_{i+1} and p_{j+1}.

(b) We construct G' from G by deleting the edges (v_i, v_{i+1}) and (v_j, v_{j+1}) and inserting new edges (v_i, v_{j+1}) and (v_j, v_{i+1}). G' consists of two cycles and is therefore disconnected.

(c) By applying the weak Extension Theorem we show that Duplicator has a winning strategy in the k-round Ehrenfeucht game on G and G'.

Let $\delta_i(x)$ be the distance of vertex $x \in p_i$ from v_i. Let $\mathcal{N}$ be such that there are at most $\mathcal{N}$ different k-types $\tau_k^{(p_i, \delta_i)}(x)$ (cf. Lemma 2). Now let n be chosen such that $n > 2(\mathcal{N}^2 + 1)(f(n) + 2)^{2(2^k + 1)}$. This is possible because $f(n) = n^{o(1)}$ and $\mathcal{N}$ does not depend on n. By Lemma 8 there are more than $\mathcal{N}^2$ edges that are 2^k-good in G and whose 2^k-neighbourhoods are pairwise disjoint. Hence there are distinct odd numbers i and j such that $\tau_k^{(p_i, \delta_i)}(v_i) = \tau_k^{(p_j, \delta_j)}(v_j)$ and $\tau_k^{(p_{i+1}, \delta_{i+1})}(v_{i+1}) = \tau_k^{(p_{j+1}, \delta_{j+1})}(v_{j+1})$. Therefore i and j fulfil (a).

Hence Duplicator has a distance respecting k-round winning strategy on p_i and p_j on one hand and on p_{i+1} and p_{j+1} on the other hand. Because all paths are pairwise disjoint, these winning strategies can be be combined to winning strategies on the two graphs consisting of $\overline{p_i}, (v_i, v_{i+1}), p_{i+1}$ and $\overline{p_i}, (v_i, v_{j+1}), p_{j+1}$ and on the two graphs consisting of $\overline{p_j}, (v_j, v_{j+1}), p_{j+1}$ and $\overline{p_j}, (v_j, v_{i+1}), p_{i+1}$.

Let G' be defined as in (b) above, by deleting the edges (v_i, v_{i+1}) and (v_j, v_{j+1}) and inserting new edges (v_i, v_{j+1}) and (v_j, v_{i+1}).

Let H_1 be the set of vertices $\{v_i, v_j, v_{i+1}, v_{j+1}\}$ and let H_1' be the respective set of vertices of G'. We have to show that conditions (i) and (ii) of Theorem 5 are fulfilled. But (i) was already shown above, and (ii) follows immediately from the construction. Hence, the statement now follows from Theorem 5. $\square$

The graph G in the proof of Theorem 9 is hamiltonian and consists of a single cycle in contrast to G'. Therefore we can conclude the following corollary.

Corollary 10. *Let $f : I\!N \to I\!N$ with $f(n) = n^{o(1)}$. Neither the class of Hamiltonian graphs nor the class of graphs that consist of a single cycle are in MonNP even in the presence of built-in relations of degree at most $f(n)$.*

References

[AF90] M. Ajtai and R. Fagin. Reachability is harder for directed than for undirected finite graphs. *Journal of Symbolic Logic*, 55(1):113–150, 1990.

[Ajt83] M. Ajtai. Σ_1^1 formulae on finite structures. *Ann. of Pure and Applied Logic*, 24:1–48, 1983.

[Cos93] S. Cosmadakis. Logical reducibility and monadic NP. In *Proc. 34th IEEE Symp. on Foundations of Computer Science*, pages 52–61, 1993.

[dR87] M. de Rougemont. Second-order and inductive definability on finite structures. *Zeitschrift für Mathematische Logik und Grundlagen der Mathematik*, 33:47–63, 1987.

[EFT92] H.-D. Ebbinghaus, J. Flum, and W. Thomas. *Einführung in die mathematische Logik*. BI, Mannheim, 3rd edition, 1992.

[Ehr61] A. Ehrenfeucht. An application of games to the completeness problem for formalized theories. *Fund. Math.*, 49:129–141, 1961.

[Fag74] R. Fagin. Generalized first–order spectra and polynomial–time recognizable sets. In R. M. Karp, editor, *Complexity of Computation, SIAM-AMS Proceedings, Vol. 7*, pages 43–73, 1974.

[Fag75] R. Fagin. Monadic generalized spectra. *Zeitschrift für Mathematische Logik und Grundlagen der Mathematik*, 21:89–96, 1975.

[Fra54] R. Fraïssé. Sur quelques classifications des systèmes de relations. *Publ. Sci. Univ. Alger. Sér. A*, 1:35–182, 1954.

[FSV93] R. Fagin, L. Stockmeyer, and M. Vardi. On monadic NP vs. monadic co-NP. In *The Proceedings of the 8th Annual IEEE Conference on Structure in Complexity Theory*, pages 19–30, 1993.

[Sch94] T. Schwentick. Graph connectivity and monadic NP. In *Proc. 35th IEEE Symp. on Foundations of Computer Science*, pages 614–622, 1994.

The Expressive Power of Clocks*

Thomas A. Henzinger, Peter W. Kopke, and Howard Wong-Toi

Computer Science Department, Cornell University, Ithaca, NY 14853

Abstract. We investigate the expressive power of timing restrictions on labeled transition systems. In particular, we show how constraints on clock variables together with a uniform liveness condition — the divergence of time — can express Büchi, Muller, Streett, Rabin, and weak and strong fairness conditions on a given labeled transition system. We then consider the effect, on both timed and time-abstract expressiveness, of varying the following parameters: time domain (discrete or dense), number of clocks, number of states, and size of constants used in timing restrictions.

1 Introduction

We study the expressive power of labeled transition systems with clocks, so-called *timed safety automata* [11]. Timed safety automata are timed automata [2] without acceptance conditions; their liveness is imposed uniformly as a progress condition on time. Timed safety automata have been used extensively for the specification and verification of real-time systems [6, 7, 11, 9]. It has been argued that with the explicit consideration of time, acceptance conditions are no longer useful abstractions to enforce liveness [8], and this paper corroborates that belief.

We look at both the time-abstract expressive power (Section 3) and the timed expressive power (Section 4) of timed safety automata. In the time-abstract view, a timed safety automaton defines a set of infinite words over the input alphabet A; in the timed view, each input symbol is labeled with a time stamp, and a timed safety automaton defines a set of infinite words over the alphabet $A \times \mathbb{T}$, where $\mathbb{T}$ is the time domain. We look at several orthogonal parameters that affect the expressiveness of timed safety automata: number of automaton states, number of clocks, size of time constants that occur in clock constraints, and time domain (discrete or dense). For instance, we examine the hierarchy obtained by fixing the number of states, and letting the number of clocks vary, and the hierarchy obtained by considering only the number of clocks.

* This research was supported in part by the National Science Foundation under grant CCR-9200794, by the United States Air Force Office of Scientific Research under contract F49620-93-1-0056, by the Defense Advanced Research Projects Agency under grant NAG2-892, and by the U.S. Army Research Office through the Mathematical Sciences Institute of Cornell University, Contract Number DAAL03-91-C-0027.

The full version of this paper is available from the Department of Computer Science, Cornell University, Ithaca, NY 14853, as Technical Report TR95-1496.

Time-abstract expressiveness. Liveness conditions are typically imposed on state-transition structures as conditions on the set of states that are visited infinitely often. The clock view, by contrast, ensures the liveness of a state-transition structure through the divergence of time under clock constraints. First, we give direct translations of generalized Büchi, Muller, Streett, and Rabin acceptance conditions, and of weak and strong fairness conditions, into clock constraints. We show that these liveness conditions can be enforced on a given state-transition structure using either just two clocks, or time constants of size at most 2. Second, we show that a single clock and the time constant 1 suffice to express all ω-regular languages. Indeed, both of these results hold for discrete and for dense time.

Over dense time, the expressive power of clocks is even greater. We show that dense-time clocks can be used to enforce any ω-regular liveness condition on a given state-transition structure. This implies that, in dense time, every ω-regular language is accepted by a timed safety automaton with a single state. The same is not true for discrete time, where, surprisingly, for any fixed state transition structure, two clocks are as expressive as any number of clocks.

Timed expressiveness. The timed language of a timed safety automaton is limit-closed [11], and therefore the timed expressiveness of timed safety automata is strictly less than that of timed automata [2]. In dense time, for any natural number k, there are timed languages that are accepted by a timed safety automaton with $k + 1$ clocks, but not by any timed safety automaton with k clocks. We thus obtain an infinite hierarchy of dense-time languages. In discrete time, the timed languages that are accepted by timed safety automata are precisely the limit-closed *timed ω-regular languages* [4, 5]. In contrast with dense time, however, a single discrete-time clock suffices to accept any limit-closed timed ω-regular language. On the other hand, for any fixed number of states, increasing the number of clocks does increase expressive power.

Two-way timed automata are studied in [3]. An infinite hierarchy of timed expressiveness is obtained, based upon the number of alternations. Clock hierarchies of a different nature are studied in [1]. Various types of timed and time-abstract observational equivalence are considered for observers with a limited number of clocks.

2   Labeled Transition Systems

A *labeled transition system* $\mathcal{A}$ is a quadruple $(S, A, \rightarrow, S_0)$, for a set S of *states*, a finite *alphabet* A of *events*, a *transition relation* $\rightarrow \subset S \times A \times S$, and a set $S_0 \subset S$ of *initial states*. We write $s \xrightarrow{a} t$ in place of $(s, a, t) \in \rightarrow$. A *run* r of $\mathcal{A}$ is an infinite sequence $s_0 \xrightarrow{a_0} s_1 \xrightarrow{a_1} s_2 \xrightarrow{a_2} \cdots$ of states $s_i \in S$ and events $a_i \in A$ such that $s_0 \in S_0$. The run r *generates* the word $a_0 a_1 a_2 \ldots$, denoted $\bar{a}$. The *language* $[\mathcal{A}] \subset A^\omega$ of the labeled transition system $\mathcal{A}$ is the set of infinite words that are generated by runs of $\mathcal{A}$.

Acceptance conditions. The labeled transition system $\mathcal{A}$ is a *finite automaton* if the state space S is finite. The liveness of finite automata is commonly enforced by acceptance conditions. The *infinite visitation set* $\inf(r)$ of the run r is the set of states that occur infinitely often along r. A *generalized Büchi condition* $\mathcal{B}$ for the labeled transition system $\mathcal{A}$ is a set of subsets of S (the Büchi acceptance condition $\mathcal{B}$ is *standard* if $|\mathcal{B}| = 1$). The run r is $\mathcal{B}$-*accepting* if for every $I \in \mathcal{B}$, $\inf(r) \cap I \neq \emptyset$. A *Muller condition* $\mathcal{M}$ is also a set of subsets of S. The run r is $\mathcal{M}$-*accepting* if $\inf(r) \in \mathcal{M}$. A *Rabin condition* $\mathcal{R}$ is a set of pairs of subsets of S. The run r is $\mathcal{R}$-*accepting* if there exists a pair $(I, J) \in \mathcal{R}$ such that $\inf(r) \cap I \neq \emptyset$ and $\inf(r) \cap J = \emptyset$. A *Streett condition* $\mathcal{S}$ is also a set of pairs of subsets of S, representing the negation of the corresponding Rabin condition: the run r is $\mathcal{S}$-*accepting* if for all $(I, J) \in \mathcal{S}$, if $\inf(r) \cap I \neq \emptyset$ then $\inf(r) \cap J \neq \emptyset$. The *Büchi (Muller; Rabin; Streett) language* $[\mathcal{A}, \mathcal{X}] \subset A^\omega$ of the labeled transition system $\mathcal{A}$ under the acceptance condition $\mathcal{X} = \mathcal{B}$ ($\mathcal{M}$; $\mathcal{R}$; $\mathcal{S}$) is the set of infinite words that are generated by $\mathcal{X}$-accepting runs of $\mathcal{A}$. The Büchi (Muller; Rabin; Streett) languages of finite automata are the ω-regular languages [12].

Fairness conditions. The labeled transition system $\mathcal{A}$ is *event-recording* if for each state $s \in S$ there is a label $a_s \in A$ such that $t \xrightarrow{a} s$ implies $a = a_s$. For every labeled transition system $\mathcal{A} = (S, A, \rightarrow, S_0)$, there is an event-recording transition system $\mathcal{A}' = (S \times A, A, \Rightarrow, S_0 \times A)$, with $(s, a) \xRightarrow{b} (t, b)$ iff $s \xrightarrow{b} t$, such that $\mathcal{A}$ and $\mathcal{A}'$ define the same language. The liveness of event-recording transition systems is commonly enforced by fairness conditions. The *enabling set* $\mathsf{enabled}(a)$ of the event a is the set of states s such that there is a successor state t with $s \xrightarrow{a} t$; the *completion set* $\mathsf{taken}(a)$ is the set of states s such that there is a predecessor state t with $t \xrightarrow{a} s$. A *weak-fairness condition* $\mathcal{F}_W$ for the labeled transition system $\mathcal{A}$ is a set of events. A run r is *weakly* $\mathcal{F}_W$-*fair* if for each $a \in \mathcal{F}_W$, infinitely many of the states of r lie in $\neg\mathsf{enabled}(a) \cup \mathsf{taken}(a)$. If $\mathcal{A}$ is a finite automaton, this corresponds to acceptance by the generalized Büchi condition that contains, for each event $a \in \mathcal{F}_W$, the set $\neg\mathsf{enabled}(a) \cup \mathsf{taken}(a)$. A *strong-fairness condition* $\mathcal{F}_S$ is also a set of events. A run r is *strongly* $\mathcal{F}_S$-*fair* if for every $a \in \mathcal{F}_S$, if infinitely many of the states of r lie in $\mathsf{enabled}(a)$, then infinitely many lie in $\mathsf{taken}(a)$. If $\mathcal{A}$ is a finite automaton, this corresponds to acceptance by the Streett condition $\mathcal{S}$ that contains, for each event $a \in \mathcal{F}_S$, the pair $(\mathsf{enabled}(a), \mathsf{taken}(a))$. The *weakly (strongly) fair language* $[\mathcal{A}, \mathcal{F}]$ of the labeled transition system $\mathcal{A}$ under the fairness condition $\mathcal{F} = \mathcal{F}_W$ ($\mathcal{F}_S$) is the set of infinite words that are generated by weakly (strongly) $\mathcal{F}$-fair runs of $\mathcal{A}$.

Timing conditions. We consider both the discrete time domain $\mathbb{T} = \mathbb{N}$ and the dense time domain $\mathbb{T} = \mathbb{R}_{\geq 0}$. Let C be a set of $\mathbb{T}$-valued variables called *clocks*. The *clock constraints* are the formulas defined by the grammar

$$\varphi ::= x \leq n \mid x \geq n \mid x \leq y + n \mid \neg\varphi \mid \varphi \wedge \varphi$$

where x and y are clocks, and n is an integer. A *closed clock constraint* is a conjunction of formulas of the form $x \leq n$ and $x \geq n$. A *clock valuation* $\nu : C \rightarrow \mathbb{T}$ is a map that assigns to each clock x a time value $\nu(x)$. The clock valuation ν, then, assigns to each clock constraint φ a truth value $\nu(\varphi)$. Given a clock valuation ν and a time delay $\delta \in \mathbb{T}$, the clock valuation $\nu + \delta$ assigns to each clock x the time value $(\nu + \delta)(x) = \nu(x) + \delta$. Given a set $X \subset C$ of clocks, the clock valuation $\nu[X := 0]$ assigns to each clock $x \in X$ the value 0, and to each clock $x \notin X$ the time value $\nu(x)$. The set of all clock valuations is denoted $\mathcal{V}_C$. It is naturally isomorphic to $\mathbb{T}^{|C|}$. Closed clock constraints specify closed subsets of $\mathcal{V}_C$. An *atomic clock command* ϕ is a pair (φ, X), where the *transition guard* φ is a clock constraint and the *reset set* X is a subset of C. The atomic clock command ϕ defines a partial function on the set $\mathcal{V}_C$ of clock valuations: if $\nu(\varphi) = true$ then $\phi(\nu) = \nu[X := 0]$; otherwise $\phi(\nu)$ is undefined. A *clock command* ψ is a finite set of atomic clock commands. The clock command ψ defines a relation on the set $\mathcal{V}_C$ of clock valuations: $(\mu, \nu) \in \psi$ iff ψ contains an atomic clock command ϕ such that $\nu = \phi(\mu)$.

A *timing condition* $\mathcal{T} = (C, \psi)$ for the labeled transition system $\mathcal{A}$ consists of a finite set C of clocks and a function ψ that assigns to each transition $s \xrightarrow{a} t$ of $\mathcal{A}$ a clock command $\psi(s, a, t)$. The timing condition $\mathcal{T}$ is *closed* if every clock constraint of every clock command $\psi(s, a, t)$ is closed. If $\mathcal{A}$ is a finite automaton, then the pair $(\mathcal{A}, \mathcal{T})$ is called a *timed safety automaton* [11].

A $\mathbb{T}$-*timing* is an infinite sequence $\overline{\delta} \in \mathbb{T}^\omega$ of time delays such that the sum $\Sigma_{i \geq 0} \delta_i$ diverges. A $\mathbb{T}$-*timed word* is a pair $(\overline{a}, \overline{\delta})$ where $\overline{a} \in A^\omega$ is an infinite word and $\overline{\delta}$ is a timing. A $\mathbb{T}$-timed word $(\overline{a}, \overline{\delta})$ is *accepted* by $(\mathcal{A}, \mathcal{T})$ if there is a run $r = s_0 \xrightarrow{a_0} s_1 \xrightarrow{a_1} \cdots$ of $\mathcal{A}$ and a sequence $\overline{\nu}$ of clock valuations such that ν_0 is the constant mapping $\lambda x.0$, and for all i, $(\nu_i + \delta_i, \nu_{i+1}) \in \psi(s_i, a_i, s_{i+1})$. In this case, the pair $(r, \overline{\nu})$ is called a *divergent execution* of $(\mathcal{A}, \mathcal{T})$. The *timed language* $[\mathcal{A}, \mathcal{T}]_{\mathbb{T}}^t$ of $(\mathcal{A}, \mathcal{T})$ under time domain $\mathbb{T}$ is the set of $\mathbb{T}$-timed words accepted by $(\mathcal{A}, \mathcal{T})$. The *untimed language* $[\mathcal{A}, \mathcal{T}]_{\mathbb{T}}$ is the set of words $\overline{a} \in A^\omega$ such that for some $\mathbb{T}$-timing $\overline{\delta}$, $(\overline{a}, \overline{\delta}) \in [\mathcal{A}, \mathcal{T}]_{\mathbb{T}}^t$.

Some previous expressiveness results. First, if the timing condition $\mathcal{T}$ is *closed*, then the choice of time domain is irrelevant, because then $[\mathcal{A}, \mathcal{T}]_{\mathbb{N}} = [\mathcal{A}, \mathcal{T}]_{\mathbb{R}_{\geq 0}}$ [10]. Second, in the discrete time domain, the set of *timed* languages accepted by timed safety automata is the class of limit-closed timed ω-regular languages [4, 5]. Third, in either time domain, the class of *untimed* languages accepted by timed safety automata is the class of ω-regular languages. This is because every timed safety automaton $\mathcal{A} = (S, A, \rightarrow, S_0)$ has a finite bisimulation [2].

Let h be the largest constant appearing in the clock constraints of $\mathcal{T}$. Define an equivalence relation $\equiv_h$ on the set $\mathcal{V}_C$ of clock valuations by $\nu \equiv_h \mu$ iff for every clock constraint φ containing no constant greater than h, $\nu(\varphi) = \mu(\varphi)$. Roughly speaking, two valuations are equivalent iff they agree on the ordering of the fractional parts of the clocks, they agree on which clocks are integers, and they agree on the the integer part of each clock with value no more than h. An equivalence class of $\equiv_h$ is called a *region*. The region of ν is denoted $\overline{\nu}$. The *region automaton* $Reg(\mathcal{A}, \mathcal{T})_{\mathbb{T}}$ for $(\mathcal{A}, \mathcal{T})$ is the finite automaton $(S \times (\mathcal{V}_C/\equiv_h), A, \to', S_0 \times \{\overline{\lambda x.0}\})$. The transition relation is defined by $((s, \overline{\mu}), a, (t, \overline{\nu})) \in \to'$ iff $s \xrightarrow{a} t$ and there exist valuations $\xi \in \overline{\mu}$ and $\zeta \in \overline{\nu}$, a duration $\delta \in \mathbb{T}$, and an atomic clock command $(\varphi, X) \in \psi(s, a, t)$ such that $(\xi + \delta)(\varphi) = true$ and $\zeta = (\xi + \delta)[X := 0]$. Let $\mathcal{B}$ be the generalized Büchi condition that requires each clock x either to be infinitely often greater than h, or both infinitely often 0 and infinitely often nonzero. The Büchi condition $\mathcal{B}$ in effect enforces time divergence on $(\mathcal{A}, \mathcal{T})$. Every divergent execution of $(\mathcal{A}, \mathcal{T})$ corresponds naturally to a run of $Reg(\mathcal{A}, \mathcal{T})_{\mathbb{T}}$ satisfying $\mathcal{B}$, and vice versa. The region automaton is the main tool for the analysis of timed safety automata [2].

3 Untimed Languages

We study the expressiveness of timing conditions with respect to untimed languages over the alphabet A.

3.1 Time Enough for Liveness

We give several ways to replace acceptance conditions on finite automata, or fairness conditions on labeled transition systems, by timing conditions. First, we decorate an existing state transition structure with timing constraints to impose a liveness condition, using as few clocks as possible. Second, we accomplish the same task using clock constraints with constants as small as possible. Third, we show that a single clock and the constant 1 suffice if we can change the state-transition structure. In this subsection, all results hold for both the discrete time domain $\mathbb{T} = \mathbb{N}$ and the dense time domain $\mathbb{T} = \mathbb{R}_{\geq 0}$.

Minimizing the number of clocks.

Theorem 3.1 *For every finite automaton $\mathcal{A}$ with state space S, and every generalized Büchi (Muller; Rabin; Streett) acceptance condition $\mathcal{X}$ for $\mathcal{A}$, there is a closed timing condition $\mathcal{T}$, using exactly two clocks and no constants greater than $|\mathcal{X}|$ ($|S||\mathcal{X}|$; $|\mathcal{X}|$; $2^{|\mathcal{X}|}|\mathcal{X}|$) in clock constraints, such that $[\mathcal{A}, \mathcal{T}]_{\mathbb{T}} = [\mathcal{A}, \mathcal{X}]$.*

Proof. Two clocks x and y can be used to store a positive integer value m by use of the invariant $|x - y| = m$. After each transition, we force one of the clocks to have the value m, and one to have value 0. To change the value from m to n we use the atomic clock commands

$$((x = m + n) \wedge (y = n), \{x\}) \text{ and } ((y = m + n) \wedge (x = n), \{y\}).$$

Notice that these commands allow at least one unit of time to pass.

We implement an acceptance condition by decomposing it into an infinite sequence of "helpful transitions." Two clocks are used to store a value that corresponds to the next required helpful transition. When this transition is taken, the value is changed to the following helpful transition. In this way, time diverges iff infinitely many helpful transitions are taken iff the acceptance condition is fulfilled. We give the proof for Streett acceptance, leaving the others to the full paper.

Let $\mathcal{X}$ be a Streett acceptance condition. A run $r = s_0 \xrightarrow{a_0} s_1 \xrightarrow{a_1} \cdots$ of $\mathcal{A}$ is accepted by $\mathcal{X}$ iff there is a subset $\mathcal{X}_{fin}$ of $\mathcal{X}$ such that

- for each $(I, J) \in \mathcal{X}_{fin}$, $\inf(r) \cap I = \emptyset$, and
- for each $(I, J) \in \mathcal{X} \setminus \mathcal{X}_{fin}$, $\inf(r) \cap J \neq \emptyset$.

Enumerate the possible values for $\mathcal{X}_{fin}$ (i.e., the subsets of $\mathcal{X}$), as $\mathcal{X}_1, \ldots, \mathcal{X}_n$. For each $\mathcal{X}_i$, enumerate the J such that $(I, J) \in \mathcal{X} \setminus \mathcal{X}_i$ for some I, as $J_{i,1}, \ldots, J_{i,m(i)}$. For $\mathcal{X}_{fin} = \mathcal{X}_i$, the helpful transitions are the transitions into the $J_{i,j}$. If entry to $J_{i,j}$ was the previous helpful transition, then the next helpful transition is defined to be entry into $J_{i,j+1 \bmod m(i)}$, where we take $\{1, \ldots, m(i)\}$ to be the range of mod $m(i)$. We store the two values i and j, corresponding respectively to the choice of $\mathcal{X}_{fin}$ and the next J to be visited.

Let $\gamma : \{(i,j) | 1 \leq i \leq n \wedge 1 \leq j \leq m(i)\} \to \{1, \ldots, n|X|\}$ be any one-to-one map. For each i, let $Bad_i = \bigcup \{I \mid \exists J.(I, J) \in \mathcal{X}_i\}$. To each transition $s \xrightarrow{a} t$ we assign four classes of atomic clock commands. First, to disallow infinitely many visits to Bad_i, we have

$$((x = 0) \wedge (y = 0), \emptyset).$$

Each run begins with some finite number of transitions of this type—a run with infinitely many converges.

Second, to guess $\mathcal{X}_i$ we have, for each i with $1 \leq i \leq n$, the atomic clock command

$$((x = \gamma(i, 1)) \wedge (y = \gamma(i, 1)), \{y\}).$$

Third, once $\mathcal{X}_i$ is chosen, we must not allow further visits to Bad_i. We must also not let time advance until the next $J_{i,j}$ is reached. For each i, j with $t \notin Bad_i$ and $t \notin J_{i,j}$, we have the atomic clock commands

$$((x = \gamma(i, j)) \wedge (y = 0), \emptyset) \text{ and } ((y = \gamma(i, j)) \wedge (x = 0), \emptyset).$$

Finally, to advance time when the next $J_{i,j}$ is met, we have for each i, j with $t \notin Bad_i$ and $t \in J_{i,j}$,

$$((x = \gamma(i, j) + \gamma(i, j + 1 \bmod m(i))) \wedge (y = \gamma(i, j + 1 \bmod m(i))), \{x\}),$$

$$((y = \gamma(i, j) + \gamma(i, j + 1 \bmod m(i))) \wedge (x = \gamma(i, j + 1 \bmod m(i))), \{y\}).$$

This completes the construction. Notice that if we did not allow time steps of duration zero, the proof would need only minor modifications. $\square$

Corollary 3.2 *For every event-recording labeled transition system $\mathcal{A}$, and every weak or strong-fairness condition $\mathcal{F}$ for $\mathcal{A}$, there is a closed timing condition $\mathcal{T}$ using exactly two clocks such that $[\mathcal{A}, \mathcal{T}]_{\mathcal{T}} = [\mathcal{A}, \mathcal{F}]$.*

Minimizing the size of constants used in clock constraints. While only two clocks are used in Theorem 3.1, the constants used in clock constraints are exponential in the number of components of the acceptance condition. If we wish to bound the constants by 2, we can get by with a logarithmic number of clocks (linear in the case of Streett acceptance).

Theorem 3.3 *Let A be a finite automaton with state space S.*

1. *For every generalized Büchi (Muller; Rabin) acceptance condition $\mathcal{X}$ for A, there is a closed timing condition T using $1 + \lceil \log_2 |\mathcal{X}| \rceil$ $(1 + \lceil \log_2 |\mathcal{X}||S| \rceil$; $1 + \lceil \log_2 |\mathcal{X}| \rceil)$ clocks, and no constants greater than 2 in clock constraints, such that $[A, T]_T = [A, \mathcal{X}]$.*

2. *For every Streett acceptance condition S for A, there is a closed timing condition T using $|S|$ clocks, and no constants greater than 1 in clock constraints, such that $[A, T]_T = [A, S]$.*

Proof sketch. A value in $\{1, ..., n\}$ may be stored in $1 + \lceil \log_2 n \rceil$ clocks by using $\lceil \log_2 n \rceil$ clocks for its binary encoding, and one additional clock to mediate increments. So for part 1, the bounds on the number of clocks result from the constants required in Theorem 3.1. For 2, use one clock $x_{(I,J)}$ for each element (I, J) of S. Allow entry to I only if $x_{(I,J)} \leq 1$, and reset $x_{(I,J)}$ only upon entry to J. □

Corollary 3.4 *For every event-recording labeled transition system A, and every weak or strong-fairness condition $\mathcal{F}$ for A, there is a closed timing condition T using $|\mathcal{F}|$ clocks, and no constants greater than 1 in clock constraints, such that $[A, \mathcal{F}] = [A, T]_T$.*

Changing the state transition structure. The previous constructions kept intact the original state transition structure. The cost of this was in the number of clocks or the size of the constants used in clock constraints. If we alter the state transition structure, then we need only one clock and the constant 1 in clock constraints. Since every ω-regular language is accepted by a finite automaton with a Streett acceptance condition with one pair, we have the following corollary to the second part of Theorem 3.3.

Corollary 3.5 *For every finite automaton A, and every generalized Büchi, Muller, Rabin, or Streett acceptance condition $\mathcal{X}$ for A, there is a timed safety automaton (C, T), with T closed, using one clock, and no constants greater than 1 in clock constraints, such that $[C, T]_T = [A, \mathcal{X}]$.*

By replicating the state space, a generalized Büchi condition may be replaced by a standard one. With this and the proof of the second part of Theorem 3.3, we can implement any fairness condition with one clock and small constants.

Theorem 3.6 *For every event-recording labeled transition system A, and every weak or strong-fairness condition $\mathcal{F}$ for A, there is a labeled transition system C and a closed timing condition T for C using one clock, and no constants greater than 1 in clock constraints, such that $[C, T]_T = [A, \mathcal{F}]$.*

3.2 Dense-Time Hierarchies

The power of dense time. Previously, we replaced liveness conditions with timing conditions. When the dense time domain $\mathbb{T} = \mathbb{R}_{\geq 0}$ is used, clocks acquire greater power. They may be used to enforce any ω-regular liveness condition on a given finite automaton.

Theorem 3.7 *For every finite automaton $\mathcal{A}$, and every ω-regular language L, there is a timing condition $\mathcal{T}$ such that $[\![\mathcal{A}, \mathcal{T}]\!]_{\mathbb{R}_{\geq 0}} = [\![\mathcal{A}]\!] \cap L$.*

Proof sketch. Let $\mathcal{A}' = (S', A, \rightarrow', S_0)$ be a finite automaton and $\mathcal{B}'$ a standard Büchi acceptance condition for $\mathcal{A}'$ such that $[\![\mathcal{A}', \mathcal{B}']\!] = L$. While performing a computation on $\mathcal{A}$, we simulate the computation of $\mathcal{A}'$ by using one clock x_s for each state s of $\mathcal{A}'$. The clock with the largest value corresponds to the current state of $\mathcal{A}'$. The current state of $\mathcal{A}'$ can be determined by the clock constraints $In_{s'}$, defined for each $s' \in S'$ by $In_{s'} = \bigwedge_{t' \in S' \setminus \{s'\}} (x_{s'} > x_{t'} \wedge x_{t'} > 0)$. When a transition is taken, all clocks but the one corresponding to the target state are reset. We use one additional clock $x_{B'}$ to enforce the acceptance condition of $\mathcal{A}'$. No transition is enabled if $x_{B'}$ is greater than one, and $x_{B'}$ is reset only when control enters $\mathcal{B}'$. The density of the time domain allows an arbitrary number of transitions in between entries into $\mathcal{B}'$. □

By applying Theorem 3.7 to the one-state automaton with language A^{ω}, we obtain the following counterpart to Corollary 3.5.

Corollary 3.8 *For every ω-regular language L, there is a timed safety automaton $(\mathcal{A}, \mathcal{T})$ with one state such that $[\![\mathcal{A}, \mathcal{T}]\!]_{\mathbb{R}_{\geq 0}} = L$.*

The state hierarchy for a fixed number of clocks. The following theorem shows that increasing the number of states, while holding the number of clocks fixed, yields an increase in expressive power.

Theorem 3.9 *For every $k, n \in \mathbb{N}$, the class of languages accepted by timed safety automata with k states and n clocks over the dense time domain $\mathbb{R}_{\geq 0}$ is properly contained in the class accepted with $k + 1$ states and n clocks.*

The clock hierarchy for a fixed number of states.

Theorem 3.10 *For every $k, n \in \mathbb{N}$, the class of languages accepted by timed safety automata with k states and n clocks over the dense time domain $\mathbb{R}_{\geq 0}$ is properly contained in the class accepted with k states and $n + 1$ clocks.*

3.3 Discrete-Time Hierarchies

The power of discrete time. Suppose $\mathcal{A}$ is a finite automaton with state space S, and $\sim$ is an equivalence relation on S. The *infinitely often change class* acceptance condition induced by $\sim$ on $\mathcal{A}$ is the generalized Büchi acceptance condition $\mathcal{B}_\sim = \{S \setminus \{t \in S \mid t \sim s\} \mid s \in S\}$. The *multiplicity* of $\sim$ is the number of elements in the largest equivalence class of $\sim$. A *k-iocc* language is a language accepted by a finite automaton with the iocc acceptance condition induced by an equivalence relation of multiplicity k. The language $(a_1 \cdots a_{k+1})^* a_1^\omega$ is $k+1$-iocc but not k-iocc. So the k-iocc languages form a strict hierarchy whose union over all k is the class of ω-regular languages. The k-iocc languages are incomparable with the class of languages defined by deterministic Büchi automata, and as far as we know, have not been studied previously.

Theorem 3.11 *For every $k \geq 1$, the set of languages defined by timed safety automata with k states over the discrete time domain N coincides with the set of k-iocc languages.*

Proof sketch. In discrete time, divergence is equivalent to infinitely many nonzero time steps, which in turn is almost equivalent to infinitely many changes of region. Modifying the region automaton by duplicating regions that are their own successors under time steps, divergence may be expressed by a k-iocc acceptance condition. To show that every k-iocc language $L = [\mathcal{A}, \mathcal{B}_\sim]$ is the language of a k-state timed safety automaton $(\mathcal{C}, \mathcal{T})$ over discrete time, use two clocks to record the $\sim$-equivalence class, as in the proof of Theorem 3.1, and use the k states of $\mathcal{C}$ to record the particular member of the equivalence class. $\square$

The state hierarchy for a variable number of clocks.

Corollary 3.12 *For every $k \in N$, there is a language accepted by a timed safety automaton with $k+1$ states over the discrete time domain N which is not accepted by any timed safety automaton with k states over N.*

Hence no finite number of states suffices to define all ω-regular languages by timed safety automata over the discrete time domain N. This is in marked contrast to the dense-time case (see Corollary 3.8), where one state suffices.

The clock hierarchy for a variable number of states collapses. While increasing the number of dense-time clocks increases the expressive power of timed safety automata, increasing the number of discrete-time clocks does not. This is because two clocks may be used to encode the region component of a discrete-time region automaton.

Theorem 3.13 *For every timed safety automaton $(\mathcal{A}, \mathcal{T})$, there is a timing condition $\mathcal{T}'$ using exactly two clocks such that $[\mathcal{A}, \mathcal{T}]_N = [\mathcal{A}, \mathcal{T}']_N$.*

4 Timed Languages

We study the expressiveness of timed automata with respect to timed languages. We provide proofs of two folk theorems: namely that increasing the number of clocks increases expressiveness in the dense time case, but not for discrete time. We also show that for both time domains increasing either the number of states with a fixed number of clocks, or vice versa, yields a strict hierarchy of expressiveness.

The dense-clock hierarchy for a variable number of states. Additional clocks in a dense time domain add expressiveness that cannot be captured by increasing the state space, since events can occur arbitrarily close together.

Theorem 4.1 *For every $n \in \mathbb{N}$, the class of timed languages accepted by timed safety automata with n clocks over the dense time domain $\mathbb{R}_{\geq 0}$ is properly contained in the class accepted by automata with $n + 1$ clocks.*

Proof sketch. Intuitively, a timed safety automaton with n clocks cannot enforce correct timing constraints over $n + 1$ concurrent events. Let $\mathcal{L}_{n+1}$ be the timed language consisting of all timed sequences $(\bar{a}, \bar{\delta})$ such that for each $i \geq 0$ there are exactly $n + 1$ a events occurring over the time interval $[i, i + 1)$, and each a event is followed by another a event exactly 1 time unit later. It is easy to construct an automaton with $n + 1$ clocks accepting this language.

On the other hand, consider any automaton for $\mathcal{L}_{n+1}$. It accepts the timed word where the i-th a occurs at time $1/i$ for $1 \leq i \leq n+1$. The timing conditions on each taken transition must imply $x = m$ for some clock x and integer m. Therefore at each time $1 + 1/i$ for $1 \leq i \leq n + 1$ there is a clock with integer value. Therefore there must be at least $n + 1$ clocks. □

The discrete-clock hierarchy for a variable number of states collapses.

Theorem 4.2 *Every timed language accepted by a timed automaton over the discrete time domain $\mathbb{N}$ is also accepted by an automaton interpreted over $\mathbb{N}$ with only one clock.*

Proof sketch. In discrete time, clock values can all be encoded in the states of an automaton, because one region may follow another only by a specific set of integral time steps. Thus, given a timed safety automaton $(\mathcal{A}, \mathcal{T})$, we may construct a timing condition $\mathcal{T}'$ over one clock x such that $[\![Reg(\mathcal{A}, \mathcal{T})_{\mathbb{N}}, \mathcal{T}']\!]_{\mathbb{N}}^t = [\![\mathcal{A}, \mathcal{T}]\!]_{\mathbb{N}}^t$, by using x to allow or disallow transitions between regions. □

The state hierarchy for a variable number of clocks. Consideration of the timed language $(a, 0)^k (b, 1)^\omega$ gives the following theorem.

Theorem 4.3 *For every $k \in \mathbb{N}$, and either time domain $\mathbb{T} = \mathbb{N}$ or $\mathbb{T} = \mathbb{R}_{\geq 0}$, the class of timed languages accepted by timed safety automata with k states over time domain $\mathbb{T}$ is properly contained in the class accepted with $k + 1$ states.*

Theorem 4.3 fails if we change the model so that time steps of duration 0 are not allowed. In this case, every timed language accepted by some timed safety automaton is accepted by one with only one state. The proof mimics that of Theorem 3.7. This is the only instance known to the authors in which the strictness of the progression of time has a bearing upon the theory.

The state hierarchy for a fixed number of clocks. We now show that for both time domains, increasing the number of states for a fixed number of clocks strictly increases expressiveness.

Theorem 4.4 *For every $k, n \in \mathbb{N}$, and either time domain $\mathbb{T} = \mathbb{N}$ or $\mathbb{T} = \mathbb{R}_{\geq 0}$, the class of timed languages accepted by timed safety automata with k states and n clocks is properly contained in the class accepted by automata with $k+1$ states and n clocks.*

The clock hierarchy for a fixed number of states.

Theorem 4.5 *For every $k, n \in \mathbb{N}$, and either time domain $\mathbb{T} = \mathbb{N}$ or $\mathbb{T} = \mathbb{R}_{\geq 0}$, the class of timed languages accepted by timed safety automata with k states and n clocks is properly contained in the class accepted by automata with k states and $n+1$ clocks.*

5 Summary

The first table summarizes the results of Section 3.1, regarding the replacement of acceptance conditions by timing conditions. Here S refers to the state space.

Acceptance condition $\mathcal{X}$	Number of Clocks	Constants
Büchi	2	$\|\mathcal{X}\|$
	$\log \|\mathcal{X}\|$	2
Muller	2	$\|\mathcal{X}\|\|S\|$
	$\log \|\mathcal{X}\|\|S\|$	2
Rabin	2	$\|\mathcal{X}\|$
	$\log \|\mathcal{X}\|$	2
Streett	2	$2^{\|\mathcal{X}\|}\|\mathcal{X}\|$
	$\|\mathcal{X}\|$	1

For $k, n \in \mathbb{N}$, and $\mathbb{T} \in \{\mathbb{N}, \mathbb{R}_{\geq 0}\}$, let $\mathcal{U}(k, n, \mathbb{T})$ be the set of untimed languages, and let $\mathcal{T}(k, n, \mathbb{T})$ be the set of timed languages, accepted by timed safety automata with k states and n clocks over the time domain $\mathbb{T}$. The following table summarizes the hierarchies for which one parameter is fixed (see Theorems 3.9, 3.10, 3.13, 4.4, 4.5, and Corollary 3.12).

$\mathcal{U}(k, n, \mathbb{N}) \subsetneq \mathcal{U}(k+1, n, \mathbb{N})$	$\mathcal{U}(k, n, \mathbb{R}_{\geq 0}) \subsetneq \mathcal{U}(k+1, n, \mathbb{R}_{\geq 0})$
$\mathcal{U}(k, 1, \mathbb{N}) \subsetneq \mathcal{U}(k, 2, \mathbb{N}) = \mathcal{U}(k, n+2, \mathbb{N})$	$\mathcal{U}(k, n, \mathbb{R}_{\geq 0}) \subsetneq \mathcal{U}(k, n+1, \mathbb{R}_{\geq 0})$
$\mathcal{T}(k, n, \mathbb{N}) \subsetneq \mathcal{T}(k+1, n, \mathbb{N})$	$\mathcal{T}(k, n, \mathbb{R}_{\geq 0}) \subsetneq \mathcal{T}(k+1, n, \mathbb{R}_{\geq 0})$
$\mathcal{T}(k, n, \mathbb{N}) \subsetneq \mathcal{T}(k, n+1, \mathbb{N})$	$\mathcal{T}(k, n, \mathbb{R}_{\geq 0}) \subsetneq \mathcal{T}(k, n+1, \mathbb{R}_{\geq 0})$

The final table refers to the clock hierarchies for a variable number of states and the state hierarchies for a variable number of clocks (see Theorems 3.11, 4.1, 4.2, 4.3 and Corollaries 3.5 and 3.8).

$\bigcup_k \mathcal{U}(k,1,\mathbb{N}) = \omega\text{-regular}$	$\bigcup_k \mathcal{U}(k,1,\mathbb{R}_{\geq 0}) = \omega\text{-regular}$
$\bigcup_n \mathcal{U}(k,n,\mathbb{N}) = k\text{-iocc}$	$\bigcup_n \mathcal{U}(1,n,\mathbb{R}_{\geq 0}) = \omega\text{-regular}$
$\bigcup_k \mathcal{T}(k,1,\mathbb{N}) = \text{timed } \omega\text{-reg.}$	$\bigcup_k \mathcal{T}(k,n,\mathbb{R}_{\geq 0}) \subsetneq \bigcup_k \mathcal{T}(k,n+1,\mathbb{R}_{\geq 0})$
$\bigcup_n \mathcal{T}(k,n,\mathbb{N}) \subsetneq \bigcup_n \mathcal{T}(k+1,n,\mathbb{N})$	$\bigcup_n \mathcal{T}(k,n,\mathbb{R}_{\geq 0}) \subsetneq \bigcup_n \mathcal{T}(k+1,n,\mathbb{R}_{\geq 0})$

Notice that the restriction of one clock and constants at most 1 does not disable acceptance of any regular language with either time domain. Similarly, the restriction of one state and constants at most 1 does not disable acceptance of any regular language with the dense time domain. However, if we make the restriction of one clock and one state, and allow constants of arbitrary size, there are indeed regular languages that are not accepted. This is implied by the strictness of the hierarchies with one fixed parameter.

References

1. R. Alur, C. Courcoubetis, and T.A. Henzinger. The observational power of clocks. In *CONCUR 94*, LNCS 836, pp. 162–177. Springer-Verlag, 1994.
2. R. Alur and D.L. Dill. A theory of timed automata. *Theoretical Computer Science*, 126:183–235, 1994.
3. R. Alur and T.A. Henzinger. Back to the future: towards a theory of timed regular languages. In *Proc. 33rd FOCS*, pp. 177–186. IEEE Computer Society Press, 1992.
4. R. Alur and T.A. Henzinger. Logics and models of real time: a survey. In *Real Time: Theory in Practice*, LNCS 600, pp. 74–106. Springer-Verlag, 1992.
5. R. Alur and T.A. Henzinger. Real-time logics: complexity and expressiveness. *Information and Computation*, 104(1):35–77, 1993.
6. R. Alur and T.A. Henzinger. Real-time system = discrete system + clock variables. In *Proc. of the First AMAST Workshop on Real-time Systems*, 1993. To appear.
7. C. Daws, A. Olivero, and S. Yovine. Verifying ET-LOTOS programs with KRONOS. In *Proceedings of FORTE '94*, 1994.
8. T.A. Henzinger. Sooner is safer than later. *Information Processing Letters*, 43:135–141, 1992.
9. T.A. Henzinger and P. Kopke. Verification methods for the divergent runs of clock systems. In *FTRTFT 94*, LNCS 863, pp. 351–372. Springer-Verlag, 1994.
10. T.A. Henzinger, Z. Manna, and A. Pnueli. What good are digital clocks? In *ICALP 92*, LNCS 623, pp. 545–558. Springer-Verlag, 1992.
11. T.A. Henzinger, X. Nicollin, J. Sifakis, and S. Yovine. Symbolic model checking for real-time systems. *Information and Computation*, 111(2):193–244, 1994.
12. W. Thomas. Automata on infinite objects. In J. van Leeuwen, editor, *Handbook of Theoretical Computer Science*, volume B, pp. 133–191. Elsevier Science Publishers (North-Holland), 1990.

GRAMMAR SYSTEMS:
A Grammatical Approach
to Distribution and Cooperation[1]

Gheorghe PĂUN
Institute of Mathematics of the Romanian Academy
PO Box 1 – 764, RO-70700 Bucharest, ROMANIA

Abstract. The title above is exactly the title of a monograph [9] published in 1994 by Gordon and Breach, London. The main aim of the present paper is to present recent notions and results not included in this monograph. One starts by introducing the basic definitions and the basic results of grammar system theory, then one surveys results related to the power of *hybrid cooperating distributed* (CD) grammar systems, to *teams* in CD grammar systems, to *parallel communicating* (PC) grammar systems with context-free and context-sensitive components. Other directions of research are also briefly mentioned, as well as a series of open problems.

1. The Basic Idea

In "classic" formal language theory, the grammars are used separately, *one* grammar produces *one* language. A grammar system is a *set* of grammars, working together, according to a specified protocol, for producing *one* language. The crucial element here is the protocol of cooperation. The aim of considering such a compound generative machinery can be many-sided: to model a real phenomenon, to increase the (generative) power of the components, to decrease the (descriptional) complexity. The cooperation protocol has to deal with all these aspects. In some sense, the theory of grammar systems is the theory of cooperation protocols; the focus is not on the generative capacity, but on the functioning of the systems, and on its influence on the generative capacity and on other specific properties.

Up to now, two main classes of grammar systems were investigated: shortly, they can be called *sequential* and *parallel*.

Of the first type are the *cooperating distributed* (CD) grammar systems: several grammars have a common sentential form, which starts from a common axiom; the component grammars work by turns, in each moment only one being enabled. The language generated in this way, by cooperation, is the language generated by the system. When a component becomes active, under what conditions it becomes again inactive, when the common string is accepted as terminal, these aspects are specified by the cooperation protocol.

In *parallel communicating* (PC) grammar systems, each component works on its own sentential form, in a synchronous manner (in each time unit, each component uses one rewriting rule). By *communication*, the components send to each other the current strings; this is done by request, using certain query

---

[1] Work supported by the Academy of Finland, project 11281

symbols. One component is designed as the master of the system and the language generated by it, using or not communications, is the language generated by the system.

2. Motivation and History

In the architecture of a CD grammar system one can recognize the structure of a blackboard model, as used in problem-solving area [36]: the common sentential form is the blackboard (the common data structure containing the current state of the problem to be solved), the component grammars are the knowledge sources contributing to solving the problem, the protocol of cooperation encodes the control on the work of the knowledge sources.

This was the explicit motivation of E. Csuhaj-Varju and J. Dassow in [7], the paper where CD grammar systems in the form we consider here were introduced. (This paper was presented at a workshop in Magdeburg already in May 1988. In the late eighties other papers by E. Csuhaj-Varju and J. Kelemen discussed the connection between the blackboard model and cooperating grammar systems, [4], [10], etc.) The syntagm "cooperating grammar systems" has been introduced by G. Rozenberg and R. Meersman in [30], with motivations related to two-level grammars. A somewhat similar idea appears already in [1], where S. Abraham has considered compound and serial grammars; this later notion is a particular type of CD grammar systems. The main aim of [1] was to increase the power of (matrix) grammars; a similar motivation have the modular grammars, of A. Atanasiu and V. Mitrana, [2], who start from time varying grammars.

While the history of CD grammar systems is not very sharply pictured, that of PC grammar systems is quite clear: these systems were introduced in [44], by L. Sântean (now Kari) and her collaborator, as a grammatical model of *parallelism* in the broad sense.

Information about the early development of CD and PC grammar systems is given in [14] and [48], respectively. The first chapter of [9] discusses in some detail many connections of CD and PC grammar systems with issues related to distribution, cooperation, parallelism in artificial intelligence, cognitive psychology, robotics, complex systems study, etc.

3. CD Grammar Systems

3.1. Definitions

A CD grammar system of degree $n, n \geq 1$, is a construct

$$\Gamma = (N, T, S, P_1, \ldots, P_n),$$

where N, T are disjoint alphabets, $S \in N$, and $P_1, \ldots, P_n$ are finite sets of rewriting rules over $N \cup T$.

The elements of N are nonterminals, those of T are terminals; $P_1, \ldots, P_n$ are called *components* of the system. Here we work with CD grammar systems having only context-free rules.

On $(N \cup T)^*$ one can define the usual one step derivation with respect to P_i, denoted by $\Longrightarrow_{P_i}$. The derivations consisting of exactly k, at most k, at least k such steps $\Longrightarrow_{P_i}$ are denoted by $\Longrightarrow_{P_i}^{=k}$, $\Longrightarrow_{P_i}^{\leq k}$, $\Longrightarrow_{P_i}^{\geq k}$, respectively. An arbitrary derivation is denoted by $\Longrightarrow_{P_i}^*$, whereas a maximal derivation is denoted by $\Longrightarrow_{P_i}^t$. Formally,

$$x \Longrightarrow_{P_i}^t y \text{ iff } x \Longrightarrow_{P_i}^{\geq 1} y \text{ and there is no } z \in (N \cup T)^* \text{ such that } y \Longrightarrow_{P_i} z.$$

Denote $D = \{*, t\} \cup \{\leq k, = k, \geq k \mid k \geq 1\}$ and $D' = \{*, t, = 1, \geq 1\} \cup \{\leq k \mid k \geq 1\}$. The language generated by the system Γ in the derivation mode $f \in D$ is

$$L_f(\Gamma) = \{w \in T^* \mid S \Longrightarrow^f_{P_{i_1}} w_1 \Longrightarrow^f_{P_{i_2}} \ldots \Longrightarrow^f_{P_{i_m}} w_m = w,$$
$$m \geq 1, 1 \leq i_j \leq n, 1 \leq j \leq m\}.$$

Five languages are associated with Γ, using the *stop conditions* in D. The components start working in an opportunistic way (nondeterministically). Various *start conditions* can be considered, too. For instance, a component can take the sentential form and rewrite it only when certain random context-like conditions are satisfied or when a more general predicate on the current sentential form is true, or according to an external control (a graph, for example, specifying the sequence of components enabling). Such variants (and many others) appear in [9]; here we confine ourselves to the basic model.

3.2. Examples

Consider the following CD grammar systems:

$$\Gamma_1 = (\{S, A, A', B, B'\}, \{a, b, c\}, S, P_1, P_2),$$
$$P_1 = \{S \rightarrow S, S \rightarrow AB, A' \rightarrow A, B' \rightarrow B\},$$
$$P_2 = \{A \rightarrow aA'b, B \rightarrow cB', A \rightarrow ab, B \rightarrow c\}.$$
$$\Gamma_2 = (\{S, A\}, \{a\}, S, P_1, P_2, P_3),$$
$$P_1 = \{S \rightarrow AA\}, \quad P_2 = \{A \rightarrow S\}, \quad P_3 = \{A \rightarrow a\}.$$
$$\Gamma_3 = (\{S, A_1, \ldots, A_k, A'_1, \ldots, A'_k\}, \{a, b\}, S, P_1, P_2), \ k \geq 1,$$
$$P_1 = \{S \rightarrow S, S \rightarrow A_1 b A_2 b \ldots b A_k b\} \cup \{A'_i \rightarrow A_i \mid 1 \leq i \leq k\},$$
$$P_2 = \{A_i \rightarrow a A'_i a, A_i \rightarrow aba \mid 1 \leq i \leq k\}.$$
$$\Gamma_4 = (\{S, S'\} \cup \{A_i, A'_i, A''_i \mid 1 \leq i \leq k\}, \{a, b\}, S, P_0, P_1, \ldots, P_{3k}), \ k \geq 2,$$
$$P_0 = \{S \rightarrow S', S' \rightarrow A_1 b A_2 b \ldots b A_k b\},$$
$$P_1 = \{A_1 \rightarrow A_1, A_1 \rightarrow a A'_1 a\},$$
$$P_{i+1} = \{A'_i \rightarrow A''_i, A_{i+1} \rightarrow a A'_{i+1} a\}, \ 1 \leq i \leq k-1,$$
$$P_{k+1} = \{A'_k \rightarrow A''_k, A''_1 \rightarrow A'_1\},$$
$$P_{k+i+1} = \{A'_i \rightarrow A_i, A''_{i+1} \rightarrow A'_{i+1}\}, \ 1 \leq i \leq k-2,$$
$$P_{2k} = \{A'_{k-1} \rightarrow A_{k-1}, A''_k \rightarrow A_k\},$$
$$P_{2k+i} = \{A_i \rightarrow A_i, A_i \rightarrow aba\}, \ 1 \leq i \leq k.$$

The reader may verify that:

$$L_f(\Gamma_1) = \{a^n b^n c^m \mid m, n \geq 1\}, \ f \in D',$$
$$L_{=2}(\Gamma_1) = L_{\geq 2}(\Gamma_1) = \{a^n b^n c^n \mid n \geq 1\},$$
$$L_{=k}(\Gamma_1) = L_{\geq k}(\Gamma_1) = \emptyset, \ k \geq 3,$$
$$L_t(\Gamma_2) = \{a^{2^n} \mid n \geq 1\},$$
$$L_{=k}(\Gamma_3) = L_{\geq k}(\Gamma_3) = \{(a^n b)^{2k} \mid n \geq 1\},$$
$$L_{=2}(\Gamma_4) = L_{\geq 2}(\Gamma_4) = \{(a^n b)^{2k} \mid n \geq 1\}.$$

3.3. Generative Capacity

We denote by $CD_n(f)$ the family of languages generated by λ-free CD grammar systems of degree at most $n, n \geq 1$, working in the derivation mode $f \in D$. When the number of components is not limited, we put ∞ instead of n. The union of all families $CD_\infty(= k), CD_\infty(\geq k), k \geq 1$, is denoted by $CD_\infty(=), CD_\infty(\geq)$, respectively. When λ-rules are allowed, we add the superscript λ: $CD_n^\lambda(f), CD_\infty^\lambda(f)$, etc.

As usual, REG, LIN, CF, CS, RE are the families in Chomsky hierarchy, MAT_{ac} (resp. MAT) is the family of languages generated by λ-free matrix grammars with (resp. without) appearance checking; $MAT_{ac}^\lambda, MAT^\lambda$ are the corresponding families generated by grammars possibly containing λ-rules, and $ET0L, EDT0L$ are the known families in L systems area.

Theorem 1.
(i) $CD_\infty(f) = CF$, for all $f \in D'$.
(ii) $CF = CD_1(f) \subset CD_2(f) \subseteq CD_r(f) \subseteq CD_\infty(f) \subseteq MAT$,
  for all $f \in \{= k, \geq k \mid k \geq 2\}, r \geq 3$.
(iii) $CD_r(= k) \subseteq CD_r(= sk)$, for all $k, r, s \geq 1$.
(iv) $CD_r(\geq k) \subseteq CD_r(\geq k+1)$, for all $r, k \geq 1$.
(v) $CD_\infty(\geq) \subseteq CD_\infty(=)$.
(vi) $CF = CD_1(t) = CD_2(t) \subset CD_3(t) = CD_\infty(t) = ET0L$.

(vii) *Except for the inclusion $CD_\infty(f) \subseteq MAT$ (which must be replaced with $CD_f^\lambda(f) \subseteq MAT^\lambda$), all the previous relations are true also for CD grammar systems allowed to use λ-rules.*

Two somewhat surprising relations are those in (vi): $CD_2(t) \subseteq CF$ and $CD_\infty(t) \subseteq CD_3(t)$. Proofs can be found in [2], [7], [9]. Because it is of interest, we present here the proof of the second inclusion above, in a form similar to that in [2] (in [7] one starts from ET0L systems, thus obtaining $ET0L \subseteq CD_3(t)$; this relation is not noticed in [2]).

Take a system $\Gamma = (N, T, S, P_1, \ldots, P_n)$ and construct $\Gamma' = (N', T, [S, 1], P_1', P_2', P_3')$, with

$$N' = \{[A, i] \mid A \in N, 0 \leq i \leq n\},$$
$$P_1' = \{[A, i] \to [w, i] \mid A \to w \in P_i, 1 \leq i \leq n\},$$
$$P_2' = \{[A, i] \to [A, i+1] \mid A \in N, 1 \leq i < n, i \text{ odd}\} \cup$$
$$\cup \{[A, n] \to [A, 0] \mid A \in N, \text{ if } n \text{ is odd}\},$$
$$P_3' = \{[A, i] \to [A, i+1] \mid A \in N, 0 \leq i < n, i \text{ even}\} \cup$$
$$\cup \{[A, n] \to [A, 1] \mid A \in N, \text{ if } n \text{ is even}\},$$

where $[w, i]$ denotes the string obtained by replacing each nonterminal $A \in N$ appearing in w by $[A, i]$ and leaving unchanged the terminals.

Then $L_t(\Gamma) = L_t(\Gamma')$. Three components suffice, but note that the first one, P_1', is as complex (as the number of productions) as the whole initial system Γ.

Let us denote by $CD_{n,m}(f)$ the family of languages generated by λ-free CD grammar systems with at most n components, each of them containing at most m productions, working in the f mode. We add the letter D in front of CD when only *deterministic* systems are used: for each component P_i, if $A \to x_1, A \to x_2$ are in P_i, then $x_1 = x_2$.

Theorem 2.

(i) $CD_{\infty,\infty}(f) = CD_{\infty,1}(f) = CF$, $f \in D'$.

(ii) $CD_{\infty,\infty}(= k) = CD_{\infty,k}(= k), CD_{\infty,\infty}(\geq k) = CD_{\infty,2k-1}(\geq k), k \geq 2$.

(iii) $CF \subset CD^\lambda_{\infty,1}(t) \subset CD^\lambda_{\infty,2}(t) \subseteq CD^\lambda_{\infty,3}(t) \subseteq CD^\lambda_{\infty,4}(t) \subseteq CD^\lambda_{\infty,5}(t) =$
$= CD^\lambda_{\infty,\infty}(t) = ET0L$.

$CF_{fin} \subset (CD^\lambda_{\infty,1}(t) = DCD^\lambda_{\infty,1}(t)) \subseteq DCD^\lambda_{\infty,2}(t) \subseteq DCD^\lambda_{\infty,3}(t) \subseteq$
$\subseteq DCD^\lambda_{\infty,4}(t) = DCD^\lambda_{\infty,\infty}(t) = EDT0L$.

(iv) $CD_{n,m}(f) \subset CD_{n+1,m}(f), CD_{n,m}(f) \subset CD_{n,m+1}(f)$, $f \in \{*, t\}$.

(CF_{fin} is the family of finite index context-free languages.) The significant relations here are $CD^\lambda_{\infty,\infty} \subseteq CD^\lambda_{\infty,5}(t)$, with its counterpart for the deterministic case (the proof can be found in [9]) and (iv) for $f = t$ (see [16]).

Open problems: Which of the inclusions not specified in Theorems 1 and 2 as proper ($\subset$) are proper ? Are the bounds 5 and 4 in Theorem 2 (iii) optimal ? They were obtained allowing the use of λ-rules. What about the case of λ-free systems ? Can the results in Theorem 2 (ii) be improved ? Are the strict inclusions in Theorem 2 (iv) true also for the derivation modes $= k, \geq k$? Are the hierarchies $CD_n(f)$, for $f \in \{= k, \geq k \mid k \geq 2\}$, infinite ? (Are systems with $n + 1$ components more powerful than systems with n components ?)

Many other problems are still open in this area (see [9]). Most of them are related to the same, technical, missing tool: counterexamples, necessary conditions (mainly for families $CD_n(f), f \in \{= k, \geq k \mid k \geq 2\}$). The examples Γ_3, Γ_4 in Section 3.2 show that the question is not trivial.

3.4. Hybrid Systems

The systems considered in the previous sections were *homogeneous*, all their components were supposed to work in the same mode. It is perhaps more realistic to consider systems consisting of components having different competences. This leads to the notion of a *hybrid CD grammar system*, which is a construct

$$\Gamma = (N, T, S, (P_1, f_1), \ldots, (P_n, f_n)), n \geq 1,$$

where $(N, T, S, P_1, \ldots, P_n)$ is a usual CD grammar system and $f_i \in D, 1 \leq i \leq n$. The language generated by Γ is

$$L(\Gamma) = \{w \in T^* \mid S \Longrightarrow^{f_{i_1}}_{P_{i_1}} w_1 \Longrightarrow^{f_{i_2}}_{P_{i_2}} \ldots \Longrightarrow^{f_{i_m}}_{P_{i_m}} w_m = w,$$
$$m \geq 1, 1 \leq i_j \leq n, 1 \leq j \leq m\}.$$

We denote by $HCD_n, n \geq 1$, the family of languages generated by λ-free hybrid CD grammar systems with at most n components; we put $n = \infty$ when no restriction is imposed on the number of components. All the results in this section hold true also for systems which are allowed to use λ-rules, but we do not explicitly state the results in this form.

Of course, $CD_n(f) \subseteq HCD_n, n \geq 1, f \in D$. Moreover, [42]

Theorem 3. *If a hybrid CD grammar system Γ of degree at least two has the following two properties, then $L(\Gamma) \in CF$:*

(1) *There is no component in Γ working in a mode $f \in \{= k, \geq k \mid k \geq 2\}$,*

(2) *There are at most two components of Γ working in the t mode.*

Conversely, there are hybrid systems not observing one of conditions (1), (2) *and generating non-context-free languages.*

The second assertion is proved by systems like Γ_1, Γ_2 in Section 4.2: one of P_1, P_2 in Γ_1 can be allowed to work in any mode in D, if the other one works in the $= k$ or in the $\geq k$ mode, then the generated language will be $\{a^n b^n c^n \mid n \geq 1\}$.

Surprisingly enough, the number of components does not induce an infinite hierarchy. More exactly, combining the results in [34] and [42], we have:

Theorem 4.
(i) $CF = HCD_1 \subset HCD_2 \subseteq HCD_3 \subseteq HCD_4 = HCD_\infty \subseteq MAT_{ac}$.
(ii) $ETOL \subset HCD_4$, $CD_\infty(=) \subset HCD_3$,
$\quad CD_\infty(=) \subseteq HCD_\infty(\textit{fin-t}) \subset (HCD_4 \cap MAT)$,

where $HCD_\infty(\textit{fin-t})$ is the family of languages generated by hybrid CD grammar systems of arbitrary degree but with a bounded number of times of using the components working in the t mode.

It is perhaps of interest to see a hybrid system (of degree 4) generating a non-ETOL language. Here is an example from [42]:

$$\Gamma = (\{S, A, B, C, X, Y\}, \{a, b, c\}, S, (P_1, t), (P_2, = 2), (P_3, t), (P_4, t)),$$
$$P_1 = \{S \to ABS, S \to ABX, C \to B, Y \to X\},$$
$$P_2 = \{X \to Y, A \to a\}, \ P_3 = \{X \to X, B \to bC\}, \ P_4 = \{X \to c, B \to c\}.$$

We obtain $L(\Gamma) = \{(ab^n c)^m c \mid 1 \leq n \leq m\}$, which is not an ETOL language (use Corollary 2.2 in [46]).

A series of problems are open in this area, too, starting with the relations not specified in Theorem 4. Moreover, denote by $HCD_n(f_1, f_2, \ldots, f_n)$ the family of languages generated by hybrid CD grammar systems $\Gamma = (N, T, S, (P_1, f_1), \ldots, (P_n, f_n))$. By the result in [34], $HCD_n(f_1, \ldots, f_n) \subseteq HCD_4(=, t, t, t)$. What one can say about families $HCD_n(f_1, \ldots, f_n)$ for $n = 2, 3, 4$? How many of them are distinct ? (Theorem 3 characterizes those of these families which equal CF; what about the others ?) Is the inclusion $ETOL \subseteq HCD_3$ a proper one ?

3.5. Teams

In the blackboard model, as well as in CD grammar systems considered so far, in each moment only one component is enabled. Removing this restriction (without however imposing to all components to work in parallel), we obtain the notion of a *team CD grammar system*, as introduced in [26]. We use here the presentation in [43]:

A CD grammar system *with* (*prescribed*) *teams* (*of variable size*) is a construct

$$\Gamma = (N, T, S, P_1, \ldots, P_n, Q_1, \ldots, Q_m), \ n, m \geq 1,$$

where $(N, T, S, P_1, \ldots, P_n)$ is a usual CD grammar system and $Q_1, \ldots, Q_m$ are subsets of $\{P_1, \ldots, P_n\}$ (called *teams*). A team $Q_i = \{P_{j_1}, \ldots, P_{j_s}\}$ is used in derivations as follows:

$$x \Longrightarrow_{Q_i} y \quad \text{iff} \quad x = x_1 A_1 x_2 A_2 \ldots x_s A_s x_{s+1}, y = x_1 y_1 x_2 y_2 \ldots x_s y_s x_{s+1},$$
$$x_l \in (N \cup T)^*, 1 \leq l \leq s+1, A_r \to y_r \in P_{j_r}, 1 \leq r \leq s.$$

Having defined the one step derivation, we can define derivations in Q_i of k steps, at most k or at least k steps, and of any number of steps, denoted by $\Longrightarrow_{Q_i}^{=k}, \Longrightarrow_{Q_i}^{\leq k}, \Longrightarrow_{Q_i}^{\geq k}, \Longrightarrow_{Q_i}^{*}$, respectively. For maximal derivations in a team Q_i we can consider three variants:

$$x \Longrightarrow_{Q_i}^{t_0} y \quad \text{iff} \quad x \Longrightarrow_{Q_i}^{\geq 1} y \text{ and there is no } z \text{ such that } y \Longrightarrow_{Q_i} z,$$

$$x \Longrightarrow_{Q_i}^{t_1} y \quad \text{iff} \quad x \Longrightarrow_{Q_i}^{\geq 1} y \text{ and for no component } P_{j_r} \in Q_i \text{ and no } z$$
$$\text{there is a derivation } y \Longrightarrow_{P_{j_r}} z,$$

$$x \Longrightarrow_{Q_i}^{t_2} y \quad \text{iff} \quad x \Longrightarrow_{Q_i}^{\geq 1} y \text{ and there is a component } P_{j_r} \in Q_i \text{ for which}$$
$$\text{no derivation } y \Longrightarrow_{P_{j_r}} z \text{ is possible.}$$

In the t_0 mode the team as a whole cannot perform any further step, in the t_1 mode no component of the team can apply any of its rules, whereas in the case of t_2 at least one component cannot rewrite any symbol of the current string. The mode t_0 is considered in [17], t_1 in [26], and t_2 in [43].

If each subset of $\{P_1, \ldots, P_n\}$ can be a team, then we say that Γ has *free teams*; when all teams have the same number of components, then we say that Γ has *teams of constant size*. For the case of teams of constant size we consider a set $W \subseteq (N \cup T)^*$ instead of an axiom $S \in N$, in order to not produce artificial counterexamples when using λ-free rules (strings of length less that s, the size of teams, cannot be generated). However, we impose the restriction that W contains only one nonterminal string.

We denote by $PT_sCD(f)$ the family of languages generated by λ-free CD grammar systems with prescribed teams of constant size s in the derivation mode $f \in \{*, t_0, t_1, t_2\} \cup \{\leq k, = k, \geq k \mid k \geq 1\}$. When dealing with free teams, the letter P is omitted. When the size of teams is not constant, we replace s with $*$. When λ-rules are allowed we add as usual the superscript λ.

Synthesizing the results in [17], [26], [43], we obtain

Theorem 5.

(i) $PT_sCD(f) = PT_*CD(f) = MAT$,

for all $s \geq 2, f \in \{*\} \cup \{\leq k, = k, \geq k \mid k \geq 1\}$.

(ii) $T_sCD(f) = PT_sCD(f) = PT_*CD(f) = T_*CD(f) = MAT_{ac}$,

for all $s \geq 2, f \in \{t_0, t_1, t_2\}$.

(iii) *All the results above hold true also for the case of using λ-rules (MAT and MAT_{ac} are then replaced by MAT^λ and MAT_{ac}^λ).*

These equalities suggest a series of interesting conclusions: teams of size two suffice for obtaining the maximal generative capacity, increasing the cooperation by the team feature increases considerably the power of CD grammar systems (remember that $ET0L \subset MAT_{ac}$), all the three variants of maximal derivation are equivalent, when λ-rules are allowed we get new characterizations of recursively enumerable languages ($MAT_{ac}^\lambda = RE$, [15]), etc.

3.6. Other Classes of CD Grammar Systems

We do not enter here into details concerning CD grammar systems with a graph control, with registers, with hypothesis languages, or paired with transducers (see details in [9]), or about CD grammar systems structured, hierarchically or by imposing an order relation on the set of components [8], [35]. We mention only the notion of a *colony*, [28], a particular case of a CD grammar system, with the components generating finite languages. We write the system in the form $\Gamma = (N = \{S_1, \ldots, S_n\}, T, w, F_1, \ldots, F_n)$, where S_i is a nonterminal associated to F_i, which, in turn, is a finite subset of $(N \cup T - \{S_i\})^*$, $1 \leq i \leq n$; $w \in (N \cup T)^*$. Rewriting S_i means to replace it with an element of F_i. Motivations for considering such devices can be found in [9], [28], results can be found in [9], [29], [40]. Most of the problems still open for CD grammar systems (including hierarchies on the number of components) are solved for colonies.

4. PC Grammar Systems

4.1. Definitions

A PC grammar system of degree n, $n \geq 1$, is a construct

$$\Gamma = (N, K, T, (P_1, S_1), \ldots, (P_n, S_n)),$$

where N, K, T are disjoint alphabets, $K = \{Q_1, \ldots, Q_n\}$, $S_i \in N$, $1 \leq i \leq n$, and $P_1, \ldots, P_n$ are finite sets of rewriting rules over $N \cup K \cup T$ such that no symbol in K appears in a left hand member of a rule.

The elements of N are nonterminals, those of T are terminals, $Q_1, \ldots, Q_n$ are called *query symbols* and they are associated in a one-to-one manner to the *components* $P_1, \ldots, P_n$. We denote $V_\Gamma = N \cup K \cup T$. Each n-tuple $(x_1, \ldots, x_n) \in (V_\Gamma^*)^n$ is called a *configuration* of Γ.

On the set of configurations we define the relation $(x_1, \ldots, x_n) \Longrightarrow_r (y_1, \ldots, y_n)$ iff either

1. $|x_i|_K = 0$ for all $1 \leq i \leq n$, and $x_i \Longrightarrow_{P_i} y_i$ or $x_i \in T^*$, $1 \leq i \leq n$, or

2. there is i such that $|x_i|_K > 0$; then for each such i we write $x_i = z_1 Q_{i_1} z_2 Q_{i_2} \ldots z_t Q_{i_t} z_{t+1}$, $z_j \in (N \cup T)^*$, $1 \leq j \leq t+1$; if $|x_{i_j}|_K = 0$, $1 \leq j \leq t$, then $y_i = z_1 x_{i_1} z_2 x_{i_2} \ldots z_t x_{i_t} z_{t+1}$ [and $y_{i_j} = S_{i_j}$, $1 \leq j \leq t$]; when, for some j, $1 \leq j \leq t$, $|x_{i_j}|_K > 0$, then $y_i = x_i$; for all i for which y_i was not specified above, we have $y_i = x_i$.

Point (1) defines a componentwise rewriting step, point (2) defines a *communication* step: each query symbol Q_{i_j} is replaced by the corresponding x_{i_j}, providing it does not contain further query symbols; in such a case, Q_{i_j} is not *satisfied*. The communication has priority over rewriting. If some query symbols are not satisfied at a communication step, they can be satisfied at the next one. If a circular query appears, then the system gets stuck. The derivation is blocked also when no query symbol is present but at least a component cannot rewrite its (nonterminal) sentential form.

The above relation $\Longrightarrow_r$ is said to be a *returning* derivation: after communicating, every component returns to its axiom.

If the symbols in brackets, [and $y_{i_j} = S_{i_j}$, $1 \leq j \leq t$], are removed, then we obtain a *non-returning* derivation: after communicating, the components continue to process the current string. We denote by $\Longrightarrow_{nr}$ such a derivation step.

The language generated by Γ in one of the modes $f \in \{r, nr\}$ is defined by

$$L_f(\Gamma) = \{w \in T^* \mid (S_1, \ldots, S_n) \Longrightarrow_f^* (w, \alpha_2, \ldots, \alpha_n), \alpha_i \in V_\Gamma^*, 2 \leq i \leq n\}.$$

The first component of the system is the *master*; its language is the language of the system.

4.2. Examples

Consider the systems

$$\Gamma_1 = (\{S_1, S_2, S_3\}, K, \{a, b\}, (P_1, S_1), (P_2, S_2), (P_3, S_3)),$$
$$P_1 = \{S_1 \to aS_1, S_1 \to aQ_2, S_2 \to bQ_3, S_3 \to c\},$$
$$P_2 = \{S_2 \to bS_2\}, \quad P_3 = \{S_3 \to cS_3\},$$
$$\Gamma_2 = (\{S_1, S_2\}, K, \{a, b\}, (P_1, S_1), (P_2, S_2)),$$
$$P_1 = \{S_1 \to S_1, S_1 \to Q_2Q_2\},$$
$$P_2 = \{S_2 \to aS_2, S_2 \to bS_2, S_2 \to a, S_2 \to b\},$$
$$\Gamma_3 = (\{S_1, S_2\}, K, \{a\}, (P_1, S_1), (P_2, S_2)),$$
$$P_1 = \{S_1 \to aQ_2, S_2 \to aQ_2, S_2 \to a\}, \quad P_2 = \{S_2 \to aS_2\},$$
$$\Gamma_4 = (\{S_1, S_2, S_3, A, B, C\}, K, \{a, b\}, (P_1, S_1), (P_2, S_2), (P_3, S_3)),$$
$$P_1 = \{S_1 \to ab, S_1 \to ba, S_1 \to S_1, S_1 \to Q_2C, C \to C, C \to bQ_2\},$$
$$P_2 = \{S_2 \to aS_2, S_2 \to a\}, \quad P_3 = \{S_3 \to A^{p-1}, A \to B\}, \ p \geq 2,$$
$$\Gamma_5 = (\{S_1, S_2, S_3, A, B\}, K, \{a\}, (P_1, S_1), (P_2, S_2), (P_3, S_3)),$$
$$P_1 = \{S_1 \to aA, S_1 \to aQ_2, B \to aA, B \to a\},$$
$$P_2 = \{S_2 \to aQ_1, A \to aQ_3\}, \quad P_3 = \{S_3 \to aQ_1, A \to aB\}.$$

We obtain:

$$L_r(\Gamma_1) = L_{nr}(\Gamma_1) = \{a^m b^{m+1} c^{m+2} \mid m \geq 1\},$$
$$L_r(\Gamma_2) = L_{nr}(\Gamma_2) = \{xx \mid x \in \{a, b\}^+\},$$
$$L_r(\Gamma_3) = \{a^{2m+1} \mid m \geq 1\},$$
$$L_{nr}(\Gamma_3) = \{a^{\frac{(m+1)(m+2)}{2}} \mid m \geq 1\},$$
$$L_r(\Gamma_4) = \{a^i ba^j \mid 1 \leq i + j \leq p\},$$
$$L_{nr}(\Gamma_4) = \{a^i ba^i \mid 1 \leq i \leq p - 1\},$$
$$L_r(\Gamma_5) = \{a^{7 \cdot 2^m - 6} \mid m \geq 1\}.$$

Γ_3 shows the important difference between the returning and the non-returning modes of work, Γ_4 contains a component, P_3, which does not produce parts of the generated strings, but limits their length.

4.3. Generative Capacity

We distinguish two classes of PC grammar systems: *centralized* and *noncentralized* (or arbitrary). A system is centralized when only P_1, the master, can introduce query symbols. (Γ_5 above shows how intricate the work of a PC grammar system can be in the non-centralized case.)

We denote by PC_nX the family of languages $L_r(\Gamma)$ for systems Γ with at most n components of type X; when only centralized systems are considered, we write CPC_nX. The corresponding families of languages generated in the non-returning mode are denoted by $NPC_nX, NCPC_nX$. When no limit on the number of components is considered, we replace n with ∞. The type X of components can be $REG, LIN, CF, CS, CS^\lambda$, with the obvious meanings.

We recall, without proofs, a series of results appearing in [9]. Denote $\Pi = \{PC, CPC, NPC, NCPC\}$.

Theorem 6.

(i) $Y_n CS^\lambda = RE$ for all n and $Y \in \Pi$.

(ii) $Y_n REG - LIN \neq \emptyset$, $Y_n LIN - CF \neq \emptyset$, $n \geq 2$, and $Y_n REG - CF \neq \emptyset$, $n \geq 3$, for all $Y \in \Pi$.

(iii) $Y_n REG - CF \neq \emptyset$, for all $n \geq 2, Y \in \{NPC, NCPC\}$.

(iv) $LIN - (CPC_\infty REG \cup NCPC_\infty REG) \neq \emptyset$.

(v) $CPC_2 REG \subset CF$, $PC_2 REG \subseteq CF$.

(vi) $CPC_\infty REG$ contains only semilinear languages.

(vii) $CPC_2 CF$ contains non-semilinear languages.

(viii) $PC_3 REG, NPC_2 REG, NCPC_2 REG$ contain one-letter non-regular languages.

(ix) If $L \subseteq V^*, L \in CPC_n REG$, then there is a constant q such that each $z \in L, |z| > q$, can be written in the form $z = x_1 y_1 x_2 y_2 \ldots x_m y_m x_{m+1}$ for $1 \leq m \leq n, y_i \neq \lambda, 1 \leq i \leq m$, and for all $k \geq 1$, $x_1 y_1^k x_2 y_2^k \ldots x_m y_m^k x_{m+1} \in L$. Consequence: $CPC_n REG \subset CPC_{n+1} REG, n \geq 1$.

(x) $PC_n REG \subset PC_{n+1} REG, n \geq 1$.

(xi) $CPC_n REG \subset CPC_n LIN \subset CPC_n CF, n \geq 1$.

Point (i) is obvious, (ii), (iii), (vii), (viii) are proved by examples (some of them are mentioned in Section 4.2), (iv) uses the linear language $\{a^n b^m cb^m a^n \mid m, n \geq 1\}$, the first relation in (v) appears already in [44], but the second one is obtained in [51] using a long construction; for (vi) it is shown that $L_r(\Gamma)$ is (the gsm image of) a matrix language of finite index, the pumping property in (ix) is proved in [27], where it is also given a direct, combinatorial, proof of (x); (xi) is obtained by combining the previous results.

Two important results were recently proved in [33], [31].

Theorem 7. (i) $NCPC_\infty CF \subseteq PC_\infty CF$, (ii) $MAT \subset PC_\infty CF$.

The first relation indicates the fact that the non-returning feature can be compensated by non-centralization. The second inclusion confirms the power of non-centralization (and, according to Theorem 1 (ii), implies $CD_\infty(f) \subset PC_\infty CF, f \in \{= k, \geq k \mid k \geq 2\}$, a first non-trivial result linking CD and PC grammar systems).

The idea of the proof of (i) is the following one: starting from a system with components $P_1, \ldots, P_n$, centralized and working in the non-returning mode, one constructs a system which contains three components, $P_{i,1}, P_{i,2}, P_{i,3}$ associated with each P_i; $P_{i,3}$ is used only at the beginning of derivations, for synchronization; $P_{i,1}, P_{i,2}$ simulate the work of P_i. When one of them uses a rule of P_i, the other gets ready to receive the current string of the partner in the case when P_0, a special new components, the master of the new system, associated with P_1, asks for the string generated by P_i. The string is sent to P_0 and at the same time to the partner in the triple $P_{i,1}, P_{i,2}, P_{i,3}$, thus saving a copy of it. In this way, although working in the returning mode, the communicated string will be processed farther. Some additional components are used for synchronizing the process and as "garbage collectors".

A similarly involved construction is used in [31] for simulating a matrix grammar (in binary normal form) by a non-centralized PC grammar system working in the returning mode. The strictness of (ii) is obtained by using the result in [22], that each one-letter matrix language is regular (from Theorem 6 (viii) we know that this is not the case for $PC_3 REG$).

Open problems: Many problems about PC grammar systems are still open. They are similar to those about CD grammar systems. Which of the inclusions not shown before to be proper are proper ? What about the hierarchies on the number of components ? For systems with regular components many questions were settled, by direct language-theoretic means or using complexity tools [3], [23], [24], [27], but very few things are known about the context-free case. The basic missing results concern counterexamples/necessary conditions.

4.4. The Context-Sensitive Case

We discusse this case separately, because it is completely settled (in a quite interesting way).

Theorem 8.
(i) $CS = Y_n CS = Y_\infty CS$, for all $n \geq 1, Y \in \{CPC, NCPC\}$.
(ii) $CS = PC_1 CS = PC_2 CS \subset PC_3 CS = PC_\infty CS = RE$.
(iii) $CS = NPC_1 CS \subset NPC_2 CS = NPC_\infty CS = RE$.

Non-centralized systems with three components working in the returning mode or with two components only, working in the non-returning mode, are powerful enough for generating any recursively enumerable language. In the centralized case, one not surpasses the power of context-sensitive grammars. Again a clear indication is obtained about the power of non-centralization.

Point (i) appears in [9], $PC_3 CS = RE$ is proved in [45], and $NPC_2 CS = RE$ in [18]. The basic idea is similar in the last two cases. For instance, starting from a language $L \subseteq V^*, L \in RE$, one takes a length-increasing grammar G for the context-sensitive language $L' \subseteq La^*b, a, b \notin V$, associated to L as in Theorem 9.9 [47], and one constructs a non-centralized PC grammar system, with two components P_1, P_2 working in the non-returning mode, as follows. For a while, P_1 is waiting (it uses rules of the form $A \to A$), whereas P_2 generates a string of the form $XDwY$, with $w \in L', X, Y, D$ being auxiliary symbols. Then also P_1 starts to produce an arbitrary string over V, using regular rules. At each step, P_2 asks for the produced string and checks whether or not the last introduced symbol is equal to the corresponding symbol in w. Only in such a case D can be moved to the right. When reaching the occurrence of a, if the derivation in P_1 is terminal, the process stops producing a terminal string, otherwise it is blocked. Therefore P_2 works like an auxiliary tape with unbounded working space, doing all the work and collecting all "garbage symbols"; this is possible due to the mode of defining the language of a PC grammar system.

Details of this construction can be found in [18]. It is interesting to note that P_1 above is a right-linear grammar never asking for the string generated by P_2; P_1 is the master, but it never starts a communication.

4.5. Non-Synchronized PC Grammar Systems

An essential feature of PC grammar systems is the synchronization of the rewriting steps: there is a universal clock which marks the time in the same way for all components and in each time unit, if no communication must be done, then each component must use one of its rules, except for components whose string is terminal. What about removing this restriction, and allowing the components to freely wait ? This is equivalent with adding rules $A \to A$ for all $A \in N$ to each component.

Let us denote by UY_nX the family of languages generated by unsynchronized PC grammar systems, corresponding to families Y_nX as above. Here are some results about these families:

Theorem 9.

(i) $UCPC_\infty X = X$, for $X \in \{REG, LIN\}$.

(ii) $UPC_2REG - REG \neq \emptyset$, $UPC_\infty REG \subseteq CF$, $UPC_2LIN - CF \neq \emptyset$, $UCPC_2CF - CF \neq \emptyset$.

(iii) $UNCPC_2REG$ contains non-semilinear languages;
$UNCPC_2CF$ contains one-letter non-regular languages.

(iv) $(CPC_\infty REG \cap NCPC_\infty REG) - UCPC_\infty CF \neq \emptyset$.

Points (i), (iv) show that removing the synchronization decreases considerably the generative capacity, but points (ii), (iii) show that unsynchronized PC grammar systems are still very powerful. For instance, for the systems

$$\Gamma_1 = (\{S_1, S_2, A, B\}, K, \{a, b\}, (P_1, S_1), (P_2, S_2)),$$
$$P_1 = \{S_1 \to bQ_2, B \to bQ_2, B \to b\} \cup \{X \to X \mid X \in \{S_1, S_2, A, B\}\},$$
$$P_2 = \{S_2 \to aS_2, S_2 \to aA, A \to aB\} \cup \{X \to X \mid X \in \{S_1, S_2, A, B\}\},$$
$$\Gamma_2 = (\{S_1, S_2, A, B\}, K, \{a, b\}, (P_1, S_1), (P_2, S_2)),$$
$$P_1 = \{S_1 \to aQ_2, S_1 \to aA, B \to bA\} \cup \{X \to X \mid X \in \{S_1, S_2, A, B\}\},$$
$$P_2 = \{S_2 \to aQ_1, A \to bB, A \to b\} \cup \{X \to X \mid X \in \{S_1, S_2, A, B\}\},$$

we obtain

$$L_{nr}(\Gamma_1) = \{(ba^n)^m b \mid m \geq 1, n \geq 2\} \text{ (non − semilinear)},$$
$$L_r(\Gamma_2) = \{a^{2m} a^{2s+1} b^{2t+1} \mid m, s, t \geq 1, s \geq t\} \text{ (non − regular)}.$$

5. Final Remarks

There are several classes of problems which we have not mentioned here. Not very much is known about closure and decidability properties of grammar systems. Some results can be found in [9], but not those in [19], [49]. A series of results are, however, known about the descriptional complexity of grammar systems of both types, mainly in comparison with context-free languages. For instance, for the measures Var (= the number of nonterminals) and $Prod$ (= the number of productions), results of the following type are true for all PC grammar systems with context-free components, as well as for CD grammar systems working in $= k, \geq k$ modes: there is a sequence $L_n, n \geq 1$, of context-free languages such that $Var_{CF}(L_n) \geq n, n \geq 1$, but $Var_{PC}(L_n) \leq p$ ($Var_{CD}(L_n) \leq p$), where p is a constant; similarly for $Prod$. Weaker results are obtained for CD grammar systems working in the t mode and for the measure $Symb$ (= the total number of symbols used in productions of a system or of a grammar, [21]).

A specific complexity measure for PC grammar systems is the number of communications in a derivation. It behaves like the index of context-free languages: it is not calculable for languages and even not for systems [38], but it gives infinite hierarchies, and has important connections to the properties of PC grammar systems [25], [37], [38], [50].

Of course, for all types of systems one can consider other classes of generative mechanisms as components, not only Chomsky grammars: L systems, pure grammars, contextual grammars, etc. A few results about systems with L systems as components can be found in [7], [9], [52].

And, last but not least: applications. Applications to the study of natural languages were started by E. Csuhaj-Varju [6] and her group. But the most promising application are the *eco-grammar systems* introduced in [12], presented first in [11], and already investigated in a series of papers [13], [32], [41], etc. In short, an eco-grammar system is a sort of superposed CD and PC grammar systems. Several *agents* live in a common *environment*. Both the agents and the environment are described by strings of symbols and evolve according to 0L systems (pure rewriting rules used in parallel). The evolution of the environment is independent of agents, but the evolution of the agents depends on the state of the environment. The agents also have *action rules*, pure rewriting rules used sequentially, for modifying locally the string-environment. Of interest is mainly the "transition space", the sequences of configurations describing the agents and the environment, but also languages can be associated to an eco-grammar system. A lot of real life-like features, characteristic to real ecosystems, can be introduced in this model: birth (by fragmentation), death, overpopulation, stagnation, pollution, hibernating, carnivorous or parasitic agents, cyclic evolution, migration, and so on and so forth. Also a wealth of theoretical problems apear. And, to close the circle, the eco-grammar systems seem to have nice applications in (theoretical) artificial intelligence, the very place where their "ancestors", the CD and the PC grammar systems were motivated (details can be found in [32]).

References

1. S. Abraham, Compound and serial grammars, *Inform. Control*, 20 (1972), 432 – 438.

2. A. Atanasiu, V. Mitrana, The modular grammars, *Intern. J. Computer Math.*, 30 (1989), 101 – 122.

3. L. Cai, The computational complexity of PCGS with regular components, *Developments in Language Theory*, Magdeburg, 1995.

4. E. Csuhaj-Varju, Some remarks on cooperating grammar systems, in *Proc. Conf. Aut. Lang. Progr. Syst.* (I. Peak, F. Gecsed, eds.), Budapest, 1986, 75 – 86.

5. E. Csuhaj-Varju, Cooperating grammar systems. Power and parameters, *LNCS* 812, Springer-Verlag, Berlin, 1994, 67 – 84.

6. E. Csuhaj-Varju, Grammar systems: a multi-agent framework for natural language generation, in [39], 63 – 78.

7. E. Csuhaj-Varju, J. Dassow, On cooperating distributed grammar systems, *J. Inform. Process. Cybern.*, *EIK*, 26 (1990), 49 – 63.

8. E. Csuhaj-Varju, J. Dassow, J. Kelemen, Gh. Păun, Stratified grammar systems, *Computers and AI*, 13 (1994), 409 – 422.

9. E. Csuhaj-Varju, J. Dassow, J. Kelemen, Gh. Păun, *Grammar Systems. A Grammatical Approach to Distribution and Cooperation*, Gordon and Breach, London, 1994.

10. E. Csuhaj-Varju, J. Kelemen, Cooperating grammar systems: a syntactical framework for the blackboard model of problem solving, in *Proc. AI and Inform. Control Syst. of Robots, 89* (I. Plander, ed.), North-Holland, 1989, 121 – 127.

11. E. Csuhaj-Varju, J. Kelemen, A. Kelemenova, Gh. Păun, Eco-grammar systems: A preview, in *Cybernetics and Systems 94* (R. Trappl, ed.), World Sci. Publ., Singapore, 1994, 941 – 949.

12. E. Csuhaj-Varju, J. Kelemen, A. Kelemenova, Gh. Păun, Eco-grammar systems: A generative model of ecosystems, *Artificial Life*, to appear.

13. E. Csuhaj-Varju, Gh. Păun, A. Salomaa, Conditional tabled eco-grammar systems versus (E)T0L systems, *JUCS*, to appear.

14. J. Dassow, J. Kelemen, Cooperating distributed grammar systems: a link between formal languages and artificial intelligence, *Bulletin of EATCS*, 45 (1991), 131 - 145.

15. J. Dassow, Gh. Păun, *Regulated Rewriting in Formal Language Theory*, Springer-Verlag, Berlin, Heidelberg, 1989.

16. J. Dassow, Gh. Păun, St. Skalla, On the size of components of cooperating grammar systems, *LNCS* 812, Springer-Verlag, Berlin, 1994, 325 - 343.

17. R. Freund, Gh. Păun, A variant of team cooperation in grammar systems, submitted, 1994.

18. R. Freund, Gh. Păun, C. M. Procopiuc, O. Procopiuc, Parallel communicating grammar systems with context-sensitive components, in [41], 166 - 174.

19. G. Georgescu, The generative power of small grammar systems, in [41], 152 - 165.

20. G. Georgescu, A sufficient condition for the regularity of PCGS languages, *J. Inform. Process. Cybern., EIK*, to appear.

21. J. Gruska, Descriptional complexity of context-free languages, *Proc. MFCS 73*, High Tatras, 1973, 71 - 84.

22. D. Hauschild, M. Jantzen, Petri nets algorithms in the theory of matrix grammars, *Acta Informatica*, 31 (1994), 719 - 728.

23. J. Hromkovic, J. Kari, L. Kari, Some hierarchies for the communicating complexity measures of cooperating grammar systems, *Theoretical Computer Sci.*, 127 (1994), 123 - 147.

24. J. Hromkovic, J. Kari, L. Kari, D. Pardubska, Two lower bounds on distributive generation of languages, *Proc. MFCS 94, LNCS* 841, Springer-Verlag, Berlin, 1994, 423 - 432.

25. C. M. Ionescu, F. L. Ţiplea, Bounded communication in parallel communicating grammar systems, *J. Inform. Process. Cybern., EIK*, 30 (1994), 97 - 110.

26. L. Kari, A. Mateescu, Gh. Păun, A. Salomaa, Teams in cooperating grammar systems, *J. Exper. Th. AI*, 1994.

27. J. Kari, L. Sântean (Kari), The impact of the number of cooperating grammars on the generative power, *Theoretical Computer Sci.*, 98 (1992), 621 - 633.

28. J. Kelemen, A. Kelemenova, A subsumption architecture for generative symbol systems, in *Cybernetics and Systems Research 92* (R. Trappl, ed.), World Sci. Publ., Singapore, 1992, 1529 - 1536.

29. A. Kelemenova, E. Csuhaj-Varju, Languages of colonies, *Theoretical Computer Sci.*, 134 (1994), 119 - 130.

30. R. Meersman, G. Rozenberg, Cooperating grammar systems, *Proc. MFCS 78, LNCS* 64, Springer-Verlag, Berlin, 1978, 364 - 374.

31. V. Mihalache, Matrix grammars versus parallel communicating grammar systems, in [39], 293 - 318.

32. V. Mihalache, General Artificial Intelligence systems as eco-grammar systems, in [41], 245 – 259.

33. V. Mihalache, On parallel communicating grammar systems with context-free components, in *Mathematical Linguistics and Related Topics* (Gh. Păun, ed.), The Publ. House of the Romanian Academy, Bucharest, 1995, 258 – 270.

34. V. Mitrana, Hybrid cooperating distributed grammar systems, *Computers and AI*, 2 (1993), 83 – 88.

35. V. Mitrana, Gh. Păun, G. Rozenberg, Structuring grammar systems by priorities and hierarchies, *Acta Cybern.*, 11 (1994), 189 – 204.

36. P. H. Nii, Blackboard systems, in *The Handbook of AI*, vol. 4 (A. Barr, P. R. Cohen, E. A. Feigenbaum, eds.), Addison-Wesley, Reading, Mass., 1989.

37. D. Pardubska, On the power of communication structure for distributive generation of languages, in *Developments in Language Theory* (G. Rozenberg, A. Salomaa, eds.), World Sci. Publ., Singapore, 1994, 90 – 101.

38. Gh. Păun, On the syntactic complexity of parallel communicating grammar systems, *Kybernetika*, 28 (1992), 155 – 166.

39. Gh. Păun (ed.), *Mathematical Aspects of Natural and Formal Languages*, World Sci. Publ., Singapore, 1994.

40. Gh. Păun, On the generative capacity of colonies, *Kybernetika*, 31 (1995), 83 – 97.

41. Gh. Păun (ed.), *Artificial Life. Grammatical Models*, The Black Sea Univ. Press, Bucharest, 1995.

42. Gh. Păun, On the generative capacity of hybrid CD grammar systems, *J. Inform. Process. Cybern.*, *EIK*, to appear.

43. Gh. Păun, G. Rozenberg, Prescribed teams of grammars, *Acta Informatica*, 31 (1994), 525 – 537.

44. Gh. Păun, L. Sântean (Kari), Parallel communicating grammar systems: the regular case, *Ann. Univ. Buc., Ser. Matem.-Inform.*, 38 (1989), 55 – 63.

45. O. Procopiuc, C. M. Ionescu, F. L. Țiplea, Parallel communicating grammar systems: the context-sensitive case, *Intern. J. Computer Math*, 49 (1993), 145 – 156.

46. G. Rozenberg, A. Salomaa, *The Mathematical Theory of L Systems*, Academic Press, New York, 1980.

47. A. Salomaa, *Formal Languages*, Academic Press, New York, 1973.

48. L. Sântean (Kari), Parallel communicating systems, *Bulletin EATCS*, 42 (1990), 160 – 171.

49. F. L. Țiplea, C. Ene, A coverability structure for parallel communicating grammar systems, *J. Inform. Process. Cybern.*, *EIK*, 29 (1993), 303 – 315.

50. F. L. Țiplea, O. Procopiuc, C. M. Procopiuc, C. Ene, On the power and complexity of parallel communicating grammar systems, in [41], 53 – 78.

51. S. Vicolov (Dumitrescu), Non-centralized parallel grammar systems, *Stud. Cercet. Matem.*, 44 (1992), 455 – 462.

52. D. Wätjen, On cooperating distributed limited 0L systems, *J. Inform. Process. Cybern. EIK*, 29 (1993), 129 – 142.

Compactness of Systems of Equations in Semigroups

T. HARJU   **J. KARHUMÄKI**[†]

Department of Mathematics, University of Turku
FIN-20500 Turku, Finland

W. PLANDOWSKI[‡]

Instytut Informatyki, Uniwersytet Warszawski
02-097 Banacha 2, Warszawa, Poland

Abstract. We consider systems $u_i = v_i$ ($i \in I$) of equations in semigroups over finite sets of variables. A semigroup S is said to satisfy the compactness property (or CP, for short), if each system of equations has an equivalent finite subsystem. It is shown that all monoids in a variety $\mathcal{V}$ satisfy CP, if and only if the finitely generated monoids in $\mathcal{V}$ satisfy the maximal condition on congruences. We also show that if a finitely generated semigroup S satisfies CP, then S is necessarily hopfian and satisfies the chain condition on idempotents. Finally, we give three simple examples (the bicyclic monoid, the free monogenic inverse semigroup and the Baumslag-Solitar group) which do not satisfy CP, and show that the above necessary conditions are not sufficient.

1. Introduction

Let $X_n = \{x_1, x_2, \ldots, x_n\}$ be a finite set of variables. We denote by X_n^+ the free semigroup generated by the set X_n. An *equation* over X_n is a pair $(u, v) \in X_n^+ \times X_n^+$ of words, which is often written as $u = v$. A *solution* of an equation (u, v) *in a semigroup* S is a morphism $\alpha: X_n^+ \to S$ such that $\alpha(u) = \alpha(v)$, i.e., such that (u, v) is in the kernel of α, $(u, v) \in ker(\alpha)$. Similarly, we let X_n^* and $X_n^{(*)}$ denote the free monoid and the free group generated by X_n, respectively. Equations $(u, v) \in X_n^* \times X_n^*$ $((u, v) \in X_n^{(*)} \times X_n^{(*)})$ and solutions $\alpha: X_n^* \to M$ $(\alpha: X_n^{(*)} \to G)$ in a monoid M and in a group G are defined similarly.

Let S be a semigroup and let $L \subseteq X_n^+ \times X_n^+$ be a *system of equations* in X_n^+. We say that L is *equivalent to a subsystem* $L' \subseteq L$, if L and L' have the same set of solutions, i.e., if $\alpha: X_n^+ \to S$ is a solution of all $(u, v) \in L'$, then α is a solution of all $(u, v) \in L$. Further, S is said to satisfy the *compactness*

[†]Supported by Academy of Finland grant 4077

[‡]Supported by the KBN grant 8 T11C 012 08

property (for systems of equations), or *CP* for short, if for all $n \geq 1$ every system $L \subseteq X_n^+ \times X_n^+$ is equivalent to one of its *finite* subsystems $T \subseteq L$.

A class C of semigroups is said to satisfy the *compactness property*, if every semigroup $S \in C$ satisfies it. We also say that a class C satisfies the compactness property *uniformly*, if for each system $L \in X_n^+ \times X_n^+$ of equations there exists a finite subsystem $T \subseteq L$ such that L is equivalent to T for all $S \in C$.

Clearly, all finite semigroups satisfy CP, but as we shall show in Section 2, the class of all finite semigroups does not satisfy CP uniformly.

As is easily seen the semigroups satisfying CP are closed under taking sub-semigroups and finite direct products, but they are not closed under morphic images. In Section 2 we prove that the semigroups (and groups) satisfying CP are not closed under infinite direct products. This result follows from the fact that the monoid $Fin(F_2)$ of all finite subsets of the 2-generator free monoid F_n is residually finite, but as shown by Lawrence [15] it does not satisfy CP.

In Section 3 we generalize a result of Albert and Lawrence [1] to varieties of monoids. We show that a variety V of monoids satisfies CP, if and only if the monoids in V satisfy the maximal condition on congruences. In particular, by Redei's Theorem, commutative semigroups and monoids satisfy the compactness property. It follows also that metabelian groups and nilpotent groups satisfy CP.

The above result for varieties does not hold for individual semigroups. Indeed, the free semigroups and the free groups satisfy CP (see, Albert and Lawrence [2], de Luca and Restivo [16]), but they do not satisfy the maximal condition on congruences (see, Section 4). In Section 4 we prove some necessary conditions for the compactness property, and we give three simple examples (the bicyclic monoid, the free monogenic inverse semigroup and the Baumslag-Solitar group) showing that these conditions are not sufficient for the compactness property. These examples also show that the maximal condition on congruences does not imply CP although this condition is sufficient in varieties of monoids.

We conclude this section with a few simple observations.

A system of equations $L \subseteq X_n^+ \times X_n^+$ is said to be *locally independent* in a semigroup S, if L is not equivalent to any of its finite proper subsystems, [14]. As a consequence, we obtain the following result, [14].

Theorem 1.1. *A semigroup S satisfies the compactness property, if and only if every locally independent system of equations is finite.*

The compactness property for systems of equations is also related to the following condition for morphisms. Let F_n be a finitely generated free semigroup and let S be a semigroup. Then S is said to satisfy the *compactness condition for morphisms*, if for all subsets $A \subseteq F_n$ there exists a finite subset $B \subseteq A$ such that for any two morphisms $\alpha, \beta : F_n \to S$, if $\alpha(u) = \beta(u)$ for all $u \in B$, then $\alpha(u) = \beta(u)$ for all $u \in A$. The problem whether or not $S = F_n$ satisfies the compactness condition for morphisms was known as Ehrenfeucht's Conjecture, and it was proved to hold in 1985 by Albert and Lawrence [2] and independently by Guba, see [19]. These proofs are based on Hilbert's Basis Theorem and on a result of Culik and Karhumäki [6] stating that for $S = F_n$ the compactness condition for morphisms is equivalent to the compactness property for systems of equations. The proof given in [6] generalizes immediately to all semigroups and groups, and hence we have the following result.

Theorem 1.2. *For any semigroup S the compactness property of systems of equations is equivalent to the compactness condition for morphisms.*

2. Closure properties

Since we consider only equations over a finite number of variables, each system of equations $L \subseteq X_n^+ \times X_n^+$ is either finite or denumerable, and hence L can be enumerated, $L = \{(u_i, v_i) \mid i = 1, 2, \ldots\}$. This implies immediately the following lemma.

Lemma 2.1. *If L does not have an equivalent finite subsystem in a semigroup, then L has an infinite subsystem $L' = \{(r_i, t_i) \mid i = 1, 2, \ldots\}$ ordered in such a way that for each j there exists a solution of the system $r_i = t_i$, $i < j$, which is not a solution of the equation $r_j = t_j$.*

The compactness property is preserved under some natural operations as shown in the next lemma, see [14].

Lemma 2.2. (1) *If a semigroup S can be embedded into a semigroup satisfying CP, then S satisfies CP. Moreover, if a semigroup S satisfies CP, then the class of subsemigroups of S satisfies CP uniformly.*

(2) *If the semigroups S_1 and S_2 satisfy CP, then so does the direct product $S_1 \times S_2$.*

Consider now a free semigroup F with an infinite number of generators. We observe that if $\alpha \colon X_n^+ \to F$ is a morphism, then the morphic image $\alpha(X_n^+)$ is a subsemigroup of a finitely generated free subsemigroup F_n of F. Since all n-generator free semigroups are isomorphic and they satisfy CP, it follows that F satisfies CP, and, moreover, each system $L \subseteq X_n^+ \times X_n^+$ of equations has a common equivalent finite subsystem for all free semigroups. The same conclusion is clearly true for all free monoids and free groups. Hence we have the following theorem.

Theorem 2.3. *The class of subsemigroups of free semigroups (respectively, sub-monoids of free monoids, and subgroups of free groups) satisfies CP uniformly.*

Notice also that a countably generated free semigroup F can be embedded into a 2-generator free semigroup F_2, and hence, by Lemma 2.2(1), if $T \subseteq L$ is an equivalent finite subsystem of $L \subseteq X_n^+ \times X_n^+$ in F_2, then T is an equivalent finite subsystem for all subsemigroups of free semigroups.

In above, the size of the finite subsystem T depends on L. Indeed, as shown in [1] for groups, for each $i \geq 1$ there is a system $L_i \subseteq X_3^{(*)} \times X_3^{(*)}$ of equations over three variables such that the size of any equivalent subsystem $T_i \subseteq L_i$ is at least i. It is not known if the same result holds for semigroups and monoids, although for these similar but weaker lower bounds are presented in [14].

The compactness property is not necessarily inherited by a morphic image (or by a quotient), simply because the 2-generator free semigroup F_2 satisfies CP, and, as we shall see, its morphic image B (the bicyclic monoid) does not.

Theorem 2.4. *The semigroups satisfying CP are not closed under morphic images.*

Next we shall prove that the semigroups satisfying CP are not closed under infinite direct products. For convenience, this and some later proofs will be given for monoids only. The restriction to monoids can be justified by the next result, where S^1 denotes the monoid, which is obtained from a given semigroup S without identity by adding an identity element 1 to S. If S is already a monoid, then we let $S^1 = S$. For the proof of the next theorem we refer to the full version of this paper [13].

Theorem 2.5. *A semigroup S satisfies CP, if and only if the monoid S^1 satisfies CP.*

First we notice that the direct products are closely related to uniformity of CP.

Lemma 2.6. *A direct product $\Pi_{i \in I} S_i$ of monoids satisfies CP, if and only if the monoids S_i ($i \in I$) satisfy CP uniformly.*

Proof. Each monoid S_j, $j \in I$, has a natural embedding into $\Pi_{i \in I} S_i$, and hence, by Lemma 2.2, if $\Pi_{i \in I} S_i$ satisfies CP, then the monoids S_j satisfy CP uniformly.

In the other direction the claim follows from the observation that for a morphism $\alpha: X_n^* \to \Pi_{i \in I} S_i$, we have $\alpha(u) \neq \alpha(v)$, if and only if for some $j \in I$, $\pi_j \alpha(u) \neq \pi_j \alpha(v)$, where $\pi_j: \Pi_{i \in I} S_i \to S_j$ is the projection onto S_j. $\square$

Recall that a monoid S is *residually finite*, if for all distinct $s, r \in S$, there exists a finite monoid R and a morphism $\alpha: S \to R$ such that $\alpha(s) \neq \alpha(r)$. It is rather straightforward to show that the free monoids are residually finite.

Lemma 2.7. *Let F_n be an n-generator free monoid. If $A \subseteq F_n$ is a finite subset of F_n, then there exists a morphism $\alpha: F_n \to S$ into a finite monoid S such that $\alpha(v) \neq \alpha(u)$ for all distinct $u, v \in A$.*

Let $Fin(S)$ denote the monoid of all nonempty *finite subsets* of the semigroup S. It was shown by Lawrence [15] that the monoid $Fin(F_2)$ does not satisfy CP. Indeed, the system L of equations

$$x_1 x_2^i x_1 = x_1 x_3^i x_1 \qquad (i \geq 1)$$

over three variables does not have an equivalent finite subsystem in $Fin(F_2)$.

Theorem 2.8. (1) *The monoids satisfying CP are not closed under infinite direct products. In fact, there is a direct product of finite monoids, which does not satisfy CP.*

(2) *The class of all finite monoids does not satisfy CP uniformly.*

Proof. Let $S = Fin(F_2)$. For the residual finiteness of S we refer again to [13].

Now, there exists a morphism $\alpha': S \to \Pi_{A \neq B} Fin(S_{AB})$ by its projections onto $Fin(S_{AB})$:

$$\pi_{AB}\alpha'(X) = \alpha'_{AB}(X) \qquad (X \in S),$$

where π_{AB} is the projection of $\Pi_{A \neq B} Fin(S_{AB})$ onto $Fin(S_{AB})$. Since we have $\alpha'_{AB}(A) \neq \alpha'_{AB}(B)$, the morphism α' is an embedding of S into the direct product $\Pi_{A \neq B} Fin(S_{AB})$. We conclude from Lemma 2.2 that the direct product $\Pi_{A \neq B} Fin(S_{AB})$ of finite monoids does not satisfy CP. Claim (2) follows now from Lemma 2.6. □

3. Compactness property for varieties of semigroups

A class $\mathcal{V}$ of monoids (resp., semigroups) is a *variety*, if it is closed under taking submonoids (resp., subsemigroups), morphic images, and arbitrary direct products. We refer to Cohn [5], Evans [8], or Neumann [21] for the theory of varieties. By Birkhoff's theorem, a variety becomes defined by a set of *identities* $u \equiv v$; these are equations $(u, v) \in X^*$ with a possibly infinite number of variables such that every morphism $\alpha: X^* \to S$ with $S \in \mathcal{V}$ is a solution of (u, v). As an example, the identity $x_1 x_2 \equiv x_2 x_1$ in X_2^* defines the variety of all commutative monoids.

A semigroup S satisfies the *maximal condition on congruences*, if each set of congruences of S has a maximal element. The following general result is easy to prove (using Zorn's lemma).

Lemma 3.1. *The following conditions are equivalent for a semigroup S.*

(1) *S satisfies the maximal condition on congruences.*

(2) *Each ascending chain $\theta_1 \subset \theta_2 \subset \ldots$ of congruences of S is finite.*

(3) *For each congruence θ of S generated by a subset $L \subseteq \theta$ there exists a finite subset $T \subseteq L$ such that T generates θ.*

Our main result for varieties of monoids generalizes a result of Albert and Lawrence [1] for varieties of groups, *cf.* [13].

Theorem 3.2. *A variety $\mathcal{V}$ of monoids satisfies CP, if and only if each finitely generated monoid $S \in \mathcal{V}$ satisfies the maximal condition on congruences.*

Redei's Theorem [23] states that the finitely generated commutative semigroups satisfy the maximal condition on congruences. For a short proof of this result, we refer to Freyd [9] or Grillet [10]. Hence we have the following corollary of Theorem 3.2 and Theorem 2.5.

Corollary 3.3. *Every commutative monoid or semigroup satisfies CP.*

It is worth noting that in Theorem 3.2 and its corollary the compactness property holds not only for finitely generated monoids, but for infinitely generated monoids as well.

Theorem 3.2 holds also for varieties of groups. In group theory congruences correspond to normal subgroups, and Theorem 3.2 can be stated as follows.

Theorem 3.4. *A variety V of groups satisfies CP, if and only if each finitely generated group of V satisfies the maximal condition on normal subgroups.*

For groups we have a stronger version of Corollary 3.3. Let $[a, b] = a^{-1}b^{-1}ab$ be the *commutator* of the elements a, b of a group. The *metabelian groups* form a (solvable) variety defined by a single identity $[[x_1, x_2], [x_3, x_4]] \equiv 1$, *i.e.*, a group G is metabelian, if and only if its second derived group is trivial. Clearly, every abelian group is metabelian. Moreover, every finitely generated metabelian group satisfies maximal condition on normal subgroups. We refer to Hall [11] for these and related results.

We also recall that a group G is *nilpotent of class n*, if in the lower central series of G we have $\gamma_{n+1}G = 1$. Hence a nontrivial abelian group is nilpotent of class one. The nilpotent groups of class at most n form a variety, since they are exactly the groups that satisfy the identity $[x_1, x_2, \ldots, x_n] \equiv 1$ for the generalized commutator over X_n. Moreover, a nilpotent group G satisfies the maximal condition on subgroups, if and only if G is finitely generated, see Hall [11] or Schenkman [24, p.200].

Corollary 3.5. *The metabelian groups and the nilpotent groups satisfy CP.*

Notice that the class of all nilpotent groups is not a variety, since it is not closed under infinite direct products. Indeed, the smallest variety that contains all nilpotent groups consists of all groups, see [24]. However, every nilpotent group belongs to a variety that satisfies CP. As will be seen in the next section, Corollary 3.5 does not extend to solvable groups.

We shall now strengthen Theorem 3.2 by showing that in a variety satisfying the compactness property, a system of equations $L \subseteq X_n^* \times X_n^*$ has an equivalent finite subsystem $T \subseteq L$ common to all $S \in V$.

Theorem 3.6. *If a variety V satisfies CP, then it satisfies CP uniformly.*

4. Semigroups without the compactness property

As shown by Lawrence [15], the monoid $Fin(F_2)$ of all finite subsets of a free monoid F_2 does not satisfy CP. In this section we show that there are even simpler monoids, semigroups and groups that do not possess CP either.

We notice that the free monoid F_n does not satisfy the maximal condition on congruences. Indeed, as an example, consider the submonoid S of the free monoid F_2 generated by the elements $a, aba, baba, baab$, where a and b are the generators of F_2. By Markov [18], S has no finite presentation. Let $S = \langle X_4 \mid u_i = v_i \ (i \in I) \rangle$ be any presentation of S, and let θ_k be the congruence of X_4^*

generated by $\{(u_i, v_i) \mid i = 1, 2, \ldots, k\}$. It follows that $\theta_1 \subset \theta_2 \subset \ldots$ is an infinite ascending chain of congruences of X_4^*.

However, all free monoids satisfy CP by Albert and Lawrence [2]. Furthermore, as shown below, the bicyclic monoid satisfies the maximal condition on congruences, but does not satisfy CP. Consequently the notions 'compactness property' and 'maximality condition' (on finitely generated subsemigroups) are incomparable in general, although they coincide on varieties.

The *bicyclic monoid* B is a 2-generator and 1-relator semigroup with the presentation $\langle a, b \mid ab = 1 \rangle$. B is isomorphic to the submonoid of the transformation semigroup $T_{\mathbf{N}}$ generated by the functions $\alpha, \beta \colon \mathbf{N} \to \mathbf{N}$:

$$\alpha(n) = \max\{0, n - 1\}, \qquad \beta(n) = n + 1,$$

see Clifford and Preston [4]. Here $\alpha\beta = 1$, the identity transformation on $\mathbf{N}$, but $\beta\alpha \neq 1$. Define $\gamma_i = \beta^i \alpha^i$, for $i \geq 0$. Hence

$$\gamma_i(n) = \begin{cases} 0 & \text{if } n \leq i, \\ n & \text{if } n > i. \end{cases}$$

Next we observe that $\gamma_i \gamma_j = \gamma_{\min\{i,j\}}$. In particular, each γ_i is an idempotent of B, i.e., $\gamma_i^2 = \gamma_i$. Consider then the system $L \subseteq X_3^+$ consisting of the equations

$$x_1^i x_2^i x_3 = x_3 \qquad (i = 1, 2, \ldots).$$

We conclude that the morphism δ_j defined by $\delta_j(x_1) = \beta$, $\delta_j(x_2) = \alpha$ and $\delta_j(x_3) = \gamma_j$, is a solution of $x_1^i x_2^i x_3 = x_3$ for all $i \leq j$, but δ_j is not a solution of $x_1^{j+1} x_2^{j+1} x_3 = x_3$. Hence the system L does not have an equivalent finite subsystem, and therefore the bicyclic monoid does not satisfy CP.

The bicyclic monoid B is an inverse semigroup, and it is simple (*i.e.*, it has no nontrivial ideals). Furthermore, for every nontrivial congruence θ, the quotient B/θ is a cyclic group, see Clifford and Preston [4]. In particular, B satisfies the maximal condition on congruences by Redei's theorem. Also, B is an example of a monoid, which does not satisfy CP, but all the proper quotients of which do satisfy CP.

If we consider the subsemigroup B_1 of B generated by the two elements $\rho = \beta^2 \alpha$ and α, we obtain a semigroup without the identity element, which does not satisfy CP either. Indeed, in B_1 we have, in the above notations, $\gamma_i = \rho^i \alpha^{i-1}$ for $i \geq 2$, and the claim follows when we consider the system of equations $x_1^i x_2^{i-1} x_3 = x_3$ for $i \geq 2$.

The set $E(S)$ of idempotents of a semigroup S can be partially ordered as follows: if $fe = e = ef$, then $e \leq f$, see [4]. We say that two elements a and b of a semigroup S form an *inverse pair*, if $a = aba$ and $b = bab$. In this case the elements ab and ba are idempotents of S. For the first claim of the next theorem we refer to Petrich [22, p.432].

Theorem 4.1. *Let S be a semigroup which contains an inverse pair a, b such that $ba < ab$. Then the subsemigroup of S generated by a and b is a bicyclic monoid with identity ab. In particular, S does not satisfy CP.*

We say that S satisfies the *chain condition on idempotents*, if each subset E_1 of $E(S)$ contains a maximal and a minimal element, *i.e.*, each chain $e_i < e_{i+1}$ ($i \in \mathbf{Z}$) of idempotents is finite. For the proof of the following result we refe to [13].

Theorem 4.2. *If a finitely generated semigroup S satisfies CP, then S satisfies the chain condition on idempotents.*

The free inverse semigroups do not satisfy the chain condition on idempotents. Indeed, the *free monogenic inverse semigroup*, which is generated by one element as an inverse semigroup, has a semigroup presentation

$$FI_1 = \langle a, b \mid a = aba, b = bab, a^m b^{m+n} a^n = b^n a^{n+m} b^n \quad (n, m \geq 1) \rangle,$$

see Petrich [22, p.427]. Here $a^n b^n$ is an idempotent for each $n \geq 1$, and $a^n b^n \cdot a^m b^m = a^n b^n = a^m b^m \cdot a^n b^n$, *i.e.*, $a^n b^n \leq a^m b^m$, for all $n \geq m$. By Theorem 4.2 we have the following result.

Theorem 4.3. *The free inverse semigroups do not satisfy CP.*

We notice also that the free inverse semigroups do not contain the bicyclic monoid.

The finitely generated free inverse semigroups are residually finite by Munn [20]. Since these semigroups do not satisfy CP, they can be used to prove Theorem 2.8 instead of the monoid $Fin(F_2)$. In particular, the free monogenic inverse semigroup FI_1 is a 2-generator semigroup, and therefore we have the following corollary.

Corollary 4.4. *The 2-generator finite (inverse) semigroups do not satisfy CP uniformly.*

Below we prove that a semigroup with the compactness property satisfies a maximal condition on restricted congruences, namely on nuclear congruences.

Following Dubreil [7] we say that a congruence θ of a semigroup S is *nuclear*, if it is induced by an endomorphism, *i.e.*, $\theta = ker(\alpha)$ for a morphism $\alpha \colon S \to S$. Hence a congruence θ of S is nuclear, if the quotient S/θ is isomorphic to a subsemigroup of S.

Lemma 4.5. *If a semigroup S satisfies CP then each sequence $\alpha_i \colon X_n^+ \to S$ of morphisms with $ker(\alpha_i) \subset ker(\alpha_{i+1})$ ($i = 1, 2, \ldots$) is finite for all n.*

Proof. Suppose α_i ($i = 1, 2, \ldots$) is an infinite sequence such that $ker(\alpha_i) \subset ker(\alpha_{i+1})$. Consider a system $L = \{(u_i, v_i) \mid i = 1, 2, \ldots\}$ of equations, where $(u_i, v_i) \in ker(\alpha_{i+1}) \setminus ker(\alpha_i)$ for each $i = 1, 2, \ldots$. Clearly, L has no equivalent finite subsystem w.r.t. to S. It follows that S does not satisfy CP. $\square$

It was shown by Harju and Karhumäki [12] that the finitely generated subsemigroups of free semigroups satisfy the maximal condition on nuclear congruences. The next theorem generalizes this result for semigroups satisfying the compactness property.

Theorem 4.6. *If a semigroup S satisfies CP then the finitely generated subsemigroups of S satisfy the maximal condition on nuclear congruences.*

Proof. Assume that S_0 is a finitely generated subsemigroup of S that has an infinite sequence $\alpha_i: S_0 \to S_0$ ($i = 1, 2, \ldots$) of endomorphisms such that $ker(\alpha_i) \subset ker(\alpha_{i+1})$ for all $i \geq 1$. Let $\mu: X_n^+ \to S_0$ be the natural morphism onto S_0. Consequently, $ker(\alpha_i\mu) \subset ker(\alpha_{i+1}\mu)$ for all $i \geq 1$, and the claim follows from Lemma 4.5. □

We conclude with one more example when the compactness property does not hold.

A semigroup S is said to be *hopfian*, if S is not isomorphic to a quotient S/θ for any nontrivial congruence θ of S, see Evans [8] or Magnus, Karrass and Solitar [17]. Equivalently, S is hopfian, if every surjective endomorphism $\alpha: S \to S$ is an automorphism.

Theorem 4.7. *If S satisfies CP then every finitely generated subsemigroup of S is hopfian.*

Proof. Assume S_0 is a non-hopfian finitely generated subsemigroup of S, and let $\alpha: S_0 \to S_0$ be a surjective endomorphism which is not injective. The nuclear congruences $\theta_i = ker(\alpha^i)$, $i = 1, 2, \ldots$, form a properly ascending chain, and hence the claim follows by Theorem 4.6. □

In particular, the finitely generated commutative semigroups are hopfian and they satisfy the chain condition on idempotents.

We notice that if a semigroup S satisfies CP, then S itself need not be hopfian (unless S is finitely generated). Indeed, the countably generated free semigroup F_∞ satisfies CP, but, as is easily seen, it is non-hopfian. For another example, consider the multiplicative group $\mathbb{C}^*$ of the nonzero complex numbers. Then $\alpha: \mathbb{C}^* \to \mathbb{C}^*$, $\alpha(c) = c^n$ ($n \geq 2$), is a surjective endomorphism of $\mathbb{C}^*$, for which $\alpha(r) = 1$ for all n-th roots of unity. In particular, α is not injective, and hence $\mathbb{C}^*$ is a non-hopfian abelian group. However, as an abelian group $\mathbb{C}^*$ satisfies CP.

We notice also that the bicyclic B monoid is hopfian, since every nontrivial quotient B/θ is a cyclic group. However, as shown above, B does not satisfy CP.

By Theorem 4.7 a finitely generated non-hopfian semigroup does not satisfy CP. Of course, the same holds for groups. The simplest non-hopfian group is the *Baumslag-Solitar group* [3], which has a group presentation $G_{BS} = \langle a, b \mid b^2a = ab^3 \rangle$, i.e., with two generators and one defining relation. Hence G_{BS} does not satisfy CP.

Indeed, the morphism $\alpha: G_{BS} \to G_{BS}$ defined by $\alpha(a) = a$ and $\alpha(b) = b^2$ is surjective, because $\alpha([a, b^{-1}]) = b$. However, as can be shown, $\alpha(u) = 1$ for $u = [a^{-1}ba, b] \neq 1$, and therefore α is not injective. Let $\beta: G_{BS} \to G_{BS}$ be defined by $\beta(a) = a$ and $\beta(b) = [a, b^{-1}]$. When we consider a and b as variables, we obtain a system of equations, $L = \{\beta^i(u) \mid i = 1, 2, \ldots\}$, which has no equivalent finite subsystem, because $\alpha^i(\beta^j(u)) = 1$ for $j < i$, but $\alpha^i(\beta^i(u)) \neq 1$.

One should notice that G_{BS} is a solvable group [3], which does not satisfy CP. Moreover, since the factors of the derived series (of length three) are abelian groups, it follows that the compactness property is not closed under group extensions.

References

[1] M.H. Albert and J. Lawrence, The descending chain condition on solution sets for systems of equations in groups, Proc. Edinburg Math. Soc. 29 (1986), 69 – 73.

[2] M.H. Albert and J. Lawrence, A proof of Ehrenfeucht's Conjecture, Theoret. Comput. Sci. 41 (1985), 121 – 123.

[3] G. Baumslag and D. Solitar, Some two-generator one-relator non-hopfian groups, Bull. Amer. Math. Soc. 68 (1962), 199 – 201.

[4] A.H. Clifford and G.B. Preston, "The Algebraic Theory of Semigroups", Vol I, Math. Surveys of the American Math. Soc.7, Providence, R.I., 1961.

[5] P.M. Cohn, "Universal Algebra", D. Reidel Publ. Co., Dordrecht, 1981.

[6] K. Culik II and J. Karhumäki, Systems of equations over a free monoid and Ehrenfeucht's Conjecture, Discrete Math. 43 (1983), 139 – 153.

[7] P. Dubreil, Sur le demi-groupe des endomorhismes d'une algébre abstraite, Lincei-Rend. Sc. fis. mat. e nat. Vol. XLVI (1969), 149 – 153.

[8] T. Evans, The lattice of semigroup varieties, Semigroup Forum 2 (1971), 1 – 43.

[9] P. Freyd, Redei's finiteness theorem for commutative semigroups, Proc. Amer. Math. Soc. 19 (1968), 1003.

[10] P.A. Grillet, A short proof of Redei's Theorem, Semigroup Forum 46 (1993), 126 – 127.

[11] P. Hall, Finiteness conditions for soluble groups, Proc. London Math. Soc. 4 (1954), 419 – 436.

[12] T. Harju and J. Karhumäki, On the defect theorem and simplifiability, Semigroup Forum 33 (1986), 199 – 217.

[13] T. Harju, J. Karhumäki and W. Plandowski, Compactness of systems of equations in semigroups, Manuscript 1994.

[14] J. Karhumäki and W. Plandowski, On the size of independent systems of equations in semigroups, Manuscript 1994.

[15] J. Lawrence, The nonexistence of finite test set for set-equivalence of finite substitutions, Bull. of the EATCS 28 (1986), 34 – 37.

[16] A. de Luca and A. Restivo, On a generalization of a conjecture of Ehrenfeucht, Bulletin of the EATCS 30 (1986), 84 – 90.

[17] W. Magnus, A. Karrass and D. Solitar, "Combinatorial Group Theory", Dover Publ., New York, 1976.

[18] Al.A. Markov, On finitely generated subsemigroups of a free semigroup, Semigroup Forum 3 (1971), 251 – 258.

[19] A.A. Muchnik and A.L. Semenov, "Jewels of Formal Languages", (Russian translation of a book by A. Salomaa), Mir, Moscow, 1986.

[20] W.D. Munn, Free inverse semigroups, *Proc. London Math. Soc.* **29** (1974), 385 – 404.

[21] H. Neumann, "Varieties of groups", Springer-Verlag, Berlin, 1967.

[22] M. Petrich, "Inverse Semigroups", John Wiley & Sons, New York, 1984.

[23] L. Redei, "The Theory of Finitely Generated Commutative Semi-Groups", Pergamon Press, Oxford, 1965.

[24] E. Schenkman, "Group Theory", D. van Nostrand Co., Princeton, 1965.

Sensing Versus Nonsensing Automata[*]

Pavol Ďuriš[1] and Zvi Galil[2]

[1] Department of Computer Science, Comenius University
842 15 Bratislava, Slovakia, duris@fmph.uniba.sk
[2] Department of Computer Science, Columbia University
New York, NY 10027, galil@cs.columbia.edu
and
Tel-Aviv University

Abstract. It is shown that one way deterministic sensing two head finite automata are more powerful than one way deterministic two head finite automata. This result solves a 14 year old open problem [2]. The proof uses two new pumping lemmas.

1 Introduction

A one way deterministic [nondeterministic] sensing k head finite automaton ($dsfa(k)$ [$nsfa(k)$]) is a one way deterministic [nondeterministic] k head finite automaton ($dfa(k)$ [$nfa(k)$]) whose heads are allowed to sense the presence of other heads on the same input position. (The concept of *sensing* was introduced by Ibarra [1].) It is well known that one way nondeterministic k head finite automata are able to recognize only regular languages over one letter alphabet for each k. On the other hand, one can easily observe that the language $\{a^{n^2} | n \geq 0\}$ can be recognized by a one way deterministic 3 head sensing finite automaton. Hence, one way sensing k head finite automata are more powerful than one way k head finite automata for each $k \geq 3$. The remaining problem has been open since it was stated in 1979 in [2].

Problem 1. Are one way deterministic [nondeterministic] sensing two head finite automata more powerful than one way deterministic [nondeterministic] two head finite automata?

The reason $nfa(k)$'s recognize only regular languages over one letter alphabet is that Ibarra showed that the languages accepted by these automata (in fact, by multihead nondeterministic pushdown automata) satisfy the semilinearity property [1] and for one letter alphabet semilinearity means regularity. Ibarra also showed that the language accepted by a $nfsa(2)$ satisfies the semilinearity property. Thus, one cannot use the simple argument above for showing that $dfsa(2)$'s (or even $nfsa(2)$'s) are more powerful than $dfa(2)$'s.

[*] This work was partially supported by NSF grant CCR-93-16209 and CISE Institutional Infrastructure Grant CDA-90-24735, and by EC cooperation Action IC 1000, ALTEC

The possible difference in computing power between sensing and nonsensing automata is of interest also in other cases. In [3] Jiang and Li showed that for any $k > 0$, a nonsensing (one-way) $dfa(k)$ cannot do string matching, settling a long standing open problem. But the problem remains open for sensing (one-way) $dfa(k)$.

Let

$$L_1 = \{uvuv | u \in \{a, b\}^+, v \in \{0, 1\}^+\},$$
$$L_2 = \{uvua^n ba^n | u \in \{a, b\}^+, v \in \{0, 1\}^+, n \geq 1\}$$

One can easily observe that $L_1 \cup L_2$ can be recognized by a $dsfa(2)$. Consequenly, the following theorem affirmatively solves Problem 1 above for deterministic automata. Unfortunately, we were not be able to solve Problem 1 also for nondeterministic automata. Note that $L_1 \cup L_2$ can be recognized by an $nfa(2)$.

Theorem 1. $L_1 \cup L_2$ *cannot be recognized by any one way deterministic two head finite automata.*

We denote the length of an input string x by $|x|$, and the cardinality of a set S by $|S|$. The main tool in our proofs is the notion of a loop. Informally, we say that an automaton is in a loop if it repeats an internal state during a computation in which each of its two heads scans the same alphabet symbol in the input, not necessarily in the same position. We next give a formal definition.

Let A be any one way deterministic two head finite automaton with input alphabet Σ and let Q be the set of internal states of A. A *configuration* of the automaton A is a triple (q, i_1, i_2), where $q \in Q$ and i_j is the position of the j-th head. A *loop* K is a triple (s, a_1, a_2), where s is a sequence $q_1, q_2, \ldots, q_p$ of internal states of A, $2 \leq p \leq |Q| + 1$, $a_1, a_2 \in \Sigma$, such that $q_1 = q_p$ and for each $i = 1, 2, \ldots, p - 1$ the following holds. If A is in internal state q_i and the j-th head scans the symbol a_j for $j = 1, 2$, then in one move A enters the internal state q_{i+1}. Let the sequence of configurations $c_1, c_2, \ldots, c_r$ be a computation of A on an input word x, where $c_m = (q_m, i_m, j_m)$ for $m = 1, 2, \ldots, r$. We will say that A performs the loop K during a computation from c_i to c_h on x, if there is an integer l, $i \leq l \leq h - p + 1$, such that A is in the internal state q_{t+1} and the j-th head scans a_j for $j = 1, 2$, when A is in the configuration c_{l+t} for $t = 0, 1, \ldots, p - 1$. The *increment* of the loop K is the pair of numbers $(i_{l+p-1} - i_l, j_{l+p-1} - j_l)$. Informally, the increment tells us, how many tape cells are crossed by heads when a loop is performed once.

Since A is deterministic automaton with $|Q|$ internal states, it is easy to see that if during $|Q| + 1$ steps one of the heads of A does not change its position and the other one traverses a block of the same symbols, then A performs a loop at least once during these steps and the loop is repeated until the head scanning the block of the same symbols leaves it. Similarly, if the strings ab^m and c^n with $m \geq |Q|$, $n \geq 2|Q|$, are located on the input tape such that one of the heads of A scans the leftmost symbol of ab^m and the other one scans the leftmost symbol of

c^n, then A performs a loop at least once within the next $2|Q|$ steps and the loop is repeated until one of the heads leaves the scanned string mentioned above.

In Section 2 we state two new pumping lemmas for dfa(2). We prove Lemma 1 and only sketch the idea of the proof of Lemma 2. The latter is quite complicated and uses Lemma 1 several times. In Section 3 we sketch the proof of Theorem 1. We give its complete proof in the Appendix.

2   Two new pumping lemmas

Lemma 1. (Pumping lemma for one stationary and one moving head.)
Let x be any string in $\{a, b, 0, 1\}^*$ and let $q, q' \in Q$ be any two internal states of A. Let i, i', k_1, k_2, n_1 and n_2 be any nonnegative integers such that $n_2 \geq |Q|$, $i' = i + k_1|Q|!$ and at least one of (a), (b) and (c) holds, where $w = xa^{n_1}ba^{n_2}$ and $w' = xa^{n_1}ba^{n_2+k_2|Q|!}$.

(a) $|x| + 1 \leq i, i' \leq |x| + n_1 - |Q|$,
(b) $i' = i \leq |w| - |Q|$ (i.e., $k_1 = 0$),
(c) $|x| + n_1 + 2 \leq i \leq |w| - |Q|$ and $|x| + n_1 + 2 \leq i' \leq |w'| - |Q|$.

Then if the j-th symbol of the string $w\$$ is the same as the j'-th symbol of the string $w'\$$ for some j, j' and if there is a computation of A on w from the configuration (q, i, j) to the configuration $(q', |w| + 1, j)$ then there is also a computation of A on w' from (q, i', j') to $(q', |w'| + 1, j')$.

Note that in (a) the first head scans the first block of a's (after x) and is far enough from its end; in (b) the first head is in the same position in both cases and is far enough from the end of w; and in (c) the first head scans the second block of a's and is far enough from the end of the corresponding input. Note also that we pump twice: (1) In (a) and (c) the first head positions can differ by a multiple of $|Q|!$ and (2) the sizes of the second blocks of a's can differ by a multiple of $|Q|!$.

Remark 1. Lemma 1 is valid also if the heads are interchanged, i.e. if the configurations (q, i, j), $(q, |w| + 1, j)$, (q', i', j'), $(q', |w'| + 1, j')$ are replaced by the configurations (q, j, i), $(q, j, |w| + 1)$, (q', j', i') and $(q', j', |w'| + 1)$, respectively.

Proof of Lemma 1. We will prove Lemma 1 only for case (a). The proofs of the other cases are similar. First note that A behaves as a one way deterministic one head finite automaton during the computation on w starting at (q, i, j), since the second head still scans the j-th symbol of the input. Thus, during the first $|Q|$ steps of the computation on w and also on w', A performs an identical loop on both inputs, since in both cases, A starts in the same state q, the first head scans a^{n_1}, the second head scans the same symbol initially and does not change its position during the computation of A on w. Thus, A works in the loop until the rightmost occurrence of b is reached by the first head. Hence, using the standard pumping technique it follows that A is in the same internal state for both inputs w and w' when the first head enters the rightmost occurrence of b, because $i' - i = k_1 |Q|!$.

Using similar reasons as above for the rest of the computations of A on w and on w', one can show that A is in the same internal state for both inputs when the first head enters \$, since $|a^{n_2 + k_2|Q|!}| - |a^{n_2}| = k_2|Q|!$. ∎

While it turned out to be relatively easy to pump twice if one of the heads is stationary, it is much more difficult to pump once when the two heads are allowed to move.

Lemma 2. (Pumping lemma for two moving heads.)
Let d be sufficiently large (see (1) in the Appendix). Then for each integer $l \geq 0$ there is a positive integer $m_l \geq l$ such that the following holds for each integer $m \geq m_l$, for each integer t, $0 \leq t \leq l$ and for each two internal states q, q' of A. Let $i_0 = 1$ and $j_0 = 1 + d + m + t$. If there is a computation of A on $w = ba^{d+m}ba^m$ from configuration $c = (q, i_0, j_0)$ to configuration $(q', |w| + 1, |w| + 1)$ then there is also a computation of A on $w' = ba^{d+m}ba^{m+(|Q|!)^2}$ from configuration c to configuration $(q', |w'| + 1, |w'| + 1)$.

Note that initially the first head scans the first b (of w or w') and the second head scans the second b (when $t = 0$) or one of the a's (the $t - 1$st) following the second b.

Remark 2. Let x be any string in $\{a, b, 0, 1\}^*$. If we replace w, w' by $w = xba^{d+m}ba^m$ and by $w' = xba^{d+m}ba^{m+(|Q|!)^2}$, respectively, then the same lemma holds also for $i_0 = |x| + 1$ and $j_0 = |x| + 1 + d + m + t$, and also for $i_0 = |x| + 1 + d + m + t$ and $j_0 = |x| + 1$.

We only sketch the proof. The proof of Lemma 2 consists of several cases. In Case 1, we can quite easily verify the conditions of Lemma 1 and fool A. Case 2 is more involved and includes three subcases that are dealt with similarly.

Recall that we chose d large enough. Given l and d, choose m large enough and consider the computation of A on w starting at the configuration c. In Case 2 the first head enters the second occurrence of b before the second head enters \$. Since w is of very simple structure, we will see that A works (with possible exception for only a few steps) in a loop K_1 with an increment (g_1, h_1) during the first phase of computation - i.e. until the first head enters the second occurrence of b. By the same reasons, A works (with possible exception for only a few steps) in a loop K_2 with an increment (g_2, h_2) during the second phase of the computation - i.e. until one of the heads (the faster one) enters \$. Our main aim is to show that the other head (the slower one) is far (at least $|Q| + (|Q| - 1)(|Q|!)^2$ tape cells) from \$ when the faster head enters \$, i.e. at the end of the second phase. To do so, we first estimate r [s], the number of times the loop K_1 [K_2] is performed during the first [the second] phase, using the fact that A works in the loop K_1 [K_2] during the first [the second] phase (with possible exception of $O(1)$ steps). The loop K_2 is performed exactly $(|Q|!)^2/g_2$ more times on $w' = wa^{(|Q|!)^2}$ than on w. We then apply Lemma 1 to complete the proof.

3 A sketch of the proof of Theorem 1

Suppose to the contrary, that there is a one way deterministic two head finite automaton A recognizing $L_1 \cup L_2$. Let Q be the set of internal states of A. Without loss of generality we may assume that A is allowed to accept the input only if both heads scan the right endmarker \$.

After setting some constants (d, m_0, p, n) we construct a special string $u = (ba^{d+p}ba^p)^{|Q|}ba^d$ and show (using a simple counting argument) that there is a string v such that A must move the heads on input $uvuv$ in such a way that one of them scans or leaves the left half of the input when the other one leaves the second occurrence of u (to be able to compare the strings following the first and the second occurrence of u). Such behavior of A enables us to know how the rest of computation looks like when we replace the suffix v by the string $a^p ba^p$ or $a^n ba^n$. There are only three cases to be considered.

In the first "extreme" case (Case I), one of the heads (the slower head) crosses only a few (less than m_0) a's of the first block a^p of the suffix $a^p ba^p$ of the input $uvua^p ba^p$, while the other one (the faster head) traverses the first $|Q|$ blocks $ba^{d+p}ba^p$ of the second occurence of u. In the second "extreme" case (Case III), one of the heads (the slower head) has crossed only a short prefix (of length at most $|uvu| - d$) of the input $uvua^n ba^n$, when the other one (the faster head) has crossed a long prefix a^h ($h > (|uvu| - d)|Q|$) of the suffix a^n of the input. The main goal in both extreme cases is to show that A works in a loop when the faster head traverses the suffix ba^p (or a^n), and moreover, the slower head does not change its position during the loop. This will enable us to fool A by a standard pumping technique by adding an extra suffix $a^{|Q|!}$ to the input (to do so, we will apply Lemma 1). To achieve the main goal is easy in the second extreme case. To do so in the first extreme case, we will first show that the behavior of A is the same when the faster head traverses the last and some former block $ba^{d+p}ba^p$ of the suffix $ua^p ba^p$. An intuitive reason for this is that the suffix $ua^p ba^p$ contains too many ($|Q| + 1$) such blocks and A is a deterministic automaton with only a few ($|Q|$) internal states. Then we will derive the main goal using the fact that the slower head crosses only a few (less than m_0) a's also when the faster head traverses the long suffix ba^p ($p = 2|Q|(m_0 + 1)$). When there is no extreme case during the computation on the input $uvua^n ba^n$ (Case II), we will fool A directly by applying Lemma 2.

References

1. O.H. Ibarra, A note on semilinear sets and bounded-reversal multihead pushdown automata , *Inform. Process. Lett.* **3** (1974) 25-28.
2. K. Inoue, I. Takanami, A. Nakamura and T. Ae, One-way simple multihead finite automata, *Theoret. Comput. Sci.* **9** (1979) 311-328.
3. T. Jiang and M. Li, k one-way heads cannot do string-matching, *Proceedings 25th ACM Symp. on Theory of Computing*, (1993) 62-70.

Appendix: The Proof of Theorem 1

Let

$$d = |Q|^2 + 7|Q|^3 + 5|Q|^5 + |Q|^3(|Q|!)^2. \tag{1}$$

Let m_0 be the integer m_l of Lemma 2 for $l = 0$ and let

$$p = 2|Q|(m_0 + 2) \quad \text{and} \quad u = (ba^{d+p}ba^p)^{|Q|}ba^d.$$

For each $y \in Y = \{0, 1\}^{\lceil \log_2 2|Q||u|^2 \rceil}$, let c^y be such a configuration (if it exists) at which both heads scan the second occurrence of u during the computation of A on the input $uyuy$; if there are several such configurations for one y, select any one of them to be c^y. Note that if there are two different strings $y, y' \in Y$ with $c^y = c^{y'}$, then since A accepts $uyuy$ and $uy'uy'$ (both in L_1) it has to accept also $uy'uy \notin L_1 \cup L_2$, a contradiction. Thus, since $|Y| \geq 2|Q||u|^2$ and the number of possible c^y's is at most $|Q||u|^2$, there is a string $v \in Y$ for which c^v does not exist. It means that during the computation of A on $uvuv$, A enters a configuration $c_1 = (q_1, i_1, j_1)$ at which one of the heads enters the second occurrence of v and the other one scans the left half of the input or it enters the right half of the input. Hence, $i_1 \leq |uv| + 1$ and $j_1 = |uvu| + 1$ or $j_1 \leq |uv| + 1$ and $i_1 = |uvu| + 1$. We will prove Theorem 1 only for the case $i_1 \leq |uv| + 1$ and $j_1 = |uvu| + 1$. (The proof for the second case is similar.)

Let m_l be from Lemma 2 for $l = (|uvu| - d)|Q| + 1$ and let

$$n = \max\{m_l, m_0, (|uvu| - d)|Q| + 1 + |Q|\}. \tag{2}$$

Consider the computation of A on the input $uvua^nba^n$. Clearly A enters c_1 on $uvua^nba^n$ exactly as it does so on $uvuv$. Let $c_2 = (q_2, i_2, j_2)$ be the first configuration reachable from c_1 on $uvua^nba^n$, at which the first head enters the rightmost occurrence of b of the second occurrence of u or the second head enters the $((|uvu| - d)|Q| + 1)$-st symbol of the suffix a^n.

We consider three cases.

Case I. $i_2 = |uvu| - d$ and $j_2 \leq |uvu| + m_0$. In this case and the next the first head enters the rightmost occurrence of b of the second occurrence of u of the input $uvua^nba^n$ when A enters c_2. Consider the computation of A on the input $uvuz$, where $z = a^pba^p$ (recall $p = 2|Q|(m_0 + 2)$). Note that the suffix uz of the input $uvuz$ is concatenation of $|Q| + 1$ strings $y = ba^{d+p}ba^p$, i.e. $uvuz = uvy^{|Q|+1}$. By (2) and the fact that $|u| > p + d$, $p \leq n$. Since $i_2 \leq j_2 \leq |uvu| + m_0 \leq |uvua^p| \leq |uvua^n|$ and c_1 precedes c_2, A enters c_1 and then c_2 on $uvuz = uvua^pba^p$ exactly as it does so on $uvua^nba^n$. Hence, A enters c_1 [c_2] on the input $uvuz = uvy^{|Q|+1} = uvy^{|Q|}ba^{d+p}ba^p$ not later than its first head enters the first occurrence of y [exactly when its first head enters the $(|Q| + 1)$-st occurrence of y] of the suffix $y^{|Q|+1}$ of the input.

For $i = 1, 2, \ldots, |Q| + 1$, let s_i denotes a computational segment starting [ending] when the first head of A enters the i-th occurrence of y [enters the

$(i+1)$-st occurrence of y or \$] in the suffix $y^{|Q|+1}$ of the input $uvy^{|Q|+1}$, and let q^i be the internal state of A when the first head enters the i-th occurrence of y in the suffix $y^{|Q|+1}$ of the input.

First we will show that behavior of A is the same during segments s_t and $s_{|Q|+1}$ for some $t \leq |Q|$. Then we will prove that A enters a loop during segment $s_{|Q|+1}$ with the following properties. The second head does not change its position during this loop and A works in this loop until end of segment $s_{|Q|+1}$. Finally we will apply Lemma 1 to fool A.

By the assumption of Case I and by the choice of c_1 and c_2, $|uvua| = j_1 \leq j_2 \leq |uvua^{m_0}| \leq |uvua^p|$. The former inequalities and the fact that each segment s_i with $1 \leq i \leq |Q|$ is a subsegment of the computation of A from c_1 to c_2 on the input $uvua^p ba^p = uvuz = uvy^{|Q|+1}$ yield that during each segment s_i with $1 \leq i \leq |Q|$, the second head is positioned between positions j_1 and j_2; i.e. it scans only a's and crosses at most m_0 of them. Let δ_i, $1 \leq i \leq |Q|$, be the number of a's crossed by the second head during segment s_i. Hence, $\delta_i \leq m_0$ for each $i = 1, 2, \ldots, |Q|$. Since $|uvua| \leq j_1 \leq j_2 + \delta_i \leq j_2 + m_0 \leq |uvua^p|$ for each $1 \leq i \leq |Q|$ and the position of the second head is somewhere between j_1 and j_2 at the end of segment $s_{|Q|}$; i.e. at the beginning of segment $s_{|Q|+1}$, there are at least δ_i a's on the input tape starting at the position of the second head at the beginning of segment $s_{|Q|+1}$.

The states of the sequence $q^1, q^2, \ldots, q^{|Q|+1}$ must periodically repeat starting at some q^h, $h \leq |Q|$, since A is a deterministic device with only $|Q|$ internal states and the first head traverses the same string y and the second head scans only a's during each segment s_i, $1 \leq i \leq |Q|$. Hence, $q^{|Q|+1} = q^t$ for some $t \leq |Q|$. But it means that the behavior of A is the same during both segments s_t and $s_{|Q|+1}$, since A is in the same internal state q^t ($= q^{|Q|+1}$) at the beginning of both segments and the same strings, that are traversed during segment s_t, occur to the right at the positions of the heads (i.e. the string y occurs on the input tape starting at the position of the first head and δ_t a's occur on the input tape starting at the position of the second head) at the beginning of segment $s_{|Q|+1}$. Thus, the second head scans only a's and crosses δ_t of them during segment $s_{|Q|+1}$. But it means that both heads of A scan only a's during a part of segment $s_{|Q|+1}$ starting at the configuration $c_3 = (q_3, |uvu| + p + 2, j_3)$ at which the first head enters the suffix a^p of the input $uvuz = uvy^{|Q|+1} = uvy^{|Q|}a^{d+p}ba^p$ and ending at the last configuration of $s_{|Q|+1}$, $c_4 = (q_4, |uvuz| + 1, j_4)$, at which the first head enters \$. Now it is easy to see that A performs a loop within the first $|Q|$ steps of the computation from c_3 to c_4 on the input mentioned above and A works in this loop until the first head leaves the suffix a^p, i.e. until A enters c_4. Moreover, the second head does not change its position during this loop, since otherwise it would change its position every time when the loop is performed, but the loop is performed at least once during each $|Q|$ steps while the first head traverses the suffix a^p, i.e. altogether at least $p/|Q| > m_0$ many times. But it contradicts the fact above that the second head crosses $\delta_t \leq m_0$ a's during segment $c_{|Q|+1}$.

Now we are ready to fool A by applying Lemma 1. Let $c_5 = (q_5, i_5, j_5)$ be the configuration of the computation from c_3 ($= (q_3, |uvu| + p + 2, j_3)$) to c_4 on input $uvua^p ba^p = uvuz = uvy^{|Q|+1}$ at which A enters the loop mentioned above within the first $|Q|$ steps. Clearly, $i_5 \leq |uvu| + p + 2 + |Q| = |uvuz| - |a^p| + 1 + |Q| \leq |uvuz| - |Q|$ (recall $p = 2|Q|(m_0 + 2)$). Since the second head does not change its position during the loop and A works in the loop until it enters c_4, it cannot change its position also during the computation from c_5 to c_4 ($= (q_4, |uvuz| + 1, j_4)$), by the choice of c_5. Hence $j_5 = j_4$. We have shown above that the second head is between positions j_1 and j_2 during each segment s_i with $1 \leq i \leq |Q|$ and the second head crosses $\delta_i \leq m_0$ tape squares during segment $s_{|Q|+1}$; consequently, $j_4 \leq j_2 + m_0$ and $j_5 = j_4 \leq j_2 + m_0 \leq |uvua^p| \leq |uvuz| - |Q|$. Clearly A enters c_5 on $uvuza^{|Q|!}$ exactly as it does so on $uvuz$. By Lemma 1 (case (b) of the version for interchanged heads, see Remark 1) there is also a computation on $uvuza^{|Q|!}$ from c_5 to $c_4' = (q_4, |uvuza^{|Q|!}| + 1, j_4)$. Since A accepts $uvuz \in L_2$, there is a computation on $uvuz$ from c_4 to an accepting configuration $(q_6, |uvuz| + 1, |uvuz| + 1)$. But, again by Lemma 1, there is a computation on $uvuza^{|Q|!}$ from c_4' to the accepting configuration $(q_6, |uvuza^{|Q|!}| + 1, |uvuza^{|Q|!}| + 1)$. But it means that A also accepts $uvuza^{|Q|!} \notin L_1 \cup L_2$, a contradiction.

Case II. $i_2 = |uvu| - d$ and $|uvu| + m_0 < j_2 \leq |uvu| + n + 1 + (|uvu| - d)|Q|$. Set $m = j_2 - |uvu|$ if $j_2 < |uvu| + n$, and set $m = n$ if $j_2 \geq |uvu| + n$. Let $w = uvua^m ba^m$ and $w' = wa^{(|Q|!)^2}$. Clearly, A enters c_2 on w and also on w' exactly as it does so on $uvua^n ba^n$. Since A accepts $w \in L_2$, there is a computation on w from c_2 to an accepting configuration $(q_3, |w| + 1, |w| + 1)$. By Lemma 2 (see Remark 2), there is also a computation on w' from c_2 to the accepting configuration $(q_3, |w'| + 1, |w'| + 1)$, i.e. A accepts $w' \notin L_1 \cup L_2$, a contradiction. Note that we apply Lemma 2 with $l = t = 0$ and $m = j_2 - |uvu| > m_0$ for the case $j_2 < |uvu| + n$, and with $l = (|uvu| - d)|Q| + 1$, $t = j_2 - |uvu| - n$ and $m = n \geq m_l$ (by (2)) for the case $j_2 \geq |uvu| + n$.

Case III. $i_2 \leq |uvu| - d$ and $j_2 = |uvu| + n + (|uvu| - d)|Q| + 2$. Let $h = (|uvu| - d)|Q| + 1$. In this case, the second head enters the h-th symbol of the suffix a^n of the input $uvua^n ba^n$ when A enters c_2. Consider the computation of A on $uvua^n ba^n$. Since the first head occupies at most $|uvu| - d$ different tape cells until A enters c_2 (by the assumption of Case III) and since it takes at least $h > (|uvu| - d)|Q|$ steps to traverse the string a^h, the first head must scan some tape cell during at least $|Q| + 1$ steps when the second head traverses the prefix a^h of the suffix a^n. But it means that A performs a loop when the second head traverses the prefix a^h of the suffix a^n (i.e. not later than A enters c_2, since $|uvua^n ba^h| \leq j_2$), during which the first head does not change its position, and clearly, A works in the loop until the second head enters \$ in configuration $c_4 = (q_4, i_4, |uvua^n ba^n| + 1)$.

Now we are ready to fool A by applying Lemma 1. Let $c_3 = (q_3, i_3, j_3)$ be the configuration at which A enters the loop for the first time. Since c_3 precedes c_2, $i_3 \leq i_2 \leq |uvu| - d \leq |uvua^n ba^n| - |Q|$ and $j_3 \leq j_2 = |uvu| + n + 1 + h \leq |uvua^n ba^n| - |Q|$, (by (2) and the assumptions of Case III). Since the first head does not change its position during the loop, it also does not change its position during the computation from c_3 to c_4. Hence $i_4 = i_3 \leq |uvua^n ba^n| - |Q|$. Since

A accepts $uvua^n ba^n \in L_2$, there is a computation on $uvua^n ba^n$ from c_4 to an accepting configuration $c_5 = (q_5, |uvua^n ba^n| + 1, |uvua^n ba^n| + 1)$. We obtain the desired contradiction as in Case I by applying Lemma 1 (case(b)) twice - for the computation from c_3 to c_4 and from c_4 to c_5, with $k_2 = 1$. Note that $uvua^n ba^{n+|Q|!} \notin L_1 \cup L_2$. This completes the proof of Theorem 1. ∎

New Upper Bounds for Generalized Intersection Searching Problems

Panayiotis Bozanis , Nectarios Kitsios , Christos Makris and Athanasios Tsakalidis

Department of Computer Engineering and Informatics, University of Patras
26500 Patras, Greece
and
Computer Technology Institute
P.O. BOX 1192, 26110 Patras, Greece
email: {bozanis, nkitsios, makris, tsak}@cti.gr

Abstract. Generalized intersection searching problems is a class of problems that constitute an extension of their standard counterparts. In such problems, we are given a set of colored objects and we want to report or count the distinct colors of the objects intersected by a query object. Many solutions have appeared for both iso-oriented and non-iso-oriented objects. We show how to improve the bounds of several generalized intersection searching problems as well as how to obtain upper bounds for some problems not treated before.

1 Introduction

A new type of problems, rich in applications and of significant theoretical interest has appeared recently ([9]), giving rise to many efficient algorithms. In these problems we are interested in preprocessing a set of colored objects (points, lines, rectangles) so that given a query object the colors of the objects intersected by the query object can be reported efficiently. For example, given a set of colored points in the plane we want to be able to report the colors of the points contained in any query range. Solutions for the standard problems yield output-insensitive solutions for their generalized counterparts. Moreover solutions for the counting versions of the standard problems, do not apply at all to the generalized problems, in which the number of the colors of the intersected objects has to be reported.

A uniform framework that yields efficient solutions for generalized problems on iso-oriented objects has been developed in [6]. Their techniques consist of geometric transformations of generalized problems into equivalent instances of some standard problems and the use of persistence as a method of imposing range restrictions to static problems. For some problems efficient solutions for a generalization of their counting versions is possible. They are called "type-2" versions and in these we are interested in reporting for each intersected color the number of the objects of that color that are intersected.

In this paper we show how to improve the bounds obtained in [6] and [7] for several generalized problems. We also give solutions for some problems not treated there. Tables 1,2 contain a summary of our results in comparison with the previous ones.

Generalized Problem		Space	Query Time	Update Time
1-D RANGE SEARCH				
dynamic counting	previous	nlogn	$\log^2 n$	$\log^2 n$
	new	n	$\log^2 n$	$\log^2 n$
static counting	previous	nlogn	logn	
	new	n	logn	
static counting (type2)	previous	nlogn	logn+i	
	new	n	logn+i	
2-D RANGE SEARCH				
static counting	previous	n^3logn	logn	
	new	n^3	logn	
dynamic reporting	previous	nlog^{3}n	$\log^2 n$ +ilogn	$\log^4 n$
	new	nlogn	$\log^2 n$ +ilogn	$\log^2 n$
static counting (type 2)	new	nlogn	$\log^2 n$ +ilogn	
ORTHOGONAL SEGM INTERSECTION				
static counting	previous	n^2logn	logn	
	new	n^2	logn	

Table 1: Isooriented objects

Generalized Problem		Space	Time (Reporting)	Time (Counting)
INTERSECTIONS OF COLORED LINES WITH A LINE SEGMENT				
VERTICAL SEGMENT				
	previous	n^2logn	logn+i	
		$n^{2-\mu/2}$logn	n^μ+i	
	new	n^2	logn+i	
		$n^{2-\mu}$logn	n^μ+i	
	new	$n^{2-\mu}\log^2 n$		$n^\mu \log^2 n$
ARBITRARY SEGMENT				
	new	n^2	logn+ilogn	
TRIANGLE STABBING				
	new	nlogn	$n^{2/3+\delta}(i+1)^{1/3-\delta}$	
		$n+\chi$	logn+i	logn
		$n^{2-\mu}$	n^μlogn+i	n^μlogn

Table 2: Non Isooriented objects

Tables 1,2: All previous bounds are obtained in [6] and [7].
Notation :
n: input size,
i: output size,
χ: number of pairwise intersecting line segments,
μ: adjustable parameter in the range $(0,1)$,
δ: arbitrarily small positive constant.

2 Intersections of Colored Lines with a line Segment

2.1 Vertical line Segment

We consider the following problem: Given n colored lines in the plane we want to preprocess them so that we can report the colors of the lines that are intersected by a vertical query segment. A solution of $O(n^2 \log n)$ (resp. $O(n^{2-\mu/2} \log n)$) space and $O(\log n + i)$ (resp. $O(n^\mu + i)$) time was given in [7]. We improve upon the above bounds while using the same approach as in [7].

Previous Approach. We will review the solution given in [7]. The arrangement of the n lines has size $O(n^2)$. For each of the $O(n^2)$ intersection points we draw vertical lines. In each of the vertical strips defined by the vertical lines the colored lines can be ordered from top to bottom. Between two neighboring strips two consecutive lines are switched in the order in each of the strips. To each strip we associate the structure of [6] for one-dimensional color range reporting, defined on the y-order of the colored lines in that strip. This structure has a query and update time of $O(\log n)$. We sweep over the arrangement and at each strip we associate a persistent version of the structure using the techniques of [5]. Between two consecutive strips $O(1)$ updates of the structure are made each requiring $O(\log n)$ time and space. So the total space needed is $O(n^2 \log n)$.

To answer a query the strip that contains the x-coordinates of the query segment is located in $O(\log n)$ time. In the same time we locate the line below the upper endpoint of the segment and the one above the lower endpoint. Then a query for the range between the lines is performed on the instance of the structure for that strip. This query reports the colors of the lines intersected by the segment in $O(\log n + i)$ time.

To tradeoff query time for space the following method is applied: we choose a parameter $s = n^\mu$, and we compute the polygonal lines described below: Let $\varepsilon_1, \varepsilon_2, ..., \varepsilon_n$ be the order of the lines in the leftmost strip. We choose the subset $\varepsilon_1, \varepsilon_{s+1}, \varepsilon_{2s+1}, ...$. Let I be the leftmost intersection of one of the lines (eg. ε_{is+1}) in the subset with another line ε. Then we replace ε_{is+1} with ε in the subset and proceed as above for the next intersection. The line segments encountered during this procedure constitute a set of $\lceil n/s \rceil$ x-monotone polygonal lines called border lines which have the following property: Any vertical line is divided by the border lines into $\lceil n/s \rceil$ intervals and each interval (except the last one) is intersected by $s-1$ lines (Fig. 1). We define vertical strips by drawing vertical lines through each vertex of a polygonal line. At any strip the region between two consecutive border lines contains a hammock (see eg. [2]) of $s-1$ original lines. We store these lines in an unordered list associated with the region between the border lines. We observe that at a vertex of a border line the list of lines for the region above the border line and on the left of the vertex differs only in one element from that on the right of the vertex. The same holds for the lists of the two regions below the border line. All other regions contain the same lines on the left and on the right strip.

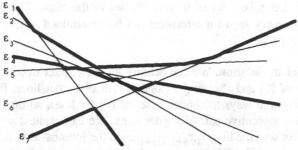

Fig. 1. A set of border lines (drawn bold) for s=2

The persistent versions of the 1-dimensional range tree are constructed for the new strips, by sweeping the arrangement as before. We make the lists of the regions persistent too at $O(1)$ space cost per strip. The size of all border lines can be bounded by $O(n^{2-\mu/2})$ as shown in [7], so the space is bounded by $O(n^{2-\mu/2}\log n)$. To answer a query we locate the strip containing the x-coordinate of the segment, as before. We locate the border line ε_u below the upper endpoint of the segment and the one ε_l above the lower endpoint. We query the persistent range structure for the interval between the two border lines in $O(\log n+i)$ time. Then we scan the list of lines in the region between ε_u and ε_{u+1} and that between ε_{l-1} and ε_l each of size $s-1=n^\mu-1$ reporting the colors of the lines that are intersected by the segment. The total time is $O(\log n+n^\mu+i)=O(n^\mu+i)$.

A Simpler Tradeoff Scheme. We first show a simpler tradeoff scheme: we divide the lines into s sets each of size $\lceil n/s \rceil$, in the following way: Let $\varepsilon_1,\varepsilon_2,...,\varepsilon_n$ be an order of the lines so that lines of the same color are in consecutive positions. Take the first $\lceil n/s \rceil$ lines as the first set, the second $\lceil n/s \rceil$ lines as the second set and so on. Note that if a color has c instances then it appears in at most $1+\lceil cs/n \rceil$ sets (if $\lceil cs/n \rceil$ >1 then $\lceil cs/n \rceil$-1 sets contain only lines of the same color). For each set of lines we use the method of [7]. So we get a $O(s(n/s)^2\log n)=O((n^2/s)\log n)$ space bound. The query time becomes $O(s\log n+i)$ since each color of more than $\lceil n/s \rceil$ instances is reported at most $1+\lceil cs/n \rceil$ times which we charge to the search time for the $\lceil cs/n \rceil$-1 sets containing only this color. For $s=n^\mu$ we get a $O(n^{2-\mu}\log n)$ space, $O(n^\mu\log n+i)$ query time algorithm, which is better than the one in [7].

For the counting problem we partition the lines into sets so that no two sets contain lines of the same color unless they contain only lines of the same color. In this way we make sure that each intersected color is counted once. It is easy to find such a partition of $\Theta(s)$ sets of size $\lceil n/s \rceil$ each. Then we only need to solve the counting problem for every set and combine the counts. By replacing the 1-d range reporting structure in the approach of [7] with its counting version [6] the counting problem for each set can be solved in $O((n/s)^2\log^2 n)$ space and $O(\log^2 n)$ query time. For $s=n^\mu$ the total space needed is $O(n^{2-\mu}\log^2 n)$ and the query time $O(n^\mu\log^2 n)$.

Lemma 2.1: Let S be a set of n colored lines in the plane. The number of colors that a vertical query segment intersects can be determined in $O(\log^2 n)$ time using $O(n^{2-\mu}\log^2 n)$ space.

A more Efficient Solution. We can achieve an even better tradeoff by returning to the approach of [7] and choosing an optimal set of border lines. First observe that there are s potential ways to choose a set of border lines, so that there are s lines between each consecutive pair of border lines. We can create s sets of border lines by choosing for set i the lines $\varepsilon_i, \varepsilon_{s+i}, \varepsilon_{2s+i}, ...$ in the leftmost strip and continuing the construction as in [7]. We can prove that no two sets of border lines contain a common segment. For suppose there was such a segment belonging both to a border line in set i_1 and to a border line in set i_2. Then the number of lines above the segment is $ks+i_1$ (for some k) since it belongs to set i_1 and $\lambda s+i_2$ (for some λ) since it belongs to i_2. This can be true only if $i_1=i_2$.

The union of the border lines of all sets is a subset of the arrangement (in fact it is the arrangement itself, that is the sets are a partition of the arrangement), so its size is $O(n^2)$. Since no segment belongs to two sets of border lines, there exists a set of size $O(n^2/s)$. We choose this set of border lines and apply the algorithm of [7]. So we get a $O((n^2/s)\log n)$ space and $O(\log n+s+i)$ query time algorithm. For $s=\log n$ we have a $O(n^2)$ space and $O(\log n+i)$ time algorithm. For $s=n^\mu$ we get an $O(n^{2-\mu}\log n)$ space, $O(n^\mu+i)$ time algorithm. The following theorem summarizes our result

Theorem 2.1: Let S be a set of n colored lines in the plane. The colors of the lines that a vertical query segment intersects can be reported in $O(n^2)$ (resp. $O(n^{2-\mu}\log n)$) space and $O(\log n+i)$ (resp. $O(n^\mu+i)$) time where i is the number of colors reported and μ an adjustable parameter in the range $(0,1)$.

It is interesting to note that we can solve the same problem using duality. To be more specific let us rotate the coordinates so that the query segment becomes horizontal instead of vertical. Let us also assume that none of the lines is vertical. Using duality each line is mapped to a point and the segment is mapped to a double wedge. It is easy to prove that for every horizontal segment the dual double wedge consists of two wedges with their common point on the y-axis. So using the terminology of [4] the problem reduces to reporting the colored points contained in two *based wedges*. With a simple modification of the method described in [4] we can reduce the problem to one dimensional searches in a set of lists, and using persistence we can achieve the same bounds as before. If there are lines that become vertical after the rotation it is easy to see that we can process them separately using the one dimensional color reporting structure of [7] and combine the answers.

2.2 Arbitrary line Segment

In this section we show how to preprocess a set S of n colored lines in the plane so that the i distinct colors of the lines that are intersected by an arbitrary query segment can be reported efficiently. The problem is equivalent to reporting the i distinct colors of n colored points that are contained in a query double wedge. Here we give a solution that uses $O(n^2)$ space and $O(\log n + i \log n)$ time where i is the output size.

In preprocessing we store the distinct colors of the input line set S at the leaves of a balanced binary tree T (in no particular order). For any node v of T let S(v) be the subset of lines of S stored at the leaves of v's subtree. At v we store a data structure D(v) in order to solve the following problem P: "does the query segment intersect any line of S(v)?". D(v) returns "yes" if the query segment intersects any line in S(v) else it returns "no".

To build D(v) we use the following approach:
The set S(v) gives rise to a planar graph G(v) called the line graph of S(v). Chazelle et al [3] show that G(v) can be set up in $O(n^2)$ time and space given S(v) as input. We assume that the line graph G(v) is represented in a suitable data structure so that in $O(\log n)$ time we can perform point location (see e.g. [12]). Then the algorithm is clear: we query G(v) with the two endpoints of the query segment. Due to the convexity of the regions the query segment intersects any line in G(v) if and only if the two endpoints belong to different regions.

So to answer the query we begin from the root and perform the query P. If the answer is "yes" then we repeat the process for the two children of the root else we stop. If we reach a leaf and the answer is "yes" we report the color of this leaf.

The correctness of the algorithm is straightforward. The space can be easily seen to be $O(n^2 \log n)$ while the query time is $O(\log n + i \log^2 n)$. An improvement to $O(\log n + i \log n)$ can be made by augmenting the tree of line graphs using fractional cascading as in [11] and the space can be reduced to $O(n^2)$ by doing a more careful building of the binary balanced tree T in such a way that the number of lines stored in each node decreases geometrically as we descend the tree.

The following theorem summarizes our result.

Theorem 2.2: The colors of the lines that an arbitrary query segment intersects can be reported in $O(n^2)$ space and $O(\log n + i \log n)$ time where i is the number of colors reported and n the number of the lines.

3 Generalized Triangle Stabbing

We are given a set of colored triangles in the plane and we want to preprocess them so that the colors of the triangles intersected by a query point can be reported efficiently. For fat triangles there is a $O(n^{3/2} \log n)$ space, $O(\log^2 n + i)$ time algorithm in [7] while no solution is known for arbitrary triangles. Here we give two algorithms with space $O(n \log n)$ (resp. $O(n^2)$) and time $O(n^{2/3+\delta}(i+1)^{1/3-\delta})$ (resp. $O(\log n + i)$).

3.1 A Space Efficient Solution

We build a balanced binary tree storing the triangles in its leaves so that all triangles of the same color appear in adjacent leaves. To each internal node we associate the triangles stored in the leaves of its subtree. We will use the "implicit point location" technique developed in [8]. This technique preprocesses n triangles in the plane in a data structure of $O(n)$ size so that we can determine in $O(n^{2/3+\delta})$ (for some δ) time whether a query point is contained in the union of the triangles. We build an instance of this structure for the set of triangles associated to each node. To answer a query we begin with the root and perform an implicit point location query. If the answer is "yes" we repeat the process for the two children of the root else stop. If a node contains only one color then we stop the recursion at the node and if the answer is "yes" report the color of the node. If we reach a leaf and the answer is "yes" we report the color of the leaf.

The correctness of the algorithm is straightforward. The space can be easily seen to be $O(n\log n)$ while a careful analysis can prove the $O(n^{2/3+\delta}(i+1)^{1/3-\delta})$ time bound.

Theorem 3.1: The generalized triangle stabbing problem for n triangles can be solved in $O(n\log n)$ space and $O(n^{2/3+\delta}(i+1)^{1/3-\delta})$ time.

3.2 A time Efficient Approach

Using more space (at most quadratic) we can achieve logarithmic time. First we compute the Union of the triangles of the same color for each color. This consists of a set of polygons possibly with holes. We compute the intersections between the polygons in $O(n+\chi)$ space and $O(n\log n+\chi)$ time ([2]), where χ is the number of pairwise segment intersections. What we get is a partition of the plane into a number of regions. We observe that the colors intersected by a point in a region are the same for all points in the same region. We store these colors in an unordered list. Furthermore lists of neighboring regions differ by at most one element. So by making the lists fully persistent we may create each list of a region by updating the list of any of its neighbors at $O(1)$ space cost. We may start from any region (the background region could be a good choice) and traverse the dual graph in a depth first manner. The construction can be completed in $O(n+\chi)$ time and space. Using a point location structure ([10, 12]) we can locate the region containing a query point in $O(\log n)$ time needing $O(n+\chi)$ space. Then in $O(i)$ time we report the colors in the list of the region. Note that in the worst case the space needed is $O(n^2)$. We can get a tradeoff between space and time by employing the simple tradeoff scheme of section 2. Then we get a worst case $O(n^{2-\mu})$ space and $O(n^{\mu}\log n+i)$ time. So we have proved the following:

Theorem 3.2: The generalized triangle stabbing problem for n triangles can be solved in $O(n^{2-\mu})$ space and $O(n^{\mu}\log n+i)$ time for $0\leq\mu<1$.

Remark 1: The counting version of the problem can be solved if we just store a count of the colors for each region, in $O(n+\chi)$ space and $O(\log n)$ time.
Remark 2: With a more careful partition of the lines into $\Theta(s)$ sets of size at most $\lceil n/s \rceil$ (as in section 2) for $s=n^{\mu}$ we can solve the counting problem too in $O(n^{2-\mu})$ space and $O(n^{\mu}\log n)$ time.

4 1-D Range type-2 Counting Problem

In this section we will show how to preprocess a set S of n colored points on the real line into a data structure of size O(n) so that for any query interval q=[a,b] the type-2 counting problem can be solved in time O(logn+i). The solution is static. The previous upper bounds for this problem consist of a O(nlogn) space, O(logn+i) query time structure proposed in [6].

The solution is simple and it consists of two Priority Search Trees and an auxiliary array A of size n. Let us name the two priority search trees PST1, PST2.

PST1 is built according to the following transformation:
For each color c, we sort the distinct points of that color by increasing x-coordinate. For each point p of color c, let pred(p) be its predecessor in the sorted order; for the leftmost point of color c, we take the predecessor to be the point -∞. We then map p to the point p'=(p,pred(p)) in the plane and associate with it the color c. Let S' be the resulting set of points. Then PST1 is a priority search tree built to answer queries of the form q'=[a,b] x (-∞,k). Let p'=(p,pred(p)) the point that is stored in a node v of PST1, and c the color of p. Then in node v in addition to p' we store a number t(p') which is the number of points of color c that lie to the right of p. It should be noted that the transformation used is the same as in [6].

PST2 is built according to the following transformation:
For each color c, we sort the distinct points of that color by increasing x-coordinate. For each point p of color c, let next(p) be its successor in the sorted order; for the rightmost point of color c, we take the successor to be the point +∞. We then map p to the point p'=(p,next(p)) in the plane and associate with it the color c. Let S' be the resulting set of points. Then PST2 is a priority search tree built to answer queries of the form q'=[a,b] x (k,+∞). Let p'=(p,next(p)) the point that is stored to a node v of PST1, and c the color of p. Then in node v in addition to p' we store a number t'(p') which is the number of points of color c that lie to the right of next(p).

To answer a query q=[l,r] we do the following steps:
a) We query PST1 with [l,r] x (-∞,l) and for each color c found we set A[c]=t(p') where p' the point stored in the node where we found c.
b) We query PST2 with [l,r] x (r,+∞) and for each color c found, we report c, and A[c]-t'(p'), where p' the point stored in the node where we found c.

The correctness of the above algorithm follows by simple geometric arguments. The space and query time can be easily bounded as stated in the following:

Theorem 4.1: The static 1-D range type-2 counting problem for a set of n colored points can be solved in O(n) space and O(logn+i) time.

5 1-D Static and Dynamic Range Counting and Applications

In [6] the generalized reporting (counting) problem was solved employing reduction to the standard grounded range reporting (counting) in two dimensions: each c-colored point on the real line is transformed to a point (p,pred(p)) in the plane, with pred(p) the predecessor of p with the same color c. So, in [l,r] there is a point of color c if and only if there is a point of color c in q'=[l,r]x(-∞,l), which is unique (if exists). Besides the auxiliary data structures that keep the correspondence between the original set S and the new set S' and have linear space and logarithmic update/query time, the authors used for the standard static (dynamic) 2-d counting a two level data structure. The points s ∈ S' are stored in non decreasing x-order at the leaves of a balanced binary search tree T (BB[á] in the dynamic case) and at each internal node v there is an array A(v) (resp. balanced binary search tree D(v)) containing the points in v's subtree in non decreasing y-order (resp. the same at leaves and at intermediate nodes a count of the number of points in their subtree). To answer a query, they determine O(logn) canonical nodes v in T such that [l,r] covers v's range but not the range of v's parent and then use A(v) (resp. D(v)) to count the number of points (that is, distinct colors) with y-coordinate less than l. This approach has O(nlogn) space and O(logn) query time in the static case and O(nlogn) space and O($\log^2 n$) query/update time in the dynamic one, determining the overall space/time complexity of the problem. But the space bounds can be improved by using instead the data structures of Chazelle ([1]) for static and dynamic 2d range counting that have the same query/update times but require only linear space. So,

Lemma 5.1: Let S be a set of n colored points on the real line. S can be preprocessed into a data structure of size O(n) such that the number of the distinctly colored points of S that are intersected by any query interval can be determined in O(logn) time in the static case and in O($\log^2 n$) time in the dynamic case with O($\log^2 n$) update time.

Immediate applications of this result can be found in the static 2d range counting and the static orthogonal segment counting problem. In [6] two of the proposed solutions use O(n^2) and O(n) instances of the static 1-d range counting data structure respectively. Thus the space complexities are O(n^3logn) and O(n^2logn) in each case whereas queries can be answered by locating in O(logn) time the correct 1-d data structure where in O(logn) time the involved colors are determined. So, employing the data structure of lemma 5.1 we have:

Corollary 5.1: (a) A set of n colored points in the plane can be preprocessed into a data structure of size O(n^3) such that the number of distinctly colored points in any axes parallel query rectangle q=[a,b]x[c,d] can be determined in O(logn) time. (b) A set S of n colored horizontal line segments in the plane can be preprocessed into a data structure of size O(n^2) such that the number of distinctly-colored segments that are intersected by a vertical query segment can be determined in time O(logn).

6 2-D Dynamic Range Reporting

Let S be a set of colored points in the plane. We consider the problem of reporting the colors of the points of S contained in an axes-parallel query rectangle, while allowing for insertions and deletions of points. A solution with $O(nlog^3n)$ space, $O(log^2n+ilogn)$ query time and $O(log^4n)$ update time was proposed in [6]. We present a simple solution with $O(nlogn)$ space, $O(log^2n)$ update time and the same query time.

Our data structure consists of a balanced tree T built on the x-coordinates of the points. The x-range of a node v of T is the range between the x-coordinate of the leftmost leaf in its subtree and the x-coordinate of the rightmost leaf. We store the points that have x-coordinate in xrange(v) in an auxiliary structure A_v associated with node v. We implement A_v as the structure of [6] for 1-dimensional range searching on the y-coordinates of the points. This structure occupies $O(n)$ space and has $O(logn+i)$ query time and $O(logn)$ update time.

To answer a query [a,b] x [c,d] we can find $O(logn)$ nodes that their x-ranges are disjoint and their union is [a,b] in $O(logn)$ time. For each such node v we query A_v with the range [c,d]. The query time is $O(logn+i)$ for each node, which sums up to a total $O(log^2n+ilogn)$ since each color in the answer may be reported once in each of the $O(logn)$ nodes. The size of the structure is $O(nlogn)$ since each point is stored in $O(logn)$ auxiliary structures and the auxiliary structures have linear size. By implementing T as a BB[α] tree we allow insertions /deletions for an $O(log^2n)$ amortized update time. By employing the techniques of Willard and Luecker [13] this time becomes worst case.

Theorem 6.1: The dynamic generalized 2D range searching problem can be solved using $O(nlogn)$ space in $O(log^2n+ilogn)$ time and $O(log^2n)$ update time.

By replacing the structure A_v, with the 1-dimensional type-2 counting structure of section 4 we solve the 2-dimensional type-2 counting problem at the same time and space. Since this structure is static the whole one is static too.

Theorem 6.2: The static type-2 range counting problem can be solved using $O(nlogn)$ space and $O(log^2n+ilogn)$ query time.

7 Conclusion and Further work

We have presented efficient solutions to several generalized interection searching problems involving objects that are either isooriented or non isooriented. Several of our bounds are improvements upon previous results in [6, 7] whereas the rest refer to solutions of new generalized problems.

Some interesting open problems are: (i) to solve the 2-d dynamic range search reporting problem with query time of the form $O(polylog(n)+i)$ and (ii) obtain dynamic data structures for intersection problems involving non isooriented objects.

References

1. B. Chazelle. A functional approach to data structures and its use in multidimensional searching. SIAM Journal on Computing 17, 3, 427-462 (1988)

2. B. Chazelle and H. Edelsbrunner. An optimal algorithm for intersecting line segments in the plane. Journal of the ACM, 39,1, 1-54 (1994)
3. B. Chazelle, L. Guibas, and D. T. Lee. The Power of Duality. In: Proceedings of the 24th IEEE Annual Symposium on Foundations of Computer Science, 1983, pp. 217-225.
4. R. Cole, C. K. Yap. Geometric Retrieval Problems. Information and Control 63, 39-57 (1984)
5. J.R. Driscoll, N. Sarnak, D.D. Sleator, and R.E. Tarjan. Making data structures persistent. Journal of Computer and System Sciences 38, 86-124 (1989)
6. P. Gupta, R. Janardan, and M. Smid. Further results on generalized intersection searching problems: counting, reporting, and dynamization. In: Proceedings of the 1993 Workshop on Algorithms and Data Structures. August 1993, pp. 361-372.
7. P. Gupta, R. Janardan, and M. Smid. Efficient Algorithms for Generalized Intersection Searching on Non-Iso-oriented Objects. In: Proceedings of the 10th ACM Symposium on Computational Geometry. 1994, pp. 369-377
8. L. Guibas, M. Overmars, and M. Sharir. Intersecting line segments, ray shooting and other applications of geometric partitioning techniques. In: Proceedings SWAT '88. Lecture Notes in Computer Science 318: Springer Verlag, Berlin: 1988, pp. 64-73
9. R. Janardan, M. Lopez. Generalized intersection searching problems. International Journal of Computational Geometry and Applications 3, 39-69 (1993)
10. D. Kirkpatrick. Optimal search in planar subdivisions. SIAM Journal on Computing 12, 1, 28-35 (1993)
11. M.S. Paterson, F.F. Yao. Point retrieval for polygons. Journal of Algorithms 7, 441-447 (1986)
12. N. Sarnak, R. E. Tarjan. Planar point location using persistent search trees. Communications of the ACM 29, 669-679 (1986)
13. D. E. Willard, G. S. Lueker. Adding range restriction capability to dynamic data structures. Journal of the ACM 32, 597-617 (1985)

OKFDDs versus OBDDs and OFDDs [*]

Bernd Becker[1], Rolf Drechsler[1], Michael Theobald[2]

[1] Dept. of Computer Science
Johann Wolfgang Goethe-University
60054 Frankfurt am Main, Germany
[2] Dept. of Computer Science
Columbia University
New York, NY 10027, USA

Abstract. *Ordered Decision Diagrams (ODDs) as a means for the representation of Boolean functions are used in many applications in CAD. Depending on the decomposition type, various classes of ODDs have been defined, the most important being the Ordered Binary Decision Diagrams (OBDDs), the Ordered Functional Decision Diagrams (OFDDs) and the Ordered Kronecker Functional Decision Diagrams (OKFDDs). In this paper we clarify the computational power of OKFDDs versus OBDDs and OFDDs from a (more) theoretical point of view. We prove several exponential gaps between specific types of ODDs. Combining these results it follows that a restriction of the OKFDD concept to subclasses, such as OBDDs and OFDDs as well, results in families of functions which lose their efficient representation.*

1 Introduction

Ordered Decision Diagrams (ODDs) as a data structure for the representation and manipulation of Boolean functions are applied in many fields of electronic design automation (see [7] for an overview). The most popular type of ODD is the *Ordered Binary Decision Diagram* (OBDD) [5]. The more recent techniques, like dynamic variable ordering by sifting [17], have made it possible to handle (some) large functions without any basic variation of the OBDD concept itself. However, sifting tends to be very time consuming and the representation of large functions remains problematic; for certain classes of functions, notably multipliers, it has been known that OBDDs will be of exponential size irrespective of the order of the variables [6]. Thus research has been geared towards variations of OBDDs. Among these variations there are those that utilize less restricted Decision Diagrams and there are other ones which augment BDDs with additional constructs. The constructs such as General BDDs [8], or pBDDs [11], IBDDs [12], XBDDs [13], and free BDDs [18] (also known as "1-time branching programs") remove the ordering constraint on BDDs at the expense of loosing the canonicity of the structure.

[*] The first and second author were supported by DFG grant Be 1176/4-2.

On the other hand there have been recent attempts at varying the nodes in OBDDs. While OBDDs are based on a recursive application of the Shannon decomposition, other decomposition types can be used to define new types of ODDs. These include Ordered Functional Decision Diagrams (OFDDs) [10], positive OFDDs (pOFDDs) [15], negative OFDDs (nOFDDs) and Ordered Kronecker Functional Decision Diagrams (OKFDDs) [9]. OFDDs, pOFDDs, nOFDDs and OKFDDs, all define canonical representations of Boolean functions. The nodes in OFDDs (pOFDDs, nOFDDs) are decomposed by (positive, negative) Davio decompositions, in OKFDDs positive, negative Davio as well as Shannon decompositions are allowed. OKFDDs thus are potentially more compact than OBDDs and OFDDs, both special kinds of OKFDDs. Recently, it has been shown [2], that OKFDDs are the most general type of ODD, that can be obtained by a variation of the decomposition type. In other words, all other decomposition types (like e.g. the equivalence decomposition [14]) are special cases of positive, negative Davio and Shannon decomposition. In this sense it is interesting and important to determine the computational power of OKFDDs versus OBDDs and OFDDs.

First promising results in this direction have been obtained in [9], where it is demonstrated that OKFDDs allow much more succinct representations of real Benchmark functions than OBDDs and pOFDDs. The relation between OBDDs and pOFDDs is studied in [3] from a more theoretical point of view. Exponential trade-offs between OBDDs and pOFDDs for a class of Boolean functions are proven. It follows that depending on the function class OBDDs and pOFDDs as well might be advantageous.

We consider trade-offs in the more general context of OKFDDs in this paper. The relation between OBDDs and OKFDDs (with a list d of decomposition types) is described by a bijective Boolean transformation τ_d, called the generalized τ-operator. We use this property to derive several results on the computational power of OKFDDs and subclasses thereof. Classes of (generalized clique) functions are given that have exponentially more concise OFDDs than OBDDs, and vice versa, independent of the ordering of the variables. In particular, the exponential gaps between OBDDs and pOBDDs as proved in [3] follow as a special case. Combining our results it can be concluded that a restriction of the OKFDD concept to subclasses, such as OBDDs, pOFDDs, nOFDDs and OFDDs as well, results in families of functions which lose their efficient representation. Thus, OKFDDs in full generality offer significant advantages in terms of representation size.

The paper is structured as follows: In Sect. 2 (ordered) Kronecker Functional Decision Diagrams and subclasses are introduced. Section 3 studies the relation between the different data structures defined in Sec. 2. Based on the generalized τ-operator ODD-sizes of specific function classes are compared in Sect. 4. We finish with a resume of the results in Sect. 5.

2 Kronecker Functional Decision Diagrams

In the following, basic definitions of Decision Diagrams and (ordered) KFDDs in particular are presented.

Definition 1. A *Decision Diagram* (DD) over $X_n := \{x_1, x_2, \ldots, x_n\}$ is a rooted directed acyclic graph $G = (V, E)$ with vertex set V containing two types of vertices, *non-terminal* and *terminal* vertices. A non-terminal vertex v is labeled with a variable from X_n, called the *decision variable* for v, and has exactly two successors denoted by $low(v), high(v) \in V$. All nodes with label x_i are denoted as *level i*. A terminal vertex v is labeled with a 0 or 1 and has no successors. The size of a DD, denoted by $|DD|$, is given by its number of non-terminal nodes.

Further structural restrictions turn out to be important.

Definition 2. A DD is *free* if each variable is encountered at most once on each path in the DD from the root to a terminal vertex. A DD is *complete* if each variable is encountered exactly once on each path in the DD from the root to a terminal vertex. A DD is *ordered* if it is free and the variables are encountered in the same order on each path in the DD from the root to a terminal vertex. Letter O will be used to denote ordered DDs.

DDs can be related to Boolean functions by using the following well-known decomposition types (given here for an arbitrary Boolean function $f : \mathbf{B}^n \to \mathbf{B}$ over the variable set X_n):

$$
\begin{aligned}
f &= \bar{x}_i f_i^0 + x_i f_i^1 & \text{Shannon } (S) \\
f &= f_i^0 \oplus x_i f_i^2 & \text{positive Davio } (pD) \\
f &= f_i^1 \oplus \bar{x}_i f_i^2 & \text{negative Davio } (nD)
\end{aligned}
$$

where f_i^0 (defined by $f_i^0(x) := f(x_1, .., x_{i-1}, 0, x_{i+1}, .., x_n)$ for $x = (x_1, x_2, \ldots, x_n) \in \mathbf{B}^n$) denotes the *cofactor* of f with respect to $x_i = 0$, f_i^1 denotes the cofactor for $x_i = 1$ and f_i^2 is defined as the Boolean difference $f_i^2 := f_i^0 \oplus f_i^1$, $\oplus$ being the Exclusive OR operation.

Decomposition types are associated to the variables in X_n with the help of a *Decomposition Type List* (DTL) $d := (d_1, \ldots, d_n)$ where $d_i \in \{S, pD, nD\}$.

Now, the Kronecker Functional Decision Diagrams can formally be defined as follows:

Definition 3. A *KFDD* over X_n is given by a DD over X_n together with a fixed DTL $d = (d_1, \ldots, d_n)$. The function $f_G^d : \mathbf{B}^n \to \mathbf{B}$ represented by a KFDD G over X_n with DTL d is defined as:

1. If G consists of a single node labeled with 0 (1), then G is a KFDD for the constant 0 (1) function.

2. If G has a root v with label x_i, then G is a KFDD for

$$\begin{cases} \overline{x}_i f_{low(v)} + x_i f_{high(v)} & : \quad d_i = S \\ f_{low(v)} \oplus x_i f_{high(v)} & : \quad d_i = pD \\ f_{low(v)} \oplus \overline{x}_i f_{high(v)} & : \quad d_i = nD \end{cases}$$

where $f_{low(v)}$ $(f_{high(v)})$ are the functions represented by the KFDD with DTL d rooted at $low(v)$ $(high(v))$.

Of course, a KFDD is an OKFDD iff the underlying DD is ordered.

A node in an OKFDD is called an *S-node* if it is expanded by Shannon decomposition. It is called a *D-node* if it is expanded by positive or negative Davio decompositions; the latter being an *nD-node* and the former a *pD-node*. Due to the decomposition type list, at every node of a fixed level i in the KFDD, the same decomposition of type d_i is applied.

The definition directly infers that KFDDs are a generalization of BDDs and FDDs: If in all levels only Shannon decomposition is applied (i.e. $d = d_S := (S, \ldots, S)$), the KFDD will be a BDD. $f_G^{BDD} := f_G^{d_S}$ then is used to denote the BDD-function of G. If only Davio decompositions are applied (i.e. $d_i \in \{pD, nD\}$ for all i), the KFDD will be an FDD and $f_G^{dFDD} := f_G^d$ denotes the FDD-function of G (for DTL d). Analogously pFDDs and nFDDs are defined for DTLs $d = d_{pD} := (pD, \ldots, pD)$ and $d = d_{nD} := (nD, \ldots, nD)$, respectively. f_G^{pFDD} and f_G^{nFDD} are the pFDD-function and nFDD-function of G, respectively. In summary, we obtain the following relation between the different classes of KFDDs:

$$\begin{array}{ll} \text{pFDDs} \subset \text{FDDs} \subset \\ \text{nFDDs} \subset \text{FDDs} \subset & \textbf{KFDDs} \\ \text{BDDs} \subset \end{array}$$

Reductions of three different types are used to reduce the size of KFDDs:

Type I: Delete a node v' whose label is identical to the label of a node v and whose successors are identical to the successors of node v and redirect the edges pointing to v' to point to v.

Type S: Delete a node v' whose two outgoing edges point to the same node v and connect the incoming edges of the deleted node to v.

Type D: Delete a node v' whose successor $high(v')$ points to the terminal 0 and connect the incoming edges of the deleted node to the successor $low(v')$.

While each node in a KFDD is a candidate for the application of reduction type I, only S-nodes (D-nodes) are candidates for the application of the reduction type S (type D). A KFDD is *reduced* iff no reductions can be applied to the KFDD. Two KFDDs G_1 and G_2 (with the same DTLs) are called *equivalent* iff G_2 results from G_1 by repeated applications of reductions and inverse reductions.

A KFDD G_2 is called the *reduction* of a KFDD G_1, iff G_1 and G_2 are equivalent and G_2 itself is reduced.

From [9], it is known that reductions can be used to define canonical representations for not only OBDDs and OFDDs, but also for OKFDDs. Furthermore, it can be concluded from [3], that reductions can be computed in linear time in the size of the OKFDD.

Example 1. An OKFDD is shown in Fig. 1, where the left outgoing edge at each node denotes $f_{low(v)}$. The OKFDD represents the function $x_1 x_2 x_4 \oplus x_1 x_2 \overline{x}_3 \oplus x_1 \overline{x}_3 \oplus \overline{x}_1 x_2 x_4$. The S-node decomposes the function into $x_2 x_4$ and $x_2 x_4 \oplus x_2 \overline{x}_3 \oplus \overline{x}_3$, respectively. The latter is in turn decomposed into $\overline{x}_3$ and $\overline{x}_3 \oplus x_4$ through the pD-node, x_2. The nD-node x_3 on the right results in x_4 and 1.

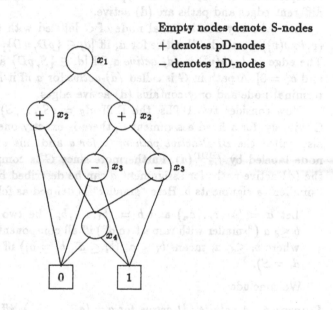

Empty nodes denote S-nodes
+ denotes pD-nodes
− denotes nD-nodes

Fig. 1. Example for OKFDD

3 Relation between BDDs, FDDs and KFDDs

In this section we investigate the relation between functions which are represented by the same DD G. Of course, changing the DTL d for G in general also changes the function represented by the DD. Nevertheless, it can be shown that the functions are related by the generalized τ-operator. In the next section we then use this relation to prove several exponential gaps between specific classes of DDs. For example, we show that there exists a class of functions which cannot be represented efficiently by OBDDs but can be by OFDDs (for a fixed DTL). This result can be extended to DTLs differing in each and every component. Furthermore, we show that there exist families of Boolean functions for which OBDD and OFDD (for a fixed DTL) representations both are exponential in size, while they have OKFDDs of polynomial size; it follows that it is advantageous to mix D- and S-nodes. In conclusion, a restriction of the OKFDD concept leads to functions which lose their efficient representation.

It is possible to establish a close relation between the functions f_G^{BDD} and f_G^d for G being a complete DD. For an intuitive approach to this problem consider an arbitrary complete DD G and a fixed DTL d. (For simplicity one may assume that G is the complete binary tree with 2^n leaves.) Then fix an assignment $a = (a_1, \ldots, a_n)$ to the variables $x_1, \ldots, x_n$. We call edges and paths *(d)-active* for a if they are considered for the computation of $f_G^d(a)$. Depending on the interpretation of a DD, i.e. the types of decompositions being valid at a node, different edges and paths are (d)-active.

Let v be any non-terminal node of G labeled with a variable x_i. The edge $(v, low(v))$ is called *(d)-active* for a, iff $[d_i \in \{pD, nD\}]$ or $[d_i = S$ and $a_i = 0]$. The edge $(v, high(v))$ is *(d)-active* a, iff $[d_i \in \{S, pD\}$ and $a_i = 1]$ or $[d_i = nD$ and $a_i = 0]$. A path in G is called *(d)-active* for a iff it leads from the root to a terminal node and only contains (d)-active edges.

Now consider two DTLs, the DTL $d_S = (S, \ldots, S)$ and a general DTL d. Obviously, for a fixed assignment a, there is exactly one (d_S)-active path in G, also called the *BDD-active path of G for a* and this path leads to a terminal node labeled by $f_G^{BDD}(a)$. Furthermore since G is complete, we conclude that the (d)-active paths for assignment a can be described by BDD-active paths for "smaller" assignments b. Here "smaller" is defined as follows:

Let $a = (a_1, \ldots, a_n)$ and $b = (b_1, \ldots, b_n)$ be two assignments. Then $b \leq_d a$ ("smaller with respect to d") iff all components i satisfy $b_i \leq_{d_i} a_i$, where $b_i \leq_{d_i} a_i$ means $b_i \leq a_i$ ($b_i \leq \overline{a_i}$, $b_i = a_i$) iff $d_i = pD$ ($d_i = nD$, $d_i = S$). [3]

We conclude:

Lemma 4. *A path is (d)-active for $a = (a_1, \ldots, a_n)$, iff it is BDD-active for an assignment $b = (b_1, \ldots, b_n)$ with $b \leq_d a$.*

Thus, $f_G^d(a)$ is defined (see Definition 3) by the parity of the number of BDD-active paths (for "smaller" assignments b) that lead to a terminal node that is marked by a 1. More precisely, $f_G^d(a) = \bigoplus_{b \leq_d a} f_G^{BDD}(b)$ and we have the relation between the functions represented by G if G is viewed as BDD and KFDD (with DTL d), respectively. We therefore define the *generalized τ-operator* [4] τ_d:

---

[3] For two elements x and y in $\mathbf{B}^n$ and $d \in \{S, pD\}^n$, it holds: $x \leq_d y \Leftrightarrow \overline{y} \leq_d \overline{x}$. Notice that this equivalence is not necessarily valid for general $d \in \{S, pD, nD\}^n$.

[4] The τ-operator τ_d for the special case $d = d_{pD} = (pD, \ldots, pD)$ was already used in [16] to analyze circuits over $\{\wedge, \oplus\}$, in [4] (as the Reed-Muller Transform) to synthesize two level circuits and in [3] to show the relation between the BDD-function and the pFDD-function.

Definition 5. Let $f \in \mathbf{B}_n$ and $d := (d_1, \ldots, d_n)$ with $d_i \in \{S, pD, nD\}$ for all i. The *generalized τ-operator* $\tau_d(f)$ is defined by

$$\tau_d(f)(x) := \bigoplus_{y \leq_d x} f(y).$$

It can easily be shown that τ_d is bijective and its inverse is given by $\tau_d^{-1}(f)(x)$ $= \bigoplus_{x \leq_d y} f(y)$. (For $d = d_{pD}$ it then follows that τ_d is identical to τ_d^{-1}.) A simple induction proof on the number of variables in G shows that the BDD-function and the KFDD-function of a complete DD G can be computed from each other through the generalized τ-operator.

Theorem 6. *For each complete DD G and each DTL d, it holds that* $f_G^d = \tau_d\left(f_G^{\mathrm{BDD}}\right)$ *and* $f_G^{\mathrm{BDD}} = \tau_d^{-1}\left(f_G^d\right)$.

It follows that the generalized τ-operator (applied to all DTLs d) provides the family of functions which are represented by the same complete DD. Furthermore, τ_d does neither depend on the variable ordering nor on the fact whether the considered complete DD is ordered or only free.

Theorem 6 is valid only for complete DDs. Fortunately, it can be shown that each free (ordered) KFDD (that is not complete) can be transformed into an equivalent complete KFDD, the size of which grows at most by a factor of $O(n)$. The idea for the transformation is simple: We only have to insert D- and S-reduced nodes by inverse reductions to make all paths complete. (For more details see [3], where the proof is given for pFDDs. The generalization to KFDDs is easy.)

4 Exponential Gaps

Combining Theorem 6 with the above mentioned transformation property, it is possible to transfer results about (free, ordered) BDDs to KFDDs and vice versa. In particular, the existence of families of functions can be proven which are *good* for OFDDs and *bad* for OBDDs and vice versa.

To do so, we need the following definitions. Consider n nodes $1, 2, \ldots, n$ and $n(n-1)/2$ Boolean variables $x_{i,j}$ $(1 \leq i < j \leq n)$. Assume that $x_{i,j} = 1$ denotes an edge between nodes i and j, whereas $x_{i,j} = 0$ indicates that the edge from i to j is missing. Then an assignment $x = (x_{1,2}, x_{1,3}, \ldots, x_{n-1,n}) \in \mathbf{B}^{n(n-1)/2}$ defines an undirected graph $G_x = (V, E_x)$ with $V = \{1, \ldots, n\}$ and $E_x = \{(i,j) | x_{i,j} = 1\}$. A group consisting of 3 nodes is called a 3-clique (or clique for simplicity), iff any two of them are connected by an edge. (The three nodes form a triangle in the graph.) Let $d = (d_{1,2}, d_{1,3}, \ldots, d_{n-1,n})$ be a DTL. Then the *polarity function* $pol_d : \mathbf{B}^{n(n-1)/2} \to \mathbf{B}^{n(n-1)/2}$ is defined by $pol_d(x_{1,2}, \ldots, x_{n-1,n}) := (x_{1,2}^d, \ldots, x_{n-1,n}^d)$ with

$$x_{i,j}^d := \begin{cases} x_{i,j} & : \quad d_{i,j} = S, d_{i,j} = pD \\ \overline{x}_{i,j} & : \quad d_{i,j} = nD \end{cases}$$

We are now ready to define the function $\oplus\text{-}cl_{n,3}^d$, the *parity-clique-function for DTL* $d = (d_{1,2}, \ldots, d_{n-1,n})$, which computes the parity of certain cliques, and the function $1\text{-}cl_{n,3}^d$, the *1-clique-function for DTL* d, which checks the existence of exactly one clique. More formally:

Definition 7. Let $d = (d_{1,2}, \ldots, d_{n-1,n})$ be a DTL.

i) $\oplus\text{-}cl_{n,3}^d : \mathbf{B}^{n(n-1)/2} \to \mathbf{B}$ is a function defined over the variables $x_{i,j}$ ($1 \leq i < j \leq n$) by: $\oplus\text{-}cl_{n,3}^d(x) = 1$, iff $G_{pol_d(x)}$ contains an odd number of cliques.

ii) $1\text{-}cl_{n,3}^d : \mathbf{B}^{n(n-1)/2} \to \mathbf{B}$ is a function defined over the variables $x_{i,j}$ ($1 \leq i < j \leq n$) by: $1\text{-}cl_{n,3}^d(x) = 1$, iff $G_{pol_d(x)}$ contains exactly three edges and these edges build a clique.

For simplicity we set $\oplus\text{-}cl_{n,3}^d =: \oplus\text{-}cl_{n,3}$ and $1\text{-}cl_{n,3}^d =: 1\text{-}cl_{n,3}$ for $d = d_{pD} = (pD, pD, \ldots, pD)$. Using these definitions we can prove by elementary computations:

Lemma 8. *Let d be any DTL with $d \in \{pD, nD\}^{n(n-1)/2}$. Then*

$$\oplus\text{-}cl_{n,3}^d = \tau_d(1\text{-}cl_{n,3})$$
$$1\text{-}cl_{n,3}^d = \tau_d(\oplus\text{-}cl_{n,3})$$

We are now ready to prove the following theorem:

Theorem 9 (Gaps between OBDDs and OFDDs). *For any order π of the variables $x_{i,j}$ and any DTL $d = (d_{1,2}, \ldots, d_{n-1,n}) \in \{pD, nD\}^{n(n-1)/2}$ it holds:*

i.) *$1\text{-}cl_{n,3}^d$ has OBDDs of size $O(n^5)$, but only free FDDs (with DTL d) of size $2^{\Omega(n^2)}$.*

ii.) *$\oplus\text{-}cl_{n,3}^d$ has OFDDs (with DTL d) of size $O(n^3)$, but only free BDDs of size $2^{\Omega(n^2)}$.*

Proof. ii.) At first, we show that the parity-clique-function $\oplus\text{-}cl_{n,3}^d$ for DTL d can be represented by an OFDD (with DTL d) of size of $O(n^3)$.

$\oplus\text{-}cl_{n,3}^d$ can be represented by the 2-level Fixed Polarity Reed-Muller Expression (FPRM)

$$\oplus\text{-}cl_{n,3}^d(x) = \bigoplus_{1 \leq i < j < k \leq n} x_{i,j}^d x_{j,k}^d x_{i,k}^d$$

The polarity of the variables in the FPRM is chosen according to the DTL d, i.e. if the variables have type pD, the polarity is positive, otherwise negative. From the fact that a reduced OFDD with DTL d for $\oplus\text{-}cl_{n,3}^d$ is a compressed representation of the FPRM - flattening of the OFDD yields the FPRM - we conclude that the number of nodes in a reduced OFDD is bounded by the number of literals in the FPRM, which on the other hand is bounded by

$3 \cdot \binom{n}{3}$ literals. (Note that this bound is independent of the variable ordering.) Thus, $\oplus\text{-}cl_{n,3}^d$ has OFDDs of size $O(n^3)$.

We now show that the parity-clique-function $\oplus\text{-}cl_{n,3}^d$ for DTL d requires exponential free BDDs. In [1] it was shown that each free BDD requires an exponential size ($2^{\Omega(n^2)}$) to represent the parity-clique-function $\oplus\text{-}cl_{n,3}^d$ for $d = d_{pD}$. We obtain a representation of $\oplus\text{-}cl_{n,3}^d$ (of $2^{\Omega(n^2)}$) for general d by exchanging the successors of every nD-node. Thus, $\oplus\text{-}cl_{n,3}^d$ is a function for that each free BDD representation has exponential size while each OFDD representation (for DTL d) has polynomial size.

i.) From Lemma 8 we know that $1\text{-}cl_{n,3}^d = \tau_d(\oplus\text{-}cl_{n,3})$, thus $1\text{-}cl_{n,3}^d$ only has free FDDs (with DTL d) of size $2^{\Omega(n^2)}$. On the other hand $\oplus\text{-}cl_{n,3}^d = \tau_d(1\text{-}cl_{n,3})$, thus $1\text{-}cl_{n,3}$ has polynomial OBDDs of size $O(n^5)$ - since τ_d operates on complete DDs, we need an additional factor of $O(n^2)$ compared to the OFDD-size of $O(n^3)$ for $\oplus\text{-}cl_{n,3}^d$ shown in part ii.) of the proof.

$\square$

It follows from the above theorem that it is advantageous to consider both OKFDDs with S-nodes and OKFDDs with D-nodes.

Above we have figured out an exponential gap between OBDDs and OFDDs with a fixed DTL d. It is an interesting question whether there also exists such an exponential gap between OFDDs with different DTLs. We now show that for OFDDs with DTL d and OFDDs with DTL $\overline{d}$ an analogous result holds. Here, the DTL $\overline{d}$ results from the DTL d by changing the decomposition types from pD to nD and vice versa.

Theorem 10 (Gaps between OFDDs with inverse DTLs). *Consider any order π of the variables $x_{i,j}$ and any DTL $d \in \{pD, nD\}^{n(n-1)/2}$. Define the DTL $\overline{d}$ by $\overline{d}_i := pD\ (nD)$ if $d_i = nD\ (pD)$. Then it holds:*

$1\text{-}cl_{n,3}^d$ has OFDDs with DTL $\overline{d}$ of size $O(n^5)$, but only free FDDs with DTL d of size $2^{\Omega(n^2)}$.

Proof. For the proof it suffices to show that $1\text{-}cl_{n,3}^d$ can be represented by OFDDs with DTL $\overline{d}$ of size $O(n^5)$. Consider a *Disjunctive Normal Form* (DNF) of $1\text{-}cl_{n,3}$ and the modified 1-clique-function $1\text{-}cl_{n,3}^d$. It is easy to see that the following holds:

$$1\text{-}cl_{n,3} = \sum_{1 \le i < j < k \le n} x_{i,j} x_{j,k} x_{i,k} \prod_{(l,m) \ne (i,j),(j,k),(i,k)} \overline{x}_{l,m}$$

Let $x^0 := x$ and $x^1 := \overline{x}$. Additionally, let $e_{i,j} := 0\ (e_{i,j} := 1)$, if $d_{i,j} = pD(nD)$. Then, it holds for the modified $1\text{-}cl_{n,3}^d$:

$$1\text{-}cl_{n,3}^d = \sum_{1 \le i < j < k \le n} x_{i,j}^{e_{i,j}} x_{j,k}^{e_{j,k}} x_{i,k}^{e_{i,k}} \prod_{(l,m) \ne (i,j),(j,k),(i,k)} x_{l,m}^{1 - e_{l,m}}$$

$$= \bigoplus_{1 \le i < j < k \le n} x_{i,j}^{e_{i,j}} x_{j,k}^{e_{j,k}} x_{i,k}^{e_{i,k}} \prod_{(l,m) \ne (i,j),(j,k),(i,k)} x_{l,m}^{1 - e_{l,m}}$$

$$= \bigoplus_{1 \le i < j < k \le n} (1 \oplus x_{i,j}^{1-e_{i,j}})(1 \oplus x_{jk}^{1-e_{jk}})(1 \oplus x_{ik}^{1-e_{ik}}) \prod_{\substack{(l,m) \neq \\ (i,j),(j,k),(i,k)}} x_{lm}^{1-e_{lm}}$$

An FPRM for the polarity defined by $\bar{d}$ results by multiplying out. Obviously, each literal in the FPRM is positive (negative), iff $\bar{d}_i = pD$ (nD). The number of and-products is $\le 8\binom{n}{3}$. Therefore the size of the FPRM is polynomial. Since the corresponding OFDD with DTL $\bar{d}$ is a compact representation of the FPRM, the size of the OFDD is bounded by the number of literals in the FPRM. Therefore, it holds: $|OFDD$ with DTL $\bar{d}| \le 8\binom{n}{3}\binom{n}{2} = O(n^5)$. □

The last two theorems are the basis for an important corollary. It states that there exist exponential gaps between the "pure" DDs, i.e. DDs containing only one decomposition type. Furthermore there is an exponential gap between DDs differing in all components of the DTL.

Corollary 11. *i.) There exists an exponential gap between each two of the "pure" DDs, i.e. OBDDs, pOFDDs and nOFDDs.*
ii.) There exists an exponential gap between two DTLs, d_1 und d_2, differing in all components.

Proof. i.) The exponential gap between OBDDs and pOFDDs as well as between OBDDs and nOFDDs follows from Theorem 9 with DTLs $d = (pD, \dots, pD)$ and $d = (nD, \dots, nD)$, respectively. The exponential gap between pOFDDs and nOFDDs follows from Theorem 10 with the DTLs $d = (pD, \dots, pD)$ and $\bar{d} = (nD, \dots, nD)$.
ii.) Notice, that there are only 6 possibilities to change a decomposition type of the DTL d_1. The proof then follows from part i.) and the fact, that one of these 6 possibilities is applied to at least $n/6$ variables.

□

Summing up the results so far, we have shown that it is advantageous to consider OKFDDs with several different DTLs. A restriction to only one DTL results in functions that lose their efficient representation. We now consider the problem, whether a restriction to a few DTLs reduces the power of OKFDDs. We show that a few DTLs are not as powerful as OKFDDs.

For this, we first prove that mixed OKFDDs, i.e. OKFDDs that contain Shannon nodes as well as Davio nodes, are more powerful than the "union" of pure OKFDDs that contain only Shannon or only Davio nodes.

Theorem 12 (Gaps for mixed OKFDDs). *Consider a fixed DTL $d = (d_1, d_2, \dots, d_{n(n-1)+2})$ with $d_i \in \{pD, nD\}$. Then the following holds:*

There exists a Boolean function f_n over $n(n-1)+2$ variables, such that each OFDD (with DTL d) and each OBDD for f_n has exponential size $2^{\Omega(n^2)}$, while there exist OKFDDs for f_n of polynomial size $O(n^5)$.

Proof. Let g_n (h_n) be a function over $n(n-1)/2$ variables that can be represented efficiently by an OBDD (OFDD with DTL d), but only has OFDDs (OBDDs) of exponential size (see Theorem 9). One may further assume that g_n and h_n depend on disjoint sets of variables. Then the function $f_n = x_1 \cdot g_n \oplus x_2 \cdot h_n$ (x_1, x_2 being new variables) can obviously be represented efficiently by an OKFDD. However, neither a small OBDD nor a small OFDD exists for f_n: Assume that there exists a small OBDD (OFDD) for f_n. Since cofactoring is an efficient operation, an efficient OBDD (OFDD) for h_n (g_n) could be obtained in contradiction to the assumption. $\square$

Combining the preceding theorem with Theorem 10, we obtain an analogous result for OBDDs and OFDDs with differing DTLs in each component. Finally we can show that a restriction to k DTLs (k constant) also leads to exponential gaps. We omit the proof for shortness of the paper.

Corollary 13. i.) *Consider a fixed DTL $d = (d_1, \ldots, d_n)$ with $d_i \in \{pD, nD\}$. Then the following holds:*
There exists a Boolean function f_n over n variables, such that each OBDD, each OFDD (with DTL d) and each OFDD (with DTL $\bar{d}$) for f_n has exponential size $2^{\Omega(n)}$, while there exist OKFDDs for f_n of polynomial size $O(n^{2.5})$.

ii.) *Consider a set $D = \{d^1, d^2, \ldots, d^k\}$ ($k \in \mathbb{N}$ constant) of DTLs $d^j = (d_1^j, d_2^j, \ldots, d_n^j)$ with $d_i^j \in \{S, pD, nD\}$. Then the following holds:*
There exists a Boolean function f_n over n variables, such that each OKFDD (with DTL d^j, $1 \le j \le k$) for f_n has exponential size $2^{\Omega(n)}$, while there exist OKFDDs for f_n of polynomial size $O(n^{2.5})$.

5 Conclusions

In this paper we presented some results to clarify the relation between OBDDs, OFDDs and OKFDDs - data structures frequently used in CAD applications. OKFDDs are a superset of OBDDs and OFDDs.

The relation between the three forms of representation was described by the generalized τ-operator. Using this operator we proved that for OBDDs and OFDDs classes of Boolean functions exist that cannot be represented efficiently, but on the other hand these functions have polynomial bounded OKFDD size. We showed for OFDDs that there is an exponential trade-off if OFDDs with decomposition type lists differing in each component are considered.

Finally it was proven that the restriction of the OKFDD concept to a fixed constant number of decomposition type lists will directly reduce the number of functions that can be represented efficiently.

References

1. M. Ajtai, L. Babai, P. Hajnal, J. Komlos, P. Pudlak, V. Rödl, E. Szemeredi, and G. Turan. Two lower bounds for branching programs. In *Symp. on Theory of Computing*, pages 30–38, 1986.

2. B. Becker and R. Drechsler. How many decomposition types do we need? In *European Design & Test Conf.*, pages 438–443, 1995.
3. B. Becker, R. Drechsler, and R. Werchner. On the relation between BDDs and FDDs. In *LATIN95, LNCS*, 1995.
4. Ph.W. Besslich and E.A. Trachtenberg. A three-valued quasi-linear transformation for logic synthesis. In C. Moraga and R. Creutzburg, editors, *Spectral Techniques: Theory and Applications*. Elsevier, North Holland, 1992.
5. R.E. Bryant. Graph - based algorithms for Boolean function manipulation. *IEEE Trans. on Comp.*, 8:677–691, 1986.
6. R.E. Bryant. On the complexity of VLSI implementations and graph representations of Boolean functions with application to integer multiplication. *IEEE Trans. on Comp.*, 40:205–213, 1991.
7. R.E. Bryant. Symbolic boolean manipulation with ordered binary decision diagrams. *ACM, Comp. Surveys*, 24:293–318, 1992.
8. J.R. Burch. Using BDDs to verify multipliers. In *Design Automation Conf.*, pages 408–412, 1991.
9. R. Drechsler, A. Sarabi, M. Theobald, B. Becker, and M.A. Perkowski. Efficient representation and manipulation of switching functions based on ordered kronecker functional decision diagrams. In *Design Automation Conf.*, pages 415–419, 1994.
10. R. Drechsler, M. Theobald, and B. Becker. Fast OFDD based minimization of fixed polarity reed-muller expressions. In *European Design Automation Conf.*, pages 2–7, 1994.
11. S. J. Friedman. *Efficient Data Structures for Boolean Function Representation.* PhD thesis, Dept. of Comput. Sciences, Princeton University, 1990.
12. J. Jain, M. Abadir, J. Bitner, D. Fussell, and J. Abraham. IBDDs: An efficient functional representation for digital circuits. In *European Conf. on Design Automation*, pages 441–446, 1992.
13. S.-W. Jeong, B. Plessier, G. Hachtel, and F. Somenzi. Extended BDD's: Trading of canonicity for structure in verification algorithms. In *Int'l Conf. on CAD*, pages 464–467, 1991.
14. U. Kebschull. Die Äquivalenz-Expansion Boolescher Funktionen. Technical report, WSI-93-15, Universität Tübingen, 1993.
15. U. Kebschull, E. Schubert, and W. Rosenstiel. Multilevel logic synthesis based on functional decision diagrams. In *European Conf. on Design Automation*, pages 43–47, 1992.
16. M.S. Paterson. On Razborov's result for bounded depth circuits over $\{\oplus, \wedge\}$. Technical report, University Warwick, 1986.
17. Richard Rudell. Dynamic variable ordering for ordered binary decision diagrams. In *Int'l Conf. on CAD*, pages 42–47, 1993.
18. I. Wegener. *The Complexity of Boolean Functions.* John Wiley & Sons Ltd., and B.G. Teubner, Stuttgart, 1987.

Bicriteria Network Design Problems

M. V. Marathe [2]   R. Ravi [3]   R. Sundaram [4]
S. S. Ravi [1]   D. J. Rosenkrantz [1]   H.B. Hunt III [1]

Abstract

We study several bicriteria network design problems phrased as follows: given an undirected graph and two minimization objectives with a budget specified on one objective, find a subgraph satisfying certain connectivity requirements that minimizes the second objective subject to the budget on the first. First, we develop a formalism for bicriteria problems and their approximations. Secondly, we use a simple parametric search technique to provide bicriteria approximation algorithms for problems with two similar criteria, where both criteria are the same measure (such as the diameter or the total cost of a tree) but differ only in the cost function under which the measure is computed. Thirdly, we present an $(O(\log n), O(\log n))$-approximation algorithm for finding a diameter-constrained minimum cost spanning tree of an undirected graph on n nodes. Finally, for the class of treewidth-bounded graphs, we provide pseudopolynomial-time algorithms for a number of bicriteria problems using dynamic programming. These pseudopolynomial-time algorithms can be converted to fully polynomial-time approximation schemes using a scaling technique.

1 Introduction

Several fundamental problems in the design of communication networks can be modeled as finding a network obeying certain connectivity constraints. In applications that arise in several real-life situations, the goal is to minimize several measures of cost associated with the network. We first develop a formalism for bicriteria problems and their approximations. A typical bicriteria problem, $(\mathcal{A}, \mathcal{B})$, is defined by identifying two minimization objectives of interest from a set of possible objectives. The problem specifies a budget value on the first objective, $\mathcal{A}$, and seeks to find a network having minimum possible value for the second objective, $\mathcal{B}$, such that this network obeys the budget on the first objective. As an example, consider the following *diameter-bounded minimum spanning tree problem* or (Diameter, Total cost) bicriteria problem: given an undirected graph $G = (V, E)$ with two different integral nonnegative weights f_e (modeling the cost) and g_e (modeling the delay) for each edge $e \in E$, and an

[1] Dept. of Computer Science, University at Albany - SUNY, Albany, NY 12222. Email:{ravi, djr, hunt}@cs.albany.edu. Supported by NSF Grants CCR 94-06611 and CCR 90-06396.

[2] Los Alamos National Laboratory P.O. Box 1663, MS M986, Los Alamos NM 87545. Email: madhav@c3.lanl.gov. Research supported by the Department of Energy under Contract W-7405-ENG-36.

[3] DIMACS, Dept. of Computer Science, Princeton University, Princeton, NJ 08544-2087. Email: ravi@cs.princeton.edu. Research supported by a DIMACS postdoctoral fellowship.

[4] Dept. of Computer Science, MIT LCS, Cambridge MA 02139. Email: koods@theory.lcs.mit.edu. Research supported by DARPA contract N0014-92-J-1799 and NSF CCR 92-12184.

integral bound B (on the total delay), find a minimum f-cost spanning tree such that the diameter of the tree under the g-costs (the maximum delay between any pair of nodes) is at most B. The following hardness results can be derived by a reduction from the partition problem [10].

Theorem 1.1 The (Diameter, Total cost), (Diameter, Diameter) and the (Total cost, Total cost) spanning tree problems are NP-hard even for series-parallel graphs.

An (α, β)-approximation algorithm is defined as a polynomial-time algorithm that produces a solution in which the first objective value is at most α times the budget, and the second objective value is at most β times the minimum for any solution obeying the budget on the first objective.

2 Summary of results and related research

There are two natural alternative ways of formulating general bicriteria problems, one where we impose the budget on the first objective and seek to minimize the second and two, where we impose the budget on the second objective and seek to minimize the first. We show that an (α, β)-approximation algorithm for one of these formulations naturally leads to a (β, α)-approximation algorithm for the other. Thus our notion of bicriteria approximations is invariant on the choice of the criterion that is budgeted in the formulation.

Arbitrary Graphs

We summarize several bicriteria spanning tree results including our new contributions in Table 1. The table contains the performance ratios for finding spanning trees under different pairs of minimization objectives. All results in the table extend to finding Steiner trees with at most a constant factor worsening in the performance ratios. We omit elaboration on these extensions. The horizontal entries denote the budgeted objective. For example the entry in row i, column j denotes the performance guarantee for the problem of minimizing objective j with a budget on the objective i. As a result of the equivalence mentioned earlier, the table is symmetric, i.e. entry (i, j) is identical to entry (j, i). For each of the problems catalogued in the table, two different costs are specified on the edges of the input undirected graph: the first objective is computed using the first cost function and the second objective, using the second cost function.

In the table, the "Diameter" objective is the maximum distance between any pair of nodes in the tree. The "Total cost" objective is the sum of the costs of all the edges in the tree. The "Degree" objective denotes the maximum degree of any node in the spanning tree; the entry (Degree, Degree) however refers to a generalization to a weighted variant based on two cost functions defined on the edges. This weighted variant of the degree objective is defined as the maximum over all nodes, of the sum of the costs of the edges incident on the node in the tree. When all edges in the graph have unit weight, this reduces to the usual notion of the maximum degree.

Cost	Degree	Diameter	Total Cost
Degree	$(O(\lg n), O(\lg n))^*$	$(O(\lg n), O(\lg n))[20]$	$(O(\lg n), O(\lg n))[18]$
Diameter	$(O(\lg n), O(\lg n))[20]$	$(1+\gamma, 1+\frac{1}{\gamma})^*$	$(O(\lg n), O(\lg n))^*$
Total Cost	$(O(\lg n), O(\lg n))[18]$	$(O(\lg n), O(\lg n))^*$	$(1+\gamma, 1+\frac{1}{\gamma})^*$

Table 1. Performance Guarantees for finding spanning trees in an arbitrary graph on n nodes. Asterisks indicate results obtained in this paper. $\gamma > 0$ is a fixed accuracy parameter.

The diagonal entries in the table follow as a corollary of the following general result proved using a parametric search algorithm.

Theorem 2.1 Let $\mathcal{P}$ denote a single criterion minimization problem defined on a graph G with costs h associated with the elements of G and let $\gamma > 0$ be a fixed accuracy parameter. Assume that there exists a ρ-approximation algorithm for $\mathcal{P}$. Then the bicriteria problem $\mathcal{P}_2$ defined on G by specifying two different costs f and g on the elements of G and the two objectives being minimizing the objective of $\mathcal{P}$ under the two different costs f and g has a $((1+\gamma)\rho, (1+\frac{1}{\gamma})\rho)$-approximation algorithm.

The diagonal entries in the table correspond to such bicriteria problems in which the two objectives are similar but only differ in the cost function on the edges under which they are computed; For such problems, we introduce a parametric search method on a hybrid cost function $h_e(\mu) = f_e + \mu g_e$ on the edges e, such that the single objective problem solved using the hybrid cost $h(\mu)$ for an appropriately chosen μ yields a good approximation for both the original objectives. For example, for the (Total cost, Total cost) problem, using Theorem 2.1 with $\rho = 1$ (using an exact algorithm to comput a MST) gives the corresponding result in our table. The result for (Diameter, Diameter) follows from known exact algorithms for minimum diameter spanning trees [7, 19]. Similarly, the result for (Degree, Degree) follows from the $O(\log n)$-approximation algorithm for the weighted degree problem in [18].

Ravi et al. [18] studied the degree-bounded minimum cost spanning tree problem. They provided an approximation algorithm with performance guarantee $(O(\log n), O(\log n))$. The problem of finding a degree-bounded minimum diameter spanning tree was studied by Ravi [20] in the context of finding good broadcast networks. He provided an approximation algorithm for the first problem with performance guarantee $(O(\log n), O(\log n))$ with an extra additive term of $O(\log^2 n)$ for the degree.

The (Diameter, Total cost) entry in Table 1 corresponds to the diameter-constrained minimum spanning tree problem introduced earlier. This problem arises naturally in the design of networks used in multicasting and multimedia applications [8, 9, 12, 14]. It is known [10] that this problem is NP-hard. In the special case when the two cost functions are identical, i.e., $f_e = g_e$ for all edges e, the diameter-bounded minimum spanning tree problem reduces to finding a spanning tree that has simultaneously small diameter (i.e., shallow) and small total cost (i.e., light), both under the same cost function. Awerbuch,

Baratz and Peleg [3] showed how to compute in polynomial-time such *shallow, light trees* while Khuller, Raghavachari and Young [13] studied an extension called *Light, approximate Shortest-path Trees* (LASTs). Kadaba and Jaffe [12] and Kompella et al. [14] considered the general diameter-bounded minimum spanning tree problem and presented heuristics without any guarantees. We present the first approximation algorithm for this problem; the performance ratios for both objectives are logarithmic.

Treewidth-bounded graphs

We also study the bicriteria problems mentioned above for the class of treewidth-bounded graphs. These graphs were introduced by Robertson and Seymour [21]. Many hard problems have exact solutions when attention is restricted to the class of treewidth-bounded graphs and much work has been done in this area (see [1, 2, 5] and the references therein). Examples of treewidth-bounded graphs include trees, series-parallel graphs and bounded-bandwidth graphs. Independently, Bern, Lawler and Wong [5] introduced the notion of decomposable graphs. Later, it was shown [2] that the class of decomposable graphs and the class of treewidth-bounded graphs are one and the same. Bicriteria network design problems restricted to treewidth-bounded graphs have been previously studied in [1, 6].

We use a dynamic programming technique to show that for any class of decomposable graphs (or treewidth-bounded graphs), there are either polynomial-time (when the problem is in P) or pseudopolynomial-time algorithms (when the problem is NP-complete) for several of these bicriteria problems. We then show how to convert these pseudopolynomial-time algorithms into fully polynomial-time approximation schemes using a general scaling technique. We summarize our results for this class of graphs in Table 2. As before, the horizontal entries denote the budgeted objective.

Using our results for the (Degree, Diameter) case along with the techniques in [20], we can obtain an $O(\frac{\log n}{\log \log n})$-approximation algorithm for the minimum broadcast time problem restricted to the class of treewidth-bounded graphs (series-parallel graphs have a treewidth of 2), improving and substantially generalizing the results of Kortsarz and Peleg [16]. We omit the details due to lack of space.

Cost Measures	Degree	Diameter	Total Cost
Degree	(open) pseudopoly	(open) pseudopoly	poly-time
Diameter	(open) pseudopoly	(weak NP-hard) pseudopoly	(weak NP-hard) pseudopoly
Total Cost	poly-time	(weak NP-hard) pseudopoly	(weak NP-hard) pseudopoly

Table 2. Bicriteria spanning tree results for treewidth-bounded graphs

Due to lack of space, the rest of the paper consists of selected proof sketches.

3 Equivalence of bicriteria formulations

We formulate a general bicriteria problem in network design as follows: Given a graph G and two integral cost functions, say c and d, defined on a class S of subgraphs of G (e.g., spanning trees of G), and a bound on the value of one of the costs (say C for the c-cost), find a subgraph in S that has cost at most C under the cost function c and the minimum possible cost under d given this restriction.[5] We call this the C-bounded minimum-d-cost subgraph problem. The alternative formulation would be to use a bound D on the d-cost of the solution and ask for a minimum-c-cost subgraph under this restriction. This alternative formulation may be termed the D-bounded minimum-c-cost subgraph problem.

Note that such bicriteria problems are meaningful only when the two criteria are "hostile" with respect to each other in that the objective of minimizing one is incompatible with that of minimizing the other. A good example of such hostile objectives are the degree and the total edge cost of a spanning tree in an unweighted graph [18]. The notion of hostility between criteria can be formalized by defining two minimization criteria to be hostile whenever the minimum value of one of the objectives is monotonically nondecreasing as the budget on the value of the other objective is decreased.

Theorem 3.1 The existence of a (ρ_c, ρ_d)-approximation algorithm for the C-bounded minimum-d-cost subgraph problem implies the existence of a (ρ_d, ρ_c)-approximation algorithm for the D-bounded minimum-c-cost subgraph problem.

The proof of the theorem uses binary search on the range of values of the c-cost with an application of the given approximation algorithm at each step of this search.

4 Parametric search for approximations of similar objective functions

We now present the approximation algorithm used to prove Theorem 2.1. We illustrate the proof of the theorem by considering the case (Total cost, Total cost) in Table 1. In this problem, we are given two costs f_e and g_e on the edges $e \in E$ of the input graph $G = (V, E)$. We are also given a budget B on the total cost of the spanning tree under g. Assume for now that the magnitude of costs f on the edges are polynomial in the size of the input graph.

Let OPT denote the minimum cost of the tree under f which obeys the restriction that its cost under g is at most B. For any tree T and cost function f, we use $COST_f(T)$ to denote the cost of T under f. To simplify the analysis, we assume that γ divides OPT. This can be enforced by scaling both the cost functions f and g by γ.

---

[5] We use the term "cost under c" or "c-cost" in this section to mean value of the objective function computed using c, and not to mean the total of all the c costs in the network.

Algorithm Two-Cost(G, f, g, B, γ)

Input: A graph $G(V, E)$ with two cost functions f and g on the edges, a budget B on the cost of the spanning tree under g and a performance requirement $\gamma > 0$.
Output: A spanning tree T of G such that the cost of T under g is no more than $(1 + \frac{1}{\gamma})B$ and the cost of T under f is no more than $(1 + \gamma)OPT$.

1   Initialize $C := 1$
2   **While** $(\text{Test}(C) = NO)$ **do**
3        $C := C + 1$
4   Output the tree T computed by **Test** as the solution.

Procedure Test(C)

1     $\mu := \frac{C}{B}$.
2   Compute a new cost function h on the edges $e \in E$ as follows: $h(e) := f(e) + \mu g(e)$.
3   Compute a minimum spanning tree T in the graph $G(V, E)$ under the cost function h.
4   If $COST_h(T) \leq (1 + \gamma)C$ then output YES else output NO.

Lemma 4.1 The function $\mathcal{R}(C) = \frac{COST_h(MST)}{C}$ as C takes increasing integral values from $1, 2, 3, \ldots$ is monotone nonincreasing.

Proof: Suppose for a contradiction that for two integral values of C, say C_1 and C_2 with $C_1 < C_2$, we have that $\mathcal{R}(C_1) < \mathcal{R}(C_2)$. Let T_1 and T_2 denote minimum spanning trees of G under h when $C = C_1$ and $C = C_2$ respectively. For $i \in \{1, 2\}$, let F_i and G_i denote the costs of the tree T_i under f and g respectively. Thus, we have that $\mathcal{R}(C_i) = \frac{F_i}{C_i} + \frac{G_i}{B}$ for $i \in \{1, 2\}$.

Consider the cost under h of the spanning tree T_1 when $C = C_2$. By the definition of F_1 and G_1, it follows that the cost of T_1 is $F_1 + \frac{G_1 \cdot C_2}{B}$. Thus the value of $\mathcal{R}(C_2)$ is at most this cost divided by C_2 which is $\frac{F_1}{C_2} + \frac{G_1}{B}$. This in turn is less than $\frac{F_1}{C_1} + \frac{G_1}{B}$, since $C_1 < C_2$. But $\frac{F_1}{C_1} + \frac{G_1}{B}$ is exactly $\mathcal{R}(C_1)$ contradicting the assumption that $\mathcal{R}(C_1) < \mathcal{R}(C_2)$. $\square$

Theorem 4.2 Let C' be the value of C when the Test procedure outputs YES. Let $T_{C'}$ denote the corresponding solution tree. Then $COST_f(T_{C'}) \leq (1 + \frac{1}{\gamma})OPT$ and $COST_g(T_{C'}) \leq (1 + \gamma)B$.

Proof: First consider the value of $COST_h(T)$ when $C = C^* = \frac{OPT}{\gamma}$ (Note that C^* is integral by assumption). This is at most $OPT + \frac{C^*B}{B} = OPT + C^* = (1+\gamma)C^*$. Thus the value of $\mathcal{R}(C^*) = \frac{COST_h(MST)}{C^*} \leq 1 + \gamma$. Since $\mathcal{R}(C^*) \leq 1 + \gamma$, the function $\mathcal{R}(C)$ is monotone nonincreasing by Lemma 4.1, and C' is the least integer such that $\mathcal{R}(C') \leq 1 + \gamma$, we have that $C' \leq C^*$. It is now easy to verify that the following inequalities hold for $COST_f(T_{C'})$ and $COST_g(T_{C'})$.
$$COST_f(T_{C'}) \leq COST_h(T_{C'}) \leq OPT + \frac{C'}{B}B \leq OPT + C^* \leq (1 + \frac{1}{\gamma})OPT.$$

$$\frac{C'}{B} COST_g(T_{C'}) \leq COST_h(T_{C'}) \leq C'(1+\gamma).$$

The second chain of inequalities implies that $COST_g(T_{C'}) \leq (1+\gamma)B$.

□

We can obtain a better running time by doing a binary search for C in **Algorithm Two-cost** thus using only $O(\log F)$ calls to the polynomial-time test procedure **Test**, where F denotes the ratio of maximum edge cost to the minimum edge cost under f. We omit the details.

5 Diameter-bounded minimum spanning trees

We now discuss our approximation algorithm for the diameter bounded minmum cost spanning tree problem with performance guarantee $(O(\log n), (\log n))$. The proof for diameter bounded minimum cost steiner tree is similar and is omitted. By an approximation preserving reduction from set cover and using the known hardness results in [4, 17], we can show that there is no polynomial-time approximation algorithm that outputs a Steiner tree of diameter at most the bound D, and cost at most R times that of the minimum cost diameter-D Steiner tree, for $R < \log k/8$, unless $NP \subseteq DTIME(n^{\log \log n})$.

We shall use the term "diameter cost" of a tree to mean the diameter of the tree under the d-costs, and the "building cost" of a tree to mean the total cost of all the edges in the tree computed using the c-costs. We shall also use the term "diameter-D path (tree)" to refer to a path (tree) of diameter cost at most D under d.

We now describe some background material that will be useful in understanding our algorithm and its analysis. Given a diameter bound D, the problem of finding a diameter-D path between two specified vertices of minimum building cost has been termed the multi-objective shortest path problem (MOSP). This problem is NP-complete and Warburton [22] presented the first fully polynomial approximation scheme (FPAS) for this problem. Hassin [11] provided an alternative FPAS for the problem without a running-time dependency on the magnitude of the costs in the problem. His algorithm runs in time $O(m(\frac{n^2}{\epsilon} \log \frac{n}{\epsilon}))$, where m and n denote the number of edges and nodes in the input graph respectively. We use the latter result in implementing our algorithm.

Next, we state a tree decomposition result from [18], and use it in the proof of the performance guarantee.

Lemma 5.1 Let T be a tree with an even number of marked nodes. Then there is a pairing $(v_1, w_1), \ldots, (v_k, w_k)$ of the marked nodes such that the $v_i - w_i$ paths in T are edge-disjoint.

A pairing of the marked nodes that minimizes the sum of the sizes of the tree-paths between the nodes paired up can be shown to obey the property in the claim above.

Overview

The algorithm begins with an empty solution subgraph where each node is in a connected component (termed a *cluster*) by itself in the solution. Assume for simplicity that the number of nodes in the input graph is a power of two. The algorithm works in $\log_2 n$ iterations where n is the number of nodes in the original graph, merging clusters in pairs during each iteration by adding edges between them. This pairing ensures that the number of iterations is as desired.

The clusters maintained by our algorithm represent node-subsets of the input graph G. However, they *do not* represent a partition of the nodes of the input graph. This is because of the way in which we merge the clusters in the algorithm. For each cluster we maintain a spanning tree on the nodes in the cluster. The spanning trees of the clusters maintained by the algorithm are not necessarily edge-disjoint as a result of our merging procedure. We sketch this procedure below. We identify a *center* in the spanning tree of each cluster. In each iteration, every cluster is paired with another cluster and merged with it by the addition of a path between their respective centers. This path may involve nodes that occur in either of the merging clusters or even nodes in other clusters currently maintained by the algorithm. However, while merging two clusters into one, we ensure that the new cluster formed has at most one copy of any node or edge.

The Algorithm

1   Initialize the set of clusters $\mathcal{C}$ to contain n singleton sets, one for each node of the input graph. For each cluster in $\mathcal{C}$, define the single node in the cluster to be the center for the cluster. Initialize the iteration count $i := 1$.

2   Repeat until there remains a single cluster in $\mathcal{C}$

3   Let the set of clusters $\mathcal{C} = \{C_1 \ldots, C_{\frac{n}{2^{i-1}}}\}$.

4   Construct a complete graph G_i as follows.

5   The node set V_i of G_i is $\{v : v$ is the center of a cluster in $\mathcal{C}\}$.

6   Between every pair of nodes v_x and v_y in V_i, include an edge (v_x, v_y) in G_i of cost equal to a $(1 + \epsilon)$-approximation of the shortest building cost of a diameter-D path between v_x and v_y in G, where ϵ is the accuracy parameter input to the algorithm. Since a FPAS is available to compute such an estimate [11], the costs of all the edges in G_i can be computed in polynomial-time.

7   Find a minimum-cost perfect matching in G_i.

8   For each edge $e = (v_r, v_s)$ in the matching

9   Let P_{rs} be the path in G represented by $e = (v_r, v_s)$. Add this path to merge the clusters C_r and C_s for which v_r and v_s were centers respectively, to form a new cluster C_{rs}, say. The node set of the cluster C_{rs} is defined as the union of the node sets C_r, C_s and the nodes in P_{rs}.

10   Define the union of the edge sets of the spanning trees for C_r and C_s and the set of edges in P_{rs} to be E_{rs}. The edges in E_{rs} form a connected

graph on C_{rs}. Choose one of v_r or v_s as the center v_{rs} of the cluster C_{rs}. Using only the diameter cost function d, find a shortest-path tree rooted at v_{rs} in the graph (C_{rs}, E_{rs}). This is the spanning tree for the cluster C_{rs}.

11   Set $C := C - \{C_r, C_s\} \cup \{C_{rs}\}$.

12     $i := i + 1$.

13   Output the spanning tree of the single cluster in C.

We prove the performance guarantee using a series of lemmas.

At each iteration, since the clusters are paired up using a perfect matching and merged to form new clusters, the number of clusters halves. Thus we have the following lemma.

Lemma 5.2 The total number of iterations of the above algorithm is $\lceil \log_2 n \rceil$.

Lemma 5.3 Let C be a cluster with center v formed at iteration i of the algorithm. Then any node u in C has a diameter-iD path to v in the spanning tree of C maintained by the algorithm.

Proof: The proof is by induction on the iteration count i. The basis when $i = 1$ is trivial. To prove the induction step, consider a cluster C_{rs} formed at iteration i (> 1) by merging two clusters C_r and C_s with centers v_r and v_s respectively. Suppose $v_{rs} = v_r$. Consider a node $u \in C_{rs}$. u is in either C_r, C_s or the path P_{rs}. In all these cases, using the inductive hypothesis and the fact that the diameter cost of path P_{rs} is at most D, it is easy to show that u has a diameter-iD path to the center $v_{rs} = v_r$ in the graph (C_{rs}, E_{rs}). Since we compute a shortest-path tree in this graph rooted at v_{rs} using the diameter costs, it follows that the path in this tree between any node and v_{rs} has diameter cost no more than iD. □

Corollary 5.4 Let C be a cluster formed at iteration i of the algorithm. Then the diameter cost of the spanning tree of C maintained by the algorithm is at most $2iD$.

Lemma 5.5 Let OPT_D be the minimum building cost of any diameter-D spanning tree of the input graph. At each iteration i of the algorithm, the cost of the minimum cost matching in G_i found in Step 7 is at most $(1 + \epsilon) \cdot OPT_D$, where $\epsilon > 0$ is the accuracy parameter input to the algorithm.

Proof: It is easy to show using a simple induction on the iteration count that, at any iteration i, the set of centers of clusters for this iteration are distinct nodes of G. Since these are exactly the nodes in G_i, we have that the graph G_i has at most one copy of any node of G. We can now apply Lemma 5.1 on the optimal diameter-D spanning tree of the input graph with the nodes of G_i marked. The lemma yields a pairing between these centers such that the pairs are connected using edge-disjoint paths in the optimal tree. Note that all these paths have diameter cost at most D since they are derived from a diameter-D tree. Furthermore, the sum of the building costs of all these paths is at most

OPT_D since they form edge-disjoint fragments of a tree of total building cost OPT_D. Thus we have identified a pairing between the nodes in G_i of "cost" at most OPT_D where the cost of a pair is the minimum building cost of a diameter-D path between its endpoints.

In constructing G_i, the cost assigned to an edge between a pair of nodes is a $(1 + \epsilon)$-approximation to the minimum building cost of a diameter-D path between these nodes in G (see Step 6). Thus between every pair identified above, there is an edge in G_i of cost at most $(1 + \epsilon)$ times the building cost of the path between them in the optimal tree. This identifies a perfect matching in G_i of cost at most $(1 + \epsilon) \cdot OPT_D$ and completes the proof. □

Note that the spanning tree finally output by the above algorithm is a subgraph of the union of all the paths added by all the matchings over all the iterations. Lemma 5.2, corollary 5.4, and lemma 5.5 prove the performance guarantees of the algorithm.

6   Treewidth-bounded Graphs

In this section we briefly discuss our ideas by describing the algorithm for solving the diameter B-bounded minimum cost spanning tree problem.

Let f be the cost function on the edges for the first objective (diameter) and g, the cost function for the second objective (total cost). Let Γ be any class of decomposable graphs. Let the maximum number of terminals associated with any graph G in Γ be k. Following [5], it is assumed that a given graph G is accompanied by a parse tree specifying how G is constructed using the rules and that the size of the parse tree is linear in the number of nodes.

Let π be a partition of the terminals of G. For every terminal i let d_i be a number in $\{1 \ldots B\}$. For every pair of terminals i and j in the same block of the partition π let d_{ij} be a number in $\{1 \ldots B\}$. Corresponding to every partition π, set $\{d_i\}$ and set $\{d_{ij}\}$ we associate a cost for G,

$Cost^{\pi}_{\{d_i\}, \{d_{ij}\}}$ = Minimum total cost under the g function of any forest containing a tree for each block of π, such that the terminal nodes occurring in each tree are exactly the members of the corresponding block of π, no pair of trees is connected, every vertex in G appears in exactly one tree, d_i is an upper bound on the maximum distance (under the f function) from i to any vertex in the same tree and d_{ij} is an upper bound the distance (under the f function) between terminals i and j in their tree. For the above defined cost, if there is no forest satisfying the required conditions the value of Cost is defined to be ∞.

Note that the number of cost values associated with any graph in Γ is $O(k^k \cdot B^{O(k^2)})$. We now show how the cost values can be computed in a bottom-up manner given the parse tree for G. To begin with, since Γ is fixed, the number of primitive graphs is finite. For a primitive graph, each cost value can be computed in constant time, since the number of forests to be examined is fixed. Now consider computing the cost values for a graph G constructed from subgraphs G_1 and G_2, where the cost values for G_1 and G_2 have already been computed. Notice that any forest realizing a particular cost value for G

decomposes into two forests, one for G_1 and one for G_2 with some cost values. Since we have maintained the best cost values for all possibilities for G_1 and G_2, we can reconstruct for each partition of the terminals of G the forest that has minimum cost value among all the forests for this partition obeying the diameter constraints. We can do this in time independent of the sizes of G_1 and G_2 because they interact only at the terminals to form G, and we have maintained all relevant information.

Hence we can generate all possible cost values for G by considering combinations of all relevant pairs of cost values for G_1 and G_2. This takes time $O(k^4)$ per combination for a total time of $O(k^{2k+4} \cdot B^{O(k^4)})$. As in [5], we assume that the size of the given parse tree for G is $O(n)$. Then the dynamic programming algorithm takes time $O(n \cdot k^{2k+4} \cdot B^{O(k^4)})$.

Acknowledgements We thank Professors S. Arnborg and H. L. Bodlaender for pointing out to us the equivalence between treewidth bounded graphs and decomposable graphs. We wish to thank A. Ramesh for bringing [15] to our attention. We also thank Dr.Vachaspati Kompella for making his other papers available to us. Finally, we thank the referees for their constructive suggestions.

References

[1] S. Arnborg, J. Lagergren and D. Seese, "Easy Problems for Tree-Decomposable Graphs," *J. Algorithms*, Vol. 12, 1991, pp. 308-340.

[2] S. Arnborg, B. Courcelle, A. Proskurowski and D. Seese, "An Algebraic Theory of Graph Problems," *J. ACM*, Vol. 12, 1993, pp. 308-340.

[3] B. Awerbuch, A. Baratz, and D. Peleg, "Cost-sensitive analysis of communication protocols," *Proc. of 9th Symp. on Principles of Distributed Computing (PODC)*, pp. 177-187, (1990).

[4] M. Bellare, S. Goldwasser, C. Lund, and A. Russell, "Efficient probabilistically checkable proofs," *Proceedings of the 25th Annual ACM Symposium on the Theory of Computing* (1993), pp. 294-304.

[5] M.W. Bern, E.L. Lawler and A.L. Wong, "Linear -Time Computation of Optimal Subgraphs of Decomposable Graphs," *J. Algorithms*, Vol. 8, 1987, pp. 216-235.

[6] H.L. Bodlaender, "Dynamic programming on graphs of bounded treewidth," *Proc. of the 15th ICALP*, LNCS Vol. 317, 1988, pp. 105-118.

[7] P. M. Camerini, and G. Galbiati, "The bounded path problem," *SIAM J. Alg. Disc., Meth.* Vol. 3, No. 4 (1982), pp. 474-484.

[8] C.-H. Chow, "On multicast path finding algorithms," *Proc. of IEEE INFO-COM '91*, pp. 1274-1283 (1991).

[9] A. Frank, L. Wittie, and A. Bernstein, "Multicast communication in network computers," *IEEE Software*, Vol. 2, No. 3, pp. 49-61 (1985).

[10] M. R. Garey and D. S. Johnson, *Computers and Intractability: A guide to the theory of NP-completeness*, W. H. Freeman, San Francisco (1979).

[11] R. Hassin, "Approximation schemes for the restricted shortest path problem," *Math. of O. R.*, Vol. 17, No. 1, pp. 36-42 (1992).

[12] B. Kadaba and J. Jaffe, "Routing to multiple destinations in computer networks," *IEEE Trans. on Comm.*, Vol. COM-31, pp. 343-351, (Mar. 1983).

[13] S. Khuller, B. Raghavachari, and N. Young, "Balancing Minimum Spanning and Shortest Path Trees," *Proc., Fourth Annual ACM-SIAM Symposium on Discrete Algorithms* (1993), pp. 243-250.

[14] V.P. Kompella, J.C. Pasquale and G.C. Polyzos, "Multicasting for multimedia applications," *Proc. of IEEE INFOCOM '92*, (May 1992).

[15] V.P. Kompella, J.C. Pasquale and G.C. Polyzos, "Multicast Routing for Multimedia Communication," *IEEE/ACM Transactions on Networking*, pp. 286-292, (1993).

[16] G. Kortsarz and D. Peleg, "Approximation algorithms for minimum time broadcast," *Proc. of the 1992 Israel Symposium on Theoretical Computer Science* LNCS 601, (1994).

[17] C. Lund and M. Yannakakis, "On the Hardness of Approximating Minimization Problems," *Proc., 25th Annual ACM Symp. on Theory of Computing*, (1993), pp. 286-293.

[18] R. Ravi, M. V. Marathe, S. S. Ravi, D. J. Rosenkrantz, and H.B. Hunt III, "Many birds with one stone: Multi-objective approximation algorithms," *Proc. of the 25th Annual ACM Symposium on the Theory of Computing* (1993), pp. 438-447.

[19] R. Ravi, R. Sundaram, M. V. Marathe, D. J. Rosenkrantz, and S. S. Ravi, "Spanning trees short or small," in *Proc. of the 5th Annual ACM-SIAM Symposium on Discrete Algorithms*, (1994), pp 546-555.

[20] R. Ravi, "Rapid Rumor Ramification: Approximating the minimum broadcast time," in *Proc. of the 25th Annual IEEE Foundations of Computer Science* (1994), pp. 202-213.

[21] N. Robertson and P. Seymour, "Graph Minors IV, Tree-width and well-quasi-ordering," *J. Combin. Theory Ser. B* 48, 227-254 (1990).

[22] A. Warburton, "Approximation of Pareto optima in multiple-objective, shortest path problems," *Oper. Res.* 35, pp. 70-79 (1987).

On Determining Optimal Strategies in Pursuit Games in the Plane

(Extended Abstract)

Ngọc-Minh Lê *

Praktische Informatik VI
FernUniversität Hagen
D-58084 Hagen, Germany

ngoc-minh.le@fernuni-hagen.de

Abstract

Pursuit games have application to robotics: the pursuer models a moving obstacle and the evader models a robot that tries to reach a goal region without colliding with the moving obstacle, at each moment the robot does not know the future trajectory of the obstacle. The motion of the pursuer and the evader is controlled by their sets of permissible velocities, called *indicatrices*. We allow indicatrices that are more general than the simple motion (i.e., velocities are bounded by an L_2-norm circle). We provide sufficient condition for a pursuit game to "have value", in this case we give optimal strategies for the pursuer and the evader. We prove that the pursuit game in which the pursuer and the evader are convex objects moving with simple motion "has value".

1 Introduction

Pursuit game, also called game of *guarding a territory*, is a basic problem in differential games. It was through its study that Isaacs initiated the theory of differential games. He stated the pursuit game as follows [11, p. 10 and p. 19].

In Figure 1, Ω is a territory which P, the pursuer, is guarding from E, the evader. Both P and E travel with simple motion with the same speed and start from the positions shown. Capture means the coincidence of P and E. The payoff is to be the distance from the point of capture, if any, to Ω, which P is to maximize and E to minimize. If E can reach Ω without being captured, he regards this outcome as best of all. How should each craft travel?

*This work was partly supported by Deutsche Forschungsgemeinschaft grant Kl 655/2-2.

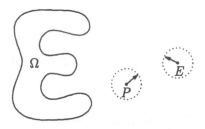

Figure 1: Pursuit game.

Isaacs proposed the following strategies:

> *Draw the bisector of P and E. Let D be the point of the bisector nearest*
> Ω. *The optimal strategies for both players decree that they travel toward*
> *D.*

Isaacs did not justify the optimality of the strategies above — he did not even define the formal concept of "strategy". The rigorous foundations for the theory of differential games were laid later by Friedman in [8] (and about the same time by others). He introduced the fundamental concepts of *strategy, value, optimal pair of strategies,* ... Friedman also considers the problem of guarding a territory [8, p. 29]. The strategy for P is simple: P always moves symmetrically with respect to the bisector of P and E (at the beginning of the game). He proves that this strategy is optimal and the game *has value*, which is the distance from D to Ω: P (resp. E) cannot increase (resp. decrease) this distance.

Reif and Tate [15] consider three-dimensional polyhedral pursuit games in which P and E are cubes and the environment is cluttered with polyhedral obstacles. The evader E wants to reach a given goal region, and the pursuer P wants to hinder E from reaching the goal by trying to capture E, i.e., collide with P. They prove that any algorithm which solves the decision problem for the polyhedral pursuit game must take at least exponential time in the worst case. This result is the first provable intractibility result for a robotics problem with complete information. They ask for exact solutions in restricted pursuit games, e.g., no obstacles.

In this paper we consider pursuit games without obstacles. We first investigate the problem of guarding a territory for point evader and point pursuer, where more general sets of permissible velocities, also called *indicatrices (of velocities)*, for P and E are allowed. Furthermore, we consider the case when P and E have convex shapes, in this case P and E are to move with simple motion (i.e., their indicatrices are equal L_2-norm circles).

For the point pursuit game where the pursuer is faster than the evader, we provide upper and lower bounds on certain numerical characteristics of the payoff of the game. Using these bounds we derive a sufficient condition for the game to "have value" and determine in that case optimal strategies for the pursuer and the evader. Similar results hold if the pursuer and the evader are equally fast.

In the case where the pursuer and the evader have convex shapes and move with simple motion we prove that the pursuit game "has value" and determine optimal strategies for the players.

In all cases above we also analyse the complexity for computing the mentioned numerical characteristics of the "value" of the game and for determining the optimal strategy of the evader (if the game "has value").

Pursuit games with indicatrices different from the L_2-norm circle — as far as we know — have not been considered before. We answer interesting questions posed by Reif and Tate [15]: they ask if exact solutions for restricted games (no obstacles) are possible. The pursuit game that we investigate here can be regarded as a *quantified* variant of such restricted games: we want to know how near the evader can approach the goal.

2 Preliminaries

We first describe some notation used in this abstract. The convex hull of a subset $X \subset \mathbf{R}^2$ is written *conv X*. If p and q are points in $\mathbf{R}^2$ then the line passing through p and q is written pq, we use $\overline{pq}$ to denote the line segment connecting p and q, we denote the length of $\overline{pq}$ by $|pq|$. The distance between two sets $X, Y \subset \mathbf{R}^2$ is written *dist* (X, Y). We say that a point $p \in X$ *realizes* *dist* (X, Y) if *dist* $(p, Y) =$ *dist* (X, Y).

2.1 Convex Distance Functions

Let C be the boundary of a convex body in $\mathbf{R}^2$ containing the origin in its interior. Using C we define a *convex distance* d_C in $\mathbf{R}^2$ as follows. Given points $p, v \in \mathbf{R}^2$, then the distance (with respect to d_C) from p to v is

$$d_C(p, v) = \frac{|pv|}{|pv'|}, \tag{1}$$

where v' denotes the intersection of the halfline from p through v with C which is centered at p. Thus, when given an indicatrix of velocities C, we can use C to define a convex distance d_C. In this case we may interpret the distance $d_C(p, v)$ as the shortest time needed to travel from p to v, provided that at each moment of time only velocities chosen from C are permissible.

Let d_C and $d_{C'}$ be two convex distances. Given two distinct points p and $q \in \mathbf{R}^2$, we define the *bisector $B(p, q)$ of p and q (with C assigned to p and C' assigned to q)* by $B(p, q) = \{v \in \mathbf{R}^2 \mid d_C(p, v) = d_{C'}(q, v)\}$. This definition of bisector generalizes the notion of bisector in the multiplicatively weighted distance and in convex distances [3]. If C and C' are indicatrices of velocities, then the bisector $B(p, q)$ consists exactly of points that can be reached by two moving point objects in the same shortest possible time, provided that they start at p and q and choose their velocity vectors only from C and C', respectively.

We call the set $D(p, q) = \{v \in \mathbf{R}^2 \mid d_C(p, v) \le d_{C'}(q, v)\}$ the *region of dominance of p over q*. This terminology is motivated by the following observation:

a moving point that starts at p using C can attain any point of its region of dominance not latter than a moving point that starts at q employing C'. Let S be a set of points in $\mathbf{R}^2$, $S = \{p, \ldots\}$, let each point of S be assigned an indicatrix. The set $R(p, S) = \bigcap_{q \in S \setminus \{p\}} D(p, q)$ is called the *Voronoi region* of p (with respect to S). A subset W of $\mathbf{R}^2$ is *star-shaped around* $p \in \mathbf{R}^2$ if for every $x \in W$ the segment $\overline{px}$ lies entirely in W. Given two indicatrices C and C', we will write $C' \prec C$ to mean that the interior domain of C' is a subset of the interior domain of C and that $C' \cap C = \emptyset$.

2.2 Object model

Besides the usual polygons in the planar linear world, we will model curvilinear objects by algebraic splinegons, which have been introduced in [7]. An *algebraic splinegon* can be described by the boundary representation consisting of a single oriented cycle of edges that bounds a planar region [4], [6]. Each edge is an algebraic curve incident to two vertices. An algebraic curve is either implicitly defined by a single polynomial equation $f(x_1, x_2) = 0$ or parametrically defined by a pair $x_1 = f_1(t)/g(t)$ and $x_2 = f_2(t)/g(t)$, where f_1, f_2, and g are polynomials. We refer to [5] for further details. To avoid a cumbersome presentation of the paper, we will always assume that all algebraic curves — e.g., edges of an algebraic splinegon — are defined implicitly. In addition, determining an algebraic curve will mean determining its implicit representation.

We assume some primitive procedures for dealing with algebraic curve segments. Examples of such primitives are: compute the intersection of a line with an algebraic curve segments, compute the intersection of two algebraic curve segments or the distance between them, determine the tangent line of a monotone curve segment from a point.

Throughout the paper we use d to denote the maximal degree of the algebraic curve segments involved. Employing methods similar to that, for instance, in [4] and [5], we can implement the primitive procedures needed to run in time $O(d^{O(1)})$.

We will need to compute the Euclidean distance between simple algebraic splinegons together with some points that realize it. Recently, Amato [2] determines the distance between simple polygons in linear time. However, to our knowledge, it is not known if the distance between an m-edge and an n-edge algebraic splinegon can be computed in time better than $O(d^{O(1)}mn)$. So let $O(d^{O(1)})\Lambda(m, n)$ denote the time needed for solving this problem. An appearance of Λ in a time bound makes explicit the dependency of the bound on an efficient distance computation. However, if one of the algebraic splinegons is *convex* then we have the following.

Lemma 1 *The distance between a convex algebraic splinegon having m edges and an algebraic splinegon having n edges (together with a pair of points that realize it) can be computed in time $O(d^{O(1)}(m + n))$.*

The proofs of the results in this extended abstract can be found in the full paper [12].

Convention. When dealing with algebraic splinegons we will usually omit, for convenience, the constant factor $d^{O(1)}$ in the Big-Oh notation for the related time bounds.

2.3 Bisectors

Throughout this section we assume that $C' \prec C$. For simplicity, we make the assumption that the interior domains bounded by C (at p) and C' (at q) are disjoint; in fact, we are allowed to shrink C and C' by an appropriate positive scale factor to ensure this assumption. Observe that if s is the scale factor then $d_{s \cdot X}(p, a) = d_X(p, a)/s$ for $X \in \{C, C'\}$.

Lemma 2 *(i) The bisector $B(p, q)$ is homeomorphic to a simple closed curve.*

(ii) The set $D(q, p)$ is star-shaped around q.

Remark. If $C' \not\prec C$ then the bisector may be disconnected.

Lemma 3 *(i) Let C and C' be convex polygons. Then the bisector $B(p, q)$ consists of line segments and can be computed in $O(|C| + |C'|)$ time.*

(ii) Let C and C' be convex algebraic splinegons. Then the bisector $B(p, q)$ is also an algebraic splinegon, and can be determined in $O(|C| + |C'|)$ time.

The following result will prove to be crucial in designing a good strategy for the pursuer in our pursuit game.

Lemma 4 *(i) Let $b \in B(p, q)$, and let $p' \in \overline{pb}$ and $q' \in \overline{qb}$ such that the segment $\overline{p'q'}$ is parallel to the segment $\overline{pq}$. Then we have conv $D(q', p') \subset$ conv $D(q, p)$.*

(ii) Let p' be any point in the segment $\overline{pq}$ with $p' \neq q$. Then we have $D(q, p') \subset D(q, p)$.

2.4 Voronoi Regions

Let S be a set of points in $\mathbf{R}^2$, i.e., $S = \{p_1, \ldots, p_M, q\}$.
Define $N := \sum_M (|C_i| + |C'|) = M|C'| + \sum_M |C_i|$.

Lemma 5 *If $C' \prec C_i$ for $i = 1, \ldots, M$ then $R(q, S)$ is star-shaped around q.*

Proof. The claim follows from Lemma 2. □

Let $\lambda_s(n)$ denote the maximum length of a Davenport-Schinzel sequence of order s composed of n symbols, see [1].

Lemma 6 *(i) Let $C_1, \ldots, C_M$ be convex polygons. Then the Voronoi region $R(q, S)$ can be determined in $O(N \log N)$ time.*

(ii) Let $C_1, \ldots, C_M$ be convex algebraic splinegons. Then the Voronoi region $R(q, S)$ is bounded by algebraic curve segments and can be computed in $O(\lambda_{k+1}(N) \log N)$ time, where k denotes the maximal number of intersections between any two curve segments of the bisectors $B(p_i, q)$ for $i = 1, \ldots, M$.

3   Optimal Strategies in Pursuit Games

To cope with the fact that differential games are played in a continuous manner, we discretize the play time interval into n steps with the size δ, and study two δ-approximating games $\{G^\delta\}$ and $\{G_\delta\}$, in which the players alternatively make their moves step-by-step. By a δ-*strategy* for the evader (resp. pursuer) we mean an n-tuple of functional mappings ("operator") for mapping any tuple of permissible controls (which are defined on the time lapses δ and model the moves of the players in the past) to a unique control that yields the next move for the evader (resp. pursuer). In the "upper" δ-game G^δ, the evader must move first; conversely, in the "lower" δ-game G_δ, the pursuer must move first.

If each player tries to do her/his best, then we call the payoff in the upper δ-game the *upper δ-value*, denoted V^δ. We define the *lower δ-value* V_δ similarly. If, for some δ-strategy pair, V^δ and V_δ converge as $\delta \to 0$ to the same value, denoted V, then we call V the *value* of the game, and say that the game *has value*. In this case, corresponding strategies are called *optimal*. We refer to [8] or [12] for more details on notation and basic results concerning differential games.

Throughout the rest of the paper Ω denotes the territory to be guarded, which is assumed to be a compact set of $\mathbf{R}^2$.

3.1   One Pursuer

We consider the pursuit game where the point pursuer P moves by choosing velocity vectors from an indicatrix C and the point evader E moves by choosing velocity vectors from an indicatrix C'. We assume that $C' \prec C$. The payoff is defined to be the distance from E to Ω when capture occurs. Denote the positions of P and E at time t by $P(t)$ and $E(t)$. The differential equation that determines the trajectories of P and E is as follows:

$$\begin{cases} \dot{P}(t) = y(t), & \text{where } y(t) \in C \\ \dot{E}(t) = z(t), & \text{where } z(t) \in C'. \end{cases}$$

The initial positions of P and E are $P(t_0) = p$ and $E(t_0) = q$. The terminal set is

$$\begin{aligned} F = \ & \{(t, x_P, x_E) \,|\, x_P = x_E, t_0 \leq t \leq T\} \cup \\ & \{(t, x_P, x_E) \,|\, x_P, x_E \in \mathbf{R}^2, t \geq T\}, \end{aligned}$$

where the deadline time T is sufficiently large and can be determined a posteriori once conditions for optimal strategies are found. Let $D(q, p)$ be the region of dominance of q over p, which is bounded by the bisector $B(p, q) = \{v \in \mathbf{R}^2 \,|\, d_C(p, v) = d_{C'}(q, v)\}$. If $\Omega \cap D(q, p) \neq \emptyset$ then E may head for an appropriate point of Ω lying in its region of dominance, and reach Ω without being intercepted by P. So we assume that $\Omega \cap D(q, p) = \emptyset$ [1]. Note that neither the

---

[1] Intersection detection for simple splinegons can be performed in linear time [6].

strategy in [8] nor the strategy in [11] for the pursuer can be generalized to solve our problem.

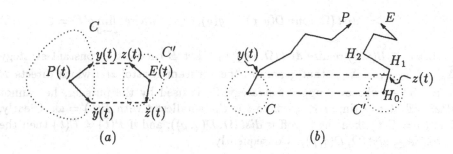

Figure 2: (a) If E plays $z(t)$, resp. $\breve{z}(t)$, then P plays $y(t)$, resp. $\breve{y}(t)$. (b) Outcome of an upper δ-game, the pursuer P uses the basic guarding rule as its upper δ-strategy.

Theorem 7 *The upper δ-value V^δ satisfies*

$$dist\,(\Omega, conv\,D(q,p)) \;\leq\; \liminf_{\delta \to 0} V^\delta \;\leq\; \limsup_{\delta \to 0} V^\delta \;\leq\; dist\,(\Omega, D(q,p)),$$

and the lower δ-value V_δ satisfies

$$dist\,(\Omega, conv\,D(q,p)) \;\leq\; \liminf_{\delta \to 0} V_\delta \;\leq\; \limsup_{\delta \to 0} V_\delta \;\leq\; dist\,(\Omega, D(q,p)).$$

Proof. • First we consider an *upper δ-game G^δ* — the evader moves first. We define an upper δ-strategy $\widetilde{\Gamma}^\delta = (\widetilde{\Gamma}^{\delta,1}, \dots, \widetilde{\Gamma}^{\delta,n})$ for the pursuer P as follows. If the evader traverses from H_0 to H_1 along a curve $H_0 H_1$ by choosing a control $z_1(\cdot) = \Delta_{\delta,1}$ then the strategy $\widetilde{\Gamma}^{\delta,1}$ generates for P a control $y_1(\cdot) = \widetilde{\Gamma}^{\delta,1}(z_1)$ determined by the rule below:

> **Basic guarding rule.** See Figure 2. If E is within the step size, then "catch" it. Otherwise consider, for each t in the corresponding time interval, the indicatrix C of the pursuer at $P(t)$. Project the control vector $z(t)$ parallel to the line pq on C. Choose control vector $y(t)$ so that the endpoint of $y(t)$ coincides with the projection of the endpoint of $z(t)$.

In the next round, if E moves from H_1 to H_2 along a curve $H_1 H_2$ by employing a control $z_2(\cdot)$ generated by a strategy $\Delta_{\delta,1}$ then the strategy $\widetilde{\Gamma}^{\delta,2}$ generates for P a control $y_2(\cdot) = \widetilde{\Gamma}^{\delta,2}(z_1, y_1, z_2)$ determined also by the basic guarding rule stated above. We define $\widetilde{\Gamma}^{\delta,3}, \dots, \widetilde{\Gamma}^{\delta,n}$ similarly.

If P chooses $\widetilde{\Gamma}^\delta$ as an upper δ-strategy then by Lemma 4(i) (to be precise, we use its infinitesimal version, which can be derived by determining the average velocities of P and E on $\overline{pp'}$ and $\overline{qq'}$ and then letting $p' \to p$ and $q' \to q$.) the trajectories resulting from $(\Delta_\delta, \widetilde{\Gamma}^\delta)$ is so that capture occurs at a point which is at most one step size away from $conv\,D(q,p)$. It follows that, for any Δ_δ, the upper δ-value satisfies

$$V^\delta \;\geq\; dist\,(\Omega, conv\,D(q,p)) - g(\delta), \qquad \text{where } \lim_{\delta \to 0} g(\delta) = 0. \tag{2}$$

Let $\omega \in B(p,q)$ realize $dist\,(\Omega, D(q,p))$. Let E choose the constant strategy $\widetilde{\Delta}_\delta = (\widetilde{\Delta}_{\delta,1}, \ldots, \widetilde{\Delta}_{\delta,n})$ that generates the constant control $\widetilde{z}(\cdot)$ which directs E to ω. Then whatever upper δ-strategy $\widetilde{\Gamma}^\delta$ is used by the pursuer, he cannot attain E at any time $t < \widetilde{t}$, where $\widetilde{t}$ is the smallest t with $E(\widetilde{t}) = \omega$. Clearly, if $P(\widetilde{t}) = E(\widetilde{t})$ then the payoff is $dist\,(\Omega, D(q,p))$, and if $P(\widetilde{t}) \neq E(\widetilde{t})$ then the payoff is $\leq dist\,(\Omega, D(q,p))$. Consequently

$$V^\delta \;\leq\; dist\,(\Omega, D(q,p)). \tag{3}$$

Now the bounds for $\liminf_{\delta \to 0} V^\delta$ and $\limsup_{\delta \to 0} V^\delta$ follows from (2) and (3).

• Finally consider a *lower δ-game G_δ* — the pursuer moves first. Define a lower δ-strategy $\widetilde{\Gamma}_\delta = (\widetilde{\Gamma}_{\delta,1}, \ldots, \widetilde{\Gamma}_{\delta,n})$ for the pursuer as follows. The strategy $\widetilde{\Gamma}_{\delta,1}$ is a constant control $y_1(\cdot)$ that makes P move linearly toward the evader. By Lemma 4(ii), the effect of this move is $D(q, P(t_1)) \subset D(q,p)$. If the evader in his turn moves from H_0 to H_1 along a curve $H_0 H_1$ by using a control $z_1(\cdot) = \Delta^{\delta,1}(y_1)$, then the strategy $\widetilde{\Gamma}_{\delta,2}$ generates for P a control $y_2 = \widetilde{\Gamma}_{\delta,2}(y_1, z_1)$ determined by the basic guarding rule defined above. The component strategies $\widetilde{\Gamma}_{\delta,3}, \ldots, \widetilde{\Gamma}_{\delta,n}$ are defined similarly. If P chooses $\widetilde{\Gamma}_\delta$ as a lower δ-strategy then by Lemma 4 the trajectories resulting from $(\widetilde{\Gamma}_\delta, \Delta^\delta)$ are so that the evader is captured at a point which is at most one step size away from $conv\,D(q,p)$. We conclude that, for any Δ^δ, the lower δ-value satisfies

$$V_\delta \;\geq\; dist\,(\Omega, conv\,D(q,p)) - h(\delta), \qquad \text{where } \lim_{\delta \to 0} h(\delta) = 0. \tag{4}$$

Let E choose the constant upper δ-strategy $\widetilde{\Delta}^\delta = (\widetilde{\Delta}^{\delta,1}, \ldots, \widetilde{\Delta}^{\delta,n})$ that makes E move linearly toward a point $\omega \in B(q, P(t_1))$ nearest to Ω, then no matter what lower δ-strategy is used by P, the pursuer position $P(t)$ cannot equal $E(t)$ for any $t_0 \leq t < \widetilde{t}$, where $\widetilde{t}$ is the smallest t satisfying $E(\widetilde{t}) = \omega$. If $P(\widetilde{t}) \neq E(\widetilde{t})$ then the resulting payoff is certainly $\leq dist\,(\Omega, D(q, P(t_0 + \delta)))$. Thus we obtain

$$V_\delta \;\leq\; dist\,(\Omega, D(q, P(t_0 + \delta))). \tag{5}$$

The claimed bounds for $\liminf_{\delta \to 0} V_\delta$ and $\limsup_{\delta \to 0} V_\delta$ follow from (4) and (5).

□

For the following, recall that $O(d^{O(1)})\Lambda(m,n)$ denotes the time needed for computing the distance between simple algebraic splinegons of size m and n, see Section 2.2. Using the preceding results, we can prove the following.

Lemma 8 *(i) Let C and C' be convex polygons, and let Ω be a simple polygon. Then, the lower bound $dist\,(\Omega, conv\,D(q,p))$ for $\liminf_{\delta \to 0} V^\delta$ and $\liminf_{\delta \to 0} V_\delta$*

can be computed in time $O(|C|+|C'|+|\Omega|)$. The upper bound $dist\,(\Omega, D(q, p))$ for $\limsup_{\delta \to 0} V^\delta$ and $\limsup_{\delta \to 0} V_\delta$ can also be found within the same time bound.

(ii) Let C, C' be convex algebraic splinegons, and let Ω be a simple algebraic splinegon. Then, the lower bound $dist\,(\Omega, conv\,D(q, p))$ for $\liminf_{\delta \to 0} V^\delta$ and $\liminf_{\delta \to 0} V_\delta$ can be computed in time $O(|C| + |C'| + |\Omega|)$. The upper bound $dist\,(\Omega, D(q, p))$ for $\limsup_{\delta \to 0} V^\delta$ and $\limsup_{\delta \to 0} V_\delta$ can be determined in time $T = O(|C| + |C'| + \Lambda)$, where Λ denotes the time needed for computing the distance between $D(q, p)$ and Ω; in particular, if either $D(q, p)$ or Ω is convex then $T = O(|C| + |C'| + |\Omega|)$.

We now continue investigating the strategy $\Gamma^* = \{\widetilde{\Gamma}_\delta\}$ for the pursuer and the strategy $\Delta^* = \{\widetilde{\Delta}_\delta\}$ for the evader, where $\widetilde{\Gamma}_\delta$ and $\widetilde{\Delta}_\delta$ are introduced in the proof of Theorem 7.

Theorem 9 *If the distance from Ω to $conv\,D(q, p)$ equals the distance from Ω to $D(q, p)$ then the game of guarding a territory has value, which is $V = dist\,(\Omega, D(q, p))$. The strategies Γ^* for P and Δ^* for E are optimal. Furthermore,*

(i) if C and C' are convex polygons, and Ω is a simple polygon, then V and the strategy Δ^* for E can be determined in time $O(|C| + |C'| + |\Omega|)$.

(ii) if C and C' are convex algebraic splinegons, and Ω is a simple algebraic splinegon, then V and the strategy Δ^* for E can be computed in time $T = O(|C| + |C'| + \Lambda)$, where $\Lambda = \Lambda(|C| + |C'|, |\Omega|)$; in particular, if either $D(q, p)$ or Ω is convex then $T = O(|C| + |C'| + |\Omega|)$.

Proof. If $dist\,(\Omega, conv\,D(q, p)) = dist\,(\Omega, D(q, p))$ then by Theorem 7 we have $\lim_{\delta \to 0} V^\delta = \lim_{\delta \to 0} V_\delta = dist\,(\Omega, D(q, p))$. It follows that $V = dist\,(\Omega, D(q, p))$. From the proof of Theorem 7 we see that the strategies Γ^* and Δ^* for P and E are optimal. Parts (i) and (ii) of the theorem follow from Lemma 8. Checking whether the sufficient condition for the existence of V is satisfied does not require more time. $\square$

3.2 Many Pursuers

We now consider a pursuit game where the pursuer is a team composed of M point pursuers $P_1, \ldots, P_M$. Assume that for $i = 1, \ldots, M$ we have $C' \prec C_i$, where C' denotes the indicatrix of the point evader E and C_i denotes the indicatrix of P_i. This game is a general variant of a game considered by Isaacs under the name *The two cutters and the fugitive ship* [11, p. 148]. Let the initial positions of $P_1, \ldots, P_M$ and E be $P_1(t_0) = p_1, \ldots, P_M(t_0) = p_M$ and $E(t_0) = q$, correspondingly.

We set $N = \sum_{i=1,\ldots,M} |C_i| + |C'| = (\sum_{i=1,\ldots,M} |C_i|) + M|C'|$ and $S = \{q, p_1, \ldots, p_M\}$. Define $\widetilde{R} = \bigcap_{i=1,\ldots,M} conv\,D(q, p_i)$. Let k denote the maximal number of intersections between any two edges of the bisectors $B(p_i, q)$ for $i = 1, \ldots, M$.

Let each pursuer P_i (with $i = 1, \ldots, M$) choose the strategy $\Gamma_i^* = \{\widetilde{\Gamma}_\delta\}$ and the evader E choose the strategy $\Delta^* = \{\widetilde{\Delta}_\delta\}$, where $\widetilde{\Gamma}_\delta$ and $\widetilde{\Delta}_\delta$ are as described in the proof of Theorem 7.

Theorem 10 *If the distance from Ω to $\widetilde{R}$ equals the distance from Ω to $R(q,S)$ then the game of guarding a territory has value, which is $V = dist\,(\Omega, R(q,S))$. The strategies Γ_i^* for P_i $(i = 1, \ldots, M)$ and Δ^* for E are optimal. Furthermore,*

(i) if $C_1, \ldots, C_M$ and C' are convex polygons, and Ω is a simple polygon, then V and the strategy Δ^* for E can be determined in time $O(N \log N + |\Omega|)$.

(ii) if $C_1, \ldots, C_M$ and C' are convex algebraic splinegons, and Ω is a simple algebraic splinegon, then the value V and the strategy Δ^* can be computed in time $T = O(\lambda_{k+1}(N) \log N + \Lambda)$, where $\Lambda = \Lambda(\lambda_{k+2}(N), |\Omega|)$. Especially, if either $R(q,S)$ or Ω is convex then $T = O(\lambda_{k+2}(N) \log N + |\Omega|)$.

3.3 Pursuer and Evader have the Same Indicatrix of Velocities

Assume now that $C = C'$. If C is strictly convex then it is not hard to see that the bisector is a bi-infinite curve that can be determined in time $O(|C|)$ by using reciprocal search as in the proof of Lemma 3. Results analogous to that in Sections 3.1 and 3.2 also hold in the case considered here. But now it may hold that $conv\, D(q,p) = \mathbf{R}^2$ (recall that if $C \prec C'$ then $conv\, D(q,p)$ is a bounded set), which makes the situation more complicated.

3.4 Pursuer and Evader Have Convex Shapes

In this section, we study a generalization of a problem that was considered by Isaacs [11, *The football players* p. 145] and Friedman [8, *Simple blocking game* p. 205]. In their problem the pursuer is a circle and the evader is a point, both move with simple motion; the territory is assumed to be a halfplane. While the solution of Isaacs is rather heuristic, the rigorous solution of Friedman makes use of heavy analytical tools ("Isaacs equation", which is a nonlinear partial differential equation). In our problem both the pursuer and the evader are allowed to be arbitrary convex objects.

Let KP and KE be the shapes of the pursuer and the evader, respectively. The payoff of the game is defined to be the distance from the shape of E to Ω when capture occurs, i.e., when the shapes of P and E collide. We fix a reference point on KP, and let KP_p denote KP located in $\mathbf{R}^2$ with its reference point at $p \in \mathbf{R}^2$. We similarly introduce the notation KE_q so that if $q = 0$ then $KE_q = KE$.

Applying the *configuration space* approach [13] we grow KP_p and Ω by KE to obtain objects in the configuration space: $KP_p \ominus KE$, $\Omega \ominus KE$, and the point q that represents KE_q, where $\ominus$ denotes the Minkowski difference, i.e., if $A, B \subset \mathbf{R}^2$ then $A \ominus B := \{a - b \,|\, a \in A \text{ and } b \in B\}$. The set $KP_p \ominus KE$ is convex and can be computed efficiently [9], [4]. Thus, in this way we have reduced the original game to a game in which the pursuer is a convex object and the evader is a point. Note that we have to grow the shape of the pursuer by the shape of the evader, and not vice versa. Note also that the distance between KE_q and Ω can be inferred from the distance between q and $\Omega \ominus KE$. It can be shown that the bisector of a convex object and a point is a bi-infinite convex differentiable curve, whose convex hull is the region of dominance of the given point. Now let $P(t)$ and $E(t)$ be the positions of the reference points of KP and

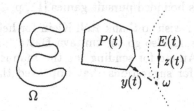

Figure 3: Illustrating the basic guarding rule for the pursuer. The dotted curve is the bisector of the convex set $KP_{P(t)} \ominus KE$ and the point $E(t)$.

KE at time t, respectively. As described above, we can equivalently consider the pursuit game for the pursuer with the shape $KP_{P(t)} \ominus KE$ (initially at p, i.e., $P(t_0) = p$) and the point evader (initially at q). Similarly as in Theorem 7, we consider the following *basic guarding rule* for the pursuer:

- **Basic guarding rule.** See Fig. 3. If E is within the step size then capture it. Otherwise consider, for each t in the corresponding time interval, the current goal point ω of E on the bisector $B(KP_{P(t)} \ominus KE, E(t))$ (if such a goal point exists). Now compute the tangent to $B(KP_{P(t)} \ominus KE, E(t))$ at ω, and choose $y(t)$ so that $y(t)$ and $z(t)$ are symmetric to that tangent. If ω does not exist — the evader "retreats" — then just choose $y(t) = z(t)$.

Let Γ^* denote the pursuer strategy which is deduced from the basic guarding rule above in a similar manner as in the proof of Theorem 7. Let Δ^* denote the evader strategy that linearly leads the evader to the point of $D(q, KP_p \ominus KE)$ nearest to Ω. The following theorem is the main result of this subsection.

Theorem 11 *The pursuit game stated above has value: V ist the distance between the set $\Omega \ominus KE$ and the set $D(q, KP_p \ominus KE)$. The strategies Γ^* for the pursuer and Δ^* for the evader are optimal. Furthermore, let KP and KE be convex algebraic splinegons.*

(i) If Ω is a convex algebraic splinegon, then the value V and the strategy Δ^* for E can be determined in time $O(|KP| + |KE| + |\Omega|)$.

(ii) If Ω is an algebraic splinegon, then V and Δ^* can be computed in time $O(|KP| + |KE| \cdot |\Omega|)$.

4 Discussion

If $dist(\Omega, conv\, D(q,p)) < dist(\Omega, D(q,p))$, we were not able to prove nor disprove whether optimal strategies for the pursuer and the evader exist, see Theorem 7.

It might be interesting to consider analogous pursuit games where the pursuer and the evader are constrained to move within a simple polygon with simple motion (von Neumann's bounded pursuit games [11, p. 270]).

Acknowledgements. I wish to thank Rolf Klein for helpful discussions about Voronoi diagrams. Thanks also go to Amitava Datta for some pointers to the literature, and Nancy Amato for sending me the manuscript [2]. I also thank an anonymous referee for suggestions that improved the presentation of this extended abstract.

References

[1] P. Agarwal, M. Sharir and P. Shor: *Sharp upper and lower bounds for the length of general Davenport-Schinzel sequences.* J. Combin. Theory, Ser. A 52(1989), pp. 228–274.

[2] N. M. Amato: *Determining the Separation of Simple Polygons.* To appear in Internat. J. Comp. Geom. Appl.

[3] F. Aurenhammer: *Voronoi Diagrams—A Survey of a Fundamental Data Structure.* ACM Computer Surveys 23(3), 1991.

[4] C. Bajaj and M.-S. Kim: *Generation of Configuration Space Obstacles: The Case of Moving Algebraic Curves.* Algorithmica 4(1989), pp. 157–172.

[5] C. Bajaj and M.-S. Kim: *Convex Hulls of Objects Bounded by Algebraic Curves.* Algorithmica 6(1991), pp. 533–553.

[6] D. B. Dobkin, D. L. Souvaine, and C. J. Van Wyk: *Decomposition and Intersection of Simple Splinegons.* Algorithmica 3(1988), pp. 473–485.

[7] D. B. Dobkin and D. L. Souvaine: *Computational Geometry in a Curved World.* Algorithmica 5(1990), pp. 421–457.

[8] A. Friedman: *Differential Games.* Wiley—Interscience, New York, 1971.

[9] L. J. Guibas and R. Seidel: *Computing Convolutions by Reciprocal Search.* Discrete Comput. Geom. 2(1987), pp. 175–193.

[10] J. Hershberger: *Finding the upper envelope of n line segments in $O(n\log n)$ time.* Information Processing Letters 33(1989), pp. 169–174.

[11] R. Isaacs: *Differential Games.* Wiley, New York, 1965.

[12] N.-M. Lê: *On Determining Optimal Strategies in Pursuit Games in the Plane.* Technical Report 172, Dep. of Comp. Science, FernUniversität Hagen, Germany.

[13] T. Lozano-Pérez: *Spatial planning: A configuration space approach.* IEEE Trans. Computers 32(2), pp. 108–120, 1983.

[14] F. P. Preparata and M. I. Shamos: *Computational Geometry.* Springer-Verlag, New York, 1985.

[15] J. H. Reif and S. R. Tate: *Continuous Alternation: The Complexity of Pursuit in Continuous Domains.* Algorithmica 10(1993), pp. 156–181.

Extension Orderings*

Albert Rubio

Universitat Politècnica de Catalunya, Pau Gargallo 5, 08028 Barcelona, Spain
E-mail: rubio@lsi.upc.es.

Abstract. In this paper we study how to extend a collection of term orderings on disjoint signatures to a single one, called an *extension ordering*, which preserves (part of) their properties. Apart of its own interest, e.g. in automated deduction, extension orderings turn out to be a new method to obtain simple and constructive proofs for modularity of termination of TRS. Three different schemes to define extension orderings are given. The first one to deal with reduction orderings, the second one to extend simplification orderings and the last one for total reduction orderings. This provides simpler and more constructive proofs for some known modularity results for (simple and total) termination of rewriting as well as some new results for rewriting modulo equational theories. Finally, our technique is applied to extend an ordering on a given signature to a new one on the signature enlarged with new symbols.

1 Introduction

General methods for proving termination are crucial for decision procedures and for using rewriting-like methods in theorem proving and programming. A well-known related problem in rewrite theory is to prove the modularity of termination of *term rewrite systems* (TRS's), that is, given two terminating rewrite systems R_1 and R_2, to show that their union $R_1 \cup R_2$ is also terminating. This property was shown to be false in general, even when the two rewrite systems do not share any function symbols [17]. Therefore many efforts have been devoted to finding sufficient conditions under which termination is modular. Most existing results impose syntactic (or abstract) conditions on (one of) the two TRS's (e.g. [16, 5, 12, 8, 14, 6]), and some other ones impose restrictions on the method used to show the termination of the involved TRS's ([10, 7]).

In this context it is quite surprising that, although usually termination of TRS's is proved by actually building a *reduction ordering* in which the rewrite relation is contained (cf. [4]), in most modularity proofs for termination such an ordering is not explicitly built, that is, in this sense these proofs are not constructive. Hence, for simplicity and to improve the understanding of the problem, it is worth to look at it from a more constructive point of view. This can give information about the weakest conditions ensuring modularity and why they are needed. As a good example for this viewpoint, *total termination* (i.e. the rewrite

---

* The author wishes to thank Hubert Comon, Jean-Pierre Jouannaud, Robert Nieuwenhuis and Aart Middeldorp for many helpful comments. Partially supported by the Esprit Working Group CCL, ref. 6028.

relation is included in some total reduction ordering) was proved modular provided that one of the TRS's is conservative in [7]. The reason for this requirement seems to be rather technical, and the proof does provide no intuition of its need. Here we prove the same result in a simple way, actually building a total ordering containing the combined system. This gives some intuition about the restriction.

Our main idea is to generalize the combination of TRS's to the combination of orderings. Instead of proving that the ordering induced by the combined rewrite system is well-founded, we build a well-founded ordering and prove that it includes the combined TRS (this technique was already used in [16]). This is done by means of *extension orderings*. By an extension ordering of a collection of orderings $\succ_1, \ldots, \succ_n$ on disjoint signatures, we mean another ordering $\succ$ s.t. $\succ_i \subseteq \succ$ for all i. Actually, an extension ordering is interesting when it preserves (part of) the properties satisfied by the initial orderings.

In section 3 we define an extension ordering $\succ_e$ on ground terms, which preserves well-foundedness of the $\succ_i$. Moreover, if every $\succ_i$ is monotonic, then $\succ_e$ fulfills a property called *quasi-monotonicity* which suffices for termination of rewriting. As expected, due to the non-modularity of termination of *disjoint union* of TRS's, $\succ_e$ cannot be lifted, in general, to terms with variables. However, lifting is possible for right-linear systems. This gives us a simple constructive proof for modularity of right-linear TRS's.

Section 4 is on simple termination (i.e. the rewrite relation is included in some simplification ordering). We define an extension ordering $\succ_s$ preserving monotonicity and the subterm property to prove the modularity of simple termination (this was first proved in a non-constructive way by [10]). In section 5 an extension ordering preserving well-foundedness, monotonicity and totality on ground terms is given, allowing us to prove the results of [7].

In section 6 these extension orderings are proved to preserve also E-compatibility for certain classes of equational theories E. Therefore we can obtain some new modularity results for rewriting modulo equational theories. This is a first example of new results that can be derived from our extension orderings approach (this problem is also studied in [6], which is, as far as we know, the only other work on this topic).

In section 7 we extend any ordering, to allow new symbols. These results can sometimes be used to prove termination of *hierarchical unions* of TRS's [3, 9]. In section 8 some possible improvements to the present results are commented and the general application of the method in termination proofs is discussed.

2 Preliminaries

We adopt the standard notations and definitions for term rewriting given in [4]. Here we recall some of them.

$|t|$ is the size (number of symbols) of t. $Vars(t)$ and $Mvars(t)$ are respectively the set and the multiset of variables of t. Operations on sets and multisets are denoted ambiguously. $t|_p$ is the subterm of t at position p; $t[s]_p$ is obtained from

t by replacing $t|_p$ by s, and $\mathcal{H}ead(t)$ is the topmost function symbol of t. $=_E$ is the congruence generated on $T(\mathcal{F})$ by the equations E.

Below let s and t (possibly indexed) be terms in $T(\mathcal{F})$ and f a function symbol in $\mathcal{F}$. A (strict partial) ordering on $T(\mathcal{F})$ (a transitive irreflexive relation) $\succ$ fulfills the *subterm property* if $f(\ldots t \ldots) \succ t$ for all f and t. It is *monotonic* if $s \succ t$ implies $f(\ldots s \ldots) \succ f(\ldots t \ldots)$. A monotonic ordering fulfilling the subterm property is a *simplification ordering* and for finite signatures it is *well-founded*: there is no infinite sequence $t_1 \succ t_2 \succ \ldots$ An ordering $\succ$ is *E-compatible* if $s' =_E s \succ t =_E t'$ implies $s' \succ t'$. Finally we say that an ordering is total on (the E-congruence classes of) $T(\mathcal{F})$ if $s \neq t$ ($s \neq_E t$) implies $s \succ t$ or $t \succ s$ for all terms $s, t \in T(\mathcal{F})$. Any well-founded monotonic ordering $\succ$ total on $T(\mathcal{F})$ is a simplification ordering.

An ordering on $T(\mathcal{F}, \mathcal{X})$ is *stable under substitutions* if $s \succ t$ implies $s\sigma \succ t\sigma$ for all terms s and t and substitutions σ. A well-founded monotonic ordering stable under substitutions is called a *reduction ordering*. An ordering $\succ$ on $T(\mathcal{F})$ can always be lifted to an ordering $\succ_v$ (or simply $\succ$) on $T(\mathcal{F}, \mathcal{X})$, which is stable under substitutions in the following way: if s and t be terms in $T(\mathcal{F}, \mathcal{X})$, then $s \succ t$ if $s\sigma \succ t\sigma$ for all substitutions σ with range in $T(\mathcal{F})$.

Let $\succ$ be a relation on terms. Then $max_{\succ}(S)$ is the set of all maximal terms in the set of terms S, i.e. the set S' s.t. $t \in S'$ iff there is no s in S such that $s \succ t$. When working modulo an equational theory E, $max_{\succ}(S)$ is any set obtained from S' as defined above by choosing one element in each E-congruence class.

Given an ordering $\succ$ on terms and a congruence relation $=$, the lexicographic extension $\succ^{lex}$ of $\succ$ wrt. $=$ for n-tuples is defined as: $\langle s_1, \ldots, s_n \rangle \succ^{lex} \langle t_1, \ldots, t_n \rangle$ iff $\exists k$ s.t. $s_1 = t_1, \ldots, s_{k-1} = t_{k-1}$ and $s_k \succ t_k$.

The extension to finite multisets of a congruence relation $=$ is defined as: $\{s_1, \ldots, s_m\} = \{t_1, \ldots, t_n\}$ iff $m = n$ and there exists some permutation π of $1 \ldots m$, s.t. $s_{\pi(i)} = t_i$ for all $i : 1 \ldots n$. The extension, wrt. a congruence $=$, of an ordering $\succ$ to finite multisets, denoted by $\twoheadrightarrow$, is defined by: $M = \{s_1, \ldots, s_m\} \twoheadrightarrow \{t_1, \ldots, t_n\} = N$ iff $M \supset N$ or $s_i \succ t_{j_1} \wedge \ldots \wedge s_i \succ t_{j_k}$ and $(M \setminus \{s_i\}) \succeq\!\!\!\succ N \setminus \{t_{j_1}, \ldots, t_{j_k}\}$ for some $i = 1 \ldots m$ and $1 \leq j_1 < \ldots < j_k \leq n$ ($k = 0 \ldots n$), where $M \succeq\!\!\!\succ N$ means $M \twoheadrightarrow N$ or $M = N$.

If $\succ$ is a well-founded ordering on terms then $\succ^{lex}$ and $\twoheadrightarrow$ are well-founded orderings on n-tuples of terms and on finite multisets of terms respectively, more precisely $\succ^{lex}$ and $\twoheadrightarrow$ preserve irreflexivity, transitivity and well-foundedness.

A term rewrite system (TRS) is a set of rules $l \rightarrow r$ where l and r are terms in $T(\mathcal{F}, \mathcal{X})$ with $Vars(l) \supseteq Vars(r)$. The union of two rewrite systems R_1 and R_2 is called a *disjoint union* (or *direct sum*), and denoted by $R_1 \uplus R_2$, if R_1 and R_2 do not share any symbol. For a given TRS R and terms s and t in $T(\mathcal{F}, \mathcal{X})$, we say that s *rewrites* into t, written $s \rightarrow_R t$ if $s|_p = l\sigma$ and $t = s[r\sigma]_p$ for some rule $l \rightarrow r \in R$, position p in s and substitution σ. Moreover, s rewrites modulo E into t, denoted by $s \rightarrow_{R/E} t$ if $s =_E s'$, $s'|_p = l\sigma$ and $t = s'[r\sigma]$, for some rule $l \rightarrow r \in R$, term s', position p in s' and substitution σ. R is *terminating* (resp. *E-terminating*) over $T(\mathcal{F})$ if $\rightarrow_R$ (resp. $\rightarrow_{R/E}$) is well-founded over $T(\mathcal{F})$. The embedding relation $\succ_{emb}$ is the smallest simplification ordering, i.e. the relation

$\rightarrow^+_{emb}$ generated by the rewrite system $emb = \{f(\ldots x \ldots) \rightarrow x \mid \forall f \in \mathcal{F}\}$.

If R is a TRS with left and right hand sides in $T(\mathcal{F}, \mathcal{X})$ then it is $(E\text{-})$terminating over $T(\mathcal{F})$ iff there is an $(E\text{-compatible})$ well-founded monotonic ordering $\succ$ s.t. $l\sigma \succ r\sigma$ for all $l \rightarrow r \in R$ and ground substitution σ with range in $T(\mathcal{F})$. Therefore if $\succ$ is an $(E\text{-compatible})$ reduction ordering it suffices to prove $l \succ r$ for all rules $l \rightarrow r$ in R. In fact, if R is $(E\text{-})$terminating then $\rightarrow^+_R$ is itself an $(E\text{-compatible})$ reduction ordering. If R is $(E\text{-})$terminating over $T(\mathcal{F})$ (and $\mathcal{F}$ contains at least one constant symbol) then R is $(E\text{-})$terminating over $T(\mathcal{F} \cup \mathcal{F}')$ for any set of function symbols $\mathcal{F}'$. This means that we can, in general, speak about termination without mentioning over which signature.

From now on we suppose that we have n relations $\succ_i$ over n sets $T(\mathcal{F}_i)$ respectively, where all $\mathcal{F}_i$ are disjoint and contain at least one constant symbol.

3 Extension of reduction orderings and modularity

In this section we extend general orderings on terms and study modularity of termination of TRS's. The ground case is considered first. Then we analyze the applicability of our ordering to terms with variables and termination of TRS's. The following mapping N_i eliminates from terms in $T(\mathcal{F})$ those function symbols that are not in $\mathcal{F}_i$ and is the basis of our extension ordering $\succ_e$.

Definition 1. N_i is the mapping from terms in $T(\mathcal{F})$ to multisets of terms in $T(\mathcal{F}_i)$ recursively defined as:

$$N_i(f(t_1,\ldots,t_m)) = \begin{cases} \{f(t'_1,\ldots,t'_m) \mid t'_j \in N_i(t_j) \cup Z_i(t_j)\} & \text{if } f \in \mathcal{F}_i \\ N_i(t_1) \cup \ldots \cup N_i(t_m) & \text{otherwise} \end{cases}$$

where 0_i is some (fixed) constant symbol in $\mathcal{F}_i$ and $Z_i(t) = \{0_i\}$ if $t \notin T(\mathcal{F}_i)$ and $Z_i(t) = \emptyset$ otherwise.

Proposition 2. For all $i, j : 1 \ldots n$ and any $t \in T(\mathcal{F}_i)$, if $j \neq i$ then $N_j(t) = \emptyset$ and if $j = i$ then $N_j(t) = \{t\}$. Furthermore, for all $s, t \in T(\mathcal{F})$ and $f \in \mathcal{F}$, if $s \notin T(\mathcal{F}_i)$ or both s and t are in $T(\mathcal{F}_i)$ then $N_i(s) = N_i(t)$ implies $N_i(f(\ldots s \ldots)) \supseteq N_i(f(\ldots t \ldots))$.

Definition 3. The relation $\succ_e$ on $T(\mathcal{F})$ extending relations $\succ_i$ for all $i : 1 \ldots n$ is defined as: $\quad s \succ_e t \quad$ iff

$$N_1(s) = N_1(t) \wedge \ldots \wedge N_{i-1}(s) = N_{i-1}(t) \wedge N_i(s) \gg_i N_i(t) \text{ for some } i : 1 \ldots n$$

Of course, the roles of all $\succ_i$ (and N_i) are symmetric here, and can be exchanged, obtaining in fact different $\succ_e$.

The following lemmas show that $\succ_e$ preserves several properties of the $\succ_i$ and extends them.

Lemma 4. $\succ_e$ extends $\succ_i$ for all $i : 1 \ldots n$.

Lemma 5. *If $\succ_i$ is an ordering for all $i : 1 \ldots n$ then $\succ_e$ is an ordering.*

Lemma 6. *If $\succ_i$ is well-founded for all $i : 1 \ldots n$ then $\succ_e$ is well-founded.*

Proof. Supose that there exists an infinite sequence $t_1 \succ_t t_2 \succ_t \ldots$. By well-foundedness of $\succ_1$ and hence of $\gg_1$, there must be some t_j in the sequence s.t. $N_1(t_j) = N_1(t_k)$ for all $k > j$. Let $t_1^1 \succ_t t_2^1 \succ_t \ldots$ the infinite subsequence starting from t_j. Repeating this reasoning n times it follows that there must be an infinite subsequence $t_1^n \succ_t t_2^n \succ_t \ldots$ s.t. $N_i(t_1^n) = N_i(t_2^n) = \ldots$ for all $i : 1 \ldots n$, which contradicts the existence of such an infinite decreasing sequence. $\square$

Although $\succ_e$ is not monotonic in general it fulfills the following property which we call *quasi-monotonicity* and which, as we will see, is enough to ensure termination of TRS's.

Lemma 7. *Let s and t be terms in $T(\mathcal{F})$. If $\succ_i$ is monotonic and $s \notin T(\mathcal{F}_i)$ or both s and t are in $T(\mathcal{F}_i)$ for all $i : 1 \ldots n$ then $s \succ_e t$ implies $f(\ldots s \ldots) \succ_e f(\ldots t \ldots)$, i.e. $\succ_e$ is* quasi-monotonic.

Proof. Assume $N_1(s) = N_1(t) \wedge \ldots \wedge N_{j-1}(s) = N_{j-1}(t) \wedge N_j(s) \gg_j N_j(t)$, for some $j : 1 \ldots n$. We have to show $f(\ldots s \ldots) \succ_e f(\ldots t \ldots)$. By proposition 2, since $t \in T(\mathcal{F}_i)$ whenever $s \in T(\mathcal{F}_i)$ for all $i : 1 \ldots n$, we have $N_1(f(\ldots s \ldots)) \succeq \succeq_1 N_1(f(\ldots t \ldots)) \wedge \ldots \wedge N_{j-1}(f(\ldots s \ldots)) \gg_{j-1} N_{j-1}(f(\ldots t \ldots))$. Assume $N_1(f(\ldots s \ldots)) = N_1(f(\ldots t \ldots)) \wedge \ldots \wedge N_{j-1}(f(\ldots s \ldots)) = N_{j-1}(f(\ldots t \ldots))$ (otherwise the lemma already holds):

- If $f \notin \mathcal{F}_j$ then $N_j(f(\ldots s \ldots)) = N_j(s) \cup X$ and $N_j(f(\ldots t \ldots)) = N_j(t) \cup X$ for some X obtained from the transformations of the other top level arguments, and therefore $N_j(f(\ldots s \ldots)) \gg_j N_j(f(\ldots t \ldots))$.
- If $f \in \mathcal{F}_j$ then we have $N_j(f(\ldots s \ldots)) = \{f(\ldots s' \ldots) \mid s' \in N_j(s) \cup Z_j(s)\}$ and $N_j(f(\ldots t \ldots)) = \{f(\ldots t' \ldots) \mid t' \in N_j(t) \cup Z_j(t)\}$. Since $t \in T(\mathcal{F}_j)$ whenever $s \in T(\mathcal{F}_j)$, either $Z_j(s) = Z_j(t) = \emptyset$ or $Z_j(s) \neq \emptyset$. Therefore, by monotonicity of $\succ_j$, $N_j(f(\ldots s \ldots)) \gg_j N_j(f(\ldots t \ldots))$.

Since in both cases we have $N_j(f(\ldots s \ldots)) \gg_j N_j(f(\ldots t \ldots))$, it holds that $f(\ldots s \ldots) \succ_e f(\ldots t \ldots)$. $\square$

Now we are going to lift $\succ_e$ to terms with variables ($s \succ_e t$ iff $s\sigma \succ_e t\sigma$ as in section 2) in order to obtain modularity results. Let R be $R_1 \uplus \ldots \uplus R_n$ and let $\succ_i$ be reduction orderings on $T(\mathcal{F}_i, \mathcal{X})$ containing the terminating rewrite relations $\to^+_{R_i}$ respectively.

A possible way to prove termination of R over $T(\mathcal{F})$ (see e.g. [2]) is to show that there exists a well-founded ordering $\succ$ on $T(\mathcal{F})$ s.t.: (1) $l\sigma \succ r\sigma$ for all rules $l \to r \in R$ and all substitutions σ with range in $T(\mathcal{F})$, and (2) $s \to_R t$ and $s \succ t$ implies $f(\ldots s \ldots) \succ f(\ldots t \ldots)$ for all $s, t \in T(\mathcal{F})$ and $f \in \mathcal{F}$. From these two properties we can derive that $s \succ t$ whenever $s \to_R t$, for all terms $s, t \in T(\mathcal{F})$, i.e. R is terminating over $T(\mathcal{F})$. Property (2) is fulfilled by $\succ_e$:

Lemma 8. *Let s and t be terms in $T(\mathcal{F})$. Then $s \to_R t$ and $s \succ_e t$ implies $f(\ldots s \ldots) \succ_e f(\ldots t \ldots)$ for all $f \in \mathcal{F}$.*

However, property (1) does not hold in general for $\succ_e$. This is not surprising since this result would imply the termination of the direct sum of any two terminating rewriting systems, which is well-known to be false, as shown by Toyama's counter example ([17]). If we apply our method to this example, indeed the resulting ordering does not fulfil the property:

Example 1. Let $\mathcal{F}_1$ and $\mathcal{F}_2$ be $\{f, 1, 0_1\}$ and $\{g, 0_2\}$ repectively and let $\succ_1$ and $\succ_2$ be the orderings generated by the TRS's $R_1 = \{f(1, 0_1, x) \to f(x, x, x)\}$ and $R_2 = \{g(x, y) \to x, g(x, y) \to y\}$ respectively. Then with the substitution $\{x \mapsto g(1, 0_1)\}$ we have:
$N_1(f(1, 0_1, g(1, 0_1))) = \{f(1, 0_1, 1), f(1, 0_1, 0_1), f(1, 0_1, 0_1)\}$ and
$N_1(f(g(1, 0_1), g(1, 0_1), g(1, 0_1))) = \{f(1, 1, 1), \ldots, f(1, 0_1, 1), f(1, 0_1, 0_1),$
$f(1, 0_1, 0_1), \ldots, f(0_1, 0_1, 0_1)\}$, which implies $f(g(1, 0_1), g(1, 0_1), g(1, 0_1)) \succ_e$
$f(1, 0_1, g(1, 0_1))$, contradicting the rule of R_1. □

In fact, the problem is that although $\succ_e$ extends each $\succ_i$ at the ground level, this becomes false at the variable level. Nevertheless, by imposing some syntactic conditions on the TRS's, we can prove that $\succ_e$ includes the rewrite relation $\to_R$ on $T(\mathcal{F})$. We consider the case where all rules $l \to r$ in R are *right linear*: all variables in $Vars(r)$ occur only once in r, i.e. r is *linear*.

Proposition 9. *Let R be a right-linear TRS over a signature $\mathcal{F}$ and let 0 be a new constant symbol (not occurring in R). Then R is terminating iff $R'' = R \cup R'$ is terminating, where $R' = \{f(\ldots 0 \ldots) \to 0 \mid f \in \mathcal{F}\}$.*

Note that this property does not hold for non right linear TRS:

Example 2. The TRS $R = \{f(x, x) \to f(g(x), h(x))\}$ is terminating. But $R \cup R'$ is not: $f(0, 0) \to_{R''} f(g(0), h(0)) \to_{R''} f(0, h(0)) \to_{R''} f(0, 0) \to''_R \ldots$ □

By proposition 9 we can consider that if R_i is right linear and terminating then there is a reduction ordering $\succ_i$ on $T(\mathcal{F}_i)$ including $\to^+_{R_i}$, where $\mathcal{F}_i$ contains a constant symbol, which is taken as 0_i, s.t. $f(\ldots 0_i \ldots) \succ_i 0_i$.

Lemma 10. *Let s and t be terms in $T(\mathcal{F}_i, \mathcal{X})$ for some $i : 1 \ldots n$. If $\succ_i$ is stable under substitutions and t is linear then $s \succ_i t$ implies $s\sigma \succ_e t\sigma$ for all substitutions σ with range in $T(\mathcal{F})$, i.e. $s \succ_e t$.*

Proof. (Hint) Use the fact that if $t \in T(\mathcal{F}_i, \mathcal{X})$ for some $i : 1 \ldots n$ and t is not a variable then $N_i(t\sigma) \supseteq \{t\sigma' \mid x\sigma' \in N_i(x\sigma) \cup Z_i(x\sigma) \wedge x \in Vars(t)\}$ for all ground substitutions σ with range in $T(\mathcal{F})$. Moreover if t is linear then $max_{\succ_i}(N_i(t\sigma)) \subseteq \{t\sigma' \mid x\sigma' \in N_i(x\sigma) \cup Z_i(x\sigma) \wedge x \in Vars(t)\}$ for all ground substitutions σ with range in $T(\mathcal{F})$. □

This lemma leads to a constructive proof of modularity of termination for direct sums of *right linear* TRS's:

Theorem 11. *If $R_1, \ldots, R_n$ are repectively right linear terminating TRS's over $T(\mathcal{F}_1), \ldots, T(\mathcal{F}_n)$ then $R = R_1 \uplus \ldots \uplus R_n$ is terminating.*

Proof. If $R_1, \ldots, R_n$ are terminating then $\rightarrow_{R_1} \ldots \rightarrow_{R_n}$ are repsctively included in some reduction orderings $\succ_1 \ldots \succ_n$. Then, by lemmas 6, 5, 8 and 10 the extension ordering $\succ_e$ of $\succ_1 \ldots \succ_2$ is a well-founded ordering s.t. $s \succ_e t$ for all ground terms $s, t \in T(\mathcal{F})$ with $s \rightarrow_R t$, i.e. $R_1 \uplus \ldots \uplus R_n$ is terminating. $\qquad\square$

4 Extension of Simplification Orderings and Modularity

Definition 12. A TRS R (with left and right hand sides in $T(\mathcal{F})$) is *simply terminating* over $T(\mathcal{F})$ iff there exists some simplification ordering $\succ$ s.t. $l\sigma \succ r\sigma$ for all $l \rightarrow r$ in R and substitutions σ with range in $T(\mathcal{F})$. We say that R is simply terminating (in general) if R is simply terminating over $T(\mathcal{F} \cup \mathcal{F}')$ for any set of new function symbols $\mathcal{F}'$.

At the end of this section we will see that if a TRS (with left and right hand sides in $T(\mathcal{F})$) is simply terminating over $T(\mathcal{F})$ then it is simply terminating in general.

We present a new extension ordering which extends any collection of simplification orderings at the ground level and at the variable level. Therefore, the modularity of simple termination follows.

Definition 13. Let S_i be a mapping from terms in $T(\mathcal{F})$ to terms in $T(\mathcal{F}_i)$ recursively defined as:

$$S_i(f(t_1, \ldots, t_m)) = \begin{cases} 0_i & \text{if } f \notin \mathcal{F}_i \\ f(S_i(t_1), \ldots, S_i(t_m)) & \text{otherwise} \end{cases}$$

where 0_i is some (fixed) constant symbol in $\mathcal{F}_i$.
Let S be a mapping from terms in $T(\mathcal{F})$ to terms in $\cup_{i=1}^n T(\mathcal{F}_i)$ defined as:

$$S(t) = S_i(t) \qquad \text{if } \mathcal{H}ead(t) \in \mathcal{F}_i$$

Let S_E be a mapping from terms in $T(\mathcal{F})$ to multisets of terms in $\cup_{i=1}^n T(\mathcal{F}_i)$ defined as:

$$S_E(t) = \{S(t') \mid t \succeq_{emb} t'\}$$

Definition 14. Let s, t be terms in $T(\mathcal{F})$. Then
$s = f(s_1, \ldots, s_p) \succ_s g(t_1, \ldots, t_q) = t$ iff

- $max_{\succ_s}(S_E(s)) \gg_u max_{\succ_s}(S_E(t))$ or
- $max_{\succ_s}(S_E(s)) = max_{\succ_s}(S_E(t))$ and $f = g$ and $\{s_1, \ldots, s_p\} \gg_s \{t_1, \ldots t_p\}$

where the relation $\succ_u$ is defined as $\cup_{i=1}^n \succ_i$.

Note that, since all $\succ_i$ are orderings on $T(\mathcal{F}_i)$, if $s \succ_u t$ then $s, t \in T(\mathcal{F}_i)$ and $s \succ_i t$ for some $i : 1 \dots n$, which implies that $\succ_u$ is an ordering whenever all $\succ_i$ also are.

Lemma 15. *Let all $\succ_i$ be orderings on $T(\mathcal{F}_i)$. Then*
(i) $\succ_s$ is an ordering on $T(\mathcal{F})$
(ii) if $\succ_i$ is monotonic and fulfills the subterm property then $\succ_s$ extends $\succ_i$
(iii) if all $\succ_i$ have the subterm property then $\succ_s$ has the subterm property on $T(\mathcal{F})$.

Lemma 16. *If $\succ_i$ are monotonic orderings on $T(\mathcal{F}_i)$ for all $i : 1 \dots n$ then $\succ_s$ is monotonic on $T(\mathcal{F})$.*

Proof. (Hint) Use the fact that $max_{\succ_s}(S_E(s)) \succeq_u max_{\succ_s}(S_E(t))$ implies $max_{\succ_s}(S_E(f(\dots s \dots))) \succeq_u max_{\succ_s}(S_E(f(\dots t \dots)))$. □

From the previous lemmas, we can derive the following theorem:

Theorem 17. *If $\succ_i$ are simplification orderings on $T(\mathcal{F}_i)$ for all $i : 1 \dots n$. Then $\succ_s$ is a simplification ordering on $T(\mathcal{F})$ extending $\succ_i$ for all $i : 1 \dots n$.*

As in the previous section now $\succ_s$ is lifted to terms with variables. Let R_i be simply terminating TRS's over $T(\mathcal{F}_i)$ for all $i : 1 \dots n$, and let R be $R_1 \uplus \dots \uplus R_n$.

Lemma 18. *Let s and t be terms in $T(\mathcal{F}_i, \mathcal{X})$ for some $i : 1 \dots n$. If $\succ_i$ is a simplification ordering and $s\sigma' \succ_i t\sigma'$ for all substitutions σ' with range in $T(\mathcal{F}_i)$, then $s\sigma \succ_s t\sigma$ for all ground substitutions with range in $T(\mathcal{F})$, i.e. $s \succ_s t$.*

Proof. (Hint) Use the fact that if s (similarly t) is not a variable then we have $max_{\succ_s}(S_E(s\sigma)) = max_{\succ_s}(\bigcup_{x \in Vars(s)} S_E(x\sigma) \cup \{s\sigma' \mid (x\sigma' \in \{S_i(s') \mid x\sigma \succeq_{emb} s'\}) \wedge x \in Vars(s)\})$ for all substitutions σ with range in $T(\mathcal{F})$. □

Theorem 19. *If $R_1, \dots, R_n$ are repectively simply terminating TRS's over the sets $T(\mathcal{F}_1), \dots, T(\mathcal{F}_n)$ then $R = R_1 \uplus \dots \uplus R_n$ is simply terminating over $T(\mathcal{F})$.*

Note that this also implies that if R is simply terminating over $T(\mathcal{F})$ then R is simply terminating over $T(\mathcal{F} \cup \mathcal{F}')$ for any $\mathcal{F}'$, and hence that simple termination is modular.

5 Extension of Total Orderings and Modularity

In this section we consider *total termination* (which implies simple termination). For our purposes, the orderings $\succ_e$ and $\succ_s$ (defined in section 3 and 4) can not be used since they do not preserve totality of the initial orderings $\succ_i$. However, it is quite easy to adapt $\succ_e$ (the transformation N_i is replaced by T_i) to obtain an extension ordering $\succ_t$, which also preserves totality (and monotonicity). Moreover, the resulting extension ordering $\succ_t$ needs weaker sufficient conditions to extend orderings on terms with variables. Hence, we obtain better results on modularity of total termination. The proofs are similar to the ones used for $\succ_e$.

Definition 20. Let T_i be a mapping from terms in $T(\mathcal{F})$ to multisets of terms in $T(\mathcal{F}_i)$ recursively defined as:

$$T_i(f(t_1, \ldots, t_m)) = \begin{cases} \{f(t'_1, \ldots, t'_m) \mid t'_j \in \{0_i\} \cup T_i(t_j)\} & \text{if } f \in \mathcal{F}_i \\ T_i(t_1) \cup \ldots \cup T_i(t_m) & \text{otherwise} \end{cases}$$

where 0_i is some (fixed) minimal wrt. $\succ_i$ constant symbol in $\mathcal{F}_i$. Note that if $\succ_i$ is a total ordering then there is only one minimal constant, and if $\succ_i$ is also monotonic then 0_i is the minimal term wrt. $\succ_i$ in $T(\mathcal{F}_i)$.

Definition 21. Let $\succ_r$ be an arbitrary relation on $T(\mathcal{F})$ and let s and t be terms in $T(\mathcal{F})$. Then $s \succ_t t$ iff

- $T_1(s) = T_1(t) \wedge \ldots \wedge T_{i-1}(s) = T_{i-1}(t) \wedge T_i(s) \gg_i T_i(t)$ for some $i : 1 \ldots n$ or
- $T_i(s) = T_i(t)$ for all $i : 1 \ldots n$ and $s \succ_r t$

Theorem 22. *If $\succ_i$ are monotonic well-founded orderings total on $T(\mathcal{F}_i)$ for all $i : 1 \ldots n$. Then $\succ_t$ is a monotonic well-founded ordering total on $T(\mathcal{F})$ extending $\succ_i$ for all $i : 1 \ldots n$.*

A term t is *conservative* wrt. another term s if $\mathcal{M}vars(s) \supseteq \mathcal{M}vars(t)$. A TRS R is conservative if r is conservative wrt. l, for all $l \to r$ in R.

Lemma 23. *Let s and t be terms in $T(\mathcal{F}_i, \mathcal{X})$ for some $i : 1 \ldots n$ and let $\succ_i$ be a monotonic ordering total on $T(\mathcal{F}_i)$. If $i = 1$ or t is conservative wrt. s then $s\sigma' \succ_i t\sigma'$ for all substitutions σ' with range in $T(\mathcal{F}_i)$ implies $s\sigma \succ_t t\sigma$ for all substitutions σ with range in $T(\mathcal{F})$, i.e. $s \succ_t t$.*

Proof. (Hint) Use the fact that due to the monotonicity and totality of the orderings $\succ_i$ we have that the maximal terms (wrt. $\succ_i$) in $T_i(s\sigma)$ and $T_i(t\sigma)$ are respectively $s\sigma'$ and $t\sigma'$ for some substitution σ', s.t. $x\sigma'$ is the maximal term (wrt. $\succ_i$) in $T_i(x\sigma)$ for all $x \in Vars(s)$. $\square$

Again, from theorem 22 and lemma 23, we derive the following result on modularity of termination:

Definition 24. A TRS R (with left and right hand sides in $T(\mathcal{F}, \mathcal{X})$) is *totally terminating* iff there exists some well-founded monotonic ordering $\succ$ total on $T(\mathcal{F})$ s.t. $l\sigma \succ r\sigma$ for all $l \to r$ in R and substitutions σ with range in $T(\mathcal{F})$.

Theorem 25. *Let $R_1, \ldots, R_n$ be totally terminating TRS's. If $R_2, \ldots, R_n$ are conservative then $R = R_1 \uplus \ldots \uplus R_n$ is totally terminating.*

6 E-compatible Extension Orderings and Modularity

Here we will study how to deal with the extension orderings when we have E-compatible orderings. We will apply the same techniques as in previous sections to obtain easily some results for certain equational theories E, more precisely for *permutative* theories:

Definition 26. An equation $s = t$, with $s = f(s_1, \ldots, s_m)$ and $t = f(t_1, \ldots, t_m)$, is *permutative* if $\{s_1, \ldots, s_m\} = \{t_1, \ldots, t_m\}$. An equational theory E is permutative if it has a presentation in which all the axioms are permutative.

In the following the same extension orderings $\succ_e$, $\succ_s$ and $\succ_t$ defined as in the previous sections, but replacing $=$ by $=_E$, allow us to prove the same results as before, but for E-termination. Note that in fact we have each $\succ_i$ being E_i-compatible for some equational theory E_i on $T(\mathcal{F}_i)$, but since we consider only disjoint signatures it is equivalent to speak about E-compatibility for all $\succ_i$ where $E = E_1 \cup \ldots \cup E_n$.

Lemma 27. *If all $\succ_i$ are E-compatible orderings then $\succ_e$, $\succ_s$ and $\succ_t$ are E-compatible orderings for any permutative theory E.*

Proof. (Hint) Show that N_i, S_i and T_i are compatible with E. ☐

All other lemmas of the previous sections can be straightforwardly adapted thanks to the E-compatibility of all orderings. This leads to the following results:

Theorem 28. *Let $R_1, \ldots, R_n$ be E-terminating TRS's over $T(\mathcal{F}_1), \ldots, T(\mathcal{F}_n)$ repectively, let R be $R_1 \uplus \ldots \uplus R_n$ and let E be a permutative theory. Then*
(i) if all R_i are right linear then R is E-terminating;
(ii) if all R_i are simply E-terminating then R is simply E-terminating;
(iii) if all R_i are totally E-terminating and $R_2, \ldots, R_n$ are conservative then R is simply E-terminating.

7 Enlarging the signature

Here we present a new extension ordering, which allows us to extend an ordering to deal with new function symbols. This extension ordering is based on a precedence on the set of function symbols. Although when dealing with terms with variables this extension ordering imposes strong syntactic conditions on the compared terms, in the case of total extensions it has a better behaviour.

These results can sometimes be applied to prove the termination of *hierarchical unions* [3, 9]: $R_1 \cup R_2$ where the defined symbols of R_2, denoted $\mathcal{D}_2$ do not occur in R_1. We can consider simply that we are enlarging the signature $\mathcal{F}_1$ with the new symbols of $\mathcal{F}_2$ which are at least $\mathcal{D}_2$. Then $R_1 \cup R_2$ is terminating if it is included in one of the extension ordering we will define below. We first show how to extend an ordering $\succ$ on $T(\mathcal{F})$, with $\mathcal{F}$ finite, to an ordering $\succ_p$ on $T(\mathcal{F}')$, where $\mathcal{F}' = \mathcal{F} \cup \mathcal{F}_0$ and $\mathcal{F}_0$ is a set of new function symbols.

Definition 29. Let P be a mapping from terms in $T(\mathcal{F}', \mathcal{X})$ to multisets of terms in $T(\mathcal{F}, \mathcal{X})$ recursively defined as:

$$P(x) = x \quad \text{if } x \text{ is a variable}$$

$$P(f(t_1, \ldots, t_m)) = \begin{cases} \{0\} & \text{if } m = 0 \text{ and } f \notin \mathcal{F} \\ \{f(t'_1, \ldots, t'_m) \mid t'_i \in P(t_i)\} & \text{if } f \in \mathcal{F} \\ P(t_1) \cup \ldots \cup P(t_m) & \text{otherwise} \end{cases}$$

where 0 is some (fixed) minimal (wrt. $\succ$) representative constant symbol in $\mathcal{F}$.

In the following definition, $\succ_{\mathcal{F}}$ is a precedence on $\mathcal{F}'$ (although, in fact, relations in $\succ_{\mathcal{F}}$ between symbols in $\mathcal{F}$ are not needed, since if $P(s) = P(t)$ and $f = \mathcal{H}ead(s) \neq \mathcal{H}ead(t) = g$ then $f \in \mathcal{F}' - \mathcal{F}$ or $g \in \mathcal{F}' - \mathcal{F}$). We also associate a status mul or lex to each function symbol. We assume that all symbols have status mul when $\succ$ is not a simplification ordering.

Definition 30. Let s and t be tems in $T(\mathcal{F}')$. Then $s = f(s_1, \ldots, s_m) \succ_p g(t_1, \ldots, t_n) = t$ iff $P(s) \succ P(t)$ or $P(s) = P(t)$ for all $i : 1 \ldots n$ and

- $s_i \succeq_p t$, for some $i : 1 \ldots m$
- $f \succ_{\mathcal{F}} g$ and $s \succ_p t_j$, for all $j : 1 \ldots n$
- $f = g$ and $Stat(f) = mul$ and $\{s_1, \ldots, s_m\} \gg_p \{t_1, \ldots, t_n\}$
- $f = g$ and $Stat(f) = lex$ and $\langle s_1, \ldots, s_m \rangle \succ_p^{lex} \langle t_1, \ldots, t_n \rangle$ and $s \succ t_i$ for all $i : 1 \ldots m$.

Lemma 31.
1. If $\succ$ is an ordering on $T(\mathcal{F})$ and $\succ_{\mathcal{F}}$ is an ordering on $\mathcal{F}'$ then $\succ_p$ is an ordering extending $\succ$ on $T(\mathcal{F}')$, preserving monotonicity, well-foundedness and the subterm property of $\succ$.
2. If $\succ$ is stable under substitutions, t is linear and $s \succ t$ then $s\sigma \succ_p t\sigma$ for all substitutions σ with range in $T(\mathcal{F}')$;
3. If $\succ$ is stable under substitutions then $\succ_p$ is stable under substitutions when comparing linear terms;
4. If $\succ$ is total on $T(\mathcal{F})$ and $Stat(f) = lex$ for all symbols $f \in \mathcal{F}'$ and $\succ_{\mathcal{F}}$ is a total precedence (in fact, as said, symbols in $\mathcal{F}$ are not need to be comparable) then $\succ_p$ is total on $T(\mathcal{F}')$;
5. If $\succ$ is a reduction ordering total on $T(\mathcal{F})$ then $\succ_p$ is a reduction ordering total on $T(\mathcal{F}')$ containing $\succ$.

Proof. (Hint) Most of the properties are proved as in previous sections. Only well-foundedness and stability under substitutions require the following additional remarks: (i) if $s, t \in T(\mathcal{F}')$ with $P(s) = P(t)$ and t is embedded in s (i.e. $s \succ_{emb} t$) then $s \succ_p t$; (ii) if $P(t) = \{t_1, \ldots, t_n\}$ then $P(t\sigma) = P(t_1\sigma) \cup \ldots \cup P(t_k\sigma)$. □

These results improve the ones given in [1], where only enlargements with new constants where considered. Finally, note that in section 5 we could have used this path extension instead of using the ordering $\succ_r$ as last component.

All results in this section can be obtained for E-compatible ordering for some equational theories E, e.g. containing associativity and commutativity axioms, provided that the new function symbols are free (or commutative) and that in the precedence the new function symbols are smaller than the old ones (in fact it is only requiered the new symbols to be smaller than the associative old ones). Then a slight modification of the ordering defined here is E-compatible (see [15] for details).

8 Improvements and conclusions

We believe that it is possible to prove most of the known modularity results for termination of TRS (and perhaps to improve them, e.g. to more general equational theories) by extension orderings with other transformations (note that in this paper we already use four different ones). Also the applicability of the method to modularity of termination of *constructor-sharing* or *composable* TRS's ([8, 14, 6, 11]) and conditional term rewrite systems ([13]) has to be further investigated. Extension orderings can also be used to prove the termination of the union of particular terminating TRS's, e.g. for $R \uplus \{g(\ldots x \ldots) \to x\}$. This gives a simple proof for the fact that adding pairing to (e.g. simply) typed lambda calculus preserves strong normalization.

References

1. L. Bachmair, N. Dershowitz, and D. Plaisted. Completion Whitout Failure. In Hassan Aït-Kaci and Maurice Nivat, editors, *Resolution of Equations in Algebraic Structures*, vol. 2: Rewriting Techniques, chapter 1, pages 1–30. Ac. Press, 1989.
2. Nachum Dershowitz. Termination of rewriting. *J. Symb. Comp.*, 3:69–116, 1987.
3. Nachum Dershowitz. Hierarchical termination, 1993. 4th Int. Workshop on Conditional Term Rewrite Systems.
4. Nachum Dershowitz and Jean-Pierre Jouannaud. Rewrite systems. In J. van Leeuwen, editor, *Handbook of Theoretical Computer Science*, vol. B: Formal Models and Semantics, chap. 6, pages 244–320. Elsevier Science Pub. B.V., 1990.
5. K. Drosten. Termersetzungssysteme. *Informatik-Fachberichte*, 210, 1989.
6. Maribel Fernandez and Jean-Pierre Jouannaud. Modularity properties of term rewriting systems revisited. Rap. de Recherche LRI-875, Univ. de Paris Sud, 1993.
7. Maria Ferreira and Hans Zantema. Total termination of term rewriting. *5th Conf. on Rewriting Tech. and App.*, LNCS 690, 213–227, 1993.
8. Bernhard Gramlich. Generalized sufficient conditions for modular termination of rewriting. 3rd Int. Conf. on Alg. and Logic Programming, LNCS 639, 1992.
9. M.R.K.Krishna Rao. Simple termination of hierarchical combinations of term rewriting systems. *11th Symp. on Th. Aspects of Comp. Sc.*, LNCS 789, 1994.
10. M. Kurihara and A. Ohuchi. Modularity of simple termination of term rewriting systems. *Journal of IPS Japan*, 31(5):633–642, 1990.
11. M. Kurihara and A. Ohuchi. Modularity of simple termination of term rewriting systems with shared constructors. *Th. Comp. Sc.*, 103:273–282, 1992.
12. Aart Middeldorp. A sufficient condition for the termination of the direct sum of term rewriting systems. In *4th Symp. on Logic in Comp. Sc.*, 396–401, 1989.
13. Aart Middeldorp. Modularity properties of conditional term rewriting systems. *Information and Computation*, 104(1):110–158, 1993.
14. Enno Ohlebusch. Modular properties of composable term rewriting systems. PhD Thesis, Universität Bielefield, Germany, 1994.
15. Albert Rubio. Automated deduction with ordering and equality constrained clauses. PhD. Thesis, Technical University of Catalonia, Barcelona, Spain, 1994.
16. Michaël Rusinowitch. On termination of the direct sum of term-rewriting systems. *Information Processing Letters*, 26(2):65–70, October 19, 1987.
17. Yoshihito Toyama. Counterexamples to termination for the direct sum of term rewriting systems. *Information Processing Letters*, 25(3):141–143, May 29, 1987.

The PushDown Method
to Optimize Chain Logic Programs

(EXTENDED ABSTRACT)

Sergio Greco,[1] Domenico Saccà[1] and Carlo Zaniolo[2]

[1] DEIS, Univ. della Calabria, 87030 Rende, Italy
{ greco, sacca }@si.deis.unical.it
[2] Computer Science Dept., Univ. of California, Los Angeles, CA 90024
zaniolo@cs.ucla.edu

Abstract. The critical problem of finding efficient implementations for recursive queries with bound arguments offers many open challenges of practical and theoretical import. We propose a novel approach that solves this problem for chain queries, i.e., for queries where bindings are propagated from arguments in the head to arguments in the tail of the rules, in a chain-like fashion. The method, called *pushdown*, is based on the fact that a chain query can have associated a context-free language and a pushdown automaton recognizing this language can be emulated by rewriting the query as a particular factorized left-linear program. The proposed method generalizes and unifies previous techniques such as the 'counting' and 'right-, left-, mixed-linear' methods. It also succeeds in reducing many non-linear programs to query-equivalent linear ones.

1 Introduction

In the last decade bottom-up evaluation of logic programming has been favored by deductive database applications over traditional top-down approaches [18]. The effectiveness of the bottom-up execution for bound queries is based on optimizations techniques [4, 6, 10, 13, 14, 16, 18] that transform the original program into an equivalent one that efficiently exploits bindings during fixpoint-based computation. These rewriting techniques give the bottom-up computation a wider applicability range than the top-down computation typical of Prolog, and have been used successfully in several deductive database prototypes. However, there still remains room for major extensions and improvements.

In this paper we shall deal with chain queries, i.e., queries where bindings are propagated from arguments in the head to arguments in the tail of the rules, in a chain-like fashion [5, 8]. For these queries, insisting on general optimization methods (e.g., the *magic-set* method [18]) does not allow to take advantage of the chain structure, thus resulting in rather inefficient query executions. Therefore, as chain queries are rather frequent in practice (e.g., graph applications), there is a need for ad-hoc optimization methods. Indeed, various specialized methods for chain queries have been proposed in the literature (e.g., in [1, 5, 8, 22]). Unfortunately, these methods do not fully exploit possible bindings. To find a method that is particularly specialized for bound chain queries, we have to go back to the *counting* method; however, this method, although proposed in the context of general queries [17], preserves the original simplicity and efficiency [4, 18] only for a subset of chain queries whose recursive rules are linear.

In this paper, we present a new method for the optimization of bound chain queries that reduces to the counting method in all cases where the latter method behaves efficiently. Our approach is based on the fact that a chain query can be associated to a context-free language and a particular pushdown automaton recognizing this language can be also used to drive the query execution, thus dramatically reducing the complexity, as confirmed by the large number of experiments carried out in [7]. The so-called *pushdown* method translates a chain query into a factorized left-linear program implementing the pushdown automaton and, therefore, it candidates for a powerful rewriting technique for a large class of practical DATALOG programs.

Besides to giving an efficient execution scheme to bound chain queries and providing an extension of the counting method, another nice property of the new method is that it introduces a unified framework for the treatment of special cases, such as the factorization of right-, left-, mixed-linear linear programs, as well as the linearization of non-linear programs. A number of specialized techniques for the above special cases are known in the literature [13, 14, 23]. Given the importance and frequency of these special situations in practical applications, novel deductive systems call for the usage of a unique method that includes all advantages of the various specialized techniques.

We point out that analogies between chain queries and context-free languages were investigated by several authors, including [5, 2, 8, 9, 19, 20]. In particular, the use of automata to compute general logic queries was first proposed by Lang [11]. The Lang's method is based on pushing facts from the database onto the stack for later use in reverse order in the proof of a goal. As the method applies to general queries, it is not very effective for chain queries; besides, it does not exploit possible bindings. Independently in [21], Vielle proposed an extension of SLD-resolution which avoids replicated computations in the evaluation of general logic queries using stacks to perform a set-oriented computation. Also this method does not take advantage from a possible chain structure but it does exploit possible bindings. The first proposal of a method that is both specialized for chain queries and based on the properties of context-free language is due to Yannakakis [22], who has proposed a dynamic programming technique implementing the method of Cocke-Younger and Kasami to recognize strings of general context-free languages [3]. This technique turns out to be efficient for unbound queries but it does not support any mechanism to reduce the search space when bindings are available.

2 Preliminaries

We shall assume that the reader is familiar with basic definitions and concepts of logic programming [12] and of the DATALOG language [18]. We next present only definitions and notations that are specific to this paper.

A (*logic*) *program* is a set of *rules* that are negation free. The *definition of a predicate symbol p* in a program P, denoted by $def(p)$, is the set of rules having

p as head predicate symbol. A predicate symbol p is called *EDB* if all rules in $def(p)$ are facts (i.e., ground rules with empty body) or *IDB* otherwise.

Given two (not necessarily distinct) predicate symbols p and q, we say that $q \leq p$ if q occurs in the body of some rule in $def(p)$ or there exists a predicate symbol r such that $q \leq r$ and $r \leq p$; then $leq(p)$ denotes the set of predicate symbols q for which $q \leq p$. We say that p is *recursive* if $p \in leq(p)$ and that p and q are *mutually recursive* if $leq(p) = leq(q)$.

A rule with p as head predicate symbol is *recursive* if p is recursive, *linear* if it is recursive and there is exactly one predicate symbol in the body that is mutually recursive with p, *left-recursive* (resp., *right-recursive*) if the first (resp., the last) predicate symbol in the body is mutually recursive with p.

A *query* Q is a pair $<G, P>$ where G is an atom, called *query-goal*, and P is a program. The *answer* to the query Q, denoted by $A(Q)$, is the set of substitutions θ for the variables in G such that $G\theta$ is derived from P. Two queries $Q = <G, P>$ and $Q' = <G', P'>$ are *equivalent* if $A(Q) = A(Q')$.

Given a DATALOG (i.e., function-symbol free) program P and a set $\mathbf{q}$ of IDB predicate symbols occurring in P, a rule of P is a $\mathbf{q}$-*chain rule* if it has the following general format:

$$p_0(X_0, Y_n) \leftarrow a_0(X_0, Y_0),\ p_1(Y_0, X_1),\ a_1(X_1, Y_1),\ p_2(Y_1, X_2), ...,$$
$$a_{n-1}(X_{n-1}, Y_{n-1}),\ p_{n-1}(Y_n, X_n),\ a_n(X_n, Y_n).$$

where $n \geq 0$, each X_i and Y_i, $0 \leq i \leq n$, are non-empty lists of distinct variables, each $a_i(X_i, Y_i)$, $0 \leq i \leq n$, is a (possibly empty) conjunction of atoms whose predicate symbols neither are in $\mathbf{q}$ nor are mutually recursive with p_0, and each p_i, $1 \leq i \leq n$, is a (not necessarily distinct) predicate symbol in $\mathbf{q}$. We require that the lists of variables are pairwise disjoint; moreover, for each i, $0 \leq i \leq n$, if $a_i(X_i, Y_i)$ is empty then $Y_i = X_i$ otherwise the variables occurring in the conjunction are all those in X_i and in Y_i plus possibly other variables that do not occur elsewhere in the rule.

If $n = 0$ —thus, r reduces to $p_0(X_0, Y_0) \leftarrow a_0(X_0, Y_0)$.— then r is called an *exit chain rule*; moreover if $a_0(X_0, Y_0)$ is the empty conjunction —thus r reduces to $p_0(X_0, X_0)$.— then r is called an *elementary chain* rule. Otherwise (i.e., $n > 0$), it is called a *recurrence chain* rule. Observe that a chain rule is linear iff it is recursive and $n = 1$; moreover, a chain rule is left-recursive or right-recursive iff $a_0(X_0, Y_0)$ or $a_n(X_n, Y_n)$, respectively, is the empty conjunction and p_1 or p_n, respectively, is mutually recursive with p_0.

A DATALOG program P is a $\mathbf{q}$-*chain program* if for each predicate symbol p in $\mathbf{q}$, every rule in $def(p)$ is $\mathbf{q}$-chain and for each two atoms $p(X, Y)$, $p(Z, W)$ occurring in the body or the head of $\mathbf{q}$-chain rules , $X = Z$ and $Y = W$ modulo renaming of the variables, thus the binding is passed through any atom of the same predicate symbol in $\mathbf{q}$ always using the same pattern.

A $\mathbf{q}$-*bound chain query* Q, is a query $<p(b, Y), P>$, where P is a $\mathbf{q}$-chain program, p is a predicate symbol in $\mathbf{q}$, and b is a bound argument.

In the next section we present a method which, given a $\mathbf{q}$-bound chain query $<p(b, Y), P>$, constructs an equivalent left-linear query. The program, so transformed can be implemented efficiently using the bottom-up least-fixpoint based computation favored by DATALOG [18]. In order to guarantee that the binding b is propagated through all q-chain rules, we shall assume that $\mathbf{q} = \{p\} \cup \mathbf{q}'$, $\mathbf{q}' \subseteq leq(p)$ and for each q in $\mathbf{q}$, every $q' \in leq(p)$ for which $q \leq q'$ is in $\mathbf{q}$ as well. Moreover, in order to restrict optimization to those portions which depend from some recursion, we shall also assume that for each q in $\mathbf{q}$, there exists at least one recursive predicate symbol q' in $\mathbf{q}$ for which $q' \leq q$.

3 The Pushdown Method

Our method, called *pushdown method* is based on the analogy of chain queries and context-free grammars [19].

Example 1. Consider the simple chain query $Q = \,<\mathrm{sg}(b, Y), P>$, on the following program P defining a non-linear same-generation program:

$$\mathrm{sg}(X_0, Y_0) \leftarrow \mathrm{a}(X_0, Y_0).$$
$$\mathrm{sg}(X_0, Y_2) \leftarrow \mathrm{b}(X_0, Y_0),\ \mathrm{sg}(Y_0, X_1),\ \mathrm{c}(X_1, Y_1),\ \mathrm{sg}(Y_1, X_2),\ \mathrm{d}(X_2, Y_2).$$

To this program there corresponds a context-free language generated by the grammar

$$G(Q) = \,<V_N, V_T, \Pi, sg>$$

where the set of non-terminal symbols V_N only includes the axiom sg, V_T is the set of terminal symbols $\{a, b, c, d\}$ and Π consists of the following production rules:

$$sg \rightarrow a$$
$$sg \rightarrow b\ sg\ c\ sg\ d$$

Note that the production rules in Π are obtained from the rules of P by dropping the arguments of the predicates and reversing the arrow.

The language $L(Q)$ generated by this grammar can be recognized by the automaton shown in Figure 1.

	b	c	d	a	ϵ
(q_0, Z_0)					$(q,\ sg\ Z_0)$
(q, sg)	$(q,\ sg\ c\ sg\ d)$			$(q,\ \epsilon)$	
(q, c)		$(q,\ \epsilon)$			
(q, d)			$(q,\ \epsilon)$		

Fig. 1. Pushdown Automaton for non-linear same generation query

This automaton can in turn be implemented by the following program $\hat{\Pi}$

$\mathrm{q}([\,\mathrm{sg}\,])$.

$\mathrm{q}([\,\mathrm{sg}, \mathrm{c}, \mathrm{sg}, \mathrm{d}\,|\,\mathrm{T}]) \leftarrow \mathrm{q}([\,\mathrm{sg}\,|\,\mathrm{T}]),\ \mathrm{b}$.

$\mathrm{q}(\mathrm{T}) \leftarrow \mathrm{q}([\,\mathrm{sg}\,|\,\mathrm{T}]),\ \mathrm{a}$.

$\mathrm{q}(\mathrm{T}) \leftarrow \mathrm{q}([\,\mathrm{c}\,|\,\mathrm{T}]),\ \mathrm{c}$.

$\mathrm{q}(\mathrm{T}) \leftarrow \mathrm{q}([\,\mathrm{d}\,|\,\mathrm{T}]),\ \mathrm{d}$.

We can now construct a program $\hat{P}$ that is query-equivalent to P by reintroducing the variables in $\hat{\Pi}$. Thus, both X and Y variables are added to the non-recursive predicates. For the recursive predicate, we add the variable Y to the occurrences of the predicate in the head, and the variable X to the occurrences of the predicate in the body. The resulting program $\hat{P}$ is:

q(b, [sg]).
q(Y, [sg, c, sg, d | T]) ← q(X, [sg | T]), b(X, Y).

q(Y, T) ← q(X, [sg | T]), a(X, Y).
q(Y, T) ← q(X, [c | T]), c(X, Y).
q(Y, T) ← q(X, [d | T]), d(X, Y).

It is easy to verify that the query $<q(Y, []), \hat{P}>$ is equivalent to the original query. Observe that the rewritten program is not any-more DATALOG. □

In general, let us consider a q-chain query $Q = <p(b, Y), P>$. Let V be the set of all predicate symbols occurring in the q-chain rules; we have that q is the set V_N of non-terminal symbols and $V_T = V - V_N$. We associate to Q the context-free language $L(Q)$ on the alphabet V_T defined by the grammar $G(Q) = <V_N, V_T, \Pi, p>$ where the production rules in Π are as follows.

For each q-chain rule r_j of the form:

$$p_0^j(X_0, Y_n) \leftarrow a_0^j(X_0, Y_0), p_1^j(Y_0, X_1), a_1^j(X_1, Y_1), ..., p_n^j(Y_{n-1}, X_n), a_n^j(X_n, Y_n)$$

with $n \geq 0$, there is the production rule:

$$p_0^j \rightarrow a_0^j \, p_1^j \, a_1^j \cdots a_{n-1}^j \, p_n^j \, a_n^j$$

The language $L(Q)$ is recognized by a two-state (q_0 and q, respectively initial and final state) pushdown automaton [15] whose transition table contains one column for each symbol in V_T plus a column for the ϵ symbol, one row for the pair (q_0, Z_0) and one row for each pair (q, v) where Z_0 is the starting pushdown symbol, and $v \in V$. (Note that, for the sake of presentation, the pushdown alphabet is not distinct from the language alphabet.) The Figure 2 reports the entry of the first row, corresponding to the start up of the pushdown consisting of entering the query goal symbol in the pushdown store, and the entries corresponding to the generic q-chain rule r_j shown above, one for a_0^j and one for each a_i^j, $1 \leq i \leq n$, that is not empty. Obviously, if the rule is an exit rule (i.e., $n = 0$), the entry corresponding to a_0^j is (q, ϵ).

	a_0^j	a_1^j	$\cdots$	a_n^j	ϵ
(q_0, Z_0)					$(q, p Z_0)$
$\cdots$					
(q, p_0^j)	$(q, p_1^j a_1^j \cdots p_n^j a_n^j)$				
(q, a_1^j)		(q, ϵ)			
$\cdots$					
(q, a_n^j)				(q, ϵ)	
$\cdots$					

Fig. 2. Pushdown Automaton recognizing $L(Q)$

Given a string $\alpha = a_{i_1}^{k_1} a_{i_2}^{k_2} \cdots a_{i_m}^{k_m}$ in V_T^*, a path spelling α on P is a sequence of $m + 1$ (not necessarily distinct) constants $b_0, b_1, b_2, ..., b_m$ such that for each

j, $1 \leq j \leq m$, $a_{i_j}^{k_j}(b_{j-1}, b_j)$ is derived from P; if $m = 0$ then the path spells the empty string ϵ [1].

It is well known that c belongs to $A(Q)$ if and only if there exists a path from b to c, spelling a string α of $L(Q)$ on P. Therefore, in order to compute $A(Q)$, it is sufficient to use the automaton of Figure 2 to recognize all paths leaving from b and spelling a string α of $L(Q)$ on P [1]. This can be easily done by a logic program $\hat{P}$ which implements the automaton. The program $\hat{P}$ can be directly constructed using all transition rules of Figure 2. In particular we use a rule for each entry in the table. The start-up of the automaton is simulated by a fact which sets both the initial node of the path spelling a string of the language and the initial state of the pushdown store. For the chain query $Q = <p(b, Y), P>$, the resulting program, $\hat{P}$ is as follows:

$$q(b, [p]).$$
$$\ldots$$

$$q(Y, [p_1^j, a_1^j, ..., p_n^j, a_n^j | T]) \leftarrow q(X, [p_0^j | T]), \mathbf{a}_0^j(X, Y).$$
$$q(Y, T) \qquad\qquad\quad \leftarrow q(X, [a_1^j | T]), \mathbf{a}_1^j(X, Y).$$
$$\ldots$$

$$q(Y, T) \qquad\qquad\quad \leftarrow q(X, [a_n^j | T]), \mathbf{a}_n^j(X, Y).$$
$$\ldots$$

The rewritten program $\hat{P}$ will be called the *pushdown-program* of the query Q; the query $\hat{Q} = <q(Y, []), \hat{P}>$ will be called the *pushdown-query* of Q. The technique for constructing pushdown-queries will be called the *pushdown method*.

Theorem 1. *Let Q be a q-chain query. Then the pushdown-query of Q is equivalent to Q.* □

We point out that a naif execution of the rewritten program can be sometime inefficient or even non-terminating for cyclic databases. Extending the approach of [10] and as shown in [7], the bottom-up execution can be efficiently done by a suitable implementation of the list corresponding to the pushdown store. This list is represented as a pair consisting of the head and a pointer to the tuple storing the tail of the list. So, each possible cyclic sequence in the pushdown store is recorded only once and, therefore, termination is guaranteed. Moreover, the simulation of the method on a large number of examples in [7] has shown very better performances than traditional methods — for instance, the query of Example 1 is executed about five times faster if hash accesses are available or, even, one order of magnitude faster if only sequential accesses are enabled.

4 Grammar Transformations to improve Pushdown

As pointed out in the previous section, the pushdown method is based on constructing a particular pushdown automaton to recognize a context-free language. In this section we demonstrate that this kind of automaton becomes more effective if the grammar of the language has a particular structure. More interestingly, we show that if the grammar does not have this structure, then the program can be rewritten so that the corresponding grammar achieves the desired structure and this rewriting is mainly done by applying the known techniques for transforming grammars, particularly to get the $LL(1)$ format [3].

Let us first consider the case of a query for which every recursive chain rule is right-linear, i.e., both right-recursive and linear. Then the associated grammar $G(Q)$ is regular right-linear and, therefore, the pushdown can be replaced by a finite state automaton. In such cases, the pushdown method continues to work efficiently as the list implementing the pushdown store always contains at most one symbol; thus the pushdown automaton actually acts as a finite state automaton. An example is given next.

Example 2. Right-linear Transitive Closure. Consider the following right-linear q-chain query $Q =<$ path(b, Y), $P>$, where q $= \{path\}$ and P is:

    path(X, Y) $\leftarrow$ arc(X, Y).
    path(X, Y) $\leftarrow$ arc(X, Z), path(Z, Y).

The pushdown query is $<$q(Y, [ ]), $\hat{P}>$ where $\hat{P}$ is:

    q(b, [path]).
    q(Y, T) $\leftarrow$ q(X, [path | T]), arc(X, Y).
    q(Y, [path | T]) $\leftarrow$ q(X, [path | T]), arc(X, Y).

It is easy to verify that T is always empty, i.e., the bottom-up execution of the pushdown query emulates a finite state automaton and implements an efficient breadth-first search algorithm.   □

Observe that if the grammar $G(Q)$ is regular left-linear then the pushdown method does not emulates a finite state automaton and, therefore, it may become rather inefficient or even non-terminating. As shown next, the problem can be removed by replacing left-recursion with right-recursion applying well-known reduction techniques for grammars [3].

Consider a q-chain query and suppose that a predicate symbol $s \in$ q is in the head of some left-recursive chain rule — we call s *left-recursive* in this case. Then, the definition $def(s)$ consists of $m > 0$ left-recursive chain rules and n chain rules that are not left-recursive — obviously $n > 0$ as well because otherwise s would be trivially unsatisfied:

$$s(X, Y) \leftarrow \alpha_i(X, Y). \qquad 1 \leq i \leq n$$
$$s(X, Y) \leftarrow s'(X, Z), \beta_i(Z, Y). \qquad 1 \leq i \leq m$$

The productions defining the symbol s in the grammar $G(Q)$ are:

$$s \rightarrow \alpha_i \qquad 1 \leq i \leq n$$
$$s \rightarrow s', \beta_i \qquad 1 \leq i \leq m$$

where α_i and β_j denote the sequences of predicate symbols appearing in $\alpha_i(X, Y)$ and $\beta_j(Z, Y)$, respectively. We can then apply the known transformations to remove left-recursion from the second group of rules for all left-recursive predicate symbols s and we accordingly rewrite the corresponding rules. It turns out that the resulting program, denoted by $can(P)$, does not contain any left-recursive q-chain — here *can* stands for *canonical format*.

Example 3. Left-Linear Transitive Closure. Assume that the program P of the query of Example 2 is as follows:

path$(X, Y) \leftarrow$ arc(X, Y).
path$(X, Y) \leftarrow$ path(X, U), arc(V, Y).

The associated grammar $G(Q)$

path $\rightarrow$ *arc* | *path arc*

is left recursive. After one step of the procedure for removing left-recursion, we obtain the right-recursive grammar

path $\rightarrow$ *arc path'*
path' $\rightarrow$ *arc path'* | ϵ

So, the program $can(P)$:

path$(X, Y) \leftarrow$ arc(X, Z), path$'(Z, Y)$.
path$'(X, X)$
path$'(X, Y) \leftarrow$ arc(X, Z), path$'(Z, Y)$.

is right-linear as the program of Example 2. $\square$

Example 4. Non-Linear Transitive Closure. Assume now that the program P of the query of Example 2 is defined as:

path$(X, Y) \leftarrow$ arc(X, Y).
path$(X, Y) \leftarrow$ path(X, U), path(V, Y).

This program is left recursive and, after the first step of the procedure for removing left-recursion, it is rewritten as:

$r_1 :$ path$(X, Y) \leftarrow$ arc(X, Z), path$'(Z, Y)$.
$r_2 :$ path$'(X, X)$.
$r_3 :$ path$'(X, Y) \leftarrow$ path(X, Z), path$'(Z, Y)$.

The second step removes left recursion from the rule r_3 that is rewritten as

$r_3 :$ path$'(X, Y) \leftarrow$ arc(X, W), path$'(W, Z)$, path$'(Z, Y)$. $\square$

Proposition 2. *Let $Q = <G, P>$ be a chain query and let $Q' = <G, can(P)>$. Then Q' is equivalent to Q.* $\square$

We now introduce a program transformation that improves the performance of the pushdown method for an interesting case of right-recursion.

Let us suppose that there exists a predicate symbol s in P such that $def(s)$ consists of a single elementary chain rule — i.e., the rule $s(X, X)$.— and $m > 0$ right-recursive chain rules of the form:

$$s(X, Y) \leftarrow \alpha_i(X, Z),\ s(Z, Y). \quad 0 \le i \le m$$

We rewrite each recursive chain rule that is in the following format:

$$s(X, Y) \leftarrow \alpha_i(X, Z),\ s(Z, W),\ s(W, Y).$$

as follows:

$$s(X,Y) \leftarrow \alpha_i(X,Z),\ s(Z,Y).$$

thus we drop one occurrence of the recursive goals at the end of the rule. Obviously, if the resulting rule has still multiple recursive goals at the end, we repeat the transformation. The program obtained after performing the above transformations for all the predicate symbols s in P is denoted by $simple(P)$.

Proposition 3. *Given a chain query $Q = <p(b,Y), P>$, Q is equivalent to $Q' = <p(b,Y),\ simple(P)>$.* □

Example 5. We have that $def(path') = \{r_2, r_3\}$ in the program $P' = can(P)$ of Example 4. The program $simple(P')$ is:

$r_1 : \mathtt{path(X,Y)} \leftarrow \mathtt{arc(X,Z),\ path'(Z,Y)}.$
$r_2 : \mathtt{path'(X,X)}$
$r_3 : \mathtt{path'(X,Y)} \leftarrow \mathtt{arc(X,U),\ path'(U,Y)}.$

Eventually, we have linearized non-linear transitive closure. □

We observe that the transformation $simple$ can be applied to a larger number of cases by applying further grammar rewriting. For instance, given the grammar:

$s \rightarrow a\, s'$
$s' \rightarrow b\, s\, s' \mid \epsilon$

we can modify it into:

$s \rightarrow a\, s'$
$s' \rightarrow b\, a\, s'\, s' \mid \epsilon$

so that we can eventually apply the transformation $simple$.

Example 6. Consider the $\{path\}$-chain query $Q = <\mathtt{path(b,Y)}, P>$ where P is defined as follows:

$\mathtt{path(X,Y)} \leftarrow \mathtt{yellow(X,Y)}.$
$\mathtt{path(X,Y)} \leftarrow \mathtt{path(X,U),\ red(U,V),\ path(V,W),\ blue(W,Z),\ path(Z,Y)}.$

We have that $can(P)$ is:

$\mathtt{path(X,Y)} \leftarrow \mathtt{yellow(X,Z),\ path'(Z,Y)}.$
$\mathtt{path'(X,X)}.$
$\mathtt{path'(X,Y)} \leftarrow \mathtt{red(X,U),\ path(U,W),\ blue(W,Z),\ path(Z,T),\ path'(T,Y)}.$

We now replace the two occurrence of path in the body of the last rule with the body of the first rule and we obtain the equivalent program P':

$\mathtt{path(X,Y)} \leftarrow \mathtt{yellow(X,Z),\ path'(Z,Y)}.$
$\mathtt{path'(X,X)}.$
$\mathtt{path'(X,Y)} \leftarrow \mathtt{red(X,U),\ yellow(U,V),\ path'(V,W),\ blue(W,Z)},$
$\qquad\qquad\quad \mathtt{yellow(Z,T),\ path'(T,S),\ path'(S,Y)}.$

By applying the transformation $simple$ to path' the last rule of P' becomes:

$\mathtt{path'(X,Y)} \leftarrow \mathtt{red(X,U),\ yellow(U,V),\ path'(V,W),\ blue(W,Z)},$
$\qquad\qquad\quad \mathtt{yellow(Z,T),\ path'(T,Y)}$ □

We now apply another transformation for the predicate symbols s for which the transformation *simple* cannot be applied because of the lack of the elementary chain rule. Let us then suppose that there exists a predicate symbol s in q such that $def(s)$ consists of $n > 0$ exit chain rules, say

$$s(X,Y) \leftarrow \beta_i(X,Y). \quad 1 \leq i \leq n$$

and $m > 0$ right-recursive chain rules of the form:

$$s(X,Y) \leftarrow \alpha_i(X,Z),\ s(Z,Y). \quad 1 \leq i \leq m$$

We rewrite the above rules as follows:

$$s(X,Y) \leftarrow s'(X,Z),\ \beta_i(Z,Y) \quad 1 \leq i \leq n$$
$$s'(X,X).$$
$$s'(X,Y) \leftarrow \alpha_i(X,Z),\ s'(Z,Y) \quad 1 \leq i \leq m$$

We now replace possible atoms in α_i having s as predicate symbol with the bodies of the rules defining s. In this way, every rule will not have two consecutive recursive predicate symbols at the end of the body. The program obtained after performing the above transformations for all the predicate symbols s in P is denoted by $simple'(P)$. As confirmed by experiments in [7], the pushdown method becomes much more efficient when applied to $simple'(P)$ rather than to P.

Proposition 4. *Given a chain query $Q = <p(b,Y), P>$, Q is equivalent to $Q' = <p(b,Y),\ simple'(P)>$.* □

Example 7. Consider the query $Q = <\text{path}(b,Y), P>$ where P is as follows:

    path(X,Y) ← yellow(X,Y).
    path(X,Y) ← red(X,V), path(V,W), path(W,Y).

We obtain that $simple'(P)$ is equal to:

    path(X,Y) ← path'(X,Z), yellow(Z,Y).
    path'(X,X).
    path'(X,Y) ← red(X,V), path'(V,W), yellow(W,T), path'(T,Y).

As discussed next, the format of $simple'(P)$ is very effective for the performance not only of the pushdown method but also of the counting method. □

5 When Pushdown reduces to Counting

In this section we describe some conditions under which the pushdown method reduces to the counting method. Actually, the counting method can be seen as a space-efficient implementation of the pushdown store. On the other hand, as the pushdown method has a larger application domain, we can conclude that the pushdown method is a powerful extension of the counting method.

Let us first observe that, given the pushdown program of a q-chain query, the pushdown store can be efficiently implemented as follows whenever it contains strings of the form $\alpha^k(\beta)^n$, with $0 \leq k \leq 1$ and $n \geq 0$. Indeed the store can be replaced by the counter n and the introduction of two new states q_α and q_β

to record whether the top symbol is α or β, respectively. This situation arises when the program consists of a number of exit chain rules and of linear right-recursive chain rules and one single linear non-left recursive chain rule. Such an implementation corresponds to applying the counting method.

Example 8. Consider the linear program defining the same-generation with the query-goal $sg(d, Y)$:

$sg(X, Y) \leftarrow c(X, Y)$.
$sg(X, Y) \leftarrow a(X, X1)$, $sg(X1, Y1)$, $b(Y1, Y)$.

The pushdown query is $<q(Y, [\,]), P'>$, where P' is:

$q(d, [sg])$.
$q(Y, [sg, b \,|\, T]) \leftarrow q(X, [sg \,|\, T])$, $a(X, Y)$.
$q(Y, T) \leftarrow \quad\quad q(X, [sg \,|\, T])$, $c(X, Y)$.
$q(Y, T) \leftarrow \quad\quad q(Y, [b \,|\, T])$, $b(X, Y)$.

Observe that the pushdown store contains strings of the form $sg(b)^n$ or of the form $(b)^n$, with $n \geq 0$. So, we replace the store with the counter n and the introduction of two new states q_{sg} and q_b to record whether the top symbol is sg or b, respectively. Therefore, the rules above can be rewritten in the following way:

$q_{sg}(d, 0)$.
$q_{sg}(Y, I) \leftarrow q_{sg}(X, J)$, $a(X, Y)$, $I = J + 1$.
$q_b(Y, I) \leftarrow \quad q_{sg}(X, I)$, $c(X, Y)$.
$q_b(Y, I) \leftarrow \quad q_b(Y, J)$, $b(X, Y)$, $I = J - 1$.

These rules are the same as those generated by the counting method. □

We now show that the above counting implementation of the pushdown store can be also done when the pushdown strings are of the form $\alpha^k (\beta\alpha)^n$ where $0 \leq k \leq 1$ and $n \geq 0$. This situation arises when the program consists of a number of exit chain rules and of recursive linear right-recursive chain rules and one single bi-linear (i.e., two recursive predicate symbols in the body) recursive chain rule that is right-recursive but not left-recursive, i.e., of the form:

$$p(X_0, Y_2) \leftarrow a_0(X_0, Y_0), p(Y_0, X_1), a_1(X_1, Y_1), p(Y_1, Y_2)$$

Example 9. Red/yellow path. Consider the query $<path(b, Y), P>$ where P is:

$path(X, X)$.
$path(X, Y) \leftarrow red(X, V)$, $path(V, W)$, $yellow(W, T)$, $path(T, Y)$.

Using the counting implementation of the pushdown store, we obtain

$q_{path}(b, 0)$
$q_{yellow}(X, I) \leftarrow q_{path}(X, I)$.
$q_{path}(Y, I + 1) \leftarrow q_{path}(X, I)$, $red(X, Y)$.
$q_{path}(Y, I - 1) \leftarrow q_{yellow}(X, I)$, $yellow(X, Y)$.

The query goal is $q_{yellow}(Y, 0)$. Observe that the above program cannot be handled by the simple counting method as first introduced in [4]; the generalized counting method of [17] is able to implement the query but introducing additional indices. So we can say that, in such cases, the pushdown method is a simplified version of the counting method. □

Aknowledgements: Work partially supported by the EC-US project "Deus ex Machina". The work of the first two authors has been also supported by the projects "Sistemi Informatici e Calcolo Parallelo" of CNR and "Metodi Formali per Basi di Dati" of MURST.

References

1. F. Afrati, S. Cosmadakis. Expressiveness of Restricted Recursive Queries. In *Proc. ACM SIGACT Symp. on Theory of Computing*, 1989, pages 113–126.
2. F. Afrati, C.H.. Papadrimitriou. The parallel complexity of simple chain queries. In *Proceedings of the Sixth ACM PODS Conf.*, 1987, pages 210–213.
3. A.V. Aho, and J.F. Ullmann. *The Theory of Parsing Translating and Compiling.* Volume 1 & 2, Prentice-Hall, 1972.
4. F. Bancilhon, D. Mayer, Y. Sagiv, and J.F. Ullman. Magic sets and other strange ways to implement logic programs. In *Proc. Fifth ACM PODS*, 1986, pages 1–15.
5. C. Beeri, P. Kanellakis, F. Bancilhon, and R. Ramakrisnhan. Bounds on the Propagation of Selection into Logic Programs, *JCSS*, Vol. 41, No. 2, 1990, pages 157–180.
6. C. Beeri and R. Ramakrisnhan. On the power of magic. *Journal of Logic Programming*, 10 (3 & 4), 1991, pages 255–299.
7. F. Buccafurri, S. Greco, and E. Spadafora. Implementation of chain queries (in Italian) Proc. 2nd Italian Conference on Advance Database Systems. 1994.
8. G. Dong, On Datalog Linearization of Chain Queries. In J.D. Ullman, editor, *Theoretical Studies in Computer Science*, Academic Press, 1991, pages 181–206.
9. G. Dong, Datalog Expressiveness of Chain Queries: Grammar Tools and Characterization. In *Proc. Eleventh ACM PODS Conf.*, 1992, pages 81–90.
10. S. Greco and C. Zaniolo, Optimization of linear logic programs using counting methods. In *Proc. of the Extending Database Technology*, 1992, pages 187–220.
11. B. Lang. Datalog Automata. In *Third Conf. on Data and Knowledge Bases*, 1988.
12. J.W. Lloyd. *Foundations of Logic Programming.* Springer-Verlag, 2nd ed., 1987.
13. J. Naughton, R. Ramakrisnhan, Y. Sagiv, and J.F. Ullman. Argument Reduction by Factoring. In *Proc. 15th Conf. on Very Large data Bases*, 1989, pages 173–182.
14. J. Naughton, R. Ramakrisnhan, Y. Sagiv, and J.F. Ullman. Efficient evaluation of right-, left-, and multi-linear rules. In *Proc. SIGMOD Conf.*, 1989, pages 235–242.
15. J.E. Hopcroft, and J.F. Ullmann. *Introduction to Automata Theory, Languages and Computation.* Addison-Wesley, 1979.
16. R. Ramakrisnhan, Y. Sagiv, J.F. Ullman, and M.Y. Vardi. Logical Query Optimization by Proof-Tree Transformation. In *JCSS*, No. 47, pages 222-248, 1993.
17. D. Saccà and C. Zaniolo, The generalized counting method of recursive logic queries for databases. *Theoretical Computer Science*, Vol. 4, No. 4, 1988, pages 187–220.
18. J.D. Ullmann. *Principles of Data and Knowledge-Base Systems.* Comp. Sc., 1989.
19. J.D. Ullmann. The Interface Between Language Theory and Database Theory. In *Theoretical Studies in Computer Science* (J.D. Ullman, ed.), Academic Press, 1991.
20. J.D. Ullmann and A. Van Gelder. Parallel Complexity of Logical Query Programs. In *Proc. 27th IEEE Symp. on Found. of Computer Science*, 1986, pages 438–454.
21. L. Vielle. Recursive Query processing: The Power of Logic. *Theor. Comp. Sc.* 1989.
22. M. Yannakakis. Graph-Theoretic Methods in Database Theory. In *Proc. Ninth ACM Symposium on Principles of Database Systems*, 1990, pages 230–242.
23. W. Zang, C.T. Yu and D. Troy. Linearization of Nonlinear Recursive Rules. *IEEE Transaction on Software Engineering*, Vol. 15, No. 9, 1989, pages 1109–1119.

Automatic Synthesis of Real Time Systems *

Jørgen H. Andersen   *Kåre J. Kristoffersen*   *Kim G. Larsen*
Jesper Niedermann
BRICS **
Department of Math. & Comp. Sc., Aalborg University

Introduction

During the last few years the area of real time systems has received a lot of attention from the research community. In particular, a variety of specification formalisms has emerged allowing real time properties to be expressed explicitly. These specification formalisms may roughly be divided into two groups, namely: real time logics (e.g. [AH89, HNSY92]) and real time process algebras (e.g. [Yi90, NRJV90]).

Central to the ongoing research has been the construction of *model–checking* algorithms; i.e. algorithms for deciding whether a given real time system satisfies a given specification. A number of model–checking algorithms exists for real timed logical specifications [ACD90] and more recently algorithms for model–checking [3] timed process algebraic specifications have been given [Cer92, LY93].

In this work, we deal with the more ambitious goal of *model–construction*: i.e. given a real time specification (logical or process algebraic) we want to automatically synthesize a real time system satisfying the specification (if such a system exists). Moreover, we consider the model–construction problem in the setting of *implicit specifications*, i.e.:

$$(A_1 \mid \ldots \mid A_n \mid X) \text{ sat } S \tag{1}$$

The requirement of (1) represents a certain stage in a top–down development of a network satisfying a given overall specification S: namely, the stage where some components $A_1 \ldots A_n$ have already been constructed, but for the completion of the development one component X remains to be constructed. We call S an implicit specification of X as it specifies the behaviour of X in a certain context.

In this paper we present a method for automatically constructing the component X (if possible) such that (1) is met. Our method is applicable to logical as well as process algebraic specifications and proceeds in two steps: First, the implicit specification S is (effectively) transformed into a *direct* specification S'

* This work has been partially supported by the European Communities under CON-CUR2, BRA 7166.
** Basic Research in Computer Science, Centre of the Danish National Research Foundation

[3] In process algebra model–checking consists in checking a suitable behavioural relationship (bisimilarity, say) between the implementation and the specification.

describing the sufficient and necessary requirement to X in order for (1) to hold; i.e.:

$$X \text{ sat } S' \quad \textit{if and only if} \quad (A_1 \mid \ldots \mid A_n \mid X) \text{ sat } S \tag{2}$$

Second, a real time system satisfying S' is generated (if possible) using a *direct* model–construction algorithm.

Our work can be seen as a real time extension of existing model–constructing algorithms for finite–state systems. For $n = 0$ the model–construction problem for (1) extends classical model–construction methods. For S a process algebraic specification the model–construction problem for (1) is a real time extension of the equation solving problem studied in [LX90a, Shi86, Par89, LQ90]. For S a logical specification our work is related to and extends the work on contexts as property transformers studied in [LX91, LS92].

Our method assumes that the network components $A_1 \ldots A_n$ and X are all regular timed agents [Yi90] or equivalently one–clock timed automata [AD94]. For reasons of clarity we have chosen to present our solution method in a somewhat simplified setting, where the notion of parallel composition is simply that of interleaving on actions and the specification language considered is a timed extension of the well–known Hennessy–Milner Logic [HM85]. In the concluding remarks suggestions for extensions will be discussed in more detail. Also, a prototype implementation of the implicit model–construction method for the full extensions has been given and is available as part of the EPSILON tool [CGL93].

1 Timed Processes and Timed Logic

Let $\mathcal{A}$ be a fixed set of actions ranged over by $a, b, c, \ldots$. We denote by $\mathbf{R}_{>0}$ the set of positive reals ranged over by $d, d_1, d_2, \ldots, d', d'', \ldots$. Similarly $\mathbf{R}$ denotes the set of non–negative reals, $\mathbf{N}$ denotes the set of natural numbers (including 0), and $\mathcal{D}$ denotes the set $\{\epsilon(d) \mid d \in \mathbf{R}_{>0}\}$. Regular timed agents are terms of the following grammar:

$$A ::= \sum_{i=1}^{n} [l_i, u_i].a_i.A_i \mid N$$

where $l_i, u_i \in \mathbf{N}$, $a_i \in \mathcal{A}$ and N ranges over a finite set of agent identifiers. For each agent identifier we assume a defining equation $N \stackrel{def}{=} A$. Intuitively, the term $\sum_{i=1}^{n} [l_i, u_i].a_i.A_i$ describes an agent which is able to perform the action a_i between the time bounds l_i and u_i after which the agent will perform according to A_i. On occasions we will use the expanded notation $[l_1, u_1].a_1.A_1 + \cdots + [l_n, u_n].a_n.A_n$ for the general summation. We shall use nil to denote the empty summation. However, we shall often omit trailing nil's; hence $[4, 5].a$ denotes the agent $[4, 5].a.nil$.

Formally, the semantics of regular timed agents are given in terms of a $\mathcal{A} \cup \mathcal{D}$ labelled transition system, where the *configurations* are pairs of the form $\langle A, v \rangle$, with $v \in \mathbf{R}$ denoting the amount by which the agent A has been delayed. The *transitions* between configurations are either delay- or action-transitions and are given by the rules (3), (4) and (5):

$$\langle A, v \rangle \xrightarrow{\epsilon(d)} \langle A, v + d \rangle \tag{3}$$

$$\langle \sum_i [l_i, u_i].a_i.A_i, v \rangle \xrightarrow{a_i} \langle A_i, 0 \rangle \text{ if } v \in [l_i, u_i] \tag{4}$$

$$\frac{\langle A, v \rangle \xrightarrow{a} \langle A', v' \rangle}{\langle N, v \rangle \xrightarrow{a} \langle A', v' \rangle} \text{ if } N \stackrel{def}{=} A \tag{5}$$

Thus, we adopt the two–phase functioning principle [NSY91] present in most real–time process algebras: i.e. the behaviour of a system is regarded as being split in two alternating phases, one where all components agree to let time progress, and one where the components compute. For an agent A the *maximum delay* $M(A)$ is defined recursively as $M(\sum_{i=1}^{n}[l_i, u_i].a_i.A_i) = \max\{u_i, M(A_i) | i = 1 \ldots n\}$, and $M(N) = M(A)$ where $N \stackrel{def}{=} A$.

Syntactically a timed network agent is a parallel composition of a number of regular timed agents; thus network agents are terms $(A_1 | \ldots | A_n)$. Behaviourally, we shall simply assume that a network agent interleaves component actions, whereas components are required to synchronize with respect to delay. Formally, a *network configuration* is a pair $\langle \overline{A}, \overline{v} \rangle$, where $\overline{A} = A_1 | \ldots | A_n$ is a network and $\overline{v} = (v_1, \ldots, v_n)$ is a delay vector, indicating how much each component of the network has been delayed. The transitions between network configurations are given by the rules (6) and (7), where for $d \in \mathbf{R}_{>0}$, $\overline{v} + d$ denotes the delay vector $(v_1 + d, \ldots, v_n + d)$, and $\overline{v}[v_i := 0]$ denotes the delay vector obtained by replacing v_i with 0 in the vector $\overline{v}$:

$$\langle \overline{A}, \overline{v} \rangle \xrightarrow{\epsilon(d)} \langle \overline{A}, \overline{v} + d \rangle \tag{6}$$

$$\frac{\langle A_i, v_i \rangle \xrightarrow{a} \langle A_i', 0 \rangle}{\langle A_1 | \cdots A_i \cdots | A_n, \overline{v} \rangle \xrightarrow{a} \langle A_1 | \cdots A_i' \cdots | A_n, \overline{v}[v_i := 0] \rangle} \tag{7}$$

Example 1. Consider the network agents $[0, 2].a | [2, 3].b$ and $nil | [2, 3].b$. Using the inference rules we can infer the following transitions from the initial network configuration:

$$A = \langle [0, 2].a | [2, 3].b, (0, 0) \rangle \xrightarrow{\epsilon(1.5)} B = \langle [0, 2].a | [2, 3].b, (1.5, 1.5) \rangle \xrightarrow{a}$$
$$C = \langle nil | [2, 3].b, (0, 1.5) \rangle \xrightarrow{\epsilon(1)} D = \langle nil | [2, 3].b, (1, 2.5) \rangle \xrightarrow{b} \langle nil | nil, (1, 0) \rangle$$

$\square$

The specification language used in this presentation is the Extended Timed Modal Logic introduced in [HLY92], here referred to as TL. The logic is an extension of the well–known Hennessy–Milner Logic [HM85], and the formulae of the logic are given by the following abstract syntax:

$$\phi ::= \text{tt} | \phi_1 \wedge \phi_2 | \neg\phi | \langle a \rangle\phi | \exists[l, u]\phi$$

We shall freely use ff as abbreviation for $\neg tt$, $\phi_1 \vee \phi_2$ for $\neg(\neg\phi_1 \wedge \neg\phi_2)$, $[a]\phi$ for $\neg\langle a\rangle\neg\phi$ and $\forall[l, u]\phi$ for $\neg\exists[l, u]\neg\phi$. For the interpretation of TL we define the satisfaction relation $\models$ between network configurations K and TL formulae ϕ inductively as follows:

$$
\begin{aligned}
&i) &&K \models tt &&\Leftrightarrow \text{true} \\
&ii) &&K \models \phi_1 \wedge \phi_2 &&\Leftrightarrow K \models \phi_1 \text{ and } K \models \phi_2 \\
&iii) &&K \models \neg\phi &&\Leftrightarrow \text{not } K \models \phi \\
&iv) &&K \models \langle a\rangle\phi &&\Leftrightarrow \exists K'.\, K \xrightarrow{a} K' \text{ and } K' \models \phi \\
&v) &&K \models \exists[l, u]\phi &&\Leftrightarrow \exists K', d.\, d \in [l, u] \text{ and } K \xrightarrow{\epsilon(d)} K' \text{ and } K' \models \phi
\end{aligned}
$$

We shall often write $\overline{A} \models \phi$ for $\langle \overline{A}, \overline{0}\rangle \models \phi$, where $\overline{0}$ is the (initial) delay vector with all components being 0. In this case we say that the network $\overline{A}$ satisfies the property ϕ.

Example 2. Consider the network $[0, 2].a \mid [2, 3].b$ from Example 1. Then it is easily seen that this network satisfies the formula $\forall[1, 2]\langle a\rangle\forall[1, 1]\langle b\rangle tt$. To see this simply observe that whenever $x \in [1, 2]$ we can infer the following transition sequence:

$$
\langle [0, 2].a \mid [2, 3].b,\, (x, x)\rangle \xrightarrow{a} \langle nil \mid [2, 3].b,\, (0, x)\rangle \xrightarrow{\epsilon(1)}
$$
$$
\langle nil \mid [2, 3].b,\, (1, x + 1)\rangle \xrightarrow{b}
$$

$\square$

2 Symbolic Processes and Model Checking

Using the by now well–known region technique of Alur and Dill [ACD90] one may obtain an algorithm for model–checking: i.e. an algorithm for deciding whether a given network agent satisfies a TL formula. The region technique provides an abstract interpretation of network agents sufficiently complete that all information necessary for model–checking with respect to TL is maintained. At the same time the abstract interpretation yields a finite–state symbolic representation of networks thus enabling standard algorithmic model–checking techniques to be applied.

For $t \in \mathbf{R}$, let $\lfloor t \rfloor \overset{def}{=} \max\{n \in \mathbf{N} \mid n \leq t\}$ denote the integral part of t, and let $\{t\} \overset{def}{=} t - \lfloor t \rfloor$ denote its fractional part. We now recall from [ACD90]:

Definition 1. Let $\overline{m} \in \mathbf{N}^n$ be a delay vector. Then $\overline{u}, \overline{v} \in \mathbf{R}^n$ are equivalent with respect to $\overline{m}$, denoted by $\overline{u} \doteq \overline{v}$ if

- For each $i = 1 \ldots n$, $u_i > m_i$ iff $v_i > m_i$,
- For each $i = 1 \ldots n$ such that $u_i \leq m_i$
  1. $\lfloor u_i \rfloor = \lfloor v_i \rfloor$
  2. $\{u_i\} = 0$ iff $\{v_i\} = 0$

– For each $i, j = 1 \ldots n$ such that $u_i \leq m_i$ and $u_j \leq m_j$, it is the case that $\{u_i\} \leq \{u_j\}$ iff $\{v_i\} \leq \{v_j\}$

Observe that $\mathbf{R}^n / \doteq$ is finite. For $\overline{v} \in \mathbf{R}^n$, we denote by $[\overline{v}]$ the equivalence class of $\overline{v}$ under $\doteq$. The equivalence classes determined by $\doteq$ are called *regions*. For model–checking with respect to TL it is important to note that integer delays of equivalent delay vectors are again equivalent. Thus, whenever $\overline{u} \doteq \overline{v}$ then $\overline{u} + n \doteq \overline{v} + n$ whenever $n \in \mathbf{N}$. Hence, we may without ambiguity write $[\overline{u}] + n$ for the region $[\overline{u} + n]$. In general, it can be shown (see e.g. [LY93]) that two equivalent delay vectors $\overline{u}$ and $\overline{v}$ go through the same future regions; i.e. $\{[\overline{u} + d] \mid d \in R\} = \{[\overline{v} + d] \mid d \in R\}$. Moreover, $\overline{u}$ and $\overline{v}$ also agree on the order in which these regions are visited according to the following notion of successor region: Let $\gamma = [\overline{v}]$ be a region. Then the *successor* region $\mathrm{succ}(\gamma)$ is the region $[\overline{v'}]$, where:

$$
v_i' = \begin{cases} v_i + \min\{1 - \{v_j\} \mid j = 1 \ldots n\} & \text{if } \forall i.\{v_i\} > 0 \\ v_i + \min\{1 - \{v_j\} \mid j = 1 \ldots n\}/2 & \text{if } \exists i.\{v_i\} = 0 \end{cases}
$$

We denote by $\mathrm{succ}^k(\gamma)$ the region obtained by applying succ k times to γ. Now, it may be shown that the future regions from a delay vector $\overline{u}$ are precisely the regions $[\overline{u}], \mathrm{succ}^1([\overline{u}]), \mathrm{succ}^2([\overline{u}]), \mathrm{succ}^3([\overline{u}]), \ldots$ and that they are visited in this order. For γ a region and n a natural number we shall denote by n_γ the unique successor number such that $\gamma + n = \mathrm{succ}^{n_\gamma}(\gamma)$. Thus, when d ranges between two integer bounds l and u the delay vector $\overline{v} + d$ resides in regions between $\mathrm{succ}^{l_{[\overline{v}]}}([\overline{v}])$ and $\mathrm{succ}^{u_{[\overline{v}]}}([\overline{v}])$. Also, as agents enable actions within integer bounds, two network configurations with identical network agent and equivalent delay vectors agree on the action transitions they can perform in the sense that if $\overline{A}$ is a network agent and $\overline{u} \doteq \overline{v}$ then if $\langle \overline{A}, \overline{u} \rangle \xrightarrow{a} \langle \overline{A'}, \overline{u'} \rangle$, then also $\langle \overline{A}, \overline{v} \rangle \xrightarrow{a} \langle \overline{A'}, \overline{v'} \rangle$ for some $\overline{v'}$ such that $\overline{v'} \doteq \overline{u'}$. Based on the above observations it can be concluded that network configurations with equivalent delay vectors satisfy the same TL formulae:

Theorem 2. *Let $\overline{A}$ be a network agent, $\overline{u}, \overline{v}$ two delay vectors, and ϕ a TL formula. Then if $\overline{u} \doteq \overline{v}$ the following holds: $\langle \overline{A}, \overline{u} \rangle \models \phi \iff \langle \overline{A}, \overline{v} \rangle \models \phi$*

To obtain the model–checking algorithm we extract a finite–state symbolic semantics of network configurations, by identifying configurations with equivalent delay vectors. Thus *symbolic* network configurations (or simply symbolic states) are pairs of the form $[\overline{A}, \gamma]$, where $\overline{A}$ is a network agent and γ is a $\doteq$–equivalence class with respect to the delay vector $\overline{m} = M(\overline{A}) = (M(A_1), \ldots, M(A_n))$. As network agents have only finitely many sub–terms [4], and there are only finitely many regions with respect to $M(\overline{A})$ it follows that the set of symbolic states reachable from $[\overline{A}, \gamma]$ is finite. The transitions between symbolic states are either

---

[4] with the usual application of unfolding in the case of recursive definition

un–quantified delay transitions (labelled χ) or action transitions and are defined by the axiom and rule in (8).

$$[\overline{A}, \gamma] \xrightarrow{\chi} [\overline{A}, \mathrm{succ}(\gamma)] \qquad \frac{\langle \overline{A}, \overline{v} \rangle \xrightarrow{a} \langle \overline{B}, \overline{u} \rangle}{[\overline{A}, [\overline{v}]] \xrightarrow{a} [\overline{B}, [\overline{u}]]} \qquad (8)$$

Moreover, symbolic transitions may be computed effectively. This rests on an effective representation of regions [5] allowing effective computation of a representative of a region as well as effective computation of the region from a delay vector. Also it suffices to consider a single representative $\overline{v}$ of γ when inferring symbolic action transitions. We may now give an alternative interpretation of TL based on the above symbolic semantics of networks.

> $i)$ $\quad [\overline{A}, \gamma] \models \mathrm{tt} \qquad \Leftrightarrow \mathrm{true}$
> $ii)$ $\quad [\overline{A}, \gamma] \models \phi_1 \wedge \phi_2 \Leftrightarrow [\overline{A}, \gamma] \models \phi_1 \mathrm{\ and\ } [\overline{A}, \gamma] \models \phi_2$
> $iii)$ $\quad [\overline{A}, \gamma] \models \neg \phi \qquad \Leftrightarrow \mathrm{not\ } [\overline{A}, \gamma] \models \phi$
> $iv)$ $\quad [\overline{A}, \gamma] \models \langle a \rangle \phi \quad \Leftrightarrow \exists [\overline{B}, \eta].\, [\overline{A}, \gamma] \xrightarrow{a} [\overline{B}, \eta] \mathrm{\ and\ } [\overline{B}, \eta] \models \phi$
> $v)$ $\quad [\overline{A}, \gamma] \models \exists [l, u] \phi \Leftrightarrow \exists l_\gamma \leq k \leq u_\gamma.\, [\overline{A}, \mathrm{succ}^k(\gamma)] \models \phi$

Clearly, due to the finite–state nature of the symbolic semantics of networks, the above symbolic interpretation is decidable using classical finite–state model–checking techniques. Moreover, the symbolic interpretation of TL is closely related to the standard interpretation as stated in the following theorem:

Theorem 3. *Let $\overline{A}$ be a network agent, $\overline{v}$ a delay vector, and ϕ a TL formula. Then the following equivalence holds:* $[\overline{A}, [\overline{v}]] \models \phi \quad \Leftrightarrow \quad \langle \overline{A}, \overline{v} \rangle \models \phi$

3 Symbolic Contexts

We want to decompose logical properties required of a network agent into necessary and sufficient properties of one of its agents. More precisely, for any given regular agents $A_1, \ldots, A_n$ and any given TL formula ϕ, we want to find a formula ψ such that the following holds:

$$(A_1 \,|\, \ldots \,|\, A_n \,|\, X) \models \phi \quad \textit{if and only if} \quad X \models \psi \qquad (9)$$

Clearly, the component property ψ will in general depend on the overall property ϕ as well as the agents $A_1, \ldots, A_n$. In the next section we shall define a *property transformer* $\mathcal{W}$, that — given the network agent $\overline{A} = (A_1 | \ldots | A_n)$ and the property ϕ — will construct a property $\psi = \mathcal{W}(\overline{A}, \phi)$ satisfying the requirement of (9).

In (9), the property ϕ expresses constraints on transitions of the complete network $(A_1 | \ldots | A_n | X)$, whereas ψ constrains transitions of the component agent

---

[5] The obvious effective representation of a region is as a linear inequation system. An alternative effective representation where each region has a canonical representation is given in [God94].

X. Thus, in order to solve the above decomposition problem we must have a way of interrelating transitions of a network with transitions of one of its components. To achieve this we provide a symbolic operational semantics of the network contexts $C = (A_1 | \ldots | A_n | [\,])$ in terms of *action transducers*. That is, a network context is semantically viewed as an object which consumes actions from its component agent and produces actions for the external environment, thus acting as an interface between the two. Obviously, we expect the new operational semantics of network contexts to be consistent with the existing operational semantics of network agents. That is, if the component agent X has an a transition, and the context $(A_1 | \ldots | A_n | [\,])$ can consume this action while producing the action b, then we expect the combined network $(A_1 | \ldots | A_n | X)$ to have a b–transition.

The idea of modelling contexts as action transducers has already been pursued for finite state systems [LX90b, LX91]. In our real–time setting we need in addition to take into account the delay of the context agents $A_1, \ldots, A_n$ as well as the delay of the component to be placed in the hole $[\,]$. However, as our transductional semantics is intended to provide the basis of an effective transformation of properties, we deal with delays in a symbolic manner using regions. Thus, formally, a symbolic $n+1$–ary network context is a pair of the form:

$$\left[(A_1 | \ldots | A_n | [\,]) \, , \, [(v_1 \ldots v_n, v)] \right]$$

Here $[(v_1 \ldots v_n, v)]$ is an $n+1$–ary region with $(v_1 \ldots v_n)$ giving delay information of $A_1, \ldots, A_n$ and v providing the delay information of the $[\,]$–component. The transductions between symbolic network contexts are given by the axioms and rule (10),(11) and (12). For γ being an $n+1$–ary region $[(v_1 \ldots v_n, v)]$, $\gamma^{\downarrow}$ denotes the unary region $[v]$ [6]. For $\bar{v} = (v_1 \ldots v_n)$ an n–ary delay vector and u a non-negative real, $\bar{v}u$ denotes the $n+1$–ary delay vector $(v_1 \ldots v_n, u)$.

$$\left[\overline{A} | [\,], \gamma \right] \xrightarrow[\chi]{\chi} \left[\overline{A} | [\,], \text{succ}(\gamma) \right] \quad \text{if } \text{succ}(\gamma)^{\downarrow} = \text{succ}(\gamma^{\downarrow}) \tag{10}$$

$$\left[\overline{A} | [\,], \gamma \right] \xrightarrow[0]{\chi} \left[\overline{A} | [\,], \text{succ}(\gamma) \right] \quad \text{if } \text{succ}(\gamma)^{\downarrow} = \gamma^{\downarrow} \tag{11}$$

$$\left[\overline{A} | [\,], [\bar{v}u] \right] \xrightarrow[a]{a} \left[\overline{A} | [\,], [\bar{v}0] \right] \qquad \frac{\langle \overline{A}, \bar{v} \rangle \xrightarrow{a} \langle \overline{B}, \overline{w} \rangle}{\left[\overline{A} | [\,], [\bar{v}u] \right] \xrightarrow[0]{a} \left[\overline{B} | [\,], [\overline{w}u] \right]} \tag{12}$$

Transductions may be inferred in two ways depending on whether the $[\,]$–component "participates" in the transduction or not. Thus for action transductions the axiom in (12) requires the $[\,]$–component to perform the a–action (after which the $[\,]$–delay is reset to 0). In the rule in (12), the a–action is performed entirely by the network context $\overline{A}$ without any involvement of the $[\,]$–component. This is modelled by a transduction using a unique 0–action (i.e. $0 \notin \mathcal{A}$). The symbolic semantics is extended in the obvious way to 0–actions by $[\overline{A}, \gamma] \xrightarrow{0} [\overline{A}', \gamma']$

---

[6] In the unary case regions are either integer points $[n, n]$, open intervals $]n, n+1[$ or open infinite intervals $]m, \infty[$, where m is the maximum delay bound.

if and only if $\overline{A}' = \overline{A}$ and $\gamma' = \gamma$. Delay transductions model progression to a successor region. The two delay transduction axioms reflect that the projected [ ]-region may either remain unchanged or change to its (unary) successor region, see axions (10) and (11). The following Lemma demonstrates that the transductional semantics of contexts does indeed provide the key to relating symbolic transitions of a network and its component:

Lemma 4. *Let* $\overline{A} = (A_1|\dots|A_n)$ *be an n–ary network agent,* X *a regular timed agent,* γ *an n+1–ary region, and let* $\alpha \in \mathcal{A}\cup\{\chi\}$. *Then* $[\overline{A}|X,\gamma] \xrightarrow{\alpha} [\overline{A}'|X',\gamma']$ *if and only if* $[\overline{A}|[\,],\gamma] \xrightarrow[\beta]{\alpha} [\overline{A}'|[\,],\gamma']$ *and* $[X,\gamma^\downarrow] \xrightarrow{\beta} [X',\gamma^\downarrow]$ *for* $\beta = \alpha$ *or* $\beta = 0$.

4 Contexts as Property Transformers

As shown in the previous Lemma 4, contexts relate *symbolic* transitions of networks with *symbolic* transitions of their components. To facilitate the transformation of logical properties we extend our logic TL with a modality explicitly concerned with symbolic delay transitions ($\xrightarrow{\chi}$). Syntactically, we add the production $\phi ::= \odot\phi$ to the syntax for formulae: We refer to the extended logic as $\mathrm{TL}^\odot$. Formulae with no occurrence of $\exists[l,u]$-modalities are called *pure* and the corresponding sublogic is refered to as $\mathrm{TL}_p^\odot$. Semantically, we interpret formulae $\odot\phi$ with respect to (standard) network configurations as well as symbolic network configurations thus extending the two existing interpretations of TL:

$$\langle \overline{A}, \overline{u} \rangle \models \odot\phi \iff \langle \overline{A}, \overline{v} \rangle \models \phi \text{ for some } \overline{v} \in \mathrm{succ}([\overline{u}])$$

$$[\overline{A}, \gamma] \models \odot\phi \iff [\overline{A}, \mathrm{succ}(\gamma)] \models \phi$$

It is straightforward to show that with these semantic definitions both Theorem 2 and Theorem 3 generalize to $\mathrm{TL}^\odot$. Furthermore, for any given network configuration, $\langle \overline{A}, \overline{v} \rangle$ the original (interval) delay modalities of TL $\exists[l,u]\phi$ can be expressed using the new $\odot$-modality as follows: $\bigvee\{\odot^k\phi \mid l_{[\overline{v}]} \leq k \leq u_{[\overline{v}]}\}$. For $C = [\overline{A}|[\,],\gamma]$ an $n+1$-ary context and ϕ a $\mathrm{TL}^\odot$-formula we now define the *transformed formula* $\mathcal{W}(C,\phi)$ as follows:

i) $\mathcal{W}(C,\mathrm{tt}) = \mathrm{tt}$ *ii)* $\mathcal{W}(C,\langle a\rangle\phi) = \bigvee_{C \xrightarrow[a]{a} C'} \langle a\rangle\mathcal{W}(C',\phi) \vee \bigvee_{C \xrightarrow[0]{a} C'} \mathcal{W}(C',\phi)$

iii) $\mathcal{W}(C,\phi_1 \wedge \phi_2) = \mathcal{W}(C,\phi_1) \wedge \mathcal{W}(C,\phi_2)$ *iv)* $\mathcal{W}(C,\neg\phi) = \neg\mathcal{W}(C,\phi)$

v) $\mathcal{W}(C,\exists[l,u]\phi) = \mathcal{W}(C, \bigvee_{k=l_\gamma}^{u_\gamma} \odot^k\phi)$ *vi)* $\mathcal{W}(C,\odot\phi) = \begin{cases} \mathcal{W}(C',\phi) & \text{if } C \xrightarrow{\chi}_0 C' \\ \odot\mathcal{W}(C',\phi) & \text{if } C \xrightarrow{\chi}_\chi C' \end{cases}$

Note that $\mathcal{W}(C,\phi)$ is always a pure $\mathrm{TL}^\odot$ formula. The following Theorem shows that the transformer $\mathcal{W}$ does indeed yield the sufficient and necessary requirement to a network component in order that the network itself satisfy a given property:

Theorem 5. Let $C = [\overline{A}|[\ ], \gamma]$ be an $n+1$-ary context, X a regular timed agent and let $\overline{v}u \in \gamma$. Then the following equivalences hold:

$$i) \quad [\overline{A}|X, \gamma] \models \phi \Leftrightarrow [X, \gamma^\downarrow] \models \mathcal{W}(C, \phi)$$
$$ii) \quad \langle \overline{A}|X, \overline{v}u \rangle \models \phi \Leftrightarrow \langle X, u \rangle \models \mathcal{W}(C, \phi)$$
$$iii) \quad [\overline{A}|X] \models \phi \Leftrightarrow X \models \mathcal{W}([\overline{A}|[\ ], [\overline{0}]], \phi)$$

Example 3. Using $\mathcal{W}$ we may now compute the necessary and sufficient requirement to the component X in order that $[0,2].a \mid X$ satisfies $\phi = \forall[1,2]\langle a\rangle\forall[1,1]\langle b\rangle$tt. After some calculations based on the transductional semantics of $[0,2].a|[\ ]$ we get the following:

$$\mathcal{W}\left(\left[\,[0,2].a|[\ ], [\overline{0}]\,\right], \phi\right) = \odot^2\left(\odot^2\langle b\rangle\text{tt} \vee \langle a\rangle \odot^2 \langle b\rangle\text{tt}\right) \wedge$$
$$\odot^3\left(\odot^2\langle b\rangle\text{tt} \vee \langle a\rangle \odot^2 \langle b\rangle\text{tt}\right) \wedge \odot^4\left(\odot^2\langle b\rangle\text{tt} \vee \langle a\rangle \odot^2 \langle b\rangle\text{tt}\right)$$

□

5   Direct Model Construction

In this section we provide an algorithm that given a pure TL$^\odot$–formula ϕ will decide whether ϕ is satisfiable by some regular agent. Moreover if ϕ is satisfiable the algorithm will construct a satisfying agent. The technique applied is based on classical tableau methods applied for modal logic (see e.g. [HC68]). To simplify this part of the presentation we use an alternative version of TL$_p^\odot$ with no negation but with all dual operators included (i.e. ff, $\vee$ and $[a]$).

Let Γ be the set of all unary regions of the form $[n,n]$ and $]n, n+1[$, where $n \in \mathbf{N}$. Then a *problem* Π is a finite subset of $\Gamma \times \text{TL}_p^\odot$. We say that a problem Π is *satisfiable* if there exists a regular timed agent X such that $[X, \gamma] \models \phi$ whenever $(\gamma, \phi) \in \Pi$. In this case we call X a *solution* to Π. It follows from the results of the previous sections that if X is a solution to an *initial* problem of the form $\{(\mathcal{O}, \phi)\}$ where $\mathcal{O} = \{0\}$ then $X \models \phi$.

A problem Π is called *simple* if whenever $(\gamma, \phi) \in \Pi$ then ϕ is of the form $\langle a\rangle\psi$ or $[a]\psi$; i.e. all outer conjunctions, disjunctions and $\odot$–modalities have been resolved. As we shall see in the following it is particularly easy to decide satisfiability of simple problems. However, we first provide a reduction mechanism for transforming problems into simple ones. The reduction relation $\leadsto$ between problems is defined as the least relation satisfying the following axioms [7]:

$$i) \quad \Pi \uplus \{(\gamma, \mathrm{tt})\} \rightsquigarrow \Pi$$

$$ii) \quad \Pi \uplus \{(\gamma, \phi_1 \wedge \phi_2)\} \rightsquigarrow \Pi \cup \{(\gamma, \phi_1)\} \cup \{(\gamma, \phi_2)\}$$

$$iii) \quad \Pi \uplus \{(\gamma, \phi_1 \vee \phi_2)\} \rightsquigarrow \Pi \cup \{(\gamma, \phi_1)\}$$

$$iv) \quad \Pi \uplus \{(\gamma, \phi_1 \vee \phi_2)\} \rightsquigarrow \Pi \cup \{(\gamma, \phi_2)\}$$

$$v) \quad \Pi \uplus \{(\gamma, \odot\phi)\} \rightsquigarrow \Pi \cup \{(\mathrm{succ}(\gamma), \phi)\}$$

As the use of $\rightsquigarrow$ always strictly decreases the total size of the formulae in Π it is clear that any reduction sequence from Π must be finite. In fact any problem determines a finite reduction tree with the leaves being the irreducible reductions of Π; i.e. Π' is an irreducible reduction of Π if $\Pi \rightsquigarrow^* \Pi'$ and $\Pi' \not\rightsquigarrow$. Now it follows directly from the semantic definition of the various operators of $\mathrm{TL}_p^\odot$ that there is a close connection between the satisfiability of a problem and its irreducible reductions:

Lemma 6. *A problem is satisfiable if and only if one of its irreducible reductions is satisfiable.*

Moreover, it is clear from the definition of $\rightsquigarrow$ that any irreducible problem is either simple or contains a pair of the form (γ, ff) in which case it is obviously not satisfiable. Thus, we are left with the problem of deciding satisfiability of simple problems. First we define for $a \in \mathcal{A}$, $\gamma \in \Gamma$ and Π a problem the *projected* problem Π_a^γ as follows: $\Pi_a^\gamma = \{(\mathcal{O}, \phi) \mid (\gamma, [a]\phi) \in \Pi\}$ From the symbolic interpretation of $[a]\phi$ it follows directly that whenever X is a solution to Π and $[X, \gamma] \xrightarrow{a} [X', \mathcal{O}]$, then X' is a solution to Π_a^γ. Moreover, when $\gamma =]n, n+1[$, X' is also a solution to $\Pi_a^{[n,n]}$ and $\Pi_a^{[n+1,n+1]}$ as regular agents enable actions within *closed* integer–bound intervals.

Theorem 7. *Let Π be a simple problem. Then Π is satisfiable if and only if whenever $([n, n], \langle a\rangle\phi) \in \Pi$ then (a) $\left\{(\mathcal{O}, \phi)\right\} \cup \Pi_a^{[n,n]}$ is satisfiable, and whenever $(]n, n+1[, \langle a\rangle\phi) \in \Pi$ then (b) $\left\{(\mathcal{O}, \phi)\right\} \cup \Pi_a^{[n,n]} \cup \Pi_a^{]n,n+1[} \cup \Pi_a^{[n+1,n+1]}$ is satisfiable.*

It now follows from the properties of $\rightsquigarrow$ and Theorem 7 that satisfiability of problems is decidable: to determine satisfiablity of a problem Π first (non–deterministically) reduce it to a simple problem Π' and then use the construction of Theorem 7. This leaves the satisfiability of the problems in (a) and (b) to be settled. However, as these problems all have strictly smaller maximum modal depth than Π' (and Π) we can apply the method recursively, with termination guaranteed.

---

[7] $\uplus$ denotes disjoint union of sets.

Concluding Remarks

The presentation of this paper has been based on a somewhat simplified setting, and we want here to comment in slightly more detail on how our results extend.

The logic TL considered may be extended with constructs for defining properties recursively. The symbolic interpretation of TL extends easily to this recursive extension, thus providing the basis for decidability of model–checking. As for transforming recursive properties the techniques given in [LX90b, LX91] can be directly applied. Our direct model–construction method extends to maximal recursively defined properties using the techniques of [JLJL93].

The notion of parallel composition considered in this paper is simply that of interleaving of actions. However, our results extend to a variety of parallel compositions via parameterization on a *synchronization function* as studied in [HL89]. Thus, we may consider *parameterized* network of the form $(A_1, \ldots, A_n)|_f$, where f is a synchronization (partial) function of type $(A \cup \{0\})^n \hookrightarrow A$. The use of the special no–action 0 enables the modelling of synchronizations where only some components participate. Also, the partiality of f enables synchronization of certain combinations of actions to be disallowed.

In this presentation we have not yet considered implicit process algebraic specifications; i.e. specifications of the form:

$$(A_1 \mid \ldots \mid A_n \mid X) \equiv B \tag{13}$$

where B is a regular timed agent and $\equiv$ is some abstracting equivalence (timed bisimilarity, say). However, (13) may easily be transformed into an equivalent logical implicit specification by using a *characteristic formula* ϕ_B for B; i.e. a formula such that $A \equiv B$ if and only if $A \models \phi_B$. Both for timed and time–abstracting bisimilarity such characteristic formulae can be effectively constructed. Future work includes extension of our method to implicit specifications for arbitrary n–clock automata.

References

[ACD90] R. Alur, C. Courcoubetis, and D. Dill. Model–checking for Real–Time Systems. In *Proceedings of Logic in Computer Science*, pages 414–425. IEEE Computer Society Press, 1990.

[AD94] R. Alur and D. Dill. Automata for Modelling Real–Time Systems. *Theoretical Computer Science*, 126(2):183–236, April 1994.

[AH89] R. Alur and T. A. Henzinger. A Really Temporal Logic. In *Proceeding of IEEE Symp. on Foundations of Computer Science*, Foundations of Computer Science, 1989.

[Cer92] Karlis Cerans. Decidability of bisimulation equivalences for parallel timer processes. In *Proceedings of CAV'92*, volume 663 of *Lecture Notes in Computer Science, Springer Verlag*, Berlin, 1992. Springer Verlag.

[CGL93] K. Cerans, J. C. Godskesen, and K. G. Larsen. Timed modal specifications — theory and tools. In *Proceedings of 5th International Conference on Computer Aided Verification*, volume 697 of *Lecture Notes in Computer Science*, 1993.

[God94]   J. C. Godskesen. *Timed Modal Specifications — A theory for verification of real–time concurrent systems*. PhD thesis, Aalborg University, 1994.

[HC68]   G. E. Hughes and M. J. Cresswell. *An Introduction to Modal Logic*. Methuen and Co., 1968.

[HL89]   H. Hüttel and K. G. Larsen. The Use of Static Constructs in a Modal Process Logic. In *Logic at Botik'89.*, volume 363 of *Lecture Notes in Computer Science*, 1989.

[HLY92]   U. Holmer, K.G. Larsen, and W. Yi. Decidability of bisimulation equivalence between regular timed processes. In *Proceedings of CAV'91*, volume 575 of *Lecture Notes in Computer Science, Springer Verlag*, Berlin, 1992.

[HM85]   M. Hennessy and R. Milner. Algebraic laws for nondeterminism and concurrency. *Journal of the Association for Computing Machinery*, pages 137–161, 1985.

[HNSY92] T. A. Henzinger, Z. Nicollin, J. Sifakis, and S. Yovine. Symbolic model checking for real-time systems. In *Logic in Computer Science*, 1992.

[JLJL93] O.H. Jensen, J.T. Lang, C. Jeppesen, and K.G. Larsen. Model construction for implicit specifications in modal logic. In *Proceedings of 4th International Conference on Concurrency Theory*, volume 715 of *Lecture Notes in Computer Science*, 1993.

[LQ90]   P. Lewis and H. Qin. Factorization of finite state machines under observational equivalence. *Lecture Notes in Computer Science, Springer Verlag*, 458, 1990.

[LS92]   K.G. Larsen and A. Skou. Compositional verification of probabilistic processes. *In Proceedings of CONCUR'92. To appear in Lecture Notes in Computer Science.*, 1992.

[LX90a]   K. G. Larsen and L. Xinxin. Equation Solving Using Modal Transition systems. *In Proceedings of Logic in Computer Science*, 1990.

[LX90b]   K.G. Larsen and L. Xinxin. Compositionality through an operational semantics of contexts. In *proceedings of ICALP'90*, volume 443 of *Lecture Notes in Computer Science*, 1990.

[LX91]   K.G. Larsen and L. Xinxin. Compositionality through an operational semantics of contexts. *Journal of Logic and Computation*, 1(6):761–795, 1991.

[LY93]   K. G. Larsen and W. Yi. Time Abstracted Bisimulation: Implicit Specifications and Decidability. In *Proceedings of MFPS*, 1993.

[NRJV90] X. Nicollin, J. L. Richierand, J.Sifakis, and J. Voiron. ATP: an algebra for timed processes,. In *Proceedings of the IFIP TC 2 Working Conference on Programming Concepts and Methods*, IFIP, 1990.

[NSY91]   X. Nicollin, J. Sifakis, and S. Yovine. From ATP to Timed Graphs and Hybrid Systems. volume 600 of *Lecture Notes in Computer Science*, 1991. In Real–Time: Theory in Practice.

[Par89]   J. Parrow. Submodule construction as equation solving in CCS. *Theoretical Computer Science*, 68, 1989.

[Shi86]   M.W. Shields. A note on the simple interface equation. Technical Report SE/079/1, University of Kent at Canterbury, Electronic Engineering Laboratories University, June 1986.

[Yi90]   W. Yi. Real–Time Behaviour of Asynchronous Agents. In *Theories of Concurrency: Unification and Extension*, volume 458 of *Lecture Notes in Computer Science*, 1990.

Self-Correcting for Function Fields of Finite Transcendental Degree

Manuel Blum[1*] and Bruno Codenotti[2**] and Peter Gemmell[3***] and Troy Shahoumian[4†]

[1] Computer Science Division, UC Berkeley, Berkeley, CA 94720, and International Computer Science Institute, Berkeley CA 94704. e-mail: blum@cs.berkeley.edu
[2] IMC-CNR, Via S.Maria, 46, 56100 - Pisa (Italy). e-mail: codenotti@iei.pi.cnr.it
[3] Sandia National Labs MS 1110, PO Box 5800, Albuquerque, NM 87185-1110. e-mail: psgemme@cs.sandia.gov
[4] Computer Science Division, UC Berkeley, Berkeley, CA 94720. e-mail: troys@cs.berkeley.edu

Abstract. We use algebraic field extension theory to find self-correctors for a broad class of functions. Many functions whose translations are contained in a function field that is a finite degree extension of a scalar field satisfy polynomial identities that can be transformed into self-correctors. These functions can be efficiently corrected in a way that is simpler and different from how the functions are actually computed. This is an essential feature of program self-correcting. Among the functions for which we present self-correctors are many rational expressions of x, e^x, and $\sin(x)$ (over the real and complex fields) as well as many rational expressions of x, g^x (g a generator) mapping the integers into a finite field and many rational expressions of $x, \log_h(x)$ (h a generator) mapping a finite field into the reals.

The new tools presented in this extended abstract will be useful to the theory of program self testing/correcting. Furthermore, they may yield new results in complexity theory. Previous work in the self-testing of polynomials had important applications in the PCP protocols that proved the hardness of approximating max-SNP problems.

1 Introduction

Recent papers have been dedicated to the issues of program checking and program self testing/correcting (see for example [Bl88], [BK89], [BLR90], [RR90], [L91], [GLRSW91], [GS92], [ABCG93]).

* Supported by NSF grant CCR-9201092.
** Part of this work was done while visiting the International Computer Science Institute, Berkeley, CA 94704. Partially supported by the ESPRIT B.R.A. Project 9072 GEPPCOM.
*** Part of this work was performed at Sandia National Laboratories and was supported by the U.S. Department of Energy under contract DE-AC04-76DP00789; part of the work was done while at UC Berkeley supported by NSF grant number CCR-9201092.
† Supported by a Fannie and John Hertz Foundation fellowship.

Self-correcting, which is due to [BLR90] and [L91], is an off-shoot of *program result checking* which is due to [Bl88] and [BK89]. A self-corrector, C, is a program which takes as input a program P which computes a function f correctly on *most* inputs and produces a program C_P which is correct on all inputs with high probability. Note that the probability is over the coin tosses of C and is *not* over the inputs. We give a rigorous definition of self-correctors in section 2.

Self-correctors have been known only for limited classes of functions. Previous work has shown self-correctors for such functions as polynomials, logarithm, exponential, sin, and (other) homomorphisms, but not for arbitrary combinations of these. This extended abstract describes self-correctors for broad classes of functions, including many algebraic combinations of the above functions. The technique that allows us to achieve these results is based on the nature of field extensions of finite transcendental degree and has been used previously in some applications to fault tolerance [V93, V91]. The reader is encouraged to see, e.g. [La65] for an excellent introduction to field extensions.

We describe self-correctors for many functions $f : \mathbf{R} \to \mathbf{F}$ such that $\{f(x + a)\}_{a \in \mathbf{R}} \subset \mathbf{K}$ and many functions $g : \mathbf{F_1} \to \mathbf{F}$ such that $\{g(ax)\}_{a \in \mathbf{F_1}} \subset \mathbf{K}$, where $\mathbf{R}$ is a ring, $\mathbf{F_1}$ is a field, and $\mathbf{K}$ is a function field that is a finite degree extension of the scalar field $\mathbf{F}$.

Expanding on Vainstein's work [V93, V91] on *polynomially checkable* functions, we demonstrate self-correctors for many functions in the function fields $\overline{\mathbf{C}[x, e^x, \sin(x)]}$, $\overline{\Re[x, e^x, \sin(x)]}$, $\overline{\mathbf{F_p}[x, g^x]}$, and $\overline{\Re[x, \log_h(x)]}$. Here, $\mathbf{C}$ refers to the complex numbers, $\Re$ to the real numbers, and $\mathbf{F_p}$ refers to the finite field of p elements, g is an element of $\mathbf{F_p}$, and h generates a finite field. These function fields contain all functions, mapping into the relevant scalar field, that can be written as the roots of polynomials whose coefficients are polynomial expressions of x, e^x, and $\sin(x)$ (or x, g^x) over the relevant scalar field. For example, $f(x) = \frac{e^x x^3}{x + \sin^3(x)}$ is an element of $\overline{\mathbf{C}[x, e^x, \sin(x)]}$ (and also has a self-corrector).

In the full paper, we discuss *approximate-self-correctors* for single- and multi-variable functions on the fields $\mathbf{C}$ and $\Re$, as well as finite fields (where we define an appropriate metric for the finite fields). An approximate-self-corrector, C, is a program which takes as input a program P which computes a function f nearly correctly (under some norm) on *most* inputs and produces a program C_P which is nearly correct on all inputs. For related work on this topic, see [RR90], [GLRSW91], [ABCG93].

This extended abstract shows that field extension theory provides us with classes of equalities can be used for function verification. Therefore, field extension theory sheds light into the relation between verification and computation. In addition, we show how to compute the *checking polynomial* efficiently by solving linear systems, and hence our approach also has a practical significance.

A related approach consists of using the notion of *functional equations* to find self-testers/correctors [RR94]. However we believe that field extension theory is more directly applicable to some central problems in complexity theory.

The rest of this extended abstract is organized as follows. In section 2 we give some preliminary definitions; in section 3 we present our results on polynomial checkability and on self correcting univariate functions. In section 4, we discuss modifying the self-correctors to work on sets that are not rings. In section 5 we show some extensions to the multivariate case. In section 6, we describe work in progress. Finally, in section 7 we present some concluding remarks.

2 Notation and Definitions

2.1 Self-Correcting

Here, we introduce basic notation to do with self-correcting and approximate self-correcting.

Notation: We use $x \in_R \mathcal{D}$ to denote that x is chosen uniformly at random from domain $\mathcal{D}$.

The way in which we quantify that our self-correctors are doing something different than the program being self-corrected is by insisting that they have the little-oh property, which intuitively says that the self-corrector must run asymptotically faster than any program computing f.

Definition 1. Suppose that we have a self-corrector for a program computing a function f and that $T(n)$ is the best known running time of any program computing f on inputs of length n. We say that the self-corrector has the **little-oh property** of its running time is $o(T(n))$, where each call to the program being tested counts as one step.

When we describe self-correctors for polynomially checkable functions in subsections 3.3 and 3.4 and section 5, the self-correctors will satisfy the little-oh property if the time required to solve for the coefficients of the checking polynomial(s) and the time to solve the resulting equation(s) for $f(x)$ is asymptotically smaller than the time needed to compute $f(x)$ directly. Therefore, whether or not the self-corrector will satisfy the little-oh property depends on the degree of the checking polynomial, the degree of the field extension, and f's computational complexity.

Definition 2. A program P ϵ-**computes** f on the domain $\mathcal{D}$ if P is correct on all but an ϵ fraction of inputs in the domain; that is:

$$\Pr_{x \in_R \mathcal{D}} [P(x) \neq f(x)] < \epsilon.$$

$P(\cdot)$ denotes the output of P on a given input.

Definition 3. An ϵ-**self-corrector** for f on a domain $\mathcal{D}$ is a randomized algorithm C such that (1) it has the little-oh property, (2) it uses program P as a black box, and (3) it has the property that for every program P such that P ϵ-computes f on $\mathcal{D}$,

$$\forall x \in \mathcal{D} \quad \Pr\left[C^P(x) = f(x)\right] \geq 2/3.$$

Here the probability is over the random coin flips of C.

When we say that C treats $\mathbf{P}$ as a black box, we mean that C does not depend on any features of $\mathbf{P}$ other than the fact that it ϵ-computes f. For example, C should work regardless of the particular algorithm employed by $\mathbf{P}$.

We now present a definition of self-correcting for functions that only *approximate* the correct output for most inputs:

Definition 4. Let $0 \leq \delta$ and $0 \leq \epsilon < 1$. We say that a program $\mathbf{P}$ (ϵ, δ)-**approximates** the function f, if $\|\mathbf{P}(x) - f(x)\| \leq \delta$ for all but a ϵ fraction of the inputs, where $\| \cdot \|$ is a norm on the range space.

Given a program $\mathbf{P}$ that computes a function f to a certain accuracy for a given fraction of inputs, an approximate self-corrector is a program that computes f on all inputs to a given accuracy with high probability, where the probability is over the random coin flips of the approximate self-corrector. The approximate self-corrector uses $\mathbf{P}$ as a subroutine. We formalize this as follows:

Definition 5. Let $0 \leq \epsilon < 1$. Let $0 \leq \delta_1 \leq \delta_2$. A $(\epsilon, \delta_1, \delta_2)$-**approximate self-corrector** for the function f is a randomized algorithm C_f having the following property: for every program $\mathbf{P}$, and security parameter β, if $\|\mathbf{P}(y) - f(y)\| \leq \delta_1$ for at least a $1 - \epsilon$ fraction of inputs y, then, for all inputs x, $\|C_f^{\mathbf{P}}(x, \beta) - f(x)\| \leq \delta_2$ holds with probability at least $1 - \beta$. This probability is over the coin tosses of C_f, and so this property holds for each and every input x. C_f must have the little-oh property.

2.2 Algebraic Definitions

We will be needing the following basic definitions:

By *function field*, we mean a field whose elements are functions.

Definition 6. Let $\mathbf{F}$ be a field. By $\mathbf{F}[T_0, \ldots T_k]$, we mean the ring of polynomials with k variables and with coefficients contained in $\mathbf{F}$.

Definition 7. Let $\mathbf{F}$ be a field. Let $\{y_j\}_{j=1}^k$ be a set of elements not contained in $\mathbf{F}$. Then the **field extension** $\mathbf{F}(y_1 \ldots y_k)$ is the smallest field containing both $\mathbf{F}$ and $\{y_j\}_{j=1}^k$.

Definition 8. Elements $\{z_j\}_{j=1}^k$ in a field extension $\mathbf{K}$ of a field $\mathbf{F}$ are **algebraically dependent** (relative to $\mathbf{F}$) if there is a polynomial P with coefficients in $\mathbf{F}$ (not all of which are zero) such that $P(z_1, \ldots, z_k) = 0$. Elements that are not algebraically dependent are **algebraically independent**.

In this extended abstract, we will be considering function fields which are extensions of scalar fields. For example, $\mathbf{C}(x)$ is the smallest field which contains the complex numbers as well as the symbol x. x is assumed to be algebraically independent of $\mathbf{C}$. We can also view $\mathbf{C}(x)$ as a set of functions from some set into the complex numbers. This set of functions is the set of all rational expressions of complex polynomials.

Definition 9. The **transcendental degree** of a field extension $\mathbf{K}$ of $\mathbf{F}$ is the maximum value n such that there exist n elements in $\mathbf{K}$ that are algebraically independent (relative to $\mathbf{F}$).

Definition 10. $\overline{\mathbf{F}}$, the **algebraic closure** of a field $\mathbf{F}$, is the smallest field that contains both $\mathbf{F}$ as well as all roots of all polynomials with coefficients from $\mathbf{F}$.

3 Polynomial Checkability and Self-Correcting Univariate Functions

3.1 Definitions of Polynomially Checkable and Linearly Checkable

Vainstein [V93] made the following definition of univariate polynomial checkability:

Definition 11. A function $f : \mathbf{C} \to \mathbf{C}$ (or $f : \Re \to \Re$) is **polynomially checkable** or **PC** if there exists an integer k, such that for any $a_1, \ldots, a_k \in \mathbf{C}$ (or $\Re$), such that $a_i \neq 0$ for $1 \leq i \leq k$ and $a_i \neq a_j$ for $i \neq j$, the functions $f_0(x) = f(x)$, $f_1(x) = f(x+a_1), \ldots, f_k(x) = f(x+a_k)$ are algebraically dependent. This means that there exists a non-trivial multi-variate polynomial $P_{a_1,\ldots,a_k} \in \mathbf{C}[T_0, \ldots T_k]$ (or $\Re[T_0, \ldots T_k]$) such that $\forall x \in \mathbf{C}($ or $\Re)$, $P_{a_1,\ldots,a_k}(f_0(x), \ldots, f_k(x)) = 0$. Such a polynomial $P_{a_1,\ldots,a_k}$ of minimum degree is called the **checking polynomial** of f (for offsets $a_1 \ldots a_k$).

We extend this definition to other functions, including those that map from domains that are rings and fields other than the complex and real numbers and those that map into other fields. We also consider multiplicative (as well as additive) offsets. We will see in subsections 3.3 and 3.4 that many polynomially checkable functions have self-correctors. Therefore, we try to include as many functions as possible in the class of PC functions, particularly functions defined on finite domains.

Definition 12. Let $\mathbf{R}$ be a ring and let $\mathbf{F}$ be a field. A function $f : \mathbf{R} \to \mathbf{F}$ is **polynomially checkable** or **PC** if there exists an integer k, such that for any $a_1 \ldots a_k \in \mathbf{R}$ such that $a_i \neq 0$ for $1 \leq i \leq k$ and $a_i \neq a_j$ for $i \neq j$, the functions $f_0(x) = f(x)$, $f_1(x) = f(x + a_1)$, $\ldots$, $f_k(x) = f(x + a_k)$ are algebraically dependent. We will say also that f is polynomially checkable if f's domain is a field and the functions $g_0(x) = f(x)$, $g_1(x) = f(a_1 x)$, $\ldots$, $g_k(x) = f(a_k x)$ are algebraically dependent for all offsets $a_1, \ldots, a_k$ meeting the above criteria. This means that there exists a non-trivial polynomial $P_{a_1,\ldots,a_k}$ with coefficients $\in \mathbf{F}$ such that $\forall x \in \mathbf{R}$, $P_{a_1,\ldots,a_k}(f_0(x), \ldots, f_k(x)) = 0$ or $\forall x \in \mathbf{R}$, $P_{a_1,\ldots,a_k}(g_0(x), \ldots, g_k(x)) = 0$. Such a polynomial of minimum degree $P_{a_1,\ldots,a_k}$ is called a **checking polynomial** of f for (additive or multiplicative) offsets $a_1 \ldots a_k$.

Vainstein also made a definition for linearly checkable functions and we present our extension here. We will show that linearly checkable functions have relatively simple self-correctors.

Definition 13. Let $\mathbf{R}$ be a ring and let $\mathbf{F}$ be a field. A function $f : \mathbf{R} \to \mathbf{F}$ is **linearly checkable** or **LC** if f is PC and furthermore, for all $a_1, \ldots, a_k \in \mathbf{R}$, the checking polynomial $P_{a_1,\ldots,a_k}$ is linear (i.e. of the form $\forall y_0 \ldots y_k \in \mathbf{F}, P_{a_1 \ldots a_k}(y_0 \ldots y_k) = c + c_0 y_0 + \ldots c_k y_k$, where $c, c_1 \ldots c_k \in \mathbf{F}$).

3.2 Functions That Are Polynomially Checkable/Linearly Checkable

In this subsection, we present examples of functions that are polynomially checkable (PC).

[V93] proved the following functions are PC:

Theorem 14. *Any function* $f : \mathbf{C} \to \mathbf{C}, f \in \overline{\mathbf{C}[x, e^x, \sin(x)]}$ *is polynomially checkable.*

We expand on his results as follows:

Lemma 15. *All functions* f *such that* $f : \mathbf{Z_{p-1}} \to \mathbf{F}_p, f \in \overline{\mathbf{F}_p[x, g^x]}$ *or* $f : \mathbf{F}_{p^k} \to \Re, f \in \overline{\Re[x, \log_h(x)]}$ *are polynomially checkable, where* g *generates* $\mathbf{F}_p^*$, h *generates* $\mathbf{F}_{p^k}^*$.

Proof. (sketch)

The proof is similar to Vainstein's proof of theorem 14.

1. For all $f \in \overline{\mathbf{F}_p[x, g^x]}$, for all $a_1, a_2 \in \mathbf{Z}_{p-1}$, we will show that $f(x), f(x + a_1), f(x+a_2)$ are algebraically dependent. For $j = 1, 2$, let $f_j(x) = f(x+a_j)$. Because $f \in \overline{\mathbf{F}_p[x, g^x]}$, $\exists$ polynomial $Q \in (\mathbf{F}_p[x, g^x])[T]$ such that $\forall x \in \mathbf{Z_{p-1}}, (Q(f))[x] = 0$. Let $Q(y) = \sum_{i=0}^{k} A_i(x) y^i$, $A_i(x) \in \mathbf{F}_p[x, g^x]$.
    For $j = 1, 2, 3$ and $i = 0, \ldots k$, define $A_{i,j}(x) = A_i(x + a_j)$. For $j = 1, 2, 3$, define $Q_j(y) = \sum_{i=0}^{k} A_{i,j}(x) y^i$. Note that for $j = 1, 2$ and for all $x \in \mathbf{Z}_{p-1}$, $Q_j(f_j)(x) = 0$. We know that $A_{i,j} \in \mathbf{F}_p[x, g^x]$. Hence, $f_1, f_2 \in \overline{\mathbf{F}_p[x, g^x]}$.
    Since $\mathbf{F}_p[x, g^x]$ has transcendental degree 2, $\overline{\mathbf{F}_p[x, g^x]}$ also has transcendental degree 2. Therefore, f, f_1, and f_2 are algebraically dependent.
2. For $f : \mathbf{F}_{p^k} \to \Re, f \in \overline{\Re[x, \log_h(x)]}$, our arguments are similar, except that we show that for all $a_1, a_2 \in \mathbf{F}_{p^k} - \{0\}$, $f(x), f(a_1 x), f(a_2 x)$ are algebraically dependent. $\qquad \square$

Vainstein proves the following theorem which describes the class of LC functions over $\Re$. His result is of particular importance because all LC functions have simple self-correctors.

Theorem 16. *1. Let* $f : \Re \to \Re$ *satisfy a linear differential equation:*

$$c_m \frac{d^m f}{dx^m} + c_{m-1} \frac{d^{m-1} f}{dx^{m-1}} + \ldots + c_0 f + c = 0$$

for some constants $c_m, \ldots, c_0, c$. *Then* f *is linearly checkable with* $k = m$.

2. Let $f : \Re \to \Re$ *be an LC function with a given* k. *Then* f *is infinitely differentiable and satisfies a linear differential equation with constant coefficients and where the highest derivative is* $\leq k$.

We now characterize those functions that are solutions to linear differential equations with constant coefficients. All of these equations would have a linear checking polynomial. If $c_0, \ldots, c_n$, α and β are arbitrary constants, then (1) $(c_n x^n + \cdots + c_0)e^{\alpha x}$, (2) $(c_n x^n + \cdots + c_0)e^{\alpha x} \sin \beta x$ and (3) $(c_n x^n +$

$\cdots + c_0)e^{\alpha x} \cos \beta x$ all are solutions to a linear differential equations. In addition, (4) the sum of any two equations that are the solutions of linear differential equations again is a solution to a linear differential equation. This gives a relatively wide range of functions that have linear checkers. For example, $x^4 + (7x^5 + 2x)e^{8x} \sin(7x + 5) + x^9 e^{8x} \cos x$ has a linear checking polynomial.

As an example, $f(x) = x^2 e^{3x}$ has a checking polynomial. It the offsets are $a_1 = 1$, $a_2 = 2$ and $a_3 = 3$, the checking polynomial is $f_0(x) - \frac{3}{e^3}f_1(x) + \frac{3}{e^6}f_2(x) - \frac{1}{e^9}f_3(x) = 0$.

Corollary 17. *Consider those functions over finite fields and the complex numbers that satisfy the same algebraic relationships as do LC functions over $\Re$. (For example, 2^x over Z_p satisfies the same algebraic constraints as does 2^x over $\Re$.) These functions are also LC with k-values (the number of offsets) less than or equal to the k-values of the corresponding functions over $\Re$.*

3.3 Self-Correctors for PC Functions with Multi-linear Checking Polynomials

We can now prove the following theorem about self-correcting PC functions that have multilinear checking polynomials. For all t, let $MS(t)$ be the time required to solve a set of t linear equations in t variables. Using LUP-decomposition, for example, this can be done in $O(t^3)$ field operations.

Theorem 18. *If $f : \mathbf{R} \to \mathbf{F}$ is a PC function contained in an extension of a field $\mathbf{F}$ such that the transcendental degree of the extension is k and the checking polynomials associated with f are multilinear, then f has a $\frac{1}{3((k+1)\binom{2k+1}{k}+k)}$- self-corrector that requires $(k+1)\binom{2k+1}{k} + k$ calls to the program and also time $O(MS(\binom{2k+1}{k}))$ to solve the resulting linear equations.*

Proof. Without loss of generality, we assume that the offsets are additive. Given program **P**, function f, and input x, the self-corrector determines $f(x)$ as follows: (1) Choose a set $\{a_i\}_{i=1}^k$ uniformly at random from **R**. (2) Compute the checking polynomial $P_{a_1,...,a_k}$. (3) Plug the values $\{\mathbf{P}(x + a_i)\}_{i=1}^k$ into $P_{a_1,...,a_k}$ for $f_1, \ldots, f_k$ and solve the resulting equation in one unknown for f_0. $\square$

We describe how to find the coefficients of a checking polynomial, $P_{a_1,...,a_k}$, via the following lemma:

Lemma 19. *Suppose f is a function from ring $\mathbf{R}$ to finite field $\mathbf{F}$ and f has a checking polynomial $P_{a_1,...,a_k}$ of maximum total degree d for fixed offsets $a_1 \ldots a_k \in \mathbf{R}$. Suppose also that we have a program $\mathbf{P}$ such that $Pr_{x \in_R \mathbf{R}}[f(x) \neq \mathbf{P}(x)] < \epsilon$. Then one can correctly compute the coefficients of $P_{a_1,...,a_k}$ with probability at least $1 - (k+1)\binom{k+d+1}{d}\epsilon$ by making $(k+1)\binom{k+d+1}{d}$ calls to $\mathbf{P}$ and by solving a system of $\binom{k+d+1}{d}$ simultaneous equations.*

Proof. (sketch) P is a polynomial in the $k+1$ variables $f_0, \ldots, f_k$. The number of different terms in P (= the number of coefficients that need to be solved for) is the same as the number of ways to choose d items from a set of $k+2$ items with replacement. Therefore, the number of coefficients of P is $\binom{(k+2+d)-1}{d} = \binom{k+d+1}{d}$.

To find the $\binom{k+d+1}{d}$ coefficients of $P_{a_1,\ldots,a_k}$, one chooses $\binom{k+d+1}{d}$ random values of x and, for each value of x, we call $\mathbf{P}$ on $x, x+a_1, \ldots, x+a_k$. One then solves the resulting system of $\binom{k+d+1}{d}$ linear equations for the coefficients. With probability at least $1-\epsilon$, the $(k+1)\binom{k+d+1}{d}$ values of computed by $\mathbf{P}$ are all correct and therefore the coefficients of $P_{a_1,\ldots,a_k}$ are also correct. $\quad\square$

In the case of LC functions, finding the coefficients requires solving only $k+2$ equations. The above theorem yields the following corollary about the self-correctability of LC functions. For a given k—k being the minimum number of offsets required to construct a checking polynomial—the checking polynomials for LC functions have, in general, a lot fewer terms than the checking polynomial for a general PC function. Thus, the program $\mathbf{P}$ needs to be called fewer times in order to determine the checking polynomial.

Corollary 20. *If f is an LC function contained in an extension of a field $\mathbf{F}$ such that the transcendental degree of the extension is k, then f has a $\frac{1}{3(k+1)(k+2)}$-self-corrector.*

Proof. (sketch) Without loss of generality, we assume that the offsets are additive. For a given x, one finds the value of $f(x)$ by computing the coefficients of the checking polynomial $P_{a_1,\ldots,a_k}$ for random $a_1,\ldots,a_k$ and then solving $P_{a_1,\ldots,a_k}(y, \mathbf{P}(x+a_1), \ldots, \mathbf{P}(x+a_k)) = 0$ for y. The vaule of y that solves $P_{a_1,\ldots,a_k}(y, \mathbf{P}(x+a_1), \ldots, \mathbf{P}(x+a_k)) = 0$ is the correct value for $f(x)$ assuming that $P_{a_1,\ldots,a_k}$ is correct and $\mathbf{P}(x+a_1), \ldots, \mathbf{P}(x+a_k)$ are correct. $\quad\square$

An immediate corollary of the theorem 16, corollary 17, and corollary 20 is:

Corollary 21. *Any real function that satisfies a linear differential equation has a self-corrector.*

Furthermore, any function that maps into the complex numbers or a (sufficiently large) finite field and that satisfies the same algebraic constraints as any one of the above real functions has a self-corrector.

3.4 Self-Correctors for Other Univariate PC Functions

If f is a PC function whose checking polynomial is not multilinear, but is of low degree, d, then we can still solve for $P_{a_1,\ldots,a_k}$'s coefficients. Let $\mathbf{P}$ be a program which usually computes f correctly. We can plug the values $\{\mathbf{P}(x+a_i)\}_{i=1}^{k}$ (or $\{\mathbf{P}(a_i x)\}_{i=1}^{k}$) in for $y_1 \ldots y_k$ of $P_{a_1,\ldots,a_k}(y_0, \ldots y_k) = 0$ and find the roots of the polynomial $P'_{a_1,\ldots,a_k}(y) = P_{a_1,\ldots,a_k}(y, \mathbf{P}(x+a_1), \ldots, \mathbf{P}(x+a_k))$. From the definition of checking polynomial, we know that one of these roots is $f(x)$.

We now define a property of PC functions that we will call *the unique root property*. We will show that this property enables these functions to be efficiently self-corrected. Intuitively, the unique root property says that the checking polynomials of the function define that function.

Property 1 *Given ring $\mathbf{R}$ and field $\mathbf{F}$, let $f : \mathbf{R} \to \mathbf{F}$ be a PC function. If f is PC with additive offsets (resp. multiplicative offsets), then we say that f satisfies the **unique root property** if for all $x \in \mathbf{R}$ and for two uniformly distributed independent random sets of $\{a_i\}_{i=1}^{k}, \{b_i\}_{i=1}^{k} \subset \mathbf{R}$, the two sets of solutions for*

$f(x)$ *found by plugging* $\{f(x+a_i)\}$ *and* $\{f(x+b_i)\}$ *(resp.* $\{\{f(a_ix)\}$ *and* $\{f(b_ix)\}$ *)* *into the checking polynomials* $P_{a_1,...,a_k}$ *and* $P_{b_1,...,b_k}$ *respectively will intersect in one point only with probability at least 7/8.*

We now show that functions that satisfy the unique root property have self-correctors.

Lemma 22. *If function f satisfies the unique root property and is contained in an extension of a field $\mathbf{F}$ such that the transcendental degree of the extension is k, then f has a $\frac{1}{6(k+1)\binom{k+d+1}{k}}$-self-corrector.*

Proof. (sketch) This self-corrector works in much the same way as the previous one of theorem 18 and corollary 20 except for the following change. We find the coefficients of two checking polynomials (for random values of $\{a_1, \ldots, a_k\}$ and $\{b_1, \ldots, b_k\}$). For fixed x, we compute and intersect the roots of $\{P'_{a_1,...,a_k}(y)\}$. If we do not know the degree of the checking polynomial, then we can find it by first assuming it is 1, then 2, etc, each time seeing if the algorithm gives a consistent answer. □

4 Self-correcting PC Functions on Finite Fixed Point Domains

We note here that, up to this point, the theorems, lemmas, and corollaries related to computing on the infinite fields $\mathbf{C}$ and $\Re$ made assumption that we can represent and compute on arbitrary elements of these fields as well as choose random elements from them. Because these may be unrealistic assumptions, we note here that the self-correctors described in the previous section can be modified to work when we are actually computing on finite fixed point subsets of these infinite fields.

In particular the self-correctors will assume that the domain on which the program in question computes well is different than the set that the self-corrector can actually correct over. Similar assumptions are made in [GLRSW91]. The details will appear in the full paper.

5 Polynomial Checkability and Self-correcting Multivariate Functions

In this section, we build on the ideas of the previous section to describe self-correctors for a range of multivariate functions. The description again uses ideas from field theory. We recount and expand on some of Vainstein's work on polynomially checkable multivariate functions and discuss applications to self-correcting.

We modify Vainstein's definitions of PC and LC multivariate functions to include arbitrary fields and multiplicative offsets.

Let n be a positive integer. Let $\mathbf{R}$ be a ring and $\mathbf{F}$ be a field. Let f be a function from $\mathbf{R}^n$ to $\mathbf{F}$. Let $S_1(f) = \{f(x+a)|x, a \in \mathbf{R}^n\}$. Let $S_2(f) = \{f(ax)|x, a \in \mathbf{R}^n\}$.

Definition 23. Function f is called a PC function if either the transcendental degree of $\mathbf{F}(S_1(f))$ or the transcendental degree of $\mathbf{F}(S_2(f))$ is finite.

Definition 24. Function f is called an LC function if either the dimension of $\mathbf{F}[S_1(f)]$ or the dimension of $\mathbf{F}[S_2(f)]$ is finite. Here, $\mathbf{F}[S_1(f)]$ (resp $\mathbf{F}[S_2(f)]$) refers to the vector space (over $\mathbf{F}$) obtained by taking all polynomial combinations of the elements of S_1 (resp S_2).

We can apply these definitions to self-correcting and prove corollaries similar to theorem 18, corollary 20, and lemma 22.

Corollary 25. *If f is a PC function contained in an extension of a field $\mathbf{F}$ such that the transcendental degree of the extension is k and the checking polynomials are multilinear, then f has a $\frac{1}{3(k+1)\binom{2k+1}{k}}$-self-corrector that requires $O(\binom{2k+1}{k})$ calls to the program and also time $O(MS(\binom{2k+1}{k}))$ to solve the resulting linear equations.*

Corollary 26. *If f is an LC function contained in an extension of a field $\mathbf{F}$ such that the transcendental degree of the extension is k, then f has a $\frac{1}{3(k+1)(k+2)}$-self-corrector.*

Corollary 27. *If function f satisfies the unique root property and is contained in an extension of a field $\mathbf{F}$ such that the transcendental degree of the extension is k, then f has a $\frac{1}{6(k+1)\binom{k+d+1}{k}}$-self-corrector.*

6 Approximate Self-Correctors

In the full paper, we will present **approximate-self-correctors** for programs which only approximate PC functions on most inputs. These approximate-self-correctors work in a very similar way to the self-correctors described in subsection 3.3. One major difference is that, because errors are introduced in the evaluation of the program, one must take steps to avoid having to solve poorly conditioned sets of linear equations when finding the coefficients of the PC polynomial.

7 Concluding Remarks

In this extended abstract we looked at correcting computations using equalities made available through field extension theory. This has allowed us to obtain self-correctors for quite a broad class of functions, both in the univariate and multivariate case. The technique is very interesting in itself because it enriches the whole area of program checking with a very powerful mathematical tool. We believe that interesting connections to complexity theoretic issues will be found.

Further work to be done includes the solution to the problems and to the conjectures described in section 3 as well as a rigorous treatment of the whole area of program checking by using the algebraic framework introduced here. In addition, we intend to work on applying these general techniques on the important area of numerical problems in control theory, where the issues of checking are of central importance. Specific topics are checking matrix functions, such as the exponential matrix.

References

[ABCG93] S. Ar, M. Blum, B. Codenotti, P. Gemmell. Checking Approximate Computations over the Reals. Proc. 25st ACM Symposium on Theory of Computing, 1993.

[Bl88] M. Blum. Designing Programs to Check their Work. ICSI TR-88-009, 1988.

[BK89] M. Blum, S. Kannan. Designing Programs that Check their Work. In *Proc. 21st ACM Symposium on Theory of Computing*, 1989.

[BLR90] M. Blum, M. Luby, R. Rubinfeld, "Self-Testing/Correcting with Applications to Numerical Problems," *Proc. 22nd ACM Symposium on Theory of Computing*, 1990.

[GLRSW91] P. Gemmell, R. Lipton, M. Sudan, R. Rubinfeld, A. Widgerson. Self-Testing/Correcting for Polynomials and for Approximate Functions. *Proc. 23rd ACM Symposium on Theory of Computing*, 1991.

[GS92] P. Gemmell, M. Sudan. Highly Resilient Correctors for Polynomials. *Information Processing Letters*, 28 Sept. 1992, vol. 43 (no 4): 169-174

[La65] S. Lang. Algebra. *Addison-Wesley Pu. Co.*, 1965.

[L91] R. Lipton. "New directions in testing", in Distributed Computing and Cryptography, DIMACS Series in Discrete Mathematics and Theoretical Computer Science, vol. 2, American Mathematical Society, 1991.

[RR90] R. Rubinfeld. A mathematical theory of Self-Checking, Self-Testing, and Self-Correcting Programs. PhD Thesis, UC Berkeley, and ICSI TR-90-054, 1990.

[RR94] R. Rubinfeld. Robust Functional Equations with Applications to Self-Testing/Correcting. Proc. 35th IEEE FOCS, 1994.

[V91] F. Vainstein. Error Detection and Correction in Numerical Computation by Algebraic Methods. Lecture Notes in Computer Science, 539, Springer Verlag, 1991, pp.456-464.

[V93] F. Vainstein. Algebraic Methods in Hardware/Software Testing. Ph.D. Thesis, Boston University Graduate School, 1993.

Measure, Category and Learning Theory

Lance Fortnow[*1], Rūsiņš Freivalds[**2], William I. Gasarch[***3], Martin Kummer[4], Stuart A. Kurtz[1], Carl Smith[†3], Frank Stephan[‡4]

[1] Department of Computer Science, University of Chicago, 1100 E. 58th St., Chicago, IL 60637, USA. {fortnow; stuart}@cs.uchicago.edu
[2] Institute of Mathematics and Computer Science, University of Latvia, Raiņa bulvāris 29, LV-1459, Riga, Latvia. rusins@mii.lu.lv
[3] Department of Computer Science, University of Maryland, College Park, MD 20742, USA. {gasarch; smith}@cs.umd.edu
[4] Institut für Logik, Komplexität und Deduktionssysteme, Universität Karlsruhe, 76128 Karlsruhe, Germany. {kummer; fstephan}@ira.uka.de

Abstract. Measure and category (or rather, their recursion theoretical counterparts) have been used in Theoretical Computer Science to make precise the intuitive notion "for most of the recursive sets." We use the notions of effective measure and category to discuss the relative sizes of inferrible sets, and their complements. We find that inferrible sets become large rather quickly in the standard hierarchies of learnability. On the other hand, the complements of the learnable sets are all large.

1 Introduction

Determining the relative size of denumerable sets, and those with cardinality $\aleph_1$, led mathematicians do develop the notions of measure and category [Oxt71]. Described in this report is an application of measure and category techniques to a branch of learning theory called inductive inference [AS83]. The models of learning used in this field have been inspired by features of human learning.

The goal of this work is to determine the relative sizes of set theoretically incomparable classes of inferrible sets of functions. The idea is to determine when, within the well studied hierarchies of identification criteria, the classes of learnable functions become "large." To learn a "large" class would require a significantly larger hypothesis space than would be required to learn a "small" class. As the hypothesis space grows, so does the amount of resources required to search through it. The point we are looking for is where the size of the hypothesis

* Supported in part by NSF Grant CCR 92-53582.
** Supported in part by Latvian Council of Science Grant 93.599 and NSF Grant 9119540.
*** Supported in part by NSF Grant 9301339.
† Supported in part by NSF Grants 9119540 and 9301339.
‡ Supported by the Deutsche Forschungsgemeinschaft (DFG) Grant Me 672/4-2.

space must take a quantum leap from small, with respect to category and/or measure, to large. We find this point exactly.

We are also interested in the sizes of the complements of the learnable classes. The idea is that if the complement of a class is small, then the class itself must be significantly larger than any class with a large complement. It turns out that unless an inference criteria is sufficiently powerful so as to be able to learn all the recursive functions, then the complement of the learnable sets of functions is large. This means that every practical learning system must be very far from general purpose.

The notions of measure and category have been studied within the context of theoretical computer science. Mehlhorn [Mel73] and Lutz [Lut92] used constructive notions of category and measure to study subrecursive degree structures.

2 Technical Preliminaries

The natural numbers are denoted by $\mathbb{N}$. The complement of a set S, with respect to $\mathbb{N}$, is denoted by $\bar{S}$. The rational numbers are denoted by Q For strings σ and τ, $\sigma \sqsubseteq \tau$ will denote the situation where σ is a prefix of τ. If σ is a proper prefix of τ, we will write $\sigma \sqsubset \tau$. This work is concerned primarily with the recursive (computable) functions as formalized in [MY78, Smi94] and many other places. $REC_{0,1}$ is the set of $\{0,1\}$-valued single argument recursive functions. The restriction of a function to a domain containing all the values less than x is written $f \upharpoonright x$. $M_0, M_1, \ldots$ is an effective list of Turing machines φ_i is the partial recursive function that is computed by M_i. For all results in this paper $\varphi_0, \varphi_1, \ldots$, could be replaced by any acceptable programming system.

This paper is also concerned with the ideas of measure and category that are defined over point sets in real interval $[0,1]$. In order to compare the learning of recursive functions with either measure or category, it is necessary to represent functions as points in the interval $[0,1]$. To do so, it will be convenient to represent a function by a sequence of values from its range. Such a representation is called a *string representation*. So, for example, the sequence $01^2 0^4 1^\infty$ represents the (total) function:

$$f(x) = \begin{cases} 0 & \text{if } x = 0 \text{ or } 3 \leq x \leq 6, \\ 1 & \text{if } 1 \leq x \leq 2 \text{ or } x \geq 7. \end{cases}$$

Every recursive function f can be represented by the $\{0,1\}$-valued function: $1^{f(0)+1} 0 \, 1^{f(1)+1} 0 \cdots$. Functions with range $\{0,1\}$ will be called *predicates*. We will use $\hat{f}$ to denote the string representation of the predicate that corresponds to f. The point in the interval $[0,1]$ associated with f is $0.\hat{f}$. For example, 0 represents the everywhere 0 predicate, and 1 corresponds to the constant 1 predicate.

The idea of *measure* comes from [Bor05, Bor14], see [Oxt71]. It is more convenient for us to use the martingale functions [Lut92, Lut93]. A function $m : \{0,1\}^* \to Q$ is a martingale if for all σ, if $m(\sigma 0) \downarrow$ or $m(\sigma 1) \downarrow$ then

1. $m(\sigma) \geq 0$;
2. $m(\sigma) \downarrow$, $m(\sigma 0) \downarrow$ and $m(\sigma 1) \downarrow$;
3. $m(\sigma) = \frac{m(\sigma 0) + m(\sigma 1)}{2}$.

Martingale functions can be viewed as betting strategies. A player starts with capital $m(\epsilon)$ and bets on the successive values of a sequence of bits. After he has seen the inital segment σ his capital is $m(\sigma)$ and he bets $m(\sigma b)/2$ that next bit is b for $b = 0, 1$. In (2) we require that the wages must sum up to the current capital $m(\sigma)$. If the outcome is b, he receives twice of his wager on σb and looses his wager on $\sigma(1 - b)$. His capital after σb is therefore $m(\sigma b)$, as intended. This gambling terminology is used in our proofs.

Let $A \subseteq \{0, 1\}^*$ be a path in the tree $2^{\{0,1\}^*}$. We say m succeeds on A if

1. For all n, $m(A[n]) \downarrow$, and
2. $\limsup_{n \to \infty} m(A[n]) \to \infty$.

We say a class C has measure zero if there is a martingale m that succeeds on all $A \in C$. We define recursive and partial recursive measure zero by requiring m to be recursive or partial recursive respectively.

Next we review an effective notion of category [Kur58, Oxt71]. We will give two definitions of effectively meager. Lisagor [Lis81] (also see Fenner [Fen91, Theorem 3.7]) showed that they are equivalent. We first need to define a Banach-Mazur game.

Definition: Let $C \subseteq \{0, 1\}^\omega$. The *Banach-Mazur game associated with C* is defined as follows. On the first move player I chooses $\sigma_1 \in \{0, 1\}^*$ and player II chooses τ_1 such that $\sigma_1 \sqsubseteq \tau_1$. In all subsequent moves the player picks an extension of the current string. If the final infinite string is in C then player I wins, else player II wins. We may also take C to be a subset of $2^{\{0,1\}^*}$ and use characteristic strings.

Definition: A set $C \subseteq REC_{0,1}$ is *effectively meager* iff there is a recursive winning strategy for player II in the Banach-Mazur game associated to C. Note that this strategy beats any strategy by player I, even nonrecursive ones.

Definition: (Due to Mehlhorn [Mel73]) A set $C \subseteq REC_{0,1}$ is *effectively meager* iff there is a sequence $\{h_k\}_{k \in \omega}$ of uniformly recursive functions $h_k : \{0, 1\}^* \to \{0, 1\}^*$ such that $\sigma \sqsubseteq h_k(\sigma)$ for all k, σ, and for every $f \in C$ there is k such that $h_k(\sigma) \not\sqsubseteq f$ for all σ. (This formalizes that C is contained in an effective union of effectively nowhere dense sets.)

The basic model of learning used in this paper was first used by philosophers interested in modeling the scientific method [Put75]. The model was studied formally by linguists who were interested in the learning of natural languages [Gol67] and cast recursion theoretically in [BB75]. An *inductive inference machine* (IIM) is an algorithmic device that inputs the graph of a recursive function and, while doing so, outputs a sequence of computer programs. Typically, an IIM M learns a recursive function f, if M, when given the graph of f as input, outputs a sequence of programs that converges to a program that computes f. The order in which the input arrives does not effect the learnability [BB75]. Each IIM M will learn an undecidable set of recursive functions denoted by $EX(M)$. EX is used to denote the collection of all such sets as M varies across IIMs. A common example of a set in EX is the set of so called self describing functions [Bar74, BB75]. For these functions, if x is the least number such that $f(x) \neq 0$, then $\varphi_x = f$.

A refinement of the notion of learning in the limit is to consider the number of times an IIM outputs a program that is different from the most recently produced program. When this happens, we say the M has made a *mind change*. If M learns f as above while making only $n \in \mathbb{N}$ mind changes, we say $f \in EX_n(M)$. Each IIM will learn, with respect to at most n mind changes, a certain set of recursive functions denoted by $EX_n(M)$. The collection of all the sets $EX_n(M)$ as M varies over IIMs is denoted by EX_n. Another commonly studied restriction on learning algorithms is to demand that they only produce as conjectures programs for total recursive functions. Such IIMs are called *Popperian* (see [CS83]). The resultant class of learnable sets of functions is denoted by PEX.

One way to expand the class of functions being learned is to allow the final conjecture output to be incorrect on some number of points. Let $a \in \mathbb{N}$. A set of functions $\mathcal{A}$ is in EX^a if there exists an IIM M such that, for all $f \in \mathcal{A}$, the final conjecture M makes about f is a partial recursive function that agrees with f except on at most a points. A set of functions $\mathcal{A}$ is in EX^* if there exists an IIM M such that, for all $f \in \mathcal{A}$, the final conjecture M makes about f is a partial recursive function that agrees with f except on some finite number of points.

One way to expand the class of functions being learned is to have more algorithms involved in the learning. The notion of *team inference* defined in [Smi82] and generalized in [OSW86]. A team n IIMs $[m, n]$ learns a functions if at least m of the team members learn the function according to the given criterion of success. Recently, team learning was shown to be equivalent to a form of learning resembling the well known "process of elimination" [FKS94]. Team learning is also strongly related to probabilistic learning [PS88, Pit89]. Some of our results pertain to the special case teams of IIMs judged according to the EX_0 criterion. Team learning for the criterion EX_0 has been investigated in [DPVW91, DKV92, DKV93, Vel89, JS90].

3 Categoricity of Inferrible Classes

Theorem 1. *The class of predicates of self describing functions is not effectively meager.*

Sketch. We describe an effective winning strategy A for the first player of the Banach-Mazur game for the self describing functions. Suppose B is an effective strategy for the second player of the game. Player A first plays a string that encodes the index of the following algorithm: After defining $f \upharpoonright n$, see what $B(f \upharpoonright n)$ is and extend f to be the same. $\square$

Theorem 2. *Every class of predicates in PEX is effectively meager.*

Proof. Suppose S is a set in PEX. Let M be an IIM serving as the witness. We describe an effective winning strategy B for the second player of the Banach-Mazur game for S. Suppose that it is B's turn to play and that σ is the finite function that has been determined by the game so far. Let x be the least value not in the domain of σ. Let $e = M(\sigma)$. Since $S \in PEX$, φ_e is a recursive function. Let $y = 1 \dot{-} \varphi_e(x)$. B plays y, e.g. the next initial segment is $\sigma \cdot y$. Clearly, the function constructed is not in S. □

4 Measurability of Inferrible Classes

Theorem 3. *Every finite class of recursive predicates has partial recursive measure zero. The class $S_0 = \{\sigma 0^\infty : \sigma \in \{0,1\}^*\}$ has recursive measure zero.*

Proof. Let $m(\sigma) = (1/2)^{x_1}(3/2)^{x_0}$ where x_i is the number of i's in σ. □

Theorem 4. *The class $S_1 = \{f | 0^e 1 \sqsubseteq f \text{ and } f = \varphi_e\}$ does not have recursive measure zero. (S_1 is often referred to as the class of self-referential predicates.)*

Proof. Let m be a martingale. Recall that, for all σ, either $m(\sigma 0) \leq m(\sigma)$ or $m(\sigma 1) < m(\sigma)$. Define by the recursion theorem the predicate $\varphi_e(x)$ as:

1. If $x < e$ then $\varphi_e(x) = 0$.
2. If $x = e$ then $\varphi_e(x) = 1$.
3. Otherwise $x > e$. Let $\sigma = \varphi_e \upharpoonright x$ be the already defined initial segment of φ_e and

$$\varphi_e(x) = \begin{cases} 0 & \text{if } m(\sigma 0) \leq m(\sigma); \\ 1 & \text{otherwise } (m(\sigma 1) < m(\sigma)). \end{cases}$$

The inductive definition guarantees $m(\sigma) \leq m(0^e 1)$ for all $\sigma \sqsubseteq \varphi_e$ of length greater than e. □

Theorem 5. *Every class of predicates in EX_0 has partial recursive measure zero.*

Proof. Let M be an inductive inference machine for some class of languages in EX_0. Define a partial recursive martingale m as follows for $\sigma \in \Sigma^*$ and $i \in \Sigma$:

- Let $m(\lambda) = 1$.
- If $M(\sigma)$ does not output a value then $m(\sigma b) = 1$ for $b = 0, 1$.
- If $M(\sigma)$ outputs a value and $\varphi_{M(\sigma)}(|\sigma|) = b$ then let $m(\sigma b) = 2m(\sigma)$ and $m(\sigma(1-b)) = 0$. □

The question arises as to whether every class in EX_1 has partial recursive measure 0. We will now see that this is false. To see this, we employ the following lemma. It is more general than is needed now; however, we shall use its full generality later.

Lemma 6. *If m is a total recursive martingale, $\sigma \in \{0,1\}^*$, and $c > m(\sigma)$, then there exists τ such that*

1. τ extends σ
2. $(\forall \eta)[\sigma \sqsubseteq \eta \sqsubseteq \tau \rightarrow m(\eta) \leq c]$.
3. $m(\tau 0) \leq c$ and $m(\tau 1) \leq c$.
4. There exists $k \leq \frac{m(\sigma)}{c - m(\sigma)}$ such that $\sigma 1^k \sqsubseteq \tau$ and $m(\sigma 1^k 0) \leq c$. Hence if $c = m(\sigma) + \frac{1}{2^n}$ then $k \leq 2^n m(\sigma)$.

Proof. Let $\eta_0 = \sigma$ and let

$$\eta_{i+1} = \begin{cases} \eta_i 0 & \text{if } m(\eta_i 1) > c; \\ \eta_i 1 & \text{if } m(\eta_i 0) > c; \\ \eta_i & \text{otherwise.} \end{cases}$$

If $\eta_{i+1} \neq \eta_i$ then $m(\eta_{i+1}) < 2m(\eta_i) - c \leq m(\eta_i) - (c - m(\sigma))$. Since a martingale maps to the positive numbers there exists a least $i \leq \frac{m(\sigma)}{c - m(\sigma)}$ such that $\eta_{i+1} = \eta_i$. Let $\tau = \eta_i$.

Conditions 1 clearly hold. Since m is a martingale and $c > m(\sigma)$ condition 2 holds (in fact $m(\eta) \leq m(\sigma)$). Since $m(\tau 0) \leq c$ and $m(\tau 1) \leq c$. condition 3 holds.

Let k be the maximal string such that $\sigma 1^k \sqsubseteq \tau$. Clearly $k \leq i \leq \frac{m(\sigma)}{c - m(\sigma)}$ and $m(\sigma 1^k) \leq m(\sigma)$. If $\tau = \sigma 1^k$ then, by condition 3, $m(\sigma 1^k 0) \leq c$. If $\sigma 1^k 0 \sqsubseteq \tau$ then by condition 2 $m(\sigma 1^k 0) \leq c$. Hence in any case $m(\sigma 1^k 0) \leq c$. $\qquad\square$

Theorem 7. *There exists a class in EX_1 that does not have partial recursive measure zero.*

Proof. A string $\sigma = 1^e 01^{a_1} 01^{a_2} 0 \cdots 01^{a_n}$ is said to have *few ones* if $0 \leq a_i < 2^{i+2}$ for $i = 1, 2, \ldots, n$. A function has *few ones* iff every of its initial segments has few ones. Let S contain the following functions:

- φ_e if φ_e is total, $1^e 0 \sqsubseteq \varphi_e$ and φ_e has few ones.
- $\sigma 1^\infty$ if σ has few ones.

S is EX_1-inferable: If $\sigma \sqsupseteq 1^e 0$ has few ones, then guess φ_e. Otherwise let $\tau \sqsubset \sigma$ denote the longest initial segment of σ with few ones and guess $\tau 1^\infty$.

Assume that S has partial recursive measure 0 via a martingale m; without loss of generality assume that $m(1^e 0) = 1$ for all e. Now define by the recursion theorem the predicate $\varphi_e(x)$ in effective stages of finite extension as follows. Initialize the function to be $1^e 0$. We now proceed by stages. Inductively assume that before stage $n + 1$, (1) the function is of the form $\sigma = 1^e 01^{a_1} 0 \cdots 1^{a_n} 0$, (2) σ has few ones, and (3) for all $\eta \sqsubseteq \sigma$, $m(\eta) \leq \sum_{i=0}^n 2^{-i}$. Let k be the value obtained by applying Lemma 6, part 4, to σ as above and $c = m(\sigma) + 1/2^n$. Extend the function by $1^k 0$ and go to the next stage. It is easy to show that the function still satisfies the inductive hypothesis.

Clearly, the function constructed is in S. Since $m(\eta) \leq \sum_{i=0}^n 2^{-i} < 2$ on all initial segments of f the martingale is less than 2. Hence, the martingale fails to make a profit on f. This is a contradiction. $\qquad\square$

Theorem 8. *Every class of predicates in $[c, d]EX_0$ has partial recursive measure zero iff $c > d/2$.*

Proof. ($\Rightarrow$) If $1 \leq c \leq d/2$ then $[1, 2]EX_0 \subseteq [c, d]EX_0$. Since the set S from the previous proof belongs to $[1, 2]EX_0$ it follows that $[c, d]EX_0$ contains a class which does not have partial recursive measure zero.

($\Leftarrow$) Suppose we are given a team of d machines which $[c, d]EX_0$-infer a class S. Consider the strings σ such that at least c of the machines output the values $e(\sigma, 1), \ldots, e(\sigma, c)$ on input σ, and no proper prefix of σ has this property. Note that this is an r.e. set of mutually incomparable strings.

For each such σ we uniformly enumerate the tree T_σ whose c branches are given by the functions $\varphi_{e(\sigma,1)}, \ldots, \varphi_{e(\sigma,c)}$. For every $f \in S$ there is such a σ, and since $c > d/2$, $f = \varphi_{e(\sigma,i)}$ for some i and f is a branch of the tree T_σ.

We define the martingale in $T = T_\sigma$ as follows: $m(\sigma) = 1$; if $m(\tau)$ is already defined, we enumerate T until either $\tau 0$ or $\tau 1$ appears. If τb appears first we bet $3/4 m(\tau)$ on τb and $1/4 m(\tau)$ on $\tau(1-b)$.

Consider any infinite branch f in T. Since T has only finitely many branching nodes there is $\eta \sqsubset f$ such that $(\forall \tau \in T)[\eta \sqsubseteq \tau \Rightarrow \tau \sqsubseteq f]$. Thus, for every n there is an initial segment $\tau \sqsubset f$ with $m(\tau) = (3/2)^n m(\eta)$. Since $m(\eta) > 0$, the martingale is unbounded on f. $\qquad\square$

We strengthen the previous theorem.

Theorem 9. *Let c, d be such that $c > \frac{d}{2}$. Every class of predicates in $[c, d]EX_0^*$ has partial recursive measure 0.*

Proof. Assume $S \in [c, d]EX_0^*$ via IIM's $M_1, \ldots, M_d$. We describe a partial recursive martingale m for S intuitively; we formalize it later. Assume that if σb ($\sigma \in \{0, 1\}^*$, $b \in \{0, 1\}$) is input then σ is an initial segment of a fixed 0-1 valued function $f \in S$. We define $m(\sigma 0)$ and $m(\sigma 1)$ simultaneously. They are how much we are betting that $f(|\sigma|) = 0$ and $f(|\sigma|) = 1$. Since we can compute $m(\tau)$ for all $\tau \sqsubseteq \sigma$ recursively we assume that we have access to all prior bets and outcomes.

If $|\{i : M_i(\sigma) \downarrow\}| < c$ then we bet on 0 and 1 equally. Once we have σ such that $|\{i : M_i(\sigma) \downarrow\}| \geq c$ we set up the following parameters.

1. e_i is the guess that the M_i makes as to the index of f. Initially it is $M_i(\sigma)$ if it exists, and undefined otherwise. It may be undefined now but become defined later.
2. k is our current guess for a bound on how many errors the φ_{e_i} that are supposed to $=^*$ compute f makes in trying to compute f. Initially $k = 0$. This parameter increases when its current guess looks infeasible.
3. E_i is the number of errors φ_{e_i} has made. Initially $E_i = 0$.
4. *POSS* is the set of all i such that we currently think $\varphi_{e_i} =^k f$ is possible. *POSS* will always be $\{i \mid E_i \leq k\}$. This set may (1) grow when more e_i's are defined, (2) shrink when a φ_e is seen to be wrong more than $k+1$ times, and (3) grow when k is increased. This set will always have at least c elements.
5. *CANCEL* is the set of indices that we have seen make at least $k+1$ errors. It will always be $\{e_i \mid E_i \geq k+1\}$.

On input σ we do the following. Look for $(x, i, b) \in N \times POSS \times \{0, 1\}$ such that $|x| \geq |\sigma|$ and $\varphi_{e_i}(x) \downarrow = b$. (Since $c > \frac{d}{2}$ and $|POSS| \geq c$ there exists $i \in POSS$ such that $\varphi_{e_i} =^* f$, hence an (x, i, b) will be found.) We will now plan to (1) bet even money on $\tau 0$ and $\tau 1$ for all $|\tau|$ such that $|\sigma| \leq |\tau| \leq x - 1$, and (2) bet $\frac{3}{4}$ that $f(x) = b$ and $\frac{1}{4}$ that $f(x) = 1 - b$. If we later find out that this prediction is wrong when we will set $E_i = E_i + 1$. If $|E_i| \geq k + 1$ then we cancel e_i and remove i from $POSS$. If $|POSS| < c$ then the value of k was to low so we set $k := k + 1$, $POSS := POSS \cup CANCEL$, and $CANCEL := \emptyset$.

We now present this algorithm formally. Since we will be calling it recursively we make each execution of the algorithm yields many parameters as well as the answer.

1. Input($\sigma 0, \sigma 1$).
2. If $\sigma = \lambda$ (the empty string) then $m(\sigma 0) = m(\sigma 1) = 1$. If $|\{i : M_i(\sigma) \downarrow\}| < c$ then set $m(\sigma 0) = m(\sigma 1) = m(\sigma)$. $POSS = \emptyset$ and all other parameters are undefined.
3. (We can assume $|\{i : M_i(\sigma) \downarrow\}| \geq c$.) Compute $m(\sigma)$. From this computation we extract two parameters: $POSS$ and (x, i, b). Intuitively, x is a number that we established a while back that we would like to bet that $f(x) = b$, since $\varphi_{e_i}(x) \downarrow = b$.
4. (This step will only execute on the shortest initial segment that causes $|\{i : M_i(\sigma) \downarrow\}| \geq c$.) If $POSS = \emptyset$ then we need to initialize several parameters.

$$e_i := \begin{cases} M_i(\sigma) & \text{if } M_i(\sigma) \downarrow \\ \text{undefined} & \text{otherwise} \end{cases}$$
$$k := 0$$
$$E_i := \begin{cases} 0 & \text{if } e_i \text{ is defined} \\ \text{undefined} & \text{otherwise} \end{cases}$$
$$POSS := \{i \mid E_i \leq k\}$$
$$CANCEL := \emptyset$$

5. If (x, i, b) is defined and $x > |\sigma|$ then set $m(\sigma 0) = m(\sigma 1) = m(\sigma)$. All parameters retain their values.
6. If (x, i, b) is defined and $x = |\sigma|$ then set $m(\sigma b) = \frac{3}{2}m(\sigma)$ and $m(\sigma(1 - b)) = \frac{1}{2}m(\sigma)$. All parameters retain their values. (We will in the very next stage see if this bet was correct and take appropriate action.)
7. If (x, i, b) is defined and $x = |\sigma| - 1$ then do the following.
   (a) If b is not the last bit of σ then set $E_i := E_i + 1$.
   (b) If $E_i > k$ then $POSS := POSS - \{i\}$ and $CANCEL := CANCEL \cup \{i\}$.
   (c) If $|POSS| < c$ then $k := k + 1$, $POSS := POSS \cup CANCEL$, and $CANCEL := \emptyset$.
   (d) Declare (x, i, b) undefined.
   (e) If $(\exists j)[(j \notin CANCEL \cup POSS) \wedge (M_j(\sigma) \downarrow)]$ then for all such j set $e_j := M_j(\sigma)$, $E_j = 0$, and $POSS := POSS \cup \{j\}$.
8. If (x, i, b) is undefined then find (by dovetailing) a triple $(x, i, b) \in N \times POSS \times \{0, 1\}$ such that $x \geq |\sigma|$ and $\varphi_{e_i}(x) \downarrow = b$. Define (x, i, b) to be the first such triple found. Execute steps 5 and 6.

We show that if $f \in S$ then this martingale wins: Let $f \in S$. Let $\sigma \sqsubseteq f,k'$, and k'' be such that the following hold.

1. Let $e_i = M_i(\sigma)$ if it exists, undefined otherwise. Let $CORR = \{i \mid \varphi_{e_i} = ^{k'} f\}$. We require $|CORR| \geq c$.
2. For all $i \in CORR$ if $\varphi_{e_i}(x) \neq f(x)$ then $x < |\sigma|$.
3. When $m(\sigma)$ is defined the value of k is k''.

We look at how the martingale behaves on inputs $\tau \sqsubseteq f$ such that $|\tau| > |\sigma|$. If the (x, i, b) picked is such that $i \in CORR$ then the martingale will win. The number of times an $i \notin CORR$ can be picked such that the martingale loses is at most $(d - c) \max\{k', k''\}$. Hence eventually the martingale will always win. $\square$

Definition 10. Let $\mathcal{A} \subseteq REC_{0,1}$. $\mathcal{A}$ is *uniformly r.e.* if there exists a recursive function g such that $\mathcal{A} = \{\lambda x.g(i, x) \mid i \geq 0\}$. $\mathcal{A}$ is *uniformly r.e.-in-X* if instead of g being recursive we have $g \leq_T X$.

Theorem 11. *Let $\mathcal{A}$ be a uniformly r.e. class of 0-1 valued recursive functions. Then $\mathcal{A}$ has recursive measure 0.*

Proof. Since $\mathcal{A}$ is uniformly r.e. there exists a recursive function g such that $\mathcal{A} = \{\lambda x.g(i, x) \mid i \geq 0\}$.

Let $m(\lambda) = 1$. Assume $m(\sigma)$ is defined. We define $m(\sigma 0)$ and $m(\sigma 1)$ simultaneously. Find the minimal $i \leq |\sigma|$ such that for all x, $|x| \leq |\sigma|$, $g(i, x) = \sigma(x)$. Let $g(i, |\sigma| + 1) = b$. Let $m(\sigma b) = \frac{3}{2}m(\sigma)$ and $m(\sigma(1 - b)) = \frac{1}{2}m(\sigma)$.

Let f be the recursive function being fed to the martingale. Let e be the minimal i such that $(\forall x)[g(i, x) = f(x)]$. The i picked in determining $m(|\sigma|)$ will reach a limit of e. Once it reaches that limit the martingale always bets more on the correct answer than the incorrect answer. Hence the theorem is established. $\square$

Corollary 12. *Every class of predicates in PEX has recursive measure zero.*

5 Complements of Inference Classes are Large

In this section, if $\mathcal{A} \subseteq REC_{0,1}$, then "$\mu(\mathcal{A}) = 0$" denotes that $\mathcal{A}$ has partial recursive measure 0.

Definition 13. An *infinite binary tree* is a mapping T from $\{0,1\}^*$ to $\{0,1\}^*$ such that (1) $\sigma \prec \tau$ implies $T(\sigma) \prec T(\tau)$, and (2) $T(\sigma 0)$ and $T(\sigma 1)$ are incompatible. Let $d_0 d_1 \cdots \in \{0,1\}^\omega$. Let $A = \lim_{n \to \infty} T(d_0 \cdots d_n)$. A is *a branch of T*. A *recursive binary tree* is a tree computed by a recursive function.

Definition 14. Let T be a recursive binary tree and let $\mathcal{C} \subseteq REC_{0,1}$. We would like to define the set of branches of T that are 'guided' by functions in $\mathcal{C}$. Formally let

$$RECB(T, \mathcal{C}) = \{f \mid (\exists \phi \in \mathcal{C})[f = \lim_{n \to \infty} T(\phi(0)\phi(1) \cdots \phi(n)]\}.$$

Definition 15. An inference class $\mathcal{I}$ is *reasonable* if, for all recursive trees T and sets $\mathcal{C} \subseteq REC_{0,1}$,

$$\mathcal{C} \in \mathcal{I} \text{ iff } RECB(T, \mathcal{C}) \in \mathcal{I}.$$

Virtually all known inference classes are reasonable. In particular $[a, b]I_c^d$, is reasonable for $I \in \{EX, BC, PEX\}$ and any choice of $a, b, c, d \in \mathbb{N} \cup \{*\}$. The inference classes dealt with in the study of learning via queries [GS92], are not reasonable.

Lemma 16. *Let $\mathcal{B}$ be a dense subset of $REC_{0,1}$ such that $\mu(\mathcal{B}) = 0$ or $\mathcal{B}$ is effectively meager. Then there exists a recursive binary tree T such that no branch of T is in $\mathcal{B}$.*

Proof. Case 1: $\mu(\mathcal{B}) = 0$. Let m be a partial recursive martingale for $\mathcal{B}$. Since $\mathcal{B}$ is dense, m is total recursive.

We define T as follows. Let $T(\lambda) = \lambda$. Assume $T(\sigma)$ is already defined. Let $c = m(T(\sigma)) + 2^{-|\sigma|}$. By Lemma 6 there exists a $\tau \sqsupseteq T(\sigma)$ such that

$$(\forall \eta)[T(\sigma) \sqsubseteq \eta \sqsubseteq \tau \to m(\eta) \leq c]$$

and

$$m(\tau 0) \leq c \text{ and } m(\tau 1) \leq c.$$

We can find such a τ by searching. Let $T(\sigma 0) = \tau 0$ and $T(\sigma 1) = \tau 1$.

We show that these conditions imply no branch of T is in $\mathcal{B}$. Let $d_1 d_2 d_3 \cdots \in \{0, 1\}^\omega$ By induction we have

$$[\sigma \sqsubseteq T(d_1 d_2 \cdots d_n)] \to \left[m(\sigma) \leq m(T(\lambda)) + \sum_{i=0}^{n} 2^{-n} \right].$$

Hence if $A = b_1 b_2 \cdots$ is any branch of T then

$$\lim_{n \to \infty} m(b_1 \cdots b_n) \leq m(T(\lambda)) + \sum_{i=0}^{\infty} 2^{-i} \leq m(T(\lambda)) + 2.$$

Since m is a martingale for $\mathcal{B}$ we have $A \notin \mathcal{B}$.

Case 2: $\mathcal{B}$ is effectively meager. Let $\{h_k\}_{k \in \omega}$ be the associated uniformly recursive functions. Let $T(\lambda) = \lambda$ and for, $b \in \{0, 1\}$ let $T(\sigma b) = h_{|\sigma|}(T(\sigma)b)$. Clearly no branch of T is in $\mathcal{B}$. $\square$

Theorem 17. *Let $\mathcal{I}$ be a reasonable inference class. If there exists $\mathcal{A} \in \mathcal{I}$ such that $\mu(REC_{0,1} - \mathcal{A}) = 0$ or $REC_{0,1} - \mathcal{A}$ is effectively meager then $REC_{0,1} \in \mathcal{I}$.*

Let $\mathcal{B} = REC_{0,1} - \mathcal{A}$. There are two cases.

Proof.

Case 1: $\mathcal{B}$ is not dense. Hence there is a σ such that $(\forall f \in REC_{0,1})[\sigma \sqsubseteq f \to f \in \mathcal{A}]$. Hence $\sigma \cdot REC_{0,1} \subseteq \mathcal{A}$, so $\sigma \cdot REC_{0,1} \in \mathcal{I}$. Since $\mathcal{I}$ is reasonable $REC_{0,1} \in \mathcal{I}$.

Case 2: $\mathcal{B}$ is dense. By Lemma 16 there exists a recursive tree T such that no branch of T is in $\mathcal{B}$. Hence every recursive branch of T is in $\mathcal{A}$, or in our notation $RECB(T, REC_{0,1}) \subseteq \mathcal{A}$. Therefore $RECB(T, REC_{0,1}) \in \mathcal{I}$. Since $\mathcal{I}$ is reasonable $REC_{0,1} \in \mathcal{I}$. $\square$

6 Conclusions

We have shown that there are sets in EX_0 that are not effectively meager, hence, even very limited learning algorithms must have very large hypothesis spaces to search through. For recursive measure, the learnable sets become large at the same place in the hierarchy of learnable classes. For partial recursive measure, the point is at EX_1.

The complements of the learnable sets are all very large with respect to both measure and category. This indicates that unless a technique is guaranteed to learn all the recursive functions, then it will fail to learn a large set of them.

Acknowledgement: We would like to thank Marcus Schäfer for proofreading.

References

[AS83]   D. Angluin and C. H. Smith. Inductive inference: Theory and methods. *Computing Surveys*, 15:237–269, 1983.

[Bar74]   J. Barzdins. Two theorems on the limiting synthesis of functions. In Barzdins, editor, *Theory of Algorithms and Programs*, volume 1, pages 82–88. Latvian State University, Riga, U.S.S.R., 1974.

[BB75]   L. Blum and M. Blum. Toward a mathematical theory of inductive inference. *Information and Control*, 28:125–155, 1975.

[Bor05]   E. Borel. *Leçons sur les fonctions de variables réeles*. Gauthier-Villars, Paris, 1905.

[Bor14]   E. Borel. *Leçons sur la théorie des fonctions*. Gauthier-Villars, Paris, 1914.

[CS83]   J. Case and C. Smith. Comparison of identification criteria for machine inductive inference. *Theoretical Computer Science*, 25(2):193–220, 1983.

[DKV92]   R. Daley, B. Kalyanasundaram, and M. Velauthapillai. Breaking the probability 1/2 barrier in fin-type learning. *Proceedings of the Fifth Annual Workshop on Computational Learning Theory*, pages 203–217, ACM Press, 1992.

[DKV93]   R. Daley, B. Kalyanasundaram, and M. Velauthapillai. Capabilities of fallible finite learning. *Proceedings of the Sixth Annual Workshop on Computational Learning Theory*, pages 199–208, ACM Press, 1993.

[DPVW91] R. Daley, L. Pitt, M. Velauthapillai, and T. Will. Relations between probabilistic and team one-shot learners. *Proceedings of the Fifth Annual Workshop on Computational Learning Theory*, pages 228–239, Morgan Kaufmann Publishers, 1992.

[Fen91]   S. Fenner. Notions of resource-bounded category and genericity. *Proceedings of the Sixth Annual Conference on Structure in Complexity Theory*, pages 196–212. IEEE Computer Soceity, 1991.

[FKS94]   R. Freivalds, M. Karpinski, and C. Smith. Co-learning of total recursive functions. *Proceedings of the Seventh Annual Workshop on Computational Learning Theory*, pages 190–197, ACM, 1994.

[Gol67]   E. M. Gold. Language identification in the limit. *Information and Control*, 10:447–474, 1967.

569

[GS92]   W. Gasarch and C. H. Smith. Learning via queries. *Journal of the ACM*, 39(3):649–675, 1992. A shorter version is in 29th FOCS conference, 1988, pp. 130-137.

[JS90]   S. Jain and A. Sharma. Finite learning by a team. *Proceedings of the Third Annual Workshop on Computational Learning Theory*, pages 163–177, Morgan Kaufmann Publishers, 1990.

[Kur58]   C. Kuratowski. *Topologie Vol. 1*, 4th edition, volume 20 of *Monografie Matematyczne*, Panstwowe Wydawnictwo Naukowe, 1958.

[Lis81]   L. Lisagor. The Banach-Mazur game. Translated Version of *Matematicheskij Sbornik*, 38:201–206, 1981.

[Lut92]   J. Lutz. Almost everywhere high nonuniform complexity. *Journal of Computer and Systems Science*, 44:226-258, 1992.

[Lut93]   J. Lutz. The qualitative structure of exponential time. In *Proceedings of the Eighth Annual Conference on Structure in Complexity Theory Conference*, pages 158–175. IEEE Computer Society, 1993.

[Mel73]   K. Mehlhorn. On the size of sets of computable functions. *Proceedings of the Fourteenth Annual Symposium on Switching and Automata Theory*, pages 190–196, IEEE Computer Soceity, 1973.

[MY78]   M. Machtey and P. Young. *An Introduction to the General Theory of Algorithms*. North-Holland, New York, 1978.

[OSW86]   D. N. Osherson, M. Stob, and S. Weinstein. Aggregating inductive expertise. *Information and Control*, 70:69–95, 1986.

[Oxt71]   J. Oxtoby. *Measure and Category*. Springer-Verlag, 1971.

[Pit89]   L. Pitt. Probabilistic inductive inference. *Journal of the ACM*, 36(2):383–433, 1989.

[PS88]   L. Pitt and C. Smith. Probability and plurality for aggregations of learning machines. *Information and Computation*, 77:77–92, 1988.

[Put75]   H. Putnam. Probability and confirmation. In *Mathematics, Matter and Method*, volume 1. Cambridge University Press, 1975.

[Smi82]   C. H. Smith. The power of pluralism for automatic program synthesis. *Journal of the ACM*, 29(4):1144–1165, 1982.

[Smi94]   C. Smith. *A Recursive Introduction to the Theory of Computation*. Springer-Verlag, 1994.

[Vel89]   M. Velauthapillai. Inductive inference with a bounded number of mind changes. *Proceedings of the Second Annual Workshop on Computational Learning Theory*, pages 200–213, Palo, Morgan Kaufmann, 1989.

A characterization of the existence of energies for neural networks

Michel Cosnard* and Eric Goles**

Laboratoire de l'Informatique du Parallélisme - CNRS
Ecole Normale Supérieure de Lyon, 69364 Lyon - France
and
Departamento de Ingeniería Matemática
Universidad de Chile - casilla 170-3, Santiago - Chile

Abstract. In this paper we give under an appropriate theoretical framework a characterization about neural networks which admit an energy. We prove that a neural network admits an energy if and only if the weight matrix verifies two conditions: the diagonal elements are non-negative and the associated incidence graph does not admit non-quasi-symmetric circuits.

1 Introduction

The energy concepts related with neural networks have been extensively studied, essentially in the framework of symmetric weight matrices. The dynamics of such networks is often characterized by an operator, called the energy of the network, which is decreasing on any orbit (sequence of iterates). The energy is a very powerful tool to study the periodic orbits (cycles) and particularly their length. In fact, Hopfield [4] proved that in the context of weights given by an Hebbian rule for associative memories, there exists such a decreasing operator $E = -\frac{1}{2}{}^t xWx$, where $W = (\omega_{ij})$ is the correlation matrix between the stored patterns. Also, it is proved in [2] and [3] that a similar quadratic operator is decreasing for neural networks defined by symmetric matrices with non negative diagonals. Later, the existence of an energy was proved for quasi-symmetric weights generalizing the symmetry property [3], [5]. Furthermore, a decreasing operator has been proved to exists for high order neural networks (i.e. with polynomial arguments and sequential iteration) under a natural generalization of symmetry [6].

In [5], Kobuchi develops a formal study in order to characterize networks which admit a decreasing operator. This characterization holds for symmetric and quasi-symmetric networks but the hypothesis concerning the weight matrix in more general situations are very difficult to handle.

 * Support by the EC Working Group NeuroCOLT and French-Chile cooperation (ECOS-94)is aknowledged
** Partially supported by FONDECYT-94, EC-Chile project in applied mathematics and French-Chile cooperation (ECOS-94)

Remark that the existence (or non-existence) of decreasing operators driving the network dynamics is extremely sensitive to small perturbations on the weights: i.e. little alterations on the weight matrix may change completely the dynamic behavior of the network [3]. On the other hand, the decreasing operators for Hopfield networks or for (quasi-)symmetric matrices are very robust in the sense that any neural network with a weight matrix derived from the initial one by local operations which preserve the (quasi-)symmetry accepts also the same kind of decreasing functional, this time acting on the new weight matrix and threshold vector.

In this context, we shall say that a neural network $\mathcal{N}$ admits an energy when there exists a decreasing functional for any network $\tilde{\mathcal{N}}$ derived from $\mathcal{N}$ by local changes in the weight matrix W and threshold vector b. It is clear that the previous definition of energy is robust in the sense that the existence of a decreasing operator remains invariant under local changes on the weights. Of course we shall see that our definition holds in the framework of (quasi) symmetric and non-negative diagonal neural networks. Hence it generalizes all the known situations in which an energy has been shown to exist.

In this framework we prove that a neural network admits an energy operator if and only if the graph associated to the weight matrix of the network does not admit a non quasi symmetric circuit. This characterization is important since neural networks with energy are used in various manners as for example for building associative memories. In fact a direct consequence of our result is that the generic class of neural networks admitting only fixed points is composed of direct acyclic graph of quasi-symmetric strongly connected components. As will be shown in a subsequent paper this implies that we can relate the Hopfield theory of associative memories to the well known feed forward layered networks and define learning algorithms for this new class of networks.

2 Definitions and notations

Let us consider a graph $G = (V, U)$, where $V = \{1, ..., n\}$ is the set of vertices and U is the set of arcs. Let $W = (w_{ij})$ be an (n, n) real matrix. We call G the incidence graph of W if

$$(j, i) \in U \iff w_{ij} \neq 0.$$

By extension W will also be called the incidence matrix of G. We say that a set of different vertices $\{i_0, ..., i_p\}$ is a circuit of G iff $p \geq 1$ and the arcs $(i_0, i_1), ..., (i_p, i_0)$ belong to U. If $w_{ii} \neq 0$, we shall say that G has a loop at node i. If $w_{ii} > 0$, the loop will be called positive.

Define now a neural network acting on the set of states $\{0, 1\}$ as the tuple $\mathcal{N} = (G, W, b, 1)$, where G is the incidence graph of W, b is a threshold vector and the local transition function is given by:

$$y_i = 1(\sum_{j=1}^{n} w_{ij} x_j - b_i), \quad x \in \{0, 1\}^n$$

where $1(u) = 1$ if $u \geq 0$ (0 otherwise).

Assume that one iterates the network sequentially, i.e. by updating the nodes one by one in a prescribed periodic update order $I = \{\{i_1\}, \{i_2\}, ..., \{i_n\}\}$, where $\{i_1, ..., i_n\} = \{1, ..., n\}$. Hence, starting with some given $x = (x_1(0), ..., x_n(0))$, the sequence of iterated configurations is called the orbit of x. Any orbit is composed of a periodic part, called the cycle, and a non periodic one, called the transient. If the sequence is periodic, we say also that the orbit is periodic. If x is invariant under the application of the complete sequence of updates, we shall say that x is a fixed point. Without loss of generality, we assume that all the thresholds are non zero. In fact if there exists $b_i = 0$, then one may easily construct an equivalent neural network (with the same dynamics) where $b_i' \neq 0$ by adding a vertex labeled $n+i$ which always remains fixed at value 1 and linked to vertex i by a weight b_i'.

Also one may assume (see [3]) that $\forall i, \forall u \in \{0, 1\}^n, h_i(u) = \sum w_{ij} u_j - b_i \neq 0$. In fact if $h_i(u) = 0$, it suffices to make a small change in the hyperplane coefficients in order to avoid this situation without modifying the dynamics of the network.

3   Energy definition

As mentionned in the introduction we shall define the energy of a neural network as a property invariant under local operations. But before introducing such operations, we first recall the definition of a decreasing operator. In the remaining we shall take a somewhat restrictive definition of a decreasing operator, in the sense that we shall assume that the operator is independent on the update order. Recall that we assume that the update order is periodic.

Definition 1. A neural network $\mathcal{N} = (G, W, b, 1)$ admits a decreasing operator if there exists a decreasing function

$$F_\mathcal{N} : \{0, 1\}^n \to \mathcal{R} \text{ such that } F_\mathcal{N}(x^{(k)}) - F_\mathcal{N}(x) < 0 \quad \forall k \text{ such that } x^{(k)} \neq x,$$

where x is the current configuration and $x^{(k)}$ is the configuration generated by the application of the update rule on node k, i.e.:

$$x_i^{(k)} = \begin{cases} x_i & \forall i \neq k \\ 1(\sum_{j=1}^{n} w_{ij} x_j - b_i) \text{ if } i = k \end{cases}.$$

The following lemma is a straightforward consequence of this definition. It has been often used for characterizing the dynamics of neural networks.

Lemma 2. If $\mathcal{N} = (G, W, b, 1)$ admits a decreasing operator, then its periodic orbits under any periodic update order are only fixed points.

Proof. Let $F_\mathcal{N}$ be the decreasing function associated to $\mathcal{N}$. Assume that $\mathcal{N}$ admits a cycle of period $T > 1$ and that the update order is $\{\{k_1\}, \{k_2\}, ..., \{k_n\}\}$. Then the following sequence exists:

$$x(0) \to x(1) = x^{(k_1)}(0) \to ... \to x(T-1) = x^{(k_{T-1})}(T-2) \to x(0)$$

From the definition of $F_\mathcal{N}$ we deduce that $F_\mathcal{N}(x(t)) \geq F_\mathcal{N}(x(t+1))$ for $t = 0, ..., T-1$. Since the period T is strictly greater than 1, one of the inequalities is strict. Hence $F_\mathcal{N}(x(0)) > F_\mathcal{N}(x(T))$. This is a contradiction since by definition $x(T) = x(0)$.

The notion of decreasing operator has been introduced in order to study the dynamics of symmetric and quasi-symmetric neural networks [3], [4]. We shall introduce more formally this last notion and recall some of the basic results.

Definition 3. A weight matrix W is quasi-symmetric if

$$\forall i, j, \exists \lambda_i, \lambda_j > 0 \text{ such that } \lambda_i w_{ij} = \lambda_j w_{ji}.$$

By extension a neural network $\mathcal{N} = (G, A, b, 1)$ is quasi-symmetric if W is quasi-symmetric.

Note that λ_i is associated to the vertex i and is the same for all w_{ij} with fixed i. Remark also that when $\lambda_i = 1, \forall i$, we get a symmetric matrix which corresponds to the well-known Hopfield neural network [4]. Hence quasi-symmetry can be interpreted as a direct generalization of symmetry. This notion has been introduced in [3] and later in [5]. In both cases and independently, the authors prove that the sequential dynamics of a quasi-symmetric neural network is regulated by a decreasing operator and hence converges to fixed points.

Proposition 4. *A quasi-symmetric neural network $\mathcal{N} = (G, W, b, 1)$ with non-negative diagonal admits a decreasing operator.*

Proof. One can verify easily that the following function satisfies the condition of the definition:

$$F(x) = -\frac{1}{2} {}^t x \Lambda W x + {}^t(\Lambda b) x$$

where Λ is the diagonal matrix of the λ_i (i.e. $\Lambda_{ii} = \lambda_i, \Lambda_{ij} = 0, \forall i \neq j$).

The following lemma proved in [3] is a characterization of quasi- symmetry which will be very useful in the following.

Lemma 5. *W is quasi-symmetric if and only if for any set of different vertices $\{i_0, ..., i_p\}$ of the incidence graph of W, we have the following relations:*

1. $w_{i_0 i_1} . w_{i_1 i_2} ... w_{i_p i_0} = w_{i_1 i_0} . w_{i_2 i_1} ... w_{i_0 i_p}$
2. $sign(w_{ij}) = sign(w_{ji}), \forall i, j$

Remark that if we consider a quasi-symmetric neural network, it will remain quasi-symmetric if we modify the coefficients of W in a symmetric manner. Hence the existence of the decreasing operator is somehow related to the whole class of quasi-symmetric neural networks. It is an important open question to know if there exist other classes of neural networks with the same property. In the remaining of this paper, we shall investigate this problem. For this we first introduce a natural formalization of the notion of energy of a neural network and then characterize those networks admitting such an energy.

Let us define a local operation on the matrix W, called *symmetric weight modification* or sw-modification for short, consisting to multiply by a real value any weight between a given couple of vertices. More precisely: given $i, j \in X, i \neq j$ and $\gamma \in \mathcal{R} \setminus \{0\}$, we define the sw-modification $[i, j, \gamma]$ on W as:

$$(W[i, j, \gamma])_{\ell\ell'} = \begin{cases} w_{\ell\ell'} & \forall (\ell, \ell') \notin \{(i, j), (j, i)\} \\ \gamma w_{ij} & \text{if } (\ell, \ell') = (i, j) \\ \gamma w_{ji} & \text{if } (\ell, \ell') = (j, i). \end{cases}$$

The sw-modification can also be defined for diagonal elements by taking $i = j$ in the previous definition, but we assume that in this case, the modification coefficient is strictly positive. In the same way, we define the sw-modification on the threshold vector; i.e for $\gamma \in \mathcal{R} \setminus \{0\}$.

$$(b[i, \gamma])_\ell = \begin{cases} b_\ell & \forall \ell \neq i \\ \gamma b_i & \ell = i. \end{cases}$$

Clearly the incidence graph of a quasi-symmetric matrix is invariant by sw-modifications. We shall say that two matrices W and $\tilde{W}$ are sw-related if there is a set of sw-modifications such that the matrix $\tilde{W}$ modified by all the sw-modifications of the set is equal to W. It is direct that the sw-relation is an equivalence relation. The corresponding equivalence classes are called sw-classes. The quotient set is closely related to the incidence graph as shown in the following.

Proposition 6. *1. Two sw-related matrices have the same incidence graph.*
2. If two quasi-symmetric matrices with non negative diagonal elements have the same incidence graph, then they are sw-related.

Proof. Both properties follow directly from the fact that the sw-modifications are done with non zero coefficients.

Definition 7. We say that $\mathcal{N} = (G, W, b, 1)$ admits an energy if and only if for any $\tilde{W}$ of the sw-class of W and any $\tilde{b}$ derived from b by sw-modifications, the neural network $\tilde{\mathcal{N}} = (G, \tilde{W}, \tilde{b}, 1)$ admits a decreasing operator.

In other words, we say that $\mathcal{N}$ admits an energy when the existence of a decreasing operator is invariant under the operations consisting to multiply by a constant the arcs between two arbitrary nodes or the threshold. It is not difficult to see that, for a fixed neural network, the existence of a decreasing operator does not imply necessarily the existence of an energy for the whole sw-class of $\mathcal{N}$. Take for instance for G an oriented circuit, and large thresholds to insure for any sequential update the convergence to the fixed point 0. However in this case, by using sw-modifications on the thresholds one may exhibit cycles with period $T > 1$. This notion is natural and robust, in the sense that when we consider a symmetric or quasi-symmetric neural network, any network in the same sw-class admits the quadratic operators introduced in proposition 2 as an energy. i.e.:

Proposition 8. *A quasi-symmetric neural network* $\mathcal{N} = (G, W, b, 1)$ *with non negative diagonal admits an energy.*

A direct consequence of the previous definitions and propositions is the following corollary characterizing the dynamical properties of a neural network admitting an energy.

Corollary 9. *If a neural network* $\mathcal{N} = (G, W, b, 1)$ *admits an energy then the periodic orbits of any derived network by sw-modifications are only fixed points.*

4   Characterizing neural networks with energy

This section is the main part of the paper. We shall characterize the class of neural networks admitting an energy. The characterization is based on a structural property of the interconnection graph G. Essentially we show that the existence of a cycle for a sequential update is strongly related with the existence of a circuit in G which does not verify the quasi-symmetry property. Recall that quasi-symmetry of a matrix can be characterized in term of the equality of the weight products of finite sets of vertices of the incidence graph or the different signs of a couple of coefficients as presented in lemma 3. In the following, we formally define a quasi-symmetric circuit.

Definition 10. A circuit $C = \{i_0, ..., i_p\}$ of the incidence graph of W is quasi-symmetric if the weights verify:

1. $w_{i_0 i_1} . w_{i_1 i_2} ... w_{i_p i_0} = w_{i_1 i_0} . w_{i_2 i_1} ... w_{i_0 i_p}$
2. $\forall (i, j) \in C, sign(w_{ij}) = sign(w_{ji})$

A circuit that does not satisfy the previous relation will be called non-quasi-symmetric (nqs-circuit for short).

   In the following proposition we shall prove that a neural network which incidence graph G is a nqs-circuit cannot admit an energy. Recall that a non-negative loop for a neural network corresponds to a non-negative diagonal coefficient of the weight matrix.

Proposition 11. *Let* $\mathcal{N} = (G, W, b, 1)$ *be such that* G *is composed of a unique circuit* $C = \{1, ..., n\}$, *eventually with non-negative loops. If the circuit is non-quasi-symmetric then, there exist a sw-related matrix* $\tilde{W}$ *and a sw-related threshold vector* $\tilde{b}$ *such that the neural network* $\tilde{\mathcal{N}} = (G, \tilde{W}, \tilde{b}, 1)$ *admits a cycle of period* $T > 1$ *for a given sequential update.*

Proof. Since $n \geq 2$, we have two cases.

Case 1: If $n = 2$ and the circuit, say $C = \{1,2\}$ is nqs, then $sign(w_{12}) \neq sign(w_{21})$. Since C is a circuit, we may assume without loss of generality that $w_{21} > 0$ and $w_{12} < 0$. In this context, there exists $\mu_1, \mu_2, \lambda \in \mathcal{R} \setminus \{0\}$ such that

1. $\mu_1 b_1 < 0$
2. $\lambda w_{12} - \mu_1 b_1 + w_{11} < 0$
3. $\lambda w_{21} - \mu_2 b_2 > 0$
4. $w_{22} - \mu_2 b_2 < 0$

Roughly speaking, it is sufficient to take $\mu_1 = -sign(b_1)$, $\mu_2 = (w_{22} + \epsilon)/b_2$, $\epsilon > 0$ and λ large enough for satisfying the previous conditions. Using μ_1, μ_2, λ for sw-modifications of the parameters of the previous network, we construct $\tilde{N} = (C, \tilde{W}, \tilde{b}, 1)$ where $\tilde{w}_{12} = \lambda w_{12}$, $\tilde{w}_{21} = \lambda w_{21}$, $\tilde{b}_1 = \mu_1 b_1$ and $\tilde{b}_2 = \mu_2 b_2$. Clearly $\tilde{N}$ updated in the order $\{1\}, \{2\}$ has the period 4 cycle given below, where the underline sign denotes the vertex which will be updated.

$$
\begin{array}{cc}
1 & 2 \\
\underline{0} & 0 \\
1 & \underline{0} \\
\underline{1} & 1 \\
0 & \underline{1} \\
\underline{0} & 0
\end{array}
$$

Case 2: Assume now that $n \geq 3$. If there is a couple of weights such that $(w_{ij} > 0, w_{ji} < 0)$ then the vertices $\{i, j\}$ form a circuit of length 2. Hence the previous construction applies by taking for any other vertex a positive threshold large enough to fix the state at 0.

Without loss of generality, we can assume that all the weigths of the circuit have the same sign, say non negative (by using if necessary sw-modifications with -1 coefficients).

Since C is nqs, then we may assume that

$$\gamma = w_{21}.w_{32}...w_{1n} > w_{n1}.w_{n-1n}...w_{12} = \beta$$

We now construct the neural network $\tilde{N}$ using sw-modifications as follows. Let $\tilde{b}_1 = \mu_1 b_1$ with $\mu_1 = w_{1n}/b_1$. Define the weights $\tilde{w}_{12} = \lambda_1 w_{12}$ and $\tilde{w}_{21} = \lambda_1 w_{21}$ such that $\tilde{w}_{12} < \tilde{b}_1 = w_{1n}$. It is sufficient to choose $\lambda_1 < \frac{w_{1n}}{w_{12}}$.

Similarily we define $\tilde{b}_2 = \mu_2 b_2 = \lambda_1 w_{21}$, by taking $\mu_2 = \lambda_1 \frac{w_{21}}{b_2}$. Further we determine λ_2 such that $\lambda_2 w_{23} - \tilde{b}_2 = \lambda_2 w_{23} - \lambda_1 w_{21} < 0$, i.e.

$$\lambda_2 < \lambda_1 \frac{w_{21}}{w_{23}}$$

By induction, we construct $n - 1$ parameters, $\lambda_1, ..., \lambda_{n-1}$ associated with the weights and n parameters $\mu_1, ..., \mu_n$ for the thresholds such that:

$$\frac{w_{n1}}{w_{nn-1}} < \lambda_{n-1} < \lambda_{n-2} \frac{w_{n-1n-2}}{w_{n-1n}} < ...$$

$$... < \lambda_1 \frac{w_{21}...w_{n-2n-3}w_{n-1n-2}}{w_{23}...w_{n-2n-1}w_{n-1n}} < \frac{w_{1n}w_{21}...w_{n-1n-2}}{w_{12}w_{23}...w_{n-1n}}$$

$$\mu_1 = \frac{w_{1n}}{b_1}, \mu_i = \lambda_{i-1}\frac{w_{ii-1}}{b_i}, \forall i = 2, ..., n$$

Since $\gamma > \beta$, the interval $L =]\frac{w_{n1}}{w_{nn-1}}, \frac{w_{1n}w_{21}...w_{n-1n-2}}{w_{12}w_{23}...w_{n-1n}}[$ is not empty, so we can determine recursively $\lambda_{n-1}, ..., \lambda_1 \in L$. Using the previous formulas, we can compute directly the values of the parameters $\mu_1, ..., \mu_n$.

Now we have to change the values of w_{ii} such that the new ones are small enough in order to not disturb the dynamics of the neural network. Let ρ_i be such that:

$$\rho_i w_{ii} + \tilde{w}_{ii-1} - \tilde{b}_i < 0, \forall w_{ii} > 0, (\rho_i = 0 \text{ if } w_{ii} = 0)$$

We define the new diagonal terms $\tilde{w}_{ii} = \rho_i w_{ii}$.

We prove that the new network admits a cycle of period $T > 1$. Consider the sequential update $\{n\}, \{n-1\}, ..., \{1\}$ and the configuration $x(0) = (0, 1, 0, ..., 0)$. The coefficients of $\tilde{W}$ and $\tilde{b}$ have been choosen so that this configuration belongs to the following cycle.

vertex	1	2	3		$n-2$	$n-1$	n
$x(0)$	0	1	0		0	0	0
$x(1)$	0	1	0		0	0	0
$x(n-3)$	0	1	0		0	0	0
$x(n-2)$	0	1	1		0	0	0
$x(n-1)$	0	0	1		0	0	0

It suffices to remark that the update by $\tilde{N}$ of the initial configuration, $x(0) = (0, 1, 0, ..., 0)$, shifts the 1 from left to right. Hence the value 1 turns, and $x(0)$ belongs to a cycle of period $T = n(n-1) > 1$.

We can now prove the main result of this paper.

Theorem 12. A neural network $\mathcal{N} = (G, W, b, 1)$ admits an energy if and only if the following two conditions are satisfied:

1. $\text{diag}W \geq 0$
2. G does not contain a nqs-circuit.

Proof. Assume that $\mathcal{N}$ admits an energy. If there exists $w_{ii} < 0$, then by sw-modifications we construct $\tilde{W}$ and $\tilde{b}$ such that $\tilde{b}_i = \mu_i b_i < 0$, $\tilde{w}_{ii} = \rho_i w_{ii}$ and $\rho_i w_{ii} - \tilde{b}_i < 0$. For all $j \neq i$, we let $\tilde{b}_j = \mu_j b_j$ such that $\tilde{b}_j$ be very large (and positive).

We can prove that the new network admits a cycle of period $T > 1$. Consider the sequential update $\{1\}, \{2\}, ..., \{n\}$ and the configuration $x(0) = (0, 0, ..., 0)$. The coefficients of $\tilde{W}$ and $\tilde{b}$ have been choosen so that this configuration belongs to the following cycle.

vertex	i	$i+1$	$i+2$	.	.	.	$i-1$
$x(0)$	$\underline{0}$	0	0	.	0	0	0
$x(1)$	1	$\underline{0}$	0	.	0	0	0
$x(2)$	1	0	$\underline{0}$	.	0	0	0
.							
$x(n)$	$\underline{1}$	0	0	.	0	0	0
$x(n+1)$	0	$\underline{0}$	0	.	0	0	0
.							
$x(2n)$	$\underline{0}$	0	0	.	0	0	0

This is a contradiction to corollary 6.

Assume now that G has a nsq-circuit, say $C = \{1, ..., p\}$. From proposition 7, we determine thresholds and weigths on C such that the configuration $x_C = (0, 1, 0, ..., 0)$ belongs to a cycle of period $T > 1$ for the sequential update order $\{p\}, \{p-1\}, ..., \{1\}$. For all the vertices of $V \setminus C$, we modify the thresholds so that the states remain equal to 0 (it suffices to take the thresholds large enough). If there exist chordal arcs inside the circuit, we multiply the weight values by a small enough coefficient in order that they have no influence on the dynamics of the cycle.

Consider the new neural network $\tilde{N} = (G, \tilde{W}, \tilde{b}, 1)$ and the following sequential update order $\{p\}, \{p-1\}, ..., \{1\}, \{p+1\}, \{p+2\}, ..., \{n\}$. The configuration $x = (x_C, x_{V-C}) = (0, 1, 0, ..., 0, 0, ..., 0)$ belongs to a cycle of period $T > 1$ in an analogous way as in proposition 7. This ends the necessary part.

Assume that N verifies the two conditions of the theorem. Let $\mathcal{C} = \{\mathcal{C}_i\}, i = 1, ..., q$ be the partition of G in strongly connected components.

If G itself is strongly connected ($q = 1$), since any circuit of G is quasi-symmetric and $diag(W) \geq 0$, N admits the energy given in proposition 5.

Assume now that G is not strongly connected. Clearly if we only consider the arcs between different connected components, we define an order relation between the components and from that we deduce a DAG (direct acyclic graph). Call $\{\mathcal{A}_i\}, i = 1, ..., q'$ (where $q' \leq q$), the level partition of this DAG. Remark that each $\mathcal{A}_i$ is the union of a subset of strongly connected components and that $\mathcal{A}_i$ only receives arcs from $\mathcal{A}_1$ to $\mathcal{A}_{i-1}$. Consider the following operator:

$$E(x) = \alpha_1(-\frac{1}{2}\sum_{i,j \in \mathcal{A}_1} \lambda_i w_{ij} x_i x_j + \sum_{i \in \mathcal{A}_1} \lambda_i b_i x_i) +$$

$$\alpha_2(-\frac{1}{2}\sum_{i,j \in \mathcal{A}_2} \lambda_i w_{ij} x_i x_j + \sum_{i \in \mathcal{A}_2} \lambda_i b_i x_i - \sum_{i \in \mathcal{A}_2, j \in \mathcal{A}_1} \lambda_i w_{ij} x_i x_j) +$$

$$... +$$

$$\alpha_{q'}(-\frac{1}{2}\sum_{i,j \in \mathcal{A}_{q'}} \lambda_i w_{ij} x_i x_j + \sum_{i \in \mathcal{A}_{q'}} \lambda_i b_i x_i - \sum_{i \in \mathcal{A}_{q'}, j \in \bigcup_1^{q'-1} \mathcal{A}_k} \lambda_i w_{ij} x_i x_j)$$

such that

$$\alpha_i \geq C\|W\|_1 \sum_{j \geq i+1} \alpha_j \text{ and } \alpha_{q'} > 0$$

where $\|W\|_1 = \sum_{i,j} |w_{ij}|$ and C is a large enough constant.

The operator $E(x)$ is built by adding the quadratic operators of each component (as in proposition 2) scaled by a coefficient and corrected by a quadratic term taking into account the connection between different levels of the DAG. Since any level has a quasi-symmetric associated matrix, it is not difficult to see that under the previous conditions $E(x)$ is a decreasing operator. Furthermore, since the DAG structure is invariant under sw-operations we conclude that the previous decreasing operator is valid for any sw-related neural network. Hence E is an energy.

A direct consequence of the previous result is the following corollary.

Corollary 13. *If $\mathcal{N} = (G, W, b, 1)$ is such that the graph G is strongly connected then $\mathcal{N}$ admits an energy if and only if $diag(W) \geq 0$ and W is quasi-symmetric.*

5 Conclusion

Under a reasonable definition of energy, we have characterized the structure of the interconnection graph for a neural network to admit an energy: essentially non negative loops and quasi symmetry. In fact the usual notion of decreasing operator, like in [4] and [5], does not permit to give a structural characterization because this notion is not robust (the existence of a decreasing operator for a fixed matrix is dramatically sensible to small changes in the weigths). This characterization is important since neural networks with energy are used in various manners as for example for building associative memories. In fact a direct consequence of our result is that the generic class of neural networks admitting only fixed points is composed of direct acyclic graph of quasi- symmetric strongly connected components. As will be shown in a subsequent paper this implies that we can relate the Hopfield theory of associative memories to the well known feed forward layered networks and define learning algorithms for this new class of networks.

For parallel iteration (all the sites are updated at the same time), analogous results can be obtained by defining decreasing operators taking into account two consecutive updates. From that we may assert that a neural network iterated in parallel accepts an energy if and only if its graph does not contain an nqs-circuit. In this case the signs of the diagonal elements are not relevant.

Similar results can be proved for high-order neural networks (with non linear arguments) updated sequentially.

References

1. Fogelman-Soulié, F., Goles, E., Weisbuch, G.: Transient length in sequential iteration of threshold functions. Disc. Appl. Math. **6**, (1983) 95–98
2. Goles, E., Fogelman-Soulié, F., Pellegrin, D.: Decreasing energy functions as a tool for studying threshold netwoks. Disc. Appl. Math. **12**, (1985) 261–277

3. Goles, E., Martinez, S.: *Neural and automata networks*. Math. and Applications, **58**, Kluwer Acad. Pub., (1990)
4. Hopfield, J.J.: Neural networks and physical systems with emergent collective computational abilities. Proc. Natl. Acad. Sci., USA, **79**, (1982) 2554–2558
5. Kobuchi, Y. : State evaluation functions and Lyapunov functions for neural networks. Neural Networks, **4**, (1991) 505–510
6. X. Xu W.T. Tsai, Construction associative memories using neural networks. Neural Networks, **3**, (1990) 301–309

Variable-Length Codes for Error Correction[1]

H. Jürgensen and S. Konstantinidis

Department of Computer Science, The University of Western Ontario
London, Ontario, Canada N6A 5B7
(Electronic Mail: helmut@uwo.ca, stavros@csd.uwo.ca)

Abstract: The notions of decodability and error correction for information transmission over general noisy channels – including channels with
deletions and insertions of symbols – are investigated with the goal of
constructing γ-decodable codes for a given channel γ, that is, codes that
correct errors, hence uniquely decode even in the presence of errors. In
contrast to most of the classical theory of error correcting codes, we consider not only block codes, but also codes containing words with different
lengths in this context. Several general conditions and examples are provided that are valid for arbitrary channels. We explain some of the new
techniques for the case of channels with deletions. In particular, we provide
a construction method for a large class of codes that are error correcting
for such channels.

1. Introduction

With very few exceptions, the classical theory of error correcting codes treats
only errors that can be modelled as symbol substitutions and only codes in
which all words have the same length, that is, block codes. Little has been
published on error correction for other types of errors – for instance errors that
can be modelled as insertions or deletions of symbols – or on the error correction
capabilities of variable-length codes, that is, codes that may contain words of
different lengths. In the sequel, the term *code* is used in this more general sense
without special mention of the word *variable-length*. In this paper we develop a
general framework in which to investigate and construct codes that can be used
for error correction for channels with all three types of errors: substitutions,
insertions, and deletions.

Block codes correcting insertions and deletions of symbols have been investigated by Levenshtein [9], [10], [11], Sellers [16], and by Varshamov and Tenengol'ts [20]. In [9], [10], and [11], in particular, elegant techniques are developed
for constructing *block* codes capable of correcting certain error situations involving insertions and deletions. Our work reported in this paper concerns using
non-block codes for such error situations.

We first propose a general approach to the decoding problem in information
transmission. We clarify the notion of decodability in the presence of errors,

---

[1] The results reported in this paper arose from current research towards the doctoral
dissertation of the second author. We gratefully acknowledge support of this work
by the Natural Sciences and Engineering Research Council of Canada under Grant
OGP0000243.

assuming that the errors conform to a given abstract noisy channel γ. We determine properties to be required of a code K in order for it to be γ-decodable. In the case of an error-free channel the conditions lead to the usual definition of unique decipherability (see [1] or [18]). Certain classes in the hierarchy of codes (see [4] and [5]) are of particular interest when considering error correction for channels with insertions and deletions, especially the class of solid codes (see [17], [3], and [6]). This results from the fact that solid codes satisfy a certain condition of *overlap-freeness* which is also important in the case of block codes correcting insertions and deletions (see [12]). We explore the effect of this condition on the error correction capability of non-block codes and provide a construction of codes that are decodable in the presence of deletions.

2. Basic Notions and Notation

We assume that the reader is familiar with the basic ideas of the theories of error-correcting block codes, information theory, and variable-length codes. For error-correction and information theory we use [13], [14], and [15] as standard references; for codes in general, the relevant background is presented in [1] and [18]. A survey of this theory is provided in [5]. In the present section, we introduce the notation to be used and a few basic notions.

The symbol $\mathbb{N}$ denotes the set of natural numbers, and let[2] $\mathbb{N}_0 = \mathbb{N} \cup 0$. For $n \in \mathbb{N}_0$, let $\mathfrak{n}$ denote the n-th ordinal number; moreover, ω denotes the infinite ordinal number corresponding to the order type of $\mathbb{N}_0$. As usual, any ordinal number $\alpha \in \{\mathfrak{n} \mid n \in \mathbb{N}_0\} \cup \{\omega\}$ can be represented by the set $\{i \mid i \in \mathbb{N}_0 \wedge 0 \leq i < \alpha\}$; in particular, $\mathfrak{o}$ is represented by $\emptyset$. We use this representation freely and without special mention in the sequel. In this paper, an *index set* is defined to be an ordinal number in the set $\mathfrak{I} = \{\mathfrak{n} \mid n \in \mathbb{N}_0\} \cup \{\omega\}$. For an index set I, an *index subinterval* of I is a subset of consecutive *indices* in I.

We denote by $|S|$ the cardinality of a set S. For a set $S_1 \times S_2$, the first and second projections are denoted by π_1 and π_2, respectively.

An *alphabet* is a non-empty, finite set of symbols. In the rest of this paper, let X be an arbitrary, but fixed alphabet with $|X| > 1$; without loss of generality we assume that X contains at least the two symbols 0 and 1. A *word over* X is a mapping $w : I_w \to X$ for some[3] $I_w \in \mathfrak{I}$. Thus, a word w is specified by its index set I_w and the symbols $w_i = w(i) \in X$ for $i \in I_w$; hence we may also write $w = (w_i)_{i \in I_w}$. Let X^∞ be the set of all words over X. A word $w \in X^\infty$ is a *finite word* if I_w is finite and it is an *infinite word* otherwise. In particular, λ denotes the *empty word*, $\lambda : \mathfrak{o} \to X$. As usual, let X^* be the set of finite words, $X^+ = X^* \setminus \lambda$, and let X^ω be the set of infinite words over X. For a word w, $|w| = |I_w|$ is the *length* of w. For $n \in \mathbb{N}_0$, let $X^n = \{w \mid w \in X^* \wedge |w| = n\}$.

---

[2] In the sequel we usually omit the set brackets from singleton sets when no confusion is possible.

[3] This slightly non-standard definition simplifies the subsequent notation. Moreover, it is expected to permit us to generalize the theory to two-sided infinite sequences with little change of notation.

For $w \in X^*$ and $v \in X^\infty$, wv is the usual concatenation of words with the first factor finite. If $w \in X^\omega$ then wv is defined to be equal to w. For $w \in X^\infty$,

$$P(w) = \{p \mid p \in X^+ \wedge w \in p(X^+ \cup X^\omega)\}$$

and

$$F(w) = \{f \mid f \in X^+ \wedge w \in X^+ f(X^+ \cup X^\omega)\}$$

are the sets of *proper prefixes* and *proper infixes* of w, respectively. When $w \in X^*$,

$$S(w) = \{s \mid s \in X^+ \wedge w \in X^+ s\}$$

is the set of *proper suffixes* of w. Let $P_\lambda(w) = P(w) \cup \lambda$ and $S_\lambda(w) = S(w) \cup \lambda$. If $K \subseteq X^\infty$ then $P(K)$ is defined to be the set $\bigcup_{w \in K} P(w)$; the sets $F(K)$, $S(K)$, $P_\lambda(K)$, and $S_\lambda(K)$ are defined analogously.

Let K be a subset of X^+. The set K is a *uniform code*[4] if $K \subseteq X^n$ for some $n \in \mathbb{N}$. The set K is an *infix code* if, for all $u \in K$, one has $(P(u) \cup S(u) \cup F(u)) \cap K = \emptyset$. The set K is a *solid code* if it is an infix code such that, for all $u, v \in K$, one has $P(u) \cap S(v) = \emptyset$. It is an *l-solid code* with $l \in \mathbb{N}$ if it is an infix code such that, for all $u, v \in K$, if $x \in P(u) \cap S(v)$ then $|x| > l$. Let $\mathcal{K}_{\text{solid}}$ and $\mathcal{K}_{l\text{-solid}}$ denote the classes of solid and l-solid codes, respectively.

Definition 2.1 Let $y \in X^\infty$ and $Y \subseteq X^\infty$. A *factorization* of y over Y is a pair (I, v) where I is an index set and $v : I \to Y$ is a mapping with the following properties where $y_i = v(i)$ for $i < |I|$:

(1) If $I = n$ with $n \in \mathbb{N}_0$ then $y = y_0 y_1 \cdots y_{n-1}$, $y_i \in X^* \cap Y$ for $i < n - 1$, and $y_{n-1} \in X^\infty \cap Y$.

(2) If $I = \omega$ then $y = y_0 y_1 \cdots$ and $y_i \in X^* \cap Y$ for all i.

3. Channels

A *channel* over X is a binary relation $\gamma \subseteq X^\infty \times X^\infty$. If $(y', y) \in \gamma$ then, upon input y, the channel could output y'. In this definition, we follow the approach common in the theory of error correcting codes, that is, we model the reasonably likely transmission errors only and omit probabilities altogether. This approach will allow us to make general statements of the form "code K corrects d errors of type τ in n consecutive bits."

A channel γ is *noiseless* if $\gamma \subseteq \{(y, y) \mid y \in X^\infty\}$; otherwise, it is *noisy*. The latter case corresponds to physical channels that introduce errors during information transmission.

For $y \in \pi_2(\gamma)$ let $\langle y \rangle_\gamma = \{y' \in X^\infty \mid (y', y) \in \gamma\}$. Moreover, for $Y \subseteq \pi_2(\gamma)$, let $\langle Y \rangle_\gamma = \bigcup_{y \in Y} \langle y \rangle_\gamma$. Thus $\langle Y \rangle_\gamma$ is the set of all possible outputs of γ when words in Y are used as inputs. A factorization (I, v') of $y' \in \pi_1(\gamma)$ over $\langle Y \rangle_\gamma$ is γ-*admissible* if there is a $y \in X^\infty$ and a factorization (I, v) of y over Y such that $(v'(i))_{i \in J} \in \langle (v(i)_{i \in J} \rangle_\gamma$ for every index subinterval J of I. In this case we

---

[4] Called *block code* in the literature on error correcting codes.

say that (I, υ') is γ-admissible for (I, υ).

We consider channels γ satisfying the following very general conditions for any set $Y \subseteq \pi_2(\gamma)$.

($\mathcal{P}_0$) *The channel preserves finiteness and infiniteness:* If $(y', y) \in \gamma$ then y' and y are both finite or both infinite.

($\mathcal{P}_1$) *Factorizations are preserved:* If $(y', y) \in \gamma$ and (I, υ) is a factorization of y over Y then there is a factorization (I, υ') of y' over $\langle Y \rangle_\gamma$ which is γ-admissible for (I, υ).

These conditions seem to be satisfied by most physical channels. In this paper we only consider channels the error behaviour of which can be modelled as a combination of three basic operations: substitution, insertion, and deletion of symbols, denoted by σ, ι, and δ, respectively. Such channels are called *SID channels* in the sequel[5]. If γ is an SID channel, then the descripition of γ expresses which type of error can occur and with which frequency. In particular, for $L \in \mathbb{N}$ and $m \in \mathbb{N}_0$ with $m < L$, $\gamma(m, L)$ denotes a channel in which at most m errors of *type* γ can occur in any consecutive L symbols.

For example, for the channel $\delta(1, 4)$, the input $10^3 1$ can result in exactly $10^3 1$, $0^3 1$, 0^3, 1001, and 10^3 as ouputs. In the case of $\sigma(1, 4)$ one can only get $10^3 1$, $0^4 1$, 0^5, $1^2 0^2 1$, 10101, $10^2 1^2$, and 10^4 assuming that $X = \{0, 1\}$. In the case of insertions we use the following convention: *Among the $L + 1$ possible positions for insertions in a string of length L the last one is not considered as belonging to this string.* Thus, using the same input, $\iota(1, 4)$ can produce exactly the following outputs: $10^3 1$, $x 10^3 1$, $x 10^3 y 1$, $x 10^3 1 y$, $1 x 0^3 1$, $1 x 0^3 1 y$, $10 x 0^2 1$, $10^2 x 0 1$, $10^3 x 1$, and $10^3 1 x$ where $x, y \in X$.

Theorem 3.1 *Every SID channel γ satisfies $\mathcal{P}_0$ and $\mathcal{P}_1$.*

In the rest of this paper we assume, without special mention, that every channel considered satisfies $\mathcal{P}_0$ and $\mathcal{P}_1$.

4. Codes and Decodability

In information transmission, the sender *encodes* the message in order that the receiver can regain the message uniquely (with probability close to 1). With this idea in mind, we formulate the notion of decodability with respect to a given channel γ. In the context of this paper, a *code* is any set[6] $K \subseteq X^+$. A *message* is any word in K^∞ and a *received message* is any word in $\langle K^\infty \rangle_\gamma$. A necessary condition for unique decodability – in the absence of noise it is also sufficient – is that, for any $w \in K^\infty$, there exist a unique factorization[7] of w over K. A code K satisfying this condition is said to be *uniquely decodable* in the sequel. Let $\mathcal{K}_{\mathrm{UD}}$ be the set of uniquely decodable codes over X.

---

[5] We state the definitions only informally as details are not needed in the context of this paper. See [7] for details where the channel algebra is analysed.

[6] In adopting this terminology we go back to the early days of coding theory when unique decipherability was not implied by the term of *code*.

[7] For the difference between unique decodability for K^∞ and K^* see [2], [8], and [19].

The decoding problem for a given channel γ and a given code K is as follows. Given any received message $w' \in \langle K^\infty \rangle_\gamma$, find $w \in K^\infty$ such that $(w', w) \in \gamma$. For w to be unique, one must have $\langle u \rangle_\gamma \cap \langle w \rangle_\gamma = \emptyset$ for any two distinct elements $u, w \in K^\infty$. Let

$$\mathcal{K}_\gamma^\infty = \{K \mid K \in \mathcal{K}_{\mathrm{UD}} \wedge \forall u, w \in K^\infty : \langle u \rangle_\gamma \cap \langle w \rangle_\gamma \neq \emptyset \rightarrow u = w\}.$$

Definition 4.1 Let γ be a channel. A code K is γ-correcting if $K \in \mathcal{K}_\gamma^\infty$.

We also consider the class

$$\mathcal{K}_\gamma = \{K \mid K \in \mathcal{K}_{\mathrm{UD}} \wedge \forall v \in K : \langle v \rangle_\gamma \cap \langle K^* \setminus v \rangle_\gamma = \emptyset\}.$$

Clearly, $\mathcal{K}_\gamma^\infty \subseteq \mathcal{K}_\gamma$. For channels with substitution errors only and when restricted to block codes, these classes are equal and capture the usual notion of error correction for a given channel.

Theorem 4.1 *Let γ be an SID channel. Given a finite code K, its membership in $\mathcal{K}_\gamma$ is decidable.*

A γ-correcting code allows for the correction of errors in any received message w' in the sense that one can identify the unique $w \in K^\infty$ for which $w' \in \langle w \rangle_\gamma$ despite the presence of errors due to the channel γ. As a consquence, there is a *decoding mapping* $s_{\gamma,K} : X^\infty \rightarrow X^*$ with the following property: For every $w' \in \langle K^\infty \rangle_\gamma$ there are $v \in K$ and $u \in K^\infty$ such that $s_{\gamma,K}(w') \in \langle v \rangle_\gamma$, $w' \in \langle vu \rangle_\gamma$, and $w' = s_{\gamma,K}(w')u'$ for some $u' \in \langle u \rangle_\gamma$. Let $d \in \mathbb{N}_0 \cup \infty$ be minimal such that, for every $w' \in \langle K^\infty \rangle_\gamma$ and every $u' \in P(w')$ with $|u'| \geq d$, one has $s_{\gamma,K}(u') = s_{\gamma,K}(w')$. Then d is the *decoding delay* of K, and K is γ-*decodable with delay* d. The decoding mapping extracts the output of the channel corresponding to the first code word in the message from the received message. In general, a decoding mapping may require unbounded information, that is, the complete and possibly infinite received message. Our notion of decoding delay differs from the one found in the literature on coding (see [1]). However, the delay is bounded in the usual sense if and only if it is bounded in our sense, albeit with possibly a different constant.

The following sufficient condition for a code in $\mathcal{K}_\gamma$ to be γ-correcting is used in the sequel.

Lemma 4.1 *Let $K \in \mathcal{K}_\gamma$ for some channel γ with $\langle K \rangle_\gamma \subseteq X^+$. K is γ-correcting if, for any $w' \in \langle K^\infty \rangle_\gamma$ and any γ-admissible factorizations (I_1, κ_1') and (I_2, κ_2') of w' over $\langle K \rangle_\gamma$, one has $\kappa_1'(0) = \kappa_2'(0)$.*

Example 4.1 Let $\gamma = \delta(1, 5)$ and $K = \{u, v\}$ where $u = 0001$ and $v = 01011$. K is in $\mathcal{K}_\gamma$, it is 1-solid, but not solid. One has $\langle v \rangle_\gamma = \{v, 1011, 0011, 0111, 0101\}$ and $\langle u \rangle_\gamma = \{u, 001, 000\}$. Using Lemma 4.1, one can show that K is γ-correcting.

Example 4.2 Let $\gamma = \delta(1, 6)$ and $K = \{u, v\}$ where $u = 0011$ and $v = 010111$.

K is a solid code. From

$$\langle u \rangle_\gamma = \{u, 011, 001\} \quad \text{and} \quad \langle v \rangle_\gamma = \{v, 10111, 00111, 01111, 01011\}$$

one finds that $K \in \mathcal{K}_\gamma$. However, K is not γ-correcting as

$$w' = (0011)(10111)^\omega = (00111)(01111)^\omega \in \langle uv^\omega \rangle_\gamma \cap \langle v^\omega \rangle_\gamma.$$

We now provide a necessary condition for a 1-solid uniform code of length L to be γ-decodable where γ is $\iota(1, L)$ or $\delta(1, L)$.

Theorem 4.2 *Let K be a uniform 1-solid code of code word length $L \geq 2$ and let $\gamma = \iota(1, L)$ or $\gamma = \delta(1, L)$. If $K \in \mathcal{K}_\gamma$ then K is γ-decodable with a delay not exceeding $L + 1$.*

Proof: First consider the case of $\gamma = \delta(1, L)$. Consider an input $w \in K^\infty$ with factorization (I, κ) over K. Let $w' \in \langle w \rangle_\gamma$ and let (I, κ') be a γ-admissible factorization of w' over $\langle K \rangle_\gamma$ for (I, κ); by $\mathcal{P}_1$ such a factorization exists. For $i \in I$, let $w_i = \kappa(i)$ and $w'_i = \kappa'(i)$. For any $l \in \mathbb{N}_0$, let p_l be the prefix of length l of w' – if it exists. As $\gamma = \delta(1, L)$, one has $w'_0 = p_{L-1}$ or $w'_0 = p_L$. We show that $w'_0 = p_L$ if and only if $p_L \in K$, that is, w'_0 is uniquely determined. By Lemma 4.1 this implies $K \in \mathcal{K}_\gamma^\infty$.

One verifies that $w'_0 = w'$ if $|w'| \leq L$. Hence, we may assume that $|w'| > L$ and, therefore, $|I| > 1$. Suppose that $p_L \in K$ and $w'_0 \neq p_L$. Then $w'_0 = p_{L-1}$ and one deletion occurred in w_0. So $p_{L-1} \in \langle w_0 \rangle_\gamma$. Also, as $p_L \in K$, $p_{L-1} \in \langle p_L \rangle_\gamma \cap \langle w_0 \rangle_\gamma$. From $K \in \mathcal{K}_\gamma$ it follows that $w_0 = p_L = b_0 \cdots b_{L-1}$ with $b_0, \ldots, b_{L-1} \in X$.

By assumption, $w'_0 \neq p_L$, hence $w'_1 \in b_{L-1} X^+$. It is, therefore, impossible that $w_1 \in b_{L-1} X^{L-1}$ as, otherwise, $b_{L-1} \in S(w_0) \cap P(w_1)$. So $w_1 = a w'_1$ with $a \in X$ deleted. Since $|p_L a| = L + 1$ the deletion in w_0 occured at b_0, that is, $w'_0 = b_1 \cdots b_{L-1}$. Also, since $w'_0 = p_{L-1}$ it follows that $b_i = b_{i+1}$ for $i = 0, \ldots, L - 2$. Therefore, $w_0 = b_0^L$ and $b_0 \in P(w_0) \cap S(w_0)$, a contradiction. This shows that $w'_0 = p_L$.

Now, suppose that $p_L \notin K$ and $w'_0 = p_L$. Then no deletion occurred in w_0 and $w_0 = w'_0 = p_L \in K$, a contradiction. Hence $w'_0 = p_{L-1}$. This shows that K is γ-decodable with a delay of at most L.

For the case of $\gamma = \iota(1, L)$ we only sketch the main steps of the proof. Using the notation of above, $w'_0 = p_L$ or $w'_0 = p_{L+1}$. One can show that w'_0 is determined correctly as follows:

(1) If $p_L \notin K$ then $w'_0 = p_{L+1}$.
(2) If $p_L \in K$ and $b_{L-1} \neq b_L$ then $w'_0 = p_L$.
(3) Otherwise, $w'_0 = p_{L+1}$.

(1) is obvious. For (2) it is sufficient to show that, if $p_L \in K$, then an insertion in w_0 implies $b_{L-1} = b_L$. For (3) one shows that, if $b_{L-1} = b_L$, then an insertion has occurred. In this case, the delay is at most $L + 1$. $\square$

Corollary 4.1 *Let K be a uniform 1-solid code of code word length $L \geq 2$ and let $\gamma = \iota(1, L)$ or $\gamma = \delta(1, L)$. It is decidable whether K is γ-correcting.*

5. Deletion-Correcting Solid Codes

In this section we investigate sufficient conditions for a solid code – not necessarily uniform – to be γ-correcting where $\gamma = \delta(1, L)$ for some fixed $L \in \mathbb{N}$. As shown in Example 4.2, not every solid code in $\mathcal{K}_\gamma$ is γ-correcting. In this section we consider only codes K satisfying the following conditions:

K is finite, $K \in \mathcal{K}_{\text{solid}} \cap \mathcal{K}_\gamma$, and $3 < |u| \leq L$ for all $u \in K$.

This assumption is made in the sequel without special mention. The following technical lemma states some simple, but useful bounds on the number of deletions possible in a given message.

Lemma 5.1 *Let $z \in X^\infty \setminus \lambda$ and let $(z', z) \in \gamma = \delta(1, L)$.*
(1) If d deletions can occur in z, $d \in \mathbb{N}_0$, then $|z| \geq (d-1)L + 1$.
(2) If $|z'| \leq L$ then at most two deletions can have occurred in z.

Suppose $K \in \mathcal{K}_\gamma \setminus \mathcal{K}_\gamma^\infty$. Then there are non-empty words $u, w \in K^\infty$, $u \neq w$, and $u' \in \langle u \rangle_\gamma \cap \langle w \rangle_\gamma$. Let (I_u, κ_u) and (I_w, κ_w) be factorizations of u and w over K, respectively, and let (I_u, κ_u') and (I_w, κ_w') be corresponding γ-admissible factorizations of u' over $\langle K \rangle_\gamma$ according to $\mathcal{P}_1$. $K \in \mathcal{K}_\gamma$ implies that $|I_u| > 1$ and $|I_w| > 1$.

If $\kappa_u'(0) = \kappa_w'(0)$ then $\kappa_u'(0) \in \langle \kappa_u(0) \rangle_\gamma \cap \langle \kappa_w(0) \rangle_\gamma$, hence $\kappa_u(0) = \kappa_w(0)$. By $\mathcal{P}_1$, and since $u \neq w$, we may omit $\kappa_u(0)$ from the beginnings of both u and w and consider the resulting suffixes only. Hence, we may assume that $\kappa_u'(0) \neq \kappa_w'(0)$. It follows that $\kappa_u'(0) \in \langle K^* P(K) \rangle_\gamma$. Thus there is $z \in K^* P(K)$ such that $\kappa_u'(0) \in \langle \kappa_u(0) \rangle_\gamma \cap \langle z \rangle_\gamma$. Generalizing this observation, we arrive at the following definition.

Definition 5.1 Let $v \in K$, $z \in K^* P(K) \cup S(K) K^* P(K) \cup F(K) \cup S(K) K^*$, and $v' \in \langle v \rangle_\gamma \cap \langle z \rangle_\gamma$. Then $[v', v, z]$ is called an *overlap* for K and γ. Such an overlap is a *prefix overlap* if $z \in K^* P(K)$. A prefix overlap is said to be *non-trivial* if $z \notin vP(K)$.

Lemma 5.2 *There is no overlap $[u', u, z]$ for K and γ satisfying*

$$z \in S_\lambda(K) K^* K^2 K^* P_\lambda(K).$$

The proof requires a careful distinction of several cases. We continue the discussion using the assumptions and the notation of above. For the prefix overlap $[\kappa_u'(0), \kappa_u(0), z]$ we have $z \in P(K) \cup KP(K)$ by Lemma 5.2. One can show that $z \in \kappa_u(0)P(K)$ implies $\kappa_u(0) = \kappa_w(0)$ and $\kappa_u'(0) = \kappa_w'(0)$ contradicting our assumption. Therefore, $z \notin \kappa_u(0)P(K)$.

Now assume that $z \in P(\kappa_w(0))$, hence $\kappa_w(0) = zs$ for some $s \in S(\kappa_w(0))$. Since $\kappa_w'(0) \in \langle \kappa_w(0) \rangle_\gamma$ one has $\kappa_w'(0) = z's'$ with $s' \in \langle s \rangle_\gamma$ and, hence, $\kappa_w'(0) = \kappa_u'(0)s'$. Moreover, $\kappa_u'(0)s' \in \langle \kappa_u(0)K^*P(K) \rangle_\gamma \subseteq \langle KK^*P(K) \rangle_\gamma$, hence, by Lemma 5.2, $\kappa'_u(0)'s' \in \langle KP(K) \rangle_\gamma$. Thus there is $y \in KP(K)$ such that $[\kappa'_w(0), \kappa_w(0), y]$ is a non-trivial prefix overlap. In the next theorem we summarize this analysis and determine the exact form of such overlaps.

Theorem 5.1 *Let $K \in \mathcal{K}_\gamma \setminus \mathcal{K}_\gamma^\infty$. Then there exists a non-trivial prefix overlap $[u', u, z]$ for K and γ with $z \in KP(K)$. Moreover, every such prefix overlap is of one of the following two forms.*

(R₁) $u' = xyqp$, $u = xbyqp$, and $z = xyqap$ with $a, b \in X$, $xy \in K$, $qap \in P(K)$, $x \neq \lambda$, $y \neq \lambda$, and $p \neq \lambda$ such that a and b are the symbols deleted from z and u, respectively.

(R₂) $u' = xyqp$, $u = xyqbp$, and $z = xayqp$ with $a, b \in X$, $xay \in K$, $qp \in P(K)$, $x \neq \lambda$, and $q \neq \lambda$ such that a and b are the symbols deleted from z and u, respectively.

Proof: Let $[u', u, z]$ be a prefix overlap of K such that $z \in KP(K) \setminus uP(K)$. Then there are $v \in K$ and $r \in P(K)$ such that $z = vr$. Also $u' = v'r'$ for some $v' \in \langle v \rangle_\gamma$ and $r' \in \langle r \rangle_\gamma$. If $r' = \lambda$ then $u' = v' \in \langle u \rangle_\gamma \cap \langle v \rangle_\gamma$, hence $u = v$ and $z = ur$, a contradiction. Hence $r' \neq \lambda$.

Now assume that $u' = u$. Then $v'r' \in K$, hence, as K is solid, $v' \neq v$ and $r' \neq r$. Thus, by Lemma 5.1, there occurred exactly two deletions in z and $|z| = |u| + 2 \leq L + 2$. Let $r = r_1 a r_2$ with $a \in X$ deleted and $r' = r_1 r_2$. If $r_2 = \lambda$ then $r_1 \in P(K) \cap S(u)$, a contradiction. Therefore, $r_2 \neq \lambda$ and $|vr_1 a| < |z|$, hence $|vr_1 a| \leq L + 1$. This implies that the two deletions occur at the endpoints of $vr_1 a$. Thus $v = bv'$ with $b \in X$ and, consequently, $v' \in S(v) \cap P(u)$, a contradiction.

Therefore, $u' \neq u$. Thus $u = xbt$ with $b \in X$ deleted and $u' = xt = v'r'$. We distinguish two cases.

Case 1: Assume $|v'| > |x|$. Then $v' = xy$ with $y \neq \lambda$ and $t = yr'$. If $r = r'$ then $r \in P(K) \cap S(u)$ which is impossible. Thus $r \neq r'$ and $r = qap$ with $a \in X$ deleted and $r' = qp$. Then $u = xbyqp$ and $u' = xyqp$. If $p = \lambda$ then $q = r' \neq \lambda$ and $q \in P(K) \cap S(u)$, a contradiction. Therefore, $p \neq \lambda$. If $v' \neq v$ then, since $z = vr$ and $r' \neq r$, there are two deletions in z which must occur at its endpoints. Then necessarily $p = \lambda$ which is impossible. Thus $v = v'$ which implies $v = xy$ and $z = xyqap$. Finally, $x \neq \lambda$ as, otherwise, $v = y \in F(u)$. Thus, the overlap has the form R₁.

Case 2: Assume that $|v'| \leq |x|$. If $v' = v$ then $v \in P(u) \cap K$ which is impossible. Therefore, $v' \neq v$ and $v = x_1 a y$ with $a \in X$ deleted and $v' = x_1 y$. Also $v' \in P(u)$. If $r \neq r'$ then two deletions occur at the endpoints of z. In this case $v = av'$ and $v' \in P(u) \cap S(v)$ which is impossible again. Thus $r = r'$. Now $|v'| \neq |x|$ as, otherwise, $t = r' = r \in P(K) \cap S(u)$. Therefore, $r = qp$ with $q \in S(x)$ and $p = t$. This implies $x = x_1 yq$, $u = x_1 yqbp$, $u' = x_1 yqp$, and $z = x_1 ayqp$ with $q \neq \lambda$ and $x_1 \neq \lambda$. Hence the overlap has the form R₂. $\square$

Consider the code of Example 4.2. It is in $\mathcal{K}_\gamma$ but not γ-correcting and there is a non-trivial prefix overlap of the form R₁. Indeed, the overlap is $[00111, 0b0111, 0011a1]$ with $b = 1$ and $a = 0$ as the deleted symbols.

Corollary 5.1 *If all distinct* $u, v \in K$ *satisfy* $v \notin \langle P(u) \rangle_\gamma$ *and* $P(u) \cap \langle v \rangle_\gamma = \emptyset$ *then* K *is* γ*-correcting.*

Proof: If an overlap of the form R_1 exists then $v \in \langle P(u) \rangle_\gamma$ for some distinct $u, v \in K$. On the other hand, the existence of an overlap of the form R_2 implies that $P(u) \cap \langle v \rangle_\gamma$ is non-empty for some distinct $u, v \in K$. $\square$

6. A Class of Deletion-Correcting Solid Codes in $0^+1^+0^+1^+$

Not much in terms of detailed structural information is known about solid codes in general. The special case of solid codes which are subsets of the set $0^+1^+0^+1^+$ is an exception to this situation. By [6], every such solid code K can be defined by a triple $(\mathbb{D}, f, g)$ where $\mathbb{D} = \{(r_1, r_2) \mid r_1, r_2 \in \mathbb{N} \wedge 0^+1^{r_1}0^{r_2}1^+ \cap K \neq \emptyset\}$ is called the *domain* of K and where $f, g \colon \mathbb{D} \to \mathbb{N}$ are mappings such that the following condition $\mathrm{Sol}(\mathbb{D}, f, g)$ holds true:

$$\forall\, r, s \in \mathbb{D}\ (f(r) > \pi_2(s) \vee g(s) > \pi_1(r)).$$

Then $K = \{0^{f(r)}1^{r_1}0^{r_2}1^{g(r)} \mid r = (r_1, r_2) \in \mathbb{D}\}$. In this section we focus on solid codes of this kind and exhibit a subclass of codes which are error correcting with respect to a channel with single deletions.

Lemma 6.1 *Let* K *be a solid code with* $K \subseteq (0^+1^+)^2$, *let* $v \in K$ *and consider the channel* $\delta_v = \delta(1, |v|)$. *Then* $\langle v \rangle_{\delta_v} \cap \langle K^\infty \setminus K \rangle_{\delta_v} = \emptyset$.

The proof of Lemma 6.1 requires several subtle case distinctions.

If K is a code and M is a finite, non-empty subset of K, let δ_M be the channel $\delta(1, \max\{|u| \mid u \in M\})$. We now consider the following property R_δ of a solid code K:

For all words $u, v \in K$, *if* $u \neq v$, *then* $\langle u \rangle_{\delta_{\{u,v\}}} \cap \langle v \rangle_{\delta_{\{u,v\}}} = \emptyset$.

By Lemma 6.1, for every finite solid code $K \subseteq 0^+1^+0^+1^+$ satisfying R_δ one has that $K \in \mathcal{K}_\delta$. For arbitrary $\mathbb{D}$ one obtains the following characterization of solid codes satisfying R_δ.

Lemma 6.2 *Let* $K \subseteq (0^+1^+)^2$ *be a solid code given by* $(\mathbb{D}, f, g)$. K *satisfies* R_δ *if and only if the following conditions hold true for all* $r = (r_1, r_2) \in \mathbb{D}$ *and for all* $k, l \in \mathbb{N}$:

(R_δ^1) If $(r_1 + 1, r_2) \in \mathbb{D}$ then $f(r) \neq f(r_1 + 1, r_2)$ or $g(r) \neq g(r_1 + 1, r_2)$.
(R_δ^2) If $(r_1, r_2 + 1) \in \mathbb{D}$ then $f(r) \neq f(r_1, r_2 + 1)$ or $g(r) \neq g(r_1, r_2 + 1)$.
(R_δ^3) If $(r_1 - 1, r_2 + 1) \in \mathbb{D}$ then $f(r) \neq f(r_1 - 1, r_2 + 1)$ or $g(r) \neq g(r_1 - 1, r_2 + 1)$.
(R_δ^4) If $(r_1 + 1, r_2 - 1) \in \mathbb{D}$ then $f(r) \neq f(r_1 + 1, r_2 - 1)$ or $g(r) \neq g(r_1 + 1, r_2 - 1)$.
(R_δ^5) If $(1, k), (l, 1) \in \mathbb{D}$ then $f(1, k) + k \neq f(l, 1)$ or $g(1, k) \neq g(l, 1) + l$.
(R_δ^6) If $(r_1 + 1, r_2) \in \mathbb{D}$ then $f(r) \neq 1 + f(r_1 + 1, r_2)$ or $g(r) \neq g(r_1 + 1, r_2)$.
(R_δ^7) If $(r_1, r_2 + 1) \in \mathbb{D}$ then $f(r) \neq 1 + f(r_1, r_2 + 1)$ or $g(r) \neq g(r_1, r_2 + 1)$.
(R_δ^8) If $(r_1 + 1, r_2) \in \mathbb{D}$ then $f(r) \neq f(r_1 + 1, r_2)$ or $g(r) \neq 1 + g(r_1 + 1, r_2)$.
(R_δ^9) If $(r_1, r_2 + 1) \in \mathbb{D}$ then $f(r) \neq f(r_1, r_2 + 1)$ or $g(r) \neq 1 + g(r_1, r_2 + 1)$.

We omit the fairly involved proof of Lemma 6.2. It requires a sequence of subtle case distinctions.

Using these lemmata and Corollary 5.1, we can now construct a class of solid codes in $(0^+1^+)^2$ which decode correctly in the presence of deletion errors. Let $\mathbb{D}_1 = (\mathbb{N} \setminus 1) \times (\mathbb{N} \setminus 1)$ and let $\mathbb{F}$ be the set of all functions $f_{x,y} : \mathbb{D}_1 \to \mathbb{N}$, for $x, y \in \mathbb{N} \setminus 1$ with $x \neq y$, such that $f_{x,y}(r_1, r_2) = xr_1 + yr_2$. For $f \in \mathbb{F}$ and $n \in \mathbb{N} \setminus 1$, let $G_f(n) = \{\pi_1(r) \mid r \in \mathbb{D}_1 \wedge f(r) \leq n\}$.

Theorem 6.1 *Let $f \in \mathbb{F}$ and $g : \mathbb{D}_1 \to \mathbb{N} \setminus 1$ such that, for all $r \in \mathbb{D}_1$, $g(r) > \max G_f(\pi_2(r))$. Let $K = \{0^{f(r)}1^{r_1}0^{r_2}1^{g(r)} \mid r = (r_1, r_2) \in \mathbb{D}_1\}$. Then K has the following properties:*

(1) K is a solid code.

(2) K satisfies R_δ.

(3) For all $u, v \in K$ with $u \neq v$ one has $P(u) \cap \langle v \rangle_{\delta_{\{u,v\}}} = \emptyset$ and $v \notin \langle P(u) \rangle_{\delta_{\{u,v\}}}$.

(4) Every finite subset M of K is δ_M-correcting. Moreover, if $L = \max\{|u| \mid u \in M\}$, then M is δ_M-decodable with delay not exceeding L.

Proof: We only outline the main steps of the proof. For (1), one shows that $\mathrm{Sol}(\mathbb{D}_1, f, g)$ holds true. To do so, one first verifies that each of the sets $G_f(n)$ is finite. Using Lemma 6.2, one proves (2).

For the proof of (3), consider $v = 0^{f(r)}1^{r_1}0^{r_2}1^{g(r)}$ in K and $p = 0^{f(s)}1^{s_1}0^{s_2}1^{g'}$ in $P(K)$ with $r \neq s$ and $g' < g(s)$. Obviously $v \neq p$. Assume $v \in \langle p \rangle_{\delta_{\{u,v\}}}$ for $u \in K$ and $p \in P(u)$. Then the deletion in p must occur in 1^{s_1} or 0^{s_2} as, otherwise, $r = s$. So $s = (r_1 + 1, r_2)$ or $s = (r_1, r_2 + 1)$ and $f(r) = f(s)$ which is impossible. Now assume $p \in \langle v \rangle_{\delta_{\{u,v\}}}$. The deletion in v must occur in 1^{r_1} or 0^{r_2} as, otherwise, $r = s$. Thus, $s = (r_1 - 1, r_2)$ or $s = (r_1, r_2 - 1)$ and $f(s) = f(r)$ which is impossible.

For the proof of (4), let M be a finite subset of K. Let $L = \max\{|u| \mid u \in M\}$. By (3), the premiss of Corollary 5.1 is true for M. Therefore, M is δ_M-correcting.

Let $w \in M^\infty$ be a non-empty message with factorization (I, κ) over M and let $w' \in \langle w \rangle_{\delta_M}$ be a received message with a corresponding γ-admissible factorization (I, κ') over $\langle M \rangle_{\delta_M}$. As before, for any $l \in \mathbb{N}$, let p_l be the prefix of length l of w' – when it exists. Let $J = \{l \mid l \in \mathbb{N}, p_l \in \langle M \rangle_{\delta_M}\}$. We show that $\kappa'(0)$ is uniquely determined as $\kappa'(0) = p_m$ with $m = \max J$. This will show that the decoding delay of M does not exceed L.

First assume that $p_l \in M$ for some $l \in J$, but $\kappa'(0) \neq p_l$. Then $\kappa'(0) = p_k$ with $k < l$ or $k > l$. Using properties (1) and (3), one proves that this is impossible. Thus $\kappa'(0) = p_l$ and this is unique as at most one prefix of w' can be in M.

Next assume that M does not contain a prefix of w' and let $m = \max J$. Clearly $m \leq L$ and $\kappa'(0) = p_l$ for some $l \in J$. Assume $l < m$. Then $p_m = p_l b_l b_{l+1} \cdots b_{m-1}$ with $b_l, \cdots, b_{m-1} \in X$. Consider $u = (a_i)_{i \in I_u} \in M$ such that $(p_m, u) \in \delta_M$. As $p_m \notin M$, p_m is obtained from u by the deletion of one symbol, a_k say, with $k \in I_u$ and $I_u = \mathbf{m} + \mathbf{1}$. If $k < l$ then $b_l \cdots b_{m-1} \in S(u)$. If $|\kappa(1)| \leq m - l$ then $\kappa(1) \in P(b_l \cdots b_{m-1}) \cup \{b_l \cdots b_{m-1}\} \subseteq F(u) \cup S(u)$ which is impossible.

Therefore, $|\kappa(1)| > m - l$ and, hence, $b_l \cdots b_{m-1} \in P(\kappa(1)) \cap S(u)$, again a contradiction. Therefore, $k \geq l$. But then $p_l \in P(u) \cap \langle \kappa(0) \rangle_{\delta_M}$ contradicting the fact that $u \neq \kappa(0)$. This proves that the only possible choice for $\kappa'(0)$ is p_m. □

7. Discussion

We have presented a theoretical framework for dealing with the decoding problem in information transmission. In general, when insertions or deletions of symbols are allowed, the decoding problem is significantly more difficult in comparison to the case of substitution-only channels where uniform codes suffice. The approach taken in this paper for channels of the form $\delta(1, L)$ can also be used for other SID channels. One can identify the overlaps of a given code and then apply restrictions similar to the one expressed in Corollary 5.1 in order to obtain γ-correcting codes. The notion of overlaps can be generalized appropriately to include products of overlaps. This method turns out to be powerful enough to provide a necessary and sufficient condition for a code to be γ-correcting. Moreover, this permits us to prove the decidability of the following question: For a given a SID channel γ and a finite code K, is $K \in \mathcal{K}_\gamma^\infty$?

Our current research involves investigating γ-correcting codes for various SID channels, as well as efficient methods for synchronization. Moreover, one of our goals is the characterization and construction of maximal solid codes since it seems that this class of codes is one of the most appropriate to use in information transmission through SID channels.

References

1. J. Berstel, D. Perrin: *Theory of Codes.* Academic Press, Orlando, 1985.
2. S. W. Golomb, B. Gordon: Codes with Bounded Synchronization Delay. *Inform. and Control* **8** (1965), 355–372.
3. H. Jürgensen, S. S. Yu: Solid Codes. *J. Inform. Process. Cybernet. (EIK)* **26** (1990), 563–574.
4. H. Jürgensen, S. S. Yu: Relations on Free Monoids, Their Independent Sets, and Codes. *Internat. J. Computer Math.* **40** (1991), 17–46.
5. H. Jürgensen, S. Konstantinidis: The Hierarchy of Codes. In Z. Esik (editor), *Fundamentals of Computation Theory, 9th International Conference, FCT '93. Lecture Notes in Comput. Sci.* **710**, 50–68. Springer-Verlag, Berlin, 1993.
6. H. Jürgensen, M. Katsura, S. Konstantinidis: Maximal Solid Codes. Manuscript, in Preparation.
7. H. Jürgensen, S. Konstantinidis: Error-Correction for Channels with Substitutions, Insertions, and Deletions (Abstract). *4th Canadian Workshop on Information Theory, 1995.*
8. V. I. Levenshtein: Certain Properties of Code Systems. *Soviet Phys. Dokl.* **6** (1962), 858–860 (English translation).

9. V. I. Levenshtein: Binary Codes Capable of Correcting Spurious Insertions and Deletions of Ones. *Problemy Peredachi Informatsii* **1** (1965), 12–25 (in Russian).

10. V. I. Levenshtein: Binary Codes Capable of Correcting Deletions, Insertions, and Reversals. *Soviet Phys. Dokl.* **10** (1966), 707–710 (English translation).

11. V. I. Levenshtein: Asymptotically Optimum Binary Code with Correction for Losses of One or Two Adjacent Bits. *Systems Theory Research* **19** (1967), 298–304 (English translation).

12. V. I. Levenshtein: Maximum Number of Words in Codes without Overlaps. *Problemy Peredachi Informatsii* **6** (1970), 88–90 (in Russian).

13. W. W. Peterson, E. J. Weldon, Jr.: *Error-Correcting Codes*. MIT Press, Cambridge, 2nd edition, 1972.

14. S. Roman: *Coding and Information Theory*. Springer-Verlag, New York, 1992.

15. M. Schwartz: *Information Transmission, Modulation, and Noise: A Unified Approach to Communication Systems*. McGraw-Hill, New York, 3rd edition, 1980.

16. F. F. Sellers, Jr.: Bit Loss and Gain Correction Code. *IRE Trans. Inform. Theory* **8** (1962), 35–38.

17. H. J. Shyr, S. S. Yu: Solid Codes and Disjunctive Domains. *Semigroup Forum* **41** (1990), 23–37.

18. H. J. Shyr: *Free Monoids and Languages*. Hon Min Book Company, Taichung, 2nd edition, 1991.

19. L. Staiger: On Infinitary Finite Length Codes. *RAIRO Inform. Théor. Appl.* **20** (1986), 483–494.

20. R. R. Varshamov, G. M. Tenengol'ts: Codes Capable of Correcting Single Asymmetric Errors. *Avtomat.i Telemekh.* **26** (1965), 288–292 (in Russian).

Graphbots: Mobility in Discrete Spaces

Samir Khuller[1]*, Ehud Rivlin[2,3] and Azriel Rosenfeld[3]**

[1] Computer Science Department and Institute for Advanced Computer Studies,
University of Maryland, College Park, MD 20742.
[2] Computer Science Department, Technion, Haifa 32000, Israel.
[3] Center for Automation Research, University of Maryland, College Park, MD 20742.

Abstract. Most previous theoretical work on motion planning has addressed the problem of path planning for geometrically simple robots in geometrically simple regions of Euclidean space (e.g., a planar region containing polygonal obstacles). In this paper we define a natural version of the motion planning problem in a graph theoretic setting. We establish conditions under which a "robot" or team of robots having a particular graph structure can move from any start location to any goal destination in a graph-structured space.

1 Introduction

Most previous theoretical studies of motion planning deal with "robots" that have simple geometries (e.g., polygonal) and that operate in simple spaces (e.g., in a planar region containing polygonal obstacles) [10]. This paper suggests that it may also be appropriate to study "robots" (or teams of robots) that operate in *discrete* spaces, i.e. in graphs, and that have discrete "geometries" (represented by subgraphs).

1.1 Mobility in a Graph

We begin by defining our model for mobility of a graph-structured robot (or team of robots) in a graph-structured space. Let G be a connected graph, which we usually refer to from now on as the "graph space", and let H be a connected subgraph of G, which we refer to from now on as the "robot" (though, as we shall see below, it might also represent, e.g., a cooperating team of (point) robots). We allow H to incrementally "move" in G; as it moves, we require it to remain isomorphic to itself. Formally, this is defined as follows: Two isomorphic subgraphs K, L of G are said to differ by a *local displacement* if corresponding nodes (under the isomorphism) are either identical or are neighbors. A *movement* or *motion* of H from S to T (which we call the start and target locations or "states") is defined by a sequence of subgraphs $S = H_0, H_1, \ldots, H_k = T$, all isomorphic to

* Research supported by NSF Research Initiation Award CCR-9307462.
** Research supported by ARPA under Contract DACA76-92-C-0009 (ARPA Order 8459).

H, such that H_i and H_{i-1} differ by a local displacement, $1 \le i \le k$. We say that H moves from S to T by a sequence of k local displacement steps. Note that a location (or "placement") of H in G is a subgraph of G which is isomorphic to H; two isomorphisms of H into G which have the same image are regarded as defining the same placement.

In this model, the graph structure of the robot or team of robots might represent physical (but not necessarily rigid) connections between the robot's parts, or communication links between the team members. It is reasonable to require that this structure must be preserved when the robot (or team) moves. Moreover, the movement should be as "continuous" as possible; "jumps" from one position to another arbitrarily distant position should not be allowed. Our definition of movement is a very good discretized description of the way a real-world team of robots might move cooperatively (e.g., [6]); we represent the members of the team by single nodes, and each time they move, we require them to reestablish their lines of communication, which define the graph structure of the team. (Note that a team of robots might be operating in a complex environment such as a mine (a network of caves and tunnels), so that the space in which the team moves might have an arbitrary topology, represented by an arbitrary graph.) Our model can also be applied to the movement of a jointed robot (with telescoping limbs that meet at universal joints) that moves on a discrete space, where neighboring joints always occupy neighboring nodes in the space.

1.2 Outline of Results

A given graph H cannot always move *freely* in a given graph space G; H can only occupy positions in which G is "big" enough to contain an isomorphic copy of H. If H has only one or two nodes, it can obviously move freely in any connected G, but bigger H's cannot move freely in an arbitrary G.

In this paper we explore two kinds of questions:

- What conditions should G satisfy so that a given graph H can move freely in G? In other words, under what conditions can we guarantee the existence of a motion between *any two* valid placements of H in G? (We shall assume that there exist at least two possible locations, since if there is only one location H cannot move at all.) We consider many different types of H's, and establish conditions on G that permit a motion between any two valid placements of H. Our proofs easily lead to efficient algorithms for planning a motion between any two locations of H.
- Given a start state S and a target state T, what is the complexity of finding a motion with the fewest local displacements from S to T (if one exists)?

We consider the first question for various types of H's. Even for very simple H's, this question turns out to be quite interesting and non-trivial. The results characterize various classes of graphs that permit free movement of various forms of H's, which we will name after various types of arthropods.[4] In Section 2.1

---

[4] A major group of segmented invertebrates having jointed legs. (Strictly speaking, not all of our examples are arthropods.)

we give necessary and sufficient conditions on G for the movement of a two-vertex path, which we call a *tick*. In Section 2.2 we give necessary and sufficient conditions on G for a three-vertex path (a *scorpion*) to be able to move freely in G, and in Section 2.3 we give sufficient conditions for a three-vertex cycle (a *jellyfish*) to move freely. These conditions do not generalize straightforwardly to paths or cycles (or cliques) that have more than three nodes; but in Section 2.4 we give a simple sufficient condition on G for a $(k+1)$-node star (a k-legged *spider*), consisting of a "central" node and k of its neighbors, to be able to move freely in G. (The study of this class of graphs is motivated by CMU's Ambler [1]. In fact, our spider can be regarded as an abstract version of a physical legged robot that has freely pivoting leg sockets and telescoping legs and that walks on stepping stones.) These proofs yield efficient algorithms for moving these types of arthropods from one location to another.

In Section 3 we show that for any *fixed* graph H, the second question can always be answered in polynomial time. Unfortunately the time complexity of the algorithm is quite high, albeit polynomial. For special kinds of graphs G, we give faster algorithms. We also prove that the problem is NP-complete when H is part of the input. This reduction is also outlined in Section 3.

1.3 Related Work

Recently we learned of related work by Papadimitriou, Raghavan, Sudan and Tamaki [9]. The spirit of their work is similar, in the sense of trying to capture a well studied geometric motion planning problem in a graph setting. In their work, it is assumed that the robot is a single vertex and the obstacles (also single vertices) are considered to be movable. The robot needs to plan a motion and has the power to request obstacles to move. (Another way to view this situation is to think of a "fleet" of robots located in the graph; one of them wants to plan a motion and may request the others to cooperate.) In our work, we have tried to model the movement of a robot that has a "shape" and that moves in the presence of fixed obstacles. Since the obstacles are fixed, we can simply delete the nodes occupied by them from the graph; thus we never need to refer to them. (Similarly, fixed obstacles in the plane can simply be treated as holes in the plane.) Clearly, the two models could be combined; one could study the problem of planning a cooperative motion for *many* robots, each of which has a certain shape.

2 Moving Robots Freely

2.1 Moving a Tick

A "tick" is modelled by a two vertex graph linked by a single edge. We leave the proof of the following proposition as an exercise for the reader.

Proposition 2.1 *A tick can move freely in any connected graph.*

Note that if G has only two vertices (and is connected) then there is a unique placement of the tick. (Both isomorphisms define the same placement.)

2.2 Moving a Scorpion

A "scorpion" is modelled by a three-vertex graph linked by two edges. We will refer to the degree-2 vertex as b (the "body") and the degree-1 vertices as the f vertices (the "feet").

Theorem 2.2 *A scorpion can move freely in G if and only if G does not contain a vertex v with two neighbors of degree 1.*

Note that we regard the two placements of the scorpion on a three-vertex, two-edge graph as identical.

Proof. If G has such a vertex v, we can place the scorpion with b on v, and the f's on the neighbors v_i of v. Any movement of the feet requires that both must move to v. If we move one foot to v, then b must leave v and thus the other foot will not be adjacent to b's new location. Thus the scorpion cannot move from this location.

To prove the converse, assume there is no such vertex v. Let the initial placement of the scorpion be at u_1, u_2, u_3 (one f on u_1, b on u_2, and the other f on u_3), and let the target location be v_1, v_2, v_3. If there is a path joining a foot of the scorpion in the initial location to a foot of the scorpion at the final location, without passing through u_2 and v_2, then the scorpion can "creep" along this path to reach its destination. Let us assume that the only path from $\{u_1, u_2, u_3\}$ to $\{v_1, v_2, v_3\}$ goes through either u_2 or v_2 (or both). Since the degree of u_1 is not 1 (the case when the degree of u_3 is not 1 is identical), u_1 has a neighbor u_0 (other than u_2). Move the scorpion as follows. Vertex f goes from u_1 to u_0, b goes to u_1 and f goes from u_3 to u_2. Now the scorpion can creep along the path to the desired destination. (At the other end a similar orientation step may be required.) The details are left to the reader.

Basically, to corner a scorpion you have to completely immobilize it at its starting location! □

2.3 Moving a Jellyfish

A "jellyfish" is modelled by a three-vertex graph linked by three edges (i.e., a clique of size 3). We will give some conditions under which free motion of a jellyfish is possible. Before discussing the details of the conditions that the graph must satisfy, we review some definitions [3].

Biconnected Graphs: A single vertex in a connected graph whose deletion disconnects the graph is called a *cut vertex* (also known as an *articulation vertex*). A graph with no cut vertices is called *biconnected*.

λ-Connected Graphs: A graph is said to be λ-connected if the deletion of *any* subset of $\lambda - 1$ vertices leaves the graph connected. We also require that G contain at least $\lambda + 1$ vertices.

A Chordal Graph is a graph in which each cycle of length at least 4 has a chord. (A chord is an edge that connects two vertices that are not adjacent on the cycle.)

Perfect Elimination Orderings: A perfect elimination ordering (peo) is a numbering of the vertices from $\{1, \ldots, n\}$, such that for each i, the higher numbered neighbors of vertex i form a clique. Thus a **peo** is represented by a sequence σ of vertices.

Theorem 2.3 (Fulkerson and Gross [5]) *A graph G has a peo if and only if G is chordal.*

We will use this characterization of chordal graphs to prove the following theorem.

Theorem 2.4 *A jellyfish can move freely in a biconnected chordal graph that has at least three vertices.*

Note that if G itself is a clique of size three, then a jellyfish has only one possible location (recall that we do not count locations obtained by permuting the vertices as distinct locations).

Before we prove the theorem, we first prove the following simple lemma.

Lemma 2.5 *Let G be a biconnected chordal graph that has at least three vertices. Consider a **peo** for G; it must be the case that any vertex i ($i = 1, \ldots, n-2$) has at least two neighbors with higher numbers.*

Proof. We will prove this by contradiction. Start deleting vertices one at a time, starting with vertex 1. Stop the deletion process when we are about to delete vertex v that has fewer than two neighbors in the current graph. There are two possible cases.

(Case 1.) *Vertex v has one neighbor u:* The other vertices adjacent to v have been deleted. We now claim that u is a cut vertex (in G) that separates v from the other vertices in C (by our assumption, v's number in the **peo** is between 1 and $n-2$; thus C has at least one vertex other than u). To prove this by contradiction, assume that P is the shortest path (in G) from v to any vertex in C that avoids u. Observe that P has at least two edges in it since v is not adjacent to any vertex in C (other than u). Consider the first vertex of P that we delete. There must be an edge between its two neighbors; this gives a shorter path from v to C, contradicting the assumption that P was the shortest path from v to some vertex in C (avoiding u)!

(Case 2.) *Vertex v has no neighbors:* Let w be the most recently deleted vertex among the neighbors of v in G. By our assumption, w had at least two neighbors when it was deleted; since these formed a clique, v must have a neighbor that has not been deleted as yet (by our assumption on w). Thus this case cannot occur.

$\square$

Proof. (Of Theorem 2.4) We now use Lemma 2.5 to prove the theorem by induction on the number of vertices of G. The base case is a chordal graph with three vertices. A biconnected graph on three vertices is a clique of size 3 and the theorem is trivially true for this graph. Consider a peo of G. Assume that we have deleted vertices $1, \ldots, i$ and have a graph G_i left. The subgraph G_i is chordal since it is an induced subgraph of G. G_i is also biconnected since it does not have any cut vertices. (If it has a cut vertex a separating it into two connected components C_1 and C_2 then the last vertex to be removed in the peo from the component that is exhausted first will have only a as its neighbor; this is a contradiction to Lemma 2.5.) Since G_i is biconnected and chordal, by the induction hypothesis, we know that the jellyfish can move freely on G_i. What happens if we add vertex i back? We have to show that for each location of a jellyfish that uses i, it can move to any location in G_i. In a placement of the jellyfish using vertex i, let the other corners of the jellyfish be at vertices x and y. W. l. o. g assume that $x < y$. If $x \leq n - 2$, then x has at least one other neighbor z with a higher number, such that y is adjacent to z (this follows from Lemma 2.5 and the definition of a peo). Thus we can move the jellyfish from i, x, y to x, z, y. Since x, y and z are all in G_i, by the induction hypothesis the jellyfish can also move to any other location in G_i. The other case to consider is when $x = n - 1$ (in this case $y = n$). Since G_i has more than three vertices, vertex $n - 2$ is distinct from vertex i. This vertex is also adjacent to both vertices n and $n - 1$. We can move the jellyfish from $i, n - 1, n$ to $n - 1, n - 2, n$. This concludes the proof. $\square$

It is easy to see that valid locations of a jellyfish cannot be separated by a cut vertex. However, we can construct graphs that are biconnected but not chordal, which allow free movement of the jellyfish. Hence chordality is sufficient, but *not* necessary.

2.4 Moving a Spider

A spider is modelled by a $(k + 1)$-vertex graph having a central vertex denoting its "body" (vertex b) linked by edges to vertices representing its "feet". We give conditions under which free motion of a spider is possible. We will refer to such a spider as a k-legged spider with feet $f_1, \ldots, f_k$. When $k = 1$, we have a 1-legged spider, which is an inchworm; this case has already been dealt with. Subsection 2.2 gave an exact characterization of the graphs a 2-legged spider (also called a scorpion) can move freely in. In the remainder of this subsection we will assume that $k \geq 3$.

Theorem 2.6 *A k-legged spider can move freely in a $(k - 1)$-connected chordal graph. (We assume that the graph has at least two distinct placements of the spider.)*

For the rest of the discussion in this subsection, we will assume that the underlying graph is $(k - 1)$-connected and chordal. Observe that Menger's theorem [3] guarantees the existence of $k - 1$ internally vertex-disjoint paths between any pair of vertices in G.

We first prove some interesting lemmas that are used in the proof of Theorem 2.6. Before we prove the lemmas, we introduce the following notation. Let

$N(v)$ denote the set of vertices that are adjacent to vertex v. Let $N(u,v)$ denote the set of vertices that are adjacent to vertices u and v. In other words, $N(u,v) = N(u) \cap N(v)$.

Lemma 2.7 (Common Neighbors Lemma) *Let G be a graph that is $(k-1)$-connected and chordal. If x and y are two adjacent vertices then $|N(x,y)| \geq k-2$.*

Proof. Consider a set of $k-1$ vertex-disjoint paths from x to y. Clearly, one path is the edge from x to y. Let P_i ($i = 1, \ldots, k-2$) be the other $k-2$ paths from x to y. Of all such paths, we can pick the set of $k-2$ paths that have the shortest total length. If the path P_i has three or more edges, then together with the edge (x,y) it creates a chordless cycle of length 4 or more, which is a contradiction to the assumption that the graph is chordal. Let P_i be $[x, u_i, y]$. Clearly, all the vertices u_i are in $N(x,y)$. This completes the proof. $\square$

Lemma 2.8 (Connectivity Lemma) *Let G be a graph that is biconnected and chordal. For any vertex x, if G_x is the graph induced by the neighbors of x then G_x is connected.*

Proof. Suppose G_x were not connected. Let $C_1, \ldots, C_\ell$ be the connected components of G_x. Since G is biconnected, for any pair of vertices we must have a path in G avoiding x that connects the pair. Of all such paths, let P be the shortest path connecting two vertices y and z such that $y \in C_i$ and $z \in C_j$ with ($i \neq j$). Since there is no edge (y, z) this path must have at least two edges. In this case, the edge (x, y) together with $P[y, z]$ and the edge (z, x) form a chordless cycle of length 4 or more; this is because x has no edges to internal vertices of P. This is a contradiction to the assumption that G_x has least two components. Hence G_x is connected. $\square$

Lemma 2.9 (Rotation Lemma) *Consider a spider located with its body b at vertex v, and its feet $f_1, \ldots, f_k$ at vertices $u_1, \ldots, u_k$. There is a legal motion that moves the k feet to any desired subset $S \subseteq N(v)$ with $|S| = k$.*

Proof. By the Connectivity Lemma, G_v is connected; let T_v be a spanning tree of G_v. Using T_v we will move the feet of the spider from their current locations to the vertices in S. Pick any foot f_i at vertex u_j. The vertices in S that currently do not have a foot on them are called "free" vertices, and those that have a foot on them are called "occupied" vertices.

Let P be a path in T_x that connects u_j with some free vertex in S. We start moving the foot f_i along the path P from u_j to the free vertex. If any vertex on the path is occupied by a foot f_p, we stop moving f_i and start moving f_p. In other words, we exchange the role of f_i and f_p. In this way we can move all the feet to vertices in S. $\square$

Observe that the above lemma lets us reduce the problem of moving a spider from one location to another, to the problem of moving the body without worrying about the placement of the feet.

Remark: The above lemma moves the feet one by one. If one wants to move the feet as quickly as possible, it is possible to compute a set of $k - 2$ shortest disjoint paths between the initial locations of the feet and the final locations. (Observe that the feet in "front" of and "behind" the body are not moved by the rotation lemma.) This works since the graph G is $(k - 2)$-connected even after we delete the vertex occupied by the body.

Proof. (Of Theorem 2.6) Assume that the start location of the spider's body is at vertex s and its target location is at vertex t. We will concentrate on moving the spider's body from vertex s to vertex t. Using the Rotation Lemma we can always move the feet from any set of neighbors of a vertex to any desired set of neighbors of the vertex.

Let $P = [s = v_0, v_1, \ldots, v_\ell = t]$ be the shortest path in G from s to t. We will show how the spider can move from v_i to v_{i+1} by a local displacement. We first apply the Rotation Lemma and move the k feet to the set $S_i = v_{i-1} \cup v_{i+1} \cup N(v_i, v_{i+1})$. (The only cases in which this is an invalid set are $i = 0$ and $i = \ell$; these will be dealt with later.)

The spider now moves the foot at v_{i+1} to v_{i+2}, and the foot at v_{i-1} to v_i. The body moves from v_i to v_{i+1}. The feet in the set $N(v_i, v_{i+1})$ do not need to move at all in this step. Observe that for the next local displacement we apply the rotation lemma to move the feet to vertices in $N(v_{i+1}, v_{i+2})$ before we can move the body again. In particular, the feet are moved from the current location to $S_{i+1} = v_i \cup v_{i+2} \cup N(v_{i+1}, V_{i+2})$.

When $i = 0$, we define v_{-1} as any vertex in $N(v_0)$ that is not in $N(v_0, v_1) \cup v_1$. When $i = \ell$, we define $v_{\ell+1}$ as any vertex in $N(v_\ell)$ that is not in $N(v_\ell, v_{\ell-1}) \cup v_{\ell-1}$. $\square$

Let n and m respectively denote the number of vertices and edges in G. Our proof also yields an efficient algorithm for finding the motion between the start and target locations. The shortest path from s to t can be found in $O(m)$ time by using BFS. At each step we can move the legs in $O(m)$ time This yields an $O(nm)$ time algorithm for moving the spider from any start location to a target location (since the body moves for at most n steps).

Remark: Using a modification of the basic approach described above, we can design an algorithm with running time $O(m)$ (assuming that k is a fixed constant). The main bottleneck in our algorithm is due to the fact that we need to apply the rotation lemma each time the body moves one step. The following idea yields a linear time algorithm.

First compute $k-1$ vertex-disjoint paths $P_1, \ldots, P_{k-1}$ between s and t. (These can be obtained by finding a flow of value k in an appropriately defined flow network [4]). Next, preprocess these paths to make them minimal, so they have no chords that provide shortcuts. This is easily accomplished by numbering each vertex on P_i by its distance from s along the path, and traversing P_i a second time, picking the neighbor with the highest label at each step.

Define $x_i (1 \leq i \leq k)$ to be the set of vertices adjacent to s on the path P_i. The spider does one rotation step to put $k - 1$ feet on the x_i vertices. Observe that $s \cup_{i=1}^{k} x_i$ form a clique since G is chordal. The spider can move its body to any neighbor x_j on one of the paths and drag the last foot behind it. We need to recompute chordless $k - 1$ vertex-disjoint paths in $O(d(x_j))$ steps, where $d(x_j)$ is the degree of vertex x_i. This can be done by examining the neighbors of x_j on the paths P_i.

The feet are once more moved onto the new set of paths and then the body moves again. Once the body reaches t we need to do one more rotation step to finish the motion. This method saves the cost of doing a rotation step each time the body moves and thus runs in $O(m)$ time.

2.5 Moving a Four-Legged Spider

We know that we can move a three-legged spider in a biconnected chordal graph; this follows from the theorem in the previous section with $k = 3$. In this section, we prove the slightly stronger result that a four-legged spider can move freely in a biconnected chordal graph. This is the best possible, since we can show by an example that a five-legged spider cannot move freely in a biconnected chordal graph.Clearly, there are two feasible locations for the five-legged spider at the two vertices of degree-5; but since these are the only two degree 5 vertices, no movement is possible between the two locations of the spider.

The proof for the four-legged spider is more complicated since the proof technique that worked for the three-legged spider in a biconnected graph does not work for four-legged spiders. (In particular, it is *not* the case that we can select a shortest path and move the body along this path, since there may be degree-3 vertices on this path.) The proof is omitted due to lack of space.

Theorem 2.10 *A four-legged spider can move freely in a biconnected chordal graph.*

3 Shortest Motion

Assume that G has n vertices and H has ℓ vertices. There are at most $O(n^\ell)$ possible locations of H in G. We construct a new graph $G' = (V', E')$ in which each vertex corresponds to a possible valid location of H in G. There is an edge in E' between $u \in V'$ and $v \in V'$, if there is a local displacement between u and v of H. Clearly, G' can be constructed in time polynomial in the size of H. In fact, by finding the shortest path in G' from the start state to the target state, we can determine the motion with the least number of local displacements in polynomial time for any fixed graph H. In fact, when the graph H is not fixed, not only is finding the shortest motion difficult, but even the problem of simply checking to see if there exists a motion between two valid placements is NP-complete, as we show next.

3.1 Moving a Jellyfish Quickly

The previous paragraph shows that we can move any arthropod from any start location to any target location in the least possible number of moves. Clearly, the complexity of the algorithm is rather high. In this subsection we show how to move a jellyfish from any start location to any target location (in a biconnected chordal graph) in a way that uses at most *twice* the number of moves of an optimal solution. This algorithm has an $O(m)$ running time.

Theorem 3.1 *In a biconnected chordal graph G we can find a a valid motion in $O(m)$ time for which the number of steps is guaranteed to be at most twice the optimal number of steps.*

Proof. Let S denote the subgraph corresponding to the initial location of the jellyfish and T the subgraph corresponding to the target location of the jellyfish. If S and T share a common edge then one step is enough to move the jellyfish from S to T.

The key idea is to determine the shortest cycle passing through any two edges of S and T. We determine a motion for the jellyfish using only the vertices on this cycle. Each time we move the jellyfish by one step, the length of the cycle decreases by one. The number of moves we make is $c - 2$ where c is the length of the cycle. After $c - 2$ moves, the current position and the target position share a common edge and only one more move is required. This means that $c - 1$ moves are sufficient.

It is easy to show, by induction on the number of moves, that if there is a solution with p moves, then there must be a cycle of length at most $2p + 2$ that passes through an edge of S and an edge of T. (Decompose the sequence of moves into the first move, followed by a sequence of $p - 1$ moves. By the induction hypothesis we can find a cycle of length $2p$; extending the cycle only uses two extra edges.) Since we have found the shortest cycle, $c \le 2p + 2$ and $c - 1 \le 2p + 1$. Hence the total number of moves is at most $2p + 1$, where p is the number of moves in an optimal solution. $\square$

3.2 NP-completeness

`Clique`(G, k) : is the problem of checking if the graph G contains a clique of size k. This problem is known to be NP-complete [7].

We can reduce the clique problem to checking to see if there exists a motion that moves $H = K_{2k}$ from a start location to a target location. Construct a new graph $G' = (V', E')$ as follows:

$$V' = V \cup \{x_1, \ldots, x_{2k}\} \cup \{y_{2k+1}, \ldots, y_{4k}\}.$$

$$E' = E \cup E_x \cup E_y \cup E''.$$

$$E_x = \{(x_i, x_j) | 1 \le i \le j \le 2k\}.$$

$$E_y = \{(y_i, y_j) | 2k + 1 \le i \le j \le 4k\}.$$

$$E'' = \{(v, x_i), (v, y_j) | v \in V, 1 \le i \le 2k, 2k + 1 \le j \le 4k\}.$$

In other words, we add two cliques to G having $2k$ vertices each, and we add edges from each vertex in each clique to all the vertices in G.

In the start state, H occupies one clique completely, and in the target state it occupies the other clique completely. If G contains a clique on k vertices, then we have two cliques in G' of $3k$ vertices each that share k vertices of G in common. We can move k vertices of H from one copy of K_{2k} to the k vertices in G forming the clique. In the next step these k vertices move to the second copy of K_{2k}, and the remaining k vertices in K_{2k} occupy the k vertices in G forming the clique. One more step completes the motion. Thus if G contains a clique of size k, the desired motion exists.

To prove the converse, notice that H cannot simultaneously occupy vertices from the two cliques of $2k$ vertices each (since there are no edges between them). Thus if G does not have a clique of size k, there is no motion from the start state to the target state.

4   Discussion and Future Work

1. In this paper we have proposed a model for motion planning in graphs. Clearly, variations of our definition could also be considered. For example, our definition allows the "exchange" of two vertices in a single step, but none of our results make use of such a motion, so that, in essence, we have disallowed it. (We did not do so explicitly to avoid cluttering our simple definitions.) In fact, in much of this paper we have used a definition of movement in which we do not distinguish between automorphic images of H. If we strictly enforce our original definition, an automorphism of H is not necessarily a local displacement, since corresponding nodes may not be neighbors; thus by this definition, a subpath of length ≥ 3 cannot move freely on a path, since it cannot reverse itself.

2. We could also consider stricter definitions of movement—for example, "rigid" movement, in which the distances *in* G between corresponding nodes of H remain the same. Specialized types of movements would also be of interest— for example, "translation", in which every node of H is required to move to a neighboring node.

3. It would also be of interest to study the free movement problem in special types of graphs. For example, it is clear that any H can move freely in G if G is a clique, a path, or a cycle. On the other hand, it is easy to see that only a path can move freely on a tree.

4. It might be of interest to study arthropods that undergo metamorphosis!.

5. An interesting open problem is to characterize the class of graphs that a *snake* can move freely in. A snake is modelled by a four-vertex graph linked by three edges. We will refer to the four vertices as a, b, c and d. Assume that the edges are $(a, b), (b, c)$ and (c, d). To go from position (a, b, c, d) to another position, say (w, x, y, z): if there is a path from either $\{a, b\}$ to $\{w, z\}$ the snake can just "creep" along this path. The problem becomes more difficult when the snake is not oriented in a position that allows it to start crawling from an endpoint. We have identified a list of conditions that allow the snake

to re-orient itself, but this list does not appear to be exhaustive. A simple characterization does not appear to be straightforward.

6. Efficient computation of shortest motions (or approximate shortest motions) remains a very interesting open problem. The main difficulty is in charging for the cost of rotations.

7. This paper has formulated a graph-theoretic framework for studying robot mobility. It would be of interest to develop graph-theoretic formulations of other types of robotic activities, such as sensing and manipulation. However, these activities do not have straightforward graph-theoretic counterparts, since a graph has no concept of direction (though it does have concepts of adjacency and distance). Thus it is not clear how to define vision-like sensory processes in a general graph, since vision would seem to require some sort of concept of a "line of sight", which presumably would involve distinguishing between nodes that lie in given directions from a given node. Similarly, it is not clear how to define the concept of one subgraph (a "gripper") "grasping" an "object" subgraph; we can require that certain nodes of the gripper be adjacent to certain nodes of the object, but there is no obvious way of distinguishing stable from unstable grasps. In future papers we plan to consider methods of augmenting a graph in order to allow direction-dependent concepts to be defined.

Acknowledgments: We would like to thank Yossi Matias, Seffi Naor, Balaji Raghavachari and Hava Siegelmann for useful discussions.

References

1. J. Bares and W. L. Whitaker, Configuration of an Autonomous Robot for Mars Exploration, In *Proc. of the World Conf. on Robotics Research*, vol. 1, pp. 37–52 (1989).

2. B. Bollobás, *Extremal Graph Theory*, Academic Press, London (1978).

3. J. A. Bondy and U. S. R. Murty, *Graph Theory with Applications*, North Holland, Amsterdam (1977).

4. T. H. Cormen, C. E. Leiserson, and R. L. Rivest, *Introduction to Algorithms*, MIT Press, Cambridge, MA (1989).

5. D. Fulkerson and O. Gross, Incidence Matrices and Interval Graphs, *Pacific Journal of Math.*, vol. 15, pp. 835–855 (1965).

6. E. G. Mettala, The OSD Tactical Unmanned Ground Vehicle Program, In *Proc. of the DARPA Image Understanding Workshop*, pp. 159–171 (1992).

7. M. R. Garey and D. S. Johnson, *Computers and Intractability: A Guide to the Theory of NP-Completeness*, Freeman, San Francisco, CA (1979).

8. P. J. M. McKerrow, *Introduction to Robotics*, Addison-Wesley, Reading, MA (1993).

9. C. H. Papadimitriou, P. Raghavan, M. Sudan and H. Tamaki, Motion Planning on a Graph, To appear in *Proc. of the 35th Foundations of Computer Science Conference* (1994) .

10. J. T. Schwartz, M. Sharir and J. H. Hopcroft, *Planning, Geometry and Complexity of Robot Motion*, Ablex Publishing Corp., Norwood, NJ (1987).

Solving Recursive Net Equations

Eike Best* and Maciej Koutny**

Abstract. This paper describes a denotational approach to the Petri net semantics of recursive expressions. A domain of nets is identified such that the solution of a given recursive equation can be found by fixpoint approximation from some suitable starting point. In turn, a suitable starting point can be found by fixpoint approximation on a derived domain of sets of trees. The paper explains the theory on a series of examples and then summarises the most important results.

Keywords: Denotational semantics, Petri nets, Recursion.

1 Introduction

Process algebras [1, 10, 11, 13, 18] are widely used as models for the specification of concurrent programs, and so are Petri nets and their extensions [4, 12, 16, 21]. In a process algebra one may study the algebraic properties of connectives related to operators known from programming languages. In particular, a general operator expressing (guarded or unguarded) recursion is part of many process algebras. Its transition system semantics is well-understood (e.g. [1]). In Petri net theory, emphasis tends to be on the partial order semantics of a concurrent system. In order to combine the two models and their individual advantages, it has frequently been proposed that a Petri net semantics be given to a process algebra [3, 8, 13, 17, 19, 22]. In this context, the question arises how the general recursion operator could be given a Petri net semantics. Although this question has been addressed in many of the papers just mentioned and also in [6, 14], denotational fixpoint methods that seem appropriate for recursion have not yet been completely satisfactorily applied in a Petri net setting (see, for example, the list of Further Research Topics given in [19]).

The purpose of the present paper is to close this gap by demonstrating the use of fixpoint methods in combination with elementary Petri nets. Our approach makes significant use of Devillers' idea to describe transition refinement in a Petri net by name trees [6]. The structure of this paper is as follows. First, we define a syntactic domain of expressions (or terms), wherein recursion can be expressed by a system of equations. Then we define a semantic domain which delineates the class of Petri nets that will be associated to expressions. This domain is equipped with a partial ordering suitable for approximating solutions

---

* Institut für Informatik, Universität Hildesheim, Marienburger Platz 22, D-31141 Hildesheim, E.Best@informatik.uni-hildesheim.de

** Department of Computing Science, University of Newcastle upon Tyne, NE1 7RU, U.K., Maciej.Koutny@newcastle.ac.uk

of recursive equations. Both domains are defined in Section 2. The remaining sections of the paper develop a theory explaining that every term equation always has at least one (countable) Petri net solution, that there may not always be a unique solution, and how to compute a solution. The approximation of a solution is based on successive refinements in the style of [6]. For this reason, we explain a general refinement operator in Section 3. In Section 4 we define Petri net equations, which serve as the semantics of term equations. The successive refinements approximating a solution of a net equation have to start with some first element (Section 5). Unless recursion is guarded, there are several possible alternatives for choosing such a starting point depending on the equation system under consideration. We examine an auxiliary domain in which this starting point can be defined as the solution of another equation. This auxiliary domain is defined in Section 6. Section 7 is devoted to reviewing some of the main results of our theory in general terms. In Section 8, finally, having shown that solutions always exist, we show that the preceding theory allows to deduce that every net equation system associated with any given term equation system has a Petri net solution, and thus the latter has a good semantics in terms of Petri net fixpoints.

2 Expressions and boxes

In this paper we deal with recursion in its most general form. Therefore, we define a process algebra and a tailor-made class of labelled Petri nets which are general with regard to recursion but minimal in other respects.

To define the process algebra, let A be a countable alphabet of action names and let $\mathcal{X}$ be a countable set of recursion variables. We use $a, b, a_i, \ldots$ as elements of A and $X, Y, X_j, \ldots$ as elements of $\mathcal{X}$. A *recursion-free expression* or *term E*, by definition, is generated by the syntax

$$E ::= a \mid X \mid E; E \mid E + E \mid E \| E,$$

where the connectives ; (sequence), + (choice) and $\|$ (disjoint parallelism) have their usual meaning. A *recursive equation system* is a countable set of equations of the form

$$\{X_j = E_j \mid j = 1, 2, \ldots\},$$

such that the X_j are distinct variables and the E_j are recursion-free expressions. Our syntactic domain is defined to be the set of all closed recursive equation systems, which will later just be called *term equations* or (if no misunderstanding is possible) *equations*, where 'closed' means that every variable occurring on the right hand side of some equation also occurs on the left hand side of the same or a different equation. An equation system will be called a (recursive) *expression* if some seed, that is, some fixed equation, is specified. This is often done in the form

$$\mu X_i.\{X_j = E_j \mid j = 1, 2, \ldots\},$$

where i is one of the indices $1, 2, \ldots$ For reasons of brevity and clarity, we define the main constructions and state the results of this paper only for the special case of one recursive equation. This is also an expression, since the (only) index is automatically fixed. All constructions and results can easily be generalised to the case of more than one equation. We indicate such generalisations at the appropriate points later in the paper. We use the following six expressions, all of which are closed, as our running examples:

$$
\begin{aligned}
\text{Example 1: } & X = (a; X; b) \\
\text{Example 2: } & X = ((a; X; b) + c) \\
\text{Example 3: } & X = (a\|X) \\
\text{Example 4: } & X = (a; X) \\
\text{Example 5: } & X = ((a\|b) + X) \\
\text{Example 6: } & X = X.
\end{aligned}
\tag{1}
$$

Intuitively, expression 1 can do any number of a-moves (including infinitely many), but never any b-move, and it never terminates properly either. Expression 2 can do whatever expression 1 could and, in addition, at any point it can make a single c-move followed by up to as many b-moves as there have been a-moves before the c-move; its execution terminates after the last b-move. In expression 3, any (finite) number of concurrent a-moves can happen, but it is not clear whether or not the execution can terminate after an infinite number of a-moves[1]. Expression 4 has the same behaviour - but obviously not the same structure - as expression 1. Example 5 expresses a choice between one of infinitely many parallel moves of a single a and a single b such that once one of the a's (or b's) is chosen, the corresponding b (or a, respectively) must be taken. Expression 6, finally, can make no move at all.

Our objective is to associate one or more elementary (possibly infinite) Petri nets canonically to each equation. We address the questions whether this is always possible and if so, how the Petri net(s) can be computed. In order to find the answers, monotonic functions defined over a suitable semantic domain will be associated to every equation. It turns out that these functions always have fixpoints amongst which we will find our solutions. If there are two or more fixpoints then different solutions can be obtained by starting their approximation from different points.

To define the semantic domain, we recall that an (arc-weighted) *net* is a triple (S, T, W) such that S and T are disjoint sets (of *places* and *transitions*, respectively) and W is a *weight function* from the set $((S \times T) \cup (T \times S))$ to the set $\mathbf{N}$ of natural numbers. A net (S, T, W) is called an *S-net* iff

$$
\forall t \in T: \left(\sum_{s \in S} W(s, t) \right) = 1 = \left(\sum_{s \in S} W(t, s) \right).
$$

A *labelled net*, for our purpose, is a four-tuple $\Sigma = (S, T, W, \lambda)$ such that (S, T, W) is a net and λ is a partial function from S to the set $\{e, x\}$ and from T to the set A. If $\lambda(s) = e$ then s is called an *entry place*. If $\lambda(s) = x$ then s is called an *exit place*. If $\lambda(s)$ is undefined then s is called an *internal place*. If

---

[1] We will return to this question later.

$\lambda(t) = a$ then, intuitively speaking, an execution of t denotes an execution of the action a. We will see later how $\lambda(t)$ being undefined is to be interpreted. If Σ is a labelled net, by convention, $^\bullet\Sigma$ and $\Sigma^\bullet$ denote the set of e-labelled places and the set of x-labelled places, respectively, of Σ. For any place $s \in S$, $^\bullet s$ $(s^\bullet)$ denotes the set of input (output, respectively) transitions of s. For a set of places $R \subseteq S$, $^\bullet R = \bigcup\{^\bullet s \mid s \in R\}$ and $R^\bullet = \bigcup\{s^\bullet \mid s \in R\}$. Similar notation is used for transitions and sets of transitions. A *box* is defined to be a labelled net such that the following hold:

$$^\bullet(^\bullet\Sigma) = \emptyset = (\Sigma^\bullet)^\bullet \ \wedge \ ^\bullet\Sigma \neq \emptyset \neq \Sigma^\bullet \ \wedge \ \forall t \in T : {}^\bullet t \neq \emptyset \neq t^\bullet.$$

A box is called *plain* or *non-operative* if $\lambda|_T$ is a function[2]. A finite box is called an *operator box* iff λ is undefined for all the transitions in T, the underlying net is the disjoint union of connected S-nets, each such S-net comprising exactly one entry place and exactly one exit place, and the net is side-condition-free[3], i.e., $\forall t \in T : {}^\bullet t \cap t^\bullet = \emptyset$. A *marking* M of (S, T, W, λ) is a function from S to the natural numbers. A marking is *safe* iff $M(S) \subseteq \{0, 1\}$. A safe marking can and will be represented as a subset of S. A *marked box* (Σ, M) is a box Σ together with a marking M. The *initially marked box corresponding to* Σ is defined to be $(\Sigma, ^\bullet\Sigma)$. A box Σ is called *safe* iff all markings in $[^\bullet\Sigma\rangle$, that is, all markings reachable in $(\Sigma, ^\bullet\Sigma)$, are safe. A safe box Σ is called *clean* iff all markings M in $[^\bullet\Sigma\rangle$ satisfy $\Sigma^\bullet \subseteq M \Rightarrow \Sigma^\bullet = M$.

Lemma: Operator boxes are safe and clean.

Proof: Directly from the S-net decomposability property.

We equip the set of plain boxes with a relation $\sqsubseteq$. A plain box $\Sigma = (S, T, W, \lambda)$ *is included in* a plain box $\Sigma' = (S', T', W', \lambda')$ (in symbols, $\Sigma \sqsubseteq \Sigma'$) iff

$$
\begin{aligned}
&S \subseteq S' \\
&T \subseteq T' \\
&\forall(x, y) \in (S \times T) \cup (T \times S) : W(x, y) = W'(x, y) \\
&\forall x \in S \cup T : \lambda(x) = \lambda'(x).
\end{aligned}
$$

Clearly, the relation $\sqsubseteq$ is a partial order. It also has least upper bounds: the places and transitions of the limit $\Sigma = \lim_{k \geq 0} \Sigma^k$ of a sequence $\Sigma^0 \sqsubseteq \Sigma^1 \sqsubseteq \Sigma^2 \ldots$ are, the unions of the places and transitions, respectively, of the boxes Σ^k in the sequence. For two elements x and y of Σ (one of them a place, the other a transition), the weight $W(x, y)$ is defined as the weight between x and y in the first Σ^k in which both of them occur (note that by the definition of $\sqsubseteq$, the weight does not change in subsequent approximations). Finally, for an element x in Σ, the label $\lambda(x)$ is defined as the label of x in the first and hence in all Σ^k containing x. For such chains we have:

Lemma: If all plain boxes Σ^k of a $\sqsubseteq$-chain $\Sigma^0 \sqsubseteq \Sigma^1 \sqsubseteq \Sigma^2 \ldots$ are safe and clean then so is the limit $\lim_{k \geq 0} \Sigma^k$.

---

[2] The meaning of the word 'non-operative' will become clear later.

[3] This restriction is added for reasons of simplicity; it can be eliminated, however, as described in [6].

Proof: From the fact that any transition sequence of the limit can be recovered as a transition sequence of some approximation Σ^k.

A class of (so to speak) rather small plain boxes is defined as follows. A box is called an *ex box* iff its set of transitions and consequently also its weight function are empty. An *ex* box thus consists only of a nonempty set of places labelled e and a nonempty set of places labelled x.

The nets associated to term equations will be taken from the semantic domain of plain boxes defined above. The relation $\sqsubseteq$ will be used to compute them by successive approximation.

3   A tree view of refinement

A key to our understanding of recursive nets is the use of labelled trees as place and transition names. Every plain name, such as $1, 2$ etc. for places and $1, 2, t, u$ etc. for transitions, will be viewed as a special tree with a single root (labelled with $1, 2, 1, 2, t, u$ etc.) which is also a leaf. Moreover, we will allow (degree-finite) trees as place or transition names, such as for example 125, which, by definition, is an abbreviation for the tree with root 1, intermediate node 2 and leaf 5^4. Let us consider the two boxes Σ_1 and Σ_2 depicted in Figure 1. They

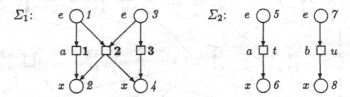

Fig. 1. A net to be refined (left) and the refining net (right)

can be written down formally in four-tuple notation as follows, where we specify the functions W_1 and W_2 by listing the pairs for which their value is 1 (all other values are implicitly assumed to be 0):

$$\Sigma_1 = (\{1, 2, 3, 4\}, \{1, 2, 3\}, \{(1, 1), (1, 2),$$
$$(3, 2), (3, 3), (1, 2), (2, 2), (2, 4), (3, 4)\},$$
$$\{(1, e), (3, e), (2, x), (4, x), (1, a)\})$$
$$\Sigma_2 = (\{5, 6, 7, 8\}, \{t, u\},$$
$$\{(5, t), (7, u), (t, 6), (u, 8)\},$$
$$\{(5, e), (7, e), (6, x), (8, x), (t, a), (u, b)\}).$$

Notice that λ_1 is undefined for transitions 2 and 3 of Σ_1. In our interpretation, this indicates that 2 and 3 may be refined by some plain box(es). Let us examine, in particular, the refinement of 2 and 3 by the same box Σ_2. Refinement of a transition by a box, in general, means that every input place of the transition

---

^{4} For brevity, we will refer to a labelled node using its label.

and every e-place of the box are combined to give a new place, and similarly with the output places of the transition and the x-places of the box, and that the new places are connected in the canonical way (see also [3, 9, 19]). The reader may check that doing this in either order in the example at hand, i.e., refining **3** after **2** or the other way round, produces exactly thê same net. This net is shown in Figure 2.

Using place and transition names in the shape of trees is a formal way of expressing the fact that refinement does not depend on the ordering in which it is done or, more precisely, of expressing simultaneous refinement. Thus when both **2** and **3** are refined by the box Σ_2, the input place shared by the two transitions, i.e., place *3*, has to be merged with the e-places *5* and *7* twice, once for transition **2** and once for transition **3**, resulting in the four upper right places of Figure 2; moreover, the input place *1* which is connected only to transition **2** but not to transition **3** has to be merged only once with *5* and *7*, resulting in the two upper left hand places of Figure 2. This merging is described by the trees that serve as e-place names in the figure, and the same is true for the x-places. The transition names are also trees describing their origins. For example, the tree name of the second transition of Figure 2 reflects the fact that this transition comes from a refinement of transition **2** of Σ_1 by the transition t of Σ_2. We shall use a linear

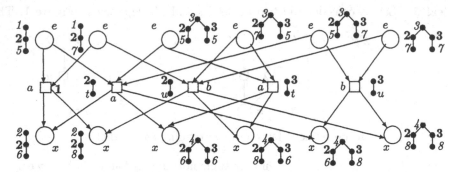

Fig. 2. A refined net

notation to express those trees. To this end, the expression $s \lhd (LT)$, where s is a node and LT is a list of trees, is a new tree where the trees of the list are appended to the root s. Thus, the set of places in Figure 2 can be written in linear form as follows:

$$\{1 \lhd (2 \lhd (5)), 1 \lhd (2 \lhd (7)), 3 \lhd (2 \lhd (5), 3 \lhd (5)), 3 \lhd (2 \lhd (7), 3 \lhd (5)), \dots\}.$$

Formally, the fact that the refinement of transition **2** commutes with the refinement of transition **3** is expressed by the fact that, for instance, $3 \lhd (2 \lhd (5), 3 \lhd (5))$ is the same tree (and hence the same place) as $3 \lhd (3 \lhd (5), 2 \lhd (5))$ resulting from different orders of refinement, and similarly for the other place names.

To complete the definition of the domain of nets in which we will be solving recursive equations, we assume that each transition name is a finite tree with nodes being transition names, and each place name is a finitely branching (possibly infinite) tree of the form $t_1 \triangleleft (t_2 \triangleleft (\ldots t_n \triangleleft (s \triangleleft (LT))))$, where $t_1, \ldots, t_n$ ($n \geq 0$) are transition names and s is a place name (c.f. Figure 2 and 4).

4   Net equations

In the following, we will denote by Σ, Σ_1 etc. plain boxes with transitions t, u, etc., and by Ω, Ω_1 etc. operator boxes with transitions $\mathbf{1}, \mathbf{2}, \mathbf{3}$ etc. Let Ω have n transitions, ordered in some arbitrary but fixed way; we will always use the ordering indicated by the names $\mathbf{1}, \mathbf{2}, \mathbf{3}$ etc. Then the equation

$$X = (\Omega; \Pi_1, \ldots, \Pi_n)$$

defines X to be a box arising from the simultaneous refinement of the n transitions $\mathbf{1}, \mathbf{2}$ etc. of Ω by the respective plain boxes Π_1, Π_2 etc. Note that this new box is plain as well, since all refining boxes Σ_i are plain and all transitions of the operator box Ω are replaced. Note also that since Ω is finite, the property that place trees are finitely branching is preserved through refinement. We now generalise this type of equation by allowing, in the list $\Pi_1, \ldots, \Pi_n$, not just plain boxes but also recursion variables $X \in \mathcal{X}$ as place-holders for other plain boxes. More precisely, let us consider a countable system of equations of the following form:

$$
\begin{aligned}
\text{line 1}: \quad & X_1 = (\Omega^1; \Pi_1^1, \Pi_2^1, \ldots) \\
\text{line 2}: \quad & X_2 = (\Omega^2; \Pi_1^2, \Pi_2^2, \ldots) \\
\text{line 3}: \quad & X_3 = (\Omega^3; \Pi_1^3, \Pi_2^3, \ldots) \\
& \ldots
\end{aligned}
\tag{2}
$$

where the left hand sides of the equations are mutually distinct recursion variables and the right hand sides are tuples such that in each line j, the number of Π_i^j's equals the number of transitions of the corresponding Ω^j, and every Π_i^j is either a plain box or a variable from the set $\{X_1, X_2, X_3, \ldots\}$. Note that since Ω^j is finite by the definition of an operator box, every line of the equation system has finite length, but the number of equations may be infinite. A *solution* of the system (2) is an assignment of a plain box Σ_j to every variable X_j such that every equation is valid if every occurrence of X_j is replaced by Σ_j.

By definition, the *dependency graph* of (2) is the digraph whose nodes are the set of variables $\{X_1, X_2, X_3, \ldots\}$ and whose arcs (X_i, X_j) indicate that X_j occurs on the right hand side of the equation having X_i as its left hand side. The equation system (2) is called *recursion free* if its dependency graph has no forward-infinite directed paths (in particular, the graph must be acyclic). In particular, if none of the right hand sides of (2) has a recursion variable then the system is recursion free since its dependency graph has no arcs at all. If the system (2) is recursion free then using simultaneous refinement a set of plain boxes $\Sigma_1, \Sigma_2, \ldots$ solving it can be calculated by executing the following procedure: starting with equations that have no variables on their right hand sides, use simultaneous refinement repeatedly until all left hand sides have been calculated. That this procedure is correct follows from König's Lemma, the finiteness of the operator boxes and the absence of infinite directed paths in the dependency graph. Note that the boxes Σ_j are unique in that case.

Lemma: If the system (2) is recursion free and if all plain boxes occurring on the right hand sides are safe and clean then the same is true for the box Σ_j obtained for the j'th equation ($j = 1, 2, \ldots$).

Proof: Operator boxes are safe and clean by a previous lemma, and the property propagates through refinement as proved in [6].

In the next section we show that the equation system (2) always has a solution, even if it is not recursion free. In the general case, this solution may not be unique.

5 Approximation of solutions

To start with, let us consider the term equation $X = (a; X; b)$ of the first example in (1). Figure 3 renders this equation as a net equation. Note how the term

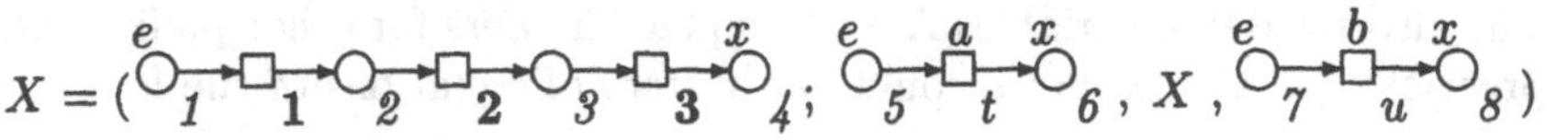

$$X = (\overset{e}{\underset{1}{\bigcirc}} \to \underset{1}{\square} \to \underset{2}{\bigcirc} \to \underset{2}{\square} \to \underset{3}{\bigcirc} \to \underset{3}{\square} \to \underset{4}{\bigcirc} ; \quad \overset{e}{\bigcirc} \to \overset{a}{\underset{t}{\square}} \to \overset{x}{\underset{6}{\bigcirc}}, \quad X, \quad \overset{e}{\underset{7}{\bigcirc}} \to \overset{b}{\underset{u}{\square}} \to \overset{x}{\underset{8}{\bigcirc}}) \quad (3)$$

Fig. 3. Equation associated to $X = (a; X; b)$

equation $X = (a; X; b)$ and the net equation (3) correspond to each other. The right hand side $(a; X; b)$ of the former is translated into an operator box with three unlabelled transitions which are to be refined by their corresponding boxes, which are specified explicitly for the transitions **1** and **3** and implicitly (as the variable X) for transition **2**. We will show later (Section 8) that every term equation system can be turned canonically into a corresponding system of net equations. For the time being, we continue to restrict our attention to only one equation, i.e., to the special case $n = 1$.

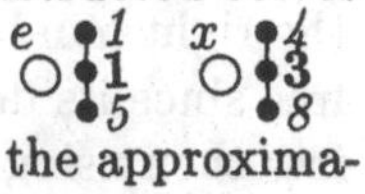

Let us examine how a solution of the net equation (3) associated with $X = (a; X; b)$ might be constructed. Suppose that there exists already an approximation Σ^0 as a starting point. Then it is a good idea to insert that approximation in the place of the X on the right hand side of equation (3). This yields (by the simultaneous refinement of the three transitions $1, 2, 3$ of the operator box of (3)) a new plain box Σ^1 which, hopefully, is a better approximation. We may then construct another approximation Σ^2 by the same principle, and so on. Again hopefully, the limit of the sequence $\Sigma^0, \Sigma^1, \ldots$ so constructed solves the equation system (3). We will see later that the ex box $\Sigma^0 = \overset{e}{\bigcirc}\overset{1}{\underset{5}{\overset{\bullet}{\square}}} \quad \overset{x}{\bigcirc}\overset{4}{\underset{8}{\overset{\bullet}{\square}}}$ (with these particular names) serves as a good starting point of the approximation of a solution of equation (3). Figure 4 shows the first three approximations resulting from this starting box, where the refinement is done according to the rules given in Section 3, except that we do not represent all name trees explicitly (the reader may easily guess the missing ones by analogy). A reason why

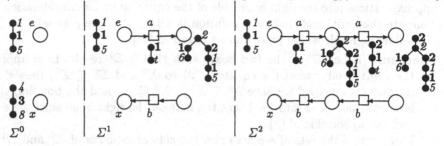

Fig. 4. Three approximations Σ^0, Σ^1 and Σ^2 of net equation (3)

this particular box Σ^0 is a good starting point is that by this choice, we get an embedding of Σ^j into Σ^{j+1} in the sense of the partial order $\sqsubseteq$. More precisely, we have $\Sigma^0 \sqsubseteq \Sigma^1 \sqsubseteq \Sigma^2 \ldots$, where $\sqsubseteq$ is the subnet relation defined at the end of Section 2. Figure 5 shows the limit of the infinite series of boxes whose first three elements are shown in Figure 4. The reader may check that this box sat-

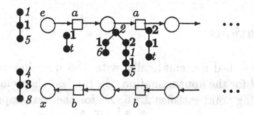

Fig. 5. A box $\Sigma_{X=(a;X;b)}$ solving the net equation (3)

isfies the equation (3) in the strictest possible sense of the word. To see this, replace the variable X on the right hand side of equation (3) by the plain box $\Sigma_{X=(a;X;b)}$ shown in Figure 5, perform the refinement of the three transitions $1, 2, 3$ according to the rules given in Section 3, and check that the result is not just isomorphic but even identical to $\Sigma_{X=(a;X;b)}$, i.e., it has the same places and the same transitions with the same connections and the same labels. Note also that, as might be expected, with the canonical initial marking $^\bullet\Sigma_{X=(a;X;b)}$ (i.e., a token on its e place), this solution has the same behaviour as the expression $X = (a; X; b)$.

It may be checked that this approach works similarly for the second example $X = ((a; X; b) + c)$, whose net equation is the same as equation (3) except that the operator box has one more transition **4** from the e-place to the x-place and the list of Σ_i^j on the right hand side has one more element with a c-labelled transition. If we use exactly the same starting box Σ^0 as before (with the same names and the same labels), the approximation yields a limit which is very much like $\Sigma_{X=(a;X;b)}$ in Figure 5, except that the places of the upper line and the places of the lower line are joined by vertical transitions labelled c.

Once an appropriate starting point Σ^0 is chosen, the approximation of a solution of a given net equation is quite automatic (we just keep refining successive approximations into the right hand side of the equation under consideration and compute the limit) and leads to a solution in all cases. More precisely, we have the following result whose proof can be found in [15]:

Theorem: (i) Let Σ^0, Σ^1 be two boxes such that if Σ^1 results from applying the right hand side of the equation (2) to Σ^0 and $\Sigma^0 \sqsubseteq \Sigma^1$, then if this iteration is repeated we have $\Sigma^0 \sqsubseteq \Sigma^1 \sqsubseteq \Sigma^2 \sqsubseteq \ldots$, and the box $\lim_{k \geq 0} \Sigma^k$ is a (safe and clean, provided that the nets on the right hand sides were safe and clean) solution of (2).

Moreover, if the sets of e-places and the sets of x-places of Σ^0 and Σ^1 are the same then they are the same in all Σ^k and hence also in the limit.

(ii) If Σ is a solution of the equation (2) then by taking an ex box Σ^0 with entry places $^\bullet\Sigma$ and exit places $\Sigma^\bullet$, and by applying the same procedure as in (i), we obtain $\Sigma^0 \sqsubseteq \Sigma^1 \sqsubseteq \Sigma^2 \sqsubseteq \ldots$ and $\Sigma = \lim_{k \geq 0} \Sigma^k$.

Note that (i) does not require Σ^0 to be an ex box. The requirement that the initial approximation is included in the next one already guarantees that the limit is a solution.

6   Tree equations

In this section we deal systematically with the question how to find a good starting point Σ^0 for the approximation. By the last theorem, we wish to ensure that such a starting point satisfies $\Sigma^0 \sqsubseteq \Sigma^1$ for the next approximation Σ^1. For instance, we may ask why the box $\ \begin{smallmatrix} e \\ \bigcirc \end{smallmatrix} \begin{smallmatrix} \bullet 1 \\ \bullet 1 \\ \bullet 5 \end{smallmatrix} \begin{smallmatrix} x \\ \bigcirc \end{smallmatrix} \begin{smallmatrix} \bullet 4 \\ \bullet 3 \\ \bullet 8 \end{smallmatrix}\ $ is so suitable a starting point for the approximation of the solutions of our first two expressions $X = (a; X; b)$ and $X = ((a; X; b) + c)$. First of all, it is a good idea always to start with an ex box, because these boxes are minimal with respect to the number of transitions they contain. It is even better (if possible) to start with an ex box containing only one e place and one x place, because these boxes are minimal with respect to $\sqsubseteq$. (Note that we have excluded the empty box[5])

Finally, by (ii) in the last theorem, we do not lose any solution by restricting our starting point only to ex boxes. Let us call this generic starting box

$$\underbrace{(P_X \cup Q_X, \emptyset, \emptyset, \{(p, e), (q, x) \mid p \in P_X, q \in Q_X\})}_{\Sigma^0}.$$

In order to exclude the empty box, the sets P_X of entry place and Q_X of exit places should both be nonempty (and preferrably singletons). They are disjoint, of course. The ex box Σ^0 is supposed to satisfy the property

$$\Sigma^0 \sqsubseteq \Sigma^1$$

---

[5] A reason for excluding the empty box is that the empty box, put in sequence with another box, would yield a non-safe net; there are also other reasons.

in order to be the starting point of an approximation of a solution of the equation under consideration. Because of this, some conditions can be imposed on the sets P_X and Q_X, essentially by performing the refinement at P_X and Q_X and limiting the required inequality to these sets. As both are sets of trees, they may be defined as solutions of an inequation ranging over the complete lattice of sets-of-trees with the subset ordering relation. In order to express such a condition, we use two operations on sets of trees: the ordinary set union, and the appending construct $\lhd$ lifted to sets of trees. More precisely, $s \lhd (LST)$ creates a set of trees from a given node s and a list of sets of trees LST. For instance, $3 \lhd (\{4,5\}, \{6\})$ equals the set of trees $\{3 \lhd (4,6), 3 \lhd (5,6)\}$, and $3 \lhd (\{4\})$ is the singleton set $\{3 \lhd (4)\}$. We also define $s \lhd (\ldots, \emptyset, \ldots) = \emptyset$.

In determining the conditions that P_X and Q_X have to satisfy, then, we simply carry out the refinement specified by the net equation under consideration, in moving from Σ^0 to Σ^1. Thus, for example, equation (3) (see Figure 3) leads to the following tree inequations for an ex box starting the approximation of $X = (a; X; b)$:

$$P_X \subseteq 1 \lhd (1 \lhd (\{5\}))$$
$$\text{and } Q_X \subseteq 4 \lhd (3 \lhd (\{8\})). \tag{4}$$

To see why, consider the e-place (for P_X) and the x-place (for Q_X) of the operator box of equation (3) and check the way in which the transitions bordering on those places (1 for P_X and 3 for Q_X) are refined by the corresponding simple boxes, viz. the first one for 1 and the third one for 3. In general, the right hand sides of such inequations can be expressed in terms of union and $\lhd$. As both operators are monotonic on the complete lattice of sets-of-trees, functions formed from them have fixpoints defined as usual. Moreover, every P_X and Q_X solving an inequation of the form (4) must be subsets, respectively, of P_X and Q_X solving the corresponding equation

$$P_X = 1 \lhd (1 \lhd (\{5\}))$$
$$\text{and } Q_X = 4 \lhd (3 \lhd (\{8\})). \tag{5}$$

In our particular case, it so happens that both the inequation and the equation have only one nonempty solution, namely the two singleton sets of trees $P_X = \{1 \lhd (1 \lhd (5))\}$ and $Q_X = \{4 \lhd (3 \lhd (8))\}$. Note that these are exactly the places of the box Σ^0 we have chosen earlier for this example. In general, if P_X and Q_X solve the equation then they determine a starting ex box to which the second part of (i) in the previous theorem can be applied, that is, they remain constant throughout the approximation of the limit of (2). Conversely, by (ii), any solution of (2) determines a solution P_X and Q_X. As a result, there is a one-to-one correspondence between the solutions of an equation on nets and the solutions of the corresponding equation on (place) trees. Therefore, in what follows we can limit ourselves to considering the equations on trees only.

If the e-places and the x-places of an operator box touch only transitions that are to be refined by a constant plain box rather than a recursion variable, such as is the case in the two examples considered so far, then the right hand sides of the equations for P_X and Q_X are always constant and finite. In this case, the equation has a unique (finite) solution for P_X and Q_X (as a consequence, the net obtained as the limit is the least solution of the net equation with respect to the subnet relation $\sqsubseteq$). More generally, the same property holds if the system of net equations under consideration is *guarded*[6].

The equations for P_X and Q_X may become more complicated for unguarded recursion. This is illustrated by our third example $X = (a\|X)$ which contains X unguardedly at both ends (that is to say, the e-labelled front end and the x-labelled rear end). The net equation shown in Figure 6 corresponds to it. Note

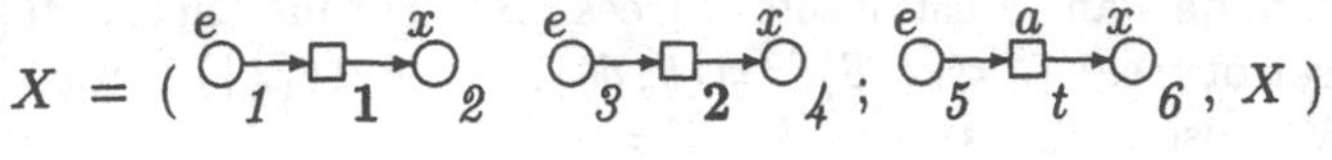

$$X = (\; \overset{e}{\underset{1}{\bigcirc}} \!\!\rightarrow\!\! \overset{}{\underset{1}{\square}} \!\!\rightarrow\!\! \overset{x}{\underset{2}{\bigcirc}} \quad \overset{e}{\underset{3}{\bigcirc}} \!\!\rightarrow\!\! \overset{}{\underset{2}{\square}} \!\!\rightarrow\!\! \overset{x}{\underset{4}{\bigcirc}} \;;\; \overset{e}{\underset{5}{\bigcirc}} \!\!\rightarrow\!\! \overset{a}{\underset{t}{\square}} \!\!\rightarrow\!\! \overset{x}{\underset{6}{\bigcirc}} \;,\; X \;) \tag{6}$$

Fig. 6. Equation associated to $X = (a\|X)$

that equation (6) indeed matches $X = (a\|X)$ in the same way as equation (3) matches $X = (a; X; b)$. We now derive a box Σ^0 which is supposed to start the approximation for this example. Again we determine two (in)equations, one for the set P_X of e-places of Σ^0 and one for the set Q_X of x-places of Σ^0. Since transition **2** of the operator box, this time, does not have a constant plain box to be replaced by, there is no constant set of e-places (or x-places) to perform the refinement with. Rather we have to re-use P_X (and Q_X, respectively) in the right hand side of the tree equation. This leads directly to the following equations:

$$\begin{aligned} P_X &= 1 \triangleleft (1 \triangleleft (\{5\})) \;\cup\; 3 \triangleleft (2 \triangleleft (P_X)) \\ Q_X &= 2 \triangleleft (1 \triangleleft (\{6\})) \;\cup\; 4 \triangleleft (2 \triangleleft (Q_X)). \end{aligned} \tag{7}$$

There are two solutions for the first of these equations. One solution contains only finite trees and the other solution contains an infinite tree as well. We call these solutions P_X^{min} and P_X^{max}, respectively, and for reasons of brevity, we write the trees in these sets in the form of sequences of node labels:

$$\begin{aligned} P_X^{min} &= \{115, 32115, 3232115, \ldots\} \\ P_X^{max} &= P_X^{min} \cup \{323232\cdots\}, \end{aligned}$$

The equation for Q_X has two corresponding solutions. Together, they determine four starting boxes $\Sigma^{0,min}$ (using P_X^{min} and Q_X^{min}), $\Sigma^{0,max/min}$ (using P_X^{max} and Q_X^{min}), $\Sigma^{0,min/max}$ (using P_X^{min} and Q_X^{max}) and $\Sigma^{0,max}$ (using P_X^{max} and Q_X^{max}) to begin the approximation of the solution of the net equation (6) for $X = a\|X$. Depending on which starting point of the approximation is chosen, different solutions of equation (6) are obtained. Two of them (intuitively, the minimal and maximal ones) are shown in Figure 7 (but again we do not represent all name trees). It is not hard to check that both boxes are indeed solutions of equation (6) in the same strict sense as before. Note that the interleaving behaviour of the two solutions is the same, but if infinitely broad behaviours are allowed then

---

[6] In general, an instance of (2) is called *guarded* iff its dependency graph contains no infinite path of the form $X^1 X^2 \ldots$ such that each X^{j+1} is refined in the equation for X^j into a transition which is connected to an e-place, or each X^{j+1} is refined in the equation for X^j into a transition which is connected to an x-place.

the first one has a terminating behaviour while the second has not (since the initial token on its isolated e-place remains there and cannot be moved to the corresponding x-place).

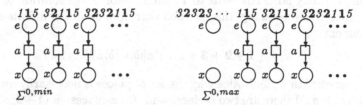

Fig. 7. Two solutions to $X = (a \| X)$

To end this section, we illustrate the technique on the three remaining examples defined in (1), albeit not in the same degree of detail as before. The fourth term equation $X = (a; X)$ corresponds to the net equation shown in Figure 8.

$$X = (\; \overset{e}{\underset{1}{\bigcirc}}\rightarrow\underset{1}{\Box}\rightarrow\underset{2}{\bigcirc}\rightarrow\underset{2}{\Box}\rightarrow\underset{3}{\overset{x}{\bigcirc}} ; \; \overset{e}{\underset{4}{\bigcirc}}\rightarrow\underset{t}{\overset{a}{\Box}}\rightarrow\overset{x}{\underset{5}{\bigcirc}} , X\;) \tag{8}$$

Fig. 8. Equation associated to $X = (a; X)$

This net equation further leads to the following tree equations:

$$\begin{aligned} P_X &= 1 \triangleleft (1 \triangleleft (\{4\})) \\ \text{and } Q_X &= 3 \triangleleft (2 \triangleleft (Q_X)). \end{aligned} \tag{9}$$

Note that the only solution for P_X is the singleton set $\{114\}$ while Q_X has no unique solution, corresponding to the fact that the expression $X = (a; X)$ is front-guarded but not rear-guarded. The minimal solution for Q_X is the empty set $\emptyset$, but the singleton set $\{32323 \cdots\}$ containing an infinite tree is also a solution. As we start the approximation with a nonempty box Σ^0, only the latter solution for Q_X is permissible as the set of x-places of Σ^0. When begun with the one-e-place one x-place box just defined, the approximation leads to a solution whose structure is similar to the box shown in Figure 5, except that the lower line is replaced by a single x-place, namely the element of Q_X.

The fifth term equation $X = ((a \| b) + X)$ corresponds to the net equation shown in Figure 9.

$$X = (\; \begin{matrix} \overset{e}{\bigcirc}1 \\ \Box 1 \quad\quad \Box 2 \\ \underset{x}{\bigcirc}2 \end{matrix} ; \; \begin{matrix} \overset{e}{\bigcirc}3 \quad \overset{e}{\bigcirc}5 \\ a\Box t \quad b\Box u \\ \underset{x}{\bigcirc}4 \quad \underset{x}{\bigcirc}6 \end{matrix} , X\;) \tag{10}$$

Fig. 9. Equation associated to $X = ((a \| b) + X)$

By the same rules as above, this leads to the following tree equations:

$$\begin{aligned} P_X &= 1 \triangleleft (1 \triangleleft (\{3\}), 2 \triangleleft (P_X)) \; \cup \; 1 \triangleleft (1 \triangleleft (\{5\}), 2 \triangleleft (P_X)) \\ Q_X &= 2 \triangleleft (1 \triangleleft (\{4\}), 2 \triangleleft (Q_X)) \; \cup \; 2 \triangleleft (1 \triangleleft (\{6\}), 2 \triangleleft (Q_X)). \end{aligned} \tag{11}$$

Note how the branching in the operator box leads to corresponding branchings in the trees. Equations (11) are interesting because, as it turns out, their largest solutions have uncountably many trees. Starting the approximation with such an uncountably large box leads to an uncountably large solution of the equation (10). Intuitively speaking, this solution corresponds to the infinite simultaneous refinement

$$(1 + 2 + 3 + \ldots \; ; \; a\|b, a\|b, a\|b, \ldots).$$

It is described in [2, 6]. Notice why the set of places is necessarily uncountable: in the box $(1; a\|b)$ there are two e-places and two x-places; in $(1+2; a\|b, a\|b)$ there are four e-places and four x-places; in $(1 + 2 + 3; a\|b, a\|b, a\|b)$ there are eight e-places and eight x-places; and so on. This uncountability can be seen keeping in mind that the right hand sides of equation (11) are recursive unions of binary trees. Nevertheless, equation (11) also has countable solutions. For instance, a singleton set of trees whose only member is shown in Figure 10 (and which is a solution of the inequation on sets-of-trees corresponding to equation (11)), leads in the limit to a countable solution for P_X. Starting the approximation with a

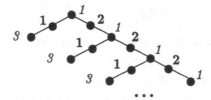

Fig. 10. Infinite tree belonging to a countable solution of equation (11)

countable solution of (11), one gets a countably infinite solution of the equation (10). Intuitively speaking, this solution corresponds to the infinite choice

$$(a\|b) + (a\|b) + (a\|b) + \ldots$$

whose box, being made of countably many finite alternatives, is countable[7].

Finally, the sixth term equation $X = X$ leads to the following net equation:

$$X = (\overset{e}{\underset{1}{\bigcirc}} \!\!\rightarrow\!\! \underset{1}{\square} \!\!\rightarrow\!\! \overset{x}{\underset{2}{\bigcirc}} \; ; \; X). \tag{12}$$

---

[7] More precisely, this box can be described as follows. There are infinitely many pairs of transitions labelled a and b, respectively; let us call them a_1, b_1 (the first pair), a_2, b_2 (the second pair) and so on. There is a countable infinity of e-places. Split this set into two halves both of which are again countably infinite. The left half is taken as preplaces of a_1, the right half as preplaces of b_1. The left half is split again into two countable infinities, and so is the right half. The left left half and the left right half are taken as preplaces of a_2, the right left half and the right right half as preplaces of b_2, and so on like a binary tree of infinities. The same is done symmetrically with the x places.

By the rules of the game, (12) leads to the following tree equations for the entry and exit places of the starting box:

$$P_X = 1 \triangleleft (1 \triangleleft (P_X))$$
$$\text{and } Q_X = 2 \triangleleft (1 \triangleleft (Q_X)). \tag{13}$$

We have $P_X = \emptyset = Q_X$ amongst the set of possible solutions, but we also have nonempty singleton solutions. In this case, the approximations never change anything, and the box solving $X = X$ has exactly one e-place, one x-place, and no transitions.

7 Main results

In this section we state the results that underpin the theory exemplified in the previous sections. Proofs can be found in [15]. We have the following theorems, (A)-(J), which we will phrase for the special case $n = 1$ of the equation system (2):

(A) The right hand side of the equation system (2) determines an operator which is monotonic w.r.t. $\sqsubseteq$ relation.

(B) The right hand sides of the tree-set equations associated to a variable X determine an operator which is monotonic on the complete lattice of sets of trees with $\subseteq$. Hence every tree-set equation has a unique minimal solution and a unique maximal solution P_X^{min} and P_X^{max} (and similarly for Q_X).

(C) P_X^{min} is always a set of finite trees (which may be countably infinite, as in example 3, or empty, as in example 5).

(D) P_X^{max} is always a nonempty set. *Proof sketch:* We need to construct a tree τ which is a member of P_X^{max}. In general, if τ satisfies $\{\tau\} \subseteq f(\{\tau\})$, where f is the operator determined by the right hand side of the tree equation for P_X, then $\tau \in P_X^{max}$. We may construct such a tree by induction, this time on the domain of trees, starting with the root of an arbitrary set on the right hand side of the tree equation. Note that this equation is always of the form $P = set \cup \ldots \cup set$, so that the construction is well-defined. *End of proof sketch for* (D). Note that the tree shown in Figure 10 is an instance of this construction.

(E) Let P be any set of trees satisfying only the $(\subseteq)$ part of the tree equations. Then $\bigcup_{i \geq 0} f^i(P)$ is a solution of the equations, where f is as in (D). Moreover, both sets lead to the same solution of (2).

(F) If P_X satisfies only the $(\subseteq)$ part of the tree equations and if Σ^0 is a box constructed from P_X and a corresponding Q_X, then starting the approximation of (2) with Σ^0, an ascending (in the $\sqsubseteq$ sense) chain of boxes whose limit solves (2) is obtained.

(G) It is always possible to start approximating a solution of (2) with a single-entry single-exit box Σ^0. *Proof:* From the proof of (D) and (E).

(H) Every equation has a countable nonempty solution. *Proof:* From (D) and the fact that the limit construction preserves countability of nets.

(J) If the net equation (2) is guarded then the associated tree equations have unique solutions and the associated solution obtained by successive approximations is unique and the least one with respect to $\sqsubseteq$.

If instead of a single equation, we are given a finite set of n equations (2) then the theory can be generalised. Instead of considering the lattice of boxes, we then consider the extended lattice of n-fold Cartesian product of the set of boxes. If we are given a countably infinite system of equations, we consider the standard extension to the corresponding function space representing the infinite Cartesian product. All constructions and results remain the same (with changes produced by the extensions).

8 Petri net semantics of recursive expressions

In this section, we summarise the constructions defined in this paper in general terms. Let a closed recursive expression

$$\mu X_i \{X_i = E_i \mid i = 1, 2, \ldots\}$$

be given. As E_i is recursion free, it is possible to represent it as one or more (auxiliary) operator boxes, even if it is deeply nested. In fact, the three operator boxes shown in Figure 11, which correspond to the three connectives of the syntax for recursion free expressions, are sufficient for that purpose. To understand

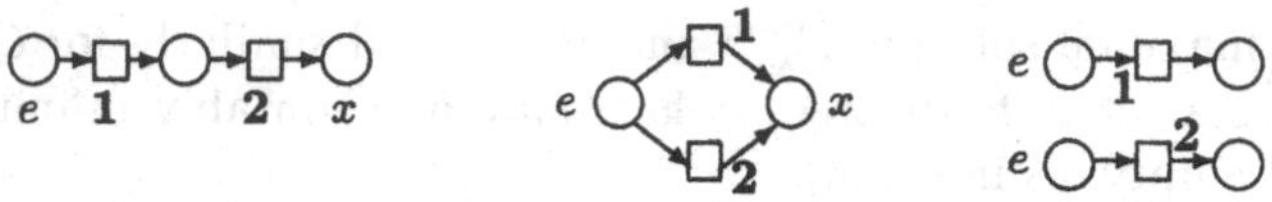

Fig. 11. Three operator boxes for ; (left), + (middle) and $\parallel$ (right)

this translation, consider the expression

$$X = ((a \parallel (b + (X; b))); (X + (b \parallel X)))$$

as an example. It can be unravelled into six equations as follows:

$$X = X_1; X_2 \quad X_1 = a \parallel X_3 \quad X_2 = X + X_4$$
$$X_3 = b + X_5 \quad X_4 = b \parallel X \quad X_5 = X; b.$$

Every such equation can be translated directly into a net equation of the form $X = (\Omega; \Sigma_1, \Sigma_2)$, where Ω is one of the operator boxes of Figure 11[8]. In this way, a net equation system (2) can be associated in a canonical way to the given recursive expression. This system can be solved using the techniques and results described earlier. Note however that a box being a solution of the net equation does not necessarily mean that it is also a good semantics of the corresponding expression. Fortunately, as it is proved in [15] that any such limit has exactly the same behaviour as the expression it comes from.

Summarising, the following overall picture emerges from our discussions:

---

[8] For simplicity, we exclude (trivial) equations of the form $X = a$ and $X = X'$.

- Any term equation has an associated net equation, and any net equation has a solution.
- Any solution of a net equation can be derived as an approximation from some suitable starting box (namely, the box consisting only of the e-places and the x-places of the solution).
- Conversely, any starting box Σ^0 satisfying $\Sigma^0 \sqsubseteq \Sigma^1$ for the next approximation Σ^1 leads to a solution.
- If recursion is guarded then both the net equation and the tree equations derived from it have unique solutions.
- Otherwise, the equations may have different solutions (cf. example 4).
- In all cases, it is possible to start the approximation with a well chosen single entry / single exit box.
- There is always a countable solution.

The fact that for expressions such as $X = (a\|X)$, more than one solution exists may be unsatisfactory, and therefore it may make sense to examine the space of possible solutions for a canonical one. While the function defined by the net equation is monotonic, it is not a function on a complete partial order, and hence the notion of minimal or maximal fixpoints (which could lead to some canonicity) do not make sense. However, they do make sense on the auxiliary lattice of sets-of-trees on which the function derived from the associated tree equations is defined. Since the minimal solution may be empty, only the maximal solution offers itself as a candidate for being 'canonical' in all cases. One may therefore take the point of view that the box approximated from a starting box that corresponds to the maximal solution of the tree equation, is 'the' semantics of the given expression; it is unique, of course. For example 5, this would mean that the uncountable box indicated earlier - rather than the countable one - is 'the' semantics of the expression $X = ((a\|b) + X)$, and for example 3, it would mean that the box shown on the r.h.s of Figure 7, rather than the box on the l.h.s., is 'the' semantics of $X = (a\|X)$. This view is taken in [6]. Here, we have demonstrated that if the uniqueness of a net associated to an expression is stressed, this view (of taking the maximal fixpoint of the tree equation) appears to be the only reasonable choice because the minimal fixpoint is not always applicable.

9 Concluding remarks

We have demonstrated how a tree-based general naming discipline for refinement leads to a technique for solving recursive Petri net equations. This technique is orthogonal to the general relabelling technique described in [5] to capture synchronisation and other transition-based operators. These two techniques can be combined with each other to yield the net semantics of process algebras featuring, beside recursion and the other control operators, also synchronisation and restriction.

Acknowledgements

This work owes to Raymond Devillers who invented and formalised the tree naming device in a more general form than needed here (applicable also to nets with side conditions), and to Javier Esparza who did a formalisation for guardedly recursive expressions [7]. Both also gave useful comments on this paper. We are also indebted to Mogens Nielsen and Ugo Montanari for comments. This work was done within the Esprit Basic Research Working Group 6067 CALIBAN.

References

1. J.Baeten, W.P.Weijland: *Process Algebra.* Cambridge Tracts in Theoretical Computer Science 18 (1990).
2. E.Best, R.Devillers, J.Esparza: General Refinement and Recursion Operators for the Petri Box Calculus. Springer-Verlag, LNCS Vol.665, 130-140 (1993).
3. E.Best, R.Devillers, J.Hall: The Petri Box Calculus: a New Causal Algebra with Multilabel Communication. Advances in Petri Nets 1992, G.Rozenberg (ed.), Springer-Verlag, Lecture Notes in Computer Science Vol. 609, 21-69 (1992).
4. E.Best, C.Fernández: *Nonsequential Processes. A Petri Net View.* Springer-Verlag, EATCS Monographs on Theoretical Computer Science Vol. 13 (1988).
5. E.Best, M.Koutny: A Refined View of the Box Algebra. Proceedings of the Petri Net Conference'95, G.De Michelis, M.Diaz (eds.), Springer-Verlag, Lecture Notes in Computer Science (1995).
6. R.Devillers: S-invariant Analysis of Petri Boxes. Technical Report LIT-273, Laboratoire d'Informatique Théorique, Université Libre de Bruxelles (1993). To appear in Acta Informatica (1995).
7. J.Esparza: Fixpoint Definition of Recursion in the Box Algebra. Memorandum, Universität Hildesheim (December 1991).
8. U.Goltz: On Representing CCS Programs by Finite Petri Nets. Proc. MFCS'88, Springer-Verlag, Lecture Notes in Computer Science Vol. 324, 339-350 (1988).
9. U.Goltz, R.van Glabbeek: Refinement of Actions in Causality Based Models. Springer-Verlag, Lecture Notes in Computer Science Vol.430, 267-300 (1989).
10. M.B.Hennessy: Algebraic Theory of Processes. The MIT Press (1988).
11. C.A.R.Hoare: *Communicating Sequential Processes.* Prentice Hall (1985).
12. R.Janicki, M.Koutny: Structure of Concurrency. Theoretical Computer Science Vol. 112, 5-52 (1993).
13. R.Janicki, P.E.Lauer: *Specification and Analysis of Concurrent Systems - the COSY Approach.* Springer-Verlag, EATCS Monographs on Theoretical Computer Science (1992).
14. A.Kiehn: Petri Net Systems and their Closure Properties. Springer-Verlag LNCS Vol.424, 306-328 (1990).
15. M.Koutny and E.Best: Operational Semantics for the Box Algebra. Draft paper (1995).
16. K.Jensen: Coloured Petri Nets. Basic Concepts. EATCS Monographs on Theoretical Computer Science, Springer-Verlag (1992).
17. P.E.Lauer: Path Expressions as Petri Nets, or Petri Nets with Fewer Tears. Technical Report MRM/70, Computing Laboratory, University of Newcastle upon Tyne (1974).

18. R.Milner: *Communication and Concurrency.* Prentice Hall (1989).
19. E.R.Olderog: *Nets, Terms and Formulas.* Cambridge Tracts in Theoretical Computer Science 23 (1991).
20. G.Plotkin: Domains. Department of Computer Science, University of Edinburgh (1983).
21. W.Reisig: *Petri Nets. An Introduction.* EATCS Monographs on Theoretical Computer Science Vol. 3, Springer-Verlag (1985).
22. D.Taubner: *Finite Representation of CCS and TCSP Programs by Automata and Petri Nets.* Springer-Verlag, Lecture Notes in Computer Science Vol. 369 (1989).

Implicit Definability and Infinitary Logic in Finite Model Theory

(Extended Abstract)

Anuj Dawar[*1], Lauri Hella[2], Phokion G. Kolaitis[**3]

[1] Dept. of Comp. Science, Univ. of Wales, Swansea, Swansea SA2 8PP, U.K.
[2] Dept. of Mathematics, P.O. Box 4, 00014 Univ. of Helsinki, Finland.
[3] Comp. and Info. Sciences, University of California, Santa Cruz, CA 95064, U.S.A.

Abstract. We study the relationship between the infinitary logic $L^\omega_{\infty\omega}$ with finitely many variables and implicit definability in effective fragments of $L^\omega_{\infty\omega}$ on finite structures. We show that fixpoint logic has strictly less expressive power than first-order implicit definability. We also establish that the separation of fixpoint logic from a certain restriction of first-order implicit definability to $L^\omega_{\infty\omega}$ is equivalent to the separation of PTIME from UP $\cap$ co-UP. Finally, we delineate the relationship between partial fixpoint logic and implicit definability in partial fixpoint logic on finite structures.

1 Introduction and Summary of Results

Finite model theory is the study of first-order logic and extensions of first-order logic on finite structures. Finite model theory has enjoyed steady growth through a continuous interaction with computational complexity, database theory, and combinatorics. In particular, the interaction between finite model theory and computational complexity has resulted in the development of descriptive complexity, an area of research whose main aim is to classify algorithmic problems according to the resources needed to express such problems in various logical formalisms. One of the most notable successes of descriptive complexity is the discovery that essentially all major complexity classes have natural characterizations in terms of logical expressibility on finite structures (cf. [Gur88, Imm89]). The prototypical result in this vein is Fagin's theorem [Fag74] which asserts that a class of finite structures is in NP if and only if it is definable by a formula of existential second-order logic. Quite often, certain logical characterizations of other major complexity classes are valid only on classes of ordered structures. Thus, Immerman [Imm86] and, independently, Vardi [Var82] showed that PTIME coincides with fixpoint logic LFP on the class of all ordered finite structures. Similarly, Vardi [Var82] established that partial fixpoint logic PFP captures PSPACE on ordered structures (cf. also [AV89]). However, on the class

---

[*] Research supported by EPSRC grant GR/H 81108
[**] Research of this author partially supported by NSF Grants INT-9024681 and CCR-9307758

of all finite structures the collection of queries expressible in fixpoint logic LFP is properly contained in PTIME, and the same is true of partial fixpoint PFP and PSPACE (cf. [CH82]). In spite of these limitations, both LFP and PFP constitute powerful extensions of first-order logic on the class of all finite structures, as they incorporate recursion and iteration mechanisms in the form of fixpoint operators. For this reason, they have been studied extensively in database theory and have become a standard for comparing and calibrating the expressive power of other database query languages (cf. [Cha88]).

A subsequent turning point in the study of the connections between logic and complexity was the paper by Abiteboul and Vianu [AV91], which established that the problem of separating fixpoint logic LFP from partial fixpoint logic PFP on the class of all finite structures is equivalent to the separation of PTIME from PSPACE. This result was further extended by Abiteboul, Vardi, and Vianu [AVV92] to results concerning a whole range of complexity classes between PTIME and EXPTIME. For this, they introduced a variety of logics equipped with suitable fixpoint operators and showed that inclusions and separations between these fixpoint logics on the class of all finite structures mirror inclusions and separations between corresponding complexity classes. In other words, each of these fixpoint logics gives rise to what is called a *relational* complexity class on all finite structures; such a class embodies the salient properties of a corresponding complexity class, but it is defined in a machine-independent and order-invariant way.

The aforementioned results in [AV91, AVV92] exploit the fact that logics with fixpoint operators can be viewed as effective fragments of the infinitary logic $L^\omega_{\infty\omega}$ with finitely many variables. During the past several years, the study of $L^\omega_{\infty\omega}$ became a focal point of research in finite model theory. One of the reasons for the interest in $L^\omega_{\infty\omega}$ is that its expressive power has an elegant characterization in terms of two-player pebble games [Bar77, Imm82]. This has made it possible to apply game-theoretic techniques in order to derive lower bounds for expressibility in $L^\omega_{\infty\omega}$, as well as to obtain positive structural results, such as 0-1 laws (cf. [KV92c]). Moreover, as demonstrated in [AV91, DLW92], there is a deeper connection between $L^\omega_{\infty\omega}$ and LFP, a connection which is behind the phenomenon that outstanding open problems in complexity theory turn out to be equivalent to separation questions between fragments of $L^\omega_{\infty\omega}$ on the class of all finite structures.

Research in finite model theory has also focused on the study of concepts from the classical model theory of first-order logic in the context of finite structures. While early work in this direction centered on the failure of many classical theorems when restricted to finite models (cf. [Gur84]), in recent years some effort has been devoted to exploring the positive aspects of this restriction. By taking as a point of departure Gurevich's [Gur84] result that Beth's definability theorem fails in the finite case, Kolaitis [Kol90] embarked on a systematic investigation of first-order implicit definability on finite structures. This investigation was carried out by studying the expressive power and computational content of the class IMP(FO) of queries Q that are members of a sequence $(Q, Q_1, \ldots, Q_n)$

of queries implicitly definable by some first-order sentence $\varphi(S, S_1, \ldots, S_n)$. In [Kol90], it was shown that LFP $\subseteq$ IMP(FO), which means that on the class of all finite structures first-order implicit definability is at least as expressive as fixpoint logic. The question of whether the inclusion LFP $\subseteq$ IMP(FO) is proper on the class of all finite structures was left as an open problem. On the other hand, it was also shown in [Kol90] that on the class of all ordered finite structures IMP(FO) coincides with UP $\cap$ co-UP, where UP is the class of NP problems computable by unambiguous Turing machines.

In the present paper, we study the interaction between infinitary logic $L^\omega_{\infty\omega}$ and implicit definability in first-order logic, fixpoint logic, and partial fixpoint logic. This investigation touches upon all directions of research in finite model theory mentioned above. Thus, we view our work as a continuation of the study of implicit definability on finite structures and as a contribution to both descriptive complexity and the model theory of $L^\omega_{\infty\omega}$. Building on recent work of Gurevich and Shelah [GS94], we show that there is a query in IMP(FO) which is not definable in $L^\omega_{\infty\omega}$. This result implies that on the class of all finite structures the inclusion LFP $\subseteq$ IMP(FO) is indeed a proper one. After this, we introduce the class IMP(FO)$|L^\omega_{\infty\omega}$ of queries that are obtained by restricting first-order implicit definition to sequences of queries all of which are in $L^\omega_{\infty\omega}$. We investigate the computational strength of the class IMP(FO)$|L^\omega_{\infty\omega}$ and demonstrate that it is a relational complexity class corresponding to UP $\cap$ co-UP. In particular, we establish that the separation of LFP from IMP(FO)$|L^\omega_{\infty\omega}$ on the class of all finite structures is equivalent to the separation of PTIME from UP $\cap$ co-UP. This result provides a new instance of the phenomenon that a complexity theoretic problem is equivalent to a problem about fragments of $L^\omega_{\infty\omega}$ on the class of all finite structures. Finally, we study implicit definability in partial fixpoint logic and determine the exact relationship between the classes PFP, IMP(PFP)$|L^\omega_{\infty\omega}$, and IMP(PFP) $\cap L^\omega_{\infty\omega}$.

2 Background Information

A *vocabulary* σ is a finite sequence $(R_1, \ldots, R_m)$ of relation symbols of arities $r_1, \ldots, r_m$, respectively. As usual, a *structure* $\mathbf{A}$ *over* σ is a tuple $\mathbf{A} = (A, R_1^{\mathbf{A}}, \ldots, R_m^{\mathbf{A}})$, where A is a non-empty set, called the *universe* of $\mathbf{A}$, and $R_i^{\mathbf{A}} \subseteq A^{r_i}$ is a relation on A of arity r_i, $1 \leq i \leq m$. From now on, we assume that all structures studied are finite.

Let k be positive integer. A *k-ary query* Q is a function which for every structure $\mathbf{A}$ returns as value a k-ary relation $Q(\mathbf{A})$ on the universe of $\mathbf{A}$ such that $Q(\mathbf{A})$ is preserved under isomorphisms, i.e., if $\mathbf{A}$ and $\mathbf{B}$ are structures and g is an isomorphism from $\mathbf{A}$ to $\mathbf{B}$, then g is also an isomorphism from $(A, Q(\mathbf{A}))$ to $(B, Q(\mathbf{B}))$.

We say that a *k-ary* query Q is *definable in a logic L* if there is a formula $\varphi(x_1, \ldots, x_k)$ of L such that for every structure $\mathbf{A}$ we have $Q(\mathbf{A}) = \{(a_1, \ldots, a_k) \in A^k : \mathbf{A} \models \varphi(a_1, \ldots, a_k)\}$.

If L is a logic, then the notation L will also be used to denote the class of queries definable in L. If L_1 and L_2 are two logics, then $L_1 \subseteq L_2$ means that the class of queries definable in L_1 is contained in the class of queries definable in L_2, while $L_1 \subset L_2$ means that the above containment is a *proper* one.

The preceding definitions relativize naturally to a class $\mathcal{C}$ of structures that is closed under isomorphisms, so that we have the concept of a *query on $\mathcal{C}$* and the notation $L(\mathcal{C})$ for the class of queries on $\mathcal{C}$ definable in L.

2.1 Fixpoint Logic and Partial Fixpoint Logic

Let $\varphi(x_1, \ldots, x_n, S)$ be a first-order formula over a vocabulary $\sigma \cup \{S\}$, where S is an n-ary relation symbol that is not in σ, and let $\mathbf{A}$ be a structure over σ. The formula φ gives rise to an operator $\Phi(S)$ from n-ary relations on the universe A of $\mathbf{A}$ to n-ary relations on A, where

$$\Phi(S) = \{(a_1, \ldots, a_n) : \mathbf{A} \models \varphi(a_1, \ldots, a_n, S)\}$$

for every n-ary relation S on A. Every such operator $\Phi(S)$ can be iterated and, thus, it gives rise to the sequence of *stages* Φ^m, $m \geq 1$, defined by the induction: $\Phi^1 = \Phi(\emptyset)$, $\Phi^{m+1} = \Phi(\Phi^m)$.

If the operator $\Phi(S)$ is *monotone* (which means that $\Phi(S_1) \subseteq \Phi(S_2)$, whenever $S_1 \subseteq S_2$), then by the Knaster-Tarski Theorem the operator $\Phi(S)$ has a *least fixpoint*, which is denoted by φ^∞. Moreover, the least fixpoint φ^∞ of $\Phi(S)$ can be obtained as the "limit" of the stages Φ^m, $m \geq 1$, that is to say for every structure $\mathbf{A}$ there is an integer m_0 such that $\varphi^\infty = \Phi^{m_0} = \Phi^m$ on $\mathbf{A}$ for every $m \geq m_0$. Monotonicity is a semantic property which follows from the syntactic property of positivity. More precisely, we say that a formula $\varphi(x_1, \ldots, x_n, S)$ is *positive in S* if every occurrence of S in φ is within an even number of negations. For such formulas, the associated operator $\Phi(S)$ is monotone and, thus, the least fixpoint φ^∞ exists.

Definition 1. *Fixpoint Logic* LFP is the extension of FO that is closed under the operations of first-order logic and the *least fixpoint rule*: if $\varphi(x_1, \ldots, x_n, S)$ is a positive formula, then $\varphi^\infty(x_1, \ldots, x_n)$ is also a formula of LFP.

If the operator Φ is not monotone, then on a given structure $\mathbf{A}$ either there is a positive integer m such that $\Phi^m = \Phi^{m+1}$ or the sequence Φ^m, $m \geq 1$, of stages cycles without ever yielding a fixpoint of Φ. If $\varphi(x_1, \ldots, x_n, S)$ is a first-order formula (not necessarily a positive one), then we define the *partial fixpoint φ^∞* of φ to be a stage Φ^m such that $\Phi^m = \Phi^{m+1}$, if such a stage exists, or the empty set $\emptyset$, otherwise.

Definition 2. *Partial Fixpoint Logic* PFP is the extension of FO that is closed under the operations of first-order logic and the *partial fixpoint rule*: if $\varphi(x_1, \ldots, x_n, S)$ is a formula, then $\varphi^\infty(x_1, \ldots, x_n)$ is also a formula of PFP.

Fixpoint logic on finite structures was first studied by Chandra and Harel [CH82]. Partial fixpoint logic was introduced by Abiteboul and Vianu [AV89], who showed that the class of PFP queries coincides with the class of queries that are expressible in first-order logic augmented with WHILE-*looping*. The latter is a powerful iteration construct investigated by Chandra and Harel [CH82].

It is easy to verify that on the class of all finite structures LFP $\subseteq$ PTIME and PFP $\subseteq$ PSPACE. The above containments are proper, since counting properties, such as "even cardinality", are not expressible in LFP or in PFP (cf. [CH82]). In contrast, on classes of *ordered* structures LFP captures PTIME and PFP captures PSPACE, where $\mathcal{C}$ is a class of *ordered* structures over a vocabulary σ if σ contains a binary relation symbol $<$ and for every $\mathbf{A}$ in $\mathcal{C}$ the relation $<^{\mathbf{A}}$ is a linear order on A. Immerman [Imm86] and Vardi [Var82] showed independently that if $\mathcal{C}$ is a class of ordered structures, then LFP($\mathcal{C}$) = PTIME on $\mathcal{C}$; moreover, Vardi [Var82] showed that PFP($\mathcal{C}$) = PSPACE on $\mathcal{C}$ (cf. also [AV89]).

Since the partial fixpoint of a positive formula coincides with its least fixpoint, it is clear that LFP $\subseteq$ PFP. Chandra and Harel [CH82] raised the problem of showing that LFP is properly contained in PFP on the class of all finite structures. Abiteboul and Vianu [AV91] established that this separation amounts to separating PTIME from PSPACE, i.e., LFP = PFP if and only if PTIME = PSPACE.

2.2 Infinitary Logics with Finitely Many Variables

Powerful extensions of first-order logic result by augmenting the syntax with infinitary connectives. In particular, $L_{\infty\omega}$ is the extension of first-order logic obtained by allowing infinite disjunctions and conjunctions, while keeping the quantifier strings finite. This logic, however, is too powerful to be of any use in finite model theory, since every class of finite structures that is closed under isomorphisms is definable by a sentence of $L_{\infty\omega}$. On the other hand, the fragments of $L_{\infty\omega}$ obtained by restricting the number of variables have turned out to be of importance in finite model theory (cf. [KV92c]).

Definition 3. Let k be a positive integer.

- The *infinitary logic* $L^k_{\infty\omega}$ *with k variables* consists of all formulas of $L_{\infty\omega}$ with at most k distinct variables.
- The infinitary logic $L^\omega_{\infty\omega}$ consists of all formulas of $L_{\infty\omega}$ with a finite number of distinct variables, i.e., $L^\omega_{\infty\omega} = \bigcup_{k=1}^\infty L^k_{\infty\omega}$.
- We write $L^k_{\omega\omega}$ for the collection of all first-order formulas with at most k variables. Thus, FO $= \bigcup_{k=1}^\infty L^k_{\omega\omega}$.

The infinitary logic $L^\omega_{\infty\omega}$ subsumes both fixpoint logic and partial fixpoint logic on finite structures, i.e., LFP $\subseteq$ PFP $\subseteq$ $L^\omega_{\infty\omega}$ (cf. [KV92b]). Moreover, on the class of all finite structures both LFP and PFP are properly contained in $L^\omega_{\infty\omega}$, since $L^\omega_{\infty\omega}$ can express non-recursive queries. Although LFP constitutes a small effective fragment of $L^\omega_{\infty\omega}$, it plays a key role in the study of the model theory of $L^\omega_{\infty\omega}$.

Definition 4. Let k be a positive integer and let **A** be a structure.

- If $(a_1, \ldots, a_k)$ is a tuple from the universe of **A**, then the k-*type* of $(a_1, \ldots, a_k)$ is the collection of all formulas ψ of $L^k_{\infty\omega}$ such that $\mathbf{A} \models \psi(a_1, \ldots, a_k)$.
- If $(a_1, \ldots, a_k)$ and $(b_1, \ldots, b_k)$ are two tuples having the same k-type, then we say that $(a_1, \ldots, a_k)$ is $L^k_{\infty\omega}$-*equivalent* to $(b_1, \ldots, b_k)$ on **A** and write $(\mathbf{A}, a_1, \ldots, a_k) \equiv^k_{\infty\omega} (\mathbf{A}, b_1, \ldots, b_k)$.

It can be shown that each k-type is equivalent to a formula of $L^k_{\infty\omega}$. Dawar et al. [DLW92] established that each k-type is actually definable by a formula of $L^k_{\omega\omega}$. Moreover, it was shown in [DLW92, KV92a] (and implicitly in [AV91]) that for each $k \geq 1$ there is a formula of fixpoint logic LFP that defines $L^k_{\infty\omega}$-equivalence. These results make use of the fact that $L^k_{\infty\omega}$-equivalence has a combinatorial characterization in terms of k-pebble games (cf. [Bar77, Imm82]).

Abiteboul and Vianu [AV91] showed that for every $k \geq 1$ there is a formula of fixpoint logic that defines a linear order $<_k$ on (the representatives of) the equivalence classes of $L^k_{\infty\omega}$-equivalence (cf. also [DLW92]). This is the main technical tool used by Abiteboul and Vianu [AV91] in establishing that the separation of LFP from PFP is equivalent to the separation of PTIME from PSPACE.

2.3 Implicit Definability

So far, we have been considering *explicit definability* of queries in a logic L. The concept of *implicit definability* in a logic L arises when the value of a query is the unique relation that satisfies a sentence of L.

Definition 5. Let L be a logic, and $(Q_1, \ldots, Q_n)$ a sequence of queries. We say that the sequence $(Q_1, \ldots, Q_n)$ is *implicitly definable in L* if there is a sentence $\varphi(P_1, \ldots, P_n)$ of L over the vocabulary $\sigma \cup \{P_1, \ldots, P_n\}$ such that for every structure **A** the sequence $(Q_1(\mathbf{A}), \ldots, Q_n(\mathbf{A}))$ is the only sequence of relations $(S_1, \ldots, S_n)$ on A such that $\mathbf{A} \models \varphi(S_1, \ldots, S_n)$.

We write IMP(L) to denote the collection of all queries Q such that $Q = Q_1$ for some sequence $(Q_1, \ldots, Q_n)$ of queries which is implicitly definable in L.

The classical Beth definability theorem (cf. [CK90]) asserts that if a query is implicitly definable in first-order logic on all *finite and infinite* structures over a vocabulary σ, then it is also first-order definable on all finite and infinite structures over σ. Gurevich [Gur84] pointed out that Beth's theorem fails on finite structures, which means that the class FO of first-order definable queries is properly contained in IMP(FO). The expressive power and computational strength of first-order implicit definability on finite structures was investigated by Kolaitis [Kol90], where it was shown that LFP $\subseteq$ IMP(FO) and IMP(FO) = IMP(LFP). Moreover, in [Kol90] it was established that if $\mathcal{C}$ is a class of ordered structures, then IMP(FO)($\mathcal{C}$) = UP $\cap$ co-UP on $\mathcal{C}$, where UP is the class of *unambiguous* NP problems introduced by Valiant [Val76]. Thus, separating LFP from IMP(FO) on ordered structures is equivalent to separating PTIME from UP $\cap$ co-UP.

In view of the above, in [Kol90] the question was raised of whether LFP is properly contained in IMP(FO) on the class of all finite structures. This question is settled in the next section.

3 Fixpoint Logic vs. First-Order Implicit Definability

Recently, Gurevich and Shelah [GS94] constructed certain finite structures, called *multipedes*, and used them to refute a conjecture of Dawar [Daw93] stating that if C is a first-order definable class of rigid finite structures, then there is a formula of fixpoint logic that defines a linear order on every member of C. In the next definition, we present a somewhat different, though essentially equivalent, variant of multipedes.

Definition 6. A *multipede* is a structure $(U, \preceq, L, H_1, H_2, \varepsilon)$ such that the following hold.

- The universe U is the disjoint union of three sets S, F, and P.
- $\preceq$ is a linear order on S.
- $L \subseteq F \times S$ is a binary relation with the following properties: for every $f \in F$ there is exactly one s such that $(f, s) \in L$; for every $s \in S$ there are exactly two elements f_1 and f_2 in F such that $(f_i, s) \in L$, $i = 1, 2$.
  If $(f_i, s) \in L$, $i = 1, 2$, then we say that f_i is a *foot* of s; we also say that s is the *segment* corresponding to f_i, $i = 1, 2$.
- There is a one-to-one correspondence between P and the powerset of S; moreover, ε is the corresponding membership relation between elements of S and elements of P.
- H_1 is a ternary relation on S such that if $(s_1, s_2, s_3) \in H_1$, then the three elements are distinct and for every permutation π of the set $\{1, 2, 3\}$ we have that $(s_{\pi(1)}, s_{\pi(2)}, s_{\pi(3)}) \in H_1$.
  We treat H_1 as a collection of three-element subsets of S.
- H_2 is a ternary relation on F satisfying the following conditions: for every triple in H_2, the triple of corresponding segments is in H_1; for every $h \in H_1$, exactly four triples are in H_2 out of the eight triples that can be formed by choosing exactly one foot of each element of h; moreover, these four triples are such that any one of them can be obtained from any other by exchanging the feet of exactly two elements of h.

A multipede is *odd* if for every non-empty subset X of S there is an $h \in H_1$ such that either $h \subseteq X$ or $|X \cap h| = 1$. Gurevich and Shelah [GS94] showed that the class of odd multipedes is first-order definable and that every odd multipede is a *rigid* structure, which means that the identity function is its only automorphism. Moreover, using a difficult probabilistic construction, Gurevich and Shelah established the following result.

Theorem 7. [GS94] *There is no formula of* $L^\omega_{\infty\omega}$ *that defines a linear order on the class of odd multipedes.*

In contrast, we can show the following:

Lemma 8. *There is a first-order formula $\varphi(\leq)$ which implicitly defines a linear order on all odd multipedes.*

Proof: *(Sketch)* We can write a first-order formula stating that:

1. all elements of S occur before any element of P and all elements of P are before any element of F in the linear order $\leq$;
2. $\leq$ coincides on S with the linear order $\preceq$, while on P it is the linear order generated from $\preceq$ by the lexicographic order on the powerset of S;
3. for every two elements f and f' of F we have that $f \leq f'$ if the segment corresponding to f in the order $\preceq$ occurs before the segment corresponding to f' in $\preceq$.

Any two linear orders that satisfy the above conditions are distinguished only by the choice of ordering the feet of individual segments. That is, for any two such orders there is a set $X \subseteq S$ such that one order is obtained from the other by exchanging the feet of exactly the segments in X.

We associate with each order $\leq$ satisfying the above conditions a subset G of H_1 defined as follows. For each $h = \{a, b, c\} \in H_1$, let (a_1, a_2), (b_1, b_2), and (c_1, c_2) be the pairs of feet of a, b, and c respectively, where $a_1 \leq a_2$, $b_1 \leq b_2$ and $c_1 \leq c_2$. We then put $h \in G$ if and only if $\{a_1, b_1, c_1\} \in H_2$.

It can be verified that two distinct choices of order cannot yield the same set G, since otherwise the set X as defined above would witness that the multipede is not odd. Now, we can write a first-order formula $\varphi(\leq)$ asserting that the set G obtained from the order $\leq$ is lexicographically minimal among all linear orders satisfying conditions 1–3 above. ∎

From Theorem 7 and Lemma 8 it follows immediately that $\text{IMP}(\text{FO}) \not\subseteq L^\omega_{\infty\omega}$. Thus, we have established the following separations.

Theorem 9. *The class of fixpoint queries is properly contained in the class of first-order implicitly definable queries on finite structures, i.e., $\text{LFP} \subset \text{IMP}(\text{FO})$. Moreover, $\text{IMP}(\text{FO}) \cap L^\omega_{\infty\omega} \subset \text{IMP}(\text{FO})$.*

Theorem 9 settles the question raised in [Kol90] about the relationship between fixpoint logic and first-order implicit definability on finite structures. On the other hand, an examination of the proof that $\text{LFP} \subseteq \text{IMP}(\text{FO})$ in [Kol90] reveals that a seemingly stronger containment holds, namely for every fixpoint query Q there is another fixpoint query Q' such that the pair (Q, Q') is first-order implicitly definable on finite structures. In other words, not only is every fixpoint query a member of a sequence which is first-order implicitly definable, but actually the other members of this sequence can be taken to be fixpoint queries (and, hence, $L^\omega_{\infty\omega}$ queries) as well. This observation motivates the following definition.

Definition 10. If L is one of the logics FO, LFP, PFP, then $\text{IMP}(L)|L^\omega_{\infty\omega}$ is the class of queries Q that are definable in $L^\omega_{\infty\omega}$ and such that there are $L^\omega_{\infty\omega}$-definable queries $Q_1, \ldots, Q_n$ and a formula φ of L which implicitly defines the sequence of queries $(Q, Q_1, \ldots, Q_n)$.

It is important to differentiate between $\text{IMP}(L)|L^\omega_{\infty\omega}$ and $\text{IMP}(L) \cap L^\omega_{\infty\omega}$. It is clear that $\text{IMP}(L)|L^\omega_{\infty\omega} \subseteq \text{IMP}(L) \cap L^\omega_{\infty\omega}$, but, as will be shown later on, this containment may be a proper one. Thus, a query Q may be both in $L^\omega_{\infty\omega}$ and in $\text{IMP}(L)$, but there may be no L-implicitly definable sequence $(Q, Q_1, \ldots, Q_n)$ of queries such that $Q_1, \ldots, Q_n$ are *all* $L^\omega_{\infty\omega}$-definable. It should also be noted that if $\mathcal{C}$ is a class of ordered structures, then $\text{IMP}(L)|L^\omega_{\infty\omega}(\mathcal{C})$ is the same as $\text{IMP}(L)(\mathcal{C})$, since in this case every query on $\mathcal{C}$ is definable in $L^\omega_{\infty\omega}$.

By combining Theorem 9 with the remarks preceding Definition 10, we obtain the following picture

$$\text{LFP} \subseteq \text{IMP}(\text{FO})|L^\omega_{\infty\omega} \subseteq \text{IMP}(\text{FO}) \cap L^\omega_{\infty\omega} \subset \text{IMP}(\text{FO}).$$

A natural question arises whether the containment of LFP in $\text{IMP}(\text{FO})|L^\omega_{\infty\omega}$ is proper. In view of the above, this can be construed as the "right" way or the "fair" way to ask if the containment of fixpoint logic in first-order implicit definability is a proper one. It should be pointed out that an answer to this new question cannot be obtained from the proof of Theorem 9, because the $\text{IMP}(\text{FO})$ query that defines a linear order on odd multipedes is not definable in $L^\omega_{\infty\omega}$. Our next result shows that the problem of separating LFP from $\text{IMP}(\text{FO})|L^\omega_{\infty\omega}$ is equivalent to a major open problem in computational complexity that we encountered earlier.

Theorem 11. $\text{LFP} = \text{IMP}(\text{FO})|L^\omega_{\infty\omega}$ *if and only if* $\text{PTIME} = \text{UP} \cap \text{co-UP}$.

Proof: (*Hint*) The proof uses the methods developed by Abiteboul and Vianu [AV91] to show that $\text{LFP} = \text{PFP}$ if and only if $\text{PTIME} = \text{PSPACE}$. More specifically, by restricting first-order implicit definability to $L^\omega_{\infty\omega}$, we can work with $L^k_{\infty\omega}$-equivalence classes and take advantage of the existence of a fixpoint definable linear order on k-types. In addition, we apply the result in [Kol90] asserting that $\text{IMP}(\text{FO}) = \text{UP} \cap \text{co-UP}$ on classes of ordered structures. ∎

Several remarks are in order now. First, it should be pointed out that results similar to Theorem 11 can be established about the relationship of $\text{IMP}(\text{FO})|L^\omega_{\infty\omega}$ to other effective fragments of $L^\omega_{\infty\omega}$. In particular, using the same technique, we can show that $\text{IMP}(\text{FO})|L^\omega_{\infty\omega} = \text{PFP}$ if and only if $\text{UP} \cap \text{co-UP} = \text{PSPACE}$. Thus, Theorem 11 and its extensions provide a new manifestation that separation problems between complexity classes are equivalent to separations of appropriate fragments of $L^\omega_{\infty\omega}$ on the class of all finite structures, even if the fragments of $L^\omega_{\infty\omega}$ do not capture the corresponding complexity classes. Moreover, these results indicate that the class $\text{IMP}(\text{FO})|L^\omega_{\infty\omega}$ should be regarded as the "relational" version of the complexity class $\text{UP} \cap \text{co-UP}$.

Although the separation of LFP from $\text{IMP}(\text{FO})|L^\omega_{\infty\omega}$ does not appear to be within reach, it turns out that we can separate LFP from $\text{IMP}(\text{FO}) \cap L^\omega_{\infty\omega}$. In order to achieve this, we need to find a query Q which is both in $L^\omega_{\infty\omega}$ and in $\text{IMP}(\text{FO})$, but whose implicit definition appears to require in a crucial way an auxiliary query that is not definable in $L^\omega_{\infty\omega}$.

Observe that if $\mathcal{C}$ is a class of structures in which there is a linear order definable in $\text{IMP}(\text{FO})$, then every query on $\mathcal{C}$ that is computable in $\text{UP} \cap \text{co-UP}$

is also definable in IMP(FO), using the linear order as an "auxiliary query". Consequently, on such a class $\mathcal{C}$ we have the containment $L_{\infty\omega}^\omega \cap \text{PTIME} \subseteq \text{IMP(FO)} \cap L_{\infty\omega}^\omega$. In [DLW92], it is proved that LFP is properly contained in $L_{\infty\omega}^\omega \cap \text{PTIME}$ by constructing an appropriate query on complete binary trees. The size of each complete binary tree is exponential in its height, but, by a result of Lindell [Lin91], the number of distinct automorphism types, and, a fortiori, the number of distinct k-types is only polynomial in the height of the tree. However, it is obvious that there is no definable linear order on complete binary trees, as they are not rigid. Nevertheless, we are able to combine the necessary properties of binary trees with those of multipedes in order to obtain structures that are rigid, but where the number of distinct k-types is logarithmic in the size of the structure. In these structures, called *tree multipedes*, each segment, instead of having two feet, is the root of a complete binary tree. Moreover, we replicate the relations H_1 and H_2 at all levels of the resulting forest.

As with multipedes, we can show that every odd tree multipede is rigid, that the class of odd tree multipedes is definable in LFP and that there is an FO formula that implicitly defines a linear order in every odd tree multipede. We can also show that for every k and n, there is an odd tree multipede of height n in which no two siblings are distinguished by a formula of $L_{\infty\omega}^\omega$. Together these facts enable us to establish the following theorem.

Theorem 12. LFP $\subset$ IMP(FO) $\cap L_{\infty\omega}^\omega$.

We conjecture that the above Theorem 12 can be strengthened to show that IMP(FO)$|L_{\infty\omega}^\omega \subset \text{IMP(FO)} \cap L_{\infty\omega}^\omega$. At present, we can obtain this separation only modulo the complexity-theoretic assumption that UP $\cap$ co-UP is properly contained in UEXP $\cap$ co-UEXP.

4 Implicit Definability in Partial Fixpoint Logic

Next, we focus on partial fixpoint logic and implicit definability in this logic. It should be clear from the definitions that the following sequence of inclusions holds:

$$\text{PFP} \subseteq \text{IMP(PFP)}|L_{\infty\omega}^\omega \subseteq \text{IMP(PFP)} \cap L_{\infty\omega}^\omega \subseteq \text{IMP(PFP)}.$$

As it turns out, it is possible to determine which of these inclusions are proper.

Theorem 13. PFP $=$ IMP(PFP)$|L_{\infty\omega}^\omega$.

Proof: (*Hint*) Abiteboul and Vianu [AV91] showed that there is a formula of LFP that defines a linear order $<_k$ on the representatives of $\equiv_{\infty\omega}^k$-equivalence classes. We use this formula to construct a formula φ of PFP such that on every structure the stages of φ cycle through all $L_{\infty\omega}^k$-definable relations. We can then check at each stage whether the relation satisfies the condition specified by the implicit definition. ∎

Theorem 13 can be viewed as a Beth definability theorem for partial fixpoint logic on finite structures. As such, it should be compared and contrasted with the earlier Theorem 11, which reveals that the analogous theorem for fixpoint logic is true if and only if UP ∩ co-UP collapses to PTIME, an unlikely complexity-theoretic eventuality. Thus, Theorems 11 and 13 indicate yet another difference between fixpoint logic and partial fixpoint logic on finite structures.

Finally, we complete the picture for partial fixpoint logic by establishing the following proper containments.

Theorem 14. $PFP \subset IMP(PFP) \cap L^\omega_{\infty\omega}$ *and* $IMP(PFP) \cap L^\omega_{\infty\omega} \subset IMP(PFP)$.

Proof: (*Hint*) The first separation is established along the lines of Theorem 12 by showing that the inclusion $PFP \subseteq L^\omega_{\infty\omega} \cap PSPACE$ is proper on the class of odd tree multipedes. The second separation follows from Theorem 9. ∎

5 Conclusions

In this paper, we investigate the relationship between the infinitary logic $L^\omega_{\infty\omega}$ and implicit definability in effective fragments of this logic on finite structures.

We were able to obtain the following sharp picture that delineates the relationship between partial fixpoint logic and implicit definability in partial fixpoint logic on finite structures:

$$PFP = IMP(PFP)|L^\omega_{\infty\omega} \subset IMP(PFP) \cap L^\omega_{\infty\omega} \subset IMP(PFP).$$

Concerning the relationship between fixpoint logic and implicit definability, we establish that $LFP \subseteq IMP(FO)|L^\omega_{\infty\omega} \subseteq IMP(FO) \cap L^\omega_{\infty\omega} \subset IMP(FO)$ and $LFP \subset IMP(FO) \cap L^\omega_{\infty\omega}$.

This picture is not as sharp as the one for partial fixpoint logic. However, we exhibited good reasons that prevent us from making it sharper, since we also demonstrated that further separations are intimately related to outstanding open problems in complexity theory, such as the separation of PTIME from UP ∩ co-UP.

References

[AV89] S. Abiteboul and V. Vianu. Fixpoint extensions of first-order logic and Datalog-like languages. In *Proc. 4th IEEE LICS*, pages 71–79, 1989.

[AV91] S. Abiteboul and V. Vianu. Generic computation and its complexity. In *Proc. 23rd ACM Symp. on Theory of Computing*, pages 209–219, 1991.

[AVV92] S. Abiteboul, M. Y. Vardi, and V. Vianu. Fixpoint logics, relational machines, and computational complexity. In *Proc. 7th IEEE Conf. on Structure in Complexity Theory*, pages 156–168, 1992.

[Bar77] J. Barwise. On Moschovakis closure ordinals. *Journal of Symbolic Logic*, 42:292–296, 1977.

[CH82]   A. Chandra and D. Harel. Structure and complexity of relational queries. *Journal of Computer and System Sciences*, 25:99–128, 1982.

[Cha88]   A. Chandra. Theory of database queries. In *Proc. 7th ACM Symp. on Principles of Database Systems*, pages 1–9, 1988.

[CK90]   C. C. Chang and H. J. Keisler. *Model Theory*. North Holland, 3rd edition, 1990.

[Daw93]   A. Dawar. *Feasible computation through model theory*. PhD thesis, University of Pennsylvania, Philadelphia, 1993.

[DLW92]   A. Dawar, S. Lindell, and S. Weinstein. *Infinitary logic and inductive definability over finite structures*. Research Report 92-20, Univ. of Pennsylvania, 1992.

[Fag74]   R. Fagin. Generalized first–order spectra and polynomial–time recognizable sets. In R. M. Karp, editor, *Complexity of Computation, SIAM-AMS Proceedings, Vol. 7*, pages 43–73, 1974.

[GS94]   Y. Gurevich and S. Shelah. On finite rigid structures. August 1994. Preprint.

[Gur84]   Y. Gurevich. Toward logic tailored for computational complexity. In M. M. Richter et al., editor, *Computation and Proof Theory, Lecture Notes in Mathematics 1104*, pages 175–216, Springer-Verlag, 1984.

[Gur88]   Y. Gurevich. Logic and the challenge of computer science. In E. Börger, editor, *Current trends in theoretical computer science*, pages 1–57, Computer Science Press, 1988.

[Imm82]   N. Immerman. Upper and lower bounds for first-order expressibility. *Journal of Computer and System Sciences*, 25:76–98, 1982.

[Imm86]   N. Immerman. Relational queries computable in polynomial time. *Information and Control*, 68:86–104, 1986.

[Imm89]   N. Immerman. Descriptive and computational complexity. In J. Hartmanis, editor, *Computational Complexity Theory, Proc. Symp. Applied Math., Vol. 38*, pages 75–91, American Mathematical Society, 1989.

[Kol90]   Ph. G. Kolaitis. Implicit definability on finite structures and unambiguous computations. In *Proc. 5th IEEE LICS*, pages 168–180, 1990.

[KV92a]   Ph. G. Kolaitis and M. Y. Vardi. Fixpoint logic vs. infinitary logic in finite-model theory. In *Proc. 7th IEEE LICS*, pages 46–57, 1992.

[KV92b]   Ph. G. Kolaitis and M. Y. Vardi. Infinitary logic and 0-1 laws. *Information and Computation*, 98:258–294, 1992.

[KV92c]   Ph. G. Kolaitis and M. Y. Vardi. Infinitary logic for computer science. In *Proc. 19th ICALP*, pages 450–473, Springer-Verlag, Lecture Notes in Computer Science 623, 1992.

[Lin91]   S. Lindell. An analysis of fixed point queries on binary trees. *Theoretical Computer Science*, 85:75–95, 1991.

[Val76]   L. Valiant. Relative complexity of checking and evaluating. *Information Processing Letters*, 5:20–23, 1976.

[Var82]   M. Y. Vardi. The complexity of relational query languages. In *Proc. 14th ACM Symp. on Theory of Computing*, pages 137–146, 1982.

The Limit of Split$_n$-Language Equivalence

Walter Vogler *

Institut für Mathematik, Universität Augsburg
D-86135 Augsburg, Germany

Keywords: concurrency, partial order semantics, action refinement

Abstract

Splitting is a simple form of action refinement that may be used to express the duration of actions. In particular, split$_n$ subdivides each action into n phases. Petri nets N_1 and N_2 are split$_n$-language equivalent, if $split_n(N_1)$ and $split_n(N_2)$ are language equivalent. It is known that these equivalences get finer and finer with increasing n.

This paper characterizes the limit of this sequence by a newly defined partial order semantics. This semantics is obtained from the interval-semiword semantics, which is fully abstract for action refinement and language equivalence, by closing it under a special swap operation. The new swap equivalence lies strictly between interval-semiword and step-sequence equivalence.

1 Introduction

Many models of concurrent systems assume that the actions which are performed by the system are instantaneous. The interleaving approach is based on this assumption: in this approach, which is traditionally followed in process algebra, the concurrent execution of two actions is regarded as being equal to performing them in any order. Hence, the behaviour of a system can be described by its language, which is the set of action sequences that the system can perform. But also in Petri net theory, where traditionally a 'true concurrency' approach is preferred, it is usually assumed that a transition firing is instantaneous.

In real life however, actions usually take time. It is often assumed that we can nevertheless work with instantaneous actions: the suggestion is to replace an action with duration by a sequence of two instantaneous actions, one denoting the start, the other the end of the original action, see e.g. [Hen88]. Hence, instead of considering the language of a system N, we first apply the operation $split_2$ to N, which splits each action a into a sequence $a_1 a_2$, and then consider the language of $split_2(N)$.

If this is a sensible treatment of durational actions, then, for each $n > 2$, splitting each action into n phases should give us the same information as splitting into two

---

*This work was partially supported by the DFG-Projekt 'Halbordnungstesten' and the ESPRIT Basic Research Working Group 6067 CALIBAN (CAusal calcuLI BAsed on Nets).

phases. To formulate this expectation mathematically, let the operation $split_n$ replace each action a by a sequence $a_1 \ldots a_n$; let us call systems N and N' $split_n$-language equivalent or simply $split_n$-equivalent, if the split systems $split_n(N)$ and $split_n(N')$ have the same language. Now, the above expectation says: if systems are $split_2$-equivalent, then they are $split_n$-equivalent for all n. Although plausible at first sight, this conjecture has turned out to be wrong. For all n, $split_{n+1}$-equivalence is strictly finer than $split_n$-equivalence [GV91]. Thus, if we want to model the durations of actions by considering them as sequences of several phases, we should work with the limit of these increasingly finer equivalences. The purpose of this paper is to characterize this limit; we will use safe Petri nets as system models, and we will give a characterization in terms of a new partial order semantics of nets, called swap interval semiwords.

Splitting is a simple form of action refinement, an operation for the hierarchical design of systems which has recently found considerable interest. A semantics supports action refinement, if it induces a congruence with respect to this operation. In this context, a specific class of partial orders is useful, so-called *interval orders*; their elements correspond to intervals of real numbers. These interval orders and their representation by suitable sequences play an essential rôle in this paper.

Processes and semiwords give well-established partial order semantics of Petri nets; interval semiwords combine the ideas of semiwords and interval orders. Interval-semiword equivalence is a congruence for action refinement, and it has the following additional property: if nets N and N' are not interval-semiword equivalent, then there exists a refinement ref such that $ref(N)$ and $ref(N')$ are not language equivalent. Hence, if we take the language as a starting point, then interval semiwords make exactly those additional distinctions between nets which are necessary to get a congruence. Expressed in a technical way, interval-semiword equivalence is *fully abstract* with respect to language equivalence and action refinement, i.e. it is the coarsest congruence for action refinement that is finer than language equivalence [Vog92]. Analogously, the purpose of this paper is to determine a fully abstract semantics with respect to language equivalence and splitting. It is clear that interval semiwords induce a congruence for splitting that is finer than language equivalence, but as we will see, it is not the coarsest such congruence – as was already conjectured in [Lar88, Vog92]. At least, interval semiwords will be the starting point for the definition of the new fully abstract semantics.

The corresponding problem on the level of bisimulations (which do not simply consider the sequences of actions that are performed, but also take into account the choices that are possible during a system run) has been solved: similarly as on the language level, $split_{n+1}$-bisimilarity is strictly finer than $split_n$-bisimilarity for all $n \in I\!N$ [GV91]. ST-bisimulation is a variant of bisimulation that corresponds to the interval idea. It is fully abstract with respect to bisimulation and general action refinement [Gla90, Vog93a], but – quite surprisingly – it is also fully abstract with respect to bisimulation and splitting [GL91]. For history-preserving bisimulation, a combination of 'true concurrency' and bisimulation, the situation is even simpler: history-preserving bisimilarity is a congruence for action refinement on systems without internal actions [GG89, BDKP91]; for systems with internal actions, it is not a congruence, but the ST-version is [Vog93a]; on this level, the $split_2$-version coincides with all $split_n$-versions, $n > 2$, and with the ST-version [Dev92].

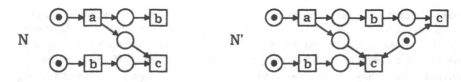

Figure 1

That the situation is different on the language level can be seen from the following example. If we want to describe the runs of the two nets in Figure 1 by partial orders, then we get essentially, that N can perform the (labelled) partial order p of Figure 2, while N' can perform q and p. These partial orders are interval orders, hence N and N' have different interval semiwords.

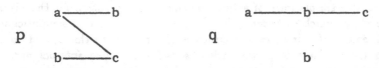

Figure 2

The split nets $split_3(N)$ and $split_3(N')$ can perform the sequence $a_1b_1b_2a_2a_3b_1b_3c_1$. From this we can deduce that N and N' can perform p: in this sequence, we see a starting, then b. Since this first b starts before the end of a, it must be independent of a; it corresponds to the b in the lower left corner of p. Similarly, the second b is independent of the first one; but it comes after a. Now, b_3 is the end of some b; the essential point is, that it must be the end of the first b, since only the first b has finished its second phase at this point. Hence, c comes after a and the first b, but it is independent of the second b.

For q, we cannot find a corresponding sequence in any $split_n(N')$. In such a sequence, we would see the lower b starting during the a-action. Hence, the upper b would start after the lower one, but it would end *before* the lower one, since c starts during the lower b. In such a situation, where one b overtakes the other, we cannot keep the two b's apart by keeping them in different phases – as we have done above for p. Hence, when some b ends, we cannot be sure which one it is. Consequently, we can only deduce that N' can perform q or p, something we already know. Thus, N and N' cannot be differentiated by looking at the languages of any of their splittings, but they are distinguished by interval-semiword equivalence. Hence, this equivalence is different from the limit of the $split_n$-equivalences.

In the above example, the presence of p 'hides' q. Here, and in similar cases, we can obtain q from p by applying a swap-operation to the two b-actions, which exchanges their successors in the partial order p. This example motivates the new semantics which characterizes the limit we are looking for: the swap interval semiwords of a net are obtained by closing the set of interval semiwords under the swap-operation.

The intention of this paper is not only to present a fully abstract semantics for language equivalence and splitting, but also to indicate how this full abstractness can be shown (details can be found in [Vog93b]). Essential for this proof is the representation of interval orders by so-called interval sequences; similar representations have also been studied in their own right, see e.g. [JK93]. In Section 2, some theorems for interval orders and their representations are given, also concerning augmentation and the swap-operation; they are needed for proving the main result, but are also interesting on their own, I believe. Section 3 shortly introduces some Petri net terminology. Our main result in Section 4 is the full abstractness of the new swap-interval-semiword semantics. We also show that the new semantics is closed under prefixes and a suitable form of augmentation; with these pleasing closure properties, we can really prove that the nets of Figure 1 are indeed split$_n$-equivalent for all $n \in I\!N$. With this in turn, we get that the new swap-equivalence lies strictly between interval-semiword and step-sequence equivalence.

2 Representing and Augmenting Interval Orders

In this section, we have a look at interval orders and their representation by sequences; a basic reference for interval orders is [Fis85, Chapter 2]. Furthermore, we study augmentations and the new swap-operation for interval orders and their representations.

A *(labelled) partial order* $p = (E, <, l)$ consists of a finite set E (of *events*), an irreflexive, transitive relation $<$ on E and a *labelling* $l : E \to \Sigma$ of the events with elements from some fixed set Σ of *actions*. We consider isomorphic partial orders as equal. If $e < e'$, e' is a *successor* of e. If $e < e'$ or $e = e'$, we write $e \leq e'$. Two events e and e' are *concurrent*, e co e', if neither $e < e'$ nor $e' < e$ (but possibly $e = e'$).

$E' \subseteq E$ is *left-closed*, if for all $e' < e \in E'$ we have $e' \in E'$. $(E', <', l')$ is a *prefix* of $(E, <, l)$, if E' is a left-closed subset of E, $<'$ equals $<$ restricted to $E' \times E'$, and l' equals l restricted E'. $(E, <, l)$ is a *(proper) augmentation* of $(E, <', l)$, if $<'$ is a (proper) subset of $<$; if in this case $(E, <', l)$ describes a system run, then $(E, <, l)$ describes essentially the same run, but includes some more ordering information about the events. If X is a set of partial orders, we call $p \in X$ *least sequential* in X if p is not a proper augmentation of any $q \in X$.

$(E, <, l)$ is an *interval order*, if for all $a, b, c, d \in E$ we have: if $a < b$ and $c < d$, then $a \leq d$ or $c \leq b$. A (proper) augmentation p of q is a *(proper) interval augmentation* of q, if p is an interval order. The name 'interval order' originates from the following.

Theorem 2.1 *[Fis85]* $(E, <, l)$ *is an interval order if and only if there are closed intervals* $[r_1(e), r_2(e)] \subseteq I\!R$, $e \in E$, *such that all events e and e' satisfy the equivalence:* $e < e'$ *if and only if* $r_2(e) < r_1(e')$. $\qquad\square$

We can think of event e as starting at time $r_1(e)$ and lasting until $r_2(e)$. Event e is smaller than event e' if the interval of e lies completely before the interval of e'.

For a finite set E let $E^{\pm} = E^{+} \,\dot\cup\, E^{-}$ be the disjoint union of two copies of E, where $E^{+} = \{e^{+} \mid e \in E\}$ and $E^{-} = \{e^{-} \mid e \in E\}$. We call e^{+} the *start* of e and e^{-} the *end* of e. A pair (w, l), where w is a sequence in $(E^{\pm})^{*}$ and $l : E \to \Sigma$, is called an *interval sequence (over E)* if for all $e \in E$:

  – w contains e^{+} once;

  – w contains e^{-} at most once, and if e^{-} occurs in w, it occurs after e^{+}.

The interval sequence (w, l) is called *closed*, if e^- occurs in w for each $e \in E$. Interval sequences (w, l) over E and (w', l') over E' are *isomorphic*, if there is some bijection $\beta : E \to E'$ such that $l(e) = l'(\beta(e))$ and w' can be obtained from w by replacing each e^+ by $\beta(e)^+$ and each e^- by $\beta(e)^-$. Since we are not so much interested in the concrete events, but in the structure of their starts and ends, we usually do not distinguish isomorphic interval sequences.

If we have an interval sequence (w, l), we can associate to each event $e \in E$ an interval $[r_1(e), r_2(e)] \subseteq \mathbb{R}$ as follows: $r_1(e)$ is the position of e^+ in w, $r_2(e)$ is the position of e^- in w or – in case that e^- does not occur in w – some value greater than the length of w. In view of Theorem 2.1, we define: an interval sequence (w, l) over E *represents* (or is a *representation* of) an interval order $(E, <, l)$ if for all $e_1, e_2 \in E$ we have $e_1 < e_2$ if and only if e_1^- occurs before e_2^+ in w.

We can think of a representation as a sequential observation where we see events start and end; thus, we observe events as intervals in time and from this we can conclude that some partial order of events occurred, namely the represented interval order.

In [Vog92], it is shown that each interval order has some closed representation. Here, we want to find out about *all* representations. To prove the next theorem, but also for later use, we define two relations $<^+$ and $<^-$ for an interval order $(E, <, l)$.

We have $e_1 <^+ e_2$ if for some e_3 we have $e_1 \ co \ e_3$ and $e_3 < e_2$. Analogously, $e_1 <^- e_2$ if for some e_3 we have $e_1 < e_3$ and $e_3 \ co \ e_2$. These relations are introduced for the following reason: if (w, l) represents $(E, <, l)$ and $e_3 < e_2$, then e_3^- occurs before e_2^+ in w; furthermore, $e_1 \ co \ e_3$ implies that e_1^+ occurs before e_3^-, since otherwise e_3^- would not occur in w or we would have $e_3 < e_1$, which are both wrong. Hence, e_1^+ must occur before e_2^+ if $e_1 <^+ e_2$. Similarly, e_1^- must occur before e_2^- or e_2^- does not occur at all, if $e_1 <^- e_2$.

Using these relations, one can show the next theorem – where the more difficult part in the proof is to show that *all* (closed) representations can be obtained.

Theorem 2.2 *Let $p = (E, <, l)$ be an interval order.*

   i) If (w, l) is a closed representation of p, then exactly all closed representations of p can be obtained from (w, l) by commuting starts and commuting ends. More precisely, define two relations on $(E^{\pm})^*$: $w_1 e_1^+ e_2^+ w_2 \equiv w_1 e_2^+ e_1^+ w_2$ and $w_1 e_1^- e_2^- w_2 \equiv w_1 e_2^- e_1^- w_2$ for all $e_1, e_2 \in E$, $w_1, w_2 \in (E^{\pm})^*$; let $\equiv^*$ be the reflexive-transitive closure of $\equiv$. Then the set of all closed representations of p consists of all (v, l) with $v \equiv^* w$.

   ii) If (w, l) is a closed representation of p, then exactly all representations of p can be obtained from (w, l) by commuting starts, commuting ends (as in i)) and deleting ends from the end of a sequence (i.e. transforming $(w' e^-, l)$ to (w', l)).

This theorem shows that interval orders are more abstract than interval sequences: several interval sequences may represent the same interval order; they differ in the ordering of starts or in the ordering of ends, but these differences are irrelevant for some purposes and are abstracted away in the interval order. 2.2 i) shows that the set of closed representations of some given p corresponds to a Mazurkiewicz trace over $E^{\pm}$, where every two starts and every two ends are independent.

One could also be interested in infinite system runs and, thus, in representing countably infinite interval orders by infinite sequences; for this, see the full version.

Some results in this paper concern all interval augmentations of some given interval order $(E, <, l)$; in order to allow an inductive proof for such a result, we must have that each interval augmentation can be reached by adding one pair at a time to $<$. For partial orders in general, this is known; the essential point here is that with the additions we stay in the class of interval orders.

Theorem 2.3 *Let $p = (E, <, l)$ be an interval order and $p' = (E, <', l)$ be a proper interval augmentation of p. Then there exists a proper interval augmentation $p'' = (E, < \cup \{(e_1, e_2)\}, l)$ of p such that p' is an augmentation of p''.*

We also have a result for the representations of the augmentations of a given interval order; the proof of this result is based on the last two theorems.

Theorem 2.4 *Let p be an interval order, (w, l) a closed representation of p. Then (v, l) is a closed representation of some interval augmentation of p if and only if v can be obtained from w by commuting starts, commuting ends as in 2.2 i) and moving starts to the end, i.e. transforming $w_1 e_2^+ e_1^- w_2$ to $w_1 e_1^- e_2^+ w_2$ where $e_1 \neq e_2$. The same holds for arbitrary representations if we additionally allow to delete ends from the end as in 2.2 ii).*

We have seen in the introduction, that p in Figure 2 has two events, say e_1 and e_2, which have the same label b and run in parallel; i.e. they are concurrent, e_1 starts before e_2 (as a formula: $e_1 <^+ e_2$) and e_1 ends before e_2 ($e_1 <^- e_2$). We have also motivated the following *swap-operation*, which exchanges the successors of two such events. This operation is not symmetric: in p, the two b-events run in parallel, and we can transform p to q; in q, we have some overtaking and *swap* is not applicable.

Definition 2.5 Let $p = (E, <, l)$ be an interval order such that for some $e_1, e_2 \in E$ we have $e_1 \, co \, e_2, e_1 <^+ e_2, e_1 <^- e_2$ and $l(e_1) = l(e_2)$. Let $q = (E, <', l)$ be defined by
$$e <' e' \quad \text{if} \quad \begin{cases} e < e' & \text{and} \quad e_1 \neq e \neq e_2 \\ e_1 < e' & \text{and} \quad e = e_2 \\ e_2 < e' & \text{and} \quad e = e_1 \end{cases}$$
Then q is obtained from p by *swapping*, denoted by $q \in swap(p)$. If some q' is obtained from p by applying this swap-operation arbitrarily often (including zero times), we write $q \in swap^*(p)$. □

This operation is actually easier to understand on the level of interval sequences, where the starts and ends of events are treated explicitly. For this reason, we will use representations in our proofs, although we are ultimately interested in system runs described as partial orders.

Definition 2.6 Let (w, l) be an interval sequence such that for some e_1, e_2 we have that $l(e_1) = l(e_2), e_1^+$ occurs before e_2^+ in w, e_2^+ before e_1^- and (provided e_2^- occurs at all) e_1^- before e_2^-. Let (v, l) be obtained from (w, l) by exchanging e_1^- and e_2^- in w. Then we say that (v, l) is obtained from (w, l) by *swapping*, denoted by $(v, l) \in swap(w, l)$. If some (v', l) is obtained from (w, l) by applying this swap-operation arbitrarily often (including zero times), we write $(v', l) \in swap^*(w, l)$. □

Obviously, (v, l) as defined above is an interval sequence. The next theorem shows the close interrelation between the swap-operation on interval orders and that on interval sequences; it is fundamental for the proofs of most of our results later on. In particular, it implies that $q \in swap^*(p)$ is indeed an interval order, since it has a representation.

Theorem 2.7 *For an interval order p, $q \in swap^*(p)$ is equivalent to each of the following:*

i) *For each representation (w, l) of p there exists some representation (v, l) of q with $(v, l) \in swap^*(w, l)$.*

ii) *For each representation (v, l) of q there exists some representation (w, l) of p with $(v, l) \in swap^*(w, l)$.*

We close this section by two results on the interplay of swapping with augmenting interval orders and with the prefix-relation, which are needed later on.

Proposition 2.8 *Let p, q and q' be interval orders such that $q \in swap^*(p)$ and q' is an augmentation of q. Then there exists an interval augmentation p' of p with $q' \in swap^*(p')$.*

Proposition 2.9 *Let p, q and q' be interval orders such that $q \in swap(p)$ and q' is a prefix of q. Then there exists a prefix p' of p such that $q' \in swap(p')$ or q' is an augmentation of p'.*

Remark: i) To see that, in 2.9, q' might indeed be a proper augmentation, consider p and q in Figure 2. Choose q' as the prefix of q containing a, one b and c. The only possible prefix p' of p consists of the c-labelled event and the two events on the left.

ii) We have seen that the operations of swapping and augmenting can just as well be performed on the level of representations as on the level of interval orders. It is interesting to note that such a correspondence does not hold for the prefix-relation. Consider again the prefix of p in Figure 2 consisting of the c-labelled event and the two events on the left, and let (v, l) be a representation of this prefix. In (v, l), we have only one start of a b-labelled event; the end of this event occurs in v, since it must occur before the start of the c-labelled event. Let (w, l) be any representation of p. Since in p there are two concurrent b-labelled events, we must have two starts of b-labelled events in (w, l) before the first end of such an event. Hence, v cannot be a prefix of w. Compare [Vog92] and the trunk-relation mentioned there. □

3 Petri Nets, Interval Semiwords and Splitting

We assume some basic knowledge of Petri nets. In this section, we explain some further notions informally, i.e. without formulas; see [Vog92] for details. We will deal with nets (place/transition-nets with initial marking) whose transitions are labelled with actions from some infinite alphabet Σ or with the empty word λ denoting an internal action. We assume that Σ contains for each $a \in \Sigma$ and $i \in IN$ also the action a_i.

General assumption All nets considered in this paper are safe and without isolated transitions.

We can extend the labelling of a net to steps and sequences in the obvious way by replacing each transition by its label and deleting internal actions. This way, we can transform firing sequences and step sequences to *action firing* and *action step sequences*. We call two nets *language* or *step-sequence equivalent* if they have the same action firing or step sequences.

Petri net theory has a long tradition of studying 'true concurrency' using partial orders. Most often a partial order semantics is given by so-called processes, where the partial order models causality. Another view is that concurrency is more than arbitrary interleaving but includes it; this idea is formalized in the partial words of [Gra81], which coincide with the semiwords of [Sta81] for the nets we consider here. Semiwords can be understood on the basis of processes as follows. (They can also be defined independently, see e.g. [Vog92] for a comparison of the definitions.) A process describes a run of a Petri net. It contains transition-labelled events, which represent transition firings, and it induces a labelled partial order on them, its *event structure*. If we change the labelling to the corresponding actions and delete events that correspond to internal actions, then we obtain the *action structure* of the process. This action structure shows the actions that occurred and their causal relations. For example, the net N' of Figure 1 has the partial orders p and q of Figure 2 and all their prefixes as action structures.

The set of *semiwords* consists of all these action structures and all their augmentations. This set is, by definition, closed under augmentation; hence, concurrency in semiwords is just seen as a possibility, since possibly concurrent transitions can also be performed sequentially. Now, an *interval semiword* is a semiword whose partial order is an interval order. The set of interval semiwords of N is denoted by $ISW(N)$. Two nets are *interval-semiword equivalent* if they have the same interval semiwords. Since the elements of an interval order correspond to intervals of real numbers, an interval semiword can be seen as a system run where each firing takes some time. Hence, $ISW(N)$ is a partial order semantics which is not so much related to causality, but rather has a temporal flavour.

Instead of the interval semiwords themselves, we can also consider their representations: an *interval word* of a net N is a representation of an interval semiword of N; the set of interval words is denoted by $IW(N)$. Interval words of Petri nets can also be defined independently of interval semiwords, see [Vog92].

In this paper, we are interested in a very simple, but natural sort of action refinement, namely in splitting. Let N be a net, $n \in \mathbb{N}$. Then $split_n(N)$, the *splitting* of N into n phases, is constructed from N as follows. Replace each transition t with label in Σ by a path consisting of transitions and places alternatingly, beginning and ending with a transition; the path contains n transitions; the first transition has the preset of t as preset, the last has the postset of t as postset; if a is the label of t, then the transitions are labelled $a_1, \ldots, a_n$; the new places are unmarked, while the initial marking stays the same on the places of N. Internal actions are not refined.

We call nets N, N' *split$_n$-language equivalent* or simply *split$_n$-equivalent* if *split$_n$(N)* and *split$_n$(N')* are language equivalent. It is not too difficult to see that split$_{n+1}$-equivalence implies split$_n$-equivalence. More surprisingly, it has been shown in [GV91] that for all $n \in \mathbb{N}$ there are nets which are split$_n$-, but not split$_{n+1}$-equivalent. Thus, for all $n \in \mathbb{N}$, split$_{n+1}$-equivalence implies split$_n$-equivalence, but not the other way round.

We only consider split-refinements where each visible action is split into the same number of phases. This is no restriction for the following reason: if all actions of N are split into at most n phases, we can obtain the language of the split net from the language of *split$_n$(N)*; namely, if action a is split into $k(a) \le n$ phases, consider only those sequences, where each $a_{k(a)}$ is immediately followed by $a_{k(a)+1} \ldots a_n$, and contract each subsequence $a_{k(a)} \ldots a_n$ to $a_{k(a)}$. Hence, split$_n$-equivalence implies equivalence with respect to arbitrary splittings into at most n phases.

4 A New Congruence for Splitting

From the general result that interval-semiword equivalence is fully abstract for language equivalence and action refinement, it is clear that, in particular, it is a congruence for splitting. In this paper, we want to determine the coarsest such congruence, i.e. one that is fully abstract with respect to splitting and language equivalence. In other words, we want to determine the limit of the sequence of the increasingly finer split$_n$-equivalences. In the introduction, we have already argued that for this purpose we are interested in a new semantics, which is defined by closing $ISW(N)$ under swapping — similarly as semiwords are obtained by closing the set of action structures under augmentation.

Definition 4.1 The *swap-interval-semiword semantics* of a net N is *swap-ISW(N)*, the set of all interval orders q for which there exists some $p \in ISW(N)$ with $q \in swap^*(p)$. If *swap-ISW(N)* = *swap-ISW(N')* for nets N and N', we call these nets *swap equivalent*. □

The proof that swap-equivalence is fully abstract is based on interval sequences. Hence, as a first step, we characterize swap-equivalence with these sequences. (The proof of this characterization is based on Theorem 2.7.)

Definition 4.2 The *swap-interval-word semantics* of a net N is the set *swap-IW(N)* of all interval sequences (v, l) for which there exists some $(w, l) \in IW(N)$ with $(v, l) \in swap^*(w, l)$. If *swap-IW(N)* = *swap-IW(N')* for nets N and N', we call these nets *swap-word equivalent*. □

Theorem 4.3 *For a net N, swap-IW(N) is the set of all representations of elements of swap-ISW(N). Hence, swap-word and swap equivalence coincide.*

Lemma 5.4.12 in [Vog92] shows how, for any action refinement *ref*, we can determine the interval words of some *ref(N)* from the interval words of N. Here, we are only interested in splitting instead of general refinement, and we are only interested in the language of the split net, not in its interval words. In Proposition 4.5 below, we adapt Lemma 5.4.12 to our setting and our slightly different notation.

Definition 4.4 Let (w, l) be an interval sequence over E, $n \geq 2$. A *concrete n-refinement* of (w, l) is a sequence $v \in (\Sigma \times E)^*$ such that:

- If (b, e) occurs in v, then $b = a_k$ for $a = l(e)$ and some $k \leq n$.
- Define a morphism $proj : (\Sigma \times E)^* \to (E^{\pm})^*$ by

$$proj(b, e) = \begin{cases} e^+ & \text{if } b = a_1 \\ e^- & \text{if } b = a_n \\ \lambda & \text{otherwise} \end{cases}$$

Then $proj(v) = w$.

- For each $e \in E$, define a morphism $proj_e : (\Sigma \times E)^* \to \Sigma^*$ by

$$proj_e(b, e') = \begin{cases} b & \text{if } e = e' \\ \lambda & \text{if } e \neq e' \end{cases}$$

Then $proj_e(v) = a_1 a_2 \ldots a_k$ for some $k \leq n$.

An *abstract n-refinement* of (w, l) is some $abs(v)$, where v is a concrete n-refinement of (w, l) and $abs : (\Sigma \times E)^* \to \Sigma^*$ is a morphism defined by $abs(b, e) = b$. □

Condition $proj(v) = w$ says that the first and last phases in v exactly match the starts and ends in w; the condition on $proj_e(v)$ shows that e is in v subdivided into a suitable number of phases. In a concrete n-refinement v, the second component tells us which a_i are phases of the same occurrence of action a. This is an information we do not have in $abs(v)$, and neither in an action firing sequence of $split_n(N)$.

Proposition 4.5 *[Vog92] Let N be a net, $n \geq 2$. Then $L(split_n(N))$ is the set of all abstract n-refinements of interval words of N.*

Besides this description for the language of a split net, the following lemma is needed.

Lemma 4.6 *Let (w, l) be an interval sequence, $(w', l) \in swap(w, l)$ and v an abstract n-refinement of (w', l) for some $n \geq 2$. Then v is an abstract n-refinement of (w, l), too.*

Now we state the main result of this paper.

Theorem 4.7   *i) Nets are swap-equivalent if and only if they are $split_n$-equivalent for all $n \in \mathbb{N}$.*

ii) Swap-equivalence is fully abstract with respect to language equivalence and splitting.

Part ii), our main result, is not hard to show from i). The 'only-if' part of i) follows quite directly from 4.5 and 4.6; the proof for the reverse implication is more difficult and completely omitted.

In order to check swap-equivalence, it seems that we have to consider all interval semiwords of a net and all interval orders that can be obtained from them by iterated swapping. If we want to check an example as the one given in the introduction, such considerations are not feasible. The following result – where ii) and iii) are shown with 2.8 and 2.9 – states some nice closure properties of the new semantics. Most importantly, these can help to check swap-equivalence.

Theorem 4.8 *i) If, for nets N_1 and N_2, $ISW(N_1) \subseteq swap\text{-}ISW(N_2)$, then we have $swap\text{-}ISW(N_1) \subseteq swap\text{-}ISW(N_2)$.*

ii) For a net N, $swap\text{-}ISW(N)$ is closed under interval augmentation, i.e. if p is an interval augmentation of $q \in swap\text{-}ISW(N)$, then $p \in swap\text{-}ISW(N)$.

iii) For a net N, $swap\text{-}ISW(N)$ is closed under taking prefixes.

To demonstrate the usefulness of this result, we apply it to the nets discussed in the introduction. Since N is contained in N', it is quite clear that $ISW(N) \subseteq ISW(N')$, hence we have $swap\text{-}ISW(N) \subseteq swap\text{-}ISW(N')$. To prove swap-equivalence of N and N' we have to show the reverse containment, and by 4.8 i) it suffices to check that all interval semiwords of N' are in $swap\text{-}ISW(N)$. In fact, it even suffices to check all least sequential interval semiwords of N' by 4.8 ii).

All interval semiwords of N' that do not involve the additional c-transition of N' are also interval semiwords of N and nothing has to be checked. Now recall that all semiwords, hence all interval semiwords, are augmentations of action structures. The only action structures of N' that contain the additional c-transition are q of Figure 2 and the sequence abc. These are interval semiwords themselves, i.e. each least sequential interval semiword that remains to be checked equals q or abc. For q we have $q \in swap^*(p)$ and $p \in ISW(N)$; hence $q \in swap\text{-}ISW(N)$. This implies with 4.8 iii) that the prefix abc of q is also in $swap\text{-}ISW(N)$, and we are done.

With the fact that N and N' are swap-equivalent, one can show:

Theorem 4.9 *Interval-semiword equivalence strictly implies swap-equivalence, which in turn strictly implies step-sequence equivalence.*

Thus, on the language level full abstractness for splitting is different from full abtractness for general action refinement. I conjecture that the same can be shown e.g. on the level of failure semantics by adapting the approach of this paper. In contrast to this, full abstractness for splitting and for general action refinement coincide on the level of bisimulation. So, which ingredient of bisimulation makes bisimulation different from the language in this comparison of splitting and action refinement? Is this ingredient characterisic for bisimulation? More precisely, is bisimulation the coarsest semantics in the linear time – branching time spectrum such that full abstractness for splitting and for general action refinement coincide?

We have seen how swap equivalence can be checked by hand for small examples using some nice closure properties of the new semantics. To decide this new equivalence in general, it seems more promising to use the (sequential) representations instead of the interval orders themselves; but so far, no decision algorithm for finite nets is known.

References

[BDKP91] E. Best, R. Devillers, A. Kiehn, and L. Pomello. Concurrent bisimulations in Petri nets. *Acta Informatica*, 28:231–264, 1991.

[Dev92]   R. Devillers. Maximality preservation and the ST-idea for action refinement. In G. Rozenberg, editor, *Advances in Petri Nets 1992*, Lect. Notes Comp. Sci. 609, 108–151. Springer, 1992.

[Fis85]   P.C. Fishburn. *Interval Orders and Interval Graphs*. J. Wiley, 1985.

[GG89]   R.J. v. Glabbeek and U. Goltz. Equivalence notions for concurrent systems and refinement of actions. In A. Kreczmar and G. Mirkowska, editors, *MFCS 89*, Lect. Notes Comp. Sci. 379, 237–248. Springer, 1989.

[GL91]   R. Gorrieri and C. Laneve. The limit of split_n-bisimulations for CCS agents. In A. Tarlecki, editor, *MFCS 91*, Lect. Notes Comp. Sci. 520, 170–180. Springer, 1991.

[Gla90]   R.J. v. Glabbeek. The refinement theorem for ST-bisimulation semantics. In M. Broy and C.B. Jones, editors, *Programming Concepts and Methods*, *Proc. IFIP Working Conference*, 27–52. Elsevier Science Publisher(North-Holland), 1990.

[Gra81]   J. Grabowski. On partial languages. *Fundamenta Informaticae*, IV.2:428–498, 1981.

[GV91]   R.J. v. Glabbeek and F. Vaandrager. The difference between splitting in n and n+1, 1991. Unpublished.

[Hen88]   M. Hennessy. Axiomatising finite concurrent processes. *SIAM J. of Computing*, 17:997–1017, 1988.

[JK93]   R. Janicki and M. Koutny. Representations of discrete interval orders and semi-orders. Technical Report 93-02, Dept. Comp. Sci. Sys., McMaster University, Hamilton, Ontario, 1993.

[Lar88]   K.S. Larsen. A fully abstract model for a process algebra with refinement. Master's thesis, Dept. Comp. Sci., Aarhus University, 1988.

[Sta81]   P.H. Starke. Processes in Petri nets. *J. Inf. Process. Cybern. EIK*, 17:389–416, 1981.

[Vog92]   W. Vogler. *Modular Construction and Partial Order Semantics of Petri Nets*. Lect. Notes Comp. Sci. 625. Springer, 1992.

[Vog93a]   W. Vogler. Bisimulation and action refinement. *Theoret. Comput. Sci.*, 114:173–200, 1993.

[Vog93b]   W. Vogler. The limit of split_n-language equivalence. Technical Report Nr. 288, Inst. f. Mathematik, Univ. Augsburg, 1993.

Divergence and Fair Testing*

V. Natarajan   Rance Cleaveland

Abstract

This paper develops a new testing-based semantic theory of processes that aims to circumvent difficulties that traditional testing/failures theories have in dealing with divergent behavior. Our framework incorporates a notion of *fairness* into the determination of when a process passes a test; we contrast this definition with existing approaches and give characterizations of the induced semantic preorders. An example highlights the utility of our results.

1   Introduction

Research into algebraic models of concurrency has focused on the use of semantic equivalences and preorders for establishing that systems meet their specifications. In such an approach to verification one formulates a specification as a system describing the required high-level behavior; a design/implementation then meets such a specification if its behavior is indistinguishable from the specification's (if one is using an equivalence) or if its behavior is in some sense better than the specification's (if one is using a preorder). A variety of different equivalences and preorders have been suggested for this purpose, with the testing/failures relations [4, 8] attracting particular attention because of their intuitive formulations in terms of the responses a system exhibits to tests. The theory underlying this framework has been extensively investigated, and several case studies attest to the utility of these relations in verification [1, 5].

One aspect of process behavior that the testing/failures framework is particularly sensitive to is *divergence*. A divergent process may engage in an infinite sequence of internal computation steps; in doing so, it ignores requests for interaction offered by its environment. As such behavior is in general undesirable, the testing/failures relations ensure that processes that may diverge are "not as good" as those that will not. However, in some cases processes deemed divergent by this framework may be viewed as unproblematic. As an example, consider a communications protocol that is designed to operate over a lossy medium. The system consisting of the protocol and this medium would exhibit a possible infinite internal computation in which the medium loses every message that the sender gives to it. In reality, however, such behavior would in all likelihood not occur; if a lossy medium is indeed capable of delivering messages, then it will certainly deliver some of the ones it is given, and the system just described ought not to be considered divergent. The traditional testing/failures framework does not admit such distinctions, however, and this fact complicates the use of the theory in reasoning about communications protocols in particular, and distributed fault-tolerant systems in general.

---

*Research supported by NSF grant CCR-9120995, ONR Young Investigator Award N00014-92-J-1582, NSF Young Investigator Award CCR-9257963, and NSF grant CCR-9402807. Authors' address: Department of Computer Science, North Carolina State University, Raleigh, NC 27695-8206, USA, e-mail: {rance, raj}@science.csc.ncsu.edu.

$$S_0 \;\Leftarrow\; send.S_0' \qquad S_0' \Leftarrow \overline{s_0}.(r_{ack_0}.S_1 + r_{ack_1}.S_0' + \tau.S_0')$$
$$S_1 \;\Leftarrow\; send.S_1' \qquad S_1' \Leftarrow \overline{s_1}.(r_{ack_1}.S_0 + r_{ack_0}.S_1' + \tau.S_1')$$

$$R_0 \;\Leftarrow\; r_0.R_0' + r_1.\overline{s_{ack_1}}.R_0 + \tau.\overline{s_{ack_1}}.R_0 \qquad R_0' \Leftarrow receive.\overline{s_{ack_0}}.R_1$$
$$R_1 \;\Leftarrow\; r_1.R_1' + r_0.\overline{s_{ack_0}}.R_1 + \tau.\overline{s_{ack_0}}.R_1 \qquad R_1' \Leftarrow receive.\overline{s_{ack_1}}.R_0$$

$$M \;\Leftarrow\; s_0.(\overline{r_0}.M + \tau.M) + s_1.(\overline{r_1}.M + \tau.M)$$
$$\qquad + s_{ack_0}.(\overline{r_{ack_0}}.M + \tau.M) + s_{ack_1}.(\overline{r_{ack_1}}.M + \tau.M)$$
$$M' \;\Leftarrow\; s_0.\overline{r_0}.M' + s_0.M' + s_1.\overline{r_1}.M' + s_1.M'$$
$$\qquad + s_{ack_0}.\overline{r_{ack_0}}.M' + s_{ack_0}.M' + s_{ack_1}.\overline{r_{ack_1}}.M' + s_{ack_1}.M'$$

$$ABP \;\Leftarrow\; (S_0|M|R_0)\backslash\{r_0,r_1,s_0,s_1,r_{ack_0},r_{ack_1},s_{ack_0},s_{ack_1}\}$$

$$ABP' \;\Leftarrow\; (S_0|M'|R_0)\backslash\{r_0,r_1,s_0,s_1,r_{ack_0},r_{ack_1},s_{ack_0},s_{ack_1}\}$$

Subscripted s and r actions denote sends to and receives from the medium.

Figure 1: The alternating bit protocol.

In this paper we develop a theory of testing that incorporates a notion of *fairness* into process testing. Intuitively, in our framework a process is deemed to "pass" a test if it may come "arbitrarily close" to success; using this notion as a basis we characterize the resulting semantic preorders and show how, as a side-effect, a natural treatment of divergence emerges that permits reasoning about systems with faulty components. The rest of the paper develops along the following lines. In the next section we offer an example designed to illustrate in a concrete manner the shortcomings of the traditional view of divergence. The section following then presents the formal definitions of traditional testing and gives our new framework, while Section 4 examines the relationship between the old and the new testing approaches. The next section offers an alternative characterization of the testing relations that do not require analyzing the behavior of processes in response to tests and re-examines the example of Section 2 in light of our new relations. The last section contains our conclusions and directions for future work as well as comparisons and contrasts with existing research on this topic.

2 Motivating Example

Traditional testing/failures semantics relates processes on the basis of their capabilities for deadlock and divergence; if two processes are equivalent then their behavior in these regards should be identical. These theories typically view divergence as "catastrophic" in the sense that once a process exhibits a capacity for infinite internal behavior, its subsequent deadlocking behavior becomes irrelevant. This insensitivity to deadlock in the face of divergence can lead to counterintuitive determinations regarding process equivalence, as the example presented in this section illustrates.

The system we consider is the Alternating Bit Protocol (ABP), a communications protocol designed to ensure reliable communication over lossy communication lines. Figure 1 contains a rendering of the ABP in CCS [13], some familiarity with which we assume on the part of the reader. The ABP consists of three entities: a sender, a receiver, and a lossy medium over which the sender and receive communicate. In order to recover from the possibility of message losses, the sender and receiver exchange messages and acknowledgements that are tagged with "sequence bits" in the following manner. When the sender is given a message to deliver, it appends its current sequence bit value i (initially $i = 0$) to the message and gives the resulting packet to the medium (by performing the action $\overline{s_i}$). It then awaits an acknowledgement from the receiver that is also tagged with i (by waiting to perform action r_{ack_i}). If it receives such an acknowledgement the sender increments its sequence bit modulo 2 and awaits for the next message to deliver. If the sender receives an acknowledgement with the wrong bit, or times out (by performing its τ action), then it resends its message. The receiver behaves in a dual fashion; it awaits a message from the medium whose sequence bit matches its current sequence bit i (initially $i = 0$), and if it receives one it delivers it (by executing its $\overline{receive}$ action), sends an acknowledgement tagged with i (by performing $\overline{s_{ack_i}}$), and increments its sequence bit modulo 2. If it receives a message with the wrong sequence bit, or if it times out, then the receiver sends an acknowledgement with sequence bit $1 - i$. In the figure, S_i and R_i represent the "start states" of the sender and receiver, respectively, when their sequence bits are i.

In this example we consider two slightly different faulty media. Medium M may lose messages given to it (by performing τ) at any point until the message is received. M', on the other hand, may only drop a message at the time it is given to it; once it decides to deliver a message, it cannot lose it. ABP represents the ABP built using medium M, while ABP' represents the ABP built using M'.

From the operational semantics of CCS, it is straightforward to establish that both ABP and ABP' are initially capable of an infinite sequence of internal τ-steps. This results from the fact that the receiver may engage in an unbounded number of time-outs, which prompt it to send acknowledgements that the medium then loses. Accordingly both are divergent, and one may establish that, with respect to testing/failures equivalence, the two systems are indistinguishable. However, ABP is incapable of deadlocking; at any state it reaches during its execution it always has the potential of engaging in future $send$ and $\overline{receive}$ actions. On the other hand, ABP' does have reachable deadlocked states; for example, after executing a $send$ followed by several τ-actions, the system may wind up in the state

$$(S_0' \,|\, \overline{r_{ack_1}}.M' \,|\, \overline{r_{ack_1}}.R_0) \backslash \{r_0, r_1, s_0, s_1, r_{ack_0}, r_{ack_1}, s_{ack_0}, s_{ack_1}\}.$$

In this state S_0' and $\overline{r_{ack_1}}.R_0$ can only advance by sending a message/acknowledgement to the medium, which in turn is waiting to deliver an acknowledgement to the sender.

Note that the fact that these systems are divergent relies on the potential inability of the medium to deliver messages given to it. Under realistic assumptions about medium behavior, however, this would not happen; some messages will eventually be delivered. Therefore, one could argue that when such "fairness" considerations are taken into account, neither ABP nor ABP' should be considered divergent. Were one able to make this determination, then one might in fact ascertain that ABP is "strictly better", with respect to the failures/testing preorders, than ABP' because of its better deadlock behavior. Unfortunately, the traditional testing/failure framework requires that ABP and ABP' be considered divergent, and hence the semantic relations are insensitive to these systems' differing potential for deadlock.

3 Processes

This section formally defines the model of processes used in this paper, reviews relevant concepts from traditional testing theory, and introduces a new approach to testing.

3.1 Processes and Tests

We model processes using *labeled transition systems*, which are defined as follows.

Definition 3.1 *A labeled transition system is a tuple $\langle P, A, \rightarrow, p_0 \rangle$ where P is a set of states, A is a set of actions containing a distinguished internal action τ, $\rightarrow \subseteq P \times A \times P$ is the transition relation, and $p_0 \in P$ is the start state.*

Intuitively, a labeled transition system encodes the operational behavior of a process. P consists of the set of states that the process may enter, and A contains the set of actions that it may perform. The relation $\rightarrow$ describes the actions available in a particular state and the states that may result when the action is executed. In what follows we write $p \xrightarrow{a} p'$ in lieu of $\langle p, a, p' \rangle \in \rightarrow$, and we also use $p \xrightarrow{a}$ to indicate that there is a p' such that $p \xrightarrow{a} p'$ and $p \not\xrightarrow{a}$ if no such p' exists. Also, $\rightarrow$ may be extended to sequences $s \in A^*$ in the obvious fashion. To simplify the presentation, we restrict our attention to *finite-sort* transition systems, namely, systems $\langle P, A, \rightarrow, p_0 \rangle$ such that A is finite. We do not, however, require that processes be finite-branching.

3.2 Traditional Testing

The traditional testing framework of [7, 8, 11] defines preorders and equivalences that relate processes on the basis of their responses to tests. In their setting a test is any process whose action set contains a distinguished *success* action w; to determine a process' response(s) to a test one must examine the computations that result when the test is "applied" to the process. These intuitions may be formalized as follows.

Definition 3.2 *Let process $\mathcal{P} = \langle P, A, \rightarrow, p_0 \rangle$ and test $\mathcal{T} = \langle T, A' \cup \{w\}, \rightarrow, t_0 \rangle$.*

1. *A configuration has form $p\|t$, where $p \in P$ and $t \in T$. $p\|t$ is successful if $t \xrightarrow{w}$.*

2. *The configuration transition relation, $p\|t \xmapsto{a} p'\|t'$, is defined as follows.*

   - $p \xrightarrow{a} p', t \xrightarrow{a} t', t \not\xrightarrow{w} \Rightarrow p\|t \xmapsto{a} p'\|t'$.
   - $p \xrightarrow{\tau} p', t \not\xrightarrow{w} \Rightarrow p\|t \xmapsto{\tau} p'\|t$.
   - $t \xrightarrow{\tau} t', t \not\xrightarrow{w} \Rightarrow p\|t \xmapsto{\tau} p\|t'$.

   We sometimes write $p\|t \xmapsto{} p'\|t'$ if $p\|t \xmapsto{a} p'\|t'$ for some $a \in A \cap A'$.

3. *A computation of $p\|t$ is a maximal (potentially infinite) sequence $p\|t \equiv p_1\|t_1 \xmapsto{a_1} p_2\|t_2 \xmapsto{a_2} \cdots$; it is strongly successful if some $p_i\|t_i$ is successful.*

4. *$p \ may_{DH} \ t$ if some computation of $p\|t$ is strongly successful. $\mathcal{P} \ may_{DH} \ \mathcal{T}$ if $p_0 \ may_{DH} \ t_0$. $p \ must_{DH} \ t$ if every computation of $p\|t$ is strongly successful. $\mathcal{P} \ must_{DH} \ \mathcal{T}$ if $p_0 \ must_{DH} \ t_0$.*

As processes and tests are in general nondeterministic, one may distinguish between the possibility of success ($\mathcal{P} \ may_{DH} \ \mathcal{T}$) and the inevitability of success ($\mathcal{P} \ must_{DH} \ \mathcal{T}$). Also, the definition of $\xmapsto{}$ departs slightly from the one given in [8] in a couple of respects. Firstly, our definition labels $\xmapsto{}$ with actions. More importantly, it also adds the premise "$t \not\xrightarrow{w}$" to all the rules used in defining the relation. The latter alteration ensures that successful configurations have no outgoing transitions; this property turns out to be technically convenient and yet does not affect the resulting semantic theory. The following preorders may now be defined.

Definition 3.3 *Let $\mathcal{P}$ and $\mathcal{P}'$ be processes.*

1. *$\mathcal{P} \sqsubseteq_{may_{DH}} \mathcal{P}'$ if for every test $\mathcal{T}$ such that $\mathcal{P} \ may_{DH} \ \mathcal{T}$, $\mathcal{P}' \ may_{DH} \ \mathcal{T}$.*

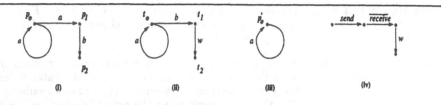

Figure 2: Sample processes and tests

2. $P \mathrel{\sqsubseteq_{must_{DH}}} P'$ *if for every test* T *such that* P *must$_{DH}$* T, P' *must$_{DH}$* T.

A number of alternative characterizations have been developed to ease the task of reasoning using these preorders. Here we present one due to De Nicola and Hennessy [8].

Definition 3.4 *Let* $P = \langle P, A, \to, p_0 \rangle$ *be a process, with* $p, p' \in P$. *Let* $s \in (A - \{\tau\})^*$ *be a sequence of visible actions.*

1. $p \overset{s}{\Rightarrow} p'$ *is defined inductively as follows:* $p \overset{\epsilon}{\Rightarrow} p'$ *if* $p(\overset{\tau}{\to})^* p'$ *and* $p \overset{as'}{\Rightarrow} p'$ *if* $p \overset{s}{\Rightarrow} \overset{a}{\to} \overset{s'}{\Rightarrow} p'$.

2. *The language,* $L(p)$, *of* p *is defined by:* $L(p) = \{ s \mid \exists p'. p \overset{s}{\Rightarrow} p' \}$. *We sometimes refer to the elements of* $L(p)$ *as traces of* p *and take* $L(P) = L(p_0)$.

3. $\Downarrow \subseteq P$ *is defined as the least predicate such that whenever* $p' \Downarrow$ *for all* p' *such that* $p \overset{\tau}{\to} p'$, *then* $p \Downarrow$.

4. $p \Downarrow s$ *is defined inductively as follows:* $p \Downarrow \epsilon$ *if* $p \Downarrow$, *and* $p \Downarrow as'$ *if* $p \Downarrow$ *and for all* p' *such that* $p \overset{a}{\Rightarrow} p'$, $p' \Downarrow s'$.

5. $S(p) = \{ a \in (A - \{\tau\}) \mid p \overset{a}{\Rightarrow} \}$.

Intuitively, $L(p)$ contains the sequences of visible actions that p may engage in, while $p \Downarrow$ holds if p is incapable of an infinite sequence of internal computation steps. $p \Downarrow s$ indicates that during any "execution" of s from p, no infinite internal computation sequences are possible. We now have the following [11].

Theorem 3.5 *Let* $P = \langle P, A, \to, p_0 \rangle$ *and* $P' = \langle P', A', \to, p'_0 \rangle$ *be processes.*

1. $P \mathrel{\sqsubseteq_{may_{DH}}} P'$ *iff* $L(p_0) \subseteq L(p'_0)$.

2. $P \mathrel{\sqsubseteq_{must_{DH}}} P'$ *iff for every* $s \in (A' - \{\tau\})^*$ *with* $p_0 \Downarrow s$, $p'_0 \Downarrow s$ *and for any* p'_1 *such that* $p'_0 \overset{s}{\Rightarrow} p'_1$, *there is a* p_1 *such that* $p_0 \overset{s}{\Rightarrow} p_1$ *and* $S(p_1) \subseteq S(p'_1)$.

3.3 Fair Testing

The definitions given above have a certain intuitive appeal: a computation is strongly successful if one observes a successful configuration at some point along the computation. However, consider the process P in Figure 2(i) and test T in Figure 2(ii). P *must$_{DH}$* T, since the computation $p_0 \| t_0 \overset{a}{\longmapsto} p_0 \| t_0 \overset{a}{\longmapsto} \cdots$ is not strongly successful. However, this computation may be viewed as "pathological" in that it "discriminates" against success: although unsuccessful, each configuration along this computation is also a starting point for the strongly successful computation $p_0 \| t_0 \overset{a}{\longmapsto} p_1 \| t_0 \overset{b}{\longmapsto} p_2 \| t_1$. We now offer an alternative definition of successful computation that aims at a different treatment of such "unfair" computations.

Definition 3.6 *Let* $\mathcal{P} = \langle P, A, \to, p_0 \rangle$ *be a process and* $\mathcal{T} = \langle T, A' \cup \{w\}, \to, t_0 \rangle$ *a test. A configuration* $c \equiv p\|t$ *is potentially successful if there is a successful configuration* c' *such that* $c \longmapsto^* c'$. *A computation is potentially successful if every configuration in it is potentially successful.*

If we replace the term "strongly successful" by "potentially successful" in Definition 3.2(4) then we arrive at definitions for *p may t*, *p must t*, *$\mathcal{P}$ may $\mathcal{T}$* and *$\mathcal{P}$ must $\mathcal{T}$*. That is, *p may t* if *p‖t* has a potentially successful computation, etc. These definitions induce the preorders $\sqsubseteq_{may}$ and $\sqsubseteq_{must}$ on processes through the obvious alterations to Definition 3.3.

Note that it is possible for a computation to be potentially successful even though it is not strongly so: the computation $p_0\|t_0 \xrightarrow{a} p_0\|t_0 \xrightarrow{a} \cdots$ in the previous example has this property. Indeed, we may show that $\mathcal{P}$ *must* $\mathcal{T}$. The following lemma shows the relationship between *may* and *must* and their traditional counterparts.

Lemma 3.7 *Let* $\mathcal{P}$ *be a process and* $\mathcal{T}$ *be a test. Then* $\mathcal{P}$ *may* $\mathcal{T}$ *iff* $\mathcal{P}$ *may$_{DH}$* $\mathcal{T}$, *while* $\mathcal{P}$ *must$_{DH}$* $\mathcal{T}$ *implies* $\mathcal{P}$ *must* $\mathcal{T}$.

The previous example shows that in general, $\mathcal{P}$ *must* $\mathcal{T}$ does not imply $\mathcal{P}$ *must$_{DH}$* $\mathcal{T}$.

4 A Topological Comparison of *must$_{DH}$* and *must*

In the remainder of this paper we give alternative characterizations of the preorders induced by *may* and *must*. Before doing so, however, we first wish to characterize precisely the distinction between the traditional notion of *must$_{DH}$* and our new notion of *must*. In general, for a given process and test the set of potentially successful computations properly contains the set of strongly successful computations. Define a computation to be *weakly successful* if it is potentially successful but not strongly successful. Our aim in this section is two-fold: to show that the set of potentially successful computations is a "reasonable" extension of the set of strongly successful ones and that the set of weakly successful computations is "small".

To formalize these notions we first define a metric space over the computations of a process and test in which weakly successful computations may be viewed as limits of sequences of strongly successful computations. We then show that the set of weakly successful computations is *nowhere dense* in the set of computations. This amounts to saying that the "volume" of the weakly successful computations is "infinitesimal".

Definition 4.1 *A metric space is a pair* $\langle X, d \rangle$, *where* X *is a non-empty set and* d, *the metric, is a mapping from* $X \times X$ *to the non-negative reals satisfying the following properties, for all* $x, y, z \in X$.

1. $d(x, y) = d(y, x)$.

2. $d(x, y) = 0$ *iff* $x = y$.

3. $d(x, z) \leq d(x, y) + d(y, z)$.

If $x_0 \in X$ *and* $\epsilon > 0$ *is real then* $B_\epsilon(x_0)$, *the open ball of radius* ϵ *centered at* x_0, *is the set* $\{ x \in X \mid d(x_0, x) < \epsilon \}$. *Set* $U \subseteq X$ *is an open set if for any* $x \in U$, *there exists* $\epsilon > 0$ *such that* $B_\epsilon(x_0) \subseteq U$. *Set* $F \subseteq X$ *is closed if* F^c, *its complement in* X, *is open. The interior,* U^0, *of a set* $U \subseteq X$ *is the largest open set contained in* U. *The closure of a set* $F \subseteq X$ *is the smallest closed set containing* F; *it will be denoted by* $\overline{F}$. *A subset* $A \subseteq X$ *is nowhere dense if* $(\overline{A})^0 = \emptyset$. *Note that a set* $U \subseteq X$ *has nonempty interior iff there is some* $x \in U$ *and a positive real* ϵ *such that* $B_\epsilon(x) \subseteq U$. *A sequence of elements* $x_1, x_2, \ldots$ *from* X *converges to* $x \in X$ *if for every* $\epsilon > 0$, $B_\epsilon(x)$ *contains all but a finite number of the* x_i; *in this case,* x *is said to be the limit of the sequence.*

The following defines two metric spaces over the computations of a process and test.

Definition 4.2 *Let* $\mathcal{P}$ *be a process with start state* p_0 *and* $\mathcal{T}$ *a test with start state* t_0, *and let* $\mathcal{C}(\mathcal{P}, \mathcal{T})$ *be the set of computations of* $p_0 \| t_0$. *If* $x \in \mathcal{C}(\mathcal{P}, \mathcal{T})$ *and* n *is a natural number then* $x\langle n \rangle$ *is the prefix of* x *of the form* $p_0 \| t_0 \overset{a_0}{\longmapsto} \cdots \overset{a_{n-2}}{\longmapsto} p_{n-1} \| t_{n-1}$; *if* x *contains fewer than* n *configurations then* $x\langle n \rangle = x$. *Let* $x, y \in \mathcal{C}(\mathcal{P}, \mathcal{T})$; *then* $d_1(x, y)$ *and* $d_2(x, y)$ *are defined as follows.* $d_1(x, y)$ *is* 1 *if* $x \neq y$ *and is* 0 *otherwise.* $d_2(x, y) = \inf\{ 2^{-n} \mid x\langle n \rangle = y\langle n \rangle \}$.

It is easy to see that $\langle \mathcal{C}(\mathcal{P}, \mathcal{T}), d_1 \rangle$ and $\langle \mathcal{C}(\mathcal{P}, \mathcal{T}), d_2 \rangle$ are metric spaces for any $\mathcal{P}$ and $\mathcal{T}$. The next lemma establishes that weakly successful computations may be seen as *limits* of sequences of strongly successful ones. Therefore, although a weakly successful computation is not strongly successful, it comes "arbitrarily close" to being so.

Lemma 4.3 *Let* $\mathcal{P}$ *be a process and* $\mathcal{T}$ *be a test, and let* $x \in \mathcal{C}(\mathcal{P}, \mathcal{T})$ *be weakly successful. Then there exists a sequence* $x_1, x_2, \ldots$ *in* $\langle \mathcal{C}(\mathcal{P}, \mathcal{T}), d_2 \rangle$ *converging to* x *such that each* x_i *is strongly successful.*

We now are able to establish the following metric-based characterizations for $must_{DH}$ and $must$. Each says in essence says that a process $must_{DH}/must$ a test exactly when "almost all" the computations (in the appropriate metric space) are successful.

Theorem 4.4 *Let* $\mathcal{P}$ *be a process and* $\mathcal{T}$ *a test, and let* $\mathcal{S} \subseteq \mathcal{C}(\mathcal{P}, \mathcal{T})$ *be the set of strongly successful computations. Define* $\mathcal{F} = \mathcal{C}(\mathcal{P}, \mathcal{T}) - \mathcal{S}$. *Then* $\mathcal{P} \; must_{DH} \; \mathcal{T}$ *iff* $\mathcal{F}$ *is nowhere dense in* $\langle \mathcal{C}, d_1 \rangle$, *and* $\mathcal{P} \; must \; \mathcal{T}$ *iff* $\mathcal{F}$ *is nowhere dense in* $\langle \mathcal{C}, d_2 \rangle$.

5 Characterizing the Preorders

This section provides alternative characterizations of $\sqsubseteq_{may}$ and $\sqsubseteq_{must}$ that are analogous to those for the traditional preorders given in Section 3. Such characterizations simplify the task of determining when two processes are related, and they may be used as a basis for developing denotational models. In view of Theorem 3.5 and Lemma 3.7 we immediately have the following.

Theorem 5.1 *Let* $\mathcal{P}$ *and* $\mathcal{P}'$ *be processes. Then* $\mathcal{P} \sqsubseteq_{may} \mathcal{P}'$ *iff* $L(\mathcal{P}) \subseteq L(\mathcal{P}')$

The remainder of this section is devoted to a characterization of $\sqsubseteq_{must}$. We approach this in two stages; we first consider the preorder induced by *finite* tests, and we then generalize this to the scenario in which arbitrary tests are allowed. Note that, in contrast with the traditional framework, infinite tests do add expressive power.

5.1 Finite Tests

Finite transition systems have a bound on the longest execution sequence they can engage in. Formally, a transition system $\mathcal{P} = \langle P, A, \rightarrow, p_0 \rangle$ is *finite* if there is a k such that for any $s \in A^*$ with $p_0 \overset{s}{\rightarrow}, |s| \leq k$. Let $\mathcal{F}$ be the class of finite tests, and define $\mathcal{P} \sqsubseteq_{must}^{\mathcal{F}} \mathcal{P}'$ to hold if for every $\mathcal{T} \in \mathcal{F}$, $\mathcal{P} \; must \; \mathcal{T}$ implies $\mathcal{P}' \; must \; \mathcal{T}$. Certainly, $\mathcal{P} \sqsubseteq_{must} \mathcal{P}'$ implies $\mathcal{P} \sqsubseteq_{must}^{\mathcal{F}} \mathcal{P}'$.

To give an alternative characterization of $\sqsubseteq_{must}^{\mathcal{F}}$, we first give the following lemma.

Lemma 5.2 *Let* $\mathcal{P} \sqsubseteq_{must}^{\mathcal{F}} \mathcal{P}'$. *Then* $L(\mathcal{P}') \subseteq L(\mathcal{P})$.

Note that this lemma stands in contrast with the analogous result for $\sqsubseteq_{must_{DH}}$, which states that if $\mathcal{P} \sqsubseteq_{must_{DH}} \mathcal{P}'$, $s \in L(\mathcal{P}')$, and the start state of $\mathcal{P}$ converges on s, then $s \in L(\mathcal{P}')$ [8]. Indeed, on the basis of this and Theorem 5.1 we have the following.

Corollary 5.3 *Let* $\mathcal{P} \sqsubseteq_{must} \mathcal{P}'$. *Then* $\mathcal{P}' \sqsubseteq_{may} \mathcal{P}$.

Using standard techniques from traditional testing, we now have the following.

Theorem 5.4 *Let* $\mathcal{P} = \langle P, A, \to, p_0 \rangle$, $\mathcal{P}' = \langle P', A', \to, p'_0 \rangle$. *Then* $\mathcal{P} \sqsubseteq^{\mathcal{F}}_{must} \mathcal{P}'$ *iff for each* s *and* p'_1 *with* $p'_0 \overset{s}{\Rightarrow} p'_1$, *there is a* p_1 *with* $p_0 \overset{s}{\Rightarrow} p_1$ *and* $S(p_1) \subseteq S(p'_1)$.

Note the similarity between this characterization and the one for $\sqsubseteq_{must_{DH}}$ given in Theorem 3.5; the difference lies in the insensitivity of $\sqsubseteq^{\mathcal{F}}_{must}$ to divergence.

5.2 Arbitrary Tests

In traditional testing theory, the preorders induced by finite tests coincide exactly with those resulting from arbitrary tests; infinite tests do not add any distinguishing power. The next example shows that this fails to be the case in our framework.

Example 5.1 *Let* $\mathcal{P}$, $\mathcal{P}'$ *and* T *be the processes and tests in Figure 2(i), (iii) and (ii), respectively. Then* $\mathcal{P} \sqsubseteq^{\mathcal{F}}_{must} \mathcal{P}'$ *but* $\mathcal{P} \not\sqsubseteq_{must} \mathcal{P}'$, *since* $\mathcal{P}$ *must* T *but* $\mathcal{P}'$ *must* T.

Although initially surprising, this fact does have an intuitive basis. In traditional testing, computations are deemed "good" (i.e. strongly successful) if, after a finite amount of time, one observes a "good" configuration. In our setting, however, computations may be successful even though no such "good" configuration is ever seen; in this case, the configurations in the computation must exhibit an ongoing capability to "become" good. Thus the infinitary behavior of a process becomes important, and as we use tests to probe a process's behavior, it follows that infinite tests add power.

We now present a characterization of $\sqsubseteq_{must}$. We begin with some definitions.

Definition 5.5 *Let* $\mathcal{P} = \langle P, A, \to, p_0 \rangle$ *be a process, with* $p, p' \in P$, *and let* $s \in (A - \{\tau\})^*$. *If* S *is a set then define* S^ω *to be the set of infinite sequences of elements of* S.

1. $L_\omega(p) = \{ \sigma \in (A - \{\tau\})^\omega \mid p \overset{\sigma}{\Rightarrow} \}$, *where* $p \overset{\sigma}{\Rightarrow}$ *is defined in the obvious fashion.*

2. $\Sigma(p) = \{ \sigma \in L_\omega(p) \mid (s \text{ is a prefix of } \sigma \wedge p \overset{s}{\Rightarrow} p') \Rightarrow \exists a \in (A - \{\tau\}): p' \overset{a}{\Rightarrow} \}$.

3. *If* $X \subseteq A^\omega \cup A^*$ *then* $s \prec X$ *if* $st \in X$ *for some* t.

4. $S(p, s) = \bigcup \{ S(q) \mid p \overset{s}{\Rightarrow} q \}$.

So $L_\omega(p)$ contains the infinite traces p may perform, while $\Sigma(p)$ consists of elements of $L_\omega(p)$ with the property that while p is "executing" any one of them it cannot reach a state in which all visible actions are disabled. Note that $s \prec X$ holds if s is a prefix of *some* element of X. We now introduce a relation on processes that we will show coincides with $\sqsubseteq_{must}$.

Definition 5.6 *Let* $\mathcal{P} = \langle P, A, \to, p_0 \rangle$ *and* $\mathcal{P}' = \langle P', A', \to, p'_0 \rangle$. *Then* $\mathcal{P} \ll_{must} \mathcal{P}'$ *if for every* $s \in (A' - \{\tau\})^*$, p'_1 *such that* $p'_0 \overset{s}{\Rightarrow} p'_1$, *and* $X \subseteq \Sigma(p'_1) \cup \{\epsilon\}$ *then:*

$$\exists s' \prec X, p_1 : p_0 \overset{ss'}{\Rightarrow} p_1 \wedge (\forall t : s't \prec X \Rightarrow S(p_1, t) \subseteq S(p'_1, s't)).$$

The definition is technically complex, but underlying intuition is not too difficult. Suppose $p_0' \Rightarrow p_1'$ and $\Sigma(p_1') = \{\sigma\}$. If $\mathcal{P} \ll_{must} \mathcal{P}'$ then (assuming $s' = \epsilon$) this condition assures the existence of some s-derivative p_1 of p_0 such that the action capabilities of p_1 after any prefix t of σ is contained in those of p_1' after t. A consequence is that if $\mathcal{P}'$ fails a test $\mathcal{T}$ because of the derivation $p_0' \Rightarrow p_1' \Rightarrow$, then surely $\mathcal{P}$ will also fail $\mathcal{T}$: the action capabilities of p_1 along σ are no larger than those of p_1' along σ, and thus if p_1' could not trigger a successful computation on $\mathcal{T}$ then neither could p_1. We now have the following; the proof may be found in the appendix.

Theorem 5.7 *Let* $\mathcal{P}, \mathcal{P}'$ *be processes. Then* $\mathcal{P} \sqsubseteq_{must} \mathcal{P}'$ *iff* $\mathcal{P} \ll_{must} \mathcal{P}'$.

A Sufficient Condition. Determining whether or not $\mathcal{P} \ll_{must} \mathcal{P}'$ can be difficult. We now give a sufficient condition that it easier to check.

Proposition 5.8 *Let* $\mathcal{P} = \langle P, A, \rightarrow, p_0 \rangle$ *and* $\mathcal{P}' = \langle P', A', \rightarrow, p_0' \rangle$. *Then* $\mathcal{P} \sqsubseteq_{must} \mathcal{P}'$ *if for every* s *and* p_1' *with* $p_0' \Rightarrow p_1'$, *there is a* p_1 *with* $p_0 \Rightarrow p_1$ *and* $L(p_1) \subseteq L(p_1')$.

The similarity between this condition and Theorem 5.4 is worth noting. From this proposition it also follows that observational equivalence [13] is finer than $\sqsubseteq_{must}$.

5.3 The ABP Revisited

We now return to the sample implementations of the Alternating Bit Protocol in Section 2 in order to examine the relationship between ABP and ABP'. Using Proposition 5.8 it is not difficult to see that $ABP' \sqsubseteq_{must} ABP$. On the other hand, $ABP \not\sqsubseteq_{must} ABP'$; to see this, consider the test $\mathcal{T}$ given in Figure 2(iv). It is straightforward to show that ABP' m$\not$st $\mathcal{T}$, since after executing a *send* this process may deadlock. However, it is the case that ABP *must* $\mathcal{T}$.

6 Conclusions

In this paper we have developed an alternative theory of process testing that incorporates a notion of "fairness" into the determination of when a process passes a test. In contrast with traditional testing theory, divergence no longer becomes catastrophic; rather, a process that is superficially capable of an infinite internal computation but is also always capable of interacting with its environment will eventually do so. We achieved this by redefining when a process must inevitably pass a test—in our setting, this holds if "almost all" of their computations are successful in a sense we made precise using simple notions from mathematical analysis. We also provided alternative characterizations of the preorders that obviate the need for reasoning about process' response to all tests and demonstrated the utility of the preorders on an example involving the Alternating Bit Protocol. In particular, we showed that two versions of the protocol that have intuitively different behavior but are equated under traditional testing theory are distinguished in our framework.

As future work, we would like to investigate denotational and algebraic issues involved in using our framework to give meaning to process algebras such as CCS. Our initial results indicate that our preorders are substitutive for typical constructs found in these languages; however, traditional fixpoint approaches to defining the meaning of recursive processes would no longer apply because of the infinitary nature of the "observables" required by our preorders. This problem also occurs in other semantic treatments of fairness, and we expect that results obtained there may be applicable. In addition, we would like to see if our alternative characterizations can be simplified, and we hope to develop algorithms for computing when two finite-state systems are related by our preorders.

Related Work. Fairness has been an active area of research for over a decade; a good overview of the subject may be found in [9]. A variety of different notions of fairness have been defined, and our definition of potentially successful may be seen as related to two of them in particular. First, define success to be *enabled* in a configuration c if there is a successful configuration reachable from c. A computation is then potentially successful if success is continuously enabled. This amounts to an assumption of *weak* fairness [9] with respect to success in that the definition of potential success implicitly assumes that if success is continuously enabled, then it is eventually "reached". For finite-state processes and tests connections may also be drawn with the notion of ∞-fairness introduced in [3]. According to this notion an execution is unfair if a transition which is enabled infinitely often by a sequence of unrestricted length occurs only finitely often. In our case, a weakly successful computation is in a certain sense ∞-unfair with respect to success transitions.

Other researchers have also given semantic models of fairness in process-algebraic settings. Following [9], in this work one may distinguish between *structural* fairness, in which notions of fairness are incorporated into the semantics of process operators [6, 10], and *observational* fairness, in one focuses on the "allowable" execution sequences [2]. Our work falls in the latter category; in what follows we briefly compare our results with those of [10] and [2], which also develop testing/failures frameworks. In [10] Hennessy gives an algebraic theory for reasoning about asynchronous communicating processes built using a fair parallel composition operator, $\|$. Divergence is still considered to be catastrophic in his model; consequently, his semantics would equate the systems ABP and ABP' considered in our example, even when the the fair parallel combination operator is used for composing the protocol's entities. On the other hand, he does offer an algebraic theory and a fully abstract denotational model. The issue of *fair abstraction* of divergent processes is addressed in [2]. Like our theory, their approach differentiates between ABP and ABP'. However, it is possible to come up with another implementation ABP'' of the Alternating Bit Protocol using a different implementation M'' of the medium entity that could induce "divergent deadlocking" states (states which are divergent and incapable of ever performing visible actions); their theory would equate ABP and ABP'' while ours would not. There are also situations in which our semantics licenses the treatment of divergence as finite delay whereas theirs does not. Like [10], their work presents proof systems that permit the verification of processes using syntactic manipulations. Koomen's Fair Abstraction Rule [12] is not valid in their semantics, although a restricted version is; neither rule holds in our framework.

References

[1] J.C.M. Baeten, editor. *Applications of Process Algebra*, vol. 17 of *Cambridge Tracts in Theoretical Computer Science*. Cambridge Press, Cambridge, 1990.

[2] J.A. Bergstra, J.W. Klop, and E.-R. Olderog. Failures without chaos: A new process semantics for fair abstraction. In M. Wirsing, editor, *Formal Description Techniques—III*, pages 77–103. North-Holland, 1987.

[3] E. Best. Fairness and conspiracies. *Inf. Proc. Letters*, 18:215–220, 1984.

[4] S.D. Brookes, C.A.R. Hoare, and A.W. Roscoe. A theory of communicating sequential processes. *Journal of the ACM*, 31(3):560–599, July 1984.

[5] R. Cleaveland, J. Parrow, and B. Steffen. The Concurrency Workbench: A semantics-based tool for the verification of finite-state systems. *ACM Trans. on Prog. Lang. and Sys.*, 15(1):36–72, January 1993.

[6] G. Costa and C. Stirling. Weak and strong fairness in CCS. *Inf. and Comp.*, 87:207–244, 1987.

[7] R. De Nicola. Extensional equivalences for transition systems. *Acta Inf.*, 1987.

[8] R. De Nicola and M.C.B. Hennessy. Testing equivalences for processes. *Theoret. Comp. Sci.*, 34:83–133, 1983.

[9] N. Francez. *Fairness*. Springer-Verlag, 1986.

[10] M.C.B. Hennessy. An algebraic theory of fair asynchronous communicating processes. *Theoret. Comp. Sci.*, 49:121–143, 1987.

[11] M.C.B. Hennessy. *Algebraic Theory of Processes*. MIT Press, Boston, 1988.

[12] C.J. Koomen. Algebraic specification and verification of communications protocols. *Science of Comp. Prog.*, 5:1–36, 1985.

[13] R. Milner. *Communication and Concurrency*. Prentice-Hall, London, 1989.

A Appendix

We begin by stating a useful lemma.

Lemma A.1 *Let* $\mathcal{P}, \mathcal{P}'$ *be such that* $\mathcal{P} <<_{must} \mathcal{P}'$. *Then* $\mathcal{P} \sqsubseteq^{\mathcal{F}}_{must} \mathcal{P}'$.

The next two lemmas together prove Theorem 5.7. In what follows, if $p_0||e_0 \overset{a_1}{\to} \cdots \overset{a_k}{\to} p_k||e_k$ is a prefix of some computation of $p_0||e_0$ then we will refer to the sequence obtained by deleting the τ's from the string $a_1 \cdots a_k$ as its *visible content*. Note that if s is the visible content of a prefix of some computation of $p||e$ then $s \in L(p) \cap L(e)$.

Lemma A.2 *Let* $\mathcal{P}$ *and* $\mathcal{P}'$ *be processes. Then* $\mathcal{P} <<_{must} \mathcal{P}'$ *implies* $\mathcal{P} \sqsubseteq_{must} \mathcal{P}'$.

Proof. Let $\mathcal{P} = \langle P, A, \to, p_0 \rangle$ and $\mathcal{P}' = \langle P', A', \to, p'_0 \rangle$, and let $\mathcal{E}$ be some test $\langle E, A'' \cup \{w\}, \to, e_0 \rangle$ such that $\mathcal{P}$ *must* $\mathcal{E}$. We must prove that $\mathcal{P}'$ *must* $\mathcal{E}$. This can be accomplished by showing any configuration $c = p'_1||e_1$ in an arbitrary computation of $p'_0||e_0$ is potentially successful. Let s be the visible content of this computation from $p'_0||e_0$ to c; then $p'_0 \overset{s}{\Rightarrow} p'_1$ and $e_0 \overset{s}{\Rightarrow} e_1$.

The proof now splits into two cases. For the first, suppose there is some configuration $c' = p'_2||e_2$ such that $c \longmapsto^* c'$ and $S(p'_2) \cap S(e_2) = \emptyset$. The fact that c is potentially successful follows from Lemma A.1 and Theorem 5.4. For the second case suppose that every $c' = p'_2||e_2$ with $c \longmapsto^* c'$ is such that $S(p'_2) \cap S(e_2) \neq \emptyset$. This means the visible content of *every* computation of c will be an element of $\Sigma(p'_1)$. Let X denote this induced nonempty subset of $\Sigma(p'_1)$. From the definition of $<<_{must}$, we can infer the existence of s' and p_1 such that $s' \prec X$ and $p_0 \overset{ss'}{\Rightarrow} p_1$. Further, if $s't \prec X$ then $S(p_1, t) \subseteq S(p'_1, s't)$. Since $s' \prec X$ $s' \in L(p'_1) \cap L(e_1)$. As $e_0 \overset{s}{\Rightarrow} e_1$, it is clear that there is some e_2 such that $e_0 \overset{ss'}{\Rightarrow} e_2$. So a prefix of a computation of $\mathcal{P}||\mathcal{E}$ with the following structure can be constructed: $p_0||e_0 \longmapsto^* p_1||e_2$ and the visible content of this part of the computation is ss'. As $\mathcal{P}$ must $\mathcal{E}$, $p_1||e_2$ must be potentially successful.

Now we claim that if $t \in L(p_1) \cap L(e_2)$ then $s't \in L(p_1') \cap L(e_1)$. The proof of this claim will be done by induction on the length of the string t. The basis case is $t = \epsilon$. Then $s' \in L(p_1') \cap L(e_1)$, since s' is the visible content of some prefix of a computation of $p_1'||e_1$. For the induction step assume $t = t'a$ for some visible action a and $t'a \in L(p_1) \cap L(e_2)$ and $s't' \in L(p_1') \cap L(e_1)$. Then $s't'$ is the visible content of some prefix of a computation of $p_1'||e_1$ and so by definition of X, $s't' \prec X$. Hence $\mathcal{S}(p_1, t')$ is a subset of $\mathcal{S}(p_1', s't')$. Because $t'a$ is an element of $L(p_1)$, a belongs to $\mathcal{S}(p_1, t')$ as well as $\mathcal{S}(p_1', s't')$. Consequently $s't'a$ belongs to the language of p_1'. Moreover $e_1 \overset{a}{\Rightarrow} e_2$ and $t'a \in L(e_2)$ implies $s't'a$ is also an element of $L(e_1)$. As $s't'a$ belongs to the languages of both p_1' and e_1, the claim follows immediately.

Since $p_1||e_2$ is potentially successful there exists some $p_2||e_3$ such that $p_1||e_2 \longmapsto^*$ $p_2||e_3$ and $e_3 \overset{w}{\to}$. Let t be the visible content of this part of the computation. Clearly $t \in L(p_1) \cap L(e_2)$, and because of the claim proved in the previous paragraph it is also the case that $s't \in L(p_1') \cap L(e_1)$. So there exists process p_3' such that $p_1' \overset{s't}{\Rightarrow} p_3'$. Since we know $e_1 \overset{s't}{\Rightarrow} e_3$ and $e_3 \overset{w}{\to}$, it follows that the configuration $c = p_1'||e_1$ is potentially successful. $\square$

Lemma A.3 *Let $\mathcal{P}$ and $\mathcal{P}'$ be processes. Then $\mathcal{P} \sqsubseteq_{must} \mathcal{P}'$ implies $\mathcal{P} \ll_{must} \mathcal{P}'$.*

Proof. Let $\mathcal{P} = \langle P, A, \to, p_0 \rangle$ and $\mathcal{P}' = \langle P', A', \to, p_0' \rangle$. Fix p_1' such that $p_0' \overset{s}{\Rightarrow} p_1'$ for some s in $(A' - \{\tau\})^*$, and let X be some subset of $\Sigma(p_1') \cup \{\epsilon\}$. We will construct a test $\mathcal{E}$ so that $\mathcal{P}'$ fails it. Since $\mathcal{P}'$ and $\mathcal{P}$ are related by $\sqsubseteq_{must}$, $\mathcal{P}$ must also fail $\mathcal{E}$. The test $\mathcal{E}$ will be constructed in such way that when $\mathcal{P}$ fails $\mathcal{E}$, it follows that $\mathcal{P}$ possesses the property given in Definition 5.6.

To define $\mathcal{E} = \langle E, A' \cup A, \to, e_0 \rangle$ let $E_1 = \{ [t] \mid t$ *is a proper prefix of s* $\}$ and $E_2 = \{ [st] \mid t \prec X \}$ and $E_3 = \{[Succ]\}$ where $Succ$ does not belong to $(A \cup A')^*$. Take $E = E_1 \cup E_2 \cup E_3$. $\to$ is the smallest subset of $E \times (A \cup A') \times E$ satisfying the following conditions: (1) $[t] \overset{a}{\to} [ta]$; (2) $e \overset{\tau}{\to} [Succ]$ if $e \in E_1$; (3) $[Succ] \overset{w}{\to} [Succ]$; and (4) $[st] \overset{a}{\to} [Succ]$ if $[st] \in E_2$ and a is some visible action belonging to $(A \cup A') - \mathcal{S}(p_1', t)$. Also take $e_0 = [\epsilon]$. Note that if $[s'] \in E_1 \cup E_2$ and $e_0 \overset{t}{\Rightarrow} [s']$ then $t = s'$. Note also that $\{ t \in (A' - \{\tau\})^* \mid [s] \overset{t}{\Rightarrow} [Succ]\} \cap L(p_1') = \emptyset$. Thus the configuration $p_1'||[s]$ is not potentially successful. As $p_0' \overset{s}{\Rightarrow} p_1'$ and $e_0 \overset{s}{\Rightarrow} [s]$, it is possible to construct a computation such that $p_0'||e_0 \longmapsto^* p_1'||[s]$. Consequently $\mathcal{P}'$ fails $\mathcal{E}$.

Since $\mathcal{P} \sqsubseteq_{must} \mathcal{P}'$, it must be the case that $\mathcal{P}$ fails $\mathcal{E}$ as well, i.e. there is a $c = p_1||e_1$ such that $p_0||e_0 \longmapsto^* c$ and c is not potentially successful. $e_1 \notin E_1$ since all the elements of E_1 have a τ-transition to $[Succ]$. So $e_1 \in E_2$ and hence has the form $[ss']$ for some $s' \prec X$. Since $e_0 \overset{t}{\Rightarrow} [ss']$ implies $t = ss'$, there exists p_1 such that $p \overset{ss'}{\Rightarrow} p_1$ and $p_1||e_1$ is *not* potentially successful. Now we must prove that for all t such that $s't \prec X, \mathcal{S}(p_1, t) \subseteq \mathcal{S}(p_1', s't)$. Assume instead the existence of t_0 such that $s't_0 \prec X$ but $\mathcal{S}(p_1, t_0)$ is not contained in $\mathcal{S}(p_1', s't_0)$. Let b be an element of $\mathcal{S}(p_1, t_0) - \mathcal{S}(p_1', s't_0)$. Obviously, $p_1 \overset{t_0 b}{\Rightarrow}$. From the construction of $\mathcal{E}$, $s't_0 \prec X$ implies $ss't_0$ is an element of E_2. Moreover $b \in (A \cup A') - \mathcal{S}(p_1', s't_0)$ implies $[ss't_0] \overset{b}{\to} [Succ]$ whence $[ss'] \overset{t_0 b}{\Rightarrow} [Succ]$. As $p_1 \overset{t_0 b}{\Rightarrow}$ and $e_1 \overset{t_0 b}{\Rightarrow} [Succ]$, we can conclude $p_1||e_1$ is potentially successful which is the desired contradiction. $\square$

Causality for Mobile Processes *

Pierpaolo Degano and Corrado Priami

Dipartimento di Informatica, Università di Pisa
Corso Italia 40, I-56125 Pisa, Italy - {degano,priami}@di.unipi.it
EXTENDED ABSTRACT

Abstract. We study causality in the π-calculus. Our notion of causality combines the dependencies given by the syntactic structure of processes with those originated by passing names. It turns out that two transitions not causally related may although occur in a fixed ordering in any computation, i.e., π-calculus may express implicitly a priority between actions. Our causality relation still induces the same partial order of transitions for all the computations that are obtained by shuffling transitions that are concurrent (= related neither by causality nor by priority). The presentation takes advantage from a parametric definition of process behaviour that highlights the essence of the topic. All the results on bisimulation based equivalences, congruences, axiomatizations and logics are taken (almost) for free from the interleaving theory.

1 Introduction

The study of the behaviour of a distributed system may benefit from knowledge on the causal relation between its events. For example, when debugging a system, it could be very expensive to examine all the observable events which precede in time a detected bug. Much simpler is to look only at the events which have influenced the bug. These are identified by a *causality* relation which traces the effects that an action has on those it causes.

In the literature we find essentially two kinds of non-interleaving semantics for calculi without name-passing like CCS [13], namely the *causal* [24, 8, 6, 19, 2] and the *local* [3, 12] semantics. The former says that an activity t is caused by another, say t', if t' is a necessary condition for the occurrence of t. The locality relation has been introduced for studying the spatial distribution of resources. The only difference between the two notions is that the causal relation operates a cross-update of the causes between the partners of communications, while locality semantics ignores completely communications. Indeed, the two coincide when no communications arise [12, 10].

Recently a few papers [22, 1, 18] begin the study of this problem in the π-calculus [16, 17]. However, many subtle aspects are to be clarified, especially related with the explicit distinction of input and output actions and with the dependencies induced by the usage of names. We assume the reader familiar with the π-calculus in the rest of the Introduction.

---

* Work partially supported by ESPRIT Basic Research Action n.8130 - LOMAPS

Following Boreale and Sangiorgi [1], we consider two kinds of dependencies: those induced by the structure of processes (called *structural*), and those originated when names are bound (here called *link*).

As links in the π-calculus are directed, i.e., the sender and the receiver are clearly identified in a communication, a better account can be given to structural causality. Consider the process $(\nu x)(a.\overline{x}y.b|c.x(z).d)$ and its computation $(\nu x)(a.\overline{x}y.b|c.x(z).d) \xrightarrow{c} (\nu x)(a.\overline{x}y.b|x(z).d) \xrightarrow{a} (\nu x)(\overline{x}y.b|x(z).d) \xrightarrow{\tau} (\nu x)(b|d)$. According to Milner [14], the effect of the communication influences only d, the residual of the receiver of y. On the contrary, b is unaffected because there is no *flow of information* from c to b. Thus, there is no *causal* relation between c and b. They are only *temporally* dependent. This shows that c has implicitly got some *priority* over b. We claim suitable for mobile processes this notion of causality, already studied for CCS in [20]. Note that priority may be ignored with an asynchronous implementation of communications, e.g., through a buffer. The sender writes a value in the buffer and leaves its residual to proceed. The receiver reads the value from the buffer and passes it to its residual. In our example, the read action xy can overlap with or even follow in time the execution of b.

Link dependencies are established when an action uses as link a name bound (through an input or an extrusion) by another. For example, in $P = (\nu a)(\overline{x}a|\overline{a}y)$ the binding of a in the first transition makes its usage possible in the second one. The output on the link a in P can occur only if a has been extruded by the left component of the parallel composition. (Indeed, the external behaviour of P coincides with the one of $(\nu a)\overline{x}a.\overline{a}y$.) Therefore, $\overline{a}y$ is *link* dependent on $\overline{x}(a)$. Analogously, in $y(a).\overline{a}x$ the behaviour of $\overline{a}x$ depends on what is read along the link y. Instead, if the extruded name is the value sent by an output (or the variable instantiated by an input) *no* causality appears due to links. E.g., in $(\nu a)(\overline{x}a|\overline{y}a)$ both components can extrude and thus they are causally independent, although *not* temporally independent. This shows a different kind of priority. It looks like the reduction of the parallel operator, via left merge, i.e. $Q|R = Q\rfloor R + R\rfloor Q$, where in the left summand the first action of Q has priority over R, and symmetrically. Consider now $(\nu a)(\overline{x}a|b(y))$. One may think that a link dependency arises between the extrusion $(\nu a)\overline{x}a$ and the bound input $b(a)$ which instantiates the name y to a. Indeed, the extrusion and the instantiation use the same symbol a to represent actually different names because $(\nu a)(Q|R)$ coincides with $(\nu a)Q|R$ if $a \notin fn(R)$, as (νa) acts as a *static* binder. Therefore, the considered process is structurally congruent to $((\nu a)\overline{x}a)|b(y)$ and the extrusion and the input are clearly independent. Instead, if we perform the free input ba after the extrusion, we establish a priority between the two transitions (see the discussion at the end of Sect. 4 and Def. 7).

Our notion of causality is time independent; of course priority is not. Indeed, given a computation, two transitions not related by either causality or priority can be executed in *any* temporal ordering.

The presentation of our causal semantics for mobile processes is based on the parametric approach introduced in [10]. More in detail, we adopt a very concrete (SOS) transition system whose transitions are labelled by encodings of their proofs. Then, we instantiate it to causal semantics through relabelling functions which maintain in the labels only the relevant information. Actually, the relabelling yields an action, as usual, and a combination of structural and link dependencies. This approach permits to use the standard definitions of bisimulation and to inherit almost for free their axiomatizations, as well as the modal characterizations of processes. More generally, one can re-use in a truly concurrent setting the theory and the tools developed in the interleaving approach.

Lack of space prevents us from reporting more examples, the proofs of our claims and our version of the causal transition system. Also, we are forced to omit a discussion on how our approach can be easily lifted to higher order mobile processes. This shows the robustness of our proposal, as little or no changes are needed. The interested reader can find more details in the full version of the paper.

2 The π-calculus

In this section we briefly recall the π-calculus [17, 16], a model of concurrent processes based on the notion of *naming*.

Definition 1. Let $\mathcal{N}$ be a countable infinite set of names ranged over by $a, b, .., x,$ $y, ..$ with $\mathcal{N} \cap \{\tau\} = \emptyset$. Processes (denoted by $P, Q, R, \ldots \in \mathcal{P}$) are built from names according to the syntax

$$P ::= 0 \mid \pi.P \mid P + P \mid P|P \mid (\nu x)P \mid [x = y]P \mid P(y_1, \ldots, y_n)$$

where π may be either $x(y)$ for *input*, or $\bar{x}y$ for *output* (where x is the *subject* and y the *object*) or τ for *silent* moves. In the sequel, the trailing 0 will be omitted.

The prefix π is the first atomic action that the process $\pi.P$ may perform. The input prefix binds the name y in the prefixed process. Intuitively, some name y is received along the link named x. The output prefix does not bind the name y which is sent along x. Silent prefixes τ denote invisible actions. Summation denotes nondeterministic choice. The operator $\mid$ describes parallel composition of processes. The operator (νx) acts as a static binder for the name x in the process P that it prefixes. In other words, x is a unique name in P different from all the external names. Finally, matching $[x = y]P$ is a choice operator: the process P is activated if $x = y$. $P(y_1, \ldots, y_n)$ is the definition of constants (in the sequel, we will write $\tilde{y}$ to denote the sequence $y_1, \ldots, y_n$).

The early operational semantics for the π-calculus is defined in the SOS style and the label of the transitions are τ for silent actions, $x(y)$ for input, $\bar{x}y$ for free output and $\bar{x}(y)$ for bound output actions. We will use μ as a metavariable for the labels of transitions (distinct from π, the metavariable for prefixes, although coincident in the first three cases below). The free (fn), bound names (bn) and names (n) of a label μ are defined in the standard way, and are extended to processes. In the sequel, we assume the *structural congruence* $\equiv$ on processes defined as the least congruence satisfying the following clauses:

- P and Q α-equivalent (they differ in the choice of bound names only) $\Rightarrow$ $P \equiv Q$,
- $(\mathcal{P}/_\equiv, +, \mathbf{0})$ is a commutative monoid,
- $[x = x]P \equiv P$,
- $(\nu x)\mathbf{0} \equiv \mathbf{0}$, $(\nu x)(\nu y)P \equiv (\nu y)(\nu x)P$ (often written $(\nu x, y)P$) and $(\nu x)P \equiv P$ if $x \notin fn(P)$.

Note that the $|$ is *not* commutative.

We call a *variant* of a transition $P \xrightarrow{\mu} Q$ a transition which only differs in that P and Q have been replaced by structurally congruent processes, and μ has been α-converted, where a name bound in μ includes Q in its scope.

Then, we consider the standard early transition system of the π-calculus generated by the *SOS* rules reported in [16].

3 Proved transition system

We enrich the labels of the standard interleaving transition system in order to encode more information, in the style of [7, 2]. In this way it is possible to derive different semantic models for the π-calculus by extracting new kinds of labels from the enriched ones, as done in [10] for *CCS*. In the next section, we apply this technique to derive a causal version of the π-calculus. We start with the definition of the enriched labels (*proof terms*). Moreover, we introduce a function (ℓ) that takes a proof term to the corresponding standard action label.

Definition 2. Let $\vartheta \in \{||_0, ||_1\}^*$. Then *proof terms* (with metavariable θ) are defined by the following syntax

$$\theta ::= \vartheta\mu \mid \vartheta\langle\vartheta_0\mu_0, \vartheta_1\mu_1\rangle$$

with $\mu_0 = x(y)$ (resp. xy) iff μ_1 is $\overline{x}(y)$ (resp. $\overline{x}y$), or vice versa.
Function ℓ is defined as $\ell(\vartheta\mu) = \mu$ and $\ell(\vartheta\langle\vartheta_0\mu_0, \vartheta_1\mu_1\rangle) = \tau$.

Our version of the early transition system for the π-calculus is reported in Tab. 1, where the symmetric rules for communications (Com_1 and $Close_1$) are omitted. Again, the transitions in (the conclusion of) each rule stand for all their variants. We call this transition system *proved*, because the labels of the transitions are encodings of (portions of) their proofs. Only the parallel structure of processes is encoded, as this suffices to derive causality.

The proved transition system differs from the standard one essentially in the rules for parallel composition and communication. The rule Par_0 (Par_1) adds to the label a tag $||_0$ ($||_1$) to record that the left (right) component is moving. The rules Com_0 and $Close_0$ have in their conclusion a pair instead of a τ to record the components which interacted. Their symmetric version Com_1 and $Close_1$ should be obvious.

The standard interleaving semantics is obtained from the proved transition system by relabelling each transition through function ℓ of Def. 2.

Often, we will write a transition $P \xrightarrow{\theta} Q$ simply as θ, when unambiguous.

Table 1. Early proved transition system of the π-calculus

$Act : \mu.P \xrightarrow{\mu} P$

$Ein : x(y).P \xrightarrow{xw} P\{w/y\}$

$Par_0 : \dfrac{P \xrightarrow{\theta} P'}{P|Q \xrightarrow{\|_0\theta} P'|Q}, bn(\ell(\theta)) \cap fn(Q) = \emptyset$

$Sum : \dfrac{P \xrightarrow{\theta} P'}{P+Q \xrightarrow{\theta} P'}$

$Par_1 : \dfrac{P \xrightarrow{\theta} P'}{Q|P \xrightarrow{\|_1\theta} Q|P'}, bn(\ell(\theta)) \cap fn(Q) = \emptyset$

$Res : \dfrac{P \xrightarrow{\theta} P'}{(\nu x)P \xrightarrow{\theta} (\nu x)P'}, x \notin n(\ell(\theta))$

$Close_0 : \dfrac{P \xrightarrow{\vartheta \bar{x}(y)} P', Q \xrightarrow{\vartheta' x(y)} Q'}{P|Q \xrightarrow{\langle\|_0\vartheta\bar{x}(y),\|_1\vartheta' x(y)\rangle} (\nu y)(P'|Q')}$

$Open : \dfrac{P \xrightarrow{\vartheta\bar{x}y} P'}{(\nu y)P \xrightarrow{\vartheta\bar{x}(y)} P'}, x \neq y$

$Com_0 : \dfrac{P \xrightarrow{\vartheta\bar{x}y} P', Q \xrightarrow{\vartheta' xy} Q'}{P|Q \xrightarrow{\langle\|_0\vartheta\bar{x}y,\|_1\vartheta' xy\rangle} P'|Q'}$

$Ide : \dfrac{P\{\tilde{y}/\tilde{x}\} \xrightarrow{\theta} P'}{Q(\tilde{y}) \xrightarrow{\theta} P'}, Q(\tilde{x}) = P$

4 Causality

Now, we define the notion of dependency of transitions that occur in a computation (a sequence of transitions in which the target state of a transition is the source state of the next one). From this, it is immediate to recover the standard representation of causality as a partial order of events. Later on, we will show that the same partial order is obtained by shuffling independent transitions in a computation. In other words, causality is a time-independent notion.

Following Boreale and Sangiorgi [1], we consider two kinds of dependencies: those induced by the structure of processes (called *structural*), and those originated when names are bound (called *link*). Our transitions will be labelled by an action, and by a combination of structural and link dependencies.

Structural dependencies coincide with the read-write ones, as defined for CCS [20]: when a communication occurs, the sender transmits its causes also to the residual of the receiver, but not vice versa. Indeed, reading a value does not causally affect the evolution of the residual of the sender (see also the *effect* of a communication as introduced by Milner in [14]).

Link dependencies are established when an action uses as link a name bound by another through an input or an extrusion.

The definition of causality between the transitions of a computation is given in three steps. Roughly, the first pertains to structural dependencies. It says that a transition labelled θ_n depends on a transition labelled θ_h if the proof part of θ_h is a prefix of the proof part of θ_n (with the tuning needed to cover communications). The underlying idea is that the two transitions have been derived using the same initial set of rules and thus they are nested in a prefix chain (or they are connected by communications in a similar way).

Definition 3. Let $P_0 \xrightarrow{\theta_0} P_1 \xrightarrow{\theta_1} \ldots \xrightarrow{\theta_n} P_{n+1}$ be a proved computation, and in the following let i, as well as j, be either 0 or 1. Then, θ_h has *structural dependency* on θ_n (written $\theta_h \sqsubseteq_{str}^1 \theta_n$) iff $0 \leq h \leq n$, and either

- $\theta_n = \vartheta\mu$, $\theta_h = \vartheta'\mu'$ and ϑ' is a prefix of ϑ; or
- $\theta_n = \vartheta\mu$, $\theta_h = \vartheta'\langle\vartheta_0'\mu_0',\vartheta_1'\mu_1'\rangle$ and $\vartheta'\vartheta_j'$ is a prefix of ϑ; or
- $\theta_n = \vartheta\langle\vartheta_0\mu_0,\vartheta_1\mu_1\rangle$, $\theta_h = \vartheta'\mu'$, ϑ' is a prefix of $\vartheta\vartheta_i$ and μ_i is an output action; or
- $\theta_n = \vartheta\langle\vartheta_0\mu_0,\vartheta_1\mu_1\rangle$, $\theta_h = \vartheta'\langle\vartheta_0'\mu_0',\vartheta_1'\mu_1'\rangle)$, $\vartheta'\vartheta_j'$ is a prefix of $\vartheta\vartheta_i$ and μ_i is an output action.

The *structural* dependencies of θ_n are obtained by reflexive and transitive closure of $\sqsubseteq_{str}^1$, i.e., $\sqsubseteq_{str} = (\sqsubseteq_{str}^1)^*$.

The last two items of Def. 3 say that a transition θ causes a communication if it causes the output component of the communication. We report an example to show the need of imposing μ_i to be an output action in the above definition. Consider the process $P_0 = (\nu b,c)((a.\bar{b}x|b(y).c(z))|\bar{c}w.d)$ and its computation $P_0 \xrightarrow{||_0||_0a}$ $P_1 \xrightarrow{||_0\langle||_0\bar{b}x,||_1bx\rangle} P_2 \xrightarrow{\langle||_0||_1cw,||_1\bar{c}w\rangle} P_3 \xrightarrow{||_1d} P_4 = (\nu b,c)((0|0)|0)$. If we ignore the condition on μ_i being an output, the following relations hold between these transitions: $||_0||_0a \sqsubseteq_{str}^1 ||_0\langle||_0\bar{b}x,||_1bx\rangle \sqsubseteq_{str}^1 \langle||_0||_1cw,||_1\bar{c}w\rangle \sqsubseteq_{str}^1 ||_1d$. In particular, the second communication inherits the reference to the transition $||_0||_0a$ through its input component. By transitive closure $||_0||_0a \sqsubseteq_{str} ||_1d$, that erroneously makes the residual of the writer (d) inherit the causes of the reader.

The second step defines link dependencies. It is simplified by noting that only extrusions do actually generate these dependencies. This is because a link dependency between an input which binds a name y and its following usage always induces also a structural dependency. In the process $P = x(y).Q$ the scope of the binding occurrence of y is Q. Since Q is guarded by $x(y)$, the prefixes in which y occurs are structurally dependent upon the input. The binding rules show that the input $x(y)$ in $P|R$ has no influence upon R. Later on we will combine structural and link dependencies, thus we may safely ignore input bindings in the following definition.

Definition 4. Let $P_0 \xrightarrow{\theta_0} P_1 \xrightarrow{\theta_1} \ldots \xrightarrow{\theta_n} P_{n+1}$ be a computation. Then, the *link* dependency of θ_n is the unique θ_h (in symbol, $\theta_h \sqsubseteq_{lnk} \theta_n$) iff h ($0 \leq h < n$) is the maximum index such that $\ell(\theta_h) = \bar{x}(a)$, and $\ell(\theta_n)$ is either $\bar{a}z$ or $\bar{a}(z)$ or az or $a(z)$.

Note that there is no need for implementing the cross inheritance of link dependencies after a communication. Indeed, if one component of a communication has the form $\vartheta \overline{x}(a)$, the link is made local to the residual of the communicating processes via (νa) (as a *Close* rule is used, if the sender performs a bound output).

Consider the computation $(\nu a)(\overline{x}(a)|b(y).y) \xrightarrow{||_0 \overline{x}(a)} (0|b(y).y) \xrightarrow{||_1 ba} (0|a) \xrightarrow{||_1 a}$ $(0|0)$, where the binder of the name a in the last transition is the input ba on which $||_1 a$ depends structurally. However, the name a has been extruded by the first transition. Hence, we establish a link dependency of $||_1 a$ on $||_0 \overline{x}(a)$.

It is now easy to relabel a computation in the proved transition system, in order to make explicit causality in our third step. All the causes of a transition t are the union of its structural dependencies, of its link dependency t', and of the set containing the link and structural causes of t'. The reason for the presence of the last set is vindicated by the following example. Consider the process $(\nu b, c)((a.\overline{b}x|b(y).\overline{y}c)|\overline{c})$. If link and structural causes are kept distinct, the only action on which $\overline{c}$ depends is $\overline{x}(c)$. However, the extrusion depends on a (via $\overline{b}x$) and so should $\overline{c}$. Thus, the transitive closure of the union of structural and link dependencies is mandatory. All dependencies of a transition are captured by the *causal* relation $\sqsubseteq = (\sqsubseteq_{str} \cup \sqsubseteq_{lnk})^*$.

We relabel each visible proved transition with a pair $ct = \langle \mu, K \rangle$ where the first component is the standard action label and the second component is the set of its causes. For simplifying the presentation, we adopt the reference mechanism of unique names for transitions introduced by Kiehn [12]. (Only some auxiliary definitions are needed to encode causes as backward pointers, as in [6]). As usual, we omit from the set of causes the self-reference (condition $h \neq k$ in Def. 5).

Definition 5. Let $\xi = P_0 \xrightarrow{\theta_0} P_1 \xrightarrow{\theta_1} \ldots \xrightarrow{\theta_n} P_{n+1}$ be a proved computation. Its associated causal computation $Ct(\xi)$ is derived by relabelling any transition θ_k as ct_k, where

$$ct_k = \begin{cases} \tau & \text{if } \ell(\theta_k) = \tau \\ \langle \ell(\theta_k), \{h \neq k | \theta_h \sqsubseteq \theta_k, \ell(\theta_h) \neq \tau\} \rangle & \text{otherwise} \end{cases}$$

5 Locality, enabling and priority

Other truly concurrent semantics can be defined easily, by slightly changing Def. 3. If one is interested in a locality model in which communications are completely ignored, also link dependencies may be discarded. Indeed, a link dependency between two actions is also structural when they occur at the same location (or one at a sub-location of the other). So, it is enough to redefine the causal relation $\sqsubseteq$ in order to ignore silent transitions as follows

$$\theta_h \sqsubseteq^1_{loc} \theta_n \text{ iff } \theta_n = \vartheta \mu, \ \theta_h = \vartheta' \mu' \text{ and } \vartheta' \text{ is a prefix of } \vartheta.$$

Then we define $\sqsubseteq_{loc} = (\sqsubseteq^1_{loc})^*$. The above definition amounts to take only the reflexive and transitive closure of the first item of Def. 3.

With the relation above, we obtain the same local semantics of [22] (for the common fragment of the language).

By deleting from the definition of $\sqsubseteq_{str}$ the conditions that μ_i is an output action one recovers a weaker notion of structural dependency ($\sqsubseteq'_{str}$, called *subject dependency* in [1]) that does not distinguish between senders and receivers. In a calculus without name passing, the latter notion coincides with the usual notion of causality [8, 6, 12].

In the π-calculus it is possible to define implicitly priorities between actions, i.e., to sequentialize in a fixed ordering transitions which are causally independent in all computations where they occur. An instance of priority comes from the distinction between input and output actions. As an example, consider the process $(\nu x)(a.\overline{x}y.b|c.x(z).d)$ of the Introduction, where c has priority over b, although the two are causally independent. A second kind of implicit priority arises from the usage of names. See the process $(\nu a)(\overline{x}a|\overline{y}a)$ also discussed in the Introduction, where the restriction (νa) acts as a sequentializer of the bound and free outputs. The former output enables the latter to be free. Priority has been studied for processes calculi, among the others, by [5, 4, 9], even if in different ways. Here, we define priority in two steps, according to its different kinds exemplified above. Since the priority due to communications is induced by the structure of processes, we call it *structural priority*.

Definition 6. Let $P_0 \xrightarrow{\theta_0} P_1 \xrightarrow{\theta_1} \ldots \xrightarrow{\theta_n} P_{n+1}$ be a proved computation, and in the following let i, as well as j, be either 0 or 1. Then, $\theta_h \ll^1_{str} \theta_n$ iff $0 \le h \le n$, and either

- $\theta_n = \vartheta\langle\vartheta_0\mu_0, \vartheta_1\mu_1\rangle$, $\theta_h = \vartheta'\mu'$, ϑ' is a prefix of $\vartheta\vartheta_i$ and μ_i is an input action; or
- $\theta_n = \vartheta\langle\vartheta_0\mu_0, \vartheta_1\mu_1\rangle$, $\theta_h = \vartheta'\langle\vartheta'_0\mu'_0, \vartheta'_1\mu'_1\rangle$, $\vartheta'\vartheta'_j$ is a prefix of $\vartheta\vartheta_i$ and μ_i is an input action.

Moreover,

$$(\theta_k \sqsubseteq_{str} \theta_h \ll^1_{str} \theta_n) \Rightarrow \theta_k \ll^1_{str} \theta_n \text{ and } (\theta_k \ll^1_{str} \theta_h \sqsubseteq_{str} \theta_n) \Rightarrow \theta_k \ll^1_{str} \theta_n.$$

The *structural* priorities of θ_n are obtained by reflexive and transitive closure, i.e., $\ll_{str} = (\ll^1_{str})^*$.

Note that the last condition of Def. 6 says that priority is preserved by causality.

We now consider priority imposed by restriction, which is due to the usage of names. We call it *object priority*, to recall that the prioritized name is an object and not a link.

Definition 7. Let $P_0 \xrightarrow{\theta_0} P_1 \xrightarrow{\theta_1} \ldots \xrightarrow{\theta_n} P_{n+1}$ be a computation. Then, the *object* priority of θ_n is the unique θ_h (in symbol, $\theta_h \ll_{obj} \theta_n$) where h $(0 \le h \le n)$ is the maximum index such that $\ell(\theta_h) = \overline{x}(a)$ and $\ell(\theta_n)$ is either $\overline{y}a$ or ya.

Finally, the *priority* relation is defined by the reflexive and transitive closure of the union of the structural and of the object priorities $\ll = (\ll_{str} \cup \ll_{obj})^*$.

The following proposition compares structural and object priority with the structural and link causality, assuming as given a computation.

Proposition 8. $\ll_{str} \cap \sqsubseteq_{str} = \{\langle \theta_h, \theta_h \rangle\}$, *and*
$\ll_{obj} \cap \sqsubseteq_{lnk} = \{\langle \theta_i, \theta_j \rangle | \theta_i \text{ is } \overline{x}(a) \text{ and } \theta_j \text{ is either } \overline{a}a \text{ or } aa\}.$

The first equation above holds because the first two items of Def. 6 are complementary to the corresponding ones of Def. 3. The second equation is proved by comparing Def. 4 and 7. Note that the two equations of the above proposition do not imply $(\ll \cap \sqsubseteq) = \{\langle \theta_h, \theta_h \rangle\} \cup \{\langle \theta_i, \theta_j \rangle | \theta_i \text{ is } \overline{x}(a) \text{ and } \theta_j \text{ is either } \overline{a}a \text{ or } aa\}.$ A counter-example is the process $(\nu a)\overline{x}a.\overline{y}a$ in which the second transition has a structural dependency on the first, and the bound output $\overline{x}(a)$ has an object priority on the free output $\overline{y}a$.

If one wants to extract from a computation a partial ordering that expresses also priorities, it suffices to take the subject dependency relation $\sqsubseteq'_{str}$ in place of $\sqsubseteq_{str}$ and to weaken the definition of $\sqsubseteq_{lnk}$ as follows. Let $\theta_h \sqsubseteq'_{lnk} \theta_n$ if $\theta_h \sqsubseteq_{lnk} \theta_n$ or if θ_n uses the name extruded by θ_h as an object of an output or of a free input. We call *enabling* the relation defined as $\preceq = (\sqsubseteq'_{str} \cup \sqsubseteq'_{lnk})^*$.

Note that enabling is not the union of causality and priority. Indeed, we have $(\ll \cup \sqsubseteq) \subset \preceq$. The strict inclusion is shown by the process $(\nu b, y)(\overline{a}.b|\overline{b}.\overline{x}y|\overline{y}z)$ where $||_0||_0\overline{a} \preceq ||_1\overline{y}z$ and neither $||_0||_0\overline{a} \ll ||_1\overline{y}z$ nor $||_0||_0\overline{a} \sqsubseteq ||_1\overline{y}z$.

Further comparisons among the introduced relations are reported in Tab. 2 at the end of the next section.

6 Concurrency

Our notion of causality is actually time-independent, as expected. We need a symmetric notion of independency between the transitions of a computation. However it cannot be the complement of causality because of priority. Roughly, two transitions are *concurrent* (for short , $\theta_h \smallsmile \theta_n$) whenever a process P can can fire them one before the other *and* vice versa. More formally, we have the following definition.

Definition 9. Let $C[\bullet, \bullet], C'[\bullet, \bullet], C_i[\bullet, \bullet]$ be contexts with two holes. If the proved transition system contains the diamond

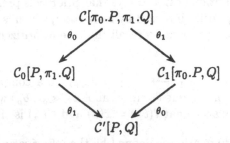

with the actions θ_0 (θ_1) originated by the same prefix π_0 (π_1), then $\theta_0 \smallsmile \theta_1$. We say that the computations $\theta_0\theta_1$ and $\theta_1\theta_0$ of Def. 9 are related by a *2-shuffling* (in symbol $\theta_0\theta_1 \asymp^2 \theta_1\theta_0$).

For the sake of simplicity, the above definition considers transitions which are not communications; to take these into account, contexts suffice with three (one communication) or with four (two communications) holes. The definition of 2-shuffling is extended to computations of any length ξ and ξ' as follows

$$\xi \asymp^2 \xi' \text{ iff } \xi = \xi_0 \theta\theta' \xi_1, \; \xi' = \xi_0 \theta'\theta \xi_1, \text{ and } \theta\theta' \asymp^2 \theta'\theta.$$

Also, let $\asymp = (\asymp^2)^*$ and say that a computation ξ is a *shuffling* of a computation ξ' if one can be obtained by repeated 2-shufflings from the other. This permits shuffling of non-adjacent transitions. Note that $\asymp$ is an equivalence relation.

The time-independency property of our notion of causality is established below. Intuitively, two proved computations differing for the order in which concurrent transitions are fired generate the same partial order of transitions. Some notation can be useful. Given a proved computation ξ, the labelled partial ordering induced by its transitions is the triple $\langle \xi, \ell, \sqsubseteq_\xi \rangle$ where the labelling function associates to each transition in ξ its standard action label μ and $\sqsubseteq_\xi$ is the causal relation on the transitions of ξ. Then, we have the following theorem.

Theorem 10. *Given two proved computations ξ and ξ' of a process, $\xi \asymp \xi'$ iff there exists a label- and ordering-preserving isomorphism between $\langle \xi, \ell, \sqsubseteq_\xi \rangle$ and $\langle \xi', \ell, \sqsubseteq_{\xi'} \rangle$.*

We end this section by comparing concurrency with the relations introduced previously. Also, let the relation $\mathcal{I}$ denote the *flow of time*. Thus, given two transitions θ_h and θ_n of a computation, we define $\theta_h \mathcal{I} \theta_n$ iff $h \leq n$. Since we represent processes as transition systems, concurrency is reduced to interleaving and nondeterminism (see Def. 9). Therefore, if $\mathcal{R}$ is any of $\sqsubseteq, \sqsubseteq_{loc}, \ll, \preceq$ and $\smile$, then $\theta_h \mathcal{R} \theta_n \Rightarrow \theta_h \mathcal{I} \theta_n$.

The relations between causality, locality, enabling, concurrency and priority are collected all together in Tab. 2. The relations indexing the rows (resp. columns) are the left (resp. right) operands of the set operators in the entries of the table. E.g., the entry in the row $\sqsubseteq_{loc}$ and in the column $\preceq$ means $\sqsubseteq_{loc} \subseteq \preceq$, while the entry in last column of the row $\ll$ means $(\ll \cap \smile) = \emptyset$ and $(\ll \cup \smile) \neq \mathcal{I}$. The entries below the main diagonal of the table are obtained from the symmetric ones (w.r.t. the diagonal).

Table 2. Comparison of dependencies and concurrency relations

	$\sqsubseteq$	$\sqsubseteq_{loc}$	$\preceq$	$\ll$	$\smile$
$\sqsubseteq$	$=$	$\supset$	$\subset$	$\cap \neq \emptyset, \cup \neq \mathcal{I}$	$\cap = \emptyset, \cup \neq \mathcal{I}$
$\sqsubseteq_{loc}$		$=$	$\subset$	$\cap \neq \emptyset, \cup \neq \mathcal{I}$	$\cap = \emptyset, \cup \neq \mathcal{I}$
$\preceq$			$=$	$\supset$	$\cap = \emptyset, \cup = \mathcal{I}$
$\ll$				$=$	$\cap = \emptyset, \cup \neq \mathcal{I}$

7 Equivalences, axiomatizations, logics and HO-calculi

The standard definitions of bisimulations for the π-calculus amount essentially to compare the labels of transitions. In other words, they compare the observable behaviour of a computational step of a system with that of another system. As the relabelling function applied to the proved transition system yields exactly (for each transition) the observable behaviour of a process, we can adopt the standard definitions without any change. Therefore, also the axiomatizations are obtained for free, as well as the modal characterizations of systems. This is due to our parametric approach that permits factorization of work, re-use of established results and separation of concerns, thus enabling one to concentrate on the relevant aspects of a topic.

We consider the early equivalence and refer the reader to [16] for its definition. Then we start studying the equivalences induced by the relations introduced in the previous sections. In the sequel, the symbol $\approx$ indexed by a dependency relation means the equivalence induced by the indexing relation. Obviously, our $\mathcal{I}$ relation induces exactly the interleaving semantics ($\approx$) of [16].

From the above and from the discussion after the definition of $\mathcal{I}$ in Sec. 6, it is immediate that the equivalences induced by any of the relations of the previous sections imply the interleaving equivalence.

All the standard hierarchies of different semantics defined on calculi without value-passing (e.g., in [10]) still hold. The bisimulation based equivalences induced by the local ($\approx_{\sqsubseteq_{loc}}$) and enabling ($\approx_{\preceq}$) semantics sketched in the previous section are incomparable with each other and with our causality equivalence ($\approx_{\sqsubseteq}$), as it already happens for CCS [12, 10, 20]. The equivalences induced by priority ($\approx_{\ll}$) and concurrency ($\approx_{\smile}$) are compared with the others below.

Theorem 11.

- $\approx_{\sqsubseteq} \neq \approx_{\sqsubseteq_{loc}}$, $\approx_{\sqsubseteq} \neq \approx_{\ll}$, $\approx_{\sqsubseteq} \neq \approx_{\preceq}$, and $\approx_{\sqsubseteq} \neq \approx_{\smile}$;
- $\approx_{\sqsubseteq_{loc}} \neq \approx_{\ll}$, $\approx_{\sqsubseteq_{loc}} \neq \approx_{\preceq}$, $\approx_{\sqsubseteq_{loc}} \neq \approx_{\smile}$;
- $\approx_{\ll} \neq \approx_{\preceq}$, $\approx_{\ll} \neq \approx_{\smile}$;
- $\approx_{\preceq} = \approx_{\smile}$.

We end this section with a couple of words on higher order processes, that permit names to represent also processes. A communication may cause processes to migrate. An example is the higher-order π-calculus [21] where a process may be communicated along a link named x. E.g., the process $\overline{x}\langle P \rangle.Q \mid x(X).X$ after the communication becomes $Q \mid P$. Since the place-holder X is *already* present in the receiver and becomes P, we can apply the machinery described so far for first-order π-calculus, without any change. Quite similar is the treatment for $CHOCS$ [23] and for the spawn operations of languages like FACILE [11] or CML [15]. This argument shows the stability of our approach to a parametric semantics of mobile processes.

References

1. M. Boreale and D. Sangiorgi. A fully abstract semantics of causality in the π-calculus. In *Proceedings of STACS'95, LNCS*. Springer Verlag, 1995.
2. G. Boudol and I. Castellani. A non-interleaving semantics for CCS based on proved transitions. *Foundamenta Informaticae*, XI(4):433–452, 1988.
3. G. Boudol, I. Castellani, M. Hennessy, and A. Kiehn. A theory of processes with localities. *Theoretical Computer Science*, 114, 1993.
4. J. Camilleri. An operational semantics for occam. *IJPP*, 18, 1989.
5. R. Cleaveland and M. Hennessy. Priorities in process algebras. In *Proceedings of LICS'89*, pages 193–202, 1989.
6. Ph. Darondeau and P. Degano. Causal trees. In *Proceedings of ICALP'89, LNCS 372*, pages 234–248. Springer-Verlag, 1989.
7. P. Degano, R. De Nicola, and U. Montanari. Partial ordering derivations for CCS. In *Proceedings of FCT, LNCS 199*, pages 520–533. Springer-Verlag, 1985.
8. P. Degano, R. De Nicola, and U. Montanari. A partial ordering semantics for CCS. *Theoretical Computer Science*, 75:223–262, 1990.
9. P. Degano, R. Gorrieri, and S. Vigna. On relating some models of concurrency. In *Proceedings of TAPSOFT'93, LNCS 668*, pages 15–30. Springer-Verlag, 1993.
10. P. Degano and C. Priami. Proved trees. In *Proceedings of ICALP'92, LNCS 623*, pages 629–640. Springer-Verlag, 1992.
11. A. Giacalone, P. Mishra, and S. Prasad. Operational and algebraic semantics for Facile: A symmetric integration of concurrent and functional programming. In *Proceedings ICALP'90, LNCS 443*, pages 765–780. Springer-Verlag, 1990.
12. A. Kiehn. Local and global causes. Technical report, TUM 342/23/91, 1991.
13. R. Milner. *Communication and Concurrency*. Prentice-Hall, London, 1989.
14. R. Milner. Action structures. TR ECS-LFCS-92-249, University of Edinburgh, 1992.
15. R. Milner, D. Berry, and D. Turner. A semantics for ML concurrency primitives. In *Proceedings of POPL'92*, 1992.
16. R. Milner, J. Parrow, and D. Walker. Modal logics for mobile processes. Technical Report ECS-LFCS-91-136, University of Edinburgh, 1991.
17. R. Milner, J. Parrow, and D. Walker. A calculus of mobile processes (I and II). *Information and Computation*, 100(1):1–77, 1992.
18. U. Montanari and M. Pistore. Concurrent semantics for the π-calculus. In *Proceedings of MFPS'95, LNCS*. Springer-Verlag, 1995.
19. V. Pratt. Modelling concurrency with partial orders. *International Journal of Parallel Programming*, 15:33–71, 1986.
20. C. Priami and D. Yankelevich. Read-write causality. In *Proceedings of MFCS'94, LNCS 841*, pages 567–576. Springer-Verlag, 1994.
21. D. Sangiorgi. *Expressing Mobility in Process Algebras: First-Order and Higher-Order Paradigms*. PhD thesis, University of Edinburgh, 1992.
22. D. Sangiorgi. Locality and non-iterleaving semantics in calculi for mobile processes. In *Proceedings of TACS'94, LNCS 789*. Springer-Verlag, 1994.
23. B. Thomsen. *Calculi for Higher Order Communicating Systems*. PhD thesis, Imperial College - University of London, 1990.
24. G. Winskel. Petri nets, algebras, morphisms and compositionality. *Information and Computation*, 72:197–238, 1987.

Internal mobility and agent-passing calculi

Davide Sangiorgi

INRIA- Sophia Antipolis, France. Email: `davide@cma.cma.fr`.

Abstract. In process calculi, *mobility* indicates the possibility of dynamic reconfigurations of the process linkage. *Name-passing* calculi like the π-calculus achieve mobility via communication of names. The names exchanged can be *internal* or *external* . Accordingly, we can distinguish between *internal* and *external* mobility. In [San94b] it is shown that the subcalculus of the π-calculus which only uses internal mobility, called πI, has a simple algebraic theory but, at the same time, is expressive enough to encode, for instance, the λ-calculus.

In this paper, we compare name-passing calculi based on internal mobility with *agent-passing* calculi, i.e., calculi where mobility is achieved via exchange of agents. By imposing bounds on the order of the types of πI and of the Higher-Order π-calculus we define a hierarchy of name-passing calculi based on internal mobility and one of agent-passing calculi. We show that there is an exact correspondence, in terms of expressiveness, between the two hierarchies. This refines and complements previous results on the comparison between name-passing and agent-passing calculi by Thomsen, Sangiorgi and Amadio.

1 Motivations

The π-calculus [MPW92] is a process algebra which originates from CCS [Mil89] and permits a natural modelling of mobility using communication of names. Previous research has shown that the π-calculus has much greater expressiveness than CCS, but also a much more complex mathematical theory. Attempting to understand the reasons for this gap, we introduced πI [San94b], a calculus intermediate between CCS and the π-calculus. πI was obtained by separating the mobility mechanisms of the π-calculus into two, namely *internal* mobility and *external* mobility. The former arises when an input meets a bound output, i.e., the output of a private name; the latter arises when an input meets a free output, i.e., the output of a known name. In πI only internal mobility is retained. A pleasant property of πI is the full symmetry between its input and output constructs. The operators of matching and mismatching, that in the π-calculus implement a form of case analysis on names and are important in the algebraic reasoning, are not needed in the theory of πI.

The theory of πI is very close to that of CCS: Alpha conversion is, essentially, the only new ingredient. But, nevertheless, πI appears to have considerable expressiveness. In [San94b] we analised the encoding of the λ-calculus into πI in detail: The encoding was challenging because all known encodings of the λ-calculus into π-calculus exploit, in an important way, the free-output construct, disallowed in πI. We also hinted that data values, agent-passing calculi, the causality and locality relations among the activities of a system can be modelled in πI much in the same way as they are in the π-calculus [MPW92, Mil93, Tho90, San92, BS94].

In this paper, we continue the study of calculi based on internal mobility. Using the typing system of πI, as inherited from the π-calculus, and imposing some constraint on it, we define the calculi $\{\pi I^n\}_{n \leq \omega}$. A calculus πI^n includes those πI processes which can be typed using types of order n or less than n, and πI^ω is the union of the πI^n's. Informally speaking, the calculi $\pi I^1, \pi I^2, \ldots, \pi I^n, \ldots \pi I^\omega, \pi I$ are distinguished by the "degree" of mobility allowed. πI^1 does not allow mobility at all and is the core of CCS. The hierarchy gives us a classification of mobility and an incremental view of the transition from CCS to π-calculus.

We shall use the above hierarchy also to understand the expressiveness of *agent-passing process calculi* (sometimes called *higher-order* process calculi). In these calculi, agents, i.e., terms of the language, can be passed as values in communications. The agent-passing paradigm is often presented in opposition to the *name-passing* paradigm, followed by the π-calculus where mobility is modelled using communication of names. An important criterion for assessing the value of the two paradigms is the expressiveness which can be achieved. Agent-passing developments of CCS are the calculi *Plain CHOCS* [Tho90], and *Strictly-Higher-Order π-calculus*; the latter, abbreviated $HO\pi^\omega$, is the fragment of the Higher-Order π-calculus [San92] which is purely higher order, i.e., no name-passing feature is present. In Plain CHOCS processes only can be exchanged. In $HO\pi^\omega$ besides processes also abstractions (i.e., functions from agents to agents) of arbitrary high order can be exchanged. Roughly, $HO\pi^\omega$ is as an extension of CCS with the constructs of the simply-typed λ-calculus. As in πI, so in $HO\pi^\omega$ we can discriminate processes according to the order of the types needed in the typing. This yields a hierarchy of agent-passing calculi $\{HO\pi^n\}_{n<\omega}$, where $HO\pi^1$ coincides with πI^1 and $HO\pi^2$ is the core of Plain CHOCS. For each $n \leq \omega$, we compare the agent-passing calculus $HO\pi^n$ with the name-passing calculus πI^{n-}; the latter is a subcalculus of πI^n whose processes respect a discipline on the input and output usage of names similar to those studied in [PS93]. We show that $HO\pi^n$ and πI^{n-} have the same expressiveness, by exhibiting faithful encodings of $HO\pi^n$ into πI^{n-} and of πI^{n-} into $HO\pi^n$. The encodings are fully abstract w.r.t. the reduction relations of the two calculi.

These results establish an exact connection between agent-passing calculi and name-passing calculi based on internal mobility, and strengthen the relevance of the latter calculi.

Related work. We are not aware of other work on isolating or classifying different forms of mobility for name-passing calculi.

Encodings of agent-passing calculi into a name-passing calculus have been studied by Thomsen [Tho90], Sangiorgi [San92] and Amadio [Ama93]. Thomsen and Amadio deal with Plain CHOCS and π-calculus; Sangiorgi with Higher-Order π-calculus and π-calculus. The encoding from $HO\pi^n$ to πI^n used in this paper is a special case of the encoding in [San92] and, when restricted to $HO\pi^2$, it is the same as the encodings in [Tho90] and [Ama93]. The works [Tho90,

San92, Ama93] show that agent-passing can be mimicked using name-passing. In this paper, we push the analysis further, in that: (1) we isolate the specific features of name-passing calculi which make the encodings possible, and (2) we investigate the opposite direction, namely the modelling of name-passing using agent-passing.

The only attempt we know at encoding a name-passing calculus into an agent-passing calculus is by Thomsen [Tho90], who gives a translation of the π-calculus into Plain CHOCS. The translation makes heavy use of a relabelling operator of Plain CHOCS which behaves as a *dynamic binder* — occurrences of names not bound can later become bound. Since we only accept *static binders*, our translation of πI^{n-} into $HO\pi^n$ is quite different from Thomsen's. The absence of relabeling is indeed what distinguishes $HO\pi^2$ from Plain CHOCS.

Due to lack of space, the presentation of this paper is kept rather informal and the most of the proofs are omitted. More details can be found in the full paper[1].

Acknowledgements. I have benefited from discussions with Robin Milner, David N. Turner and David Walker. This research has been supported by the Esprit BRA project 6454 "CONFER".

2 Background: πI and internal mobility

What distinguishes π-calculus from CCS is *mobility*, that is, the possibility that the communication linkage among processes changes at run-time. In the π-calculus there are two mechanisms to achieve mobility: One is through the output of a bound (i.e., private) name, the other is through the output of a free name. We distinguish the two forms of mobility as *internal* and *external*. πI is the subcalculus of π-calculus where only internal mobility is allowed. In this section, we review the syntax and the semantics of the polyadic πI. Symbols $x, y, z, \ldots$ range over the infinite set of names, and P, Q and R over processes. We assume a set of constants, ranged over by D, each of which has a non-negative *arity*.

Definition 1 (πI). The grammar of πI processes is:

$$P ::= \sum_{i \in I} \alpha_i . P_i \mid P \mid P \mid \nu x P \mid D\langle \tilde{y} \rangle$$
$$\alpha ::= \tau \mid x(\tilde{y}) \mid \overline{x}(\tilde{y}) .$$

where each constant D has a unique defining equation of the form $D \stackrel{\text{def}}{=} (\tilde{y})P$.

The process constructs (sum, parallel composition, restriction and constants) are the familiar constructs of CCS and π-calculus. In a sum, when $I = \emptyset$ we get the inactive process, written **0**. Sometimes the trailing **0** is omitted in $\alpha . \mathbf{0}$. The three forms of prefixes correspond to the silent, input and bound output prefixes of the π-calculus. In Definition 1, all tuples $\tilde{y}$ have pairwise different names. When $\tilde{y}$ is empty, we omit the surrounding parenthesis. Binding and

---

[1] To appear as an INRIA Technical Report; also available via anonymous ftp at `cma.cma.fr` as the file `pub/papers/davide/piI.ps.Z`.

alpha conversion are defined as in the π-calculus. We write $P =_{\text{alpha}} Q$ if P and Q are alpha convertible; $\text{fn}(P), \text{bn}(P)$ (resp. $\text{fn}(\alpha), \text{bn}(\alpha)$) are the *free names* and the *bound names* of P (resp. α). In a constant definition $D \stackrel{\text{def}}{=} (\widetilde{y})P$ it must be that $\text{fn}(P) \subseteq \widetilde{y}$.

A key feature of πI is the full symmetry between its input and output constructs. The basic theory of πI, namely operational semantics, bisimilarity, axiomatisation, construction of canonical normal forms, is developed in [San94b]. Alpha conversion represents the main difference w.r.t. the theory of CCS. An exception to this is the appearance of restrictions in the communication rule for πI. Below, we review a few notions needed later.

2.1. Operational semantics and bisimilarity

We write $\overline{\alpha}$, if $\alpha \neq \tau$, for the complementary of α; that is, if $\alpha = x(\widetilde{y})$ then $\overline{\alpha} = \overline{x}(\widetilde{y})$, if $\alpha = \overline{x}(\widetilde{y})$ then $\overline{\alpha} = x(\widetilde{y})$. Table 1 contains the set of the transition rules for πI. We have omitted the symmetric of rule PAR.

$$\text{ALP:} \quad \frac{P =_{\text{alpha}} P' \quad P' \stackrel{\alpha}{\longrightarrow} P''}{P \stackrel{\alpha}{\longrightarrow} P''} \qquad \text{PRE:} \quad \alpha.P \stackrel{\alpha}{\longrightarrow} P$$

$$\text{PAR:} \quad \frac{P \stackrel{\alpha}{\longrightarrow} P'}{P \mid Q \stackrel{\alpha}{\longrightarrow} P' \mid Q} \; \text{bn}(\alpha) \cap \text{fn}(Q) = \emptyset \qquad \text{RES:} \quad \frac{P \stackrel{\alpha}{\longrightarrow} P'}{\nu x P \stackrel{\alpha}{\longrightarrow} \nu x P'} \; x \notin \text{bn}(\alpha) \cup \text{fn}(\alpha)$$

$$\text{COM:} \quad \frac{P \stackrel{\alpha}{\longrightarrow} P' \quad Q \stackrel{\overline{\alpha}}{\longrightarrow} Q'}{P \mid Q \stackrel{\tau}{\longrightarrow} \nu \widetilde{x}(P' \mid Q')} \; \widetilde{x} = \text{bn}(\alpha) \qquad \text{SUM:} \quad \frac{P_i \stackrel{\alpha}{\longrightarrow} P_i', \, i \in I}{\sum_{i \in I} P_i \stackrel{\alpha}{\longrightarrow} P_i'}$$

$$\frac{P \stackrel{\alpha}{\longrightarrow} P'}{D\langle \widetilde{x} \rangle \stackrel{\alpha}{\longrightarrow} P'} \text{if } D \stackrel{\text{def}}{=} (\widetilde{y})Q \text{ and } (\widetilde{y})Q =_{\text{alpha}} (\widetilde{x})P$$

Table 1. The transition system for πI

We call a transition $P \stackrel{\tau}{\longrightarrow} P'$ a *reduction*.

Weak transitions are defined in the usual way. Relation $\Longrightarrow$ is the reflexive and transitive closure of $\stackrel{\tau}{\longrightarrow}$; relation $\stackrel{\alpha}{\Longrightarrow}$ is $\Longrightarrow \stackrel{\alpha}{\longrightarrow} \Longrightarrow$; relation $\stackrel{\widehat{\alpha}}{\Longrightarrow}$ is $\Longrightarrow \stackrel{\alpha}{\Longrightarrow}$ if $\alpha \neq \tau$, is $\Longrightarrow$ otherwise.

Definition 2 (πI bisimilarity). Bisimilarity, written $\approx$, is the largest symmetric relation on processes s.t. $P \approx Q$ and $P \stackrel{\alpha}{\Longrightarrow} P'$ with $\text{bn}(\alpha) \cap \text{fn}(Q) = \emptyset$ imply that there is Q' s.t. $Q \stackrel{\widehat{\alpha}}{\Longrightarrow} Q'$ and $P' \approx Q'$.

In πI, alpha conversion is the only form of name instantiation needed. Alpha conversion is a "harmless" form of substitution, because it does not modify the equalities between names. This explains why, by contrast with π-calculus, in πI bisimilarity is preserved by name instantiations and there is no proliferation of bisimulations (like the late, early or open of the π-calculus).

2.2. Replication and πI$^\omega$

Some presentations of the π-calculus have the replication operator in place of recursion. A replication $! P$ stands for an infinite number of copies of P in parallel. Its transition rule says that if $P \mid !P \xrightarrow{\alpha} P'$, then $!P \xrightarrow{\alpha} P'$. The comparison between replication and recursion is interesting. These operators are notational devices to represent syntactically-infinite objects. Replication yields infinity in width (for instance, $!\alpha.P$ stands for $\alpha.P \mid \ldots \mid \alpha.P$). Recursion, by contrast, can also capture infinity in depth: For instance, if $D \stackrel{\text{def}}{=} (a)\bar{a}(b)$, then $D\langle a_1 \rangle$ stands for $\bar{a}_1(a_2).\bar{a}_2(a_3).\ldots\bar{a}_n(a_{n+1}).\ldots$.

Milner [Mil93] showed that in the π-calculus replication and recursion yield the same expressive power, provided that the number of recursive definitions is finite. We prove in Section 4 that in πI recursion is strictly more powerful. We call πI^ω the language defined as πI but with replication in place of constants.

3 Typing

Having polyadicity, one needs to impose some discipline on names so to avoid run-time arity mismatchings in interactions, as for $x(y).P \mid \bar{x}(y,z).Q$. In the π-calculus, this discipline is achieved by means of a *typing system* (in the literature sometimes called *sorting system*). A typing allows us to specify the arity of a name and, recursively, of the names carried by that name. The same formal systems can be used for the typing of πI processes. We follow the *by-structure* presentation of the typing system, as in [PS93]. Our language for types is:

$$S ::= (\widetilde{S}) \mid X \mid \mu X : S.$$

S and T range over types. X is a type-variable; $\mu X : S$ is a recursive type. Equality between two types S and T, written $S \asymp T$, means that their infinite unfoldings yield the same tree. There are efficient algorithms for checking whether two recursive types denote the same tree; see for instance [Cou83, PS93].

We write $x : S$ and $D : S$ if name x and constant D have type S. Intuitively, $x : (S_1, \ldots, S_n)$ means that x carries n-uples of names whose i-th component has type S_i; similarly, $D : (S_1, \ldots, S_n)$ means that D accepts n-uples of names as parameters, and the i-th name has type S_i.

Definition 3. A *typing* is finite set of assignments of types to names and constants:

$$\Gamma ::= \emptyset \mid \Gamma, x : S \mid \Gamma, D : S.$$

Names and constants appearing in a typing Γ are always taken to be pairwise distinct. $\Gamma[x]$ (resp. $\Gamma[D]$) is the type assigned to x (resp. D) by Γ.

Definition 4. A process P in πI or in πI^ω is *well-typed for* Γ if $\Gamma \vdash P$ can be inferred from the rules of Table 2. P is *well-typed* if there is Γ s.t. P is well-typed for Γ.

Reasoning by transition induction, one can prove a subject reduction property for the type system.

$$\frac{\Gamma[x] \asymp (\widetilde{S}) \quad \Gamma, \widetilde{y}:\widetilde{S} \vdash P}{\Gamma \vdash x(\widetilde{y}).P,\ \Gamma \vdash \overline{x}(\widetilde{y}).P} \qquad \frac{\Gamma \vdash P}{\Gamma \vdash \tau.P} \qquad \frac{\Gamma \vdash P \quad \Gamma \vdash Q}{\Gamma \vdash P \mid Q} \qquad \frac{\Gamma \vdash P_i,\ i \in I}{\Gamma \vdash \sum_{i \in I} P_i}$$

$$\frac{\Gamma, x:S \vdash P,\ \text{for some } S}{\Gamma \vdash \nu x\, P} \qquad \frac{\Gamma \vdash P}{\Gamma \vdash !P} \qquad \frac{\Gamma[D] \asymp (\Gamma[\widetilde{x}]) \quad \widetilde{y}:\Gamma[\widetilde{x}] \vdash P}{\Gamma \vdash D(\widetilde{x})}\ \text{if } D \overset{\text{def}}{=} (\widetilde{y})P$$

Table 2. The typing rules for the operators of πI and πI$^\omega$

---

4 A hierarchy of calculi based on internal mobility

4.1. Non-recursive and order-bounded types Dropping recursion, the language of types simply becomes $S ::= (\widetilde{S})$. We call them *non-recursive types*, and we call a typing that only uses non-recursive types a *non-recursive typing*.

The *order* of a non-recursive type S is the maximal level of bracket nesting in its definition. For instance, () has order 1 and $((), (()))$ has order 3.

If only non-recursive types are used, it makes sense to concentrate on the processes of the language πI$^\omega$ (i.e., the recursion-free processes, Section 2.2): As we shall see in Section 4, the confinement to non-recursive types does not affect the typability of processes in πI$^\omega$, whereas it affects that of processes in πI. We can discriminate processes according to the order of the types needed in the typing. We use ω to denote the first ordinal limit. A non-recursive typing which does not include assignments to constants is a πI$^\omega$ typing.

Definition 5 (calculi $\{\pi\text{I}^n\}_{n<\omega}$). A process $P \in \pi$I$^\omega$ is in πI^n, $n < \omega$, if, for some πI$^\omega$ typing Γ, there is a derivation proof of $\Gamma \vdash P$ in which all types used have order n or less than n.

That is, the typability of processes in πI^n can be established utilising types of order at most n. Thus πI^1 represents the core of CCS, for in πI^1 names can only be used for pure synchronisation. πI^2 includes processes like $x(y, z).(\overline{y} \mid z)$ and $y.x(z).\overline{y}.z$ where if a name carries another name, then the latter can only be used for pure synchronisation. Informally, let us say that a name *depends on* another name if the latter carries the former; for instance, in $x(y).\overline{y}(z).z.\mathbf{0}$, name y depends on x and z depends on y. A *dependency chain* is a sequence $x_1, \ldots, x_n$ of names s.t. x_{i+1} depends on x_i, for all $1 < i < n$. Then the processes in πI^n are those which have dependency chains among names of length at most n; for instance, process $x(y).\overline{y}(z).z.\mathbf{0}$ is in πI^m, for all $m \geq 3$.

Dependency chains are important w.r.t. mobility. If a process has dependency chains of length n at most, then also its traces (i.e., the sequences of actions that the process can perform) have dependency chains of length n at most. In a trace, a dependency between names indicates the creation of a link — hence the creation of mobility. (For instance, if P can perform the action $\overline{x}(z)$, then an interaction in which this action is consumed creates a new link, called z in P.) Similarly, in a trace a dependency chain of length n indicates $n-1$ nested creations of links. Therefore, if a process P, simulating a process Q, has to reproduce the mobility that Q creates, then the dependency chains in traces of P should be at least as long as those in traces of Q (they could be longer: the creation of a new link by Q might be simulated in more than one step by P).

Theorem 6. 1. *There is a trace of a process in πI^n, $n < \omega$, with a dependency chain of length n.*
  2. *No trace of a process in πI^n, $n < \omega$, has a dependency chain of length $n + 1$ or greater than $n + 1$.* □

Theorems 6 shows that the calculi $\{\pi I^n\}_n$ form a strict hierarchy of expressiveness classes: Processes in πI^{n+1} exhibit a "higher degree" of mobility than processes in πI^n. For future investigations, we would like to see if there are stronger formulations of this non-expressiveness result which did not require to explicitly take into account dependency chains.

4.2. Recursion versus replication Since replication is a special case of recursion, every process in πI^ω can be simulated by a process in πI.

Lemma 7. *If $P \in \pi I^\omega$ and well-typed, then also $P \in \pi I^n$, for some $n < \omega$.* □

Corollary 8. *Suppose $P \in \pi I^\omega$. Then there is $n < \omega$ s.t. no traces of P have a dependency chain of length $n + 1$ or greater than $n + 1$.* □

On the other hand, using recursion we can define a process like $D\langle x_1 \rangle$, for $D \stackrel{\text{def}}{=} (x)\overline{x}(y). D\langle y \rangle$, which has traces with unbounded dependency chains.

The typability of process $D\langle x_1 \rangle$ requires recursive types. We expect that recursion and replication become interdefinable if only non-recursive types are allowed, and even if a bound on the order of types is imposed.

5 Agent-passing calculi

In *agent-passing* process calculi (sometimes called *higher-order* process calculi), agents, i.e., terms of language, can be passed around.

For our study of agent-passing process calculi we use the *Higher-Order π-calculus*, a development of the π-calculus introduced in [San92]. Since we want to compare purely-agent-passing calculi with purely-name-passing calculi, we disallow the name-passing features of the Higher-Order π-calculus, namely communication of names and abstraction on names. We call the resulting calculus the *Strictly-Higher-Order π-calculus*, briefly HOπ^ω. We review its syntax in the next subsection; we refer to [San92] for more details.

5.1. The Strictly-Higher-Order π-calculus A HOπ^ω *agent* (or *term*) can be a *process* or an *abstraction*, i.e. a parametrised process. In the following, P and Q stand for processes, F and G for abstractions, A for agents. X, Y range over the set of variables; as in λ-calculus, a variable is supposed to be instantiated with a term. The grammar of HOπ^ω combines familiar CCS-like process constructs with the λ-calculus constructs — abstraction, application and variable.

$$\text{(processes)} \quad P ::= \sum_{i \in I} \alpha_i . P_i \mid P \mid P \mid \nu x P \mid \ !P \mid X\langle \tilde{A} \rangle$$

$$\text{(prefixes)} \quad \alpha ::= \tau \mid x(\tilde{X}) \mid \overline{x}\langle \tilde{A} \rangle$$

$$\text{(agents)} \quad A ::= P \mid F$$

$$\text{(abstractions)} \quad F ::= (\tilde{X})P \mid (\tilde{X})X\langle \tilde{A} \rangle$$

where in the last production, namely $F ::= (\tilde{X})X\langle\tilde{A}\rangle$, expression $X\langle\tilde{A}\rangle$ represents a partial application — i.e., $\tilde{A}$ does not include all arguments that X requires.

Variable application $X\langle\tilde{A}\rangle$ is needed to provide an abstraction received as an input with the appropriate arguments. An abstraction is an agent which takes some arguments before becoming a process. The typical form of an abstraction is $(\tilde{X})P$, where $(\tilde{X})$ represents the formal parameters. For instance, $F \overset{\text{def}}{=} (Y)(P|Y)$ abstracts on a process variable; F takes a process and runs it in parallel with P. We can also abstract on abstraction variables, as in $G \overset{\text{def}}{=} (X)(P \mid X\langle Q\rangle)$; then F applied to G yields $P \mid P \mid Q$. The machinery can be iterated, progressively increasing the order of the resulting abstraction.

An abstraction $(\tilde{X})A$ and an input prefix $x(\tilde{X}).A$ bind all free occurrences of variables $\tilde{X}$ in A. An agent is *open* if it may have free variables in it; *closed* otherwise. Abstraction has the highest precedence among the operators, application the lowest. The notations introduced for the π-calculus, regarding substitutions, tuples, brackets, etc., are extended to $\text{HO}\pi^\omega$ in the obvious way.

We shall only consider well-typed $\text{HO}\pi^\omega$ terms. We ascribe types to $\text{HO}\pi^\omega$ expressions following the type assignment of the simply-typed λ-calculus. The process-type () is our only basic type. We adopt a bracket-nesting notation for functional types to have the same language for types used in Section 4, namely $S ::= (\tilde{S})$. A term of type $S = (S_1, \ldots, S_n)$ takes a sequence of terms of type $S_1, \ldots, S_n$ as arguments before becoming a process. A $HO\pi^\omega$ *typing* is a finite sequence of assignments of types to names and variables:

$$\Gamma ::= \emptyset \mid \Gamma, x : S \mid \Gamma, X : S$$

Judgements are of the form $\Gamma \vdash A : S$, to be read as "under assumptions Γ, agent A has type S". The typing rules are the expected ones. For instance, the rules for abstraction and and output prefix are:

$$\frac{\Gamma, \tilde{X} : \tilde{S}' \vdash A : (\tilde{S})\,, \text{for some } \tilde{S}'}{\Gamma \vdash (\tilde{X})A : (\tilde{S}', \tilde{S})} \qquad \frac{\Gamma[x] = (\tilde{S}) \quad \Gamma \vdash \tilde{A} : \tilde{S} \quad \Gamma \vdash A_1 : ()}{\Gamma \vdash \overline{x}\langle\tilde{A}\rangle. A_1 : ()}$$

If $\Gamma \vdash A : S$ holds, for some S, then we say that A is *well-typed for* Γ. Agent A is *well-typed* if there is Γ s.t. A is well-typed for Γ.

We omit the transitional rules of the calculus, see the full paper or [San92].

5.2. A hierarchy of agent-passing process calculi Similarly to what we did for πI^ω, so from $\text{HO}\pi^\omega$ we define a hierarchy of calculi using the order of their typings.

Definition 9 (calculi $\{\text{HO}\pi^n\}_{n<\omega}$). An agent $A \in \text{HO}\pi^\omega$ is in $\text{HO}\pi^n$, $n < \omega$ if, for some typing Γ and type S, there is a derivation proof for $\Gamma \vdash A : S$ in which all types used have order n or less than n.

In $\text{HO}\pi^1$ no value is exchanged in communications; hence $\text{HO}\pi^1$ coincides with πI^1 and is the core of CCS. In $\text{HO}\pi^2$ only processes can be passed as values in communications; $\text{HO}\pi^2$ is the core of Thomsen's Plain CHOCS [Tho90].

6 Comparison between agent-passing calculi and name-passing calculi

In this section, we let n range over $\{1, 2, \ldots, n \ldots, \} \cup \{\omega\}$. We compare the expressiveness of the calculi $\{HO\pi^n\}_n$ with that of the calculi $\{\pi I^n\}_n$. It turns out that πI^n is slightly more powerful than $HO\pi^n$. To obtain an exact correspondence, we cut down the class πI^n, by imposing a few syntactic conditions on the usage of names in processes. The resulting calculus is called πI^{n-}.

Given a πI^ω process P, we say that an occurrence of a name in P is a *name-variable* if such occurrence is bound by an input prefix of P. For instance, the name-variables have been underlined in $a(b).\,(\underline{\overline{b}}(c).\,c(d) \mid \underline{\overline{b}}(e).\,e(f).\,\underline{f})$.

Definition 10 (calculi $\{\pi I^{n-}\}_{n \leq \omega}$). We call πI^{n-}, $n \leq \omega$, the class of processes in πI^n which satisfy the following syntactic constraint: For any subterm Q of a process in πI^{n-} it holds that

1. if $Q = x(\widetilde{y}).\,R$, then any $y \in \widetilde{y}$ appears free in R only in output position;
2. if $Q = \overline{x}(\widetilde{y}).\,R$, then any $y \in \widetilde{y}$ appears free in R only in input position;
3. if $Q = \overline{x}.R$ and x is a name-variable, then $R = 0$.

Conditions 1 and 2 could also be described using a typing system similar to that proposed in [PS93], where types also carry informations about the input/output usage of names. The results for $\{\pi I^n\}_n$ in Section 4 can be adapted to $\{\pi I^{n-}\}_n$ to prove that also these calculi form a hierarchy in expressiveness.

6.1. From $HO\pi^n$ to πI^{n-}   To translate $HO\pi^n$ into πI^{n-}, we can use the compilation in [San92] from the Higher-Order π-calculus to the π-calculus, since when restricted to $HO\pi^n$ processes, it defines a compilation $\{\![\,]\!\}$ into πI^{n-}. Moreover, the correctness of the compilation in [San92] implies that of $\{\![\,]\!\}$. For the definition of the compilation and other properties about it, we refer to [San92] or the full version of this paper.

In the compilation, the communication of an agent A is translated as the communication of a private name which acts as a pointer to (the translation of) A and which the recipient can use to trigger a copy of (the translation of) A.

Example 1 (from $HO\pi^2$ to πI^{2-}). Let $R \stackrel{\text{def}}{=} \overline{v}.0$ and $P \stackrel{\text{def}}{=} \overline{w}\langle R \rangle.\,0 \mid w(X).\,X$. It holds that $P \stackrel{\tau}{\longrightarrow} R$ (we garbage-collect 0 processes). Translating P we get $\overline{w}(y).\,!\,y.R \mid w(x).\,\overline{x}$ and, using simple algebraic laws, we can infer $\{\![P]\!\} \stackrel{\tau}{\longrightarrow} \nu y\,(\,!\,y.R \mid \overline{y}) \approx R$. We have $P \in HO\pi^2$ and $\{\![P]\!\} \in \pi I^{2-}$.

The computation by a process $\{\![P]\!\}$ may require more reductions than the corresponding computation by P. But if we ignore internal work, then P and $\{\![P]\!\}$ have the "same" behaviour.

Theorem 11 (full abstraction for $\{\![\,]\!\}$ on reductions). For all $P \in HO\pi^n$:

1. If $P \stackrel{\tau}{\longrightarrow} P'$, then $\{\![P]\!\} \stackrel{\tau}{\longrightarrow} \approx \{\![P']\!\}$;

2. the converse, i.e., if $\{\![P]\!\} \xrightarrow{\tau} P''$, then there is P' s.t. $P \xrightarrow{\tau} P'$ and $P'' \approx \{\![P']\!\}$. $\qquad\qquad\qquad\qquad\qquad\qquad\qquad\qquad\qquad\qquad\qquad\qquad$ □

6.2. From πI^{n-} to $HO\pi^n$ The translation $\{\![\,]\!\}$ from $HO\pi^n$ to πI^{n-}, in the previous section, used *name-pointers* to model the communication of agents. The translation $[\,]$ from πI^{n-} to $HO\pi^n$, in this section, uses simple *agent-continuations* to model the communication of private names, in the following way. Suppose that a process of πI^{n-} sends a name y, and that the recipient uses y to send another name z and then becomes the process P. In the translation, the communication of y is replaced by the communication of a continuation which has two parameters. The recipient instantiates the first parameter with the continuation for z and the second parameter with (the translation of) P. Continuations for names whose type has order 1, i.e., names used for pure synchronisation, have one parameter only (since by condition 3 in the definition of πI^{n-}, such names can only prefix the $\mathbf{0}$ process — that is, the process called P above in this case is always $\mathbf{0}$).

Example 2 (from πI^{2-} to $HO\pi^2$). If $P \stackrel{\text{def}}{=} \overline{x}(y).\,y.\,\mathbf{0} \mid x(y).\,\overline{y}.\,\mathbf{0}$, then we have

$$P \xrightarrow{\tau} \nu y\,(y.\,\mathbf{0} \mid \overline{y}.\,\mathbf{0}) \stackrel{\text{def}}{=} P_1 \xrightarrow{\tau} \nu y\,(\mathbf{0} \mid \mathbf{0}) \stackrel{\text{def}}{=} P_2\,.$$

If $\mathrm{Cont}_y \stackrel{\text{def}}{=} \overline{y}.\,\mathbf{0}$, then the translation of P is $\nu y\,y\,\overline{x}\langle\mathrm{Cont}_y\rangle.\,y.\,\mathbf{0} \mid x(Y).\,Y$ and:

$$[P] \xrightarrow{\tau} \nu y\,(y.\mathbf{0} \mid \mathrm{Cont}_y) = \nu y\,(y.\mathbf{0} \mid \overline{y}.\mathbf{0}) = [P_1] \xrightarrow{\tau} \nu y\,(\mathbf{0} \mid \mathbf{0}) = [P_2]\,.$$

Example 3 (from πI^{3-} to $HO\pi^3$). Let $P \stackrel{\text{def}}{=} \overline{x}(y).\,y(z).\,\overline{z}.\,\mathbf{0} \mid x(y).\,\overline{y}(z).\,z.\,\mathbf{0}$; then:

$$
\begin{aligned}
P &\xrightarrow{\tau} \nu y\,(y(z).\,\overline{z}.\,\mathbf{0} \mid \overline{y}(z).\,z.\,\mathbf{0}) &&\stackrel{\text{def}}{=} P_1 \\
&\xrightarrow{\tau} (\nu y, z)(\overline{z}.\,\mathbf{0} \mid z.\,\mathbf{0}) &&\stackrel{\text{def}}{=} P_2 \\
&\xrightarrow{\tau} (\nu y, z)(\mathbf{0} \mid \mathbf{0}) &&\stackrel{\text{def}}{=} P_3\,.
\end{aligned}
$$

If $\mathrm{Cont}_y \stackrel{\text{def}}{=} (W, U)\overline{y}\langle W\rangle.\,U$ and $\mathrm{Cont}_z \stackrel{\text{def}}{=} z.\,\mathbf{0}$, then the translation of P is

$$[P] \stackrel{\text{def}}{=} \nu y\,y\,\overline{x}\langle\mathrm{Cont}_y\rangle.\,y(Z).\,Z \mid x(Y).\,\nu z\,z\,Y\langle\mathrm{Cont}_z, z.\mathbf{0}\rangle\,.$$

We have
$$
\begin{aligned}
[P] &\xrightarrow{\tau} \nu y\,(y(Z).\,Z \mid \nu z\,z\,\overline{y}\langle\mathrm{Cont}_z\rangle.\,z.\,\mathbf{0}) = [P_1] \\
&\xrightarrow{\tau} (\nu y, z)(\overline{z}.\mathbf{0} \mid z.\mathbf{0}) = [P_2] \\
&\xrightarrow{\tau} (\nu y, z)(\mathbf{0} \mid \mathbf{0}) = [P_3]\,.
\end{aligned}
$$

Note that there is a one-to-one match between actions of P and of $[P]$. In the definition of the encoding $[\,]$, the difference between names and name-variables is important: πI^{n-} names are mapped onto $HO\pi^n$ names, whereas πI^{n-} name-variables are mapped onto $HO\pi^n$ variables. The encoding is in Table 3. To ease readability, it is only defined on the subclass of πI^{n-} processes whose names have at most arity one; the generalisation to the calculus with arbitrary arities is straightforward. In the table, in the definition of Cont_y, order[y] is the order of the type assigned to y in a correct typing derivation for the original source process. The encoding is parametrised over a finite set of names, ranged over by V. Occurrences of names in this set have to be treated as name-variables in the translation. We abbreviate $V \cup \{y\}$ as $V \cup y$, $V - \{y\}$ as $V - y$, and $[P]_{\emptyset}$ as $[P]$.

Let Cont_y be the agent $\overline{y}.0$ if $\text{order}[y] = 1$, and $(W, Z)\overline{y}\langle W \rangle.Z$ if $\text{order}[y] > 1$, where variable W has order equal to $\text{order}[y] - 1$, and variable Z has order 1. We then define $[\ ]_V$ thus:

$$[\overline{x}(y).P]_V \stackrel{\text{def}}{=} \begin{cases} \nu y\, \overline{x}\langle \text{Cont}_y \rangle.[P]_{V-y} & \text{if } x \notin V \\ \nu y\, X\langle \text{Cont}_y, [P]_{V-y}\rangle & \text{if } x \in V \end{cases} \qquad [\overline{x}.P]_V \stackrel{\text{def}}{=} \begin{cases} \overline{x}.[P]_V & \text{if } x \notin V \\ X & \text{if } x \in V\, (*) \end{cases}$$

$$[x(y).P]_V \stackrel{\text{def}}{=} x(Y).[P]_{V \cup y} \qquad [x.P]_V \stackrel{\text{def}}{=} x.[P]_V \qquad [\tau.P]_V \stackrel{\text{def}}{=} \tau.[P]_V$$

$$[P \mid Q]_V \stackrel{\text{def}}{=} [P]_V \mid [Q]_V \qquad [\nu x\, P]_V \stackrel{\text{def}}{=} \nu x [P]_V \qquad [\textstyle\sum_{i \in I} P_i]_V \stackrel{\text{def}}{=} \textstyle\sum_{i \in I} [P_i]_V$$

$$[\,! P]_V \stackrel{\text{def}}{=} \,! [P]_V$$

(*) note that by condition 3 of definition of πI^{n-}, if $x \in V$ then $P = 0$.

Table 3. The encoding $[\ ]$ from πI^{n-} to $\text{HO}\pi^n$, $n \leq \omega$.

We extend $[\ ]_V$ to typing as follows: If Γ is a πI^ω typing, then $[\Gamma]_V$ is the $\text{HO}\pi^\omega$ typing obtained from Γ by adding the variable assignments $X : S$, for all name assignments $x : S$ in Γ s.t. $x \in V$.

Proposition 12. *If $P \in \pi I^{n-}$ and is well-typed for Γ, then $[P]_V \in \text{HO}\pi^n$ and is well-typed for $[\Gamma]_V$.* $\qquad\qquad\square$

Theorem 13 (full abstraction for $[\ ]$ on reductions). *For all $P \in \pi I^{n-}$:*

1. *If $P \stackrel{\tau}{\longrightarrow} P'$, then $[P] \stackrel{\tau}{\longrightarrow} [P']$;*
2. *The converse, i.e., if $[P] \stackrel{\tau}{\longrightarrow} P''$, then there is P' s.t. $P \stackrel{\tau}{\longrightarrow} P'$ and $P'' = [P']$.* $\qquad\qquad\square$

7 Conclusions and future work

The work in this paper leads to a classification of name-passing process calculi according to the "degree" of mobility permitted: π-calculus permits both internal and external mobility; πI permits internal mobility; πI^ω permits internal but not recursive mobility; πI^n, $n < \omega$, permits internal mobility of order n at most; πI^1 — the core of CCS — does not permit mobility at all.

This scale can be used for comparing calculi as well as processes. We used it in Section 6 to study the expressiveness of various agent-passing calculi. As another example, in [BS94] the modelling of the *locality* relation only utilises internal mobility of order 3, whereas the modelling of the *causality* relation requires at least internal mobility of order 4; this reflects the fact that causality is a more sophisticated relation than locality.

We have also presented a hierarchy of agent-passing process calculi: In $HO\pi^\omega$ agents of arbitrary order can be communicated; in $HO\pi^n$, $n < \omega$, agents of order n at most can be communicated. Roughly, $HO\pi^1$ coincides with πI^1 and CCS, and $HO\pi^2$ with Thomsen's Plain CHOCS. We have proved that there is a strong connection, in terms of expressiveness, between this hierarchy of agent-passing calculi and the hierarchy of name-passing calculi $\pi I^1, \pi I^2, \ldots, \pi I^\omega$, i.e., the calculi using internal and non-recursive mobility. Rather surprising perhaps, process-passing allows us to achieve only the expressiveness of πI^2 (indeed, something even less than this, namely πI^{2-}), which represents just *the first step* in the hierarchy from CCS to πI or π-calculus.

These are results of *relative* expressiveness. Further work is needed, both to complete the comparison among the above-mentioned calculi, and to understand their *absolute* expressiveness. We are particularly interested in the expressiveness of πI, which we expect to be rather close to that of the π-calculus.

Another topic for future research is how to increase the expressiveness of agent-passing process calculi. The most powerful agent-passing process calculus considered in this paper is $HO\pi^\omega$; we have seen that its expressiveness is not greater than that of πI^ω. To increase the expressiveness of $HO\pi^\omega$ — so as to get closer to that of πI — one might add recursive types to $HO\pi^\omega$. This extension, however, is non-trivial because it breaks important properties of $HO\pi^\omega$ like termination of substitutions.

References

[Ama93] R. Amadio. On the reduction of CHOCS bisimulation to π-calculus bisimulation. *Proc. CONCUR '93*, LNCS 715. Springer Verlag, 1993.

[BS94] M. Boreale and D. Sangiorgi. A fully abstract semantics for causality in the π-calculus. Proc. STACS '95, LNCS 900, Springer Verlag 1995.

[Cou83] B. Courcelle. Fundamental properties of infinite trees. *Theoretical Computer Science*, 25:95–169, 1983.

[Mil89] R. Milner. *Communication and Concurrency*. Prentice Hall, 1989.

[Mil93] R. Milner. The polyadic π-calculus: a tutorial. In *Logic and Algebra of Specification*, Springer Verlag, 1993.

[MPW92] R. Milner, J. Parrow, and D. Walker. A calculus of mobile processes, (Parts I and II). *Information and Computation*, 100:1–77, 1992.

[PS93] B. Pierce and D. Sangiorgi. Typing and subtyping for mobile processes. In *8th LICS Conf.*. IEEE Computer Society Press, 1993.

[San92] D. Sangiorgi. *Expressing Mobility in Process Algebras: First-Order and Higher-Order Paradigms*. PhD thesis CST–99–93, Edinburgh, 1992.

[San94b] D. Sangiorgi. πI: A symetric calculus based on internal mobility. To appear in Proc. ICALP '95, LNCS.

[Tho90] B. Thomsen. *Calculi for Higher Order Communicating Systems*. PhD thesis, Department of Computing, Imperial College, 1990.

Author Index

K. Ambos-Spies	384	M. J. Golin	256
J. H. Andersen	535	S. Greco	523
V. Auletta	232	R. Grossi	111
J. L. Balcázar	208	T. Hagerup	268
F. Bao	147	T. Harju	1,444
F. Bauernöppel	220	L. Hella	624
B. Becker	475	T. A. Henzinger	324, 417
A. M. Ben-Amram	360	H. B. Hunt III	487
J.-C. Bermond	135	Y. Igarashi	147
M. Bertol	27	H. Jürgensen	581
E. Best	605	J. Karhumäki	444
M. Blum	547	J. Kari	51
C. Blundo	171	M. Karpinski	183
H. L. Bodlaender	87, 268	S. Khuller	593
P. Bozanis	464	N. Kitsios	464
S. Chaudhuri	244	J. Köbler	196
Z.-Z. Chen	99	P. G. Kolaitis	624
R. Cleaveland	648	S. Konstantinidis	581
B. Codenotti	547	P. W. Kopke	417
D. G. Corneil	292	M. Koutny	605
M. Cosnard	570	E. Kranakis	220
K. Culik	51	K. J. Kristoffersen	535
A. Dawar	624	D. Krizanc	220
A. De Santis	171	W. Kuich	39
P. Degano	660	M. Kummer	558
V. Diekert	15	S. A. Kurtz	558
R. Drechsler	475	K. G. Larsen	535
P. Ďuriš	455	N.-M. Lê	499
A. Ehrenfeucht	1	S. Leonardi	303
Z. Ésik	27	M. Lipponen	63
S. A. Fenner	393	L.-M. Lopez	336
B. de Fluiter	87	J. H. Lutz	393
L. Fortnow	558	A. Maheshwari	220
R. Freivalds	183, 558	C. Makris	464
Z. Galil	360, 455	M. V. Marathe	487
L. Gargano	135	A. Marchetti-Spaccamela	303
W. I. Gasarch	558	E. Mayordomo	393
P. Gastin	15	Y. Métivier	75
P. Gemmell	547	P. Narbel	336
R. Giancarlo	111	V. Natarajan	648
E. Goles	570	J. Niedermann	535

S. Nikoletseas	159	A. Rubio	511
M. Noy	220	D. Saccà	523
S. Olariu	292	J.-R. Sack	220
D. Parente	232	D. Sangiorgi	672
A. Parra	123	P. Scheffler	123
G. Păun	429	T. Schwentick	405
G. Persiano	171, 232	T. Shahoumian	547
H. Petersen	315	C. Smith	558
A. Pietracaprina	372	P. Spirakis	159
J.-E. Pin	348	F. Stephan	558
W. Plandowski	444	L. Stewart	292
C. Priami	660	R. Sundaram	487
G. Pucci	372	M. Theobald	475
M. Rauch Henzinger	280	A. Tsakalidis	464
R. Ravi	487	J. Urrutia	220
S. S. Ravi	487	U. Vaccaro	135, 171
J. Reif	159	W. Vogler	636
A. A. Rescigno	135	P.-A. Wacrenier	75
G. Richomme	75	O. Watanabe	196
E. Rivlin	593	P. Weil	348
A. Rosenfeld	593	H. Wong-Toi	417
D. J. Rosenkrantz	487	M. Yung	159
G. Rote	256	C. Zaniolo	523
G. Rozenberg	1	C. D. Zaroliagis	244

Springer-Verlag
and the Environment

We at Springer-Verlag firmly believe that an international science publisher has a special obligation to the environment, and our corporate policies consistently reflect this conviction.

We also expect our business partners – paper mills, printers, packaging manufacturers, etc. – to commit themselves to using environmentally friendly materials and production processes.

The paper in this book is made from low- or no-chlorine pulp and is acid free, in conformance with international standards for paper permanency.

Lecture Notes in Computer Science

For information about Vols. 1–871
please contact your bookseller or Springer-Verlag

Vol. 872: S Arikawa, K. P. Jantke (Eds.), Algorithmic Learning Theory. Proceedings, 1994. XIV, 575 pages. 1994.

Vol. 873: M. Naftalin, T. Denvir, M. Bertran (Eds.), FME '94: Industrial Benefit of Formal Methods. Proceedings, 1994. XI, 723 pages. 1994.

Vol. 874: A. Borning (Ed.), Principles and Practice of Constraint Programming. Proceedings, 1994. IX, 361 pages. 1994.

Vol. 875: D. Gollmann (Ed.), Computer Security – ESORICS 94. Proceedings, 1994. XI, 469 pages. 1994.

Vol. 876: B. Blumenthal, J. Gornostaev, C. Unger (Eds.), Human-Computer Interaction. Proceedings, 1994. IX, 239 pages. 1994.

Vol. 877: L. M. Adleman, M.-D. Huang (Eds.), Algorithmic Number Theory. Proceedings, 1994. IX, 323 pages. 1994.

Vol. 878: T. Ishida; Parallel, Distributed and Multiagent Production Systems. XVII, 166 pages. 1994. (Subseries LNAI).

Vol. 879: J. Dongarra, J. Waśniewski (Eds.), Parallel Scientific Computing. Proceedings, 1994. XI, 566 pages. 1994.

Vol. 880: P. S. Thiagarajan (Ed.), Foundations of Software Technology and Theoretical Computer Science. Proceedings, 1994. XI, 451 pages. 1994.

Vol. 881: P. Loucopoulos (Ed.), Entity-Relationship Approach – ER'94. Proceedings, 1994. XIII, 579 pages. 1994.

Vol. 882: D. Hutchison, A. Danthine, H. Leopold, G. Coulson (Eds.), Multimedia Transport and Teleservices. Proceedings, 1994. XI, 380 pages. 1994.

Vol. 883: L. Fribourg, F. Turini (Eds.), Logic Program Synthesis and Transformation – Meta-Programming in Logic. Proceedings, 1994. IX, 451 pages. 1994.

Vol. 884: J. Nievergelt, T. Roos, H.-J. Schek, P. Widmayer (Eds.), IGIS '94: Geographic Information Systems. Proceedings, 1994. VIII, 292 pages. 19944.

Vol. 885: R. C. Veltkamp, Closed Objects Boundaries from Scattered Points. VIII, 144 pages. 1994.

Vol. 886: M. M. Veloso, Planning and Learning by Analogical Reasoning. XIII, 181 pages. 1994. (Subseries LNAI).

Vol. 887: M. Toussaint (Ed.), Ada in Europe. Proceedings, 1994. XII, 521 pages. 1994.

Vol. 888: S. A. Andersson (Ed.), Analysis of Dynamical and Cognitive Systems. Proceedings, 1993. VII, 260 pages. 1995.

Vol. 889: H. P. Lubich, Towards a CSCW Framework for Scientific Cooperation in Europe. X, 268 pages. 1995.

Vol. 890: M. J. Wooldridge, N. R. Jennings (Eds.), Intelligent Agents. Proceedings, 1994. VIII, 407 pages. 1995. (Subseries LNAI).

Vol. 891: C. Lewerentz, T. Lindner (Eds.), Formal Development of Reactive Systems. XI, 394 pages. 1995.

Vol. 892: K. Pingali, U. Banerjee, D. Gelernter, A. Nicolau, D. Padua (Eds.), Languages and Compilers for Parallel Computing. Proceedings, 1994. XI, 496 pages. 1995.

Vol. 893: G. Gottlob, M. Y. Vardi (Eds.), Database Theory – ICDT '95. Proceedings, 1995. XI, 454 pages. 1995.

Vol. 894: R. Tamassia, I. G. Tollis (Eds.), Graph Drawing. Proceedings, 1994. X, 471 pages. 1995.

Vol. 895: R. L. Ibrahim (Ed.), Software Engineering Education. Proceedings, 1995. XII, 449 pages. 1995.

Vol. 896: R. N. Taylor, J. Coutaz (Eds.), Software Engineering and Human-Computer Interaction. Proceedings, 1994. X, 281 pages. 1995.

Vol. 897: M. Fisher, R. Owens (Eds.), Executable Modal and Temporal Logics. Proceedings, 1993. VII, 180 pages. 1995. (Subseries LNAI).

Vol. 898: P. Steffens (Ed.), Machine Translation and the Lexicon. Proceedings, 1993. X, 251 pages. 1995. (Subseries LNAI).

Vol. 899: W. Banzhaf, F. H. Eeckman (Eds.), Evolution and Biocomputation. VII, 277 pages. 1995.

Vol. 900: E. W. Mayr, C. Puech (Eds.), STACS 95. Proceedings, 1995. XIII, 654 pages. 1995.

Vol. 901: R. Kumar, T. Kropf (Eds.), Theorem Provers in Circuit Design. Proceedings, 1994. VIII, 303 pages. 1995.

Vol. 902: M. Dezani-Ciancaglini, G. Plotkin (Eds.), Typed Lambda Calculi and Applications. Proceedings, 1995. VIII, 443 pages. 1995.

Vol. 903: E. W. Mayr, G. Schmidt, G. Tinhofer (Eds.), Graph-Theoretic Concepts in Computer Science. Proceedings, 1994. IX, 414 pages. 1995.

Vol. 904: P. Vitányi (Ed.), Computational Learning Theory. EuroCOLT'95. Proceedings, 1995. XVII, 415 pages. 1995. (Subseries LNAI).

Vol. 905: N. Ayache (Ed.), Computer Vision, Virtual Reality and Robotics in Medicine. Proceedings, 1995. XIV,

Vol. 906: E. Astesiano, G. Reggio, A. Tarlecki (Eds.), Recent Trends in Data Type Specification. Proceedings, 1995. VIII, 523 pages. 1995.

Vol. 907: T. Ito, A. Yonezawa (Eds.), Theory and Practice of Parallel Programming. Proceedings, 1995. VIII, 485 pages. 1995.

Vol. 908: J. R. Rao Extensions of the UNITY Methodology: Compositionality, Fairness and Probability in Parallelism. XI, 178 pages. 1995.

Vol. 909: H. Comon, J.-P. Jouannaud (Eds.), Term Rewriting. Proceedings, 1993. VIII, 221 pages. 1995.

Vol. 910: A. Podelski (Ed.), Constraint Programming: Basics and Trends. Proceedings, 1995. XI, 315 pages. 1995.

Vol. 911: R. Baeza-Yates, E. Goles, P. V. Poblete (Eds.), LATIN '95: Theoretical Informatics. Proceedings, 1995. IX, 525 pages. 1995.

Vol. 912: N. Lavrac, S. Wrobel (Eds.), Machine Learning: ECML – 95. Proceedings, 1995. XI, 370 pages. 1995. (Subseries LNAI).

Vol. 913: W. Schäfer (Ed.), Software Process Technology. Proceedings, 1995. IX, 261 pages. 1995.

Vol. 914: J. Hsiang (Ed.), Rewriting Techniques and Applications. Proceedings, 1995. XII, 473 pages. 1995.

Vol. 915: P. D. Mosses, M. Nielsen, M. I. Schwartzbach (Eds.), TAPSOFT '95: Theory and Practice of Software Development. Proceedings, 1995. XV, 810 pages. 1995.

Vol. 916: N. R. Adam, B. K. Bhargava, Y. Yesha (Eds.), Digital Libraries. Proceedings, 1994. XIII, 321 pages. 1995.

Vol. 917: J. Pieprzyk, R. Safavi-Naini (Eds.), Advances in Cryptology - ASIACRYPT '94. Proceedings, 1994. XII, 431 pages. 1995.

Vol. 918: P. Baumgartner, R. Hähnle, J. Posegga (Eds.), Theorem Proving with Analytic Tableaux and Related Methods. Proceedings, 1995. X, 352 pages. 1995. (Subseries LNAI).

Vol. 919: B. Hertzberger, G. Serazzi (Eds.), High-Performance Computing and Networking. Proceedings, 1995. XXIV, 957 pages. 1995.

Vol. 920: E. Balas, J. Clausen (Eds.), Integer Programming and Combinatorial Optimization. Proceedings, 1995. IX, 436 pages. 1995.

Vol. 921: L. C. Guillou, J.-J. Quisquater (Eds.), Advances in Cryptology – EUROCRYPT '95. Proceedings, 1995. XIV, 417 pages. 1995.

Vol. 922: H. Dörr, Efficient Graph Rewriting and Its Implementation. IX, 266 pages. 1995.

Vol. 923: M. Meyer (Ed.), Constraint Processing. IV, 289 pages. 1995.

Vol. 924: P. Ciancarini, O. Nierstrasz, A. Yonezawa (Eds.), Object-Based Models and Languages for Concurrent Systems. Proceedings, 1994. VII, 193 pages. 1995.

Vol. 925: J. Jeuring, E. Meijer (Eds.), Advanced Functional Programming. Proceedings, 1995. VII, 331 pages. 1995.

Vol. 926: P. Nesi (Ed.), Objective Software Quality. Proceedings, 1995. VIII, 249 pages. 1995.

Vol. 927: J. Dix, L. Moniz Pereira, T. C. Przymusinski (Eds.), Non-Monotonic Extensions of Logic Programming. Proceedings, 1994. IX, 229 pages. 1995. (Subseries LNAI).

Vol. 928: V.W. Marek, A. Nerode, M. Truszczynski (Eds.), Logic Programming and Nonmonotonic Reasoning. Proceedings, 1995. VIII, 417 pages. 1995. (Subseries LNAI).

Vol. 929: F. Morán, A. Moreno, J.J. Merelo, P. Chacón (Eds.), Advances in Artificial Life. Proceedings, 1995. XIII, 960 pages. 1995 (Subseries LNAI).

Vol. 930: J. Mira, F. Sandoval (Eds.), From Natural to Artificial Neural Computation. Proceedings, 1995. XVIII, 1150 pages. 1995.

Vol. 931: P.J. Braspenning, F. Thuijsman, A.J.M.M. Weijters (Eds.), Artificial Neural Networks. IX, 295 pages. 1995.

Vol. 932: J. Iivari, K. Lyytinen, M. Rossi (Eds.), Advanced Information Systems Engineering. Proceedings, 1995. XI, 388 pages. 1995.

Vol. 933: L. Pacholski, J. Tiuryn (Eds.), Computer Science Logic. Proceedings, 1994. IX, 543 pages. 1995.

Vol. 934: P. Barahona, M. Stefanelli, J. Wyatt (Eds.), Artificial Intelligence in Medicine. Proceedings, 1995. XI, 449 pages. 1995. (Subseries LNAI).

Vol. 935: G. De Michelis, M. Diaz (Eds.), Application and Theory of Petri Nets 1995. Proceedings, 1995. VIII, 511 pages. 1995.

Vol. 936: V.S. Alagar, M. Nivat (Eds.), Algebraic Methodology and Software Technology. Proceedings, 1995. XIV, 591 pages. 1995.

Vol. 937: Z. Galil, E. Ukkonen (Eds.), Combinatorial Pattern Matching. Proceedings, 1995. VIII, 409 pages. 1995.

Vol. 938: K.P. Birman, F. Mattern, A. Schiper (Eds.), Theory and Practice in Distributed Systems. Proceedings, 1994. X, 263 pages. 1995.

Vol. 939: P. Wolper (Ed.), Computer Aided Verification. Proceedings, 1995. X, 451 pages. 1995.

Vol. 940: C. Goble, J. Keane (Eds.), Advances in Databases. Proceedings, 1995. X, 277 pages. 1995.

Vol. 941: M. Cadoli, Tractable Reasoning in Artificial Intelligence. XVII, 247 pages. 1995. (Subseries LNAI).

Vol. 942: G. Böckle, Exploitation of Fine-Grain Parallelism. IX, 188 pages. 1995.

Vol. 943: W. Klas, M. Schrefl, Metaclasses and Their Application. IX, 201 pages. 1995.

Vol. 944: Z. Fülöp, F. Gécseg (Eds.), Automata, Languages and Programming. Proceedings, 1995. XIII, 686 pages. 1995.

Vol. 945: B. Bouchon-Meunier, R.R. Yager, L.A. Zadeh (Eds.), Advances in Intelligent Computing - IPMU '94. Proceedings, 1994. XII, 628 pages.1995.

Vol. 946: C. Froidevaux, J. Kohlas (Eds.), Symbolic and Quantitative Approaches to Reasoning and Uncertainty. Proceedings, 1995. X, 420 pages. 1995. (Subseries LNAI).

Vol. 947: B. Möller (Ed.), Mathematics of Program Construction. Proceedings, 1995. VIII, 472 pages. 1995.

Vol. 948: G. Cohen, M. Giusti, T. Mora (Eds.), Applied Algebra, Algebraic Algorithms and Error-Correcting Codes. Proceedings, 1995. XI, 485 pages. 1995.